Lecture Notes in Networks and Systems

Volume 335

The series "Lecture Notes in Networks and Systems" publishes the latest developments in Networks and Systems—quickly, informally and with high quality. Original research reported in proceedings and post-proceedings represents the core of LNNS.

Volumes published in LNNS embrace all aspects and subfields of, as well as new challenges in, Networks and Systems.

The series contains proceedings and edited volumes in systems and networks, spanning the areas of Cyber-Physical Systems, Autonomous Systems, Sensor Networks, Control Systems, Energy Systems, Automotive Systems, Biological Systems, Vehicular Networking and Connected Vehicles, Aerospace Systems, Automation, Manufacturing, Smart Grids, Nonlinear Systems, Power Systems, Robotics, Social Systems, Economic Systems and other. Of particular value to both the contributors and the readership are the short publication timeframe and the world-wide distribution and exposure which enable both a wide and rapid dissemination of research output.

The series covers the theory, applications, and perspectives on the state of the art and future developments relevant to systems and networks, decision making, control, complex processes and related areas, as embedded in the fields of interdisciplinary and applied sciences, engineering, computer science, physics, economics, social, and life sciences, as well as the paradigms and methodologies behind them.

Indexed by SCOPUS, INSPEC, WTI Frankfurt eG, zbMATH, SCImago.

All books published in the series are submitted for consideration in Web of Science.

For proposals from Asia please contact Aninda Bose (aninda.bose@springer.com).

More information about this series at https://link.springer.com/bookseries/15179

Andre Batako · Anna Burduk ·
Kanisius Karyono · Xun Chen ·
Ryszard Wyczółkowski
Editors

Advances in Manufacturing Processes, Intelligent Methods and Systems in Production Engineering

Progress in Application of Intelligent Methods and Systems in Production Engineering

Springer

Editors
Andre Batako
Liverpool John Moores University
Liverpool, UK

Kanisius Karyono
School of Civil Engineering
and Built Environment
Liverpool John Moores University
Liverpool, UK

Ryszard Wyczółkowski
Faculty of Mechanical Engineering
Silesian University of Technology
Gliwice, Poland

Anna Burduk
Wroclaw University of Science
and Technology
Wrocław, Poland

Xun Chen
General Engineering Research Institute
Liverpool John Moores University
Liverpool, UK

ISSN 2367-3370 ISSN 2367-3389 (electronic)
Lecture Notes in Networks and Systems
ISBN 978-3-030-90531-6 ISBN 978-3-030-90532-3 (eBook)
https://doi.org/10.1007/978-3-030-90532-3

This Springer imprint is published by the registered company Springer Nature Switzerland AG
The registered company address is: Gewerbestrasse 11, 6330 Cham, Switzerland

Preface

The GCMM is a major biennial international congress for widespread dissemination of knowledge in manufacturing, management, production engineering, maintenance predictive technologies and Industry 4.x. Recent events held in Australia (2018), China (2016), India (2014), New Zealand (2012) and Thailand (2010) were highly successful and attracted delegates from over 20 countries.

GCMM2021 was special, for it was the first time, this global congress was held in Europe, in the UK, to celebrate the 30/31st anniversary of GCMM and to share new attainments in research and industrial practice between academic scholars and industry practitioners. This congress has always attracted high-quality research papers and continues to serve as a central hub where the world's knowledge of manufacturing research is fused with industry strategy in a globalised market.

GCMM2021 continued this tradition, though held online, it attracted over 200 scholars and higher regarded professionals from academia and industry around the world, USA, Australia, New Zealand, India, China, Southeast Asia, Africa and Europe. Regardless on the restrictive conditions imposed by the COVID, GCMM2021 has attracted scholars from over 16 countries. It attracted an equal number of male and female keynotes speakers that shared their experience with the young researchers and the challenges faced by women in academia and engineering. A critical topic like ***"Women in engineering in academia: Are we there yet?"*** was expounded by Executive Dean of the Faculty of Engineering at Queensland University of Technology, Prof Anna Deletic, exploring the global challenge faced by women in various areas of human activities.

GCMM2021 was organised and hosted jointly by Liverpool John Moores University and Wroclaw University of Science and Technology along with its **ISPEM** (Intelligent Systems for Production Engineering Maintenance).

GCMM2021 has received a good support from its sponsors, namely the Vellore Institute of Technology (India), Zhengzhou University, Xi' An university of Science and Technology (China) and Liverpool John Moores University (UK).

GCMM2021 benefited from the support of several associates: IAAST, (China), the IET (UK), QUT (Australia), Ford Motors Company (Liverpool) and Kawasaki (Liverpool) and Brunel University, UK.

GCMM operates within a broad range of manufacturing sciences and covered five main streams, i.e. global manufacturing challenges, intelligent systems in production engineering and maintenance, management of production process, advances in materials and manufacturing and Industry 4.0, with, over forty topics. However, this issue is a collection of technical and research papers dedicated to intelligent methods in production engineering, providing new solutions for innovative factories of the future, advances in production and maintenance from the viewpoint of Industry 4.0. Therefore, this volume discusses the applications of intelligent systems, methods and tools in production engineering, maintenance, logistics, along with elements of information systems and product development.

This collection of new developments and advancement in intelligent control and optimisation system for production engineering will serve as a good companion to scholars, manufacturing companies, RTO to improve the efficiency of production systems. It forms an excellent basis for the development of intelligent manufacturing system for Industry 4.0 digital and distributed manufacturing and factories for future.

The papers presented in this volume were selected after meticulous review of three independent and international reviews, the reviewers had no acquaintance with the authors, and papers were sent to different countries avoiding the country of its origin. All papers were scored, and the reviewers gave their levels of confidence. In some doubtful cases such as borderline or contradicting reviews, and, where the level of confidence of the reviewers was low, papers were resent to two additional reviewers making a total of five reviews. With this stringent review process, over sixty per cent of the submitted papers were rejected, and this ensured the quality of the work presented here.

GCMM2021 organising committee scrutinised further all the papers and awarded four prizes for the best technical papers, the best presenter and the best poster.

GCMM2021 expresses its deep gratitude to all Sponsors for their generous financial contribution that help to run the conference online, to all the associates for their support, to the keynote speakers for their time and the high quality of the material presented and experience shared.

Special gratitude is expressed to Chancellor of VIT Honourable Dr. Viswanathan, to Vice Chancellor of LJMU, Prof. Ian G Campbell and to Pro-vice Chancellor of LJMU, Prof. Keith George, to the LJMU corporate communication team, to the Conference Partners International—CPI, and, to all the scholars for their contributions.

July 2021

Anna Burduk
Andre D. L. Batako
Xun Chen
Kanisius Karyono
Ryszard Wyczókowski

GCMM2021 Hosts

LIVERPOOL JOHN MOORES UNIVERSITY

Wrocław University of Science and Technology

Conference Organisers

Dr Andre Batako : LJMU
Prof. Anna Burduk : WUST
Mr Kanisius Karyono : LJMU

GCMM2021 Sponsors

LIVERPOOL JOHN MOORES UNIVERSITY

VIT Vellore Institute of Technology

GCMM2021 Associates

国际应用科学与技术协会

IET The Institution of Engineering and Technology

QUT Queensland University of Technology

FORD MOTOR COMPANY HALEWOOD TRANSMISSION

Kawasaki Powering your potential

Organisation

Organising Committee

Patrons

G. Viswanathan (Chancellor)	VIT, India

Chairs/Co-chairs

Prasad K. D. V. Yarlagadda	QUT, Australia
Yong Ming Zhu	Zhengzhou Univ., China
Anthony Xavior	VIT Univ., India
Anna Burduk	WUST, Poland
Mike Riley	Liverpool John Moores Univ., UK
Andre D. L. Batako	Liverpool John Moores Univ. Conference Chair

Secretaries

Xun Chen	Liverpool John Moores Univ.
Martin Sharp	Liverpool John Moores Univ.

United Kingdom

Liverpool John Moores University

Jin Wang
Vijay Karunamoothei
Paulo Lisboa
Guang-Ming Zhang
James Ren
Jun Ren
David Harvey
Musa Bashir

Rafid Al-Khaddar
Ibijoke Idowu
Martin Sharp
Badr Abdullah
Tahsin Opoz
Alex Teng
Mohammad Ateeq

Other Universities

Cheng Kai	Brunel University
Vadim Silberschimidt	Loughborough Univ.
Alison McMillan	Glyndŵr Univ.
Abdul-Majeed Mahamadu	Univ. West of England
Henry Abanda	Oxford Brookes Univ.
Arivazhagan Anbalagan	Birmingham Univ.

Poland

Wrocław University of Science and Technology

Joanna Helman
Kamil Krot
Tomasz Chlebus
Anna Woźna
Maria Rosienkiewicz
Dagmara Łapczyńska
Paweł Krowicki
Kamil Musiał
Tomasz Bêdza
Małgorzata Rusiñska
Joanna Kochañska

Silesian University of Technology

Cezary Grabowik
Damian Krenczyk
Aleksander Gwiazda
Ryszard Wyczółkowski
Krzysztof Kalinowski

GCMM Board

President

G. Viswanathan (Chancellor) VIT Univ.

Secretary

Prasad K. D. V.	Yarlagadda, OAM, QUT

Board Members

Ayo Olofinjana	Univ. of Sunshine Coast, Australia
C. Brito (President)	CRES, Brazil
H. ElMaraghy	Univ. of Windsor, Canada
Yongming Zhu	Zhengzhou Univ., China
David Wang	IAASET
Shouqin Zhou	CIMC Co., Ltd.
H. Elwany	Alexandria Univ., Egypt
S. Tichkiewitch	Domaine Univ., France
K. Adjallah	ENIM, Loraine Univ.
Ing. P. Gümpel	Fachochschule Konstanz, Germany
Krishnanand Lanka	NIT, India
Ravichandran	CEO, Lucas TVS
L. Gopalakrishnan	PSG & Sons Charities
Walter C. K. Wong	Berjaya Univ., Malaysia
C. Acosta	Universidad de las America-Pueba, Mexico
E. De Bruijn	Univ. of Twente, Netherlands
Sarat Singamneni	AUT, New Zealand
Mark P. Taylor	Univ. of Auckland
Leszek A. Dobrzanski	Science Centre ASKLEPIOS, Gliwice, Poland
Korneliusz Pylak	Lublin Univ. of Tech.
Seeram Ramakrishna	NUS, Singapore, Singapore
Janez Kopac	Univ. of Ljubijana, Slovenia
E. Tovar	Universidad Politechnica de Madrid, Spain
Y.T. Im	KAIST, South Korea
Ill-Soo Kim	Mokpo National Univ.
T. Schelker	Swiss College of Agriculture, Switzerland
S. Butdee	King Mongkutś Inst. of Tech., Thailand
Batako D. L. Andre	Liverpool John Moores Univ., UK
K. S. Taraman	Lawrence Technological Univ., USA
Lal Tummala	San Diego State Univ.

International Scientific Committee

Kondo Adjallah	Université de Lorraine, France
Baghal-Wadji Alireza	Univ. Cape Town, South Africa
Andrzej Ambroziak	Wrocław Univ. of Science and Technology, Poland
Dario Antonelli	Politecnico di Torino, Italy
Katarzyna Antosz	Rzeszow Univ. of Technology, Poland
Zbigniew Banaszak	Koszalin Univ. of Technology, Poland

Josef Basl	Univ. of West Bochemia, Czech Republic
Ryszard Błażej	Wrocław Univ. of Science and Technology, Poland
Wojciech Bożejko	Wrocław Univ. of Science and Technology, Poland
Milan Brandt	RMIT Univ., Australia
Anna Burduk	Wrocław Univ. of Science and Technology, Poland
S. Butdee	King Mongkut's Inst. of Tech., Thailand
Yan-Lin Cai	Zhengzhou Univ., China
Elegbede	Charles A.O. Ariane Group, France
Edward Chlebus	Wrocław Univ. of Science and Technology, Poland
Tomasz Chlebus	Wrocław Univ. of Science and Technology, Poland
Mariusz Cholewa	Wroclaw Univ. of Science and Technology, Poland
Melany M. Ciampi	ING-PAED IGIP, Brazil
Grzegorz Ćwikła	Silesian Univ. of Technology, Poland
Jacek Diakun	Poznan Univ. of Technology, Poland
Albert H. Ding	Harbin Inst. of Tech. China
Ewa Dostatni	Poznan Univ. of Technology, Poland
Luboslav Dulina	Univ. of Zilina, Slovakia
Bogdan Dybała	Wrocław Univ. of Science and Technology, Poland
A. Elaya Erumal	Anna Univ., India
Zhang Feihu	Harbin Institute of Technology. China
Kristofer Gamstedt	Uppsala Univ., Sweden
Arkadiusz Gola	Lublin Univ. of Technology, Poland
Filip Górski	Poznan Univ. of Technology, Poland
Cezary Grabowik	Silesian Univ. of Technology, Poland
Marta Grabowska	Poznan Univ. of Technology, Poland
Milan Gregor	Univ. of Zilina, Slovakia
Katarzyna Grzybowska	Poznan Univ. of Technology, Poland
Manoj Gupta	NUS, Singapore
B. Gurumoorthy	IISc, Bengaluru, India
Aleksander Gwiazda	Silesian Univ. of Technology, Poland
Adam Hamrol	Poznan Univ. of Technology, Poland
Joanna Helman	Wroclaw Univ. of Science and Technology, Poland
Małgorzata Jasiulewicz-Kaczmarek	Poznan Univ. of Technology, Poland
P. Jeyapandiarajan	VIT, India
Ma Jiang	Shenzhen Univ., China

Contents

Optimization of Packaging Processes for Special Orders Using Process Activity Mapping

Jacob Søe Mikkelsen, Justina Lydekaityte, and Torben Tambo(✉)

Department of Business Development and Technology, Aarhus University, Birk Centerpark 15, 7400 Herning, Denmark

201610644@post.au.dk, {justina,torbento}@btech.au.dk

Abstract. Packaging has a significant impact on the performance of logistic systems, therefore it is necessary to optimize the packaging process in order to enhance efficiency and add more value to a product and customer. This paper investigates the current packaging processes for special orders at one of stainless steel drainage systems manufacturers in Denmark. The aim of this research is to examine the packaging process for special orders and identify valuable and wasteful activities. The Process Activity Mapping has been selected to determine and eliminate waste. The analysis of the packaging process was carried out through direct non-participatory observations that listed 22 steps with recorded and calculated area, machine, distance moved, time spend, and people involved. Based on the analysis, only 5 out of 22 steps were identified as value-adding, 6 as necessary non-adding value and the rest as non-adding value. The majority of non-adding value and necessary non-adding value operations were related to transportation, waiting, inappropriate processing and unnecessary motion. Based on the results, the majority of wasteful activities was observed in the shrink wrapping station, where the propane torch is used to shrink the foil. Wrapping process requires highly intensive manual work that also triggers exceeded noise exposure and poor working postures. As a result, the research suggests a set of recommendations to establish a more optimized, efficient and ergonomic flow of packaging processes, including the rearrangement of activities, changes in the plant layout, and wrapping process automation.

Keywords: Packaging process optimization · Process Activity Mapping · Value-adding activities

1 Introduction

Nowadays, efficiency and competitiveness are the critical challenges for the manufacturing companies in the global market [1]. Companies are constantly concerned about how to provide high-quality goods and services in the effortless, fast and cost-efficient manner. Consequently, there is a need to enhance the efficiency of the production line to reduce lead time and downtime rates, as well as to ensure high levels of customer satisfaction with their products [2]. According to the authors, the continuous improvement and cost reduction in manufacturing are induced by (i) the identification of what type of

A. Batako et al. (Eds.): GCMM 2021, LNNS 335, pp. 1–14, 2022.
https://doi.org/10.1007/978-3-030-90532-3_1

activities can increase the value to a product and a customer, and (ii) the identification and elimination of waste [2]. This concept is widely used to improve business responsiveness to be able to win the competition and meet customer demands [1]. Unidentified waste results in adverse consequences for the company including increased production costs for additional handling and space of the inventory, increased rate of redundancy operations due to over-processing, or a growing pile of work between workstations because of delays [2]. As a result, reference [3] states that the identification and systematic elimination of waste in the process should be a priority.

This study investigates the final part of the manufacturing – the packaging process for logistics. Overall, the packaging is referred as a "coordinated system of preparing goods for safe, secure, efficient and effective handling" [4]. Although the packaging is usually not concerned as a part of the product or the logistic process, it has a significant impact on the performance of logistic systems including such activities as transportation, warehousing, inventory, and handling [4]. Furthermore, the ability of packaging to integrate a commercial, environmental, and logistical vision allows increasing the business competitiveness [5].

As a result, this research aims to analyze the packaging process for special orders by listing every activity the selected product is involved in and identifying valuable and wasteful operations. In order to identify and eliminate waste, the Process Activity Mapping has been adapted from [6] that matches the most common wastes occurring in the studied company. Further analysis includes a set of recommendations to eliminate, rearrange, or simplify activities to reduce waste and increase efficiency.

2 Literature Review

2.1 Packaging and Wrapping

Packaging is considered as an integral and inseparable part of the majority of product-related companies' business model [7]. In general, packaging is intended to facilitate handling, communicate product information, and protect goods in the supply chain until they reach the end consumer [8]. There are three different levels of packaging's hierarchic structure: primary, secondary, and tertiary package that differ from one another based on the number of layers from being in direct contact with the packaged good [5]. This research investigates the process of tertiary packaging, which assembles a number of primary or secondary packages on a pallet or roll container [5].

Tertiary packaging is used to transport goods from their point of origin to their final destination [10]. Therefore, the protection and functional performance are significantly essential for logistic processes, where the package or pallet has to withstand robust physical handling along with the main logistic activities of transportation, distribution, inventory, and storage [11]. Consequently, the packaging process has to include operations and materials that ensure mechanical strength to prevent damages from the external factors including temperature, compression, vibration, mechanical shock and similar [12, 13]. Furthermore, the majority of shipping providers have mandatory compliances that organizations have to follow in order to ensure safe, secure, efficient and effective distribution. As a result, the packaging for logistics has to encompass a set of operations before being ready for shipping. In order to maintain stability, provide product protection, and increase handling efficiency, products are typically distributed on pallets in unitized loads that are built utilizing stretch wrap, stretch hood (or hood-stretching), thermal shrink wrap, strapping or banding [14].

Currently, the investigated company in this study employs thermal shrink-wrapping. Reference [15] defines shrink-wrapping as the process of stabilizing product loads by covering a unit load/pallet with a heat-sensitive plastic material, such as low-density polyethylene (LDPE), and then applying heat to that material to shrink it around the load adjusting to the shape of the load and secure the product to the pallet. The shrink-wrapping process varies from manual wrapping and shrinking with a hand torch to fully automatic machines [9]. On the one hand, shrink-wrapping protects a load from dirt, moisture, dust and other contaminants, provides great vertical strength, and shrinks tightly around any shape of the product [15]. On the other hand, this method is cost-consuming due to a foil price and the heat required in the process [9].

According to [14], the most common method of load unitization is stretch wrapping, where a stretchable plastic film, most commonly a linear low-density polyethylene (LLDPE), is wrapped around the pallet. More specifically, stretch film is "applied under tension and elastic recovery conforms the film to the load" [14]. Stretch wrapping can be divided into three automation levels: manual, semi-automatic and automatic [9]. Furthermore, there are three types of stretch wrapping systems: turntable, rotary arm and orbital [14]. This method has lower stability than thermal shrink wrapping or hood stretching, but the load containment can be increased by additional wraps, heavier gauge film and increased post-stretch [9, 14]. Moreover, stretch wrapped pallets are water-resistant only to some degree [9].

2.2 Mapping Tools and Types of Waste

The effectiveness and efficiency of the production system have a direct effect on both the performance and profitability of a manufacturing company [16]. The latter is related to the company's ability to optimize the output in comparison to the input used [16]. According to [17], in order to be able to optimize a process, the first step is to understand the current process and thereby to identify where the process has flaws and needs to be optimized. One way to understand the processes entirely is to map intercompany and intracompany operations and categorize them based on their value-adding/non-value adding nature [17]. The value-adding processes not only increase the worth of the final product or service but also make them more valuable to the end customer [6]. Therefore, the goal is to transform the supply chain/value chain into a value stream, where all processes are adding value to the product.

In the literature [6], there are different mapping tools to map processes in the supply chain. Each mapping tool is defined by wastes and their level of correlation and usefulness (low to high) [6]. Consequently, the decision of which of the mapping tools to select is based on the determination of the most relevant wastes to the process. According to reference [3], waste is referred to such process activities that "do not contribute with, or add value to, the transformation of a product into its final form as sold to the customer". Reference [1] adds that waste consumes time, space, and resources without any contribution to satisfying the needs of customers.

Taiichi Ohno has introduced the seven wastes commonly occurring in the production processes and named them Muda [18]. Muda enables to understand the production system better and helps to identify the most critical and problematic areas [19]. More specifically, Muda is a classification of diverse wastes that unfavorably tend to increase the cost of the production activities, and is categorized into seven wastes:

- Overproduction. It is one of the most detrimental wastes in enterprises that prevents the smooth flow of goods or services and often results in excessive lead and storage times. Furthermore, overproduction might restrain quality and productivity due to the late detection of defects [6].
- Waiting. When time is used ineffectively, the waste of waiting occurs. In a production facility, waiting happens when a product is not moved or being worked on and affects both goods and workers [6].
- Transport. Overall, every movement of a product can be considered as a waste. Excessive movement can cause damage and deterioration of goods. However, movement of goods is often a necessary activity, and therefore companies tend to minimize movement rather than eliminate [6]. The plant layout is usually the main cause of excess transport [1].
- Inappropriate processing. This waste befalls when complex solutions are found to simple processes and problems. The over complexity prompts workers to overproduce in order to recover from the large investment of the complex machine and often results in a poor production layout leading to excessive transport and scarce communication [6].
- Unnecessary inventory. It tends to conceal the production struggles for a company and prevents from rapid identification of problems. Furthermore, unnecessary inventory requires larger storage and production space that not only leads to high costs but also to poor communication that results in a lower value stream [6].
- Unnecessary motion. It is related to the ergonomic-human factor in the production processes. All unnecessary bends, pick-ups or stretches triggered by poorly organized work stations should be avoided in order to not to harm workers, lessen productivity and quality [1, 6].
- Defect. It is a direct cost for the company because the produced goods have no value [6]. Defective or rejected finished goods can impede production and be in need of expensive rework [1].

The described wastes are tools that help researchers or practitioners to identify waste in individual value streams [6]. Once identified, the waste has to be either removed or reduced to a minimum. The elimination of waste can decrease the production cost, reduce the number of work in process, increase resources utilization, and thereby improve the productivity, i.e. operational efficiency, of a company [16].

2.3 Classification of Value

The systematic selection of wastes begins with the categorization that divides wastes and operations occurring in manufacturing and production into three types [6, 16]:

- Non-adding value (NV);
- Necessary non-adding value (NNV); and
- Value-adding (VA).

The non-adding value category includes redundant and unnecessary procedures that append no value to a product, and thus they need to be eliminated completely [6], for instance, waiting time between operations, stacking intermediate products, or double handling [20].

The necessary non-adding value category consists of wasteful but imperative activities to the production process to complete the product or service, e.g. walking long distances to pick up parts, unpackaging deliveries, and inspection [3]. The elimination of such practices often results in major changes in the operating system that usually can not be implemented immediately or within the economic sense, e.g. the establishment of a new production layout or new arrangement with suppliers to deliver unpacked goods [21].

Finally, value-adding activities comprise operations that provide benefits to the product and customer. It usually involves the conversion or processing of raw materials or part-manufactured products through the use of manual labor such as subassembly of parts, painting and forging raw materials [20].

NV and NNV activities end in a higher product/service cost and production delays since they are time- and resource-consuming [20]. Therefore, companies often strive to eliminate the NV and NNV waste activities to come closest to complete value-adding production.

2.4 Process Activity Mapping

The process activity mapping (PAM) has its origin in industrial engineering and strives to eliminate waste, inconsistencies and irrationalities to provide high-quality goods and services in the effortless, fast and cost-efficient manner. The PAM is a technical approach, a step-by-step analysis of the activities making up the process on the production floor [16, 22]. The key concept of this tool is to map every stage of activity occurring in transportation, inspection, operations, delay, and storage, and then group them into three clusters regarding their value stream [16]. PAM aims to eliminate the non-value adding activities and reduce or remove necessary non-value adding activities, and focus only on the particular operations that add value to the products and services [20]. More specifically, PAM provides the average time and resources requirement for each activity in a particular process [22] that can be transformed into a cost for each specific operation [3]. Process activity mapping tool is useful for creating greater efficiency within a sequence of operations that can result in the reduced manufacturing process lead time [3]. This tool consists of five stages [20, 22]:

First. Study of process flow;
Second. Identification of existing waste;
Third. Consideration whether the process can be rearranged to be more efficient;
Fourth. Consideration about better flow patterns, involving flow layout or transport routing; and

Fifth. Consideration whether every process is really necessary and what would happen if it was removed.

3 Methodology

This research employs a case study conducted in a stainless steel drainage systems manufacturing company, located in Vildbjerg, Denmark. Data collection used both quantitative and qualitative sources to enable triangulation to increase the robustness of the findings. In order to collect data, one of the authors spent 1–2 days a week within a three-month period in a company to understand, observe and analyze the current packaging process. Several research activities were carried out, including interviews, literature review, observation exercise and process activity mapping. Semi-structured interviews were conducted with a number of members of staff at the business, operational and functional levels to gather information about the current processes, internal and external compliances and requirements, and to determine potential issues. It was decided to investigate the packaging process, particularly for special orders. The selected order no. 670-18394-03 goes through every stage of the packaging process without skipping any operations and is of the most common orders handled by the packaging operators on a daily basis.

Based on the information provided by the Continuous Improvement Manager and Project Owner at the studied company, the most commonly experienced wastes within the packaging operations of special orders are transportation, waiting, inappropriate processing and unnecessary motion. In order for a manufacturing company to add value to a product and customer consistently by identifying and eliminating waste, a mapping tool has been adapted from [6], who demonstrated the relationship between diverse techniques and various types of waste. As a result, the Process Activity Mapping matched the most common wastes occurring in the studied company, and therefore it was selected for further analysis.

The stages of PAM were performed chronologically. First, a preliminary analysis of the processes was carried out through direct non-participatory observations to list every activity the product is involved in. It resulted in 22 steps categorized into five categories based on the activity type, i.e. Operation (O), Transport (T), Inspection (I), Storage (S), Delay (D). Simple process flow was made to illustrate all steps in Fig. 1. Then each step was analyzed more thoroughly. The area, where the step took place, and the machine, used to carry out the specific activity, were recorded together with the calculations of the distance moved, time taken, and a number of people involved (see Table 1). Regarding the distance, it was recorded when the products were moving, e.g. in step 2, the staining pallet is transported to the packaging area that is 50 m away from the staining station, whereas steps 9 and 10 are carried out at the same place, therefore there is no movement of products. Also, it was decided not to record waiting times, because they highly differ from one order to another due to workers' occupation, delays and similar. Moreover, the main objective of this research was to identify valuable and wasteful activities rather than the precise timing of the steps. Furthermore, unstructured follow-up interviews were also conducted with operational workers after the observation exercises to clarify any uncertainties and specifications.

Once the PAM was completed, the analyzed steps were divided into non-adding value, necessary non-adding value and value-adding processes depending on the wastes occurring in packaging operations. Further analysis included a set of recommendations to eliminate, rearrange, or simplify activities to reduce waste and increase efficiency. In order to provide for detailed recommendations, several additional measurements were also taken during observations, including noise exposure and Ovaka Working Postures Analysing System (OWPA) provided by [23].

4 Results

4.1 The Process Flow

In the company, there is a wide range of standard and special products being packaged every day. In general, such factors as steps, flows and distances remain the same throughout the entire packaging process for special orders. However, the amount of manual labor and the amount of time spent in each process step highly depend on the size and shape of the product as well as on the size of the order. As a result, for this research, it was chosen to analyze one of the most frequently packaged special orders that packaging operators are working with daily. It is a medium-size order no. 670-18394-03. It consists of six medium-sized drains packaged on Euro pallets. All six drains are placed on one standard Euro pallet in three layers with two drains in each of it (L1.2 × W0.8 × H1.8).

In general, the optimization of any process requires an in-depth understanding of the current processes [6]. Therefore the very first step in the Process Activity Mapping is to study the process flow. The studied company's packaging process consists of four main blocks (staining, packaging pallets, wrapping pallets, and shipping) with intermediate steps. The packaging process flow for special orders is presented in Fig. 1. As seen from the illustration, once the products are welded, they are collected in a staining container and stained to improve their corrosion resistance. Then, a staining container is moved to the packaging area, where products await operator to be inspected for quality control, manually stained (if needed), transferred either to the backfilling area on the work trolley or onto packaging pallets (if products do not need backfilling). These steps are repeated until the packaging pallet is loaded to its maximum capacity or contains all the items from the customer order. For this particular order, steps from 4 to 9 in the Process Activity Mapping have to be repeated six times. Later on, items are strapped to pallets to secure the products during transportation.

Strapped pallets are first moved to the waiting area for shrink wrapping and only after an indefinite period of time (depending on the worker occupancy) are transported to the shrink wrapping station. Here pallets are inspected to make sure that all products ordered are on the pallet including standard products that are packaged in cardboard boxes and, if needed, can be placed in gaps between special orders on the pallet. The wrapping process consists of several steps. Firstly, the foil is stretched over the floor, and the pallet is placed on top of it. Then, the wrapping operator attaches the foil to the pallet with a manual staple gun. Lastly, the foil is shrunk around the pallet manually using a propane torch.

In the final series of steps, the transport declaration/label is attached to the pallet, and it is transferred to the shipping area, where the pallet is stored before the transportation to the customer.

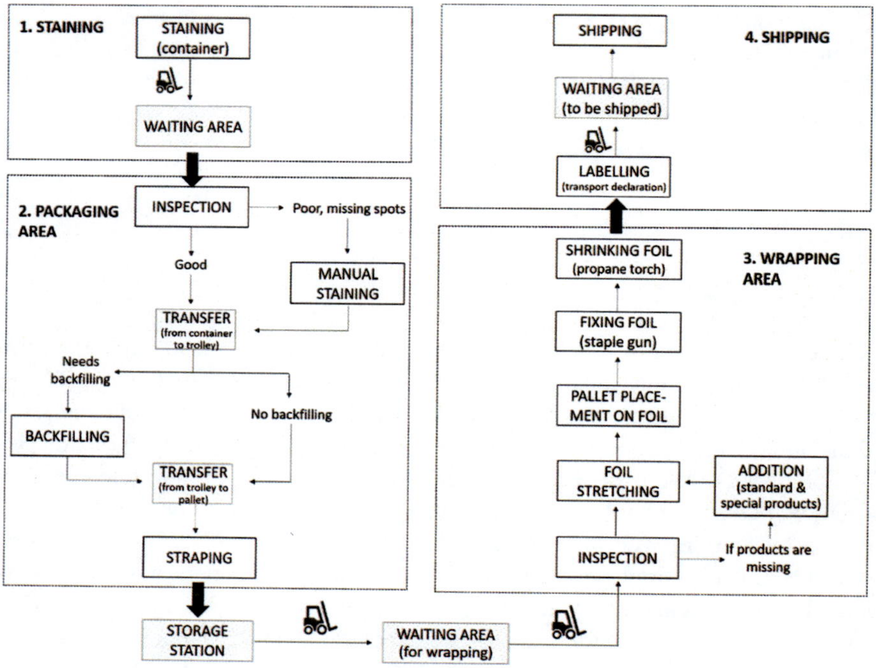

Fig. 1. Packaging process flow for special orders

4.2 The Process Activity Mapping

In order to carry out a more comprehensive and thorough analysis, the Process Activity Mapping tool was chosen provided by [6] to divide all activities occurring in the packaging process into smaller steps, determine the time and resources needed for each step, and finally categorize them based on their value stream (NV, NNV, or VA).

As stated in the literature section, operations and processes can be categorized into non-adding value, necessary non-adding value and value-adding processes [6]. Classifying the process steps into these categories provides a more comprehensive insight into the packaging process and where it can be optimized.

4.3 Non-value Adding Activities

Based on the analysis, only 5 out of 22 steps were identified as value-adding, 6 as necessary non-adding value and the rest as non-adding value (see Table 2). In fact, half of all packaging activities add no value to the product, and thereby to the customer. The majority of non-adding value operations were related to transfer from one working station to another and waiting times/storage:

- **Transportation.** In the packaging process, pallets have to be transported multiple times from one area to another. For instance, when goods are strapped to pallets, they can not be shrink wrapped right away at the same place because of the backfilling

Table 1. The process activity mapping of packaging processes

No.	Packaging Process Activity	Flow	Machine	Area	Dist (m)	Time (min)	People	Operation	Transport	Inspection	Storage	Delay	Value
1	Staining	O	Stainplant	Staining				X					VA
2	Transport to packaging area	T	Forklift	In between	50	01:00	1		X				NV
3	Waiting for packaging	S		Packaging							X	X	NV
4	Inspection of staining quality	I		Packaging		01:00	1			X			VA
5	Transport stained products to work trolley	T	Hand/Crane	Packaging	10	01:00	2		X				NV
6	Transport to backfilling area	T	Trolley	Packaging	10	00:30	1		X				NV
7	Backfilling	O	Pistol	Backfilling		02:00	1	X					VA
8	Transfer from work trolley to pallet	T	Hand	Backfilling	15	00:25	2		X				NV
9	Attachment of O-rings sealings to products	O	Hand	Packaging		01:00	1	X					VA
10	Fastening products to pallet	O	Pallet strapper	Packaging		03:00	1	X					NNV
11	Waiting for transport/storage	S		Packaging							X	X	NV
12	Transport to waiting area for foiling	T	Forklift	In between	150	02:00	1		X				NV
13	Waiting for foiling/storage	S		Stockarea							X	X	NV
14	Transport to foiling area	T	Forklift	In between	10	00:30	1		X				NV
15	Inspection of orders for missing items	I		Foil Wrapping		01:00	1			X			VA
16	Stretching foil over the floor	O	Foilroll	Foil Wrapping		01:30	1	X					NNV
17	Pallet placement on stretched foil	O	Forklift	Foil Wrapping	10	00:20	1		X				NNV
18	Fixing foil to product	O	Staple gun	Foil Wrapping		01:30	1	X					NNV
19	Shrinking foil	O	Torch	Foil Wrapping		04:00	1	X					NNV
20	Attachment of transport declaration	I		Foil Wrapping		00:30	1			X			NNV
21	Transport to shipping area	T	Forklift	In between	50	01:00	1		X				NV
22	Waiting for shipping	S		Shipping							X		NV

station with various chemicals situated nearby. Therefore the stripped pallets have to be moved to another area for shrink wrapping. Consequently, various waiting areas are designated in the facility, where the movement of pallets occurs frequently.

– **Waiting.** Multiple waiting times occur during the packaging processes. First, it happens when the product is put on hold to be packaged and strapped to pallets. Products from different orders are stained together, and therefore they are placed on the same staining pallet. Each order is handled separately, and therefore the timing of the product to be moved from staining pallet to packaging pallet differs. Second, when the product is strapped to the packaging pallet, this pallet is placed in the waiting area to be transported to the wrapping area. Third, after being transported to the wrapping station, it is placed again in a waiting area before being foiled. Finally, the finished pallet is put to storage in preparation for shipping.

4.4 Necessary Non-adding Value Activities

The necessary but non-adding value activities have been mostly observed during operations in the wrapping area (steps 16–20). Here the inappropriate processing and unnecessary motion are the main occurring wastes that are intertwined with one another. The inappropriate processing provokes unnecessary motion to happen:

– **Inappropriate processing.** It mainly occurs when the open fire is used to shrink the foil to the pallets. The propane torch open flame causes the risk of workers to burn themselves, therefore they have to be extremely cautious and wear proper clothing to avoid such risk. Moreover, the usage of the propane torch stipulates noise exposure. Maximum and average noise exposures were measured, respectively 93.6 dB(a) and 89.1 dB(a), beside the worker while using the torch. The Danish Working Environment Service recommends wearing hearing protection when noise exposure exceeds 80 dB(a). However, during the observation, none of the workers wore hearing protection while using a propane torch.

- **Unnecessary motion.** Overall, the entire shrink wrapping process requires highly intensive manual work that involves poor working postures. Particularly in the wrapping area, a forward inclination of head, neck and trunk were observed. The postures during fixing a foil to a product with a staple gun and shrinking it later with a propane torch require the operator body to be twisted, giving an asymmetrical posture. Furthermore, wrapping and shrinking involve multiple awkward postures, including forward and backward bending, twisting, kneeling, stooping, and squatting. During the observation exercise, working postures were analyzed using OWPA provided by [23]. Analyzed postures were classified as class 3 and class 4, indicating that they need to consider in the near future, otherwise the deformations will occur and the workers' muscular health might be damaged.

Table 2. The results of process activity mapping

Value	Operation (Quantity)	Transport (Quantity)	Inspection (Quantity)	Storage (Quantity)	Total	Percentage (%)
Value-adding	3	–	2	–	5	22.73
Non-adding value	–	7	–	4	11	50.00
Necessary non-adding value	5	–	1	–	6	27.27
Total					22	100.00

5 Discussion

Once the study of the process flow and the identification of existing waste are carried out, the next three stages of the Process Activity Mapping is considerations whether the process can be rearranged, improved by better flow patterns, or removed [20].

5.1 The Third Stage: Rearrangement

It would be the most optimal solution if all the working stations, i.e. staining, backfilling, packaging, and wrapping, were located in succession one after another, e.g. employing a conveyor belt, with the final stop at the finished good inventory. Reduced storage and movement would eliminate the majority of non-adding value steps in regards to transportation and waiting. However, such rearrangement would not be feasible due to several reasons. First of all, the layout of the plant (too many locations). Second, due to a high variation in products' size and shape, the size of the pallets can vary from a standard Euro pallet (0.8 × 1.2 m) to long custom-built pallets with the maximum length of 6.2 m. Consequently, it would be complicated to make it compatible with the conveyor belt. Furthermore, the staining station contains not only special but also standard orders that are handled differently. Finally, the backfilling and wrapping stations have to be situated within appropriate distance due to the epoxy and other chemicals used in for the backfilling.

On the other hand, a few rearrangements could be made to improve the overall process. For instance, steps 15–18 could be divided and performed in the early stages of the process. The inspection, foil stretching and pallet placement (steps 15–17) would take place during the transfer from work trolley to a pallet (step 8). Likewise, the foil fixing (step 18) would be done after the fastening products to a pallet (step 10). These changes would allow to eliminate step 12.

5.2 The Fourth Stage: Layout and Transport

Based on the results, transportation is mainly caused by the plant layout, where different working stations are placed in great distances from one another. During this study, the improvements were suggested for the relocation of the working stations. As seen in Fig. 2, the packaging and backfilling areas have been transferred closer to the wrapping area. According to the observations in PAM (Table 1), the distance from the staining plant to the packaging area is 50 m, therefore for a single pallet, 100 m in total (forth and back) would be saved. As a result, the relocation would reduce the extra lead time of unnecessary movement, and, in turn, resolve in savings for transportation and time spent to transport the products. However, the improved layout does not reduce the number of steps necessary to complete the packaging process, nor eliminates any non-adding value or necessary non-adding value activities.

5.3 The Fifth Stage: Elimination

The fifth stage of PAM considers whether each process is essential and what would happen if it was removed. In regards to shipping, there are mandatory compliances to secure goods to pallets properly depending on the shipping method, i.e. air shipping, container, or truck. The first two methods require goods to be both strapped and wrapped, therefore any of the steps of PAM can not be removed from the packaging process. However, wrapping is not mandatory when transporting with trucks. On the one hand, the elimination of the wrapping process would take out the steps 12–20, resulting in cost savings in materials and labor. On the other hand, it would increase the probability of damaging the product during transportation that would result in more frequent customer complaints and returns.

Another improvement that could be considered is process automation. Current packaging processes involve poor working postures, especially in the foiling area, where the forward inclination of head, neck and trunk were observed that can result in muscular health damage. According to [24], poor working postures are physical factors that cause

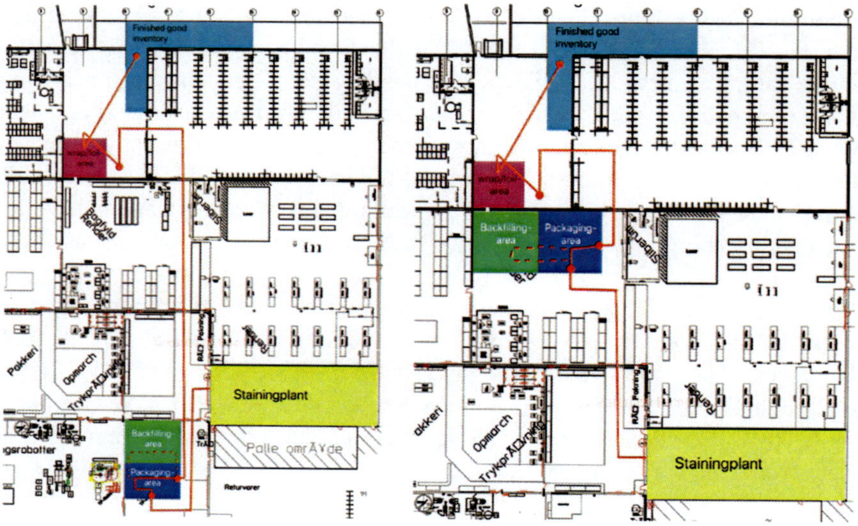

Fig. 2. The layout of the packaging process for special orders: before (left); after (right). Packaging area (dark blue); Backfilling area (green); Wrapping area (purple); Finished good inventory (light blue)

occupational musculoskeletal injuries. Furthermore, the employee might not experience any infirmities over short periods, but prolonged or long-term durations of noise exposure could lead to noise-induced hearing loss [25]. Therefore automation of the process could be a potential solution, especially where it is utilized for ergonomic reasons when technologies take over tasks that may be harmful to perform for humans [26]. Historically, automation has proven to be an efficient way to fulfill a higher need not only for ergonomics but also for product quality, product variation, and cost-efficiency [27, 28]. As a result, a solution of automatic wrapping machine would be able to address the current ergonomic-, human labor-, transport-related concerns, as well as to eliminate necessary not-adding value activities (steps 16–19) and wastes occurring in wrapping/foiling station that are performed manually. An automated horizontal stretch wrapping machine could be adjustable to the current requirements at the company's packaging process.

6 Conclusion

To sum up, the study has investigated the packaging process for a selected special order's tertiary packaging for logistic purposes employing the Process Activity Mapping and direct non-participatory observation exercise. PAM resulted in 22 activities the selected order is involved in. Based on the analysis, only 5 out of 22 steps were identified as value-adding, 6 as necessary non-adding value and the rest as non-adding value. In fact, half of all packaging activities add no value to the product, and thereby to the customer. The majority of non-adding value and necessary non-adding value operations were related to transportation, waiting, inappropriate processing and unnecessary motion. Based on the research results, besides the excessive transport and storage, the most concerning area is the shrink wrapping station, where inappropriate processing provokes unnecessary motion to happen. Overall, the entire shrink wrapping process requires highly intensive manual work that triggers excessive poise exposure and poor working postures. Therefore, it needs urgent attention from the production department in order to avoid irreversible consequences. The suggested wrapping process automation would remove necessary non-adding value activities from the process and contribute to the packaging process optimization.

The purpose of the PAM is to provide suggestions for improvement for a short period of time. Therefore, for future research, it would be beneficial to develop a simulation model in order to test the recommended potential improvement and to measure the profits gained in changing the state of the process. For instance, to carry out a comparison of lead time before and after the packaging process improvement. This study is limited to examine only packaging for special orders, therefore the extent to which the results can be applied for standard orders and to other manufacturing companies is restricted.

To conclude, the identification of waste is an essential activity to be performed in order to understand where and what waste occurs in the production process. Focusing on value-adding activities could be a great strategy for achieving competitive advantage that also would result in created value for customers.

References

1. Zahrotun, N., Taufiq, I.: Lean manufacturing: waste reduction using value stream mapping. In: E3S Web of Conferences, vol. 73, p. 6. EDP Sciences (2018)
2. Amrina, E., Lubis, A.: Minimizing waste using lean manufacturing: a case in cement production. In: 2017 4th International Conference on Industrial Engineering and Applications (ICIEA), pp. 71–75. IEEE (2017)
3. Marriott, B., Garza-Reyes, J.A., Soriano-Meier, H., Antony, J.: An integrated methodology to prioritise improvement initiatives in low volume-high integrity product manufacturing organisations. J. Manuf. Technol. Manag. **24**(2), 197–217 (2013)
4. Shi, J., Shi, X., Li, D., Du, Q.: Risk evaluation for sustainable packaging logistics solution: a quantitative method and case study. In: 2015 International Conference on Transportation Information and Safety (ICTIS), pp. 720–728. IEEE (2015)
5. García-Arca, J., Prado-Prado, J.C., Garrido, A.: Packaging logistics: promoting sustainable efficiency in supply chains. Int. J. Phys. Distrib. Logist. Manag. **44**(4), 325–346 (2014)
6. Hines, P., Rich, N.: The seven value stream mapping tools. Int. J. Oper. Prod. Manag. **17**(1), 46–64 (1997)
7. Chan, F.T.S., Chan, H.K., Choy, K.: A systematic approach to manufacturing packaging logistics. Int. J. Adv. Manuf. Technol. **29**(9), 1088–1101 (2006)
8. Lindh, H., Williams, H., Olsson, A., Wikström, F.: Elucidating the indirect contributions of packaging to sustainable development: a terminology of packaging functions and features. Packag. Technol. Sci. **29**(4–5), 225–246 (2016)
9. Kristiansen, U.S.: Tips for choosing a pallet-wrapping method. Powder Bulk Eng. **17**(5), 37–44 (2003)
10. Verghese, K., Lewis, H.: Environmental innovation in industrial packaging: a supply chain approach. Int. J. Prod. Res. **45**(18–19), 4381–4401 (2007)
11. Rundh, B.: The multi-faceted dimension of packaging: marketing logistic or marketing tool? Br. Food J. **107**(9), 670–684 (2005)

12. Mumani, A., Stone, R.: State of the art of user packaging interaction (UPI). Packag. Technol. Sci. **31**(6), 401–419 (2018)
13. Olsmats C.: Packaging and Supply Chain Management. Position Paper, World Packaging Organization WPO (2017)
14. Singh, J., Cernokus, E., Saha, K., Roy, S.: The effect of stretch wrap prestretch on unitized load containment. Packag. Technol. Sci. **27**(12), 944–961 (2014)
15. Smith, J.D.: The Warehouse Management Handbook. Tompkins Press, Releigh (1998)
16. Amrina, E., Andryan, R.: Assessing wastes in rubber production using lean manufacturing: a case study. In: 2019 IEEE 6th International Conference on Industrial Engineering and Applications (ICIEA), pp. 328–332. IEEE (2019)
17. Hines, P., Rich, N., Esain, A.E.: Value stream mapping: a distribution industry application. Benchmark. Int. J. Marts **6**(1), 60–77 (1999)
18. Ohno, T.: Toyota Production System: Beyond Large-Scale Production. Taylor & Francis Group, Abingdon (1988)
19. Abdulmouti, H.: The role of Kaizen (continuous improvement) in improving companies' performance: a case study. In: 2015 International Conference on Industrial Engineering and Operations Management (IEOM), p. 6. IEEE (2015)
20. Moinuddin, K., Collins, T., Bansal, A.: Process activity mapping-activity-based costing for semiconductor enterprises. J. Cost Manag. **21**(2), 29–33 (2007)
21. Jain, V., Benyoucef, L., Bennett, D., Seth, D., Seth, N., Goel, D.: Application of value stream mapping (VSM) for minimization of wastes in the processing side of supply chain of cottonseed oil industry in Indian context. J. Manuf. Technol. Manag. **19**(4), 529–550 (2008)
22. Zuting, K.R., Mohapatra, P., Daultani, Y., Tiwari, M.K.: A synchronized strategy to minimize vehicle dispatching time: a real example of steel industry. Adv. Manuf. **2**(4), 333–343 (2014). https://doi.org/10.1007/s40436-014-0082-1
23. Karhu, O., Kansi, P., Kuorinka, I.: Correcting working postures in industry: a practical metods for analysis. Appl. Ergon. **8**(4), 199–201 (1977)
24. Lee, T.H., Han, C.S.: Analysis of working postures at a construction site using the OWAS method. Int. J. Occup. Saf. Ergon. **19**(2), 245–250 (2013)
25. Le, T.N., Straatman, L.V., Lea, J., Westerberg, B.: Current insights in noise-induced hearing loss: a literature review of the underlying mechanism, pathophysiology, asymmetry, and management options. J. Otolaryngol.-Head Neck Surgery **46**(1), 15 (2017)
26. Frohm, J.: Levels of Automation in Production Systems. Chalmers University of Technology, Göteborg (2008)
27. Micheler, S., Goh, Y.M., Lohse, N.: Influencing factors for implementing automation in manufacturing businesses–a literature review. In: Proceedings of the 14th International Conference on Manufacturing Research (ICMR), p. 7. Loughborough University, Loughborough (2016)
28. Salmi, A., David, P., Blanco, E., Summers, J.: Deciding the level of automation during the design of assembly systems: literature review of decision methods and a new approach proposal. In: Proceedings of the 45th International Conference on Computers & Industrial Engineering (CIE45), Metz/France, p. 8 (2015)

The Construction of a Crude Oil Tanker Ship: A Life Cycle Assessment Review

Georgios Papantoniou and Eduardo Blanco-Davis(✉)

Liverpool Logistics, Offshore and Marine (LOOM) Research Institute, Liverpool John Moores University, 3 Byrom Street, Liverpool L3 3AF, UK
E.E.BlancoDavis@ljmu.ac.uk

Abstract. Marine shipping encompasses the most significant commercial transportation service around the globe; this last evidences the high sustainability of the sector. The current environmental needs and regulations, which require further reduction of Green House Gas emissions, demand engineers and specialists to take a closer look and action on the subject. This specific project presents a Life Cycle Assessment, analysing significant processes of the construction of a VLCC tanker vessel. More significantly, this paper includes two scenarios of construction; the comparison and the results are discussed. The first construction case consists of the Gas Metal Arc Welding (GMAW) technique, which is commonly used in the industry. The second case showcases Laser Arc-Hybrid Welding (LAHW), a more technologically advanced technique. The analysis also includes the emission rates produced from auxiliary machinery and indoor and outdoor transportation of the material in the shipyard. Results and graphs include the measured Global Warming Potential (GWP) in value of CO_2 equivalent emissions. Using specialised LCA software, a good approximation of the total efficiency and pollution of the two scenarios is revealed. Although the LAHW can produce more accurate welding in less operational time, GMAW has shown to be superior in terms of emissions produced and energy consumption.

Keywords: LCA · Ship construction · Shipyard · Marine engineering ship emissions

1 Introduction

The maritime industry consists of the most exceptional commercial transportation service around the globe, given that it carries 90% of the world's trade [1]. This non-stop service connects all the continents, producing a worldwide commercial network. As the Secretary-General Ban Ki-moon stated: "Everybody in the world benefits from shipping, yet few people realise it" [2].

Due to the environmental damage caused throughout the last centuries, science in cooperation with industries is striving to reduce emissions and wastes. Significant regulations and limits have already been set in different industrial sectors. Albeit the efficiency of maritime is the highest in any form of transportation mode, it comes with an environmental toll.

A. Batako et al. (Eds.): GCMM 2021, LNNS 335, pp. 15–32, 2022.
https://doi.org/10.1007/978-3-030-90532-3_2

Based on scientific surveys and literature, exhaust emissions from ships contribute to the phenomenon of climate change [3]. The use of lower quality fuel may be more economically efficient for shipping companies; however, it is not compatible with the new regulations and the global eco-friendly drive. From 1 January 2020, the limit for sulphur in fuel oil used onboard ships operating outside designated emission control areas is reduced to 0.50% m/m (mass-by-mass) [4].

Furthermore, the design and construction of ship structures are also evolving based on today's reality, forecasting a more environmentally sustainable industry. To obtain the finished product, the vessel needs to go through various phases. The decisions made in the early design phase, will most likely have a significant environmental impact on the vessel's life cycle.

The construction of a marine vessel can be a long and complicated process. Along with the utilisation phase will determine the overall efficiency of the product; more significantly the economic and environmental costs during its lifespan.

In recent years, Life Cycle Assessment (LCA) studies have been introduced to the marine sector [5]. The use of LCAs proved to be a great tool to forecast the impacts of systems, products, and/or services. The methodology includes a detailed analysis of the energy and material used throughout a system or product's lifecycle [6]. It takes into account the processes associated with the extraction of raw materials, the construction, the utilisation and the maintenance of the product, until the final disposal.

The work herein includes a Life Cycle Assessment study of a tanker ship during its construction phase. It aims at presenting major processes that comprise the vessel's construction, and the amount of emissions and waste flows produced at each stage. By using specialised LCA software (GaBi), the efficiency of the different processes is analysed [7]. For this LCA study, environmental efficiency will be determined mainly as a factor related to the production of CO_2.

2 Literature Review

2.1 Global Climate Change

The greenhouse effect is the phenomenon where the atmosphere retains thermal energy and contributes to the rise of temperature of the surface shell. Earth receives energy from the sun in the form of radiation. One part of that radiation is reflected into space. The other part endures in the atmosphere and gets absorbed by the greenhouse gases. The principal gases, which contribute to the greenhouse effect, are water vapour (H_2O), carbon dioxide (CO_2), methane (CH_4), ozone (O_3) and nitrous oxide (N_2O) [8]. These five gases produce a unique energy balance. Any increase or decrease at any of the five concentrations results in affecting this equilibrium. A further consequence is a change in temperature or climate.

The French mathematician-astronomer Josef Fourier (in 1838), and the Swedish chemist Svante Arrhenius (in 1896), were the first whom examined and demonstrated this effect. Arrhenius conceived the theory that the very fast-evolving industries of the 19th century, which emit carbon oxides and other wastes to the atmosphere, have similar climate outcomes with major volcanic eruptions. In early 1896, he published two reports based on the influence of carbonic acid (CO_2) in the air and temperature of the ground [9].

Since 2000, the amount of CO_2 generated by human activities has expanded by 3% annually, commencing Earth's ecosystem rapid climate change [10]. The Intergovernmental Panel on Climate Change (IPCC) prognosticates a mean temperature inclination of 7.5 °F over the following century. The results of a small change in temperature as predicted will be the lengthening of the frost-free season, more droughts and heatwaves, stronger hurricanes, sea-level rise, and the de-icing of the Arctic [11].

2.2 Environmental Impact of Ships

Exhaust emissions from ships create carbon dioxide (CO_2), nitrogen oxides (NOx), carbon monoxide (CO), sulphur dioxide (SO_2), volatile organic compounds (VOC), black carbon (BC), and particular organic matter (POM). All these aerial emissions can easily be transferred worldwide through the atmosphere and the sea, and from the sea to the coasts. The global marine fleet is responsible for 2–3% of CO_2, 10–15% of NOx and 4–9% of SOx total anthropogenic global emissions [12]. These types of emissions are related to premature death, health and breathing problems, acid rain, and climate change [13].

The most common kind of engine used in commercial shipping is the marine diesel engine. The required energy is produced by internal combustion. Even in the most efficient combustion cases, hydrocarbon fuels will emit some form of carbon compound, i.e. CO_2 or CO. Marine diesel engines offer powerful and high torque propulsion drives, which make for a highly efficient propulsion option to sizeable cargo laden vessels.

Despite the conspicuous exhaust emissions from the operation, ships raise further environmental concerns in other departments. Although not always directly visible, the construction and repair of vessels also produce an additional environmental impact.

Shipyards are large spaces, which are open to the atmosphere and usually have direct access to river or coastal water. The potential aftermath of construction and repair processes is that, many hazardous deposits can easily be transferred to air and waterways. The processing of raw materials and the different shipbuilding phases anticipate improvements as to sustainable energy. Also, the fact that the most prominent shipbuilding industries are located in countries where environmental regulations are significantly loose, cannot go unnoticed. Environmental impact of shipbuilding is not easily detected. Nevertheless, in a life cycle analysis of a vessel, environmental consequences and waste can be recognised through all of the construction processes.

2.3 Life Cycle Assessment

The Life Cycle Assessment is a methodological tool, which provides us with detailed information about a specific product during its lifespan. This cradle to grave analysis was initially introduced in the 70s and made an essential effect on the preliminary design idea and production of products. It contains data as the full inventory collection of the raw materials, from the production until the final disposal of them. The life cycle analysis can give a valuation of all the environmentally related problems at different life phases [14].

The introduction of LCA in modern society proved very valuable. The analyses showed that logical thinking does not always offer the most sustainable solutions. For example, the paper over plastic bags was not that apparent superior in terms of using less energy and materials.

2.4 Life Cycle Assessment in the Maritime Industry

LCA is a tool that became more popular in the maritime sector during recent years. As described earlier, the need for a more environmentally sustainable solution has also affected the shipping industry. In the last years, many attempts were made in order to measure the sustainability of these vast structures that carry most of the total commercial trade.

The work herein is based on the LCA methodology. Other sources explaining the methodology with detail include Baumann and Tillman, Curran, and Hauschild et al. [15–17]. Additionally, more specific LCAs to the shipping industry have been tackled by Blanco-Davis and Zhou [3], Bengtsson [18], Fet et al. [19], Blanco-Davis et al. [20] and Shama among others [21].

Shama's LCA of ships is addressed to clarify energy consumption and environmental impacts specific to the industry. Particular emphasis is placed on the methods commonly used to reduce energy consumption and environmental effects over the three stages of ship construction, ship operation, and ship scrapping (see Fig. 1). The rationalisation of materials used in shipbuilding and ship scrapping is emphasised while a definite waste management hierarchy is presented [21].

The encompassing nature of lifecycle thinking constitutes one of the main obstacles for the implementation of the concept into the industry, as described in the Organisation for Economic Co-operation and Development (OECD) report of Environmental and Climate Change Issues in the Shipbuilding Industry [22]. This last is due to the methodology insisting that companies look beyond conventional organisational boundaries when considering the environmental impacts of their activities. On the same OECD report, it is suggested that the shipbuilding industry should begin taking action on climate change issues by moving toward a lifecycle environmental approach. The entry point for the industry would be the adoption of the carbon footprint concept. This last would include the greenhouse gas emissions that can be measured and accounted for in terms of the ship, the shipyard, and the shipbuilding industry as a whole [22].

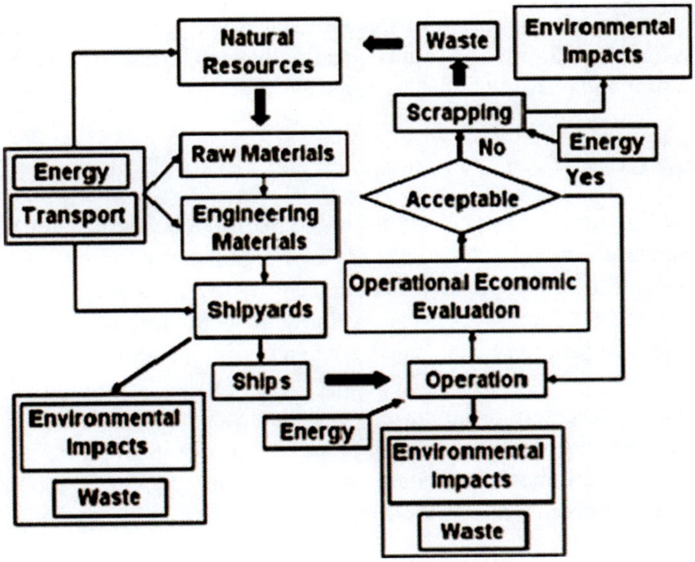

Fig. 1. Ship's life cycle [9]

3 Adopted Approach

3.1 Methodology

The main scope of this study is to find the most environmentally efficient construction techniques and processes for a crude oil VLCC tanker ship. This LCA will gather and analyse data from emissions produced in the construction phase, giving particular importance to CO_2 production. Use of a specialised LCA software has been made to quantify energy inputs and waste outputs.

As far as this report is concerned, two cases will be further analysed and discussed. Both examples include the construction of the same VLCC tanker vessel. The difference between the two cases is on the welding processes that are used. As described in the literature review section, the LCA is conducted for the lifespan of a product service or even for a specific process. An example showing the fundamental concept of LCA can be seen in Fig. 2.

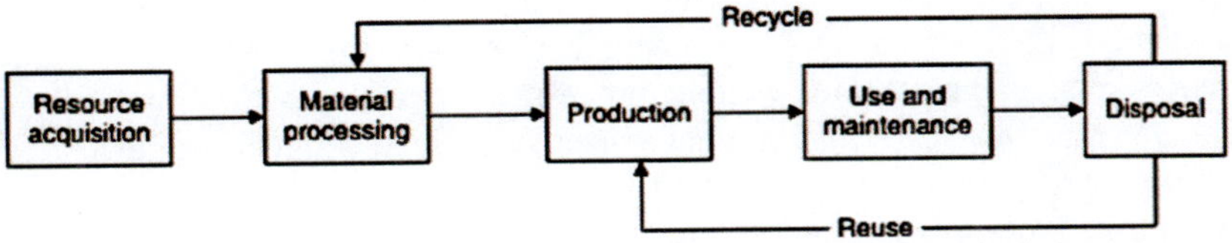

Fig. 2. Generic lifecycle stages (arrows represent transportation) [23].

The basic concept of an LCA includes the goal and scope of the study, the lifecycle inventory, the impact analysis on the environment and the final discussion of the results

[6]. After compiling the inventory of the assessment and all input and output data, a thorough discussion follows. The results acquired have to be analysed in terms of their weight of impact and possible future outcomes.

3.2 The Life Cycle Model of a Ship

Regardless of the limited lifecycle studies in the maritime industry, compared to other industrial sectors, the methodology does not vary significantly for the case of a tanker ship. The lifespan of a VLCC depends on many factors, but the average period of operation is about 20–25 years [24]. That period starts with the design and construction, continues with the operation/maintenance, and ends with the scrapping of the ship [14, 19].

Taking into consideration the current environmental problematic, it is acknowledged that the operational phase of a ship produces most of its lifecycle emissions [3]. The International Maritime Organization and other environmental organizations report the importance of operational pollution and try to decrease it by updating the related regulations.

Construction and scrapping comprise the beginning and the end of the utilization period, respectively. OECD, with its report of 2010, highlights the large number of processes happening in shipyards that have a significant environmental impact. The extraction of the different required raw materials, along with their manufacturing processes, are also a part of the design/construction phase of a ship's lifecycle, and should therefore, be included in the assessment. Moreover, the recycling or disposal, after the dismantling, also consist of sub phases that have to be taken into consideration. Finally, yet importantly, another factor taken into account should be the transportation of the material between the different phases of the lifecycle. An example would be the transfer of resources and supplies to the construction site or shipyard.

3.3 Designing the Ship's LCA Model (Cradle to Gate)

The LCA study is a cradle-to-gate case, and it includes only the construction phase of the selected VLCC vessel. Energy consumption and emission production are monitored for specific processes. These are mainly at the delivery of the raw materials and supplies, the transportation system within the shipyard, and some of the welding processes. The reason behind this initial assumption is because these processes are found to be repeated often for the construction of various vessels –and therefore a common denominator–, whereas one of the main objectives of the study is to gauge the potential environmental difference amongst the specified welding processes.

The construction phase may not have been researched as much as other lifecycle phases in existing literature, but consists of a very crucial stage as it will encompass major manufacturing processes such as the material cutting, forming, welding and transporting processes. All of these are operations producing environmental emissions [3]. Construction is also one of the ship's lifecycle sub phases that is significantly energy-intensive. Most of the energy consumed is covered by electricity, the production of which has its own environmental penalties.

Additionally, a large number of toxic materials, fumes, and fluids are also produced daily [22]. Due to the large scale of these vessels, some of the construction activities cannot be accomplished in indoor spaces. The activities in ship repair and new-buildings with the greatest direct environmental concern according to OECD include: the metal cutting, grinding and welding, the surface treatment operations, and the noise [22].

The lifecycle of a ship, based on the terminology, would also contain the individual lifecycle of each material that is used for the accomplishment of the construction. VLCC vessels are constructed by several materials, which provide the structural integrity needed for open sea conditions. However, among these, the primary material required is steel. The hull, machinery, and outfitting equipment are some of the steel parts that cover the major percentage of the ship's total mass. By adopting the idea that the vessel is mainly a steel structure, this specific LCA could be aimed at disregarding materials found in fewer percentages than steel, i.e. plastics, composites, wood, and etcetera. Eventually, the main flow of the total steel processing can be introduced in the modelling.

3.4 Shipyard Design and Layout

The shipbuilding area consists of a substantial investment, which takes a long time to plan and build. One of the early stages of the construction is the design of the layout, which can be adjusted to suit different types of ships. This step contains the final position of each compartment. The stage of the pre-construction design is crucial as it directly relates the total efficiency of the shipyard with regards to the timeline of any construction project.

Studies, such as Song and Woo [25] have discussed and analysed different yard layouts. The initial survey associated with a facility and/or factory layout has progressed because of the request for maximising the production efficiency of large-scale automated plants. The presented layout model has been introduced by the author and consists of a prototype. However, the average layout and size of the facilities are based on measurements of existing medium-large size shipyards using the Google Maps tool. The data on transportation (cranes and transporters) is also based on existing industry machinery (engines, payloads, and speeds).

Figure 3 presents the layout of the shipyard used for the work herein. The plan includes eleven main areas. At these compartments, the main sub phases of the vessel's construction take place. Table 1 includes the dimensions of each section-area. The components of the shipyard are in line with the building process:

1. The first area is the Delivery dock, where, the delivered materials, in this instance grade steel plates, are discharged from a cargo ship. This section encompasses the dock, a Tier 1 (T1) plate-unloading crane, as well as a heavy load, ground transporter. The crane stacks the transporter, which will take the steel plates to the next field.
2. The second region is the steel Plate stock. This department is where the unloaded plates are stocked until they are needed by the manufacturing line and finally, the ship assembly. This indoor space encompasses two plate handling (T2), ceiling cranes, which assist in transferring the plates throughout the available space.

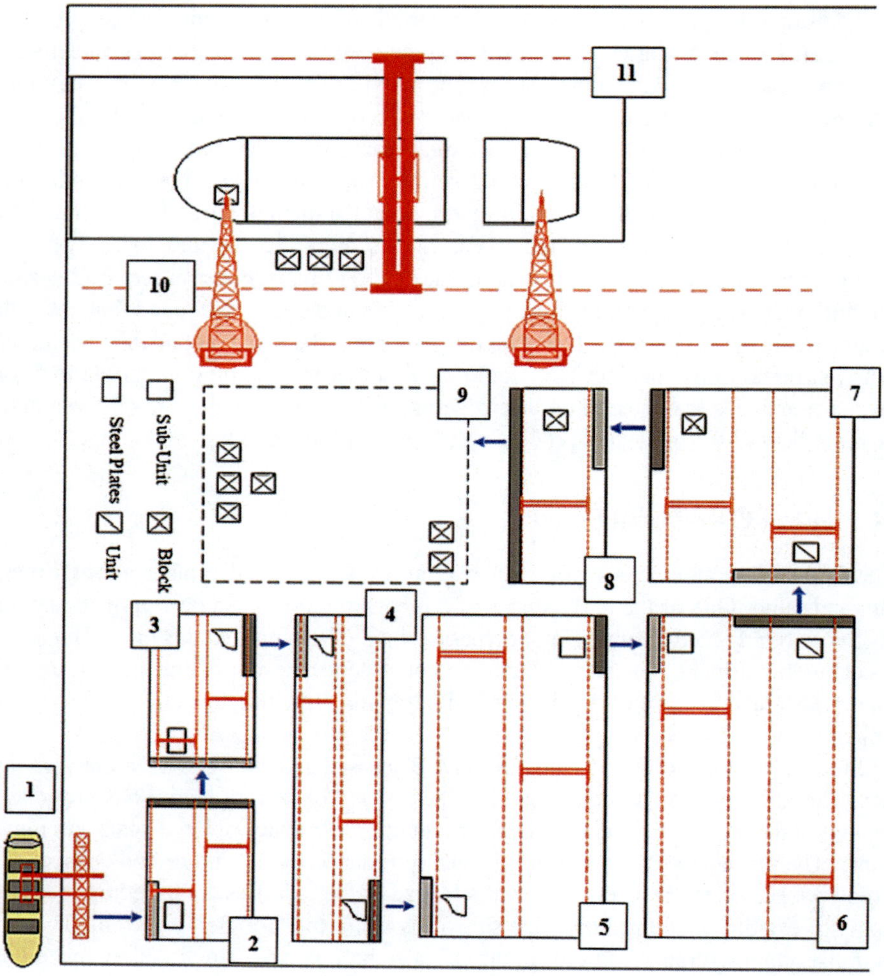

Fig. 3. Shipyard layout

3. The next location in line is the plate Cutting facility. It is linked with the beginning of the steel material modification. In this infrastructure, automatic cutting machines are fed steel plates from the stock. The plates are delivered to the cutting facility by a transporter, and then positioned out by two plate handling (T2) cranes.
4. The following indoor area is the Fabrication facility. Different machinery will add different geometry to the cut plates. The forming machines are loaded with the cut plates by two plate handling (T2) cranes. The now bent, or differently shaped plates are then transferred to the next facility.
5. The fifth, more largely space is the Sub unit assembly. It consists of the start of the leading manufacture line. That is observed because processes such as grinding and welding take place. The unit contains two block handling (T3) cranes. The

products of this process, sub units, have higher mass and justify the significant weight capacity of these cranes.

6. The subsequent stage is the Unit assembly. This indoor space's purpose is the construction of the ship's units, and it is usually one of the most space abundant sections in the yard. This facility further contains two block handling cranes, which can transport heavy materials to different welding spaces, and units once they are formed.
7. Succeeding the unit construction stands the Block assembly facility, equipped with two block handling cranes. The blocks created from the units after being welded will be carried to the next space by transporters. These are capable of lifting significant structural weight components.
8. The Paint shop is located after the Block assembly, and contains one block handling crane for arranging the constructed blocks into position for suitable paint operations. Moreover, it is the final indoor facility for this proposed shipyard model.
9. The Outfitting area is usually an outdoor space where all other systems, machinery and pipping, are added to the painted blocks.
10. The Pre-Erection site is a space where the constructed blocks are placed before the final erection. Within this section, two single boom (T4) cranes (triangular, thin beam structures) and one goliath gantry (T5) crane (thick double beam structure) operate on rail tracks (between label 9,10 and 11). This massive machinery is used for the erection and positioning of the blocks within the Dry-Dock (DD) assembly. Moreover, one of the two single boom cranes is further used for the transportation of the outfitted blocks to the Pre-Erection site.

Table 1. Dimensions of compartments.

No	Area	Dimensions (m)
1	Delivery dock	N/A
2	Stock	100 × 70
3	Cutting	80 × 70
4	Forming	80 × 200
5	Sub-assembly	300 × 200
6	Unit-assembly	400 × 200
7	Block-assembly	250 × 250
8	Paint-shop	200 × 80
9	Outfitting	260 × 300
10	Pre-erection	N/A
11	Dry-dock	180 × 350

3.5 Transportation System

For the production of this VLCC vessel, material follows an assumed route through the available transportation system. The route includes cranes located in the facilities and ground transporters, covering the distances between the facilities. During the first phase, the steel plates are unloaded by a Plate Handling crane and more significantly a T1 crane. The cranes are named according to their type. Table 2 illustrates the types of cranes and their particulars.

Table 2. Particulars of crane types.

Crane tier	Payload	No. of cranes	Consumption (l/h)
T1	75 t	1	94
T2	30 t	6	42.6
T3	350 t	6	420
T4	350 t	2	420
T5	2800 t	1	500

On the other hand, only one type of transporter is used in the shipyard. Its consumption is about 64 L per hour and its payload capability at 216 tons. The distance that they cover is represented by the blue arrows, in Fig. 3. The distance from each site to another is 60 m while only the first route (from facility 1 to 2) is 40 m. It is essential to highlight that transportation after the block assembly is different. The block products have large masses, which sometimes are over the payload limits of the transporters.

For this reason, two combined ground transporters are used in order to transport the blocks. Moreover, the blocks located at the outfitting section are transferred to the PE site by one of the single boom cranes.

The transporter operates in a single trip from point A –exit point of the previous process, to point B– entrance point of next process. On the other hand, the crane T1, T2, T3 and T5 operate a lifting motion followed by a gantry ground motion and finally, a descending motion. The T4 crane operates the same action but instead of a gantry ground motion, executes a circular movement. In Fig. 3, the cranes and their axis of movement is represented in red. All machinery is operated at close to maximum payload.

3.6 Welding Processes

The steel plates used for the construction of the vessel, proceed through welding procedures. The main welding processes occur at the phase of subassembly, unit assembly, and block assembly. Secondary welding also occurs at the stages of outfitting, PE and final assembly. The instalment of the pipping arrangement and machinery systems also require welding in order to secure their position. Furthermore, at the PE stage, blocks are welded together, creating modules which will be directly transferred and positioned within the dry dock.

For all previously referred processes, an approximate percentage of the additional material has been introduced. The two scenarios of this analysis included welding procedures for the Gas Metal Arc Welding (GMAW) and the Automatic Laser Hybrid-Arc Welding (LAHW).

Sproesser [26] illustrated a comparison of the welding process for a thick metal plate by conducting an LCA. The LAHW proved to be the most efficient process due to the high performance in tandem with low overall weld volume.

The GMAW includes a metal arc and a consumable electrode with internally added shielding gas. It belongs to the category of wire feed welding. The system consists of a welding gun, the electrode, an electrode feed unit, the shielding gas, cables and hoses, and the power supply. The gun contains two hoses with the electrode feed and the shielding gas and an electrical current conductor. The shielding gas is a mixture of CO_2 and argon (Ar) at a capacity of 18% and 82%, respectively [26].

This welding process is widely used in the industry as it offers large scales of flexibility when automated technologies apply it. The GMAW used for this construction operates with spray arc transfer. This technique achieves significant higher deposition rates [26].

The LAHW was introduced by W. Steen of the Imperial College of Science and Technology in the 1970s [27]. In comparison with other GMAW, LAHW attains higher welding speeds, effectively reducing the construction time. Another essential feature is the accuracy of that method. The weld is produced in excellent form with a minimal number of passes and less material consumption. These two characteristics lead to resource savings, a fact of great importance for lifecycle utilization.

The laser beam is used for the first interaction with the surface, as it can achieve deep penetration or otherwise known as 'keyholes' into the base material. Following the laser beam, a GMAW torch is used. The structure of which was described earlier. It provides the melting material while giving the final composition of the weld. The weld cools down and finally stabilises.

In the specific report, only the consumption of the techniques was taken into account. The utilized material and energy usage and emission production data from Sproesser [26] were taken into account for the calculations. The main power feed of the welding machinery is electricity. For this reason, the emissions produced are linked with the amount of electrical power produced and provided to processes.

3.7 Global Warming Potential

One aspect of this report consists the characterization of the emissions produced as for their Global Warming Potential (GWP) in 100 years. This impact category is measured in kilograms of CO_2. More significantly, each emission potential contribution in the global warming effect, is defined in equivalent units of the same substance; carbon dioxide (CO_2eq.) In addition, the residence time of these emissions to the atmosphere are also considered to the calculation, in this case for 100 years. In cases where one substance as CO_2 is produced more than others in quantity, it may have less effect in the pollution phenomenon. For example, 1 kg of CO is equivalent with 3 kg of CO_2 emission. More significant information regarding the GWP factors can accessed at the review of Blanco-Davis and Zhou [28], JRC [29] and IPCC [30].

4 The Flow Cases

The study contains two scenarios for the construction. For each case, a basic flow of the procedure is used, exactly as described at the construction line reference. All processes have the same inputs and outputs except those that include welding procedures (and are therefore, a common denominator of both).

The first case includes the LAHW processes. The laser welding technique is used for the assembly of the steel plates while for the second flow case, the GMAW process is used. The flow of the construction consists of the main, the auxiliary, and the secondary phases. More specifically, the main process is a welding process, an auxiliary process is a transporter or crane, and a secondary process could be the supply of electrical power or fuel.

The inputs and outputs of each process have been set following the calculations made based on the consumption and payload capabilities of each machinery. The estimations of the inputs-outputs of LAHW and GMAW are based on the Sproesser et al. report [26]. Their welding experiments of thick steel plates introduced approximate values such as the energy consumption (kWh) per specific mass of wire feed (g). Based on the same logic, information of the welding fumes production has also been attained.

Included in the flow of the construction of both cases, several mass addition processes exist. Systems such as the pipping arrangement and the machinery installations are includes in the entire steel structure. The approximate mass of these systems is presented in Table 3. The information used for the calculation of these values is based on Papanikolaou [31]. Further values for the estimation of the welded mass can be accessed in Table A.1.

Table 3. Weight of additional systems.

Deadweight (DWT)	Lightweight (W_L)	Outfitting (W_{OT})	Machinery (M_W)
317,000 t	54,112 t	5,950 t	3,576 t

5 Results and Discussion

The results of an LCA can provide a good approximation, not only for the specific project but also for future scenarios. As described in previous sections, the difference between the two cases relies on the welding processes. The data presented in Fig. 4 compiles the emission production for the auxiliary processes. The results in Fig. 5 include emission production from the welding processes.

The first diagram of Fig. 4(a) presents the GWP in 100 years from all crane systems used in the shipyard. The results are the outcome of the specific routes in which they operate, and the subjected payload. The global warming emissions are presented in equivalent units of CO_2. Clearly, the T5 crane produces the majority of the GWP rates. Following, the next significant amount is produced by T4 cranes. Both types of cranes operate in the erection process. The mass needed to be carried out to the dry dock is the maximum related to the other sub phases, a fact which evidences the high energy consumption of these cranes, and subsequently their high emission production.

The ground transporters results, representing the GWP (CO_2 eq.), are presented in Fig. 4(b). Similar to the cranes, more mass is added after each phase. In other words, each transporter produces more emissions than the previous.

Moreover, the carrier (to stock) contains the lowest CO_2 mass because the actual distance covered from the unloading dock to the stock of the steel plates was set to 40 m, in controversy with the 60 m between the other sites.

The first diagram of Fig. 5 represents the Global Warming Potential in 100 years of the laser welding processes. The main power feed of the welding machinery is electricity. For this reason, the emissions produced are linked with the amount of electrical power produced and provided to the processes.

As illustrated by the graph, the electricity feed of the subassembly process includes the highest amount of emissions. The subassembly is the first phase of the construction line containing welding. The delivered material, in this case, the steel plates, goes through welding procedures in order to shape sub-units needed for the next phase. The large surfaces of the plates consume a great mass of welded steel that drives electricity consumption to the highest rates.

The emissions produced from each GMAW in Fig. 5(b), follow the same weight sequence as at the laser welding technique. The first stages of the construction require more welding, concluding to more electricity consumption. As can be observed in Fig. 5(a), the LAHW method has a greater amount of total emissions than GMAW. The laser beam consumes much more electricity, and that is translated to emission generation from the production of the supplied energy.

Additional remarks were made for the production of welding fumes from both processes. Based on Sproesser et al. [26] the laser welding process produces fewer fumes than GMAW. This fact is valid for all phases of the construction line containing welding (Fig. 6). The welding fume has not been defined for its substances but consists of emission with great environmental pollution potential. The particulars of its impact are directly related to the welding material. More significantly, it consists of vapours and particles of the feed material. Most of the steels used in the maritime industry are alloys of carbon steel. The primary type of carbon steel consists of carbon, chromium, molybdenum, iron and silicon [32].

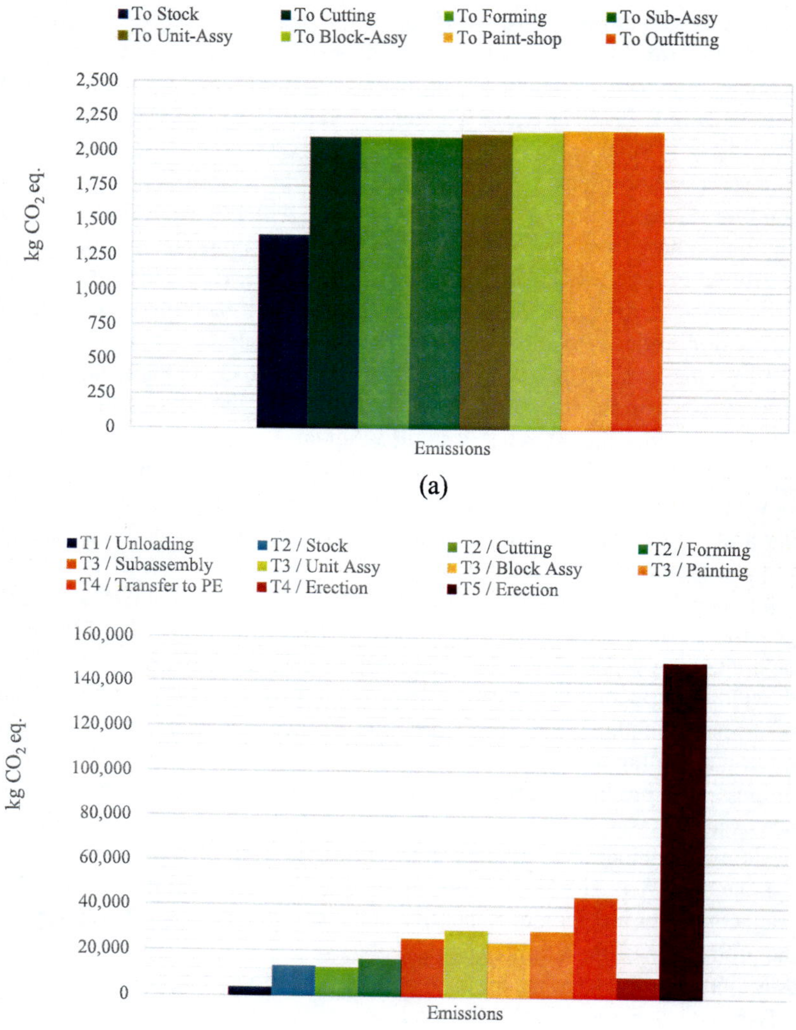

Fig. 4. Cranes (a) and transporters (b) – Global Warming Potential 100 years.

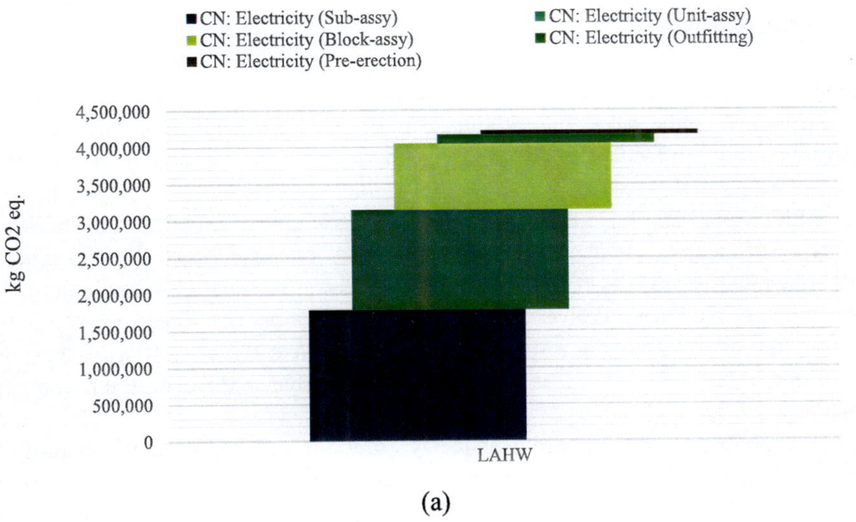

Fig. 5. LAHW (a) and GMAW (b) welding emissions – GWP 100 years.

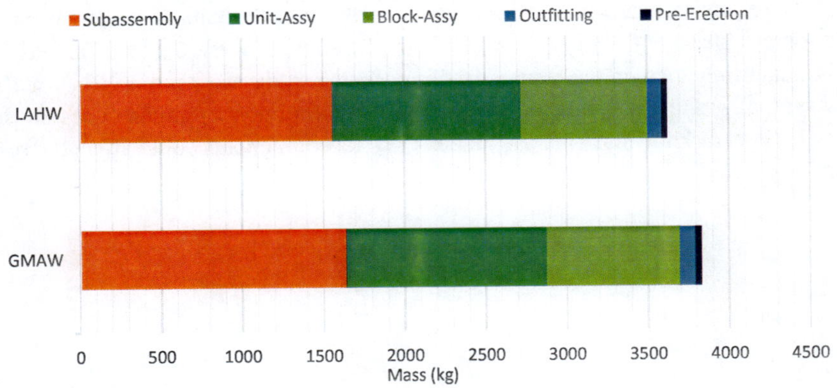

Fig. 6. Production of welding fumes

6 Conclusion

The general data-result, including the transportation and lifting machinery, give an approximation of the emissions produced and more specifically, the carbon dioxide equivalents through the GWP. The cranes with the most considerable payload capabilities operate in lower speeds, but with heavier loads and require a corresponding high consumption compared to all other cranes. It is essential to highlight that these specific types of cranes include a basic industrial diesel generator. More significant studies have shown that more advanced power generating systems, such as hybrid-electric, can reduce the emissions and deliver the required amount of energy. These approximations are important for future purposes, as a way to improve the re-design of the layout and/or operations, when it comes to finding ways to enhance the overall environmental efficiency.

The specific case of shipyard uses heavy ground transporters. Their flexibility and heavy load capabilities are the most significant advantages for the connection of material flow through the construction process. As only one type of transporter was used for this scenario, the emission production was about the same for all operations.

The two similar welding processes, the LAHW and the GMAW showed some exciting results. Although many studies have proved that laser welding is one of the most efficient welding techniques, for this specific project, the outcome illustrates the opposite. It is quite clear that laser welding produces fewer fumes than GMAW. This last indicates the environmental efficiency of the process, but more variables should be considered for a more wholesome utilisation. The energy required for LAHW is greater in terms of electricity used than that required for the alternative method, GMAW. If non-renewable resources produce the electricity used for this process, then the sustainability factor changes. In that case, laser welding produces more emissions than gas metal arc welding.

In future research work, more precise processes inputs could lead to more accurate results. The results and conclusion of this study consist of a product based on the specific type of shipyard and shipbuilding process. A factor of error could exist, as the cranes and transporters carried loads near to their maximum capabilities, in order to operate the least routes possible, leading to fewer emissions. Moreover, other auxiliary processes which were not included may offer better insights into the energy consumption and emission production of the system as a whole. Last but not least, the basic material has a significant influence on the results. In this case, the study was accomplished with the basic type of grade steel. In the maritime industry, the most common types are from AISI 316 steel or known as "marine grade" [33]. Its components could affect the results differently.

Appendix A

Table A.1. Weld mass considered for the calculations.

Phase	Product steel (t)	Weld	Weld (t)
Sub-assy	40071	1.00%	400.71
Unit assy	40374	0.75%	302.805
Block assy	40577	0.50%	202.885
Outfitting	46527	0.05%	23.2635
PE	46527	0.02%	9.3054
Erection	50103	0.05%	25.0515

References

1. International Chamber of Shipping: ICS-Shipping and World Trade. Ics-shipping.org. (2017). Accessed 23 Jan 2020
2. United Nations: Maritime Transport Is 'Backbone of Global Trade and the Global Economy', Says SecretaryGeneral in Message for International Day. Meetings Coverage and Press Releases. Un.org. (2016). Accessed 5 Oct 2019
3. Blanco-Davis, E., Zhou, P.: Life Cycle Assessment as a complementary utility to regulatory measures of shipping energy efficiency. Ocean Eng. **128**, 94–104 (2016). Accessed 24 Oct 2019
4. IMO: Sulphur 2020 – cutting sulphur oxide emissions. Imo.org. International Maritime Organisation (2019)
5. Chatzinikolaou, S., Ventikos, N.: Applications of life cycle assessment in shipping. In: 2nd International Symposium on Naval Architecture and Maritime, Istanbul, Turkey (2014)
6. ISO 14044:2006: Environmental management - Life cycle assessment - Requirements and guidelines. International Standard (2006)
7. Thinkstep: Life Cycle Assessment LCA Software (2020). www.gabi-software.com
8. US EPA: Overview of Greenhouse Gases. US EPA. United States Environmental Protection Agency (2018). Accessed 10 Apr 2020
9. Crawford, E.: Royal Swedish Academy of Sciences Arrhenius' 1896 model of the greenhouse effect in context. Source Ambio **26**(1), 6–11 (1997). Accessed 24 Oct 2019
10. Woolf, D., Amonette, J.E., Street-Perrott, F.A., Lehmann, J., Joseph, S.: Sustainable biochar to mitigate global climate change. Nature Commun. **1**(1) (2010). Accessed 24 Oct 2019
11. Jackson, R.: Global Climate Change: Effects. Climate Change: Vital Signs of the Planet (2018). Accessed 4 Dec 2018
12. Endresen, Ø., Eide, M.: The environmental impacts of increased international maritime shipping - Past trends and future perspectives [Internet] (2008). Accessed 24 Oct 2019
13. Nunez, C.: Greenhouse gases, explained. Nationalgeographic.com (2019). Accessed 2020
14. Blanco-Davis, E., Zhou, P.: LCA as a tool to aid in the selection of retrofitting alternatives. Ocean Eng. **77**, 33–41 (2014)
15. Curran, M.A.: A Brief History of Life-Cycle Assessment, May 2006

16. Baumann, H., Tillman, A.-M.: The Hitch Hiker's Guide to LCA. Studentlitteratur, Lund (2004)
17. Hauschild, M.Z., Rosenbaum, R.K., Olsen, S.I. (eds.): Life Cycle Assessment. Gewerbestrasse 11, vol. 6330. Springer, Cham (2018). https://doi.org/10.1007/978-3-319-56475-3
18. Bengtsson, S.: Life Cycle Assessment of Present and Future Marine Fuels. [Department of Shipping and Marine Technology, Chalmers University of Technology, Gothenburg, Sweden] (2011)
19. Fet, A.M., Aspen, D.M., Ellingsen, H.: Systems engineering as a holistic approach to life cycle designs. Ocean Eng. **62** (2013)
20. Blanco-Davis, E., del Castillo, F., Zhou, P.: Fouling release coating application as an environmentally efficient retrofit: a case study of a ferry-type ship. Int. J. Life Cycle Assess. **19**(10), 1705–1715 (2014). https://doi.org/10.1007/s11367-014-0780-8
21. Shama, M.A.: Life Cycle Assessment of Ships. Maritime Transportation and Exploitation of Ocean and Coastal Resources. Taylor & Francis Group, London (2005). Accessed 28 Feb 2020
22. OECD: Environmental and Climate Change Issues in the Shipbuilding Industry. OECD Council Working Party in Shipbuilding (WP6), p. 10–2, November 2010
23. Curran, M.A.: Life Cycle Assessment - an overview | ScienceDirect Topics. Sciencedirect.com (2017). Accessed 23 Aug 2019
24. Kim, D.K., et al.: Effect of corrosion on the ultimate strength of double hull oil tankers - Part I: stiffened panels. Struct. Eng. Mech. **42**(4), 507–530 (2012)
25. Song, Y.J., Woo, J.H.: New shipyard layout design for the preliminary phase & case study for the green field project. Int. J. Naval Archit. Ocean Eng. **5**, 132–46 (2013)
26. Sproesser, G., Chang, Y.-J., Pittner, A., Finkbeiner, M., Rethmeier, M.: Life Cycle Assessment of welding technologies for thick metal plate welds. J. Cleaner Prod. **108**, 46–53 (2015). Accessed 3 Mar 2020
27. Acherjee, B.: Hybrid laser arc welding: state-of-art review. Opt. Laser Technol. **99**, 60–71 (2018). Accessed 5 Jan 2020
28. Blanco-Davis, E., Zhou, P.: Life Cycle Assessment as a complementary utility to regulatory measures of shipping energy efficiency. Ocean Eng. **128**, 94–104 (2016)
29. JRC: Framework and requirements for Life Cycle Impact Assessment models and indicators, ILCD Handbook, 1st edn. European Commission, Joint Research Centre, Institute for Environment and Sustainability, Italy (2010)
30. IPCC 2007: Direct Global Warming Potentials, IPCC Fourth Assessment Report: Climate Change (2007)
31. Papanikolaou, A.: Ship Design. Springer, Dordrecht (2016). https://doi.org/10.1007/978-94-017-8751-2
32. Michaud, D.: Iron & Steel Composition. Mineral Processing & Metallurgy (2016). Accessed 13 Mar 2020
33. Nový, F., Zatkalíková, V., Bokůvka, O., Miková, K.: Gigacycle fatigue endurance of marine grade stainless steels with corrosion pits. Period. Polytech. Transp. Eng. **41**(2), 99 (2013)

Evaluation of Palm Oil-Nano Particle Cutting Fluids on Turning of Inconel 718

Sit Lu Fung Derek[1], Ezutah Udoncy Olugu[1](✉), Cik Suhana Binti Hassan[1], V. Karunamoothei[2], and Kah Hou Teng[1,3]

[1] Faculty of Engineering, Technology and Built Environment, UCSI University, Kuala Lumpur, Malaysia
olugu@ucsiuniversity.edu.my
[2] Clean Growth UK, Faculty of Engineering and Technology, Liverpool John Moores University, Liverpool L3 3AF, UK
[3] Built Environment and Sustainable Technology Institution (BEST), Faculty of Engineering and Technology, Liverpool John Moores University, Liverpool L3 3AF, UK

Abstract. It is an established fact that mineral based cutting fluids facilitates the quality of machining operation. Nevertheless, the negative impacts inherent in these conventional cutting fluids such as environmental pollution and adverse effects on operators' health have become major source of concern. Sequel to these concerns, there has been continuous effort at finding alternatives to these mineral based cutting fluids. In this study, diamond, copper oxide and iron oxide were selected as additives to virgin palm oil as nano-based cutting fluids applied in the turning of Inconel 718 under minimum quantity lubrication (MQL) method. The performance of each nano-based cutting fluids was evaluated in terms of machining characteristics such as cutting temperature, chip morphology and chip reduction coefficient as well as tool wear. The results obtained from the experiment were statistically analyzed using Signal to Noise (S/N) ratio technique. S/N ratio was used to determine which cutting parameter had significant effect on machining characteristics during the turning operation. The results showed that cutting speed was the most influential control factor on cutting temperature while feed rate was the most influential control factor on chip reduction coefficient. The results also indicated that the nano-diamond based palm oil outperformed nano-copper oxide and nano-iron oxide-based palm oil in terms of cutting temperature, chip reduction coefficient as well as tool wear. It is expected that these findings would contribute to the industrial application of nanosized sustainable lubricants. The study culminates with recommendations for further studies.

Keywords: Inconel 718 · Lathe turning · Cutting fluid · Vegetable oil · Nano-particles

1 Introduction

Machining is one of the most vital manufacturing processes involving material removal aimed at producing desired parts shape or component. As a result of the friction generated due to the relative motion between the work piece and the cutting tool, high

A. Batako et al. (Eds.): GCMM 2021, LNNS 335, pp. 33–45, 2022.
https://doi.org/10.1007/978-3-030-90532-3_3

temperature is generated which negatively affects the cutting tool and work piece material. Conventional cutting fluids are well applied in the manufacturing industry to reduce the friction and heat generated during the machining process especially on the shear zones of tool-workpiece interface. It is estimated that manufacturing companies consume approximately 38 million metric tonnes of lubricants in machining per annum [1]. This amount of lubricants is projected to rise every 10 years at a rate of 1.2% [2]. Due to the complex composition of these cutting fluids, it has further been established that their use has some negative effects on the environment and human health during its use as well as disposal [3]. According to Mannekote et. al. [4] and Ozcelik et al. [5], it is estimated that about 50% of the mineral based lubricants enter the environment due to volatility, spills, and improper disposal. It is further asserted that 95% of lost lubricants are mineral based lubricants which are in most cases not biodegradable, which may end up polluting waters and causing other environmental problems [6, 7].

Inconel 718 has gained recognition for its applications in aircraft gas turbine, nuclear power plant, high temperature fasteners and reciprocating engines such as exhaust valves hot plugs, turbochargers and valve seat inserts [8]. In fact, nickel-based superalloy is known as difficult-to-cut materials as compared to other materials. This is due to the low thermal conductivity of nickel-based alloys where their cutting temperature will be elevated up to 1200 °C [9]. This will lead to high tool wear on the cutting insert. Furthermore, the high strength of Inconel 718 will require high cutting forces and hence vibrations will be generated which will affect the tool, workpiece and machine tool during machining operation. In order to enhance the quality of the process and products, extreme precaution must be taken in the selection of a proper tooling and coating materials, method of machining, types of cutting fluids, and cutting parameters to be controlled to prolong the tool life [10].

The machining of difficult-to-cut materials requires the application of cutting fluids to improve the quality of machining process. Cutting fluids act as coolants to dissipate the heat produced between workpiece and tool chip [11]. This will increase the lifespan of the tool. It also acts as a cushion between the workpiece and tool as a layer of lubrication is provided [12]. Vegetable oil has been identified as the sustainable alternative to mineral based cutting fluid [5, 6].

One of the plausible approaches to improve the sustainability and efficiency of these vegetable oil is to apply MQL on nanoparticles-based vegetable oil as cutting fluids [13–15]. Various studies have demonstrated that nanotechnology developments signal advent of new cutting fluid for manufacturing processes [1]. Examples include Tran Minh Duc et al. [16] in the evaluation of MQL method by using aluminium oxide (Al_2O_3) and molybdenum disulphide (MoS_2) nano particle fluids in the turning of 90CrSi steel. The application of molybdenum disulphide (MoS_2) nano cutting fluid under MQL method was able to reduce the cutting forces Fx and Fz while the application of aluminium oxide (Al2O3) nano cutting fluid under MQL method was more suitable to be used to reduce thrust force Fy. Moreover, the application of aluminium oxide (Al_2O_3) soybean based nano fluids under MQL method recorded the lowest value of surface roughness as compared to other cutting fluids. Furthermore, MoS_2 nanoparticles have been observed to improve the quality of machined surfaces in aluminium alloy [17].

Amrita et al. [18] evaluated the machining performance in terms of cutting force when different emulsifier-based nano cutting fluids were used during turning of AISI 1040 steel rod with uncoated cemented carbide tool. The nano cutting fluids were used during turning operation to compare the cutting forces with dry machining and conventional cutting fluid under MQL technique. They stated that the overall cutting forces had been reduced when nano cutting fluids were used during machining operation and nano molybdenum disulphide (MoS_2) recorded with the least cutting forces required during machining operation as compared to other cutting fluids.

Based on the afore-presented reviews, it can be concluded that the application of MQL method with nanofluids is a good alternative which improves the performance of machining. Hence, the current study aims at evaluating the performance of nano-sized diamond, copper, and iron oxide mixed with palm oil delivered by MQL during the turning of Inconel 718 with ceramic inserts. Various machining characteristics such as cutting temperature, chip morphology and chip reduction coefficient as well as tool wear were evaluated. The results obtained from experiment were statistically analysed using Signal to Noise (S/N) ratio technique.

2 Methodology and Experimental Setup

This section explains the experimental setup adopted during the turning of Inconel 718 using three types of nano-based palm oil under MQL conditions. The complete details of the workpiece materials, tool materials, properties of nanoparticles and palm oil, preparation of nanofluids, equipment and design of experiment are discussed in the ensuing sections.

2.1 Details of Workpiece and Cutting Tool

In this work, the turning operation was conducted on a nickel-based alloy which is Inconel 718. This alloy is widely used for aircraft gas turbine, nuclear power plant and reciprocating engines. The diameter and length of the material used was 50 mm and 200 mm respectively. The chemical composition and mechanical properties of Inconel 718 are as shown in Table 1 and Table 2. Furthermore, the cutting tool used during the turning of Inconel 718 was ceramic insert having a serial code of TNGA 160408 A65 with triangular geometry. The insert was fixed on a ETJNR-1616K16 type tool holder as it is used to support the insert during the turning operation. The machine tool was an Optimum lathe machine (D360 $\times$ 1000 DPA) with a maximum capacity of 1800 rpm and power drive motor rating of 2.5 kW.

2.2 Method of Lubrication

Minimum quantity lubrication (MQL) method was adopted in this experiment due to its benefits and effectiveness over the conventional method. The MQL supplies the nanofluids with a flow rate of 100 mL per hour.

Table 1. Chemical composition of Inconel 718.

Element	Content (%)	Element	Content (%)	Element	Content (%)
Al	0.49	Mg	0.0001	Ti	1.08
C	0.027	Si	0.030	Cr	18.02
P	0.005	Ca	0.0005	Ni	54.11
Bi	<0.0001	Mn	0.040	Nb	5.08
Fe	18.06	S	0.0002	Cu	0.030
Se	<0.0001	Co	0.10	Pb	<0.0002
B	0.004	Mo	2.91	Ta	0.01

Table 2. Mechanical properties of Inconel 718.

Hardness HRC	Hardness HB	Yield strength 0.2% (MPa)	Tensile strength (MPa)	Elongation AL = 4D (%)
47/48	425/423	1199	1411	22.0

2.3 Nanofluids Preparation

The current work applied three types of nanoparticles having an average size of less than 50 nm. The nanoparticles selected in this work were nano-diamond (Diamond), nano-copper oxide (CuO) as well as nano-iron oxide (Fe_2O_3) respectively. The 0.5 wt.% of each of these nanoparticles was mixed with 100 ml of palm oil respectively. In order to obtain the homogenous dispersion of nanoparticles, the three different mixtures were placed in an ultrasonic bath and then was sonicated for 30 min. After ultrasonication, the nano fluids were considered ready to be used as cutting fluid for machining operation. The physicochemical properties of palm oil and the physical properties of nanoparticles are as shown in Table 3 and Table 4 respectively.

2.4 Machining Characteristic Measurements

In this work, different machining characteristics were evaluated for each of the of nanofluids. These includes cutting temperature, chip morphology and chip reduction coefficient as well as tool wear. For the measurement of cutting temperature, digital infrared thermometer was used to measure the temperature at the cutting zone after each run of experiment.

The thickness of the chips collected from the experiment were measured using Vernier caliper. OLYMPUS BX 51 microscope was used to observe te surface using a 50x and 1500x zoom microscopic lens. The microscopic images were captured using analySIS software. A scale of 500 μm was used to measure the flank wear of the insert in order to compare which nanofluid showed the least wear pattern on the inserts.

Table 3. The physicochemical properties of palm oil.

Properties	Value
Density (g/cm^2)	0.8875
Viscosity at 40 °C (cst)	45.9
Viscosity at 100 °C (cst)	9.4
Viscosity index	195
Iodine value (gI^2/100 g)	52.5
Flash point (°C)	314
Fire point (°C)	341
Pour point (°C)	5

Table 4. Physical properties of nanoparticles.

Properties	Diamond	Copper oxide	Iron oxide
Purity (%)	99	99.9	99.5
Colour	Grey	Black	Red brown
Particle size (nm)	<50	<50	<50
Relative density (g/cm^3)	3.05–3.30	6.4	5.25
Thermal conductivity (W/(m · K))	50	33	2.2

2.5 Process Parameters and Design of Experiment

The process parameters can be categorized as input and output parameters. In this work, types of oils, cutting speed, feed rate and depth of cut were identified as the input parameters. The output parameters considered were cutting temperature, chip morphology, chip reduction coefficient and tool wear. Besides that, Taguchi design of experiments was adopted in the study. The experimental layout for the machining input parameters adopted the L9 orthogonal array design. Taguchi design of experiment categorized variables into two different types such as control factors as well as noise factors. Adopting Taguchi method enabled more parameters to be studied with a smaller number of experiments. Hence, it saved time and cost. The machining parameters with their respective levels are presented in Table 5. Moreover, the results obtained from the experiment were statistically analyzed by smaller-the-better Signal to Noise (S/N) ratio method. The purpose of using S/N ratio method was to determine which cutting parameter had significant effect on the machining characteristics [19].

3 Results and Discussion

In this section, the results obtained from the experiment in terms of cutting temperature, chip morphology, chip reduction coefficient and tool wear were discussed. The analysis using S/N ratio method was included as well.

3.1 Cutting Temperature

Figure 1 presents the cutting parameters corresponding to the S/N ratio values at different level. The types of oil were indicated by N-D, N-CuO and N-Fe2O3 referred to the three types of nano-based oil such as nano-diamond based palm oil, nano-copper oxide based palm oil as well as nano-iron oxide based palm oil respectively. Hence, it can be clearly seen in Fig. 1 that 108 rpm is the best cutting speed to get the lowest cutting temperature in this study as it recorded with the lowest SN ratio of −32.68. It may also be inferred that using nano-diamond based palm oil as cutting fluid, 108 rpm cutting speed, 0.1 mm/rev feed rate and 0.6 mm depth of cut is the best cutting condition for achieving the minimum cutting temperature in this study. The cutting temperature increased based on the increase in cutting speed and types of oil used. It can be observed from Table 6 that cutting speed has the highest rank among the 4 parameters followed by types of oil, feed rate and depth of cut respectively. The computed values of Taguchi Signal to Noise (S/N) ratio shows that the delta value of cutting speed recorded with the value of 1.36 and this indicated that cutting speed is the influential control factor which affected the cutting temperature during this machining operation due to the smallest S/N ratio value. It can be proven by the previous research studies by Singh and Pal [20]. As the cutting speed increases, the major portion of chips is removed from cutting tool edge and it will eventually generate high friction between the tool chip interfaces. As a result, large amount of heat will be generated at the tool chip interfaces and this will increase the cutting temperature during the machining operation. High production machining will be constrained by the elevated temperature. Furthermore, the heat dissipated on the workpiece material could alter the metallurgical properties.

In addition, the types of oil also have a significant effect on cutting temperature as this parameter has the second highest rank among the 4 parameters as shown in Table 6. Nano-diamond based palm oil outperformed the other two oils as it recorded the lowest S/N ratio of −32.87. This indicates that nano-diamond based palm oil produced the lowest cutting temperature as compared to nano-copper oxide-based and nano-iron oxide-based palm oil. This is due to the physical properties of nano-diamond as it has relatively higher thermal conductivity as compared to the other two nano-based oils. The higher thermal conductivity of nano-diamond based palm oil as stated beforehand helps to dissipate heat at the primary shear zone. This is because it provides high wettability as well as better heat dissipation which leads to lower cutting temperature at the tool chip interfaces. Furthermore, the size of nano-diamond particle is much smaller as compared to nano-copper oxide particle and nano-iron oxide particle and hence it will be much easier to penetrate the cutting zone [21].

Table 5. Machining parameter with their respective levels.

No.	Oil type	Cutting speed (RPM)	Feed rate (rev/mm)	Depth of cut (mm)	a_1	a_2	λ	S/N ratio
i.	N-Diamond oil	108	0.10	0.4	0.10	0.23	2.30	−7.2346
ii.	√	140	0.15	0.6	0.15	0.27	1.80	−5.1055
iii.	√	205	0.20	0.8	0.20	0.30	1.50	−3.5218
iv.	N-CuO oil	108	0.15	0.8	0.15	0.33	2.20	−6.8485
v.	√	140	0.20	0.4	0.20	0.39	1.95	−5.8007
vi.	√	205	0.10	0.6	0.10	0.29	2.90	−9.2480
Vii.	N-Fe_2O_3 oil	108	0.20	0.6	0.20	0.35	1.75	−4.8608
viii.	√	140	0.10	0.8	0.10	0.32	3.20	−10.1030
ix.	√	205	0.15	0.4	0.15	0.38	2.53	−8.0624

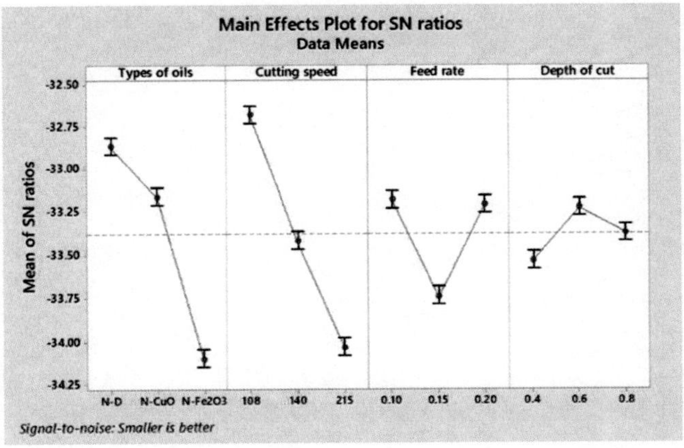

Fig. 1. Main effects plot for S/N ratios of cutting temperature.

Table 6. Response table for signal to noise ratios of cutting temperature.

Level	Types of oils	Cutting speed	Feed rate	Depth of cut
1	−32.87	−32.68	−33.18	−33.54
2	−33.17	−33.42	−33.74	−33.23
3	−34.10	−34.04	−33.21	−33.37
Delta	1.24	1.36	0.56	0.31
Rank	2	1	3	4

3.2 Observation of Chip Morphology

The chip morphology can be considered as one of the significant aspects in evaluating the machining performance. The behavior and nature of chips usually depend on the types of oil used as well as the cutting parameters controlled during the machining operation. In this study, the observation of chip formation was conducted during the turning of Inconel 718. The chip was collected after every run of experiment and the thickness of the chip was measured using Vernier caliper. Different types of chips were formed with respect to the different level of cutting parameters and types of oil used. The chips were mainly continuous with tubular and helical shape. This demonstrates stable cutting conditions with smooth flow of chips as the metal is plastically deformed without fracture in front of the cutting edge of the tool. In addition, discontinuous or segmental chips were formed. This kind of chips is synonymous with reduced friction force between the tool and chips. This will result in good surface finish. The illustration of chips formed during the machining operation is shown in Fig. 2 and Fig. 3.

Chip Reduction Coefficient. In this study, chip reduction coefficient was used as the technique to study and analyze the behavior of chip formation regarding the degree of deformation. According to Anurag et al. [22] and Das et al. [23], the thickness of the produced chip (a2) will be always be greater than the thickness of uncut chip (a1).

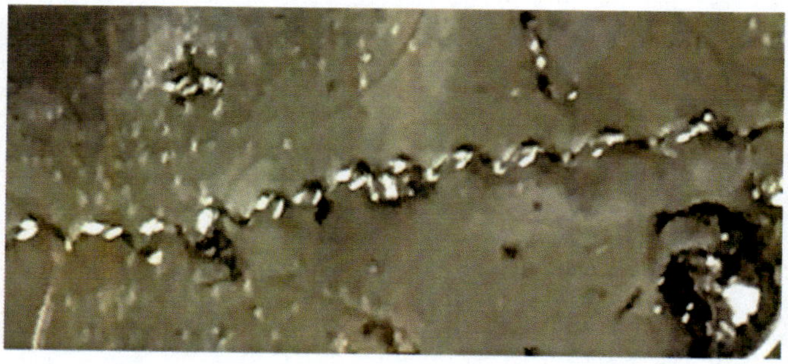

Fig. 2. Continuous chips with tubular and helical shape.

The thickness of uncut chip (a1) is also known as the feed. Hence, the ratio of the thickness of produced chip to the thickness of uncut chip is known as chip reduction coefficient (λ). This was calculated using the equation shown in Eq. (1).

$$\lambda = \frac{a_2}{a_1} \tag{1}$$

The graph shown in Fig. 4 represents the cutting parameters corresponding to the S/N ratio values at different levels. It can be seen from Fig. 4 that that feed rate was the most influential control parameter on the chip reduction coefficient as it has the lowest S/N ratio value recorded with a value of −4.728. It is further noticed that feed rate is inversely proportional to the chip reduction coefficient. The chip reduction coefficient decreases with an increment in feed rate. Hence, it can be inferred that the optimal cutting parameter

Fig. 3. Discontinuous or segmental chips.

set for minimum chip reduction coefficient would be using nano-diamond based palm oil as cutting fluid with a cutting speed of 108 rpm, feed rate of 0.2 mm/rev and depth of cut of 0.6 mm. It was further demonstrated that feed rate was the most influential control parameter for chip reduction coefficient as it has the highest rank among the other cutting parameters as shown in Table 7. The computed values of Taguchi Signal to Noise (S/N) ratio showed that the delta value of feed rate recorded with the value of 4.134. This indicated that feed rate is the most influential control parameter on the chip reduction coefficient due to the smallest S/N ratio value. In addition, it can be observed that nano-diamond based palm oil outperformed the other two nano-based palm oils in achieving the lower chip reduction coefficient. The nano-diamond based palm oil recorded the lowest S/N ratio value of −5.287 and this indicates that nano-diamond based palm oil was the best oil for lowest chip reduction coefficient in this study.

3.3 Microscopic Analysis of Tool Wear

In this study, flank wear was observed in the form of notching during the turning of Inconel 718. The measurements of the flank wear on three different inserts with respect to the types of nano-based oils used are as shown in Fig. 5. The measurement of flank wear was indicated by the white line drawn on the flank faces.

Microscopic analysis was performed to evaluate the tool wear under effect of different lubricant. As reported in Fig. 5, it is clearly seen that the tool used under the effect of nano-diamond additives based palm oil lubricant has reported the least tool wear which is 0.3333 mm compared to nano-copper oxide additives based palm oil (1.05784 mm) and nano-iron oxide additives based palm oil (1.42046 mm) under similar working condition. A clearer illustration of the tool wear is presented in Table 8.

According to the results shown in Table 8, the percentage of wear of the tool (insert) when nano-diamond based palm oil was used is 17.14%. This is relatively lower than the percentage of tool wear when nano-copper oxide-based palm oil and nano-iron

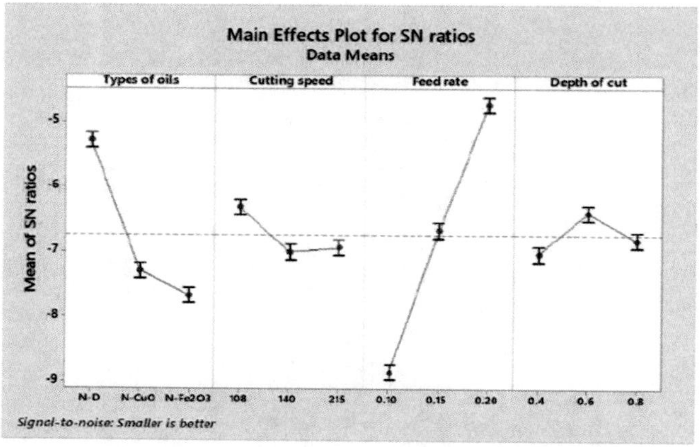

Fig. 4. Main effect plot for S/N ratios of chip reduction coefficient.

Table 7. Signal to noise ratios of chip reduction coefficient.

Level	Types of oils	Cutting speed	Feed rate	Depth of cut
1	−5.287	−6.315	−8.862	−7.033
2	−7.299	−7.003	−6.672	−6.405
3	−7.675	−6.944	−4.728	−6.824
Delta	2.388	0.688	4.134	0.628
Rank	2	3	1	4

Table 8. Results of tool wear and the percentage of tool wear

Inserts & oil	Original thickness (mm)	Tool wear (mm)	Final thickness of insert (mm)	Percentage of tool wear (%)
Diamond	1.945	0.33333	1.61167	17.14
CuO oil	1.945	1.05784	0.88716	54.39
Fe_2O_3	1.945	1.42046	0.52454	73.03

oxide-based palm oil were used (value of 54.39% and 73.03% respectively). This can be directly related to the tool life of an insert where the low percentage of tool wear leads to higher lifespan of an insert. In a nutshell, it can be summarized that the types of oil used during the machining operation have significant impact on the tool wear. It can also be concluded that low wear values and less wear pattern were observed on the rake face of insert when nano-diamond based palm oil was used as compared to when nano-copper oxide based palm oil and nano-iron oxide based palm oil were used during the turning operation. The flank wear on rake face of insert when nano-diamond based palm oil was used is relatively smaller as compared to the flank wear observed on rake face of insert when nano-copper oxide and nano-iron oxide based palm oil were used during the machining operation. This can be explained by the physical properties of nano-diamond as it has relatively higher thermal conductivity as compared to nano-copper oxide and nano-iron oxide. The high thermal conductivity of nano-diamond has better cooling effect and hence it helps to dissipate heat at the cutting zone more efficiently. As a result, less friction was developed at the tool-workpiece interface. Hence, less tool wear and the lifespan of a cutting tool could be improved. Furthermore, less notching was observed on the rake face of insert when nano-diamond based palm oil was used while high notching and thermal cracking were observed on the rake face of inserts when nano-copper oxide and nano-iron oxide based palm oil were used respectively. As a result, these wear patterns that appeared on the insert caused a large area to be removed from the rake face.

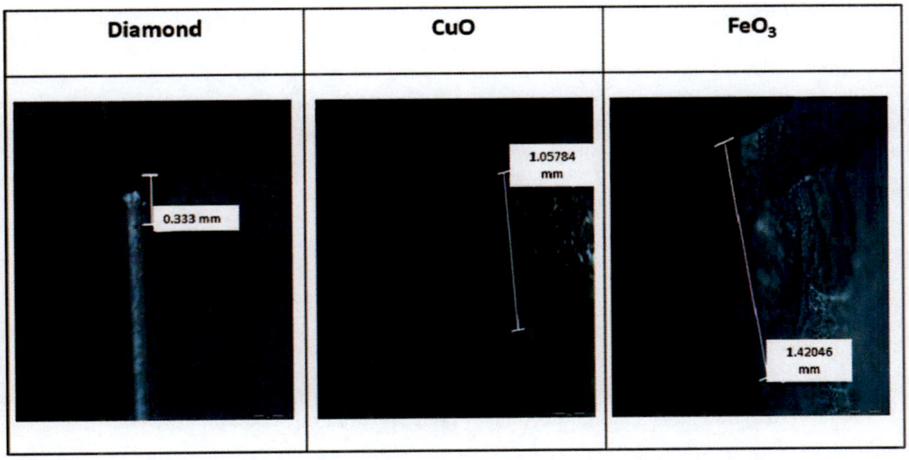

Fig. 5. Flank wear measured when nano-based palm oil was used

4 Conclusion

This study investigated the performance of nano-diamond, nano-copper oxide and nano-iron oxide based palm oil as cutting fluids on the turning of Inconel 718. Taguchi method with orthogonal array of L9 was adopted as the design of experiment in this study. Signal to Noise (S/N) ratio smaller the better was used to analyze the results obtained. Nano-diamond based palm oil performed the best in terms of cutting temperature as it recorded with the lowest generated S/N ratio of −32.87. Cutting speed was identified as the most influential control parameter which affected the cutting temperature due to the smallest S/N ratio value of −32.68. The optimal parameter setting for achieving minimum cutting temperature was denoted as using nano-diamond based palm oil as cutting fluid, 108 rpm cutting speed, 0.1 mm/rev feed rate and 0.6 mm depth of cut. Nano-diamond based palm oil outperformed the other nanofluids in achieving lower chip reduction coefficient as it recorded with the lowest S/N ratio of −5.287. Feed rate was the most influential control factor affecting the chip reduction coefficient as it had the lowest S/N ratio value recorded with a value of −4.728. The optimal parameter setting for minimum chip reduction coefficient would be using nano-diamond based palm oil as cutting fluid, cutting speed of 108 rpm, feed rate of 0.2 mm/rev and depth of cut of 0.6 mm. Nano-diamond based palm oil outperformed the other two nanofluids as the flank wear measured on the insert when nano-diamond based palm oil was used during the turning operation was the least, followed by nano-copper oxide based palm oil and nano-iron oxide based palm oil. The measurements of flank wear for nano-diamond based palm oil, nano-copper oxide-based palm oil and nano-iron oxide-based palm oil are 0.33333 mm, 1.05784 mm and 1.42046 mm respectively. Hence, the study showed that diamond nano particle is the most compatible on palm oil. Further study is suggested to investigate the performance of these nanoparticles on other vegetable oil.

Acknowledgements. The authors would like to express the utmost gratitude to UCSI University Centre of Excellence for Research, Value Innovation and Entrepreneurship (CERVIE) for funding this project under the Pioneer Scientist Incentive Fund (Proj-2019-in-FETBE-062).The authors gratefully acknowledge Faculty of Engineering, Technology and Built Environment, UCSI University Malaysia and Faculty of Engineering and Technology, Liverpool John Moores University for support to conduct this research work.

References

1. Hegab, H., Darras, B., Kishawy, H.A.: Sustainability assessment of machining with nano-cutting fluids. Proc. Manuf. **26**(1), 245–254 (2018)
2. Alotaibi, J.G., Yousif, B.F., Yusaf, T.F.: Brass, aluminium, and mild steel sliding against stainless steel under dry conditions. Int. J. Precis. Technol. **5**(2), 85–96 (2015)
3. Sharma, V.S., Singh, G., Sørby, K.: A review on minimum quantity lubrication for machining processes. Mater. Manuf. Processes **30**(8), 935–953 (2015)
4. Mannekotte, J., Kailas, S.V., Kathyayini, N.: Environmentally friendly functional fluids from renewable and sustainable sources. J. Renew. Sustain. Energy **1**(1), 1787–1801 (2017)
5. Ozcelik, B., Kuram, E., Cetin, M.H., Demirbas, E.: Experimental investigations of vegetable based cutting fluids with extreme pressure during turning of AISI 304l. Tribol. Int. **44**(12), 1864–1871 (2011)
6. Majak, D., Olugu, E.U., Lawal, S.A.: Analysis of the effect of sustainable lubricants in the turning of AISI 304 stainless steel. Proc. Manuf. **43**(2020), 495–502 (2020)
7. Rapeti, P., Pasam, V.K., Gurram, K.R., Revuru, R.S.: Performance evaluation of vegetable oil based nano cutting fluids in machining using grey relational analysis-a step towards sustainable manufacturing. J. Clean. Prod. **172**(1), 2862–2875 (2018)
8. Mall, V.K., Kumar, P., Singh, B.: A review of optimization of surface roughness of Inconel 718 in end milling using Taguchi method **4**(12 part 3), 103–109 (2014)
9. M'Saoubi, R., Outeiro, J.C., Chandrasekaran, H., Dillon, O.W., Jr., Jawahir, I.S.: A review of surface integrity in machining and its impact on functional performance and life of machined products. Int. J. Sustain. Manuf. **1**, 203–236 (2018)
10. Dudzinski, D., Devillez, A., Moufki, A., Larrouquere, D., Zerrouki, V., Vigneau, J.: A review of developments towards dry and high speed machining of Inconel 718 alloy. Int. J. Mach. Tools Manuf. **44**, 439–456 (2004)

11. Kumar, B.S., Padmanabhan, G., Krishna, P.V.: Experimental investigations of vegetable oil based cutting fluids with extreme pressure additive in machining of AISI 1040 steel. Manuf. Sci. Technol. **3**(1), 1–9 (2015)
12. Lawal, S.A., Choudhury, I.A., Sadiq, I.O., Oyewole, A.: Vegetable-oil based metalworking fluids research developments for machining processes: survey, applications and challenges. Manuf. Rev. **1**(22), 1–12 (2014)
13. Padmini, R., Krishna, P.V., Rao, G.K.M.: Effectiveness of vegetable oil based nanofluids as potential cutting fluids in turning AISI 1040 steel. Tribol. Int. **94**, 490–501 (2016)
14. Khandekar, S., Sankar, M.R., Agnihotri, V., Ramkumar, J.: Nano-Cutting fluid for enhancement of metal cutting performance. Mater. Manuf. Processes **27**(9), 963–967 (2012)
15. Li, B., et al.: Heat transfer performance of MQL grinding with different nanofluids for Ni-based alloys using vegetable oil. J. Clean. Prod. **154**(2017), 1–11 (2017)
16. Duc, T.M., Long, T.T., Chien, T.Q.: Performance evaluation of MQL parameters using Al_2O_3 and MoS_2 nanofluids in hard turning of 90CrSi steel. Lubricants **7**(40), 1–17 (2019)
17. Rahmati, B., Sarhan, A.A.D., Sayuti, M.: Morphology of surface generated by end milling AL6061-T6 using molybdenum disulphide (MoS2) nanolubrication in end milling machining. J. Clean. Prod. **2013**, 685–691 (2013)
18. Amrita, M., Shariq, S.A.: Experimental investigation on application of emulsifier oil based nano cutting fluids in metal cutting process. Proc. Eng. **97**, 115–124 (2014)
19. Rahul, D., Pretesh, J.: Application of Taguchi-based design of experiments for industrial chemical processes, statistical approaches with emphasis on design of experiments applied to chemical processes. Intech Open, 138–155 (2018)
20. Raghvendra Singh, M.K.P.: Experimental investigation of nano cutting fluids as sustainable cooling technique in turning operation. Int. J. Manag. Res. **6**(4), 414–421 (2016)
21. Yuan, S., Hou, X., Wang, L., Chen, B.: Experimental investigation on the compatibility of nanoparticles with vegetable oils for nanofluid minimum quantity lubrication machining. Tribol. Lett. **66**(3), 1 (2018). https://doi.org/10.1007/s11249-018-1059-1
22. Anurag, K.R., Joshi, K.K., Das, R.K.: Analysis of chip reduction coefficient in turning of Ti-6Al-4V ELI. IOP Conf. Ser. Mater. Sci. Eng. **390** (2018). The 3rd International Conference on Materials and Manufacturing Engineering 2018, Tamilnadu, India, 8–9 March 2018
23. Das, A., Patel, S.K., Das, S.R.: Performance comparison of vegetable oil based nanofluids towards machinability improvement in hard turning of HSLA steel using minimum quantity lubrication. Mech. Ind. **20**(5), 506 (2019)

Design and Fabrication of Laboratory Scale Carbon Dioxide Controlling Chamber for Leafy Vegetables

Chun Hau Kua[1], Norhafizah Amal Aisyah Yunos[2], Rafidah Petra[2], Fethma M. Nor[3,4], and Denni Kurniawan[1,5](✉)

[1] Mechanical Engineering Programme Area, Universiti Teknologi Brunei, Gadong BE1410, Brunei Darussalam
denni.kurniawan@utb.edu.bn
[2] Electrical and Electronic Engineering Programme Area, Universiti Teknologi Brunei, Gadong BE1410, Brunei Darussalam
[3] Department of Mechanical Engineering, Curtin University, 98009 Miri, Malaysia
[4] School of Design, Universiti Teknologi Brunei, Gadong BE1410, Brunei Darussalam
[5] Centre for Research on Agri-Food Science and Technology, Universiti Teknologi Brunei, Gadong BE1410, Brunei Darussalam

Abstract. Plants are found to have an increase in yield production of 32% when the carbon dioxide concentration is doubled from 330 ppm to 660 ppm. For greenhouse cultivation, the recommended carbon dioxide concentration level for optimum production is about 700 ppm to 900 ppm. This project designs and develops a CO_2 controlling chamber (closed system) to allow studies based on the effect of elevated CO_2 on leafy vegetables' growth and production yield. A CO_2 production system by fermentation as an external source for the chamber is also being designed and fabricated. Conceptual design was first done based on product design specification. Afterwards, fabrication of the designed CO_2 chamber was done. Monitoring system was installed to provide support in monitoring and regulating the CO_2 concentration in the chamber. The retaining capability of the chamber was being tested under three different levels of concentration (727 ppm, 854 ppm and 1011 ppm). The chamber was found to be properly sealed for all concentrations with less than 1% of CO_2 leakage. As for CO_2 generation, testing was done under three different concentrations of sucrose (12.5%, 25% and 50%) by fermentation. The results showed that 25% sucrose concentration was able to maintain CO_2 concentration above 1000 ppm for a duration of 30 h and still remained above 900 ppm until 120 h.

Keywords: Carbon dioxide · Chamber · Design · Leafy vegetable · Fermentation

1 Introduction

While public attention has focused on the effects of carbon dioxide (CO_2) on climate change, increasing CO_2 concentrations is also likely to have a profound direct impact

A. Batako et al. (Eds.): GCMM 2021, LNNS 335, pp. 46–56, 2022.
https://doi.org/10.1007/978-3-030-90532-3_4

on plant growth, physiology and chemistry [1]. The effects are the result of CO_2 being essential to the metabolism of plants. Plants take up atmospheric CO_2 as photosynthetic organism, thus chemically reducing carbon. This not only reflects the acquisition of accumulated chemical energy for the plant, but also provides the organic molecules that constitute the structure of a plant with carbon skeletons. Currently, there are difficulties in identifying the best practice for a carbon dioxide controlling chamber as there are limited studies on this. In addition, there are also insufficient guidelines in determining the best possible approach for the fabrication of a carbon dioxide controlling chamber. Currently, there are two types of carbon dioxide chambers for plant growth: open/field-based system and closed/growth-chamber based system [2]. The advantages and disadvantages of open/field-based and closed/growth chamber-based system are specified in Table 1. In this project, the closed/growth chamber-based system is selected as its advantages overweighs the open/field system. One of the main criteria for this selection is the ability to allow simple control of environmental conditions, mainly CO_2 concentration in the chamber. In addition to that, the volume of CO_2 present in the chamber can be easily maintained using regulators, valves and CO_2 sensor as it is in an enclosed system. In an open system, the difficulty in maintaining a certain level of CO_2 concentration is higher as there is a tendency for the pumped CO_2 gas to escape because of its open top. Apart from that, the open system has a major disadvantage which is the exposure to air pollution. This will result in inaccurate readings as the plant's response may be affected by some of the air pollutants, giving wrong readings.

Instead of procuring commercially available carbon dioxide in a tank, generation of carbon dioxide is preferred. There are multiple methods of generating carbon dioxide. Some of the common methods include animal houses, peatland soils, fungi, and fermentation. Animal houses of cattle, pigs, and poultry can generate CO_2 from respiration [3]. The recorded CO_2 production can range from 0.17 to 0.2 $m^3h^{-1}hpu^{-1}$, depending on the animal's respiration quotient, which is affected by animal activity, body mass, and energy metabolism [3–5]. Peatland soil is a representation of the storage of carbon since large amounts of organic matter are present in it. The water-saturated soil contributes to slow decay since the bacteria that break down organic materials have insufficient oxygen, producing methane (CH_4). CO_2 emission from peatland soil depends on the aerobic and anaerobic production rates of peat and root respiration and the transport of the gas to the surface of the peat. In three different types of peatlands (bog, fen and swamp), the microbial processes leading to the production of carbon dioxide are influenced by two factors—temperature and moisture content [6]. Overall, the CO_2 emission from peatland soils can range from 0.6 to 13.9 g m^{-2} d^{-1}. The highest carbon dioxide emission comes from bog peatland at 23 °C and 40 cm water table. Carbon dioxide can also be generated by fungi which decomposes plant biomass or animal metabolism byproduct. A study on the CO_2 production by dead wood fungi which is from their mycelial respiration reported that with different treatment at different duration can produce around 0.52 g CO_2 kg^{-1} h^{-1} [7]. As for fermentation, baker's yeast (*Saccharomyces cerevisiae*) metabolism of sugars can be used. There has been a study on the effect of different sucrose concentration on the CO_2 production [8]. Fermentation of sucrose recorded the highest CO_2 emission of 8,003 ppm after 21 h, although the concentration decreased to below 1,000 ppm at 100 h and further decreased to below 700 ppm when after 110 h [8].

Table 1. Advantages and disadvantages of open and closed system.

System	Advantage	Disadvantage
Open/Field-based system	More resemblance to natural environment Simple to design and construct	Less precise control of environmental condition Limits the type of research that can be performed More expensive Exposure to air pollution Difficulty in maintaining the volume of CO_2 present in the chamber
Closed/Growth chamber-based system	Simpler control of environmental conditions (CO_2, temperature, light, and relative humidity) Eliminates natural and unpredictable fluctuations Cheap, simple to design and fabricate Easier to maintain the volume of CO_2 present in the chamber	Continuous external supply of carbon dioxide gas is required

These carbon dioxide generation methods are assessed by Pugh matrix approach in Table 2. Fermentation method shows the highest total score which is contributed by high scores in availability, cost, efficiency, and manufacturability. In local settings, the materials required for the fermentation method which are sucrose solution and Baker's yeast are low cost and readily available. Fermentation method can generate approximately 1000 ppm of CO_2 concentration for a duration of 5 days with only 11 g of Baker's yeast [8]. After 5 days, the solution can be easily replaced with a new one and the process continues. In addition to that, the fabrication process for this fermentation method is also simple which consists of two water bottles connected together using tube connectors with one bottle being the CO_2 generator and the other being gas separator. Some reasons why the other methods are low scored in the assessment include the followings. Animal house method is costly to construct and difficult to maintain as animals are involved despite large amount of CO_2 that can be generated. Decay of peatland soils releases an unwanted greenhouse gas which is methane (CH_4) that may affect the plant's response giving inaccurate results. Lastly, CO_2 generation from mycelial respiration of fungi is unreliable because high CO_2 flux happens only during the early stage of fungi growth. At later stages, the CO_2 flux decreases.

This current project aims to design and fabricate a customised laboratory scale CO_2 chamber capable of containing leafy vegetables in a closed system. An external CO_2 generation by using fermentation of sucrose by yeast is also fabricated as an input to the chamber. The chamber is evaluated in terms of CO_2 retaining capability, while the performance of the CO_2 gas production system is in term of sustaining CO_2 concentration at 900 ppm as long as possible.

Table 2. Concept evaluation of CO_2 generation methods using Pugh matrix.

Criteria	Weightage (%)	Animal house	Peatland soils	Fungi	Fermentation
Availability	15	8	10	10	12
Low cost	25	15	22	22	20
Efficiency	25	22	15	13	20
Simplicity	10	5	7	7	8
Reliability	10	8	6	5	8
Manufacturability	15	8	10	10	12
Total weightage (100%)		66	70	67	80

2 Methodology

A product design specification was first determined based on the stated objectives and local settings, in order to keep the cost of the project to be minimal. The product design specification of the laboratory scale carbon dioxide controlling chamber and generator is as shown in Table 3.

Table 3. Product design specification for the CO_2 chamber.

Category	Specification
Functions	Self generates CO_2 Allows to perform study on the effects of elevated CO_2 in leafy vegetables
Features	Simple, easy to fabricate Light weight Portable Visible/transparent
Materials	Acrylics Aluminium frame CO_2 generator kit
Maintenance requirements	Replacement of sucrose solutions if the fermentation process ended
Product lifespan	Minimum of 3 years
Fabrication cost	Under BND 200

Considering the product design specification, the conceptual design for the carbon dioxide controlling chamber and CO_2 generator were constructed using a CAD software. This design consists of a combination of a carbon dioxide controlling chamber connected to a CO_2 generator. The components of this conceptual design include a CO_2 generator (using fermentation), bubble counter (gas separator), regulator/control valve,

CO_2 controlling chamber, and chamber lid. The overall conceptual design is displayed in Fig. 1.

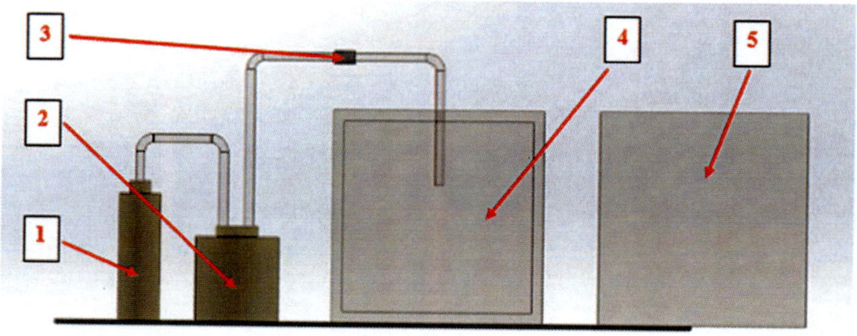

Fig. 1. Front view of CO_2 controlling chamber with CO_2 generator. (Note: (1) CO_2 generator (fermentation), (2) bubble counter (gas separator), (3) regulator/control valve, (4) CO_2 controlling chamber, and (5) chamber lid).

Further, the specific dimensions for the components of the CO_2 controlling chamber with CO_2 generator are stated in Table 4.

Table 4. Specific dimensions of different components.

Components	Dimensions
CO_2 controlling chamber	0.5 m (l) × 0.5 m (w) × 0.5 m (h)
Yeast CO_2 generator	20 cm (d) × 35 cm (h)
Gas separator	25 cm (d) × 25 cm (h)
Rubber tubing	3 mm (d)
Chamber lid	2 cm (t) × 0.5 m (l) × 0.5 m (w)

t = thickness, w = width, l = length, h = height, d = diameter

Based on the conceptual design, the final design of the carbon dioxide controlling chamber was made and fabricated along with the CO_2 generator. The fabricated carbon dioxide controlling chamber is shown in Fig. 2 and CO_2 generator in Fig. 3. The carbon dioxide chamber is made of acrylic sheets cut into specific dimensions (0.5 m × 0.5 m × 0.5 m) with a thickness of 5mm and mounted to aluminium frames using silicone based glue. To achieve the objective of retaining the carbon dioxide in the chamber, 8 individual locking hinges were installed with 2 on each side of the chamber and then sealed off with a layer of styrofoam tape. A door hinge is also attached at the middle of the chamber lid to allow opening and closing the lid at ease.

Figure 3 shows the carbon dioxide generator. Carbon dioxide generated from the first bottle enters the second bottle through a connecting pipe. The purpose of the water inside the second bottle is to filter off the unwanted gases produced during fermentation process

Fig. 2. Carbon dioxide controlling chamber.

so that only CO_2 will enter the controlling chamber. The amount of CO_2 entering the chamber is controlled by a solenoid valve. The solenoid valve is programmed to allow switching on or off through a mobile application, Blynk. The list of components and its individual function are stated in Table 5.

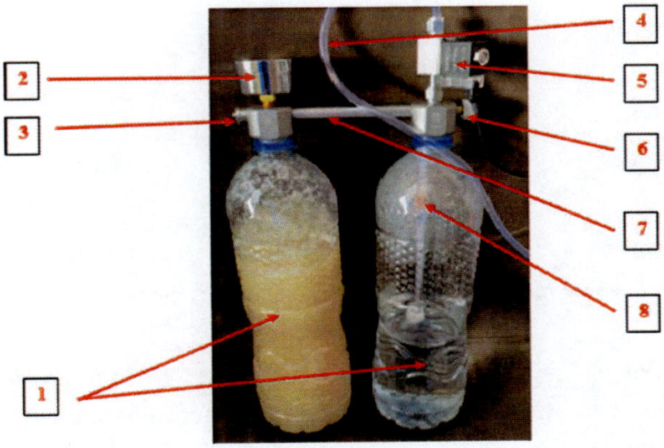

Fig. 3. CO_2 generator using fermentation method. (Note: component listing (1 to 8) is as stated in Table 5).

The final design of the customised CO_2 controlling chamber and CO_2 generator is shown in Fig. 4. Additional features such as sensors were installed in this chamber to monitor different parameters of the plant inside the chamber. Such parameters include CO_2 level, soil moisture, pH level, humidity, air temperature, and pressure. The values of these parameters are displayed on the LCD panel display located on top of the chamber lid. The different types of sensors are listed in Table 6.

The performance of the CO_2 controlling chamber and generator requires measurements to be taken in a controlled-field. The variable required to validate the experiment were mainly carbon dioxide concentration level. That said, other variables such soil moisture, air temperature, humidity, air pressure and soil pH were also able to be measured. The monitoring system was tested by attaching the specific components to its

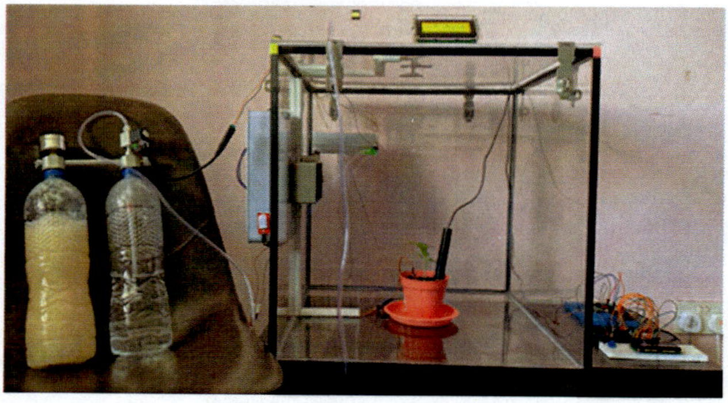

Fig. 4. Assembled CO_2 controlling chamber and generator

respective position to the plant. The monitoring system was an Arduino hardware built on an I/O board and several development environments that implements the scripting language. The software used for this system was 'Arduino IDE' and the language script was 'C language'.

Table 5. List of components.

No.	Component	Quantity	Function	
1	Water bottles	2	Left	Container for fermentation
			Right	Storage for CO_2 generated
2	Pressure gauge	1	Displays the pressure at glance	
3	Safety relieve valve	1	Protects the pressurised bottle against over pressure	
4	Rubber tubing	2 m	Transports the carbon dioxide	
5	Solenoid valve (DC12V)	1	Opens or closes the valve using electrical current	
6	Micrometer valve	1	Adjusts air output manually (back up)	
7	Aluminium connector	20 cm	A hollow tube to connect the bottles and allow the transfer of CO_2	
8	Check valve	1	Acts as a one-way valve to ensure that there is no backflow of CO_2	

Arduino Uno used was an open-source microcontroller board based on the Microchip ATmega328P microcontroller. The board has a number of digital and analog input/output (I/O) pins that can be used to connect to different expansion boards (shields) and other circuits. NodeMCU used is a low-cost open source IoT platform. It is an open source firmware for which open source prototyping board designs are available. The firmware uses LUA scripting language but C scripting language was also applicable. With the application of this Wi-Fi module, readings measured from the sensor were visualised

Table 6. List of sensors

Sensor	Model
Carbon dioxide	MQ-135
Soil moisture	HL-69
Air temperature and humidity	DHT22
Air pressure	BMP280
pH	SEN0249

and recorded through an IoT analytic platform 'ThingSpeak' or a mobile application called 'Blynk'. The dashboard for both ThingSpeak and Blynk was configured to show graphs of different measured variables.

To measure the level of carbon dioxide concentration present in the controlling chamber, a carbon dioxide sensor was used. The carbon dioxide sensor was attached on the wall of the chamber and was connected to a circuit. With a control valve, the CO_2 level can be constantly monitored and regulated depending on the type of testing required by the student.

For the experiment, the materials used for the fermentation are water, cane sugar, Baker's yeast, and baking soda. All ingredients were sourced locally. The mixture of ingredient in the fermentation container is mentioned in Table 7. The solution was stirred well and left to stand for 1 h before performing the experiment. In the storage container, 600 ml of water was added. Three sucrose concentrations of 12.5%, 25% and 50% were prepared. The experiment was performed twice for each concentration. The amount of CO_2 produced by fermentation was measured at 30-s intervals for 5 days.

Table 7. Ingredients of solution mixture for bottle A

Ingredients	Quantity	
Lukewarm water	1 L	
Sugar	12.5% sucrose concentration	125 g
	25% sucrose concentration	250 g
	50% sucrose concentration	500 g
Baker's yeast	11 g	
Baking soda	5 g	

For testing the chamber's CO_2 retaining capability, fermentation of 50% sucrose concentration by yeast was used. The CO_2 produced by the generator was injected into the chamber until it reached the required concentration level. The chamber is then left for 12 h before the amount of CO_2 in the chamber was measured again. The difference in both readings becomes the amount of CO_2 leakage (Eq. 1),

$$\% \; of \; CO_2 \; leakage = \frac{C_{Be} - C_{Aft}}{C_{Be}} \times 100\% \tag{1}$$

where C_{Be} = CO_2 concentration (before) and C_{Aft} = CO_2 concentration (after).

3 Results and Discussion

3.1 CO_2 Retaining Capability

The retaining capability of the chamber can be determined by the percentage of CO_2 leakage. The results compiled in Table 8 showed that the enclosed chamber fulfilled the objective of retaining CO_2 inside the chamber. It has been proven that even with higher concentration, the chamber was still able to maintain the level of CO_2 presents in the chamber without showing much leakage even after a duration of 12 h. that the CO_2 leakage from the chamber is very minimal, to be around 1% for all CO_2 concentrations. In fact, this 1% leakage might also be due to the precision limit of the sensor. Therefore, this customised chamber is considered to be properly sealed.

Table 8. CO_2 retaining capability of the chamber.

	CO_2 concentration (ppm)		Percentage of CO_2 leakage (%)
	Before	After	
Lower concentration	727	720	0.96
Medium concentration	854	848	0.70
Higher concentration	1011	1002	0.89

3.2 CO_2 Production at Different Sucrose Concentration

According to the results shown in Fig. 5, S. cerevisiae with 12.5% sucrose concentration was only capable of generating CO_2 at maximum CO_2 concentration of 1005 ppm at 30th hour and remained above 1000 ppm from the 24th to 40th hour, a duration of only 16 h. This level of CO_2 continued to decrease until it reaches 900 ppm at 75 h and approximately 820 ppm at 120 h.

At 25% sucrose concentration, the fermentation was able to produce CO_2 at maximum CO_2 concentration of 1075 ppm. The concentration of CO_2 remained above 1000 ppm for 30 h starting from 15th hour until 46th hour. This rate of CO_2 production continues to decrease but still remained above 900 ppm until 120 h.

At sucrose concentration of 50%, the fermentation managed to produce CO_2 at maximum concentration of CO_2 of 1150 ppm at 26th hour and remained above 1000 ppm for a duration of 34 h from 9th to 43rd hour. However, this rate of CO_2 produced later

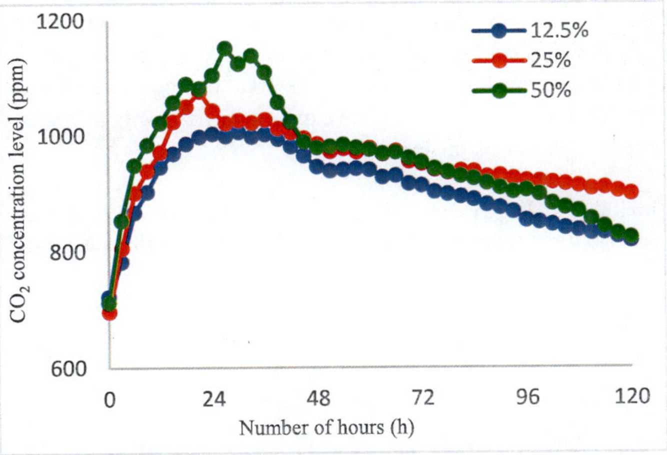

Fig. 5. CO_2 concentration from fermentation of sucrose at 12.5%, 25%, and 50% concentrations for 120 h.

continue to decrease gradually to approximately 900 ppm at 100th hour and 820 ppm after 120 h.

By comparing the results of three different sucrose concentrations, it was clearly seen that S. cerevisiae with 50% sucrose concentration recorded the highest level of CO_2 production at 1150 ppm. The reason is because highest sucrose concentration generated highest amount of ethanol, a by-product of fermentation. Therefore, they also produced the fastest rate of fermentation [9]. With a faster rate of fermentation, more CO_2 were produced. S. cerevisiae with 25% sucrose concentration showed the longest CO_2 production as it remained to be above 900 ppm almost throughout the experiment (120 h), compared to 12.5% and 50% sucrose. Although the yeast with 50% sucrose concentration showed the highest CO_2 production, it could not be sustained for a longer period of time. With a higher rate of CO_2 production, the yeast grows and metabolises sugar faster, accumulating more alcohol. This accumulation of alcohol becomes toxic and eventually kill the cells [10]. Another reason might be due to the active yeast activity which causes the solution mixture to break out of the CO_2 generator bottle into the filter bottle which in turn decreases the CO_2 production. Overall, the fermentation with 25% sucrose concentration was found to be the most efficient for CO_2 generation purpose.

4 Conclusion

The objective of designing a carbon dioxide controlling chamber with carbon dioxide generation has been achieved in this project. Furthermore, the controlling mechanism was also defined by having a solenoid valve connected to an IoT analytic platform which enables remote regulation and monitoring of the CO_2 flow rate. In term of functionality, the retaining capability of the customised chamber was proven by the 1% CO_2 leakage. Out of the three sucrose concentrations (12.5%, 25% and 50%) being tested, it was found out that sucrose concentration of 25% was the best solution in terms of CO_2 generation by fermentation. This customised carbon dioxide chamber is ready for use with other systems to study the effects of elevated CO_2 on the growth rate and yield production of horticultural plants. For this purpose, the sample plant can be placed inside the chamber for testing. The CO_2 required for the testing can be supplied using the yeast fermentation approach demonstrated in this project.

Acknowledgements. Funding by Universiti Teknologi Brunei, Brunei Darussalam through Internal Grant Projects 2020 and sponsorship by Sarawak Convention Bureau, Malaysia are gratefully acknowledged.

References

1. Smith, H.: Light quality, photoperception, and plant strategy. Annu. Rev. Plant Physiol. **33**, 481–518 (1982)
2. Van Iersel, M.W., Bugbee, B.: A multiple chamber, semicontinous, crop carbon dioxide exchange system: design, calibration and data interpretation. J. Am. Soc. Hortic. Sci. **125**, 86–92 (2000)
3. Pedersen, S.: Carbon dioxide production in animal houses: a literature review. Agric. Eng. Int. CIGR Ejournal **10**, BC 08 008 (2008)
4. Brouwer, E.: Report of sub-committee on constants and factors. In: Proceedings of the 3rd Symposium on Energy Metabolism of Farm Animals **11**, 441–443 (1965)
5. Jentsch, W., Piatkowski, B., Derno, M.: Relationship between carbon dioxide production and performance in cattle and pigs. Archiv Tierzucht **52**, 485–496 (2009)
6. Moore, T.R., Dalva, M.: The influence of temperature and water table position on carbon dioxide and methane emissions from laboratory columns of peatland soils. J. Soil Sci. **44**, 651–664 (1993)
7. Pavlík, M., Fleischer, P., Fleischer, P., Pavlík, M., Šuleková, M.: Evaluation of the carbon dioxide production by fungi under different growing conditions. Curr. Microbiol. **77**(9), 2374–2384 (2020). https://doi.org/10.1007/s00284-020-02033-z
8. Azzami, A.M.M., Yaapar, M.N., Jusoh, M., Zulkifli, N.A.: A preliminary study on the carbon dioxide (CO_2) production from *Saccharomyces cerevisiae* fermentation activity. Trans. Malaysian Soc. Plant Physiol. **26**, 238–241 (2019)
9. Bryan, A., Hart, C., Wise, M., Roberts, B.: Glucose concentrations effect on rate of fermentation in yeast. J. Undergraduate Biol. Lab. Invest. **1** (2018)
10. Mohagheghi, A., Tucker, M., Grohmann, K., Wyman, C.: High solids simultaneous saccharification and fermentation of pretreated wheat straw to ethanol. Appl. Biochem. Biotechnol. **33**, 67–81 (1992)

Total Quality Management with Safety Conformity: Shaping of Working Environment by Use of the ISO 45001 Guidelines

Adam Górny(✉)

Faculty of Engineering Management, Poznań University of Technology, Prof. J. Rychlewski Street No 2 (Room 357), 60-965 Poznań, Poland
adam.gorny@put.poznan.pl

Abstract. The concept of TQM is a vision of an organization seeking to adapt to new environments by extending the scope of its management to new areas. One such area is occupational health and safety and, in particular, working environment. Such an environment is viewed as one of the factors that impact the performance of tasks in business organizations. The paper addresses the potential of incorporating the guidelines prescribed in the ISO 45001 standard into the TQM philosophy. It points to the option of expanding the scope of the TQM model used, as developed by Oakland, by adding new issues, particularly those relating to the working environment.

Compliance with the ISO 45001 standard allows organizations to align their measures with other fields of management. Such conformity can also be seen as a universal approach practicable regardless of the industry in which an organization operates or its size, ownership structure and business purpose. The main aim of the study is refer to the existing conditions that determine the effectiveness of the above-mentioned activities. The paper takes into account the literature information and the author's own experience related to his functioning as an auditor of occupational health and safety management systems and as a the Euro-Ergonomist (CREE).

Although the following is by no means an exhaustive discussion of the issue, the author have highlighted a vital aspect of efforts taken to enable persons responsible for specific actions to better complete their jobs. To achieve that, organizations need to scrutinize their internal processes. By examining the issues in question against the backdrop of internal processes and the labour force, organizations end up viewing their workers as internal target groups. By providing the workers with an environment that is free of negative strains and that reduces the adverse biological impacts of work, lowering the incidence of adverse events and sick leave rates, organizations can significantly increase job satisfaction. Meanwhile, with improved work quality and efficiency, businesses also reduce their operating expenses.

Keywords: Total Quality Management · Safety · Work environment · ISO 45001 · Occupational hazards and risk

A. Batako et al. (Eds.): GCMM 2021, LNNS 335, pp. 57–70, 2022.
https://doi.org/10.1007/978-3-030-90532-3_5

1 Introduction

The concept of TQM can be seen as multidimensional integrated management of an organization's various fields of operation. Such management relies on the application of rules, practices and techniques aimed at increasing efficiency and quality with a view to meeting customer requirements and expectations [1]. Improvements and their effects are achieved by securing the involvement of all members of the organization in efforts designed to benefit its stakeholders. TQM becomes an improvement driver that inspires the continuous betterment of operations. Its use entails raising awareness and ensuring across-the-board engagement throughout the organization [2, 3]. In the consequence, it improves the effectiveness of systemic management, and allows for more effective achievement of the goals indicated in the adopted policy.

As for the essence of the issue at hand, which is viewed with reference to factors for the effective conduct of processes, it should be noted that a working environment that provides safety for all workers in the organization is an indispensable part of the requirements discussed here. Workers in a process-focused organization are seen as customers for internal processes. Thus, organizations need to account for their needs and expectations from the very outset of process performance and provide workers with a friendly environment for their process-related work. All this enables organizations to achieve their desired results [4–6]. Any individuals operating in such a system must be empowered to influence the transformations taking place around them. Only then can one speak of informed engagement in the change effort [3].

By taking ISO 45001 requirements on board, organizations become more effective in deploying their measures, especially where TQM is extended to other safety improvements achieved by properly shaping environments of process. However, it requires a proper connection of the existing conditions and support the attention to the determinants of effective implementation of tasks. The key potential benefits to be derived from adopting the ISO standard include reductions in workplace incident rates (and especially those of severe incidents), the management's stronger commitment to proactively improving OHS performance and creating a robust health and safety culture in which workers are encouraged to play an active role in promoting their own OHS and to report incidents and non-conformities [7]. In such an approach, workers are viewed as a resource whose proper utilization will boost an organization's ability to accomplish the outcomes it is pursuing. The working environment should also be seen as a resource [4]. Its conformity with worker requirements and expectations is a precondition for the proper completion of tasks. Failures to provide workers with an adequate working environment may prevent organizations from achieving quality criteria in their processes.

As suggested by TQM guidelines, workers will perform better if they are familiar with and understand the goals of their company and the purposes of the measures it takes [8]. TQM will only ensure efficient performance of processes if adequate conditions are provided for meeting requirements. Process environments and indicators need to be interlinked to achieve the desired results. In addition, better conformity with one set of requirements must not result in the neglect of any others [9]. Arbitrary standards, worker evaluations and other verification tools are not the correct approach to increase production and worker efficiency. To accomplish all of the desired results, TQM must account for all aspects of an organization's activities [8]. By bringing all such aspects together, organizations can be certain to achieve their goals, no matter what industry [10, 11]. The approach must incorporate guidelines for managing the working environment, especially where such guidelines affect the ability to complete tasks. Special consideration should also be given to those parts of the environment in which safety shortcomings result from unacceptable working conditions having deleterious effect on workers. Given their significance for an organization's ability to achieve the desired results, such factors deserve special consideration in process management.

2 Guidelines for Managing Occupational Safety Management Systemically in Conformity with ISO 45001

The purported purpose of management in any area of operations is to utilize the available resources with a view to generating specified benefits. The foregoing applies also to occupational health and safety management that duly accounts for the group at which its measures are aimed and surroundings of the performed tasks.

At company level, occupational health and safety management applies to such aspects of organizational management as involve the development and pursuit of a safety and health policy covering workers and any specific safety objectives formulated on its basis [12, 13]. Such management extends to fields traditionally associated with systemic management (such as organizational structures and responsibilities within organizations, resource planning and procurement, rules of conduct, recordkeeping, information flows, purchasing, and subcontracting) and fields of essential importance for occupational safety. They include rules for responding to hazards, contingency planning guidelines and rules of emergency rescue and victim assistance [14]. In view of the nature of occupational safety measures, which mainly target company workers, one may posit that the key aim of occupational safety management in business organizations is to facilitate such measures, which provide an assurance of the efficient performance of workers. This, in turn, can be seen as a sine qua non condition for enabling workers to carry out their jobs [15]. To achieve these objectives, organizations need to create a working environment that fulfills worker needs, as commonly outlined in laws as well as the recommendations and standards that complement them.

The organizations that adopt the requirements laid down in the international ISO 45001 standard may be said to contribute to a more efficient management of their business and to provide a worker-friendly working environment [12, 16]. ISO 45001 has been drafted to help [17, 18]:

- Increase the capacity of companies to assess hazards and take measures to prevent the adverse impact of such hazards on workers and company operations,
- Account for the impact of the increasingly complex environments of organizations,
- Ensure an integrated approach to occupational safety management and integration with other management systems.

In order to improve management effectiveness, these requirements must be integrated into the PDCA improvement loops. It allows you to carry out tasks efficiently, based on the current analysis of the order and conditions. The proposed solution relies on the use of the iterative PDCA method comprised of planning actions in response to identified risks and opportunities as well as processes critical for delivering desired results, implementing a methodology consistent with the action plan and assessing actions to determine their compliance with policy objectives. The key purpose behind such steps is to secure the delivery of the desired result [16]. The measure that is recommended to be used in order to determine the performance of tasks is risk assessment. The recommended way of finding the conditions that are appropriate for completing the task is through risk assessment. To conduct the assessment, an organization has to identify the impact of all areas of uncertainty on the outcomes of company operations and evaluate its ability to achieve and verify the desired results through its actions so as to detect disruptions in the achievement of such desired outcomes as are in keeping with the maturity of its management and as dictate the manner in which tasks are performed [12]. Taking into account all the conditions related to the functioning of the enterprise is a necessary condition for achieving the intended effects, which are the result of improvement.

Solutions based on systemic management guidelines enable organizations to ensure safety at the required level by following the continuous improvement loop [12]. The solutions adopted in line with PDCA guidelines create the capacity to establish appropriate conditions for the functioning of workers in the working environment [19]. This is vital for any organization facing numerous risks and hazards in its operations. For such an organization, mitigating adverse working-environment impacts on worker health and safety and facilitating the performance of work by workers constitutes an essential part of the requirements. Once put in place, proper working conditions substantially reduce the strains experienced by workers in the workplace. Improvements in line with ISO 45001 guidelines allow businesses to identify the need to meet the desired safety standards and determines the scope of efficiency improvements, helping them to accomplish their intended business objectives [20]. Operating efficiency can be achieved by focusing on various aspects of management. The desired end result is for actions to be both effective and efficient.

In the course of improvement measures, proper note should be taken of preconditions critical for the capability to deploy specific solutions. Such preconditions are [21]:

- Proposing improvement measures that realistically account for existing constraints on time, budgets, technologies and procedures,
- Adopting solutions that are commensurate with the needs that prompt their use,
- Ensuring the solutions lastingly prevent the recurrence of problems,
- Having the solutions accepted by all of the organization's persons and business units that are either directly or indirectly impacted by related decisions,
- Having solutions whose cost in a financial and broader sense is commensurate with that to the organization resulting from their deployment and their environmental impact.

Note that ISO 45001 places considerable emphasis on the environment in which the organization operates and pursues its goals. A systemic approach to occupational management compels managers to continuously track impacts on the local community (e.g. the population of the town in which the organization does its business). To ensure consistency with systemic requirements, a broad view of the problem is required. An organization may not limit itself to merely assessing direct hazards and their impacts on the local community and to conducting its actions in operational terms only. It is essential to engage in cooperation with the local community and to consult all stakeholders [20]. For that to happen, organizations need to examine their options for effectively implementing solutions so as to ensure they will successfully satisfy all preconditions for effective occupational safety management and ultimately further their own development.

3 Essential Considerations for Incorporating ISO 45001 Requirements into the TQM Approach

Total Quality Management can well be described as a form of organizational management geared towards achieving long-term success. Such success is measured by customer satisfaction and the benefits provided to all stakeholders. Hence, any improvements that make an organization more efficient should be adjusted for the specific nature of its business and the specific challenges it faces. A reference to relevant requirements allows the organization to give priority to the expectations of workers seen as internal customers for ongoing processes. This can be treated as a necessary condition for improving the efficiency of systemic management.

In an effort to ensure effective management in conformity with TQM guidelines, proper account needs to be taken of all factors contributing to a company's success. Such factors include resources, which in turn include a working environment that enables the organization to accomplish its objectives. Equally important are mission-critical factors traditionally associated with the business organizations that pursue their goals systemically. These are leadership, customer satisfaction, process efficiency, good relationships with the suppliers of goods and services, motivation to act and competent staff. All such factors must be embraced as a prerequisite for the successful delivery of desired results in the pursuit of the company's objectives, regardless of industry.

The incorporation of working environment management into the applied TQM model can be guided on the set of guidelines proposed by Oakland [22]. Regardless of the field in which working environment requirements are adopted, once preconditions critical for the effective completion of tasks have been satisfied, the effective operation of the new structure will invariably require a focus on the efficient fulfillment of customer and supplier needs and expectations. This result can be delivered by ensuring that:

- Tasks are carried out by professional (properly trained) workers who rely on the synergies of teamwork,
- All persons involved in the completion of tasks follow the adopted principles of systemic management that are based on tried and tested methodologies,
- In its effort to improve the effectiveness and efficiency of task performance, the organization relies on improvement methods and tools that enable it to identify challenges and deliver improvement measures,
- The adopted methods enable the company to act in an orderly and duplicable manner on the basis of data and information collected with the use of adequate tools. The organization needs to recognize that its success hinges on employing a very simple tool set that will allow it to collect and process data and information on various aspects of business management.

A synergy among all these areas can only be achieved in an environment that is characterized by a proper culture, proper commitments, adequate capabilities and the will to communicate both among the workers themselves and between the workers and the local community in which the organization is embedded. A crucial part of such requirements are issues having to do with the workplace, the conformity with which ensures the delivery of desired results. To that end, the workplace should be designed to facilitate the efficient conduct of the processes that contribute to increasing customer and supplier satisfaction. These are conditions explicitly equated with ensuring the effectiveness of the measures implemented.

The nature of links among all such fields of action is shown in Fig. 1.

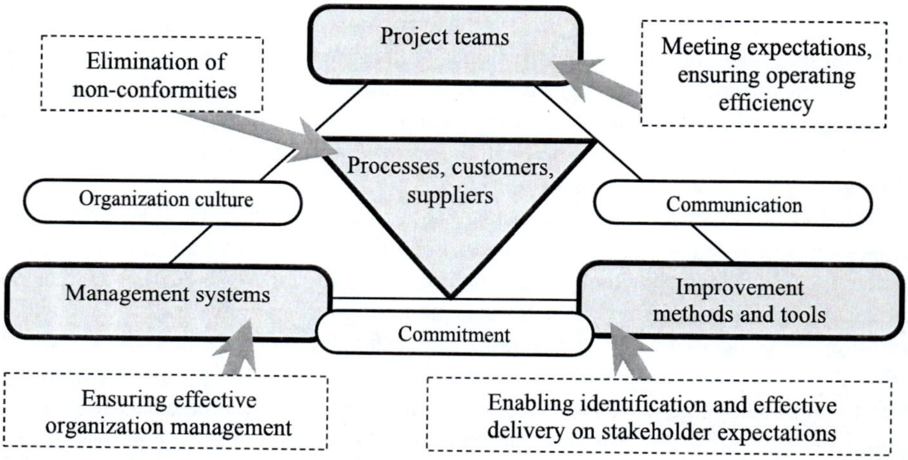

Fig. 1. The Oakland TQM model created in conformity with working environment guidelines. Source: Author's research based on [22, 23].

The measures to be undertaken in conformity with ISO 45001 to further efforts to satisfy the requirements of the organizational model proposed in TQM guidelines are presented in Table 1.

In all such efforts, it is vital to embrace various aspects of the working environment, which should guide the manner in which tasks are conducted and ultimately facilitate the functioning of workers and mitigate any adverse impacts on their health and lives. Working environment parameters may therefore be seen as a test of the environment's alignment with worker expectations. Such parameters are equally important for the operation of business as quality, efficiency, financial and performance indicators, etc. Neither should one overlook the effects of environmental factors on company operation in fields critical for the delivery of results.

The primary objective behind the solutions discussed here is to prevent losses caused by companies failing to deliver in areas that are essential for meeting stakeholder needs and expectations. Responsibility for such prevention falls squarely on company management. Their efforts are then furthered, either directly or indirectly, by all workers across the organization. Specifically, such efforts include [24, 25]:

Table 1. Measures taken in keeping with ISO 45001 guidelines.

The area of requirements in line with the TQM model	Characteristics of undertaken measures
Management systems	Ensure effective organization management by: – Incorporating working environment considerations into the requirement matrix to enable the organization to deliver on its commitments, – Referring to business principles and rules to prevent the manifestations of any non-conformities identified in the organization's environment, – Embracing guidelines for the safe performance of tasks to enable the organization to achieve its aims and desired results with respect to all significant stakeholders, – Incorporating factors for the safe performance of tasks into the overall management matrix, – Securing such environmental and other resources as are necessary for timely delivery on commitments
Processes, customers, suppliers	The requirement to eliminate all non-conformities imposed by external parties is met in day-to-day operations by: – Embracing working environment principles and requirements in the organization's strategy, – Setting goals based on guidelines designed to secure a range of benefits including, most importantly, those that affect internal processes, – Eliminating process-performance-related impediments to the delivery of benefits to customers and to the extension of measure scopes to cover suppliers and sub-suppliers, – Adopting non-conformity identification processes and incorporating non-conformity identification capacities into every process in the organization, – Planning and overseeing the conditions in which processes are conducted, – Recognizing the findings of conformity assessments and the findings of assessments of measures taken to increase their effectiveness, – Linking all cost and loss reduction actions with the organization's business processes

(*continued*)

Table 1. (*continued*)

The area of requirements in line with the TQM model	Characteristics of undertaken measures
Project teams	In order to meet the expectations of all parties involved in the performance of tasks that follow existing processes and to ensure their operating efficiency, organizations must: – Make a commitment to ensuring safe operation, – Respect systemic management guidelines covering the identification of risks and opportunities for the development of safe operations, – Ensure the active involvement of workers in shaping and meeting expectations in keeping with working environment guidelines, – Make efforts to help improve the functioning of workers in their specific environments, – Recognize the importance of working conditions and the related measures for their efficient operation
Improvement methods and tools	Stakeholder expectations are identified and met by: – Applying methods and tools helpful in defining, effectively assessing and deploying measures in keeping with stakeholder expectations, – Employing process improvement methods and tools that cover various aspects of the working environment
The desired results are achieved by way of measures that:	
– Culture	– Promote a culture supportive of efforts to develop worker-friendly working conditions
– Communication	– Ensure communication with all parties involved in the work process, – Ensure transfers of knowledge on the organization's operations and on guidelines for creating a working environment
– Commitment	– Involve the management committing to solutions that account for the environmental aspects of creating and managing working conditions, – Encourage proactive efforts to develop the working environment

Source: Own work based on ISO 45001 with proper account taken of [5, 8, 12, 17, 22]

- Establishing a policy of improvement that will allow the organization to identify goals whose delivery requires it to create a worker-friendly working environment that will enable workers to achieve the desired results,
- Taking improvement measures as part of a continuous improvement effort that recognizes both existing non-conformities and the adopted integration goals,
- Prioritizing the need for the continuous improvement of measures while mitigating the adverse impacts of the working environment on workers,
- Unifying the management systems in place to ensure that improvements across various areas of the company's operations are pursued on the basis of the same or twin models underpinned by the same guidelines,
- Proactively managing the working environment to either mitigate or fully eliminate its adverse impact on the ability to carry out tasks and on worker performance,
- Promoting health education and training that forms part of and affects the scope of the continuous improvement effort,
- Adopting solutions that effectively eliminate any existing hazards and strains and reflect the specific nature of their impacts and links with ongoing integration,
- Ensuring preparedness to adequately respond to departures from the desired state by adhering to relevant contingency plans to the extent possible,
- Ensuring conformity of actions with the relevant laws and regulations that the organization has committed to respect,
- Basing the measures taken on findings of risk assessments recognized as indicative of the extent to which specific non-conformities compromise the organization's ability to benefit from efficient operation.

It should be noted that the above guidelines cannot be considered as equivalent to systemic management guidelines in every area of company operations and in every organization regardless of the nature of its business. The reason for this is that specific requirements come into play and need to be considered in many fields of management [16].

4 Benefits of Incorporating Working Environment Guidelines into the TQM Concept

As technological advances continue, manufacturing systems grow more complex requiring more effective interactions among their components. It is therefore essential to pay particular attention to creating conditions that will enable human operators to function properly in specific technological and social environments. The adoption of Total Quality Management helps organizations to satisfy worker needs and improve working environments while embracing specific aspects of systemic management. One such potential aspect is the systemic management of occupational health and safety [21], which helps workers perform their duties properly and effectively.

TQM may be viewed as a way to eliminate business risks, in particular those pertaining to relations with customers, beneficiaries and their direct suppliers [9]. This is of particular importance for companies whose options for improving their business are highly limited. In such circumstances, methods commonly associated with quality engineering help improve effectiveness. Quality tools can be used to collect data that is essential for ensuring proper improvements and the proper application of methods that support management. In particular, management tools help [26–29]:

- Group existing situations by appropriate grouping criteria,
- Monitor changes as they unfold,
- Identify relationships between analyzed factors,
- Rank risks and strain factors and their underlying causes,
- Determine the capacity to undertake improvements and identify the required preconditions that need to be met to achieve desired benefits,
- Visualize the course of improvement activities to support a more accurate evaluation.

The improvements of working conditions that result from adopting systemic occupational health and safety management principles allow organizations to modify worker behaviors on safety and create a new safety culture [30, 31]. Such behaviors and culture are critical for achieving lasting outcomes such as the effective performance of tasks and the elimination of non-conformities in the functioning of workers. However, the process of securing the desired results cannot be external to the organization [22]. It requires the engagement of both the management and rank-and-file employees and their representatives, including persons responsible for managing the working environment. These activities are results from taking into account the guidelines of systemic management. The incorporation of systemic requirements into the overall structure of actions increases the effectiveness of TQM projects. This is done primarily by enabling organizations to recognize certain facts, confirm their suppositions regarding the nature of the challenges they face, carry out fair assessments of their status, identify issues and their underlying causes, and define adequate responses that will allow them to accomplish their objectives. All in all, such measures allow organizations to effectively address and resolve problems as they arise. The structure of the above activities implementation is presented in Fig. 2.

Business managers are increasingly aware of the benefits of deploying working environment management systems based on management models that are analogous to other systems, especially those subject to certification. The vast majority of the measures they take conform to quality paradigm guidelines while furthering such ambitious total management models as Total Safety Management [32]. By embracing the requirements of ISO 45001 in improving the working environment, organizations make their systemic actions more efficient and unified. This applies in particular to organizations whose prior systemic efforts focused on other areas of management. The existing and potential non-conformities that guide improvement measures based on changes observed in the environment can be resolved by methods closely resembling those employed in other fields of operation. In such circumstances, working environment characteristics should be seen as determinants of actions that benefit workers in the form of improved working conditions.

By incorporating systemic principles of working conditions management into Total Quality Management systems, organizations stand to streamline their operations. One way towards that goal requires the adoption of basic risk assessment management techniques and the formalization of worker responsibilities for safety and safety training.

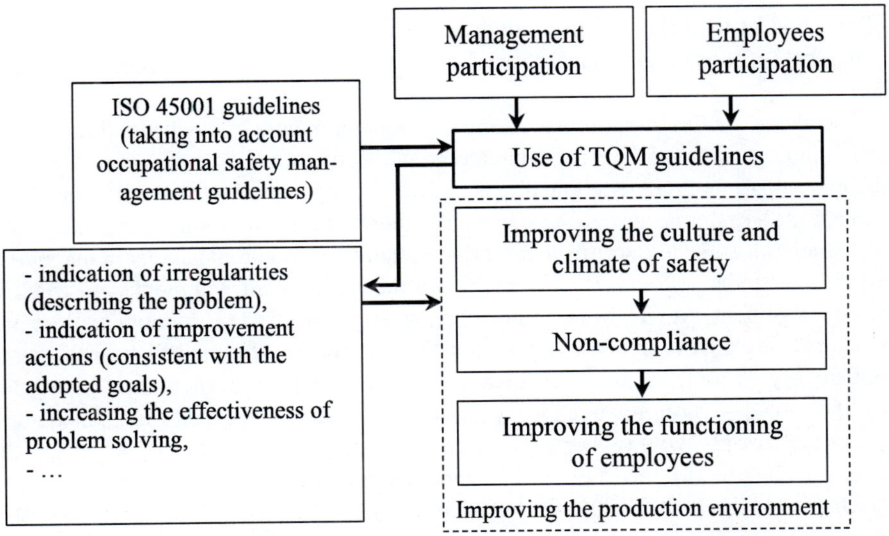

Fig. 2. The structure of the activities used to improve the production environment.

Businesses should also recognize the need to actively raise the awareness of persons involved in the performance of tasks. Surveys show clearly that Total Quality Management has the potential to promote safety management systems even where well-tested basic practices are not being followed [14, 32]. By combining the potential of ISO 45001 with ISO 45001 guidelines, organizations can make their efforts to achieve desired results even more efficient [12]. Ultimately, organizations gain greater capacity to:

– Integrate safety management with other management systems to make the overall management structure more uniform,
– Respond to more complex environments which they need to control to improve their operations,
– Assess hazards in their working environments and the impacts of such hazards on their ability to perform tasks and ultimately meet customer expectations,
– Modify their operations for better operational quality and capacity.

The benefits result from effective improvement measures undertaken in the following five basic fields:

– The identification of internal and external challenges in business operations and specifically in the environment in which tasks are performed,
– The elimination or mitigation of risks accordingly to the nature of existing hazards and strains and to the findings of assessments of the capacity to adopt specific improvements,
– Effective efforts to achieve conformity with requirements laid down in any relevant laws and regulations that the organization has committed to respect,
– Preparedness for an instant response to failures, which requires the prior identification of relevant hazards and simulated testing of emergency response measures,

– Extension of the measures in place to all business units and companies that operate on behalf of the company and in which systemic requirements are being adopted.

The above applies in particular to organizations seeking to move beyond traditional occupational health and safety management that is commonly seen as limited to minimizing occupational accidents and diseases. Such management can hardly be considered to constitute a systemic approach to managing the working environment and to measurably benefit all stakeholders. To incorporate systemic occupational safety management into their systems, especially with respect to the working environment, and to align such management with TQM principles, organizations need to profoundly explore any safety affecting factors and embrace the tenets of modern pro-humanistic business management. By doing that, they will extend their improvements to such areas associated with the working environment as impact task performance in various ways. Such areas include the spatial layout of workplaces, the organizational efficiency of processes, the use of human-friendly forms of work organization, and their psychological and social aspects [4, 14, 19, 31, 33]. This puts businesses in a position to create organizations that are open to market challenges and view humans as a foundation for their development. As they do so, businesses will benefit from the growing dedication of their workers to the performance of tasks.

5 Summary

In order to effectively implement improvement measures, organizations need to employ such tools and methods as will adequately enable them to identify, diagnose and effectively remove challenges as they arise. This applies to the working environment and, similarly, to other areas of company operations. Adverse impacts of factors in the working environment commonly result from a disconnect between the characteristics of the working environment and criteria adopted to assess working conditions and align tasks with the capacities of human operators. To resolve the issue, it is essential that all workers throughout the organization engage in advancing improvements. Once they do, the organization will achieve benefits that become evident in assessments of operational capacities.

It was indicated, based on the conducted literature review, that taking into account the TQM guidelines is necessary to ensure effective improvement. In addition, the rank and significance of the suggested activities were indicated by referring them to the guidelines of systemic management of work safety. Once organizations conform to TQM guidelines in their performance of tasks, the adoption of ISO 45001 will produce efficiency gains in such performance, increasing their chances of satisfying requirements, especially in the management of occupational safety and the working environment. The use of systemic requirements in tandem with adequate infrastructure and training should enable companies to accurately identify issues related to the performance of work, undertake improvement measures and ultimately ensure the effective operation of their business. Solutions compliant with ISO 45001 help improve safety while reducing operating expenses and mitigating potential losses that may result from failures to establish a proper manufacturing environment. However, given considerable similarities between the provisions of ISO 45001 and the rules of other areas of systemic management, such benefits may additionally arise in other areas, such as business environment management. In addition, human-friendly forms of work organization are designed to adjust the working environment to human needs, effectively reduce strains and increase job satisfaction. An ergonomically advanced working environment will help support work efficiency, reduce the biological cost of work, minimize the number and cost of flaws, reduce the number of human errors, curb sick leaves and boost the enjoyment of interacting with technical equipment. Verification of the indicated concept and its guidelines requires relying on real data and information from enterprises operating under market conditions. The potential achievement of compliance with the assumptions should be treated as a confirmation of the assumed conditions.

References

1. Bajpai, P.: Total quality management. In: Biermann's Handbook of Pulp and Paper, vol. 2: Paper and Board Making, pp. 443–454. Elsevier, Amsterdam (2018)
2. Kiran, D.R.: Total Quality Management. Key Concepts and Case Studies. Butterworth-Heinemann, Oxford (2017)
3. Iqbal, A., Asrar-ul-Haq, M.: Establishing relationship between TQM practices and employee performance: the mediating role of change readiness. Int. J. Prod. Econ. **203**, 62–68 (2018)
4. Górny, A.: Man as internal customer for working environment improvements. Proc. Manuf. **3**, 4700–4707 (2015)
5. Górny, A.: The element of work environment in the improvement process of quality management system structure. In: Karowowski, W., Salvendy, G. (eds.) Advances in Human Factors, Ergonomics and Safety in Manufacturing and Service Industries, pp. 599–606. Taylor and Francis Group, Boca Raton (2011)
6. Rae, A., Provan, D.: Safety work versus the safety of work. Saf. Sci. **111**, 119–127 (2019)
7. Ewertowski, T.: Just culture as a useful tool for the organizations in the context of ISO 45001:2018 standard implementation. In: Pacholski, L., et al. (eds.) Human Factors in Contemporary Organizations, pp. 95–103. DEStech Publications, Lancaster (2019)
8. Prajogo, D.I., Sohal, A.S.: The relationship between organization strategy, total quality management (TQM), and organization performance - the mediating role of TQM. Eur. J. Oper. Res. **168**(1), 35–50 (2006)
9. Tisca, I.A., Cornu, G., Diaconu, N., Dumitrescu, C.D.: Diagnosis risk and efficiency in the implementation of TQM in small and medium enterprises. Proc. Econ. Financ. **26**, 215–218 (2015)
10. Pereira-Moliner, J., Claver-Cortés, E., Molina-Azorín, J.F., Tarí, J.J.: Quality management, environmental management and firm performance: direct and mediating effects in the hotel industry. J. Clean. Prod. **37**, 82–92 (2012)
11. Mrugalska, B.: Lean and ergonomics competencies: knowledge and applications. In: Ahram, T., Karwowski, W., Pickl, S., Taiar, R. (eds.) IHSED 2019. AISC, vol. 1026, pp. 654–660. Springer, Cham (2020). https://doi.org/10.1007/978-3-030-27928-8_100
12. Górny, A.: Occupational health and safety management in the international condition (consistent with objectives the ISO 45001 standard). Modern Manage. Rev. **22**(4), 73–88 (2015)
13. Li, Y., Guldenmund, F.W.: Safety management systems: a broad overview of the literature. Saf. Sci. **103**, 94–123 (2018)
14. Zymonik, Z., Hamrol, A., Grudowski, P.: Zarządzanie jakością i bezpieczeństwem, Polskie Wydawnictwo Ekonomiczne, Warszawa (2013). (in Polish)

15. Choe, S., Seo, W., Kang, Y.: Inter- and intra-organizational safety management practice differences in the construction industry. Saf. Sci. **128**, 104778 (2020)
16. Morgado, L., Silva, F.J.G., Fonseca, L.M.: Mapping occupational health and safety management systems in Portugal: outlook for ISO 45001:2018 adoption. Proc. Manuf. **38**, 755–764 (2019)
17. ISO/DIS 45001: Understanding the new international standard for occupational health & safety. Mapping Guide, BSi Group, 03 April 2020. http://www.bsigroup.com/en-GB/ohsas-18001-occupational-health-and-safety/ISO-45001
18. ISO 45001:2018: Occupational health and safety management systems. Requirements with guidance to use. International Organization for Standardization, Geneva (2018)
19. Sławińska, M.: Ergonomic Engineering of Technological Devices. Publishing House of Poznan University of Technology, Poznan (2019)
20. Kazutaka, K.: Roles of participatory action-oriented programs in promoting safety and health at work. Saf. Health Work **3**(3), 155–165 (2012)
21. Gauthereau, V., Bergman, B.: Safety from a TQM perspective. In: Cottam, M.P., Pape, R.P., Harvey, D.W. (eds.) Proceedings of Conference on Foresight and Precaution Location, Foresight and Precaution, pp. 1237–1238. European Safety & Reliabil Assoc., Edinburgh (2000)
22. Oakland, J.S. (ed.): Oakland on Quality Management. Elsevier, Amsterdam (2004)
23. Górny, A., Sadłowska-Wrzesińska, J.: Ergonomics aspects in occupational risk management. In: Arezes, P., et al. (eds.) SHO 2016, Occupational Safety and Hygiene, pp. 102–104. Portuguese Society of Occupational Safety and Hygiene (SPOSHO), Guimarães (2016)
24. Górny, A.: Use of quality management principles in the shaping of work environment. In: Stephanidis, C. (ed.) HCI 2015, Part II. CCIS, vol. 529, pp. 136–142. Springer, Cham (2015). https://doi.org/10.1007/978-3-319-21383-5_23
25. Occupational Health and Safety Management. BS OHSAS 18001 moving to ISO 45001. Rapport of International Register of Certificated Auditors, Charted Quality Institute, London (2014)
26. McCormick, K.: Quality. Butterworth-Heinemann, Oxford (2002)
27. Starzyńska, B.: Practical applications of quality tools in Polish manufacturing companies. Organizacija **47**(3), 153–164 (2014)
28. Škulj, G., Vrabič, R., Butala, P.: Experimental study of work system networking in production environment. CIRP Ann. **63**(1), 401–404 (2014)
29. Woźniak, K. (ed.): Contemporary Tools for Improving Organization Management Systems. Mfiles.pl, Cracow (2012). (in Polish)
30. Frick, K.: Worker influence on voluntary OHS management systems – a review of its ends and means. Saf. Sci. **49**(7), 974–987 (2011)
31. Gabryelewicz, I., Sadłowska-Wrzesińska, J., Kowal, E.: Evaluation of safety climate level in a production facility. Proc. Manuf. **3**, 6211–6218 (2015)
32. Álvarez-Santos, J., Miguel-Dávila, J.-Á., Herrera, L., Nieto, M.: Safety management system in TQM environments. Saf. Sci. **101**, 135–143 (2018)
33. Debnath, A.K., Blackman, R., Haworth, N.: Common hazards and their mitigating measures in work zones: a qualitative study of worker perceptions **72**, 293–301 (2015)

First Principle Calculation and Data Analysis of Anisotropy in Elastic Properties of Cubic Carbides TiC, VC and NbC

C. Tang[1](✉), J. Guo[1], B. Li[1], O. Kostenevych[2], L. Wang[3], G. Rothwell[3], and J. Ren[1]

[1] School of Engineering, Liverpool John Moores University, Liverpool L3 3AF, UK
c.tangliverpool@gmail.com

[2] The Experimental Design-Technological Office of the E.O. Paton Electric Welding Institute of the National Academy of Sciences of Ukraine, Kyiv, Ukraine

[3] School of Engineering, University of Bolton, Bolton BL3 5AB, UK

Abstract. In this work, first principle calculation is applied to establish systematic ground state elastic property data of three typical cubic monocarbides (Titanium carbide (TiC), Vanadium carbide (VC) and Niobium carbide (NbC)). The bulk modulus, Young's modulus, shear modulus, and Poisson's ratio are determined from the Voigt-Reuss-Hill approximation method and compared with other published data. The micro-Vickers hardness is also predicted, the hardness data is compared to published data. The anisotropy in elastic properties of the carbides is studied through calculating the universal elastic anisotropic index; percent anisotropy in compressibility and shear as well as the shear anisotropic factors on the specific crystal planes. A python program is developed to integrate first principle calculation in Materials Studio data and establish the directional anisotropy in elastic properties in 3D space and projection on key crystal planes. 3D constructions and plane projection of the elastic properties based on the maximum, minimum and averaged values approaches is used to present and compare the anisotropy feature of the carbides. The data for E, G and Poisson's ratio is comparatively analysed and discussed, and future direction in developing systematic data through physical modelling is discussed.

Keywords: First principle · Anisotropy · Cubic carbides

Nomenclature

A_{comp}: Percent anisotropy in compressibility
A_{Shear}: Percent anisotropy in shear
A_U: Universal elastic anisotropic index
C_{ij}: Stiffness tensor
E: Young's Modulus
G: Shear modulus
H_v: Vickers hardness
K: Bulk modulus
S_{ij}: Compliance tensor
v: Poisson's ratio

A. Batako et al. (Eds.): GCMM 2021, LNNS 335, pp. 71–85, 2022.
https://doi.org/10.1007/978-3-030-90532-3_6

1 Introduction

The ground state structural, elastic, mechanical property data of different carbides (e.g. Titanium carbide (TiC), Vanadium carbide (VC) and Niobium carbide (NbC)TiC, VC and NbC) are important for material development and processing. Carbides are widely used in key engineering ceramics and manufacturing tools. Carbides also exist as secondary particles in steels, the structure and properties of which may affects the stiffness, strength, hardness, toughness as well as the grain structures and manufacturability of the material. This is particularly relevant for materials with complex alloying systems such as stainless steels and welded structures. An in-depth understanding of the crystal structures, the mechanical and physical properties of these phases are important for the prediction of the deformation and performance in service. Detailed data of these high melting point phase also relevant to other processes such as nucleation and growth of metallic phases as well as other complex carbides (e.g. Ferrite, Austenite, M7C3) [1–3].

First principle calculation is increasingly used in predicting crystal structure, electronic, physical and elastic properties of ceramics (e.g. carbides, nitrides, and multicomponent carbides). First principle calculation is a method to calculate physical properties directly from basic physical quantities (such as the mass and charge, and coulomb force of an electron, etc.) based on the principle of quantum mechanics. It naturally links multidisciplinary research areas and data in physics, chemistry, materials and processing such as steel making, and welding. Typical relevant data from first principles calculation covers elastic constants, physical properties, thermal properties, interface between different phases, electrochemical, oxygen reduction reaction, corrosion, etc. [4–6]. The computerized calculation process and subsequent large data analysis generate a rich source of materials data for establishing more detailed understanding of material parameters and their correlations. For analysis of mechanical properties, the prediction of the elastic properties involves mathematical approximation combining the Voigt, Reuss and Hill bounds of bulk and shear moduli [7–9]. In addition, the data from first principle calculation and subsequent analysis also provide effective means to estimate the properties such as Vickers hardness based on the ground state elastic parameters/properties [10, 11]. As materials informatics is increasingly being used in engineering analysis and design of complex materials systems, data from physical modelling also offers new opportunities in further understanding of the correlation between different sets of properties with more details. In addition, by integrating first principle calculation with data analysis, it is possible to establish data for enhancing the understanding and visualization of the property difference in different crystal planes or directions. The systematic data in anisotropy is often challenging, or in some cases very difficult, to be established through pure experimental approaches in particular for secondary particles in steel and welded structures.

In this work as part of a main project in developing data led approach in advanced steels and welded structures, first principle calculation is used to establish systematic data of three typical cubic binary monocarbides (TiC, VC and NbC). The key elastic constants are obtained from the first principle calculation. The bulk modulus, Young's modulus, shear modulus, and Poisson's ratio are determined from the Voigt-Reuss-Hill approximation method and compared with other published data. The micro Vickers hardness is also predicted and compared to published data. The anisotropy in elastic properties of the carbides is studied through calculating the universal elastic anisotropic index, percent anisotropy in compressibility and shear as well as the shear anisotropic factors on the specific crystal's planes. A python-based program is developed and used to integrate first principle calculation (using Materials Studio CASTEP) and data processing to establish the elastic properties in 3D space and projection on key crystal planes. The plane projection of the elastic properties on the (100) plane based on the maximum, minimum and averaged values approaches is used to present and compare the anisotropy feature of the carbides. The data for E, G and Poisson's ratio is comparatively analysed and discussed, and future direction in developing systematic data through physical modelling is discussed.

2 Computational Method and Data Integration

Figure 1 shows the crystal structures of carbides TiC, VC and NbC. TiC is a binary carbides of group VI metal, while the VC and NbC are binary carbides of Group V metal. These carbides are widely used in making key engineering ceramics. In addition, as secondary phases, their structures and properties are also critical for high alloy steels and welded structures. It affects the strength, hardness and thermal stability, as well as the microstructures (e.g. the nucleation of ferrite/austenite). These carbides were also reported to have refinement functions on other more complex multi-components carbides systems such as M(Fe, Cr)7C3 carbides [12]. All these three carbides follow a cubic structure. The carbide for Ti has only one main form (TiC) while Niobium and Vanadium may form carbide of different formulas and compositions (NBC and NB2C for Nb; V8C7, V6C5, V4C3, V5C3, V2C for V) [13, 14]. The results presented in this paper is focused on data for the simple form of the carbides and anisotropy in the elastic properties analysed through general anisotropy index and directional anisotropy through 3D surface configuration and plane projections.

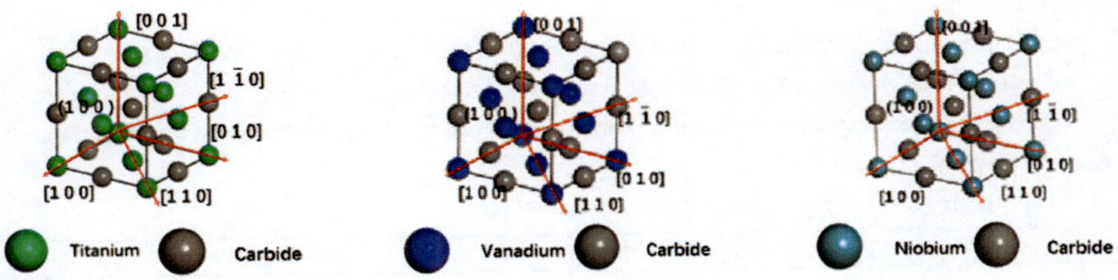

Fig. 1. The crystal structure of (a) Titanium carbide (TiC), Vanadium carbide (VC) and Niobium carbide (NbC).

Figure 2 shows the data flow and main works integrating the calculation in Materials studio CASTEP, ground state property estimation and anisotropy analysis. Materials Studio CASTEP is an ab initio quantum mechanical program that was developed based density functional theory (DFT) to simulate the properties of solids as well as characteristics of interfaces and surfaces. First principle calculation is able to calculate physical properties directly from basic physical quantities (such as the mass and charge, Coulomb force of an electron, etc.) based on the principle of quantum mechanics. As shown in Fig. 2, the main functions in the materials studios for calculating the mechanical properties is the geometry optimization and elastic constants prediction, this procedure can optimize the cell size (Lattice parameters) by identifying stable structure with lower total energy. The Generalized gradient approximation (GGA) with Perdew-Burke-Ernzerhof function (PBE) was applied by CASTEP code package. The main data output from the process is the lattice parameters including the lattice constants and the angles. Then the elastic stiffness constants are determined and output in a matrix form. As shown in Fig. 2, in the second stage, a python program is developed to calculate the compliance tensors (Sij), which is the derivative of strain with respect to stress. Then the ground state elastic properties are calculated based on the Voigt–Reuss–Hill (VRH) equations [15]. The theoretical hardness of the carbides are also determined from the bulk and shear modulus. The general anisotropic index of the crystal is calculated, which represents the difference in the values of K, G and E. The data is a generalized property without considering the actual anisotropy at directions. In the third part, a more complex program and data analysis is performed to plot the elastic properties in spherical coordinates to visualize the data of K, G, E and v in 3D, which provides a mean to analyze the directional anisotropy in E, G and Poisson's ratio and their projections on key crystal planes. This provide a more detailed way to quantify and visualize the anisotropy in different elastic properties.

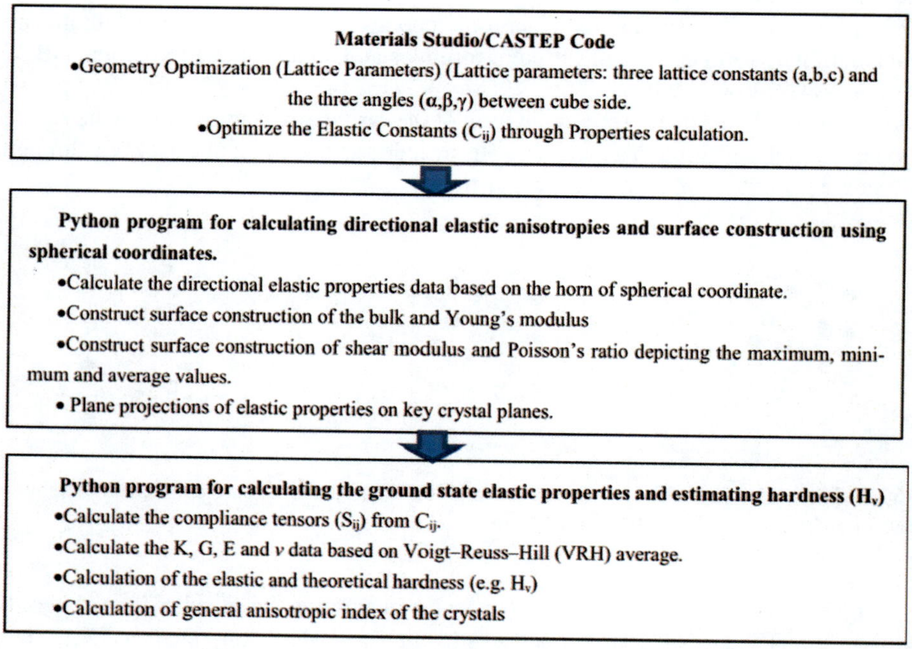

Fig. 2. Main programs integrating materials studio and data.

3 Results and Analysis

3.1 Elastic Constants and Ground Elastic Properties

The elastic constant and structures optimization were performed in Material Studio. The generalized gradient approximation (GGA) with Perdew-Burke-Ernzerhof function (PBE) was applied by CASTEP code package [15–17]. The key calculation parameters (e.g., the cut-off energy and the k-points) was selected based on systematic convergence analysis. In the test, the cut-off energy and the k-points values were systematically changed until the variation of the energy becomes negligible. The final cut-off energy used is 520 eV. The SCF tolerance were set as 1e−6 eV/atom, and for the Brillouin-zone sampling, Monkhorst-Pack mesh was set as $10 \times 10 \times 10$ k-points. One key output data used to calculate the ground state elastic properties are the elastic constants from the CASTEP. It involves calculating stress strain response of the crystal under several different strain patterns based on Hooke's law. Due to symmetry of a cubic crystal, the related elastic constants can be expressed as in Eq. (1):

$$\begin{pmatrix} \sigma_1 \\ \sigma_2 \\ \sigma_3 \\ \tau_4 \\ \tau_5 \\ \tau_6 \end{pmatrix} = \begin{pmatrix} C_{11} & C_{12} & C_{12} & & & \\ C_{11} & C_{11} & C_{12} & & & \\ C_{12} & C_{12} & C_{11} & & & \\ & & & C_{44} & & \\ & & & & C_{44} & \\ & & & & & C_{44} \end{pmatrix} \begin{pmatrix} \varepsilon_1 \\ \varepsilon_2 \\ \varepsilon_3 \\ \gamma_4 \\ \gamma_5 \\ \gamma_6 \end{pmatrix} \quad (1)$$

The relationship ($\sigma_{i=} C_{ij}\ \varepsilon_j$) is for small stresses. Where σ is the normal stress, τ is the shear stress, ε is the normal strain, γ is the shear strain. The elastic properties of solid material are calculated from stiffness constant C_{ij} (obtained from the first principle calculation) and elastic Compliance Constants S_{ij} [18]. The compliance constants S_{ij} can be determined as the inverse matrix of the elastic stiffness matrix C_{ij} ($[S_{ij}] = [C_{ij}] -$ 1). As for simple cubic lattice symmetry, there are three independent variables in $[C_{ij}]$ and $[S_{ij}]$, "C_{11} C_{12}, C_{44}" and "S_{11}, S_{12}, S_{44}", respectively. The Voigt-Reuss-Hill (VRH) approximation is an effective way for converting anisotropic single-crystal elastic constants into effective isotropic (polycrystalline) elastic moduli [19]. The Voigt approach sets the upper bond, the Reuss approach sets the lower bound of the elastic properties while the Hill values takes the average between the Voigt and Reuss predictions [7–9]. The approach provides a mean to calculate the average properties, the calculation also produce data showing potential difference in properties through comparing the upper and lower limit of the materials properties. Details of the theoretical background can be found in [7–9]. The bulk modulus can be calculated from the elastic constants C_{ij} as:

$$K_v = 2(C_{11} + C_{12} + C_{33}/2 + 2C_{13})/9 \quad (2)$$

$$K_R = 1/[(S_{11} + S_{22} + S_{33}) - 2(S_{12} + S_{23} + S_{31})] \quad (3)$$

$$K_H = (K_V + K_R)/2 \quad (4)$$

The effective shear modulus can be written as:

$$G_V = [(C_{11} + C_{22} + C_{33}) - (C_{12} + C_{23} + C_{31}) + 3(C_{44} + C_{55} + C_{66})]/15 \quad (5)$$

$$G_R = 15/[4(S_{11} + S_{22} + S_{33}) - 4(S_{12} + S_{23} + S_{31}) + 3(S_{44} + S_{55} + S_{66})] \quad (6)$$

$$G_H = (G_V + G_R)/2 \quad (7)$$

The most common equation to estimate the Young's Modulus was based on the Hill bulk modulus (K) and shear modulus (G) with the following equation:

$$E_H = 9K_H G_H/(3K_H + G_H) \quad (8)$$

The effective Poisson's ratio of the crystal can be obtained from bulk modulus and shear modulus. When the Hill model is used, it can be written as:

$$v_H = (3K_H - 2G_H)/(6K_H + 2G_H) \quad (9)$$

Table 1. Calculated values of bulk modulus (K), Young's modulus (E), shear modulus (G), Poisson's ratio (v) based on the Reuss, Voigt and Hill models. (The unit for K, E, G is GPa)

	K_H	K_v	K_R	E_H	E_V	E_R	G_H	G_V	G_R	v_H	v_H	v_H
TiC	252	252	252	427	428	427	175	176	175	0.217	0.217	017
VC	306	306	306	501	504	498	204	205	203	0.227	0.226	0.228
NbC	300	300	300	484	493	476	197	201	192	0.23	0.226	0.235

Table 1 lists bulk modulus (K), Young's modulus (E), shear modulus (G), Poisson's ratio (v) based on the Voigt–Reuss–Hill (VRH) models (Eqs. 2–9). The araged values (Hill model) are comparable to published works [20–26]. As shown in the data, the bulk modulus of the crystals determined by the Vioget, Reuss and Hill method are identical, the VC and NbC has much higher compressibility than TiC based on the values of K. The E_v and E_R is slightly different for TiC and VC, but there is a clear difference in the Young's modus of NbC, the Reuss model give a value of 493 GPa and the Reuss model give a much lower value of 476 GPa. There is limited difference between the data for the shear modulus between the upper (G_V) and lower bounds (G_R) for each of the carbides. The calculated values for the Poisson's ratio (v_V, v_R, v_H) are also similar for each material. Figure 3(a) presents the calculated bulk modulus, Young's modulus, shear modulus and Poisson's ratio (average values based on the Hill model). The E and G values for VC is slightly higher than that for NbC and TiC. The bulk modulus of NbC is comparable to that of VC, both are higher than the K value of TiC. The Poisson ratio for the carbides are all comparable between the model calculations with the v(H) being slightly higher. The predicted data of the modulus shows a good agreement wh the published experimental and other theoretical data, for example, TiC [20, 21, 27]; VC

[4, 23]; NbC [24–26]. Other data sources used included Matweb (http://www.matweb.com), CRC Materials Science and Engineering Handbook. Further details can be found in [28]. The Reuss, Voigt and Hill data also provide a mean to estimate the hardness (Eqs. 10 and 11) from the bulk and shear moduli. The data for these carbides is shown in Fig. 3(b) together with the range of published data. The different colour (yellow and green) represents the maximum and minimum among the reported values over different sources. It illustrates that, the predicted Vickers hardness values from both models are close to average of published data based on a mix of reference including material data base [e.g., 22, 28, 29].

Tickers hardness can be calculated following formula for theoretical hardness prediction from the bulk modulus (K) and shear modulus (G) [10, 11].

$$\text{Tian's Model: } H_v = 0.92\left(\frac{G}{K}\right)^{1.137} G^{0.708} \tag{10}$$

$$\text{Chen's Model: } H_v = 2\left(\left(\frac{G}{K}\right)^{0.585} G\right) - 3 \tag{11}$$

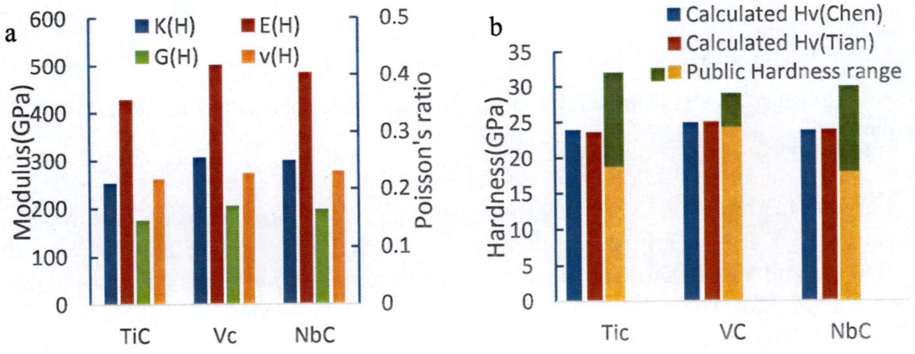

Fig. 3. (a) Predicted ground state elastic properties (B, E, G, v); (b) Vickers hardness of the carbides.

3.2 Anisotropy in the Ground State Elastic Properties

For ceramics, the anisotropies in elastic properties can influence the formation and propagation of micro cracks and reduce the mechanical durability. The hardness and wear resistance are also related to the anisotropy of the crystal [30]. The anisotropy in crystals is normally represented by anisotropic indexes such as the universal elastic anisotropic index (A_U) [31], percent anisotropy in compressibility and shear (A_{comp} and A_{shear}) [19], and the shear anisotropic factors on the (100), (010), and (001) pnes (A1, A2 and A3) of cubic crystals [32, 33]. The equations are:

$$A_U = 5G_v/G_R + K_V/K_R \tag{12}$$

$$A_{comp} = (K_V - K_R)/(K_V + K_R) \tag{13}$$

$$A_{shear} = (G_V - G_R)/(G_V + G_R) \tag{14}$$

For cubic structure:

$$A1 = A2 = A3 = 4C_{44}/(C_{11} + C_{33} - 2C_{13}) \tag{15}$$

For an elastically isotropic solid, $A_U = A_{Shear} = 0$; $A_1 = 1$. Larger values for A_U and A_1 indicate more elastic anisotropy.

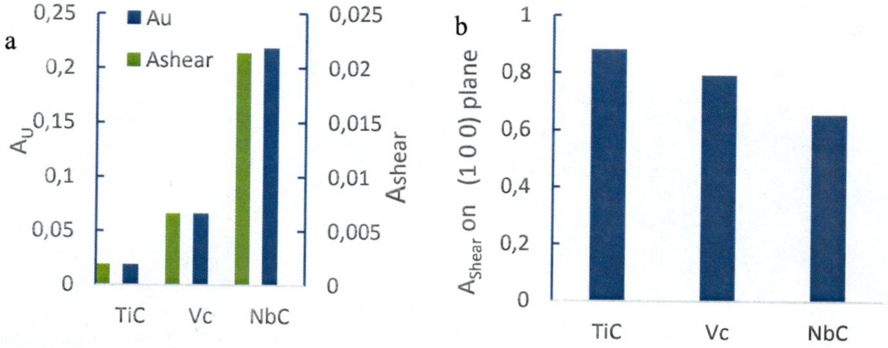

Fig. 4. (a) Universal elastic anisotropic index; (b) Shear anisotropic factor on the (1 0 0) of the carbides.

Figure 4a&b plots universal elastic anisotropic index (A_U), percent anisotropy in shear (A_{shear}). It is clearly shown that the AU for TiC and VC is relatively low with TiC (0.01) and VC (0.066), both are close to isotropic conditions, but NbC has a higher A_U value of 0.21. The values for A_{shear} of the three carbides are all very low with TiC (0.001), VC (0.006) and NBC (0.021). The shear anisotropic factors on the (100) plane is represented by the A1 values (Fig. 3(b)). The values are lower than 1 for all the carbides, indicating anisotropic characteristics on the (100) plane. The A_1 value for TiC is closer to 1, suggesting low anisotropy, VC, NbC shows a much stronger anisotropy with A_1 value of about 0.8 and 0.65, respectively. The data shows that the percent anisotropy in shear seems to be a much clearer indicator for anisotropy for the carbides. Both A_U and A_{shear} are generalised value, data for directional anisotropy is required to establish more detailed data and present the anisotropy in a better visualizable way for both materials research, development, teaching and training.

3.3 Directional Anisotropy Data of Elastic Properties Based on Surface Construction and Plane Projection

In order to establish and visualise anisotropy of elastic properties in further details, a python program is developed to calculate the 3D directional distribution of the E, G and Poisson's ratio in spherical coordinate system. The process involves calculating the directional dependence of K, E, G and v in spherical coordinate to establish the 3D surface construction [15]. Details of the theory can be found in [34] and details of the implementation in Python can be found in [28].

In general, an elastically isotropic solid has a spherical 3D surface construction of elastic parameter. A non-spherical 3D surface construction indicates an elastically anisotropic solid. The more deviation the 3D surface construction from a sphere reflects a higher degree of elastic anisotropy is [17, 35]. On a crystal plane (e.g. (100) plane), the degree of anisotropy can be represented by the closeness of the isoline to a circular shape. For cubic structure, the [010], [001] are identical anisotropy. So, the data on (100) is representative to the anisotropy feature including maximum and minimum values. Figure 5 shows the surface construction and plane projection of the bulk moduli. For all the carbides, the surface plot of the bulk modulus is spherical, indicating isotropic distribution. This corresponds to the isotropic parameter of compression being '0' (Eq. 13).

Figure 6 shows 3D surface constructions of Young's moduli for the carbides, which is the main indication of stiffness. The Young's Modulus of TiC has a near-spherical 3D graph, the projection on (100) is approximately circular indicating limited anisotropy. The Young's modulus for VC is deviated away more from near spherical, while the contour for NbC shows significant anisotropy. The plane projection on [001] shows more clearly the directional anisotropy. The data indicate that TiC has limited anisotropy while NbC showed clear visible anisotropy with a high value at [011] direction.

Figure 7 shows the 3D surface constructions and (100) plane projection of the shear moduli of TiC, VC and NbC. Different from the data for the Young's modulus in Fig. 6, there are two surface plots (two sets of data), one is based on the maximum value (in red) and the other one is based on the minimum values (in green). This reflects the complex nature of the plane and direction of the shear modulus [31]. The 3D construction of G for TiC is close to a spherical shape indicating isotropic, the surface for VC shows some increased anisotropy as the shape of the 3D surface is deviated from a perfect spherical shape. The shear modulus of NbC shows a stronger anisotropic, but the scale of anisotropic is not as significant as that for the Young's Modulus. The plane projections of shear modulus show the anisotropy more clearly as shown in Figs. 7(d, e, f) based on the isolines of the maximum values (in red); minimum values (in green line) and an isoline representing the average value (in blue). This approach could indicate more clearly the nature and scale of anisotropy. The isolines on the plane projection shows clearly that the higher G values are at the [011] direction.

Figure 8 shows 3D surface constructions and plane projection of Poisson's ratio of TiC, VC and NbC. The same colour code and approach is used for presenting the anisotropy in the Poisson's ratio. In general, all the three carbides showed clear anisotropy in the Poisson's ratio, a much higher Poisson's ratio exist in [011] direction.

Figure 9 plots the ratio between the maximum value and the minimum value of E, G, and v. The data shows a much stronger difference in the data for the Poisson ratio than that for E and G for all the carbides. The scale of anisotropy increased from TiC, VC then NbC. The results showed that 3D construction and plane projection data provide more detailed information than the generalized anisotropy indexes from the Reuss, Voigt and Hill models estimation. The work highlighted the anisotropy nature of the cubic structure carbides in particular in the Poisson's ratio. In general, the anisotropy in these monocarbides is much lower than that of more complex system with the same

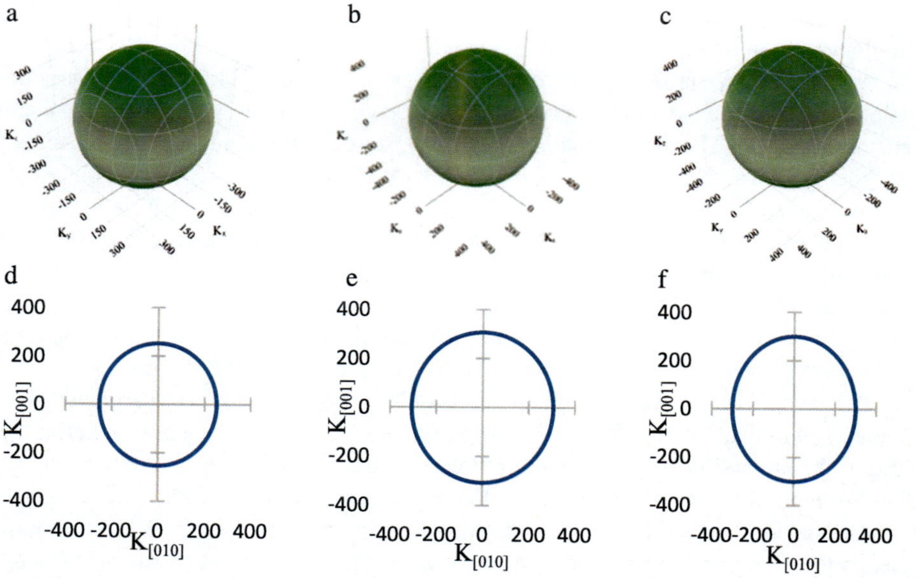

Fig. 5. Surface construction (a) TiC; (b) VC; (c) NbC and plane projection (d) TiC; (e) VC; (f) NbC of Bulk modulus of the carbides.

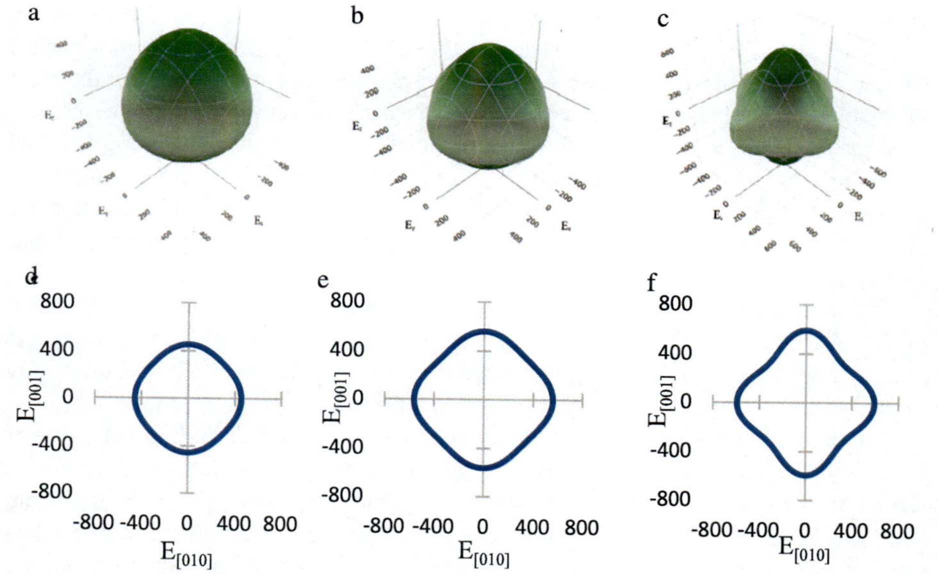

Fig. 6. Surface construction (a) TiC; (b) VC; (c) NbC and plane projection (d) TiC; (e) VC; (f) NbC of Young's modulus of the carbides.

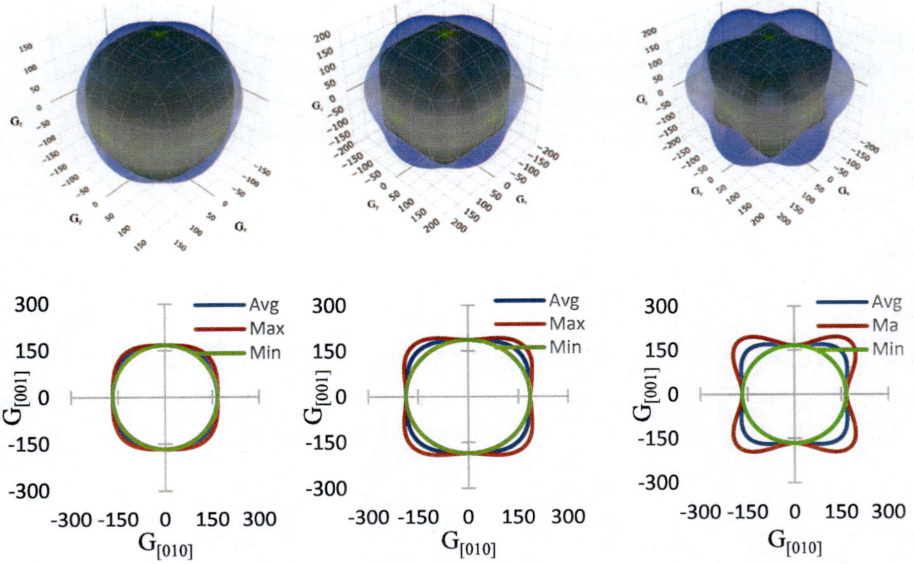

Fig. 7. Surface construction (a) TiC; (b) VC; (c) NbC and plane projection (d) TiC; (e) VC; (f) NbC of Shear modulus of the carbides.

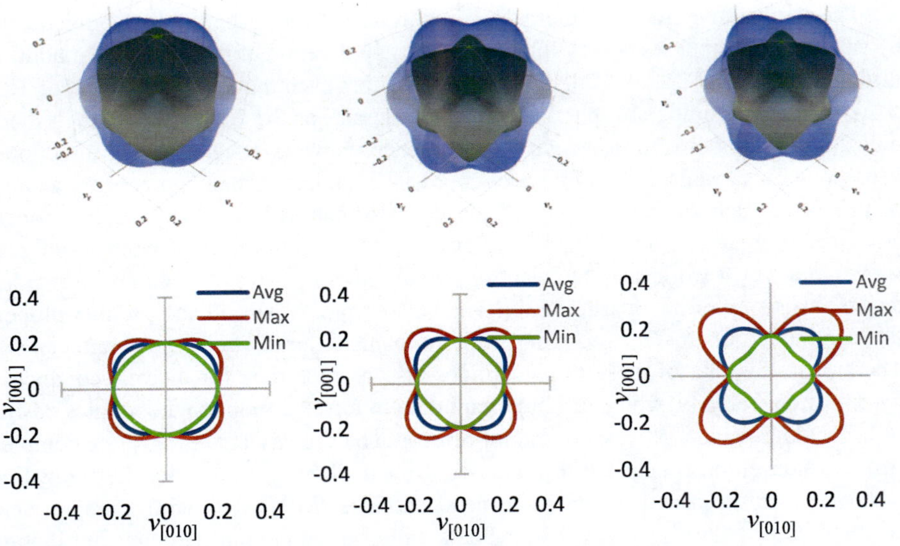

Fig. 8. Surface construction (a) TiC; (b) VC; (c) NbC and plane projection (d) TiC; (e) VC; (f) NbC of Poisson's ratio of the carbides

chemical element system. For example, the A_U for V_2C is as high as over 4 [17]. The integrated approach combining the first principle calculation with python program for anisotropy analysis in particular 3D surface construction and plane projection will help with representing the elastic properties and establishing systematic data at larger scale between different groups of carbides, stoichiometric and non-stoichiometric compounds.

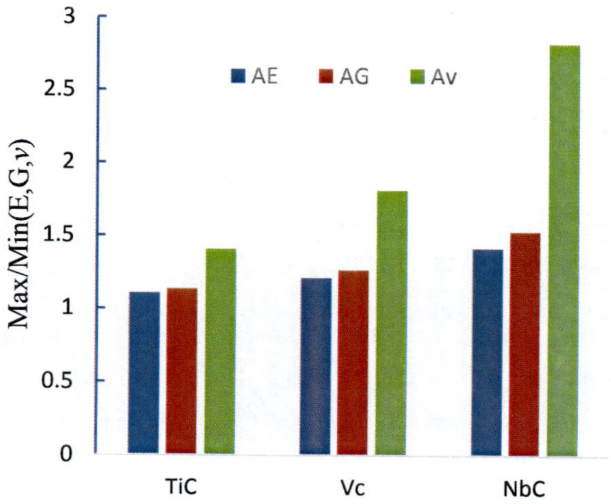

Fig. 9. Anisotropy in elastic properties (E, G, v) based on the ratio between maximum and minimum property value (Max/Min).

The comparative studies of these three important carbides in particular the anisotropy of mechanical properties is very important in the application of materials and manufacture processes [3–5, 17, 36–45]. The work shows that even though the hardness of TiC, VC and NbC are comparable but there is clear difference on the directional Young's modulus and the Poisson's ratio, which may influence the wear, toughness and directional behaviors. Elastic anisotropy of crystals could have fundamental influences over a range of processes such as development of plastic deformation in crystals and microscale cracking in ceramics [38]. It is also relevant to the development of nanoparticles as shown in a recent work that the anisotropy of elastic properties of niobium carbide led to the anisotropy of nanoparticle [39]. The anisotropies in elastic and thermal properties can induce microcracks to reduce the mechanical durability of the ceramics [17]. The anisotropy data of these three carbides also provide data for thermal conductivity and wave velocity, which are important data to modern functional ceramics design [40]. In addition, data processing framework could be used as a base for future complex structures development such as diffusion-controlled alloying strategy for the realization of covalent multi-principal transition metal carbides (MPTMCs) with a single face-centered cubic phase [43]. The detailed data will also help to quantify the contribution of the anisotropy to materials deformation under complex loading conditions and bring insight to the scatter of property data from different sources. These carbides have similar effective hardness, but there is widely reported difference in hardness anisotropy and fracture toughness (K_{IC}) [37] in both powder or precipitation forms. A future work is to link the anisotropy features of the elastic properties to hardness anisotropy and fracture toughness.

The TiC, VC and NbC are the most common precipitates in many steels group and welded structures, which directly influence the mechanical properties, wear resistance, etc. Systematic data of the ground state structure and properties of these carbides are key for comparative understanding of the precipitation strengthening and weldability of steels such as high manganese steels, stainless and multicomponent carbides [1, 3, 4, 12, 13, 28, 36, 37]. The automatic data program and anisotropy in the elastic properties provide an effective mean to analyse the coherent, semi-coherent and incoherent interface, which are crucial for comparing the effectiveness of the precipitation of TiC, VC and NbC on dislocations and in ferrite (e.g. Ti–Nb–Mo–V steels [41]). The ground state data is also relevant to hydrogen trapping of NbC and VC and TiC precipitates in high strength steels [42]. The program developed has been extended to systematically establish the potential effect of intrinsic defects, complexes defects [46] and factors affecting the phase nucleation process (e.g. lattice misfit) between carbides (TiC/NbC; NbC/M_7C_3) [28], which is key to alloy design and microstructure control in steels and welding [3, 12, 28, 44–46].

4 Conclusion

This paper presents the development of key procedure and data of first principle studies of three carbides (TiC, VC and NbC). A python-based program is developed integrating first principle calculation (Materials Studio) data processing to establish detailed 3D presentation of the elastic properties and their projection on key crystal planes. The bulk modulus, Young's modulus, shear modulus, and Poisson's ratio of the carbides are compared together with micro Vickers hardness. The theoretical Vickers hardness with different analysis models are comparable between the three carbides, while there is different level of anisotropy in elastic properties. A much stronger anisotropy in the Poisson's ratio is identified, and NbC shows a higher anisotropic feature than TiC and VC. Link between the data from first principle modelling and other properties, including effects of defects, hardness anisotropy, fracture toughness, effectiveness of precipitation, is discussed with reference to future developments.

Acknowledgements. This work has received support from the European Union's Horizon 2020 research and innovation program under the Marie Skłodowska-Curie grant agreement No. 793114 and No. 823786.

References

1. Yu, W., Li, J., Shi, C., Zhu, Q.: Effect of titanium on the microstructure and mechanical properties of high-carbon martensitic stainless steel 8Cr13MoV. Metals **6**(8), 193 (2016)
2. Birosca, S.: Process-structure-property relationships. In: Birosca, S. (ed.) Metals (2018)
3. Xiong, H., Zhang, H., Zhang, H., Gan, L.: First principle calculation of NbC precipitation competition between TiC particle and ferrite matrix. J. Wuhan Univ. Technol.-Mater. Sci. Ed. **33**(5), 1076–1081 (2018). https://doi.org/10.1007/s11595-018-1937-2

4. Sun, Z., Ahuja, R., Lowther, J.: Mechanical properties of vanadium carbide and a ternary vanadium tungsten carbide. Solid State Commun. **150**(15–16), 697–700 (2010)
5. Cuppari, M., Santos, S.: Physical properties of the NbC carbide. Metals **6**(10), 250 (2016)
6. Wen, X., Wang, Y., Zhao, J.: Negatively charged boron nitride nanosheets as a potential metal-free electrocatalyst for the oxygen reduction reaction: a computational study. New J. Chem. **42**(15), 12838–12844 (2018)
7. Voigt, W.: Handbook of Crystal Physics. Taubner, Leipzig (1928)
8. Hill, R.: The elastic behaviour of a crystalline aggregate. Proc. Phys. Soc. **65**, 349 (1952)
9. Reuss, A.: Calculation of the flow limits of mixed crystals on the basis of the plasticity of monocrystals. Z. Angew. Math. Mech. **9**, 49–58 (1929)
10. Chen, X., Niu, H., Li, D., Li, Y.: Modeling hardness of polycrystalline materials and bulk metallic glasses. Intermetallics **19**(9), 1275–1281 (2011)
11. Tian, Y., Xu, B., Zhao, Z.: Microscopic theory of hardness and design of novel superhard crystals. Int. J. Refract. Metal Hard Mater. **33**, 93–106 (2012)
12. Chong, X., Hua, M., Wu, P., et al.: Tailoring the anisotropic mechanical proeprties of hexagonal M7C3 multialloying. Acta Mater. **169**, 193–208 (2019)
13. Hamblyn, S.M.L., Reuben, B.G.: Use of radio-frequency plasma in chemical synthesis. Adv. Inorg. Chem. Radiochem. **17**, 89–114 (1975)
14. Wu, L., Yao, T., Wang, Y., et al.: Understanding the mechanical properties of vanadium carbides: nano-indentation measurement and first-principles calculations. J. Alloys Compd. **548**, 60–64 (2013)
15. Liu, Y., Chong, X., Jiang, Y., Feng, J.: Stability, electronic structures, and mechanical properties of Fe–Mn–Al system from first-principles calculations. Chin. Phys. B **26**(3), 037102 (2017)
16. Wang, B., Ma, B., Song, W., Fu, Z., Lu, Z.: First-principles calculations of structural, electronic, magnetic and elastic properties of Mo_2FeB_2 under high pressure. Roy. Soc. Open Sci. **5**(7), 172247 (2018)
17. Bao, L., Qu, D., Kong, Z., Duan, Y.: Anisotropies in elastic properties and thermal conductivities of trigonal TM2C (TM = V, Nb, Ta) carbides. Solid State Sci. **98**, 106027 (2019)
18. Nye, J.F.: Physical Properties of Crystals. Oxford University Press, Oxford (1985)
19. Chung, D.H., Buessem, W.R., Vahldiek, F.W., Mersol, S.A. (eds.): Anisotropy in Single Crystal Refractory Compound. Plenum, New York (1968)
20. Gilman, J., Roberts, B.: Elastic constants of TiC and TiB_2. J. Appl. Phys. **32**(7), 1405 (1961)
21. Kim, J., Kang, S.: Elastic and thermo-physical properties of TiC, TiN, and their intermediate composition alloys using ab initio calculations. J. Alloy Compd. **528**, 20–27 (2012)
22. Yang, Y., Lu, H., Yu, C., Chen, J.: First-principles calculations of mechanical properties of TiC and TiN. J. Alloy Compd. **485**(1–2), 542–547 (2009)
23. Liu, H., Zhu, J., Liu, Y., Lai, Z.: First-principles study on the mechanical properties of vanadium carbides VC and V_4C_3. Mater. Lett. **62**(17–18), 3084–3086 (2008)
24. Brown, H., Armstrong, P., Kempter, C.: Elastic properties of some polycrystalline transition-metal monocarbides. J. Chem. Phys. **45**(2), 547–549 (1966)
25. Brenton, R., Saunders, C., Kempter, C.: Elastic properties and thermal expansion of Niobium mono-carbide to high temperatures. J. Less Common Metals **19**(3), 273–278 (1969)
26. Gao, X., Jiang, Y., Liu, Y., Zhou, R., Feng, J.: Stability and elastic properties of Nb_xC_y compounds. Chin. Phys. B **23**(9), 097704 (2014)
27. Török, E., Perry, A.J., Chollet, L., Sproul, W.D.: Young's modulus of TiN, TiC, ZrN and HfN. Thin Solid Films **153**(26), 37–43 (1987)
28. Tang, C.: Ph.D. thesis, Liverpool John Moores University, UK (2020)

29. Edström, D., Sangiovanni, D., Hultman, L., Petrov, I., Greene, J., Chirita, V.: Elastic properties and plastic deformation of TiC-and VC-based pseudobinary alloys. Acta Mater. **144**, 376–385 (2018)
30. Ren, X.J., Hooper, R.M., Griffiths, C., Henshall, J.L.: Indentation-size effects in single-crystal MgO. Philos. Mag. A **82**(10), 2113–2120 (2002)
31. Ranganathan, S.I., Ostoja-Starzewski, M.: Universal elastic anisotropy index. Phys. Rev. Lett. **101**(5), 055504 (2008)
32. Ravindran, P., Fast, L., Korzhavyi, P.A., Johansson, B.: Density functional theory for calculation of elastic properties of orthorhombic crystals: application to $TiSi_2$. J. Appl. Phys. **84**(9), 4891 (1998)
33. Liu, X., Feng, Q., Tang, B., et al.: First-principles calculations of mechanical and thermodynamic properties of tetragonal $Be_{12}Ti$. RSC Adv. **9**(10), 5302–5312 (2019)
34. Luan, X., Qin, H., Liu, F., Dai, Z., Yi, Y., Li, Q.: The mechanical properties and elastic anisotropies of cubic Ni_3Al from first principles calculations. Curr. Comput.-Aided Drug Des. **8**, 307 (2018)
35. Sun, S., Fu, H., Lin, J., Guo, G., Lei, Y., Wang, R.: The stability, mechanical properties, electronic structures and thermodynamic properties of (Ti, Nb)C compounds by first-principles calculations. J. Mater. Res. **33**(4), 495–506 (2017)
36. Park, G., Jeong, S., Lee, C.: Fusion weldabilities of advanced high manganese steels: a review. Met. Mater. Int. **27**(7), 2046–2058 (2020). https://doi.org/10.1007/s12540-020-00706-9
37. Casellas, D., Caro, J., Molas, P.S.J., Valls, I.: Fracture toughness of carbides in tool steels evaluated by nanoindentation. Acta Mater. **55**(13), 4277–4286 (2007)
38. Kube, C.M.: Elastic anisotropy of crystals. AIP Adv. **6**, 095209 (2016)
39. Gusev, A.I.: Anisotropy of microstructure and elastic properties of niobium carbide nanopowders. Solid State Sci. **100**, 106092 (2020)
40. Luis, N.C.: Editorial, Surface engineering, special issue: coating production and surface engineering of carbide-based composites **36**(11), 1129–1131 (2020)
41. Jia, Z., Misra, R.D.K., O'Malley, R., Jansto, S.J.: Fine-scale precipitation and mechanical properties of thin slab processed titanium–niobium bearing high strength steels. Mater. Sci. Eng. A **528**(22), 7077–7083 (2011)
42. Wei, F.G., Hara, T., Tsuzaki, K.: Nano-precipitates design with hydrogen trapping character in high steel. In: Proceedings of the 2008 International Hydrogen Conference - Effects of Hydrogen on Materials, pp. 448–455 (2008)
43. Peng, C., Gao, X., Wang, M., et al.: Diffusion-controlled alloying of single-phase multi-principal transition metal carbides with high toughness and low thermal diffusivity. Appl. Phys. Lett. **114**(1), 011905 (2019)
44. Dong, J., Hou, D., Li, J.-Y., Huang, R.: Investigation of NbC/TiC heterogeneous nucleation interface by first-principles and experimental methods. Metals **9**(12), 1265 (2019). https://doi.org/10.3390/met9121265
45. Guo, J., Feng, Y., Tang, C., Ren, X.J.: Intrinsic defects, Mo-related defects, and complexes in transition-metal carbide VC: a first-principles study. J. Am. Ceram. Soc. **103**, 7226–7239 (2020)
46. Liu, S., Wang, Z., Guo, J., Shi, Z., Ren, X., Yang, Q.: Thermal-mismatch-stress induced deformation zone on primary-carbide/austenite interface. Mater. Lett. **248**, 55–59 (2019)

Grinding of Titanium Alloys (Ti-6Al-4V) Using Vibration Assisted Machining

Elsiddig Eldaw Ibrahim, Heisum Ewad, Xun Chen, and Andre D. L. Batako(✉)

General Engineering Research Institute, Liverpool John Moores University, Byrom Street, Liverpool L3 3AF, UK
a.d.batako@ljmu.ac.uk

Abstract. The machinability of titanium alloys (Ti-6Al-4V) have been reviewed in many publications recently special grinding and mill-ing as these are among the key processes for the production of compressor and turbine blades for jet engine, where high heat resistance materials are preferred. Titanium alloys differ from other metallic materials, because of its low density, high strength to weight ratio, low thermal conductivity and high resistance to corrosion and erosion. These aerospace materials have been under investigation for a possible replacement for nickel based super alloys in aircraft engines components because of weight issue. Since it has been difficult to be machined through the traditional grinding methods, vibration assisted machining technique (VAM) was introduced in the perpendicular direction of the grinding wheel to reduce cutting forces, energy consumption, wheel wear and to improve material removal rate, surface roughness and to extend the grinding tool life. Experiments were carried out under wet conditions, employing aluminium oxide (Al_2O_3) grinding wheel instead of super abrasive grinding wheels due to work-piece surface damage issues. The experimental results showed that implementing (VAM) in the axial direction of the spindle has reduced grinding forces by 40% and increased the materials removal rate by 25%. Consequently, energy consumption has been reduced and surface quality has been improved at specific boundary conditions (100 Hz vibration frequency, 130 μm vibration displacement, 17 m/s wheel speed and 250 mm/s feed rate) where burnout, chatter and crack disappeared.

Keywords: Titanium alloys · Vibration · Resonance grinding · Cutting forces

1 Introduction

1.1 A Subsection Sample

Grinding is one of the crucial technologies to achieve the desired surface quality of the components. Recently, progress in materials science has led to the development of new materials especially in the aerospace industry, where high heat resistance materials are preferred, such as Ti-6Al-4V and gamma titanium aluminide (γ-TiAl). These materials differ significantly from other metallic materials, because of its low density, high strength to weight ratio, low thermal conductivity and poor machinability.

A. Batako et al. (Eds.): GCMM 2021, LNNS 335, pp. 86–94, 2022.
https://doi.org/10.1007/978-3-030-90532-3_7

These aerospace materials, although it have been used heavily in the manufacturing of aircraft engines components, yet they are difficult to be machined through the traditional grinding methods. Through over the years, vibration assisted grinding was introduced to reduce cutting forces, improve material removal rate and surface roughness quality. However, vibration is also used in various technological processes to improve the performance of the machines by intelligently exploiting the synergy of the oscillations. Vibration assisted machining is a technique in which a certain frequency of vibration is applied to the cutting tool or the workpiece (besides the original relative motion between these two) to achieve better cutting performance, [1, 7].

The basic element of vibration assisted machining (VAM) is the oscillation that can be generated using series of methods; however, piezo-actuators are most commonly used, [2]. It has been reported by a number of researchers that machining with vibration assistance has positive effects on the machining performance as whole. The purpose of this study is to investigate the advantages of vibration assisted grinding technique when it introduced in the axial direction of the grinding wheel to reduce cutting forces during the grinding of Titanium alloy (Ti-6Al-4V). Hence, grinding of titanium differs significantly from that of other metallic materials, primarily due to its poor machinability. However, titanium and titanium alloys are among the most difficult-to machine materials, because of their low thermal conductivity and high chemical reactivity which resulted in poor surface quality.

Figure 1 shows the distribution of titanium alloys such as Ti-6Al-4V, which contribute up to 50–60% to the total production of titanium alloys [3]. These alloys are characterised by high tensile strength around 1100 Mpa at 20 °C combined with low density (~4 g/cm^3) along with a good corrosion resistance. These properties explain why titanium in used in the manufacturing of commercial aircraft engine (33% of engine weight) [4].

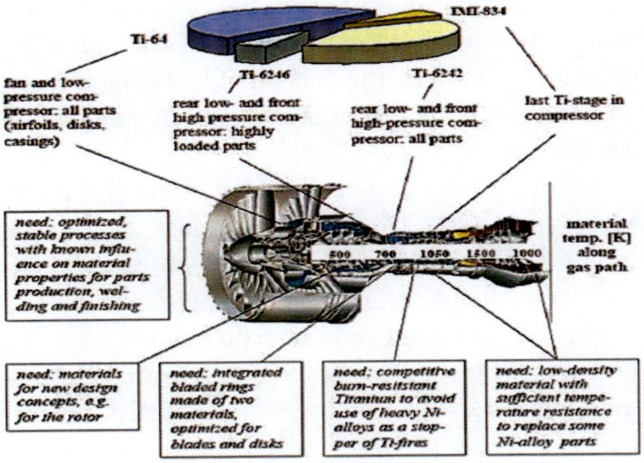

Fig. 1. Titanium in Aero-engines (J. Esslinger) [3]

One can see in Fig. 1 the material usage in a Rolls-Royce aero engine, particularly the usage of titanium in a cooler parts of the engine (fan and compressor) and nickel based super alloys in the hotter parts of the engine (combustion chamber and turbine). Since there are several constraints restricting the usage of titanium in the hot section of the aero-engine, such as the ability to operate at temperatures above ~350 °C and alloy burn resistance, the nickel super alloys became the preferred choice for the hotter parts of the gas turbine because of its durability and thermal stability. Therefore, nickel-based alloys such as Inconel 718 are used to accommodate temperatures of 600 °C in the high pressure compressor (HPC) and even higher gas flow temperatures in the combustor, [4]. Titanium aluminide is the latest generation of thermos-mechanically produced duplex γ-TiAl alloys such as Ti-45Al-8Nb-0.2Cat.%, which has a tensile strength above 900 MPa, is able to operate at temperatures of up to 800 °C [5, 6, 8]. Therefore, this higher temperature resistance allows these alloys to be considered for replacement of nickel alloys as a favourite material of aero-engine components including Low Pressure (LPC) and High Pressure (HPC) Compressor blades and stators as well as Low Pressure Turbine (LPT) blades [4, 10, 11]. However, weight saving and the requirements for engines with high trust, higher turbine temperatures and fewer compressor and turbine stages, are the main cause of the demand for new materials that secure improvement in fuel economy.

The above aforementioned materials need to be ground before undergo any manufacturing process to produce jet engine components. Hence, all previous and current studies to grind these materials were focused on vibration applied in the direction of grinding to create slots lead to ease the cooling system, while this study investigated the possibility of vibration imposed in the axial direction of the grinding wheel (perpendicular) which has not been done before. This new system allow to vibrates the workpiece in a direction that perpendicular to the grinding direction by employing a piezo electric-actuator, where 100 Hz vibration frequency was applied to the system to generate 130 μm displacement to match low excitation of (4 V) to secure lower power consumption. The frequency and the amplitude are controlled by the closed loop PID controller shown in Fig. 4 below [9, 13, 14] which was not possible to be achieved in ultrasonic vibration. However, in this study low vibration frequency was introduced with variation of amplitude up to 130 μm, instead of the currently used ultrasonic vibration approach to machine brittle aerospace materials, with much higher imposed oscillations that could reach up to 100 kHz to secure very limited displacements. The reason why because it has been proven that the bigger the amplitude the better the surface quality might be achieved of the ground product, [12].

Aerospace materials are categorised in two different groups; brittle such as titanium alloys and hard such as nickel alloys. To grind these materials a grinding wheel selection procedure should take place to specify the correct wheel speeds.

It has been reported that the operator has to use half to one third of the conventional operating wheel speeds to get the best results with titanium. Also, to secure good surface finish the input parameters of the experiment should be correctly selected, such as cutting depth, feed rate and diameter of the grinding wheel. Hence, for the task of grinding Ti-6Al-4V alloys, a surface grinding machine was chosen to match the application of vibration assisted in the axial direction of the machine spindle, at the same time an aluminium oxide grinding wheel has also been selected together with cooling system to tackle the problem of the excessive heat flux that has been generated in the contact zone during the grinding process of titanium alloys, [15–17].

2 Experiment's Setup and Methodology

Titanium (Ti-6Al-4V) workpieces with the following dimensions: 60 mm length × 8 mm width × 20 mm height were fabricated for the experimental work. A conventional surface grinding machine (ABWOOD 5025), with analogue pneumatic controls for the automatic traverse cycles, was used to grinding the workpieces with and without the application of vibration. The process was lubricated with cutting fluid dilute at 5% emulsion HySol-X.

Figure 2 illustrates the self-contained vibrating stage mounted on the grinding machine tool. Figure 3 depicts the actual arrangement of the devices during the experiment, where the frequency and the amplitude of the vibration were specified in the closed-loop system, which determined the performance of the piezoelectric actuator using the signal from the data acquisition system. Also all the data signals were recorded through the data acquisition system using LABVIEW software, and stored on a PC for analysis. Hence, the machining performance of the materials was studied in terms of four factors such as tool life, cutting forces, power consumption and surface finish.

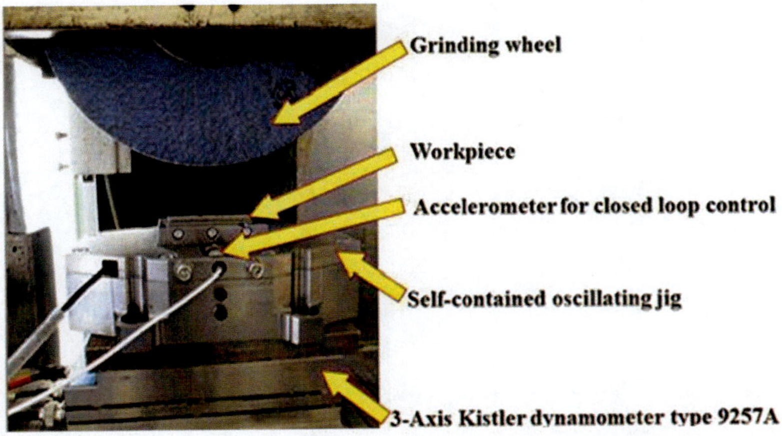

Fig. 2. Experimental set up with the actual oscillating jig

Table 1 shows the experiment's parameters used to grind Titanium alloys (T6-Al-4V) by employing aluminium oxide grinding wheel using cooling system The experimental work was carried out with and without vibration assistance, and an Aluminium Oxide grinding wheel was used to grind workpieces of Titanium (Ti-6Al-4V) alloys. However, many researchers pointed that the surface of titanium alloys can be damaged easily and cracked during machining, because of the heat generated within the grinding contact zone. Likewise, in this experiment, during the machining of the Ti-6Al-4V samples a similar damage was observed and this is shown in Fig. 5b.

The cutting fluid was used to induce cooling effect and to improve the quality of the workpiece surface finish along with an improved material removal rate. This investigations mainly focused on grinding forces, power consumption reduction and the specific grinding energy improvement.

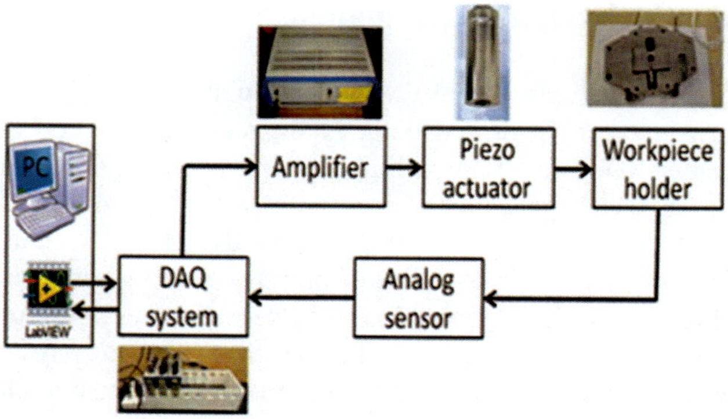

Fig. 3. Configuration of closed loop control system

Table 1. Experiment setup parameters.

Grinding parameter	Value
Grinding wheel type	Al_2O_3 (OVU33 A602HH 10VB)
Wheel speed (V_s)	17 m/s
Wheel diameter	208 mm
Work speed (V_w)	250 mm/s
Grinding condition	Wet (coolant)
Wheel feed	Traverse
Workpiece material	Ti-6Al-4V
Vibration frequency	100 Hz
Vibration amplitude	130 μm
Applied depth of cut	15–45 μm

3 Results and Discussions

3.1 Surface Roughness

The Taguchi experiment design was used to optimize a d reduce the number of grinding tests in the experimental study. A Taylor Hobson surface profile machine was used to measure surface roughness, along with Bruker contactless measurement machine.

Figure 4 depicts the surface roughness of samples ground with and without vibration. It is observed in Fig. 4b that there is a decrease in surface roughness vibration assisted grinding compared with the specimen without vibration. This shows that there is a good improvement in surface quality due to the application of vibration assisted machining. The superimposed vibration led to better process performance in terms of quality of the workpiece surface quality/finish with about 25% improvement with reference to the process without the application of vibration. However, as the depth of cuts increased, some burnout on the workpiece surface was noticed at depths from 25 μm and above. There was no burnout observed when grinding at low wheel speed (17 m/s) and moderate feed rate of 15 m/min. Meanwhile, Figs. 4a and 4b show the surface roughness profile obtained using GFM surface area measuring machine. The results reflected a clear and smooth pattern of the surface roughness with vibration assisted grinding for 20 μm depth of cut. It has been also proved that by applying vibration assisted the surface roughness Ra is slightly better than the quality achieved by conventional grinding without vibration (25% improvement).

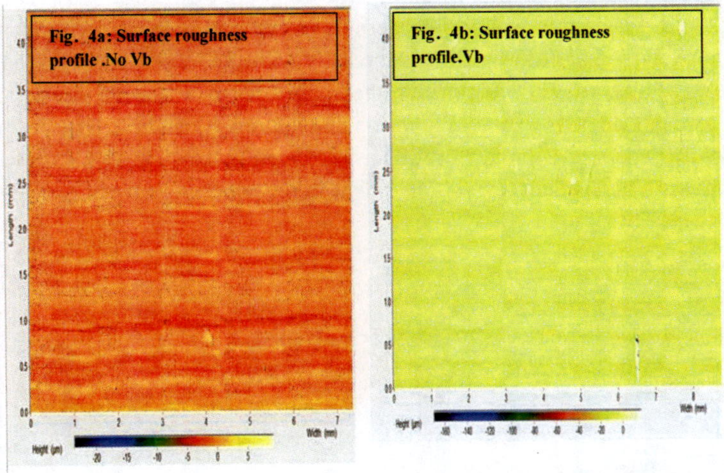

Fig. 4. Ground surface roughness by Bruker: a) No vibration, b) with vibration

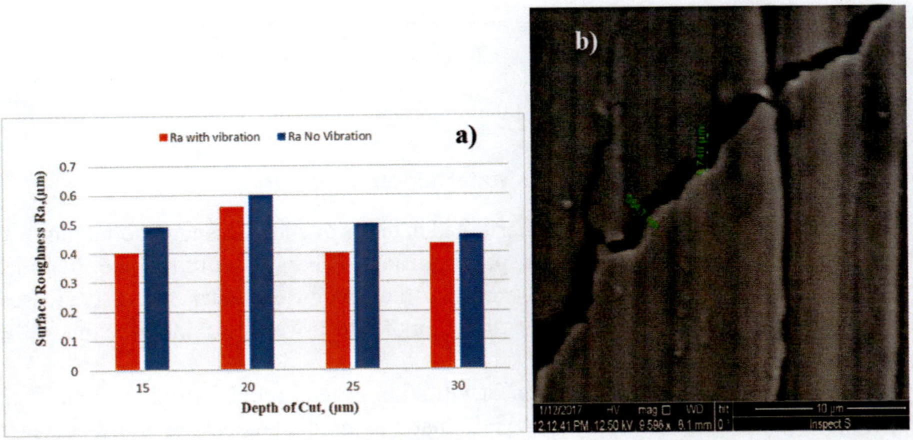

Fig. 5. Ground surface profiles: a) Taylor Hobson contact measurement; b) thermal cracks SEM

Figure 5a illustrates the direct contact measurement using stylus on the Taylor Hobson machine, where, the vibration assistance provided relatively better surface roughness. In Fig. 5b displays the surface damage caused by intensive heat concentration in the cutting zone.

3.2 Grinding Forces

The experimental work included the measurement of the cutting forces using a 3-axis dynamometer. Figure 6a and 6b represent the relationship between grinding cutting forces in the normal direction *(Fn)* and in the tangential direction *(Ft)*. These forces were measured during the grinding process in both cases with and without vibration. The results showed that cutting forces increased with the increase of applied depth of cut. There were a decrease in forces with the application of vibration assisted, with reference to forces values that measures in the conventional grinding. The cutting forces measured with the application of vibration assisted grinding of Ti6Al-4V showed significant decrease, which reflects an improvement in the process performance. On average forces decreased by around 40–45%.

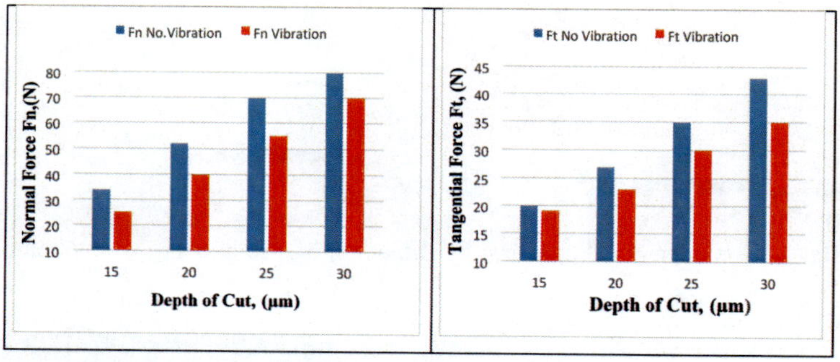

Fig. 6. Cutting force as a function of applied depth of cuts

3.3 Specific Grinding Energy and Material Removal Rate

The process performance was also determined in terms of energies and material removal rated. Figure 7a below illustrates the specific grinding energy as a function of material removal rate. It is equally observed here that the specific grinding energy decreased due to the superimposed vibration the direction of which was the axial direction of the grinding wheel rotation.

In this experiment the application of vibration at feed rate of 15 m/min and with low grinding wheel speed provided the best material removal rate along with a decrease in cutting forces. On the other hand, one of the disadvantages of grinding at low feed rate at high wheel speed is the risk of thermal damage for the work piece surface due to excessive rubbing and friction leading to burn out.

Figure 7b shows the amount of material ground off the workpieces in wear test. The figure also supports that the application of vibration in grinding Ti6Al4V secured

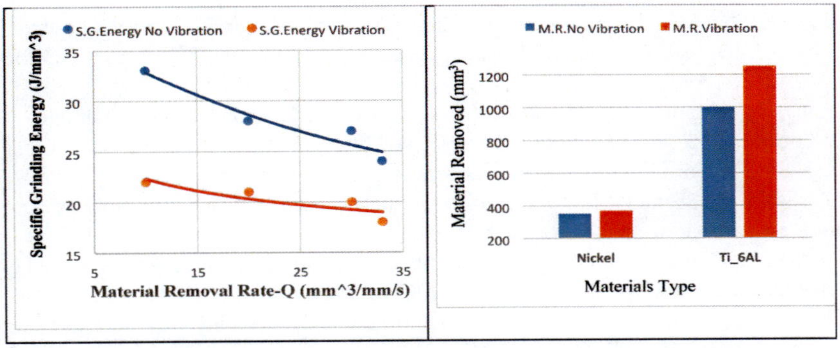

Fig. 7. Process performance: a) Specific grinding energy; b) Material removed

higher material removal rate with reference to grinding in conventional mode with no vibration. Vibration assisted grinding also secured low tool wear (35%). It also proved that vibration assisted grinding prevented wheel loading and extended tool life by (30%) due to the reduced duration of cutting tool contact with the workpiece.

4 Conclusion

This experimental work was carried out to investigate the benefits of superimposed vibration in a perpendicular direction to the machine spindle when grinding Ti-6Al-4V. A fixed low frequency of 100 Hz was the main factor to identify suitable boundary conditions such as wheel speed, feed rate and applied depth of cut at 120 μm displacement. The experiment results proved that vibration assisted reduces cutting forces by 40–45% consequently the power consumption decreased, while surface quality was improved by 25% and the material removal rate also improved by 35%. The application of Vibration secured low tool wear (35%), and prevented wheel loading and extended tool life by (30%). The reason why vibration assisted improved surface roughness was the lapping effect of the cutting tool over the cutting area enhancing finished surface quality.

References

1. Kumar, M.N., Subbu, K.S., Vamsi, K.P., Venugopal, A.: Vibration assisted conventional and advanced machining. In: 12th Global Congress on Manufacturing and Management, GCMM (2014)
2. Li, K.-M., Hu, Y.-M., Yang, Z.-Y., Chen, M.-Y.: Experimental study on vibration-assisted grinding. J. Manuf. Sci. Eng. **134**(4), 041009 (2012)
3. Nithyanandam, J., Laldas, S., Palanikumar, K.: Surface roughness optimisation in machning of titanium alloy (Ti-6al-4V). Adv. Mater. Res. Acad. J. **984–985**, 42 (2014)
4. Ibrahim, G.A., Haron, C.C., Ghani, J.A., Arshad, H.: Taguchi optimization method for surface roughness and material removal rate in turning of Ti-6Al-4V. Int. Rev. Mech. Eng. **4**(3), 216 (2010)
5. Brehl, D.: Review of vibration assisted machining. Precis. Eng. **32**, 153–173 (2008)
6. Esslinger, J.: Titanium in aero-engines. In: Proceedings of the 10th World Conference, Munich, Germany (2003)

7. Rugg, D.: The current status of titanium alloy use in aero-engines. In: Proceedings of the 10th World Conference on Titanium, Hamburg, Germany, vol. IV, pp. 2727–2735 (2003). ISBN: 3-257-30306-5
8. Voice, W.E., Henderson, M., Sheton, E.F.J., Wu, X.: Gamma titanium aluminide TNB. Intermetalics **13**, 959–964 (2005). ISSN: 0966-9795
9. Keith, D., Aspinwall, D.: Turbo-machinery components manufacture (2013)
10. Moriwaki, T., Shamoto, E.: Ultraprecision diamond turning of stainless steel by applying ultrasonic vibration. Ann. CIRP **40**, 559–562 (1991)
11. Teicher, U., Ghosh, A.: On the grinding ability of Titanium alloy by brazed type monolayered superabrasive grinding wheels. Int. J. Mach. Tools Manuf. **46**(6), 620–622 (2006)
12. Tsiakoumis, V.: An investigation into vibration assisted machining - application to surface grinding processes, Ph.D. thesis, LJMU (2011)
13. Van Arkel, A.E., de Boer, J.H.: Prepatation of pure titanium, zirconium, hafnium and thorium metal. Zeitschrift fur anorganich & Allgemeine Chemie **48**, 345–350 (1925)
14. Zareen, A.R., Wong, Y.S.: High speed machining of aerospace alloy Ti-6Al-4V. In: Society for Advancement of Material and Process Engineering Conference (2001)
15. Wang, X., Zhou, M., Gan, J.G.-K., Ngoi, B.: Theoretical and experimental studies of ultraprecision machining of brittle materials with ultrasonic vibration. Int. J. Adv. Manuf. Technol. **20**(2), 99–102 (2002). https://doi.org/10.1007/s001700200130
16. Ewad, H.: Development of a closed loop control system for vibration assisted grinding. Ph.D. thesis, LJMU (2016)
17. Alharbi, W.N.: Development of a closed loop control system for vibratory milling. Ph.D. thesis, LJMU (2018)

Simulation of Bacterial Motion Under Flow Inside Micro Channel Using CFD and DPM

S. W. M. A. I. Senevirathne, J. Hasan, A. Mathew, M. Woodruff, and P. K. D. V. Yarlagadda(✉)

Faculty of Engineering, Queensland University of Technology, Brisbane, QLD 4000, Australia
y.prasad@qut.edu.au

Abstract. Discrete Phase Method (DPM) is coupled in this study with Computational Fluid Dynamics (CFD) to simulate motion of rod-shaped motile bacteria under flow conditions inside a microfluidic device where a micro scale chamber is connected to same scale inlet and outlet channels. Here, bacterial cells are represented by spherical solid particles with equivalent volume of a typical bacterial cell, and physio-chemical interactions between cells and solid surface are omitted. Bacterial suspension is assumed as a Newtonian fluid in a laminar flow. Particle injection mass flowrate was selected to emulate OD_{600} 0.1 bacterial cell concentration and lowered up to 10% of initial concentration. Particle diffusion pattern was affected by the fluid velocity (within the range 0.05–0.005 $m.s^{-1}$), but the pattern remined similar for particle mass flowrates varying from 8–80 $\times$ 10^{-9} $kg.s^{-1}$. Particle mass concentration long the flow direction was varied with fluid velocity but not affected by the injection concentration. At 0.005 $m.s^{-1}$ fluid velocity, maximum particle mass concentration varied with a linear relationship with particle injection mass flowrate, and the variation at 0.05 $m.s^{-1}$ fluid velocity too showed a linear relationship with injection mass flowrate but with higher gradient. This is a new finding on *P. aeruginosa* cell motion and diffusion inside microchannels.

Keywords: P. *aeruginosa* · CFD · DPM · Bacterial motion · Microfluidic devices

1 Introduction

Computational fluid dynamics (CFD) simulations are extensively used in engineering to study wide variety of applications ranging from flow inside pipes, fuel aerosol flow inside engines, and aerodynamics of vehicles. CFD is also used by researchers to visualise the flow characteristics in micro-scale [1–3] where analytical solutions are not adequate to picture wall shear stress induced or fluid velocity field. Finite volume method is used to analyse micro scale fluid flows such as interstitial fluid flow [4], flow in side microfluidic biomedical devices [5], microfluidic flow cells [2, 6]. In CFD simulation of bacterial suspension flow, properties of water is assumed for the properties of bacterial suspension with assumptions of Newtonian fluid and no-slip condition at the boundary [2, 3, 7, 8]. Therefore, CFD simulation methods are applicable and appropriate boundary conditions

A. Batako et al. (Eds.): GCMM 2021, LNNS 335, pp. 95–108, 2022.
https://doi.org/10.1007/978-3-030-90532-3_8

can be established for simulation of bacterial suspension under flow. A CFD simulation model can be invaluable for predicting bacterial adhesion onto different substrates, by avoiding or minimising *in vitro* experiments and much more resource consuming *in vivo* experiments.

CFD simulation has been done by Moreira et al. [7] on a parallel plate flow cell (PPFC) and shown that sufficient length of flow channel causes the flow to be fully developed before entering the chamber. The results of this study indicate that geometry of the inlet port does not affect the flow regime and therefore can be omitted for the purpose of CFD simulation. Ponmozhi et al. have conducted a CFD on fluid flow in a microfluidic device which has complicated flow channels by simplifying the channel geometry, and simulating the flow in the area of interest for wall shear stress in the fluid volume [2]. Wiklund et al. conducted CFD simulation to study the drag force exerted by fluid flow on rod shaped bacterium [9]. However, this CFD model had an error of 12–15% compared to empirical data as pointed by the authors. Nevertheless, this work shows that CFD approximations are valid with bacterial cells represented as solid objects, for the purpose of hydrodynamics studies. CFD couple with discrete element method (DEM) has been used by Jayathilake et al. to model bacterial twitching motion in fluid flows [10]. Simplification of bacterial cells in flowing suspension as solid particles in a liquid suspension with neglecting physical and biological interactions of bacterial cells is also reported [11]. Therefore, CFD methods can be used to visualise fluid flow characteristics and bacterial motion in micro scale fluid flow.

The domain of bacteria is classified as motile and non-motile, based on the ability of self-motion. *Pseudomonas aeruginosa* is a pathogenic bacteria that is the primary cause of pulmonary infections [12, 13]. It is Gram-negative, rod-shaped, motile bacteria that moves using its extracellular appendages such as flagella [14]. *P. aeruginosa* has several types of motion such as, swimming, swarming, and twitching [15]. Flagellated bacteria uses polar flagellum as a rotor to swim in liquid environments [16]. In addition to passive motion in a flowing fluid, *P. aeruginosa* does have active motions too.

Numerous studies are reported on quantification of size and motion of *P. aeruginosa.* A number of literature reported dimensions of *P. aeruginosa*, diameter (Ø) in the range of 0.5–1.0 μm and length in the range of 1.0–5.0 μm [17, 18]. Several studies in literature were found to report velocities of motion of *P. aeruginosa.* Maximum velocity of *P. aeruginosa* reported as 71.0 $\mu m.s^{-1}$ [19] and 73.0 $\mu m.s^{-1}$ [18]. Average velocity of the same bacteria is reported to be 56.8 $\mu m.s^{-1}$ [19]. Direct measurement of rotational speed of *P. aeruginosa* was not found in literature. *E. coli* is also a rod-shaped, gram-negative bacteria. This bacteria is reported to have rotational motion without tethering into solid surface with angular velocity of 50 $rad.s^{-1}$ by Dominick et al. [20]. Lauga stated, a flagellated bacteria cell has a rotational mobility of 10 Hz [21] which translates to 62.8 $rad.s^{-1}$. Cell mass density of *P. aeruginosa* was not found in literature. However, several studies reported density of bacteria in general to be 1.1 $g.cm^{-3}$ [22–24]. Bacterial cell mass, size, and movements of *P. aeruginosa* bacteria can be established with these literatures.

In this study, a CFD simulation coupled with discrete phase method (DPM) was used to examine the bacterial cell motion inside microchannels of a newly developed microfluidic device designed to study bacterial motion. In this simulation, rod shaped bacteria were approximated by solid spherical particles with no physiochemical interaction between the bacterial cells and surface. Suspension fluid flowrate and particle inoculum concentrations were varied in this simulation and particle mass concentrations inside the fluid volume was studied. This study reports a new finding on bacterial cell diffusion inside microchannels with varying suspension flowrate.

2 Methodology

2.1 Microchannel

The newly designed microfluidic device has a microfluidic chamber attached with an inlet and outlet as shown in Fig. 1 (half-section of the microchamber and channels are shown here). This device was developed following the principle of PPFC. The fluid volume inside the actual microfluidic device was simplified for simulation. Its' microchamber has a height of 0.2 mm, width of 3.0 mm and length of 10.0 mm, while the two microchannels i.e. inlet and outlet are 2.0 mm wide, 0.2 mm high and 10.0 mm long. Flow of bacterial suspension through the device is in X+ direction.

2.2 Flow Regime

Fluid Flow

The bacterial suspension fluid was assumed as a Newtonian fluid, and laminar flow was used for the bacterial suspension flow. The flow inside the channel was verified to be in laminar region by analysing Reynolds number for the flow and streamlines. Simulation was conducted for bacterial suspension at 37 °C to emulate human body temperature. Density and viscosity of bacterial suspension fluid at 37 °C was assumed to be that of water the at the same temperature.

Representation of Bacteria as Spherical Particles

Bacterial cells are simplified as spherical particles with no interaction with the surface. Propulsion of cells in terms of swimming is added to the model. As *P. aeruginosa* is a rod-shaped bacterium, it is represented as a sphere with equivalent volume in the simulations. Mean diameter and length were calculated by averaging values from literature, as Ø = 0.75 μm, and L = 3.0 μm. Therefore, the mean volume of bacterial cell was taken as 1.325 μm^3. Diameter of a sphere with a volume of 1.325 μm^3 is 1.363 μm, and this value was taken as the particle diameter for representing bacterial cell in the simulation.

P. aeruginosa swimming velocity was calculated by taking the average of swimming velocities reported in the literature. Since angular velocity of *P. aeruginosa* was not found on literature, that of another rod-shaped Gram-negative bacterium *E. coli*, was taken as an approximation. Calculation of Re_p for bacterial motion was done using Eq. (1).

$$Re_p = \frac{\rho U_0 r}{\mu} \tag{1}$$

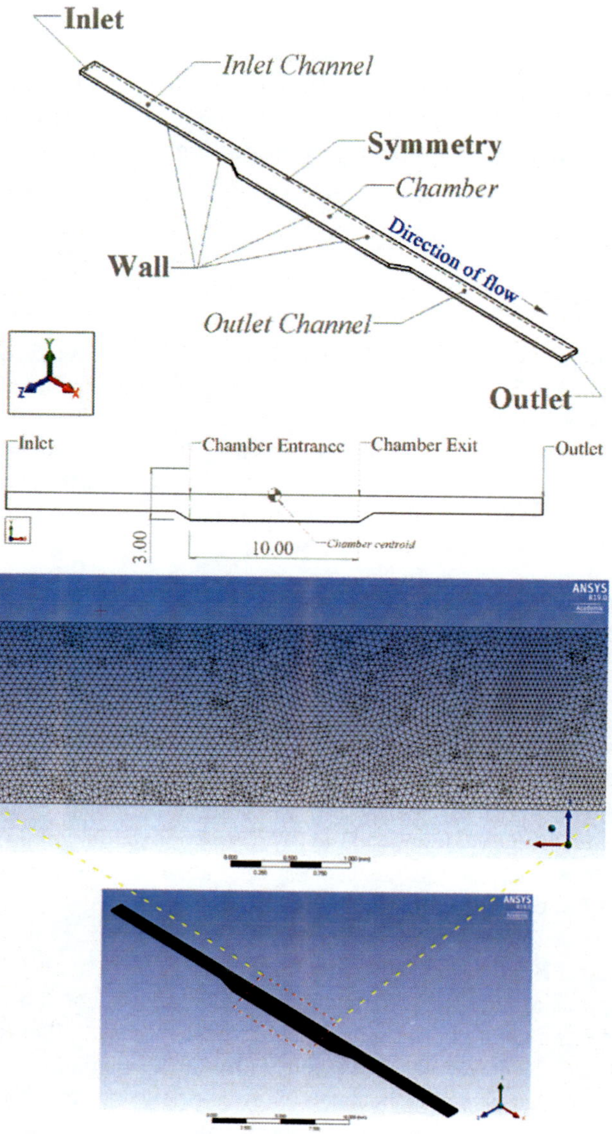

Fig. 1. Fluid volume and boundaries (top two images), and mesh (bottom two images) for particle flow simulations. (Dimensions are in mm)

In this case bacteria swimming velocity was taken as velocity relative to the fluid. τ_r is given by Eq. (2). Since $Re_p < 1$ Stokes' law is applicable. Drag coefficient in creeping flow around a spherical particle [25] is given by Eq. (3).

$$\tau_r = \frac{\rho_P D_P^2 24}{18\mu C_D Re} \quad (2)$$

$$C_D = \frac{24}{Re_p} \quad (3)$$

$$Re_p = \frac{\rho D_P |u_P - u|}{\mu},$$

$$\text{and } C_D = \frac{24}{Re_p}$$

$$\therefore \tau_r = \frac{\rho_P D_P^2}{18\mu}$$

Bacterial Cell Concentration and Particle-Injection Mass Flowrate

Mass flowrate of the particles at the injection was calculated based on the mass flowrate of bacterial cells in suspension with turbidity measured by optical density OD_{600} 0.1 flown at the different fluid velocities. Volume of a bacterium was calculated using dimensions of *P. aeruginosa* from literature assuming a perfect cylindrical shape. Mass of a bacterium was calculated using cell mass density of bacteria reported in literature. *P. aeruginosa* colony forming unit number has a linear relationship with OD_{600} as reported by Don-Ju et al. [26]. Bacterial mass flowrate was calculated using Eq. (4) and (5) for a bacterial suspension, and the values are compared. Bacterial mass flow rate is calculated using Eq. (6).

$$Y(CFU/ml) = 2.0 \times 10^8 \times (OD_{600}) + 4.0 \times 10^6 \quad (4)$$

$$Y(mg/ml) = 2.0087 \times (OD_{600}) + 0.0764 \quad (5)$$

$$\dot{m} = Y \times Q \quad (6)$$

2.3 Analytical Model for Fluid and Particle Flow in PPFC

Sedimentation of bacteria can occur based on the fluid flow characteristics. Bacteria are transported in liquid medium, and the gravitational effect on bacterial cells can be influential if the diffusive transport is higher than advective transport. Pe defines the ratio between advective transport and diffusive transport. Peclet number for a PPFC is derived by Smoluchowski-Levich approximation [27] given in Eq. (7) was used.

$$Pe = \frac{3 v_m r^3}{2 b^2 D} \quad (7)$$

Stokes' law makes the following assumptions for the behavior of a particle in a fluid; laminar flow, spherical particles, homogeneous material, smooth surfaces in the particle, and the particles do not interfere with each other, which are valid for this case. Therefore, Stokes-Einstein equation for diffusion of spherical particles through a liquid with low Reynolds number [28] given by Eq. (8) was used for this study.

$$D = \frac{k_B T}{6\pi \mu r} \tag{8}$$

$$\tau = \frac{6.Q.\mu}{W.H^3}(H - 2y) \tag{9}$$

2.4 Simulation of Bacterial Flow Inside the Microfluidic Device

Finite Element Method (FEM) is used for this analysis. Ansys Fluent DPM model is based on Euler-Lagrange approach. Its fluid phase is treated as continuum and a large number of particles are tracked in disperse phase. Since the volume fraction of bacteria is only 0.3%, DPM is applicable. First, the fluid flow without particle injection was studied using laminar flow model in CFD. Since the fluid volume is symmetric about XY plane, half of the volume is used in the CFD simulations as shown in Fig. 1. Adaptive size function is used for meshing, and a mesh dependence study was conducted to obtain mesh independent FEA model. Adaptive mesh with 0.01 mm element size was found to be the mesh-independent element size. Face sizing and mesh-inflation was used to refine the mesh. The model had a total of 945,365 elements. Convergence criteria was defined as residuals $< 10^{-5}$ and error of maximum velocity at outlet $< 10^{-5}$.

Particles were introduced to the flow at inlet as surface injections. Four levels of particle mass concentrations based on the bacterial mass flowrates were used. Particle mass flowrate was increased from 10% of calculated from bacterial mass flowrate ($\dot{m}$), to 50%, 75%, and 100%. Two levels of mean fluid velocities (low = 0.005 $m.s^{-1}$ and high = 0.05 $m.s^{-1}$) at the centroid of the chamber portion of the fluid volume were used. Average particle mass concentrations on four planes along the direction of flow was probed from the simulations, as shown in Fig. 1. No-slip condition at the MFD channel walls is assumed. Bacterial cell volume calculated as a perfect cylinder, is 1.325×10^{-18} m^3/cell. Therefore, cell mass calculated taking density of bacterium is as 1.1 $g.cm^{-3}$ is 1.458×10^{-15} kg/cell. According to Eq. (4), *P. aeruginosa* cell count with OD_{600} of 0.1 is 2.4×10^7 CFU/ml. However, this value contradicts with another experimental quantification which stated that OD_{600} of 0.1 of *P. aeruginosa* in PBS solution is 10×10^7 CFU/ml [29]. Equation (5) specifies the relationship between optical density of bacterial suspension and bacterial mass. According to Eq. (5) the bacterial cell density in the suspension,

$$Y = 0.277 \text{ mg/ml}.$$

From Eq. (6),
$\dot{m} = 8.3 \times 10^{-9}$ $kg.s^{-1}$.

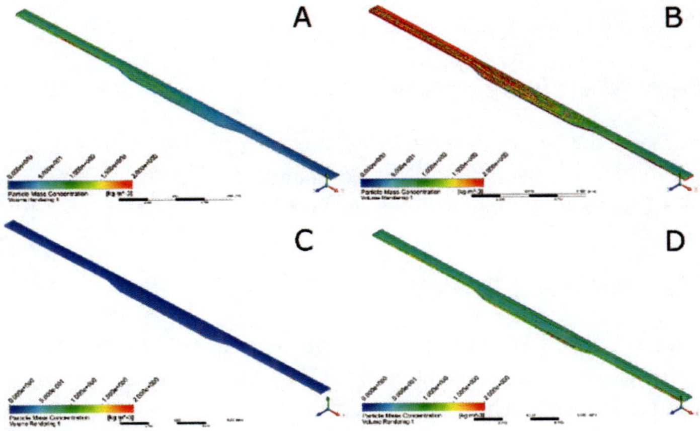

Fig. 2. Fluid velocity contours at entrance to the chamber (A), on a plane through the midpoint of the chamber (B), and exit of the chamber (C), on a plane perpendicular to the flow direction. Velocity streamlines (D) of the entire fluid volume and (E) at the mid area of the streamline.

For each of the two fluid velocities, four particle mass flowrates were taken with 10%, 50%, 75%, and 100% of calculated bacterial cell flowrate of 80×10^{-9} kg.s^{-1} leading to a total of eight states for simulation.

Selection of Simulation Model and Parameters

Particle relaxation time was calculated using Eq. (2) to determine the particle time step size in the simulations. The representative particles are micron scale, and it was assumed the particle surfaces are smooth. Therefore, spherical drag law is used in the simulation. The particle to fluid density ratio is greater than unity, therefore, the virtual mass and pressure gradient forces are applied in the model. Particle rotation is enabled, and appropriate drag and lift laws were selected.

Angular velocity of bacteria was taken as $\Omega = 50$ rad/s, and hence particle rotational velocity defined accordingly. Rotational drag law followed was Dennis et al. [30], and the Magnus lift law was Oesterle-Bui-Dinh [31]. Particles are assumed to be injected from the entire inlet surface of the fluid volume. Particles are assumed as spherical inert particles with material density equal to bacterial density reported in [22–24]. Injection is assumed to be perpendicular to injection surface, and velocity with reference to the fluid is taken as 6×10^{-5} m.s^{-1} in all 3 principal directions. *Escape* condition was assumed for inlet and outlet, so that the particles reaching to surface can pass-through. *Trap* condition was assumed for walls of the fluid volume, indicating adhesion of particle to the surface when it reaches the surface. Adhesion forces were not modelled, and the walls were assumed to be smooth.

3 Results of Particle Flow

3.1 Theoretical Wall Shear Stresses

Wall shear stress at the centroid of the PPFC chamber bottom and top faces, under the low and high fluid velocities was calculated using Eq. (9). The wall shear stress was 0.105 Pa and 1.063 Pa under low and high mean fluid velocity, respectively.

3.2 Fluid Flow Simulation Inside the MFD

Fluid flow was simulated at both low and high flowrates. Velocity contours at three different locations of the fluid chamber and velocity streamlines are shown in Fig. 2(A–C) and (D–E) respectively. These confirmed the fluid flow remained laminar at both flowrates.

3.3 Motion of Bacterial Cells

For the bacterial suspension: $\rho = 993.2$ kg.m^{-3}, $\mu = 0.000692$ Pa.s. For the particles: $U_0 = 6.00 \times 10^{-5}$ m.s^{-1}, and $r = 6.81 \times 10^{-7}$ m. Therefore, from Eq. (1),

$$Re_p = 1.17 \times 10^{-4}$$

Particle relaxation time calculated from Eq. (2),

$$\tau_r = 1.6382 \times 10^{-7} \text{ s}$$

Diffusion coefficient of bacterial cells under flow is calculated from Eq. (8) by taking $k_B = 1.380649 \times 10^{-23}$ J.K^{-1}, T = 310.15 K, $\mu = 0.000692$ Pa.s, r = 6.81×10^{-7} m.

$$D = 4.82 \times 10^{-13} \text{m}^2.\text{s}^{-1}$$

Peclet Number for the flow inside microchannel is calculated for low and high fluid flowrates using Eq. (7) by taking $b = 0.1$ mm, $v_m = 0.005$, and 0.05 m.s^{-1}, substituting values to the equation. Calculated Peclet numbers for the low and high flow velocity are 0.491 and 4.914 respectively.

3.4 Particle Flow Simulation

In this simulation, particles were injected with 6×10^{-5} m.s^{-1} velocity relative to fluid in X, Y, and Z directions from inlet surface with 50 rad.s^{-1} angular velocity about X axis (principal flow axis), with rough wall assumption at the wall boundaries. Wall shear from the simulation at the substrate centroid was in well agreement with theoretical calculations and does not affected by the particle flow. All tracked particles were escaped from the fluid volume predicting no trapping of particles.

3.5 Particle Mass Concentration in the Fluid Volume

Particle mass concentration was visualised using volume rendering as shown in Fig. 2. Low fluid velocity simulations showed a larger variation in particle mass concentration in the direction of flow, while the high velocity fluid flow showed a subtle variation in particle mass concentration. Figure 2(A) and (B) shows the 10% and 100% particle mass flowrates at low fluid velocity. At this velocity, the 3D field of particle mass concentration shows higher concentration upstream and lower concentration down the stream. As shown in Fig. 2(C) and (D), the particle mass concentration field with higher velocity did not cause major difference for both mass flowrates.

3.6 Variation of Particle Mass Concentration Along Flow Direction

Since the 3D particle mass concentration field plots indicated variation in mass concentration variation along the flow direction, average concentrations were taken on several planes at four separate locations along the flow direction. Average particle mass concentrations on the planes through inlet, chamber entrance, chamber exit, and outlet are varied in two distinct patterns for low and high fluid velocities, irrespective of injection particle mass flowrates. Figure 3 shows the variations in relative particle mass concentration at chamber-entrance, chamber-exit, and outlet, with respect to inlet, for varying fluid velocity (FV), and particle mass flowrate (PMF).

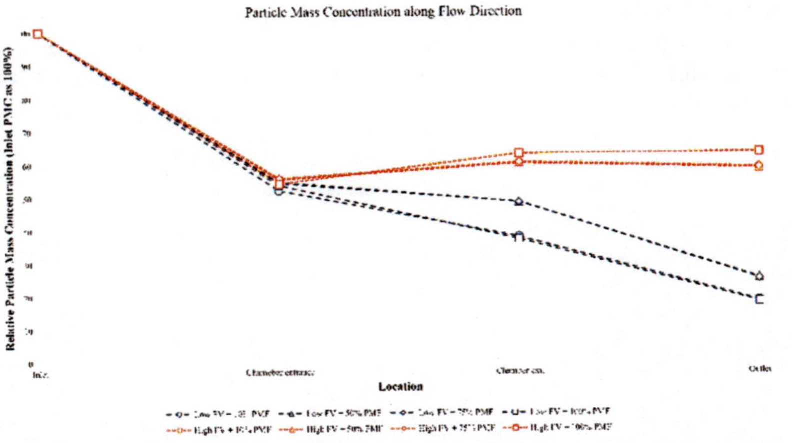

Fig. 3. Variation in average particle mass concentration on a plane across each location along the direction of flow

The maximum particle mass concentration probed in the fluid volume is varied with fluid velocity and particle mass flowrate. The variation in maximum particle mass concentration had a linear relationship with mass flowrate percentage. Regression models were fitted based on adjusted coefficient of determination (R^2) for the two fluid flowrates. Figure 4 shows the maximum particle mass concentration in fluid volume at the two-fluid velocity levels with four mass flowrates, along with the regression lines. At both fluid velocity levels with different particle mass flowrates, decrease in particle mass concentration from the inlet to chamber entrance was within the range of 44%–48%. Apart from that, percentage-wise change in particle mass concentrations were not similar between the two fluid velocities. However, the percentage-wise changes of particle mass concentration with the same velocity under different particle mass flowrates were similar. For low fluid velocity, differences between chamber ends were within −11%~31%, and chamber exit to outlet were within −48%~52%. Ranges for the same with high velocity, was 8–16% and −3%–0% respectively. Low velocity concentration profiles were with negative slopes in all 3 regions, but the at high velocity, the slope was positive in chamber-entrance to -exit region.

3.7 Variation in Maximum Particle Mass Concentration in the Fluid Volume

The maximum particle mass concentration probed in the fluid volume is varied with fluid velocity and particle mass flowrate. The variation in maximum particle mass concentration had a linear relationship with mass flowrate percentage. Regression models were fitted based on adjusted coefficient of determination (R2) for the two fluid flowrates. Figure 4 shows the maximum particle mass concentration in fluid volume at the two-fluid velocity levels with four mass flowrates, along with the regression lines. At both fluid velocity levels with different particle mass flowrates, decrease in particle mass concentration from the inlet to chamber entrance was within the range of 44%–48%. Apart from that, percentage-wise change in particle mass concentrations were not similar between the two fluid velocities. However, the percentage-wise changes of particle mass concentration with the same velocity under different particle mass flowrates were similar. For low fluid velocity, differences between chamber ends were within −11%~31%, and chamber exit to outlet were within −48%~−52%. Ranges for the same with high

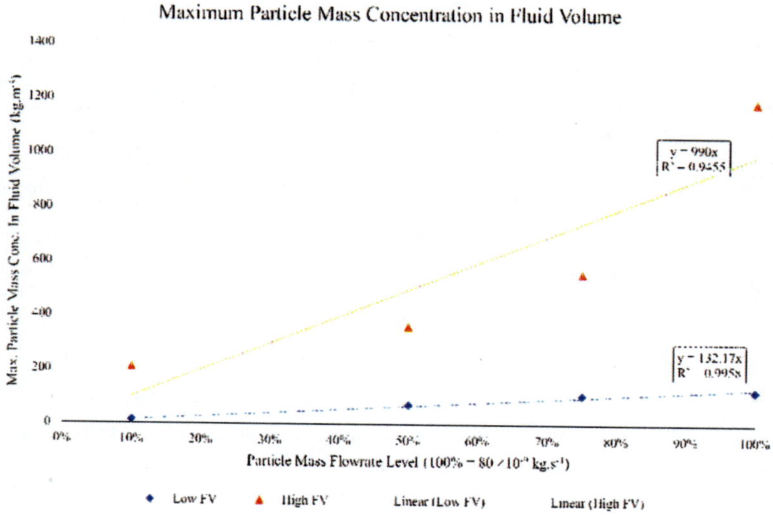

Fig. 4. Variation in Maximum Particle Mass concentration inside the fluid volume with fluid velocity and particle mass flowrate

velocity, was 8–16% and −3%~0% respectively. Low velocity concentration profiles were with negative slopes in all 3 regions, but the at high velocity, the slope was positive in chamber-entrance to -exit region.

4 Discussion

Equivalent sphere radius for *P. aeruginosa* cell was calculated in this work, and the value obtained by calculation is comparable with previous studies in literature as radius of *P. aeruginosa* cell is assumed to be 0.68 μm by [32] for the purpose of simplification of cell geometry for computer simulations.

Diffusion coefficient was calculated for the representative spherical particles in 37 °C water. This calculated diffusion coefficient agrees with experimental observation for bacteria found in literature. Diffusion coefficient of cocci-shaped *S. aureus* is reported as 3.31×10^{-13} $m^2.s^{-1}$ [33] and rod-shaped *P. fluorescens* as 3.75×10^{-13} $m^2.s^{-1}$ [34]. *S. aureus* is of spherical shape and has Ø 1 μm [35], and *P. fluorescens* is of rod-shape and has Ø 0.5 μm and length 2.0–2.5 μm [36]. Therefore, based on the particle size, the comparison of diffusion coefficient is valid.

For mean fluid velocity 0.005 $m.s^{-1}$ the Peclet number for *P. aeruginosa* is less than 1. This indicates that, the diffusive transport of bacterial cells represented by particles is greater than advective transport. Therefore, sedimentation of bacterial cells is predicted to have a low flowrate. However, for 0.05 $m.s^{-1}$ Peclet number is 4.914, suggesting advection is dominant. Therefore, sedimentation of cells inside the channel is unlikely. At low velocity particle mass concentration gradient in X direction is observed, and on the other hand at high fluid velocity mass concentration is more uniform through the length of the channel. As the Peclet number analysis predicted, the advective mass transport is dominant at high velocity, whereas diffusive transport is dominant at low velocity.

Particle mass concentration from the inlet to chamber entrance was reduced for both the fluid velocities. Effective cross-sectional area increases by 50% from inlet to chamber entrance, and as a result, particles diffuses. This appears as a reduction in particle mass concentration. With higher flowrate, particle stagnation inside the chamber is less and hence the particle mass concentration gradient is not apparent along the principal axis from chamber entrance to outlet. However, lower flowrate results in more particle stagnation, inside the chamber. As shown with Peclet number calculations, sedimentation is likely to occur with the lower flowrate.

Despite diffusive transport of particles is dominant at the lower fluid velocity, particle trapping was not observed in simulations. Interactions between particles and solid surface are not modelled in these simulations and therefore, trapping of particles at the walls would not occur.

Interactions between particle-particle, and particle-surface, are overlooked in this simulation. Expanding the simulation model to capture those effects is suggested as future work.

5 Conclusions

Low flow velocity in this device allows the bacterial cells to diffuse, while the high flow velocity causes higher advection of cells. Variation of highest bacterial cell concentration in the fluid volume with increasing inoculum concentration depends on suspension flowrate. At low flowrates, highest cell concentration inside the fluid volume increases in a linearly with increasing bacterial inoculum concentration, while at high flowrates, the same increases linearly but with a higher gradient.

Lower fluid flowrates can cause increased cell settlement inside the fluid chamber. This could eventually clog the device as cell grow and proliferate.

Acknowledgements. Authors wishes to acknowledge support from, Institute of Health and Biomedical Innovation (IHBI), Faculty of Engineering, Queensland University of Technology (QUT), Australia, Advance Queensland Industry Research Fellowship, Konica Minolta and Bionics Queensland. The first author is a lecturer from University of Moratuwa, Sri Lanka, currently attached to QUT. Funding was received for this work from AHEAD project (Grant: AHEAD/PhD/R2/ENG/TECH/161), University Grants Commission of Sri Lanka.

References

1. Salta, M., Capretto, L., Carugo, D., Wharton, J.A., Stokes, K.R.: Life under flow: a novel microfluidic device for the assessment of anti-biofilm technologies. Biomicrofluidics **7**(6), 064118 (2013). https://doi.org/10.1063/1.4850796
2. Ponmozhi, J., Moreira, J.M.R., Mergulhão, F.J., Campos, J.B.L.M., Miranda, J.M.: Fabrication and hydrodynamic characterization of a microfluidic device for cell adhesion tests in polymeric surfaces. Micromachines **10**(5), 303 (2019). https://doi.org/10.3390/mi10050303
3. Moreira, J.M.R., Araújo, J.D.P., Miranda, J.M., Simões, M., Melo, L.F., Mergulhão, F.J.: The effects of surface properties on Escherichia coli adhesion are modulated by shear stress. Colloids Surf. B Biointerfaces **123**, 1–7 (2014). https://doi.org/10.1016/j.colsurfb.2014.08.016
4. Yao, W., Li, Y., Ding, G.: Interstitial fluid flow: the mechanical environment of cells and foundation of meridians. Evid.-Based Complement. Altern. Med. **2012**, 1–9 (2012). https://doi.org/10.1155/2012/853516
5. Salek, M.M., Martinuzzi, R.J.: Numerical simulation of fluid flow and hydrodynamic analysis in commonly used biomedical devices in biofilm studies. In: Numerical Simulations - Examples and Applications in Computational Fluid Dynamics (2010). https://doi.org/10.5772/13117
6. Schlegel, C., et al.: Analyzing the influence of microstructured surfaces on the lactic acid production of Lactobacillus delbrueckii lactis in a flow-through cell system. Eng. Life Sci. **17**(8), 865–873 (2017). https://doi.org/10.1002/elsc.201700045
7. Moreira, J.M.R., Ponmozhi, J., Campos, J.B.L.M., Miranda, J.M., Mergulhão, F.J.: Micro- and macro-flow systems to study Escherichia coli adhesion to biomedical materials. Chem. Eng. Sci. **126**, 440–445 (2015). https://doi.org/10.1016/j.ces.2014.12.054
8. Westerwalbesloh, C., et al.: Modeling and CFD simulation of nutrient distribution in picoliter bioreactors for bacterial growth studies on single-cell level. Lab Chip **15**(21), 4177–4186 (2015). https://doi.org/10.1039/c5lc00646e

9. Wiklund, K., Zhang, H., Stangner, T., Singh, B., Bullitt, E., Andersson, M.: A drag force interpolation model for capsule-shaped cells in fluid flows near a surface. Microbiology **164**(4), 483–494 (2018). https://doi.org/10.1099/mic.0.000624
10. Jayathilake, P.G., Li, B., Zuliani, P., Curtis, T., Chen, J.: Modelling bacterial twitching in fluid flows: a CFD-DEM approach. Sci. Rep. **9**(1), 2–5 (2019). https://doi.org/10.1038/s41598-019-51101-3
11. Sjollema, J., Busscher, H.J., Weerkamp, A.H.: Deposition of oral streptococci and polystyrene latices onto glass in a parallel plate flow cell. Biofouling **1**(2), 101–112 (1988). https://doi.org/10.1080/08927018809378100
12. Kong, D., Lin, W., Pan, Y., Zhang, K.: Swimming motion of rod-shaped magnetotactic bacteria: the effects of shape and growing magnetic moment. Front. Microbiol. **5**, 1–11 (2014). https://doi.org/10.3389/fmicb.2014.00008
13. Lyczak, J.B., Cannon, C.L., Pier, G.B.: Establishment of Pseudomonas aeruginosa infection: lessons from a versatile opportunist. Microbes Infect. **2**(9), 1051–1060 (2000). https://doi.org/10.1016/S1286-4579(00)01259-4
14. Yang, A., Tang, W.S., Si, T., Tang, J.X.: Influence of physical effects on the swarming motility of Pseudomonas aeruginosa. Biophys. J. **112**(7), 1462–1471 (2017). https://doi.org/10.1016/j.bpj.2017.02.019
15. Toutain, C.M., Zegans, M.E., O'Toole, G.A.: Evidence for two flagellar stators and their role in the motility of Pseudomonas aeruginosa. J. Bacteriol. **187**(2), 771–777 (2005). https://doi.org/10.1128/JB.187.2.771-777.2005
16. Terashima, H., Kojima, S., Homma, M.: Chapter 2 flagellar motility in bacteria, vol. 270, pp. 39–85. Academic Press (2008). https://doi.org/10.1016/S1937-6448(08)01402-0
17. Vater, S.M., et al.: Swimming behavior of Pseudomonas aeruginosa studied by holographic 3D tracking. PLoS One **9**(1) (2014). https://doi.org/10.1371/journal.pone.0087765
18. Greenberg, E.P., Canale-Parola, E.: Motility of flagellated bacteria in viscous environments. J. Bacteriol. **132**(1), 356–358 (1977). https://doi.org/10.1128/jb.132.1.356-358.1977
19. Shigematsu, M., Meno, Y., Misumi, H., Amako, K.: The measurement of swimming velocity of vibrio cholerae and Pseudomonas aeruginosa using the video tracking method. Microbiol. Immunol. **39**(10), 741–744 (1995). https://doi.org/10.1111/j.1348-0421.1995.tb03260.x
20. Dominick, C.N., Wu, X.L.: Rotating bacteria on solid surfaces without tethering. Biophys. J. **115**(3), 588–594 (2018). https://doi.org/10.1016/j.bpj.2018.06.020
21. Lauga, E.: Bacterial hydrodynamics. Ann. Rev. Fluid Mech. **48**(1), 105–130 (2016). https://doi.org/10.1146/annurev-fluid-122414-034606
22. Bratbak, G., Dundas, I.: Bacterial dry matter content and biomass estimations. Appl. Environ. Microbiol. **48**(4), 755–757 (1984)
23. Average density - bacteria - BNID 102239 n.d. https://bionumbers.hms.harvard.edu/bionumber.aspx?id=102239. Accessed 6 Nov 2019
24. Loferer-Krößbacher, M., Klima, J., Psenner, R.: Determination of bacterial cell dry mass by transmission electron microscopy and densitometric image analysis. Appl. Environ. Microbiol. **64**(2), 688–694 (1998). https://doi.org/10.1128/aem.64.2.688-694.1998
25. Hensley, Z.D., Papavassiliou, D.V.: Drag coefficient correction for spherical and nonspherical particles suspended in square microducts. Ind. Eng. Chem. Res. **53**(25), 10465–10474 (2014). https://doi.org/10.1021/ie5007646
26. Kim, D.: Relation of microbial biomass to counting units for Pseudomonas aeruginosa. Afr. J. Microbiol. Res. **6**(21), 4620–4622 (2012). https://doi.org/10.5897/ajmr10.902
27. Adamczyk, Z., Van De Ven, T.G.M.: Deposition of particles under external forces in laminar flow through parallel-plate and cylindrical channels. J. Colloid Interface Sci. **80**(2), 340–356 (1981). https://doi.org/10.1016/0021-9797(81)90193-4

28. Midelet, J., El-Sagheer, A.H., Brown, T., Kanaras, A.G., Werts, M.H.V.: The sedimentation of colloidal nanoparticles in solution and its study using quantitative digital photography. Part. Part. Syst. Charact. **34**(10) (2017). https://doi.org/10.1002/ppsc.201700095
29. Song, C., Li, H., Zhang, Y., Yu, J.: Effects of Pseudomonas aeruginosa and Streptococcus mitis mixed infection on TLR4-mediated immune response in acute pneumonia mouse model. BMC Microbiol. **17**(1) (2017). Article number: 82. https://doi.org/10.1186/s12866-017-0999-1
30. Dennis, S.C.R., Singh, S.N., Ingham, D.B.: The steady flow due to a rotating sphere at low and moderate Reynolds numbers. J. Fluid Mech. **101**(2), 257–279 (1980). https://doi.org/10.1017/S0022112080001656
31. Oesterlé, B., Bui Dinh, T.: Experiments on the lift of a spinning sphere in a range of intermediate Reynolds numbers. Exp. Fluids **25**(1), 16–22 (1998). https://doi.org/10.1007/s003480050203
32. Ghanbari, A., Dehghany, J., Schwebs, T., Müsken, M., Häussler, S., Meyer-Hermann, M.: Inoculation density and nutrient level determine the formation of mushroom-shaped structures in Pseudomonas aeruginosa biofilms. Sci. Rep. **6**, 1–12 (2016). https://doi.org/10.1038/srep32097
33. Li, J., Busscher, H.J., Norde, W., Sjollema, J.: Analysis of the contribution of sedimentation to bacterial mass transport in a parallel plate flow chamber. Colloids Surf. B Biointerfaces **84**(1), 76–81 (2011). https://doi.org/10.1016/j.colsurfb.2010.12.018
34. Rijnaarts, H.H.M., Norde, W., Bouwer, E.J., Lyklema, J., Zehnder, A.J.B.: Bacterial adhesion under static and dynamic conditions. Appl. Environ. Microbiol. **59**(10), 3255–3265 (1993)
35. Foster, T. Staphylococcus. In: Baron, S. (ed.) Medical Microbiology, 4th edn (1996)
36. Martínez-García, P.M., et al.: Complete genome sequence of Pseudomonas fluorescens strain PICF7, an indigenous root endophyte from olive (Olea europaea L.) and effective biocontrol agent against Verticillium dahliae. Stand. Genomic. Sci. **10**, 1–7 (2015). https://doi.org/10.1186/1944-3277-10-10

Influence of Structural Parameters on the Wear Resistance of ADI During Dry Sliding Friction

N. Khidasheli[1(✉)], G. Gordeziani[1], S. Gvazava[1,2], G. Tavadze[1], R. Tabidze[1], and A. D. L. Batako[2]

[1] Ferdinand Tavadze Metallurgy and Materials Science Institute, Mindeli Street 10, Tbilisi, Georgia
khidly@gtu.ge

[2] Liverpool John Moores University, Byrom Street, Liverpool L3 3AF, UK
a.d.batako@ljmu.ac.uk

Abstract. The work presented in this paper studied the tribological characteristics (wear resistant, friction coefficient) of austempered ductile iron (ADI) with different combinations of structural composition (upper bainite, lower bainite, retained austenite) in dry sliding friction. A range of structural states of the metal matrix was obtained by changing the regimes of isothermal quenching of a high-strength cast iron. The tribological tests were carried out using two sets of isothermal quenched cast irons. After austenitization at 900 °C for 60 min, the specimens from the first group were isothermal quenched at the 300 °C temperature and the specimens from the second set – at 400 °C.

The investigations showed that the isothermal quenching increases the friction coefficient of high strength cast irons. The friction coefficient was found to be in the range from 0.4 to 0.55 for cast irons depending on the structures of the metal matrix. The quenched cast irons having lower bainite demonstrate higher wear resistance in dry friction conditions. The dependence of wear resistance on the amount of retained austenite in isothermal quenched cast irons has a nonlinear characteristic and reaches its maximum value when the content of retained austenite was about 15–22%. The boron micro-additives allowed to reduce the friction coefficient of ADI and increase their wear resistance by 1.5–1.7 times.

Keywords: ADI · Austenitization · Isothermal quenching · Wear-resistance · Tribological characteristics

1 Introduction

According to numerous data, hardened high-strength bainitic cast irons, by combination of characteristics, remain the most promising material for the manufacture of gears and brake discs in modern brake systems [1–5]. The efficiency of application the bainitic cast irons can be increased owing to the rational structuring of their metal matrix taking into account conditions of exploitation. This is due to the possibility of varying the structural state and functional properties of these materials within wide limits through the change the phase composition of the components [6–9]. Depending on the parameters of the

A. Batako et al. (Eds.): GCMM 2021, LNNS 335, pp. 109–116, 2022.
https://doi.org/10.1007/978-3-030-90532-3_9

technological process receiving the structure of isothermally quenched cast irons may include various amounts of upper and lower bainite, metastable and stabilized retained austenite, cementite and carbides of alloying elements [10]. The above listed phases significantly differ by the level of physical-mechanical parameters and differently affect the formation of final properties of bainitic cast irons, particularly, on the wear resistance under various types of loading. Friction coefficient and intensity of surface wearing to a large extent depends on the structure of the contacting places. To reduce adhesion, the sectors of contacting surfaces should be heterogeneous in structure, optimal in terms of occupied areas and mechanical properties. Due to development of new compositions alloys and changing their exploitational conditions, the problem of optimizing the microstructure of bainitic cast irons remain actual.

In the present work, are studied the tribological properties of basic and micro-alloyed high strength cast irons with various type of bainitic matrix under conditions of dry sliding friction.

The purpose of this work was to study the influence of the structural components of alloyed cast irons on their tribo-technical properties under conditions of self-lubrication of friction in a wide range of specific loads.

2 Experimental

The experimental melting of cast iron was carried out in a 50-kg-capacity crucible using an electrical induction furnace of intermediate frequency. The furnace charge consisted of High-Purity Pig Iron: 3.62% C, 2.00% Si, 0.35% Mn, 0.003% S, 0.06% P and up to 10% steel scrap. After melting at 1350 °C, the liquid metal was held during 5 min followed by spheroidization of graphite inclusions by treatment with metallic Mg vapor, through graphite equipment. Then, the liquid cast iron was micro-alloyed by Boron at the 1440 °C temperature by using the Fe17%B ligature. The Y block ingots of high strength cast irons were poured into sand and graphite forms with steel-pouring ladle.

Table 1 shows the standard chemical composition of high strength cast iron which was studied in the presented work. The silicon content was maintained within 2.2–2.4%, since according to data of various studies [11, 12], the maximum crack resistance of high strength cast iron was achieved in such a way.

Table 1. Chemical composition of experimented ductile irons (wt, %).

High strength cast iron	Chemical composition of high strength cast iron, %						
	C	Si	Mn	S	P	B	Mg
Unalloyed	3.45	2.30	0.40	0.004	0.06	–	0.040
By Boron alloyed	3.52	2.40	0.35	0.003	0.06	0.024	0.045

To obtain a various structure of the metal matrix, the high-strength cast irons were undergone to different regimes of heat treatment. Two groups of high-strength cast irons (unalloyed and micro-alloyed by boron) were investigated, which were isothermally quenching in melted metal bath (Pb-Sb-Sn) for different durations of times under the temperatures: 300 °C and 400 °C, after austenitization at 900 °C during 60 min for receiving the various amount (~10, ~20 and ~35%) of retained austenite in the metallic structure of high-strength cast irons.

The microstructure and phase composition were studied using a metallographic microscope, X-ray diffraction analysis and scanning electron microscopy. Optical microscopy gives a qualitative picture of the present phases. Quantitative study was carried out by X-Ray diffraction technique. X-ray diffractometer (model ДРОН-4) with Cu Kα radiation ($\lambda = 1.54$ Å) was used as the source.

Hardness was measured for all the heat treatment conditions before and after wear test. Hardness values were taken on a Rockwell hardness testing machine in Rc scale for a load of 150 kg.

Dry sliding wear tests were carried out using a conventional tribometer: model CMЦ-2–"pin-on-disk", capable of maintaining a constant unidirectional, sliding velocity between the pin and disk. Wear tests were performed at 196 N normal load, 2,3 m/s sliding speed and 3000 m sliding distance in dry running conditions. The quenched carbon steel (0.9% carbon content) disk with 62 HRC hardness was used as a counter-body. During each test, the evolution of the friction coefficient was recorded. For tribological characteristic, wear properties were determined based on the study of worn out surfaces and wear-induced loss amounts.

3 Results and Discussion

In the present work, by variation of heat treatment technological parameters of high strength cast iron, unalloyed and micro-alloyed by Boron and with 3.4% carbon and 2.2% silicon content have been received the experimental specimens with different amounts (10, 20, 35%) of retained austenite in the upper and lower bainitic matrix. A typical microstructure is shown in Fig. 1.

The results of metallographic and X-ray diffraction analysis allow us to conclude that microalloying by Boron (0.03%) of the liquid metal increases the dispersion and disorientation of the structural elements of a bainitic matrix.

Iron borides are formed in the structure and ultra-dispersion (<1 μm) inclusions – presumably of boron carbides and nitrides are observed in the thin section plane. As a result of microalloying, the hardness of specimens with the structure of upper bainite increases to 45–48 HRC and at 300 °C temperature of isothermally quenched specimens to 57–60 HRC (Table 2).

The higher hardness of Ductile Irons microalloyed by boron and with the low-bainite structure is achieved due to formation of fine-grained matrix structure, strengthened by the dispersion particles of boron carbides and nitrides.

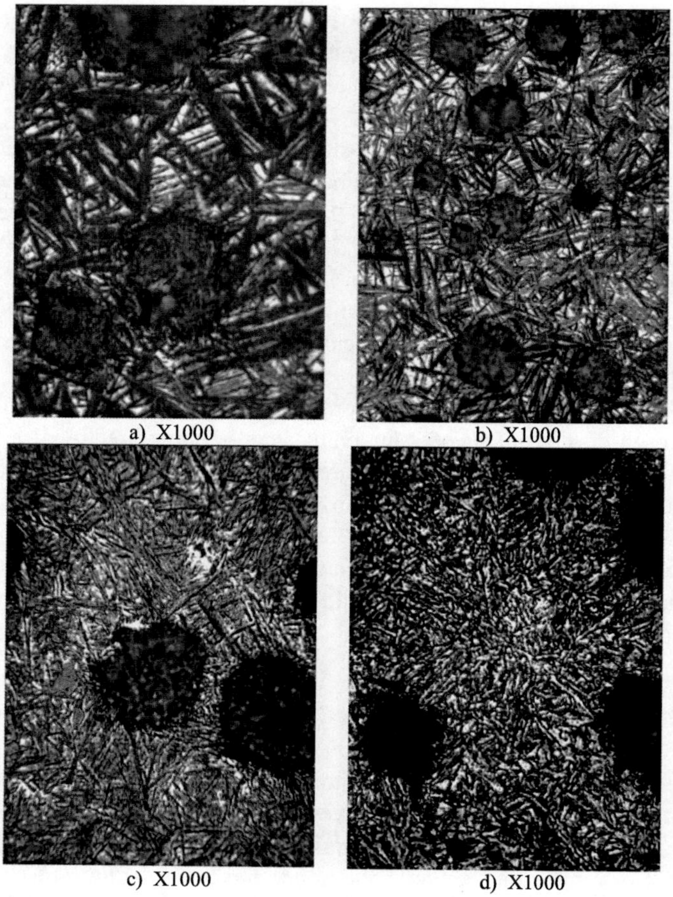

Fig. 1. Typical microstructure received by isothermal quenching at 300 °C (a, b) and at 400 °C temperature (c, d). a, c – unalloyed cast iron; b, d – by Boron micro-alloyed

Table 2. Hardness of experimented austempered ductile irons (HRC)

N	Ductile iron	Hardness of Ductile Iron (HRC) with different Metllic Matrix (10% retained austenite)	
		Upper bainit	Lower bainit
1	Unalloyed	32; 34; 35	50; 47; 48
2	By Boron alloyed	45; 48; 47	57; 59; 60

An analysis of received data from wear tests let us do preliminary estimation of working ability of high-strength bainitic cast irons with various combinations of phases of the metal matrix under conditions of dry sliding friction.

In the Fig. 2 and 3 is shown the changes of the steady-state friction coefficient, of the studied high-strength cast irons depending on their structure. As it seems from presented data the friction coefficient is changed in the range from 0.4 to 0.57 and an increase in the amount of retained austenite is accompanied by an increase in the values of this parameter. The microalloying of high strength cast iron by Boron causes decreasing the friction coefficient. Insignificant fluctuation of the friction coefficient during testing indicates the structural stability of experimental bainitic cast irons under the studied regimes of dry friction.

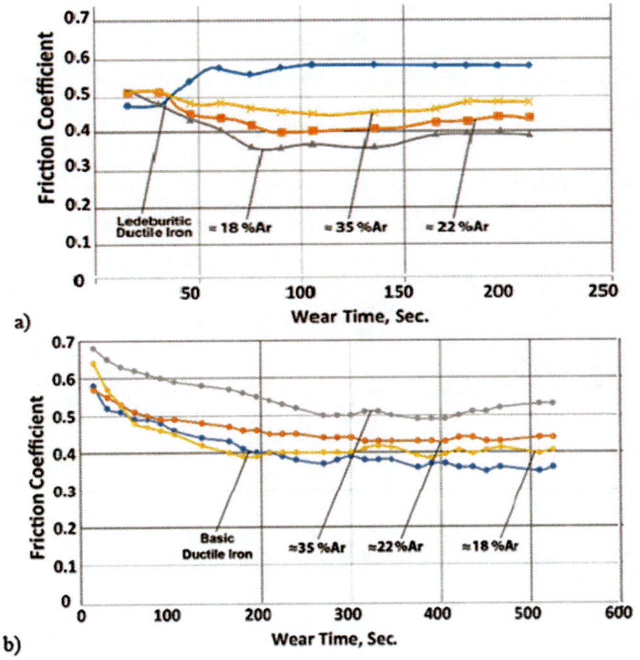

Fig. 2. Effect of retained austenite amount on the friction coefficient of unalloyed and microalloyed by Boron high strength cast iron isothermal quenched at 300 °C: a – unalloyed and b – microalloyed by Boron

During the experiments, it was determined that with an increasing in the volume fraction of retained austenite from 10 to 35% in the structure, the wear rate of bainitic cast iron first decreases, reaches to minimum at 20–22% Ar, and then begins to grow continuously (Fig. 4).

This regularity is more clearly appeared in cast iron with the structure of lower bainite. Among the studied, isothermally quenched at 300 °C temperature and microalloyed by boron high-strength cast irons have the maximum wear resistance. The investigation of friction surfaces shows (Fig. 5), that with an excessive content of retained austenite in the cast iron structure, the surface layers undergo of elastic-plastic deformation effected axial loads in the zone of friction contact.

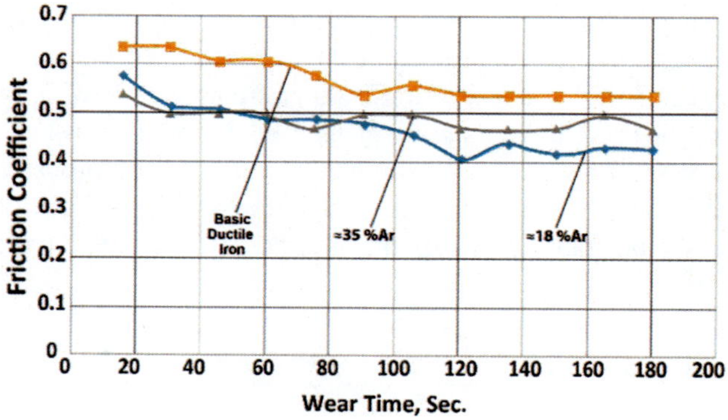

Fig. 3. Effect of retained austenite amount on the friction coefficient of unalloyed high strength cast iron isothermal quenched at 400 °C

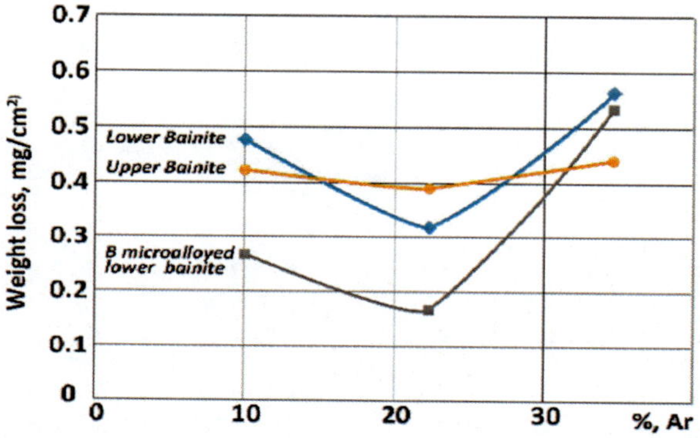

Fig. 4. Effect of retained austenite on the wear resistance of high strength cast iron with different structure

A thin austenitic interlayer is formed on the surface of the cast iron specimen, which partially covers graphite inclusions. Thus, the contact area with graphite inclusions is reduced and they are less effective as solid lubricants.

Multiple cyclic deformation of the surface layer causes fatigue failure of the austenitic interlayer. The deformational interaction of frictional surfaces with insufficient containing of retained austenite in the structure, causes brittle fracture of bainitic cast iron.

Microalloyed by boron in high-strength cast irons are generated borides, nitrides, carbides and as a result, a more multiphase structure of the metal matrix is formed. In the friction zone, the number of supporting particles are increased, the stiffness of contaction is grown, and a more favorable distribution of contact loads is provided, which increases the functioning resource of the material.

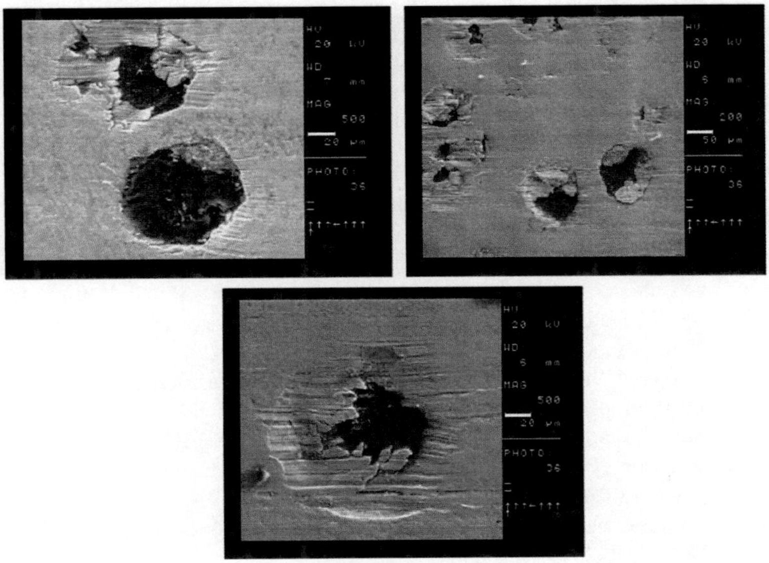

Fig. 5. SEM images of wear surfaces of bainitic ductile iron

4 Conclusions

The influence of boron microadditions on the hardness and tribological properties of ductile iron was studied.

As a result of investigations, it is determined, that the maximum wear resistance of bainitic high-strength cast irons (ADI) achieved only with a balanced relation of structural components.

In high strength cast irons microalloyed by boron more fine-dispersional bainitic metallic matrix is forming, which has been additionally hardened by boron carbides and nitrides, as a result of which the wear resistance of these alloys increases 1.6–1.7 times.

Under conditions of dry friction by sliding the high-strength cast irons with a structure of lower bainite are characterised with highest wear resistance. In this case the optimal amount of retained austenite is 10–15%.

The friction coefficient of high strength cast irons microalloyed by boron (0.03%), depending on the amount of retained austenite changes within the range of 0.38–0.49.

Based on the received results, can by concluded, that bainitic ductile irons microalloyed by boron can be recommended as a base material for ring gears, braking systems and vehicles transmission components.

References

1. Concli, F.: Austempered Ductile Iron (ADI) for gears: contact and bending fatigue behavior. Procedia Struct. Integrity **8**, 14–23 (2018)
2. Gorla, C., Conrado, E., Rosa, F.: Contact and bending fatigue behaviour of austempered ductile iron gears. Proc. Inst. Mech. Eng. Part C J. Mech. Eng. Sci. **232**(6), 998–1008 (2017)

3. Qi, Z., Jianhua, L., Rizwanulhaque, S., Jing, Z., Yang, X., Baojian, L.: The evaluation of bending fatigue testing of austempered ductile iron spur gear. In: ASME 2019 International Design Engineering Technical Conferences and Computers and Information in Engineering Conference, Anaheim, California, USA, 18–21 August 2019 (2019)
4. Zammit, A., Bonnici, M., Mhaede, M., Wan, R., Wagner, L.: Shot peening of austempered ductile iron gears. Surf. Eng. **33**(9), 679–686 (2017)
5. Myszka, D., Skołek, E., Wieczorek, A.: Manufacture of toothed elements in nanoausferritic ductile iron. Arch. Metall. Mater. **59**(3), 1217–1221 (2014)
6. Goergen, F., Mevissen, D., Masaggia, S., Veneri, E., Brimmers, J., Brecher, C.: Contact fatigue strength of austempered ductile iron (ADI) in gear applications. Metals **10**(9), 1147 (2020)
7. Li, W., Yang, X., Wang, S., Xiao, J., Hou, Q.: Comprehensive analysis on the performance and material of automobile brake discs. Metals **10**(3), 377 (2020)
8. Yu, I., Barber, G., Weiwei, C., Bingxu, W.: Improved wear resistance of austempered gray cast iron using shot-peening treatment. WCX SAE World Congress Experience (2020)
9. Wang, B., Pan, Y., Liu, Y., Barber, G.C., Qiu, F., Hu, M.: Wear behavior of composite strengthened gray cast iron by austempering and laser hardening treatment. J. Mater. Res. Technol. **9**(2), 2037–2043 (2020)
10. Wen, F., et al.: The role of bainite in wear and friction behavior of austempered ductile iron. Materials **12**(5), 767 (2019)
11. Yasinskyi, O., Bublikov, V., Berchuk, D., Zelena, L., Yasinska, O.: Influence of modifying method and silicon content on ductile cast iron structure and mechanical properties Metal and casting of Ukraine **6**, 265 (2015)
12. Ribeiroa, B.C.M., Rochab, F.M., Andradea, B.M., Lopesa, W., Corrêaa, E.C.S.: Influence of different concentrations of silicon, copper and tin in the microstructure and in the mechanical properties of compacted graphite iron. Mat. Res. **23**(2) (2020)

Prioritization of Drivers for Lean and Green Manufacturing Using Fuzzy Analytic Hierarchy Process (FAHP) Approach

Sarita Prasad(✉), Neelakanteswara A. Rao, Krishnanand Lanka, and Vijaya Kumar Manupati

Department of Mechanical Engineering,
National Institute of Technology Warangal, Warangal 506004, India
saritap@student.nitw.ac.in

Abstract. The change in customer perception towards sustainable products and recent government policies have enforced the manufacturing industries to find a way towards best practices towards production to fulfill the customers' expectations and needs. In this paper, the integrated lean and green manufacturing practices have been proposed that can help organizations by reducing waste emissions, costs, and negative impacts on the environment. To fulfill the organizations' need, drivers for lean and green manufacturing has been identified from the literature as well as with the help of experts' opinions in the related field. A total of 16 drivers are identified in this study and also classified into four main drivers, namely economic, social, environmental, and organizational drivers. Later, with the multi-criterion decision-making (MCDM) based fuzzy analytical hierarchy process (AHP) analysis approach these drivers are ranked. The study reveals that the desire to reduce cost, government support, adequate technology development, competitiveness, and top management initiatives for the adoption of green strategies are the five most important drivers for lean and green implementation in Indian manufacturing industries. Finally, the paper concludes with some managerial implications of the study, which will help the management in enterprises and policymakers in government to focus on the main drivers to implement lean and green practices in various firms.

Keywords: Lean manufacturing · Green manufacturing · Lean and green manufacturing · Drivers · Fuzzy AHP

1 Introduction

Immense pressure due to competition in the global and local markets has fetched the attention from the various organizations to adopt the latest techniques to balance the gap between customers and manufacturers [1]. With the awareness of the environmental impact, government policies, and carbon emissions, firms are moving towards innovating techniques to achieve sustainability in the supply chain [2]. Many customers at the global and local levels are focusing on standard norms, rules, and regulations to

A. Batako et al. (Eds.): GCMM 2021, LNNS 335, pp. 117–134, 2022.
https://doi.org/10.1007/978-3-030-90532-3_10

qualify at environmental terms. Hence, in companies, managers are focusing on adopting new advanced techniques, methods, and procedures, which has less impact on the environment [3]. Among many techniques and strategies, lean and green manufacturing is an integrated approach which focuses on reducing and eliminating waste activities with the application of environmentally friendly practices and achieve financial profits through cost reduction [4, 5]. This integrated system helps the firms to move towards sustainability through environmental strategies. The integration between lean and green manufacturing has been explored by various researchers in the recent few years. The integrated approach will help the firms to be sustainable through the improvement in the social, economic, and environmental way [4, 6, 7]. Orji [2] prioritized the drivers for lean led approaches with sustainability in a Chinese manufacturing organization. Micheli [8] analyzed the effect of moderator drivers (such as firm size, ISO 14001 certification, past performance, etc.) on green drivers and practices belongs to different manufacturing industries. The study focuses on drivers for green manufacturing, specifically for Chinese manufacturing industries. Luo [9] analyzed the synergies between lean and green manufacturing and the factors which can affect the implementation of lean and green manufacturing in Malaysian manufacturing industries. In developed countries, organizations have a road map which helps the organization to implement the lean and green practices in their firms [6]. However, in developing countries like India, the adaption and implementation of these techniques are still in the beginning phase [6, 10]. There is a various factor which motivates the organization for the adoption of lean and green manufacturing. These motivation factors are also known as 'drivers'. In this regard, this research aims to prioritize the drivers for lean and green manufacturing which help in the implementation of these techniques in Indian manufacturing industries. Initially, with the help of experts' opinions and literature review, a total of sixteen drivers have identified for further study. Fuzzy AHP has used for the ranking and prioritizing the drivers for lean-green manufacturing.

In summary, there is a need for the identification of lean and green drivers for Indian manufacturing as it will help the firm to achieve economic, social, and environmental sustainability. The ranking and prioritization of drivers will help the policymakers, government agencies, and high authorities in various firms to focus and facilitates the major lean and green drivers with limited resources. The framework developed in this study can help multiple manufacturing industries situated in other developing countries for the implementation of lean and green manufacturing. The main objective of this study is the identification and prioritization of drivers for lean and green manufacturing implementation in Indian industries. An extensive literature review has been conducted to analyze the contribution of other researchers in the identification of drivers. The paper has been organized into seven sections, including the introduction. Section 2 presents the literature review to identify the drivers for lean and green manufacturing and the significance of ranking lean and green drivers. Section 3 describes the methodology adopted in the study. Section 4 describes the implementation of the proposed methodology for drivers' prioritization with the Fuzzy AHP approach. Section 5 presents the results and discussion of outcomes. Section 6 presents the managerial implications of the study. Section 7 concludes the study with future scope and limitations.

2 Literature Review

Many studies have focused on the integration approach of lean and green manufacturing as it is beneficial for operational and environmental efficiency [6, 11–13]. Pampanelli [14] proposed a hybrid system of lean and green manufacturing which will help to manage the challenges in manufacturing industries. Mittal [15] analyzed the impact of lean and green practices on the footwear manufacturing company performance and also the prioritization of improvement in the system. The study conducted by farias has used an easy to track matrix called carbon value efficiency matrix, which integrates the lean and green matrices to show the implementation of these techniques in a practical manner [16]. Ng et al. [17] conducted a study for the factors which influence the integrated approach of lean and sustainable manufacturing systems in Indian automotive industries. The study [6] has developed a systematic model to analyze the relationship among determinants to implement lean-green manufacturing in SMEs. The study [18] discussed the synergies between lean practice and green practice. Campos [18] developed a model for drivers that can help to implement the hybrid lean sustainable system in the manufacturing industry. The study [4] analyzed the impact of lean-green practices on the organization and environmental efficiency and found that ISO 14001 is the most important factor for its implementation. This study also analyzed the influence of integrated lean and green approaches on the performance of small and medium industries. Gaikwad [19] conducted a study to analyze the framework, drivers and barriers for lean, green and six sigma manufacturing strategies. Bag [20] has analyzed the interrelationship among drivers for green procurement and its positive impact on the triple bottom line. Jaiswal [21] conducted a study for the ranking of lean–green drivers for SMEs. Researchers [22] analyzed the application of lean and green thinking on product-service systems. PSS system has developed to manage business activities with environmental sustainability.

Table 1. Drivers for lean and green manufacturing

Drivers	Description
Desire to reduce cost	Reduction in resources consumption and energy consumption results in cost reduction and financial benefits
Competitiveness	Due to market competition, organizations introducing several new and innovative strategies which creates new market opportunities
To increase market share	Organizations can acquire more market share via competition over prices/cost by the adaption of advanced green technologies
Government support/government assistance programs/government funded economic incentives	Support by the government in terms of assisting in skill development. The collaboration of government with industries to provide incentives schemes

(*continued*)

Table 1. (*continued*)

Drivers	Description
Public pressure - local communities, politicians, NGOs, media, community groups	Pressure from various social communities to reduce industrial emissions such as gas emissions, wastewater and solid waste which are harmful to the environment
Green brand image/company image	The respective environmental image in the view of society, as the reputation, can play a significant role in the company's growth
Environmental awareness of customers	Customer concerned nature over environmental friendly products can create pressure on the manufacturer
ISO 14001 certification	ISO 14001 certification regulation can pressure the organizations to implement the environmental management system in their manufacturing strategies to reduce the negative effect on environmental performance
Potential use of energy resources	Effective and optimized consumption of resources
Reusing/recycling materials and packaging	Green supply chain supports the reusing and recycling of material. It also encourages firms to use recyclable packaging materials
Reverse logistics	Reverse logistics focuses on the product design in such a way that the product dismantling will be easy at the end of the life of products
Green design/green purchasing/green innovation	Adaption of green strategies can help organizations to be environmental friendly
Top management initiatives for the adoption of green strategies/top management commitment	Top management in organizations is highly committed to producing products which enhance environmental performance
Stakeholders pressure	Stakeholders' concern towards the environment can pressure the manufacturer to adopt green practices
Environmental collaboration with suppliers	Supplier concern over environmental performance will help the firm to produce green products
Adequate technology development	Adaption of the latest energy-efficient and resource-efficient technology in the organization to increase operational and environmental performance

From the literature review, it is evident that various researchers in the past have analyzed the relationship between lean and green manufacturing. However, the study which focuses on the drivers for the integrated approach of lean and green manufacturing is significantly less. Few researchers have examined the integrated lean and green manufacturing approach for Indian industries. Still, their main focus was mainly on the application of lean and green practices on their operation and environmental activities. Hence, this study attempts to fill the gap by analyzing and ranking the drivers for lean and green manufacturing in Indian industries. These drivers have further ranked using the fuzzy AHP approach.

3 Methodology

For the prioritization and ranking of drivers, this study has divided into two phases. In the first phase, a thorough literature review has done to identify the drivers for lean and green manufacturing. These drivers have further consulted with the experts' team to finalize the main drivers. The drivers selected for the study has shown in Table 1. In the next step, fuzzy AHP has used for ranking these drivers. In the first phase of the literature review, articles have selected from various publishes such as Taylor & Francis, Science Directs, Emerald, Inderscience, Elsevier etc. The drivers have identified from the literature and further consulted to professionals for their opinions and suggestions. For this purpose, a group of eight-members team (six members belong to industries and two academic professionals) have consulted. The expert's team has helped in the formation of a hierarchical structure, which contains the main drivers and its sub drivers (see Table 2). The experts participated in the study having experience of more than 15 years in the industry. In the next phase, the fuzzy AHP method has used for the ranking of lean and green drivers. Fuzzy AHP can be considered as the most accepted technique for ranking and prioritizing factors. Fuzzy set theory was developed by Zadeh [23]. The fuzzy AHP method is based on fuzzy set theory. This method can analyze vague data and provide an informative solution. This method is used to handle the ambiguity and imprecision of judgment made by humans while using linguistic scale. Fuzzy AHP technique can use to deal with real-life problems. Chang, [24] first developed the triangular number in his study. Triangular fuzzy number (TFN) as the name represents have three digits, namely p, q and r where $r > q > p$. In this case, p denotes the smallest value, q denotes the middle value and r denotes the largest value. In this study, the author utilized the TFN scale of importance as shown in Table 3 [fuzzy AHP same scale].

Table 2. Classifications of drivers

Drivers	Sub-drivers	References
Economic drivers (D1)	Desire to reduce cost (D11)	[6, 22, 32, 38]
	Competitiveness (D12)	[8, 10, 26, 33, 35, 38–40]
	To increase market share (D13)	[2, 8, 10, 11, 18, 21, 22, 33, 35, 37, 38, 40, 42, 43]

(continued)

Table 2. (*continued*)

Drivers	Sub-drivers	References
Social drivers (D2)	Government support/government assistance programs/government funded economic incentives (D21)	[2, 8, 10, 11, 18, 21, 22, 33, 35, 37, 38, 40, 42, 43]
	Public pressure - local communities, politicians, NGOs, media, Community groups (D22)	[6, 8, 10, 26, 35]
	Green brand image/company image (D23)	[6, 8, 10, 11, 22, 26, 33]
Environmental drivers (D3)	Environmental awareness of customers (D31)	[2, 8, 10, 11, 19, 21, 22, 33, 35, 38, 40–42]
	ISO 14001 certification (D32)	[19, 26, 31, 33, 35, 40–42]
	Potential use of energy resources (D33)	[2, 11, 22, 26, 33, 41, 42]
	Reusing and recycling materials and packaging (D34)	[19, 26, 31, 33, 40–42]
	Reverse logistics (D35)	[26, 31, 33, 41, 42]
	Green design/green purchasing/green innovation (D36)	[11, 19, 26, 31, 41, 42]
Organizational drivers (D4)	Top management initiatives for adoption of green strategies/Top management commitment (D41)	[2, 6, 10, 11, 18, 21, 32–34, 36, 40]
	Stakeholders pressure (D42)	[2, 11, 22, 26, 33, 35, 38]
	Environmental collaboration with suppliers (D43)	[19, 21, 22, 33, 35, 42]
	Adequate technology development(D44)	[2, 8, 10, 21]

Each TFN has a linear depiction to its both left and right side so, the membership function can be represented as Eq. (1)

$$\mu\left(z/\tilde{M}\right) = \begin{Bmatrix} 0, & z < l \\ (z-p)/(q-p), & p \le z \le q \\ (r-z)/(r-q), & q \le z \le r \\ 0, & z > r \end{Bmatrix} \quad (1)$$

For the assignment of weight in fuzzy AHP, the author utilized the extent analysis method proposed by change [24]. The steps of this method are described below:

Let assume Z is a set of object

$$\text{Where it can be represented as } Z = \{z_1, z_2, \ldots\ldots, z_n\}$$
$$\text{V is a goal set, } V = \{v_1, v_2, \ldots\ldots v_n\}$$

where each object is considered and extent analysis for each goal g_i is calculated. So, the extent analysis values for each object can be obtained with these representations:

$$M_{gi}^1, M_{gi}^2, M_{gi}^3 \ldots\ldots M_{gi}^m;\ i = 1, 2, 3 \ldots\ldots, n;$$
$$and\ M_{gi}^j (j = 1, 2, 3 \ldots\ldots, m)$$

The extent methodology proposed by change, [24] is explained below:

(1) Fuzzy synthetic extent value can be represented as

$$Si = \sum_{j=1}^{m} M_{gi}^j \otimes \left[\sum_{i=1}^{n}\sum_{j=1}^{m} M_{gi}^j\right]^{-1} \quad (2)$$

The first step to calculate S_i value is the fuzzy addition operation of a matrix to obtain $\sum_{j=1}^{m} M_{gi}^j$ values. The process of operation is shown below:

$$\sum_{j=1}^{m} M_{gi}^j = \left(\sum_{i=1}^{m} l_i, \sum_{i=1}^{m} m_i, \sum_{i=1}^{m} u_i\right) \quad (3)$$

To acquire $\left[\sum_{i=1}^{n}\sum_{j=1}^{m} M_{gi}^j\right]^{-1}$, the addition of $M_{gi}^j (j = 1, 2, 3 \ldots\ldots, m)$ values is to be calculated, such that

$$\sum_{i=1}^{n}\sum_{j=1}^{m} = \left(\sum_{i=1}^{m} l_i, \sum_{i=1}^{m} m_i, \sum_{i=1}^{m} u_i\right) \quad (4)$$

In the next step, the inverse of the matrix is shown below:

$$\left[\sum_{i=1}^{n}\sum_{j=1}^{m} M_{gi}^j\right]^{-1} = \left(\frac{1}{\sum_{i=1}^{n} u_i}, \frac{1}{\sum_{i=1}^{n} m_i}, \frac{1}{\sum_{i=1}^{n} l_i}\right) \quad (5)$$

(2) In the next step, the degree of possibility is calculated for $M_2 = (l, m, u) \geq M_1 = (p, q, r)$

And can be defined as

$$V(M_2 \geq M_1) = \sup_{y \geq x}\left[\min\left(\mu_{M_1}(x), \left(\mu_{M_2}y\right)\right)\right] \quad (6)$$

This equation can be equivalently presented as

$$V(M_2 \geq M_1) = hgt(M_2 \cap M_2) = V(M_2 \geq M_1) = \left\{\begin{array}{ll} 1, & if\ q \geq m \\ 0, & if\ l \geq r \\ \frac{(l-r)}{(q-r)-(m-1)}, & otherwise \end{array}\right\} \quad (7)$$

The above equation is shown in Fig. 1, where d is the point of intersection between μ_{M_1} and μ_{M_2}. To compare M_1 and M_2, the values of $(M_1 \geq M_2)$ and $(M_2 \geq M_1)$ should be known.

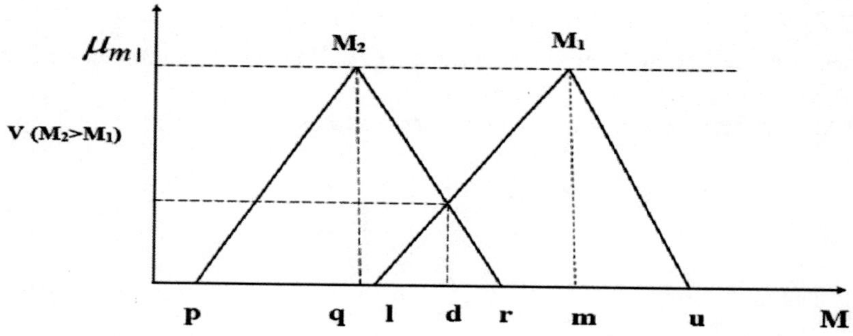

Fig. 1. Intersection points of M1 and M2

(1) Degree of possibility for a convex fuzzy number to be greater than convey fuzzy number M_i, where i = 1, 2, ..., k can be explained as

$$\begin{aligned} &V(M \geq M_1, M_{2......}M_k) \\ &= V[(M \geq M_1)\ and\ (M \geq M_2)\ and\ \ldots\ldots(M \geq M_k)] \\ &= \min V(M \geq M_i) \end{aligned} \quad (8)$$

Now assume that,

$$\begin{aligned} d'(A_i) &= \min V(S_i \geq S_k) \\ Here, k &= 1, 2, 3 \cdots\cdots, n;\ k \neq i. \end{aligned} \quad (9)$$

Therefore, the weight vector can be defined as

$$W' = \left(d'(A_1), d'(A_2), \ldots\ldots, d'(A_n)\right)^T \quad (10)$$

Where $A_i(i = 1, 2, \ldots\ldots, n)$.

(2) So, normalized weight are represented as

$$W = (d(A_1), d(A_2), \ldots\ldots, d(A_n))^T \quad (11)$$

Where W is a non-fuzzy number.

The steps followed in this study are described below [25]:

Step 1 - Identification of drivers for lean and green manufacturing from the literature review and experts' opinions. Divide the drivers in the form of sub driver to develop a hierarchical structure (see Fig. 2).
Step 2 - Develop a pairwise comparison matrix based on customer response among the drivers and their sub drivers.
Step 3 - Check for consistency (see Table 4 for random index value) for validation of results.
Step 4 - Conversion of fuzzy numbers into crisp numbers through defuzzification methods.
Step 5 - Determine the weight criteria for all the matrices.
Step 6 - Calculate the global weight of each driver through arithmetic operations.

Table 3. Fuzzy linguistic scale [fuzzy AHP same scale]

Linguistic variables	Triangular fuzzy scale	Triangular fuzzy reciprocal scale
Just equal (JE)	(1, 1, 1)	(1, 1, 1)
Equally important (EI)	(1/2, 1/1, 3/2)	(2/3, 1/1, 2/1)
Weakly more important (WMI)	(1/1, 3/2, 2/1)	(1/2, 2/3, 1/1)
Strongly more important (SMI)	(3/2, 2/1, 5/2)	(2/5, 1/2, 2/3)
Very strongly more important (VSMI)	(2/1, 5/2, 3/1)	(1/3, 2/5, 1/2)
Absolutely more important (AMI)	(5/2, 3/1, 7/2)	(2/7, 1/3, 2/5)

Steps followed for consistency check in pairwise comparison matrix.

1. Calculate the value of λmax or eigenvalues for each matrix of order n.
2. Compute the consistency index for each matrix by the formula given below:

$$\mathrm{CI} = (\lambda\mathrm{max} - \mathrm{n})/(\mathrm{n} - 1) \quad (12)$$

3. The formula can calculate consistency ratio

$$\mathrm{CR} = \mathrm{CI}/\mathrm{RI} \quad (13)$$

Where RI is called as random index and can be seen in Table 4.

Table 4. Random consistency index

Size (n)	1	2	3	4	5	6	7	8
RI	0	0	0.52	0.89	1.11	1.25	1.35	1.40

4 Lean and Green Driver Ranking Through Fuzzy AHP

Fuzzy AHP approach has used in this study to rank the driver for lean and green manufacturing. For this purpose, the following steps have used for prioritization.

4.1 Driver Identification and Categorization

In this study, a total number of 16 drivers are identified from the literature survey and later finalized through discussion with the experts' team. These drivers are later divided into sub-drivers categories to develop a hierarchical structure for the execution of the process (see Fig. 2). The drivers are categorized into four main drivers, which are economic drivers, environmental, social and organizational drivers.

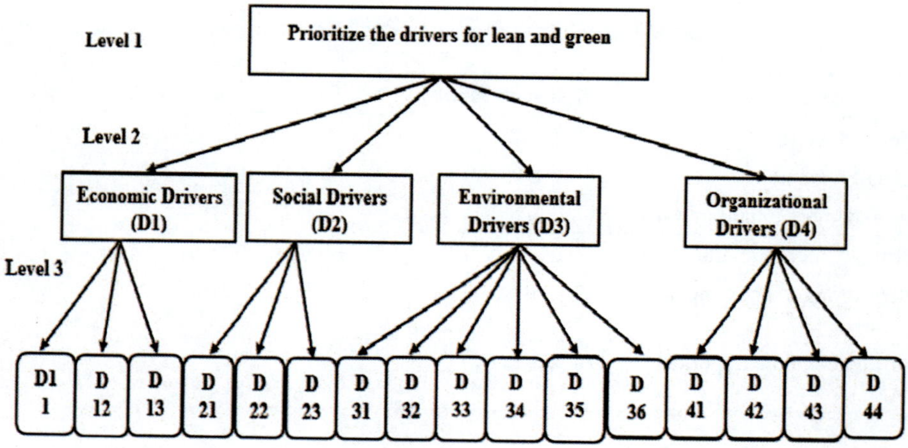

Fig. 2. Hierarchical level of drivers for lean and green manufacturing

4.2 Development of Pairwise Comparison Matrix

A pairwise comparison matrix is constructed among the drivers and sub-drivers with the help of experts (see Table 5, 6, 7, 8 and 9).

These matrices are developed with the help of fuzzy linguistic scale (see Table 3).

Table 5. Pairwise comparison of main drivers

Attributes	D1	D2	D3	D4
D1	1, 1, 1	5/2, 3, 7/2	1, 3/2, 2	1/2, 1, 3/2
D2	2/7, 1/3, 2/5	1, 1, 1	2, 5/2, 3	3/2, 2, 5/2
D3	1/2, 2/3, 1	1/3, 2/5, 1/2	1, 1, 1	2/5, 1/2, 2/3
D4	2/3, 1, 2	2/5, 1/2, 2/3	3/2, 2, 5/2	1, 1, 1

CR value for the above matrix is 0.084.

According to Sun [27], all matrix CR values should be less than or equal to 0.1. Hence, in this case, 0.084 < 0.1, so the matrix is consistent.

Table 6. Pairwise comparison of sub-drivers of economic drivers

Attributes	D11	D12	D13
D11	1, 1, 1	3/2, 2/1, 5/2	2/1, 5/2, 3/1
D12	2/5, 1/2, 2/3	1, 1, 1	3/2, 2/1, 5/2
D13	1/3, 2/5, 1/2	2/5, 1/2, 2/3	1, 1, 1

CR value for the above matrix is 0.023 < 0.1.

Table 7. Pairwise comparison of sub-drivers of social drivers

Attributes	D21	D22	D23
D21	1, 1, 1	2/1, 5/2, 3/1	3/2, 2/1, 5/2
D22	1/3, 2/5, 1/2	1, 1, 1	2/5, 1/2, 2/3
D23	2/5, 1/2, 2/3	3/2, 2/1, 5/2	1, 1, 1

CR value for the above matrix is 0.023 < 0.1.

Table 8. Pairwise comparison of sub-drivers of environmental drivers

Attributes	D31	D32	D33	D34	D35	D36
D31	1, 1, 1	1, 3/2, 2	2, 5/2, 3	3/2, 2, 5/2	5/2, 3, 7/2	1, 3/2, 2
D32	1/2, 2/3, 1	1, 1, 1	1/2, 2/3, 1	2/5, 1/2, 2/3	1/3, 2/5, 1/2	2/3, 1, 2
D33	1/3, 2/5, 1/2	1, 3/2, 2	1, 1, 1	1/2, 2/3, 1	2/5, 1/2, 2/3	2/5, 1/2, 2/3
D34	2/5, 1/2, 2/3	3/2, 2, 5/2	1, 3/2, 2	1, 1, 1	2/5, 1/2, 2/3	1/3, 2/5, 1/2
D35	2/7, 1/3, 2/5	2, 5/2, 3	3/2, 2, 5/2	1, 3/2, 2	1, 1, 1	1/2, 2/3, 1
D36	1/2, 2/3, 1	1/2, 1, 3/2	3/2, 2, 5/2	2, 5/2, 3	1, 3/2, 2	1, 1, 1

CR value for the above matrix is 0.065 < 0.1.

Table 9. Pairwise comparison of sub-drivers of organizational drivers

Attributes	D41	D42	D43	D44
D41	1, 1, 1	3/2, 2, 5/2	1, 3/2, 2	2/5, 1/2, 2/3
D42	2/5, 1/2, 2/3	1, 1, 1	2/5, 1/2, 2/3	2/7, 1/3, 2/5
D43	1/2, 2/3, 1	3/2, 2, 5/2	1, 1, 1	1/3, 2/5, 1/2
D44	3/2, 2, 5/2	5/2, 3, 7/2	2, 5/2, 3	1, 1, 1

CR value for the above matrix is $0.011 < 0.1$.

4.3 Consistency Checking and Weight Calculation for Each Matrix

Consistency check has conducted for each driver and sub driver matrix to prove the validation of results. After the checking of consistency, every matrix has evaluated for weight criteria (see Table 10).

4.4 Calculate the Global Weight

Global weight for each criterion is calculated by multiplying the weight criteria of each driver to weight criteria of corresponding sub-drivers (see Table 10).

Table 10. Local and global weight of all drivers and sub-drivers

Driver	Weight	CR	Sub-driver	Weight	Global weight	Rank
D1	0.4484	0.084	D11	0.8058	0.3613	1
			D12	0.1942	0.0871	4
			D13	0.018	0.0008	13
D2	0.2117	0.023	D21	0.806	0.1707	2
			D22	0.018	0.0038	15
			D23	0.194	0.0410	7
D3	0.1390	0.023	D31	0.3279	0.0456	6
			D32	0.1259	0.0175	10
			D33	0.0163	0.0023	16
			D34	0.0951	0.0132	11
			D35	0.2013	0.0280	9
			D36	0.2331	0.0324	8
D4	0.2008	0.065	D41	0.2357	0.0474	5
			D42	0.032	0.0064	14
			D43	0.0412	0.0083	12
			D44	0.7230	0.1453	3

5 Results and Discussion

In this study, the literature survey and expert's opinions are used to identify the sixteen main drivers for lean and green manufacturing. These drivers are grouped under four categories, namely economic drivers, social drivers, environmental drivers and organizational drivers. For the prioritization of these drivers, fuzzy AHP has implemented. The final results in terms of ranking are shown in Table 10. Among the four main drivers, economic drivers are found out to be more significant drivers with the weight of 0.4484. It represents that around 45 per cent of drivers which can encourage the organization to implement lean and green practices belongs to economic drivers. So with these drivers organizations can achieve profit with reduce cost as well as reduce the negative impact on environmental performance. Social drivers also play a significant role in the lean and green manufacturing implementation, as it contributes 0.2117 weight in the study. Organizational drivers also play a significant role with the weight of 0.2008. As the organizational drivers are directly linked to the critical decision making in the companies, it should be implied at the beginning stage of the process in the company.

Among the economic drivers, the desire to reduce cost (D11) holds the highest rank with the global weight of 0.3613, as shown in Table 10. The cost of the product is an important concern for the manufactures as well as for the customers. In the global market, price strategy plays an important role for organizations. Lean and green manufacturing approach not only considered the cost associated with products but also focuses on the cost associated with administration, capital investment, sales and environment-related techniques. The study [22] identified the reduction in manufacturing cost as a major driver for lean and green manufacturing. Researcher [29] has also analyzed that reduction in cost is the most important driver for lean manufacturing. Government Support (D21) drivers hold the second rank among all the drivers. Organizations are responsible for the various types of industrial waste emissions in terms of gases, liquid and solid. These wastes are having a negative impact on the environment. The government can provide support to organizations through assistance and can provide various incentive schemes to encourage the industries to adopt new green techniques. The government can also introduce such policies, rules and regulations which will help the organizations to control the amount of emissions produced by various industries. The study [30] shows that strict policies and regulations by the government will strongly encourage the organization to adopt green techniques in the system. A study [9] also analyzed the regulatory compliance from the government that will ensure the manufacturing industries to be sustainable [6]. Adequate technology development (D44) ranked in the third position among the drivers. Manufacturing companies need to upgrade their system by adopting and installing advanced energy-efficient technologies. The understanding of the new technologies will motivate the various industries to implement these in their manufacturing process. The study [8] analyzed that advance energy-efficient technology can help organizations to reduce it's emissions and will help the industry in effective utilization of resources. Competitiveness (D12) ranked as the fourth most important driver for the lean green system. Green products are generally getting more demand from external stakeholders. The global market becomes more competitive with the adoption of green techniques and forced several other industries to move towards the green and lean-approach. Researcher [21] found the competitiveness is an important driver for the organization to adopt lean and green practices to improve the performance of the firm and compete in the global market. Top management initiatives for adoption of green strategies/top management commitment (D41), which ranked fifth position, was also another important driver for lean and green manufacturing. The positive attitude of senior-level management towards the adaption of new techniques is necessary to implement green and lean practices in the firm. The study [6] found that top management commitment is the most critical driver for lean and green manufacturing implementation. Top management will influence the decision-making process towards the adaption of green and lean related approaches in the system. Environmental awareness of customers (D31), ranked sixth position among all drivers. The study found that environmental awareness among customers will create a rapid shift in the global market to adopt new green techniques and produce environmentally friendly products. Green brand image/Company image (D23) holds the seventh position in the ranking. Company image can be seen as an indirect driver for the organization because it requires the firms to take responsibility towards environmental protection [8]. Green design/green purchasing/green innovation (D36), reverse logistics (D35), ISO 14001 certification (D32) and reusing and recycling materials and packaging (D34) holds eight, ninth, tenth and eleventh positions respectively. A study [29] suggested that green design, green purchasing and green innovation are critical strategies for green manufacturing. It will help in the evaluation of companies towards sustainability. Certification of ISO 14001 also influences the companies to adopt the green practice. Environmental collaboration with suppliers (D43), the potential use of energy resources (D33), to increase market share (D13), public pressure (D22) and stakeholders pressure (D42) are placed at the twelfth, thirteen, fourteen, fifteen and sixteenth position respectively. These drivers are seen to be the least important drivers for lean and green manufacturing, as it appears in the bottom position in the ranking. The study [30] found that the collaboration of suppliers in environmental-related activity will help the firms to transform the whole supply chain into the green supply chain. Public pressure by the local communities, NGOs and media come under social drivers. Public views and opinions on any industry will help to maintain a reputation by a company [29]. In developed countries, stakeholders are more concerned about the environmental benefits than the developing countries, as they are more aware of the benefits and significance of green techniques [21].

6 Managerial Implications

A fuzzy-based decision methodology has used to rank and prioritize the drivers for lean and green manufacturing. The decision-makers in organizations can use this ranking as a road map to understand and assess their organization activities for greenness and leanness. This study will also provide flexibility to industrial practitioners in decision making on the implementation of lean and green manufacturing practices. It will guide the management to prioritize their most influential drivers towards the adaption of these strategies in the early stage of production. The outcome of the study will also help researchers and practitioners. Researchers can expand and compare the research outcome with other modelling techniques and practitioners can analyze the implementation benefits of lean and green manufacturing. Adaption of lean and green practices will help the organization to become socially responsible, compliance with the environmental regulation to achieve economic growth.

7 Conclusions

Environmental related issues play a significant role in decision making in the manufacturing organization [25]. High consumption of natural resources and environmental awareness among customers will help the manufacturing activities to become sustainable. The adaption and implementation of lean and green practices are a challenging task for managers as it requires high skill of management and financial support. This can be achieved by the support from the government, incentive schemes for financial support, and government assistance programs for skill development. This study shows the ranking of drivers for lean and green manufacturing for Indian manufacturing industries. This study provides a brief view of drivers which can help the firm in the implementation of lean and green manufacturing. In this study, sixteen drivers have identified from the literature survey and with the experts' opinions. These drivers are the desire to reduce cost, competitor green strategies, to increase market share, government support, public pressure, green brand image/company image, environmental awareness of customers, ISO 14001 certification, the potential use of energy resources, reusing and recycling materials and packaging, reverse logistics, green design/green purchasing/green innovation, top management initiatives for adoption of green strategies/top management commitment, stakeholders pressure, environmental collaboration with suppliers and adequate technology development. These drivers are further classified into four main drivers, namely economic drivers, social drivers, environmental drivers, and organizational drivers. These drivers are ranked with a fuzzy AHP approach to determine the most critical drivers for lean and green manufacturing. The fuzzy AHP approach has been used in this process as it will help to develop a model that is more close to real-life conditions. According to the outcome of the study, the desire to reduce cost, government support, adequate technology development, competitiveness, and top management initiatives for the adoption of green strategies are the top five drivers for lean and green implementation. The desire to reduce costs found as a significant driver for firms to adopt lean and green practices in their system. This study helps the managers to select the main drivers for lean and green practice with the weight-based ranking. For the adaption of these practices, adequate funding and high skilled management is required. These can achieve with the help of government schemes to provide funds and their assistance programs to enhance the skill set of various organizations workers. The government can also promote the usage of highly efficient technology to achieve high resource efficiency, which will also help to reduce the carbon emissions and wastewater by the industries. Government rules and regulations will mandate enterprises to reduce their emission and to become competitive globally.

The ranking of drivers will help the practitioners to implement lean and green practices in their organizations. This study will provide valuable input for researchers to further expand the work in a more comprehensive and structured way. However, this study is also having some limitations as the drivers identified in the study are based on literature review, while a survey approach can help the researchers to identify more realistic drivers which have not been reported in previous studies. The experts who participated in this study belong to manufacturing companies, but it can be applied to other industries such as the service and food industries. This study is generally based on Indian manufacturing industries, whereas the outcome of the study can be vary based on other countries. The rules and regulations, working conditions, and legislation can vary based on national government policies. The present study can be further extended with the application of various other multi-criteria decision-making methods (MCDM) such as a fuzzy technique for order of preference by similarity to ideal solution (TOPSIS), etc. and with modelling techniques such as structural equation modelling.

References

1. Yadav, G., Desai, T.N.: A fuzzy AHP approach to prioritize the barriers of integrated Lean Six Sigma. Int. J. Qual. Reliab. Manag. **34**(8), 1167–1185 (2017)
2. Orji, J., Liu, S.: A dynamic perspective on the key drivers of innovation-led lean approaches to achieve sustainability in manufacturing supply chain. Int. J. Prod. Econ. **219**, 480–496 (2020)
3. Garza-Reyes, J.A., Jacques, G.W., Lim, M.K., Kumar, V., Rocha-Lona, L.: Lean and green – synergies, differences, limitations, and the need for Six Sigma. In: Grabot, B., Vallespir, B., Gomes, S., Bouras, A., Kiritsis, D. (eds.) APMS 2014. IAICT, vol. 439, pp. 71–81. Springer, Heidelberg (2014). https://doi.org/10.1007/978-3-662-44736-9_9
4. Thanki, S., Govindan, K., Thakkar, J.: An investigation on lean-green implementation practices in Indian SMEs using analytical hierarchy process (AHP) approach. J. Clean. Prod **135**, 284–298 (2016)
5. Diaz-Elsayed, N., Jondral, A., Greinacher, S., Dornfeld, D., Lanza, G.: Assessment of lean and green strategies by simulation of manufacturing systems in discrete production environments. CIRP Ann. **62**(1), 475–478 (2013)
6. Gandhi, N.S., Thanki, S.J., Thakkar, J.J.: Ranking of drivers for integrated lean-green manufacturing for Indian manufacturing SMEs. J. Clean. Prod. **171**, 675–689 (2018)
7. Verrier, B., Rose, B., Caillaud, E., Remita, H.: Combining organizational performance with sustainable development issues: the Lean and Green project benchmarking repository. J. Clean. Prod. **85**, 83–93 (2014)
8. Micheli, G.J., Cagno, E., Mustillo, G., Trianni, A.: Green supply chain management drivers, practices and performance: A comprehensive study on the moderators. J. Clean. Prod. **259**, 121024 (2020)
9. Luo, Y., Jie, X., Li, X., Yao, L.: Ranking Chinese SMEs green manufacturing drivers using a novel hybrid multi-criterion decision-making model. Sustainability **10**(8), 2661 (2018)
10. Chiet, C.W., Ching, N.T., Huat, S.L., Fathi, M., Tzuu, T.J.: The integration of lean and green manufacturing for Malaysian manufacturers: a literature review to explore the synergies between lean and green model. In: IOP Conference Series: Earth and Environmental Science, vol. 268, no, 1, p. 012066 (2019)
11. Thanki, S.J., Thakkar, J.: Status of lean manufacturing practices in Indian industries and government initiatives: a pilot study. J. Manuf. Technol. Manag. **25**(5), 655–675 (2014)
12. Garza-Reyes, J.A.: Lean and green - a systematic review of the state of the art literature. J. Clean. Prod. **102**, 18–29 (2015)
13. Kurdve, M., Zackrisson, M., Wiktorsson, M., Harlin, U.: Lean and green integration into production system models experiences from Swedish industry. J. Clean. Prod. **85**, 180–190 (2014)
14. Pampanelli, A.B., Found, P., Bernardes, A.M.: A Lean & Green Model for a production cell. J. Clean. Prod. **85**, 19–30 (2014)

15. Mittal, V.K., Sindhwani, R., Kapur, P.K.: Two-way assessment of barriers to Lean-Green Manufacturing System: insights from India. Int. J. Syst. Assur. Eng. Manag. **7**(4), 400–407 (2016). https://doi.org/10.1007/s13198-016-0461-z
16. Farias, L.M.S., Santos, L.C., Gohr, C.F., Rocha, L.O.: An ANP-based approach for lean and green performance assessment. Resour. Conser. Recycl. **143**, 77–89 (2019)
17. Ng, R., Low, J.S.C., Song, B.: Integrating and implementing Lean and Green practices based on proposition of Carbon-Value Efficiency metric. J. Clean. Prod. **95**, 242–255 (2015)
18. Campos, L.M.S., Vazquez-Brust, D.A.: Lean and green synergies in supply chain management. Supply Chain Manag. Int. J. **21**(5), 627–641 (2016). https://doi.org/10.1108/SCM-03-2016-0101
19. Gaikwad, L., Sunnapwar, V.: An integrated Lean, Green and Six Sigma strategies. Total Qual. Manag. **32**(2), 201–225 (2020)
20. Bag, S.: Identification of green procurement drivers and their interrelationship using total interpretive structural modelling. Vision **21**(2), 129–142 (2017)
21. Jaiswal, P., Kumar, A.: Assessment of drivers to implement integrated lean green manufacturing system in Indian SMEs through IF-TOPSIS approach. Int. J. Manag. Decis. Making **17**(2), 224–243 (2018)
22. Li, Q., Found, P.: Lean and green supply chain for the product-services system (PSS): the literature review and a conceptual framework. Procedia CIRP **47**, 162–167 (2016)
23. Zadeh, L.A.: Fuzzy sets. Inf. Control **8**(3), 338–353 (1965)
24. Chang, D.-Y.: Applications of the extent analysis method on fuzzy AHP. Eur. J. Oper. Res. **95**(3), 649–655 (1996)
25. Govindan, K., Diabat, A., Shankar, K.M.: Analyzing the drivers of green manufacturing with fuzzy approach. J. Clean. Prod. **96**, 182–193 (2015)
26. Sun, C.C.: A performance evaluation model by integrating fuzzy AHP and fuzzy TOPSIS methods. Expert Syst. Appl. **37**(12), 7745–7754 (2010)
27. Zhou, B.: Lean principles, practices, and impacts: a study on small and medium-sized enterprises (SMEs). Ann. Oper. Res. **241**(1–2), 457–474 (2012). https://doi.org/10.1007/s10479-012-1177-3
28. Govindan, K., Azevedo, S.G., Carvalho, H., Cruz-Machado, V.: Lean, green and resilient practices influence on supply chain performance: interpretive structural modeling approach. Int. J. Environ. Sci. Technol. **12**(1), 15–34 (2013). https://doi.org/10.1007/s13762-013-0409-7
29. Zhou, Y., Xu, L., Muhammad Shaikh, G.: Evaluating and prioritizing the green supply chain management practices in Pakistan: based on delphi and fuzzy AHP approach. Symmetry **11**, 1346 (2019)
30. Tseng, M.L.: Green supply chain management with linguistic preferences and incomplete information. Appl. Soft Comput. J. **11**, 4894–4903 (2011)
31. Siegel, R., Antony, J., Garza-Reyes, J.A., Cherrafi, A., Lameijer, B.: Integrated green lean approach and sustainability for SMEs: from literature review to a conceptual framework. J. Clean. Prod. **240**, 118205 (2019)
32. Seth, D., Rehman, M.A.A., Shrivastava, R.L.: Green manufacturing drivers and their relationships for small and medium (SME) and large industries. J. Clean. Prod. **198**, 1381–1405 (2018)
33. AlManei, M., Salonitis, K., Xu, Y.: Lean implementation frameworks: the challenges for SMEs. Procedia Cirp **63**, 750–755 (2017)
34. Dhull, S., Narwal, M.: Drivers and barriers in green supply chain management adaptation: a state-of-art review. Uncertain Supply Chain Manag **4**(1), 61–76 (2016)
35. Wong, W.P., Wong, K.Y.: Synergizing an ecosphere of lean for sustainable operations. J. Clean. Prod. **85**, 51–66 (2014)
36. Kasim, A., Ismail, A.: Environmentally friendly practices among restaurants: drivers and barriers to change. J. Sustain. Tour. **20**(4), 551–570 (2012)

37. Mollenkopf, D., Stolze, H., Tate, W.L., Ueltschy, M.: Green, lean, and global supply chains. Int. J. Phys. Distrib. Logist. Manag. **40**(1–2), 14–41 (2010)
38. Vanalle, R.M., Ganga, G.M.D., Godinho Filho, M., Lucato, W.C.: Green supply chain management: an investigation of pressures, practices, and performance within the Brazilian automotive supply chain. J. Clean. Prod. **151**, 250–259 (2017)
39. Giunipero, L.C., Hooker, R.E., Denslow, D.: Purchasing and supply management sustainability: drivers and barriers. J. Purch. Supply Manag. **18**(4), 258–269 (2012)
40. Salonitis, K., Tsinopoulos, C.: Drivers and barriers of lean implementation in the Greek manufacturing sector. Procedia Cirp **57**, 189–194 (2016)
41. Diabat, A., Govindan, K.: An analysis of the drivers affecting the implementation of green supply chain management. Resour. Conserv. Recycl. **55**(6), 659–667 (2011)
42. Liu, J.Y., Low, S.P., He, X.: Green practices in the Chinese building industry: drivers and impediments. J. Tech. Manag. China **7**(1), 50–63 (2012)
43. Jasti, N.V.K., Kodali, R.: An empirical study for implementation of lean principles in Indian manufacturing industry. Benchmarking Int. J. **23**(1), 183–207 (2016)

Influence of the Nanostructure Parameters of Wear-Resistant Coatings on the Properties of Carbide Tools

Alexey Vereschaka[1,2(✉)], Marina Volosova[1], Nikolay Sitnikov[3], Filipp Milovich[4], Nikolay Andreev[4], Jury Bublikov[2], and Catherine Sotova[1]

[1] Moscow State Technological University STANKIN, Vadkovsky per. 1, Moscow 127994, Russia
dr.a.veres@yandex.ru
[2] ICTI RAS, Vadkovsky per. 18-1A, Moscow 127055, Russia
[3] National Research Nuclear University MEPhI (Moscow Engineering Physics Institute), Kashirskoeshosse 31, Moscow 115409, Russia
[4] National University of Science and Technology "MISIS", Leninsky prospect 4, Moscow 119049, Russia

Abstract. The article deals with the properties of the Ti-TiN-(Ti,Cr,Al)N multilayer composite coatings with a nanolayer structure of their wear-resistant layers and the nanolayer period λ within a range from 10 to 900 nm. The studies were focused on the coating nanostructures and the influence of the nanolayer period λ on the performance properties of coated cutting tools during the turning of American Iron and Steel Institute (AISI) 420 (230 HB) martensitic stainless steel. The investigation found that the coatings with the nanolayer period $\lambda = 10$–25 nm demonstrated the best resistance to the crater formation on the rake face of cutting tools and to flank wear. The highest wear resistance was demonstrated by a coated tool with the nanolayer period $\lambda = 10$ nm.

Keywords: Metal-cutting tool · Wear-resistant coatings · Nanostructures · Tool life · Heat resistance

1 Introduction

Metal-cutting remains the most common way to produce parts for various purposes. At the same time, the cutting tool remains a weak link and a limiting factor in increasing the productivity of cutting processing. The use of wear-resistant coatings is an effective way to improve the performance properties of metal-cutting tools [1–5]. In turn, the key ways to increase the characteristics of wear-resistant coatings is to complicate their structure (nanostructured coatings) and elemental composition (multi-element coatings) [2, 6–8]. Coatings of a new generation have a combination of properties such as hardness and plasticity at the same time, which makes it equally resistant to wear and brittle fracture [2, 6]. Moreover, such coatings also have increased heat resistance, resistance to oxidative and diffusion processes [7, 8]. Thus, the use of these coatings allows not only to increase

A. Batako et al. (Eds.): GCMM 2021, LNNS 335, pp. 135–144, 2022.
https://doi.org/10.1007/978-3-030-90532-3_11

the tool life and reliability of the metal-cutting tool, but also to increase the productivity of the cutting process (in particular, due to the use of higher cutting speeds) [1–4, 9].

Coatings with nanolayer structures have a number of noticeable advantages in comparison with single-layer coatings due to the more efficient performance of the barrier functions with regard to diffusion and spinodal decomposition [10–21]. Another important property of coatings with nanolayer structures is their good ability to resist crack formation and brittle fracture [9, 22, 23]. The challenge of determining the optimal thickness of nanolayers, which is essential when choosing the optimal coating parameters, was considered in [9, 20, 22, 23]. In particular, some studies note that the value of the nanolayer period λ does not significantly affect the hardness and elastic modulus [23–26], but influences the adhesion to the substrate, the coefficient of friction (COF) [23–33], and wear resistance [23, 26, 31, 33]. Other studies argue that a decrease in the nanolayer period λ leads to an increase in the hardness [27, 32] and the resistance to cohesive fracture [27]. The nanolayer structure of the coating improves its resistance to oxidation at high temperatures [29]. As the nanolayer period λ decreases, the dominant wear mechanism changes from oxidation to molecular adhesion [23]. As shown below, with a decrease in the nanolayer period λ, plastic yield in the coating is increasingly caused by rotation of nanocrystals and grain boundary sliding rather than by dislocation displacement [24]. In [25, 28], tests also detected lower residual stresses in the coatings with nanolayer structures. In [26], the studies have found that the structure of a nanolayer coating may retain nanocrystalline layers with columnar structures, although a superlattice structure may also form [32]. Coatings with nanolayer structures are also characterized by reduced surface roughness [33]. The (Ti,Cr,Al)N coating has long been successfully used to increase the tool life and reliability of metal-cutting tools [1, 2, 4, 6, 16]; however, the parameters of the (Ti,Cr,Al)N coating may be further enhanced by application of a nanolayer structure with an optimal nanolayer thickness and also by application of a three-layer architecture of the coating with three functional layers [17, 34]:

- Adhesion layer (providing the maximum adhesion strength between a coating and a substrate and also performing the function of leveling the substrate surface by filling microdefects, i.e. cracks, irregularities, and potholes);
- Transition layer (providing the barrier functions with regard to diffusion and oxidation and transition functions between a ductile and fairly soft adhesion layer and an extremely hard wear-resistant layer); and
- Wear-resistant layer (providing the maximum hardness and wear resistance in combination with high resistance to crack formation and brittle fracture).

An example of such coating is considered below.

2 Materials and Methods

The Ti-TiN-(Ti,Cr,Al)N coatings were deposited using the modernized VIT-2M (VIT-IDTI RAS) unit, with three cathode systems providing a microdroplet phase filtration and a precision turntable [34–44]. The filtered cathodic vacuum arc deposition (FCVAD) technology [34–37] was applied using broad beam sources of fast molecules with segmented cold cathodes and emissive grids [35–37] and glow discharge with electrostatic confinement of electrons in a chamber bombarded by fast electrons [37–44]. The deposition process took place with the following parameters: $I_{Al} = 160$ A, $I_{Ti} = 55$ A, $I_{Cr} = 70$ A, U = −800 DC, $P_N = 0.36$ Pa. The table rotation frequency was n = 0.07–7 rpm. The relationship between the properties of the coating nanostructure and the table rotation frequency n are presented in Table 1.

Table 1. Relationship between the properties of the coating nanostructure and the table rotation frequency.

	K0	K1	K2	K3	K4	K5	K6	K7	K8
Table rotation frequency *n*, rpm	0.07	0.25	0.5	1	1.5	2	3	5	7
Nanolayer period λ, nm	900	300	160	70	50	35	25	15	10
Number of nanolayers	3	7	13	33	41	54	85	131	215

The cutting tests were carried out under the conditions of dry turning of AISI 420 (230 HB) steel in the following modes: f = 0.25 rpm, ap = 1.0 mm, and vc = 200 m/min. The (WC + 5%TiC + 10%Co) carbide cutting inserts under SNUN ISO 1832:2012 were used, with the following geometric cutting parameters: γ = −7°, α = 7°, λ = 0, r = 0.4 mm. The cutting carbide inserts with the coatings under study were compared with each other and with the (Ti,Al)N commercial coating (single-layer coating with a thickness of about 4 μm). Flank wear was measured after every 3 min of cutting. Five experiments were conducted for tools with each type of coatings and for an uncoated tool, and after that the arithmetic mean value of the flank wear was determined. $VB_{max} = 0.3$ mm was assumed as the limiting flank wear value.

AISI 420 is martensitic stainless steel, hardened by heat treatment and characterized by high wear resistance. This steel demonstrates good ductility (delivery condition), high impact resistance, and good corrosion and heat resistance. Chromium carbides formed during the heat treatment contribute to additional dispersive hardening of steel structure and thus increase its hardness and wear resistance.

The hardness (HV) of coatings was determined by measuring the indentation at low loads according to the method of Oliver and Pharr [43], which was carried out on a micro-indentometer micro-hardness tester (CSM Instruments) at a fixed load of 10 mN (exposure 20 min).

A JEM 2100 (JEOL, Japan) high-resolution (HR) transmission electron microscope (TEM) and an FEI Quanta 600 FEG scanning electron microscope (SEM) were used for the microstructural studies. The thickness of the coating and its layers was measured on a transverse section when examined with SEM.

3 Results and Discussion

Figure 1 illustrates the structure of the coatings under study. Coating K0, deposited at the extremely low table rotation frequency (only 3 revolutions for the entire deposition process), has the very high thickness value of the nanolayer period λ equal to about 900 nm. Coatings K0, K1, and K2 cannot be formally designated as "nanolayer coatings," since their nanolayer period λ exceeds 100 nm. Despite the above, the concept of "nanolayer" will also be applied to them in order to unify the designation. As can be noticed, coatings K0, K1, and K2 have rather complicated structures due to the alternation of nanolayers with high (light in contrast) and low (dark in contrast) content of Al. It should be noted that the dark areas of the coating also contain Al, but of the lower concentration. The total coating thickness is 3500 ± 200 nm, with the thickness of an adhesion layer of about 30–50 nm and the thickness of a transition layer of about 800 ± 100 nm.

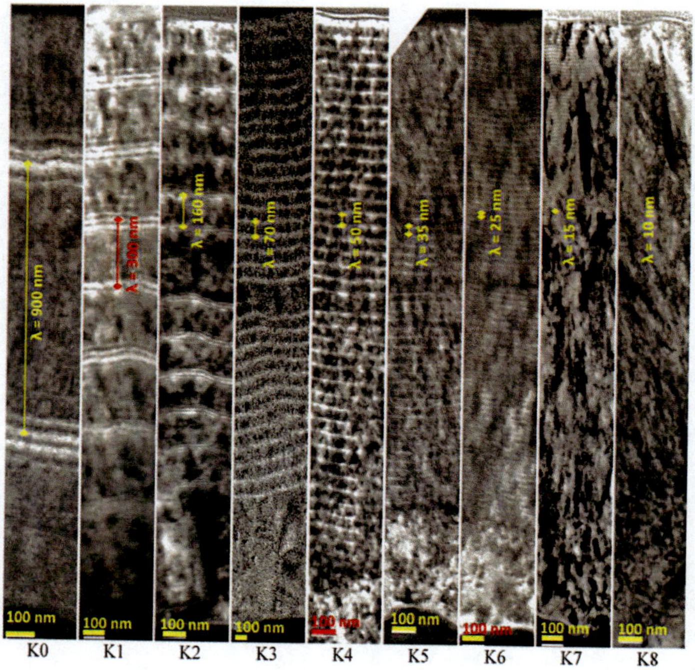

Fig. 1. Structures of the studied coatings over layer thickness.

Figure 2 depicts in more detail the structures of the nanolayers in the coatings under study. As can be noticed, each nanolayer with a high Al content consists of several variable-thickness subnanolayers, formed due to the planetary rotation of samples in the chamber. In particular, the sample with coating K0 consists of 4 subnanolayers with the high Al content, with the subnanolayer thicknesses within a range from 10 to 30 nm. With an increase in the table rotation frequency, the subnanolayer thicknesses decrease correspondingly; however, the minimum thickness of a subnanolayer possibly detected in the coating structure is about 1 nm. The studies of the chemical compositions of the coatings reveal that while in the dark areas, the Al content stays within 6–8 at%, then in the light areas, the Al content is significantly higher – 25–35 at%. An increase in the Al content enhances the hardness of the coating and reduces its ductility [44], and therefore the nanolayer structure under study is in fact a nanocomposite with alternating harder and more ductile nanolayers. The specific features of the crystalline structures of a transition and a wear-resistant layer of the coating are considered on the examples of coatings K2 and K4 (Fig. 3). In particular, Fig. 3b illustrates that the crystal sizes in the TiN wear-resistant layer are significantly larger than in the (Ti,Cr,Al)N wear-resistant layer. Such a difference can be explained by the fact that subnanolayers with the high Al content inhibit the crystal growth (Fig. 3b), thus allowing the formation of a crystalline structure with crystal sizes not exceeding the thickness of a single nanolayer.

The hardness values of the studied coatings are presented in Table 2. Based on the obtained data, it can be argued that the coatings under study are characterized by fairly close hardness, and that is in line with the data in [17, 19].

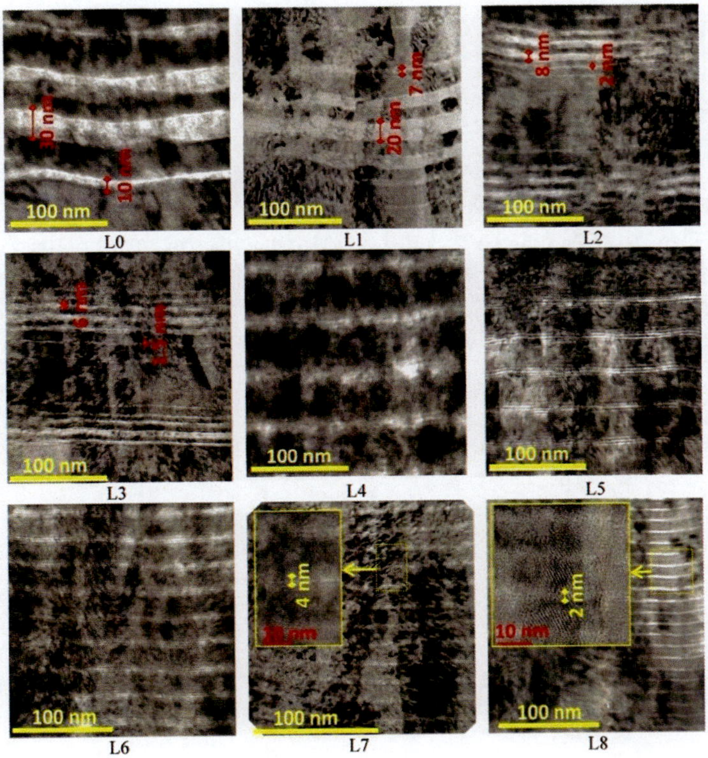

Fig. 2. Nanolayer structures of the studied coatings.

The results of the cutting tests conducted using cutting tools with the coatings under study during the turning of AISI 420 steel (Fig. 4) find that during first 13 min of cutting, all the samples demonstrate fairly close wear. In particular, flank wear is 0.21 mm for a sample with coating K8 (the lowest wear) and 0.26 mm for a sample with the

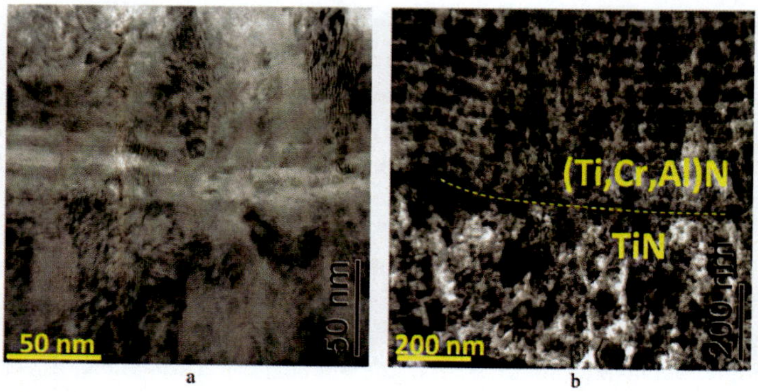

Fig. 3. Specifics of the crystalline structures of coatings (a) K2 and (b) K4.

Table 2. Hardness of the studied coatings.

	K0	K1	K2	K3	K4	K5	K6	K7	K8
$HV_{0.01}$, GPa	29.01 ± 0.92	29.87 ± 1.03	28.81 ± 0.75	27.96 ± 0.94	28.53 ± 1.12	29.06 ± 0.74	29.5 ± 0.94	29.7 ± 0.43	30.6 ± 0.34

(Ti,Al)N coating (the highest wear). However, after 13 min of cutting, intensive wear of the samples with coatings K0-K4 and (Ti,Al)N begins. Meanwhile, the wear rate of the samples with coatings K5–K8 is significantly lower. In particular, while after 19 min of cutting, the flank wear noticeably exceeds 0.4 mm for the samples with coatings K0–K4 and (Ti,Al)N, then for the samples with coatings K4–K8, the flank wear stays within 0.27–0.3 mm. The best wear resistance under the above cutting conditions is demonstrated by the sample with coating K8.

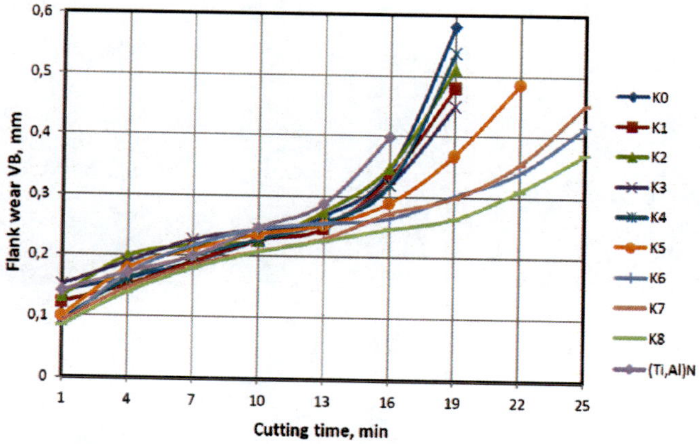

Fig. 4. Ratio of flank wear to cutting time during the turning of AISI 420 steel using the tools with the studied coatings.

Figure 5 depicts the wear patterns on the rake faces of the coated tools. As can be noticed, all samples exhibit signs of crater formation at the 13th minute of cutting. However, those signs are most weakly expressed on the samples with coatings K7 and K8. After 16 min of cutting, the samples with coatings K1–K5 demonstrate clear wear craters. Meanwhile, no wear craters are detected on the samples with coatings K7 and K8, and there are only areas with increased wear with preserved coating layers.

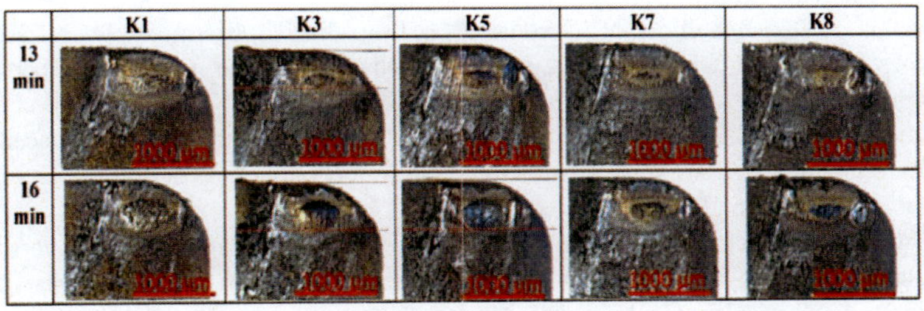

Fig. 5. Wear patterns on the contact pads on the tool rake face.

4 Conclusion

The analysis of the properties of a series of samples with the Ti-TiN-(Ti,Cr,Al)N nanolayer coatings of the same composition, but various nanolayer thicknesses, leads to the following conclusions:

- By changing the turntable rotation frequency, it is possible to control the nanolayer thickness with sufficient accuracy.
- The thicknesses of the nanolayers affect the cutting properties of the tools during the turning of AISI 420 steel, and the coating with the minimum nanolayer thickness provides the best wear resistance.
- The coatings under study demonstrate fairly close wear rates during the first 13 min of cutting, but then the coatings with the nanolayer thicknesses within a range of 10–25 nm show significantly higher wear resistance compared to the coatings with the nanolayer thicknesses of 50–900 nm.
- A possible reason for the good resistance of the coatings with the nanolayer thicknesses within a range of 10–25 nm to wear on the rake face of the tools may be their better resistance to diffusion and oxidation wear and to the phenomenon of spinodal decomposition due to the enhancement of the barrier properties of the internanolayer interfaces.

Acknowledgments. The study was supported by a grant from the Russian Science Foundation (project No. 21-19-00612). The study used the equipment from the Centre for collective use of Moscow State Technological University STANKIN (agreement No. 075-15-2021-695, 26/07/2021). The coating structure was investigated using the equipment of the Centre for collective use of scientific equipment “Material Science and Metallurgy”, purchased with the financial support of the Ministry of Science and Higher Education of the Russian Federation (GK 075-15-2021-696).

References

1. Kalss, W., Reiter, A., Derflinger, V., Gey, C., Endrino, J.L.: Modern coatings in high performance cutting applications. Int. J. Refract. Hard. Met. **24**(5), 399–404 (2006)
2. Bobzin, K.: High-performance coatings for cutting tools. CIRP J. Manuf. Sci. Technol. **18**, 1–9 (2017)
3. Vereschaka, A.S.: Working Capacity of the Cutting Tool with Wear Resistant Coatings. Mashinostroenie, Moscow (1993)
4. Bouzakis, K.D., Michailidis, N., Skordaris, G., Bouzakis, E., Biermann, D., M'Saoubi, R.: Cutting with coated tools: coating technologies, characterization methods and performance optimization. CIRP Ann. Manuf. Technol. **61**, 703–723 (2012)
5. Tkadletz, M., Schalk, N., Daniel, R., Keckes, J., Czettl, C., Mitterer, C.: Advanced characterization methods for wear resistant hard coatings: a review on recent progress. Surf. Coat. Technol. **285**, 31–46 (2016)
6. Vereschaka, A., et al.: Investigation of the influence of the thickness of nanolayers in wear-resistant layers of Ti-TiN-(Ti,Cr,Al)N coating on destruction in the cutting and wear of carbide cutting tools. Surf. Coat. Technol. **385**, 125402 (2020)
7. Grigoriev, S., et al.: Investigation of multicomponent nanolayer coatings based on nitrides of Cr, Mo, Zr, Nb, and Al. Surf. Coat. Technol. **401,** 126258 (2020)
8. Vereschaka, A.A., Grigoriev, S.N., Sitnikov, N.N., Oganyan, G.V., Batako, A.: Working efficiency of cutting tools with multilayer nano-structured Ti-TiCN-(Ti, Al)CN and Ti-TiCN-(Ti, Al, Cr)CN coatings: analysis of cutting properties, wear mechanism and diffusion processes. Surf. Coat. Technol. **332**, 198–213 (2017)
9. Vereschaka, A., et al.: Investigation of wear dynamics for cutting tools with multilayer composite nanostructured coatings in turning constructional steel. Wear **420–421**, 17–37 (2019)
10. Baker, M.A., Klose, S., Rebholz, C., Leyland, A., Matthews, A.: Evaluating the microstructure and performance of nanocomposite PVD TiAlBN coatings. Surf. Coat. Technol. **151–152**, 338–343 (2002)
11. Singh, J., Wolfe, D.E.: Nanostructured component fabrication by electron beam-physical vapor deposition. J. Mater. Eng. Perform. **14**(4), 448–459 (2005)
12. Kuzin, V., Grigor'ev, S., Volosova, M.: Effect of a TiC coating on the stress-strain state of a plate of a high-density nitride ceramic under nonsteady thermoelastic conditions, Refract. Ind. Ceram. **54,** 376–380 (2014)
13. Skordaris, G., et al.: Brittleness and fatigue effect of mono- and multi-layer PVD films on the cutting performance of coated cemented carbide inserts. CIRP Ann. Manuf. Technol. **63**, 93–96 (2014)
14. Zhang, S., Sun, D., Yongqing, F., Hejun, D.: Recent advances of superhard nanocomposite coatings: a review. Surf. Coat. Technol. **167**, 113–119 (2003)
15. Leyland, A., Matthews, A.: On the significance of the H/E ratio in wear control: a nanocomposite coating approach to optimized tribological behavior. Wear **246**, 1–11 (2000)
16. Alexey, V.: Development of assisted filtered cathodic vacuum arc deposition of nano-dispersed multi-layered composite coatings on cutting tools. Key Eng. Mater. **581**, 62–67 (2014)
17. Vereschaka, A.A., et al.: Effect of adhesion and the wear-resistant layer thickness ratio on mechanical and performance properties of ZrN-(Zr, Al, Si)N coatings. Surf. Coat. Technol. **357**, 218–234 (2019)
18. Vereschaka, A., et al.: Effect of adhesion and tribological properties of modified composite nano-structured multi-layer nitride coatings on WC-Co tools life. Tribol. Int. **128**, 313–327 (2018)

19. Vereschaka, A., Tabakov, V., Grigoriev, S., Sitnikov, N., Andreev, N., Milovich, F.: Investigation of wear and diffusion processes on rake faces of carbide inserts with Ti-TiN-(Ti, Al, Si)N composite nanostructured coating. Wear **416–417**, 72–80 (2018)
20. Hovsepian, P., et al.: TiAlN based nanoscale multilayer coatings designed to adapt their tribological properties at elevated temperatures. Thin Solid Films **485**, 160–168 (2005)
21. Antonov, M., Hussainova, I., Sergejev, F., Kulu, P., Gregor, A.: Assessment of gradient and nanogradient PVD coatings behavior under erosive, abrasive and impact wear conditions. Wear **267**(5–8), 898–906 (2009)
22. Vereschaka, A.A., et al.: Nano-scale multi-layered coatings for improved efficiency of ceramic cutting tools. Int. J. Adv. Manuf. Technol. **90**(1—4), 27–43 (2017)
23. Contreras, E., Galindez, Y., Rodas, M.A., Bejarano, G., Gómez, M.A.: CrVN/TiN nanoscale multilayer coatings deposited by DC unbalanced magnetron sputtering. Surf. Coat. Technol. **332**, 214–222 (2017)
24. Bobzin, K., Brögelmann, T., Kruppe, N.C., Arghavani, M., Mayer, J., Weirich, T.E.: Plastic deformation behavior of nanostructured CrN/AlN multilayer coatings deposited by hybrid dcMS/HPPMS. Surf. Coat. Technol. **332**, 253–261 (2017)
25. Hovsepian, P., Ehiasarian, A.P., Deeming, A., Schimpf, C.: Novel TiAlCN/VCN nanoscale multilayer PVD coatings deposited by the combined high-power impulse magnetron sputtering/unbalanced magnetron sputtering (HIPIMS/UBM) technology. Vacuum **82**, 1312–1317 (2008)
26. Zhang, Z.G., et al.: Microstructures and tribological properties of CrN/ZrN nanoscale multilayer coatings. Appl. Surf. Sci. **255**, 4020–4026 (2009)
27. Araujo, J.A., Araujo, G.M., Souza, R.M., Tschiptschin, A.P.: Effect of periodicity on hardness and scratch resistance of CrN/NbN nanoscale multilayer coating deposited by cathodic arc technique. Wear **330–331**, 469–477 (2015)
28. Araujo, J.A., Giorjão, R.A.R., Bettini, J., Souza, R.M., Tschiptschin, A.P.: Modeling intrinsic residual stresses built-up during growth of nanostructured multilayer NbN/CrN coatings. Surf. Coat. Technol. **308**, 264–272 (2016)
29. Kamath, G., Ehiasarian, A.P., Purandare, Y., Hovsepian, P.: Tribological and oxidation behaviour of TiAlCN/VCN nanoscale multilayer coating deposited by the combined HIPIMS/(HIPIMS-UBM) technique. Surf. Coat. Technol. **205**, 2823–2829 (2011)
30. Hovsepian, P., Ehiasarian, A.P., Braun, R., Walker, J., Du, H.: Novel CrAlYN/CrN nanoscale multilayer PVD coatings produced by the combined high power impulse magnetron sputtering/unbalanced magnetron sputtering technique for environmental protection of γ-TiAl alloys. Surf. Coat. Technol. **204**, 2702–2708 (2010)
31. AL-Bukhaiti, M.A., Al-Hatab, K.A., Tillmann, W., Hoffmann, F., Sprute T.: Tribological and mechanical properties of Ti/TiAlN/TiAlCN nanoscalemultilayer PVD coatings deposited on AISI H11 hot work tool steel. Appl. Surf. Sci. **318,** 180–190 (2014)
32. Yanga, Y., Yeob, I., Parkc, S., Oh, Y.: Microstructure and phase forming behavior of chrome based multi-composition nitride coatings with nanolayer structures. Ceram. Int. **40**, 11567–11573 (2014)
33. Wang, M., Toihara, T., Sakurai, M., Kurosaka, W., Miyake, S.: Surface morphology and tribological properties of DC sputtered nanoscale multilayered TiAlN/CNx coatings. Tribol. Int. **73**, 36–46 (2014)
34. Vereshchaka, A.A., Vereshchaka, A.S., Mgaloblishvili, O., Morgan, M.N., Batako, A.D.: Nano-scale multilayered-composite coatings for the cutting tools. Int. J. Adv. Manuf. Technol. **72**(1–4), 303–317 (2014)
35. Fominski, V.Yu., Grigoriev, S.N., Celis, J.P., Romanov, R.I., Oshurko, V.B.: Structure and mechanical properties of W–Se–C/diamond-like carbon and W–Se/diamond-like carbon bilayer coatings prepared by pulsed laser deposition. Thin Solid Films **520**(21), 6476–6483 (2012)

36. Fominski, V.Yu., Grigoriev, S.N., Gnedovets, A.G., Romanov, R.I.: Pulsed laser deposition of composite Mo-Se-Ni-C coatings using standard and shadow mask configuration. Surf. Coat. Technol. **206**(24), 5046–5054 (2012)
37. Metel, A.S., Grigoriev, S.N., Melnik, Y.A., Prudnikov, V.V.: Glow discharge with electrostatic confinement of electrons in a chamber bombarded by fast electrons. Plasma Phys Rep. **37**, 628–637 (2011)
38. Grigoriev, S.N., Melnik, Y.A., Metel, A.S., Panin, V.V., Prudnikov, V.V.: Broad beam source of fast atoms produced as a result of charge exchange collisions of ions accelerated between two plasmas. Instrum. Exp. Tech. **52**, 602–608 (2009)
39. Grigoriev, S., Melnik, Y., Metel, A.: Broad fast neutral molecule beam sources for industrial-scale beam-assisted deposition. Surf. Coat. Technol. **156**(1–3), 44–49 (2002)
40. Grigoriev, S.N., Melnik, Y.A., Metel, A.S., Panin, V.V.: A compact vapor source of conductive target material sputtered by 3-keV Ions at 0.05-Pa pressure. Instrum. Exp. Tech. **52,** 731–737 (2009)
41. Metel, A.S., Grigoriev, S.N., Melnik, Y.A., Bolbukov, V.P.: Broad beam sources of fast molecules with segmented cold cathodes and emissive grids. Instrum. Exp. Tech. **55**, 122–130 (2012)
42. Grigoriev, S.N., Metel', A.S., Fedorov, S.V.: Modification of the structure and properties of high-speed steel by combined vacuum-plasma treatment. Met. Sci. Heat Treat. **54**(1–2), 8–12 (2012)
43. Grigoriev, S., Metel, A.: Plasma- and beam- assisted deposition methods. Nato Sci. Ser. II Math. Phys. Chem. **155**, 147–154 (2004)
44. Vereschaka, A.A., Grigoriev, S.N., Vereschaka, A.S., Popov, A.Y., Batako, A.D.: Nano-scale multilayered composite coatings for cutting tools operating under heavy cutting conditions. Procedia CIRP **14**, 239–244 (2014)
45. Oliver, W.C., Pharr, G.M.J.: An improved technique for determining hardness and elastic modulus using load and displacement sensing indentation. J. Mater. Res. **7**, 1564–1583 (1992)
46. Chen, L., Du, Y., Wang, A.J., Wang, S.Q., Zhou, S.Z.: Effect of Al content on microstructure and mechanical properties of Ti-Al-Si-N nanocomposite coatings. Int. J. Refract. Hard. Met. **27**(4), 718–721 (2009)

Critical Reflections on 25 Years of Using Prefabrication and Modularisation in UK Construction

Neil Pickavance(✉)

School of Civil Engineering and Built Environment, Liverpool John Moores University, Liverpool, UK
n.m.pickavance@2017.ljmu.ac.uk

Abstract. Over 25 years the author undertook, in their role as construction planner at Tier 1 UK contractors, planning, programme management, and progress monitoring of offsite manufactured, prefabricated, and modularised elements on new-build construction projects between £0.5M and £100.0M. Based on empirical evidence, this study draws on personal experience in an industry grappling with modern methods of construction between 1995 and the present day, including involvement in the AMEC Education schools programme in the early 2000s to development of a prefabrication and modular process framework for BAM Construction (North West) almost 20 years later. A considerable lack of expertise is apparent within Government, associated bodies, and professional institutes, when it comes to applying knowledge and learning to the productivity problem, and offsite construction in particular. With a clear aim to provide a reflective account of P&M efficacy to explore value in programme delivery, the paper draws authentic experience into a research environment to allow an informed contribution to the topical conversation around offsite manufacturing, prefabrication, and modularisation, and debates its reputation as a panacea for construction's productivity challenge. The study critically reviews 35 examples of offsite construction, including precast frames, structural timber frames, prefabricated steel roof trusses, pre-assembled brickwork panels, prefabricated mechanical and electrical service modules (corridors and risers), hotel bathroom pods, fully modular offices, unitised glazed facades, package pumping stations, and attenuation tanks. Offsite manufacturing, prefabrication, and modularisation are used with some degree of success in the construction industry, however inefficiencies are common due to haphazard allocation plus a tendency to utilise on sections of work that possess programme float. The presumption that offsite, prefabrication, and modular is always good all of the time is flawed, creates a false sense of project success, and denies the chance to honestly appraise and maximise its potential.

Keywords: Offsite · Prefabrication · Modularisation · Programme · Reflective

1 Introduction

This paper demonstrates that the use of prefabrication and modularisation in UK construction is flawed due to being consistently incorrectly allocated to elements of the

A. Batako et al. (Eds.): GCMM 2021, LNNS 335, pp. 145–156, 2022.
https://doi.org/10.1007/978-3-030-90532-3_12

project that possess programme float. Prefabrication and modularisation (P&M) in construction can be a method for shortening of the build programme. However, its selection, specification, and application can be haphazard and poorly thought out by contractors, consultants, clients, and developers. Decisions on P&M are influenced by various parties to the contract. These can be main contractors seeking productivity betterment, subcontractors selling their own product, or consultants seeking efficiency in the design. Factory quality, reduced site labour, and safer working are seen as further side-benefits. This investigation reflects on 25 years of empirical evidence.

Advocation of prefabrication and modularisation is wide-ranging, found in UK Government-sponsored reports (Latham 1994) [11], (Wolstenholme et al. 2009) [19], (Farmer 2016) [2] and echoed by UK built environment institutes: Chartered Institute of Building (CIOB), Royal Institution of Chartered Surveyors (RICS), Royal Institute of British Architect (RIBA) and the Institute of Civil Engineers (ICE). There is belief that P&M is always good all of the time, a presumption that is flawed, poorly evidenced, and hinders P&M value optimisation. P&M selection requires consideration on a project-by-project basis. Blanket adoption, emboldened by subcontractor narrative and supplier promotion, can deliver flawed and inefficient programmes, failing clients and contractors alike. This paper collates empirical evidence of P&M efficacy from a 25-year period, comparing programme performance against critical path and non-critical schedules. Drawing on first-hand experience at tier 1 contractors this reflective study has the following aim:

To demonstrate that P&M success in shortening overall programme durations is contingent on their allocation on the project critical path.

Literature on prefabrication and modularisation shows little understanding of this programme sensitivity of P&M allocation. There appears a view, consistent with "…always good, all of the time", that P&M never fails to deliver programme improvements. Outwith the primary data here, no empirical evidence was found to support this assertion.

2 Background

2.1 Historical Overview of Offsite Construction

Pre-assembled site elements have been used since the 1940s, with prefabricated housing as per the Housing (Temporary Accommodation) Act 1944 [10]. Pre-dating this are timber windows, trusses, and doorsets. Reconstituted stone was used from the 19th century, while tunnelling saw segmental pieces, manufactured above ground and transported to the workface. Foundry-engineered cast iron was first used in a major capacity at Iron Bridge, Shropshire, UK, completed January 1781 (Trinder 1979) [17]. Locally, Ditherington Flax Mill, was the world's first iron-framed building on completion in 1797 (Trinder 1992) [18]. In the 20th century, increased demand in offsite manufacturing was driven post-war tower block schemes in UK urban areas.

2.2 Modern Methods of Construction: A Brief Definition

Contemporary P&M is broader in scope than iron, stone, prefabricated housing, and tunnelling. Modern methods of construction (MMC) include package pumping stations, precast concrete frames, structural timber frames, prefabricated roof trusses, unitised curtain walling, bathroom pods, steel sheet piles, precast concrete retaining walls, steel footbridges, plastic service chambers, facing brickwork cladding panels, precast concrete stairs, and attenuation tanks.

Images of examples used in this study are provided below. Figure 1(a) illustrates a precast concrete frame under construction at Ravensbury Primary School in Clayton, east Manchester in 2002; Fig. 1(b) shows mechanical and electrical (M&E) service modules fitted to the corridor of Lesmahagow High School in South Lanarkshire, Scotland in 2006; Fig. 1(c) is of a 51-tonne prefabricated steel roof truss being prepared for installation into the building at Wigan Joint Service Centre in 2010; and, Fig. 1(d) is of timber structural insulated panel system (SIPS) fitted to the concrete frame of the Orchard Hotel at the University of Nottingham in 2012. First Section.

Fig. 1. (a) Precast frame; (b) M&E modules; (c) roof truss; (d) SIPS panels.

3 Government and Industry Reports

3.1 Guidance, Policy, and Target Setting

P&M benefits go beyond programme shortening and includes quality, labour reduction, and safer working. Efficiencies are promoted widely by professional institutes such as the CIOB (Green [6]) and the ICE (Goodier et al. [5]). Industry bodies such as Constructing Excellence and UK Research and Innovation Council, via their Transforming Construction challenge, heavily promote offsite construction. Construction industry reports over the last thirty years (examples in Table 1) maintain a view that productivity improvements should be attainable via prefabrication and modularisation.

Table 1. Construction reports recommending use of offsite construction.

Report	1990s	2000s	2010s
Constructing the Team	1994		
Rethinking Construction	1998		
Modernising Construction		2001	
Never Waste a Good Crisis		2009	
Construction 2025			2013
Modernise or Die			2016
Reinventing construction through a productivity revolution			2017
Industrial Strategy: Building a Britain Fit for the future			2017
Industrial Strategy: Construction Sector Deal			2018

Constructing the Team (Latham (1994) [11] alludes to programme saving benefits from P&M, providing examples of case studies, although not all from the UK. In the 1990s and 2000s, Rethinking Construction (Egan 1998) [1], Modernising Construction (National Audit Office 2001) [13], and Never Waste a Good Crisis (Wolstenholme et al. 2009) [19], suggest much-needed improvements could be unlocked via offsite manufacturing. Limited investigation into P&M of optimising programme performance exists in these reports, and their contribution to the narrative consists of being advisory and establishing targets.

Construction 2025 (HM Government 2013) [7] and Modernise or Die (Farmer 2016) [2] establish their own targets with no recommendations to support the challenge. The Industrial Strategy reports: Building a Britain fit for the future (HM Government 2017) [8] and Construction Sector Deal (HM Government 2018) [9] follow the format of promotion of offsite construction but offering no empirical supporting evidence. In these reports, offsite is placed in a wider context with other proposed modern methods of working and technologies to address productivity issues.

4 Literature

4.1 Offsite Manufacturing and Construction

Egan (1998) [1] considered that greater use of prefabrication and standardisation would greatly enhance the productivity of the construction industry. Specifically, the efficiency of construction has been frequently questioned, particularly by Egan (1998) [1], who thought that "within five years, the construction industry should deliver its products to its customers in the same way as the best consumer-led manufacturing and service industries". The inefficiencies of traditional construction in the UK have been explored and are documented in government reports and related literature (Latham 1994 [11]; Egan 1998 [1]; Woodhuysen and Abley 2004 [20]).

Construction in the UK is characterised by low quality, high costs, a poor safety, adversarial nature, low investment in research and development, poor training, serviced by companies operating nationally and locally with low capital investment (Egan 1998) [1]. In general, offsite construction is a modern method incorporating both prefabrication and preassembly or prefabrication of modules (Gibb 1999) [3] usually in a factory remote from the project while groundworks, substructures, foundations, and suchlike are prepared on site. Offsite construction is increasingly gaining momentum in the UK construction industry, driven by the need for expedient delivery, technologically advanced approaches as well as government backing and influence (Siebert et al. 2018 [14]). However, in the UK it is thought that no more than 2% of construction work has any significant adoption of prefabrication (Steinhardt and Manley 2016 [16]).

Another significant benefit of adopting prefabrication is the reduction of onsite construction time due to the opportunity to simultaneously construct elements in the factory while the site-based work is progressing (Smith 2010) [15]. Goodier and Gibb (2007) [4] noted that longer lead times awaiting P&M elements could potentially delay the beginning of a project; to reduce this, the whole design and construction process needs to be better aligned. Goodier et al (2019) [15], in the proceedings of The Institute of Civil Engineers (Volume 172), provides awareness of the need to employ a decision-support tool when determining a project's suitability for offsite construction. The study notes that offsite and modular are increasing in engineering construction, and current decision-support tools are currently limited or outdated. To address this lack of process in construction, this paper presents a stage process for optimisation of prefabrication and modularisation for ensuring P&M is used on the project critical path.

5 Research Method

This study revisited a personal archive of as-built construction records spanning 25 years, reviewing extensive project documentation and combining with recollections of episodic memory. Sustained exposure to P&M over 25 years at UK contractors has come from project involvement in planning and programme management, pre-construction planning, and responsibility for progress monitoring. Construction reviews, subcontractor meetings, and post-handover reviews provided opportunities to regularly assess, and gather data on, the programme performance of projects. However, limitations of this research are the timescales involved and recollections.

The 35 P&M elements are tabulated chronologically, with the sector type noted and the main year of installation (Table 2). Central to the study has been identifying whether or not the P&M element was on the project critical path, or not on the critical path, and thereby possessing float. These two conditions are identified in the table with Yes/No markings and any shortening to the overall construction programme is noted. Where shortening occurred, a percentage is provided in the table right-hand column to indicate the overall project saving compared to an equivalent traditional solution. Figure 2 illustrates the crucial characteristic that the study looked for with each element of P&M, which is placement on the project critical path.

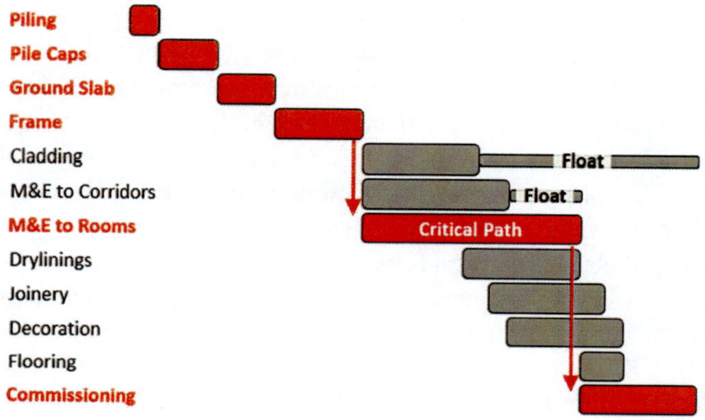

Fig. 2. Critical path potential for shortening overall programmes.

Diversity of the research sample is demonstrated in Figs. 3(a) and 3(b). The first image is the precasting facility at CV Buchan in Middlewich, Cheshire in June 2000, in preparation for new-build secondary school projects at Alsager and Wilmslow in Cheshire. The second image shows unitised glazing being installed to the No.1 Spinningfields office façade at Spinningfields, central Manchester in September 2016.

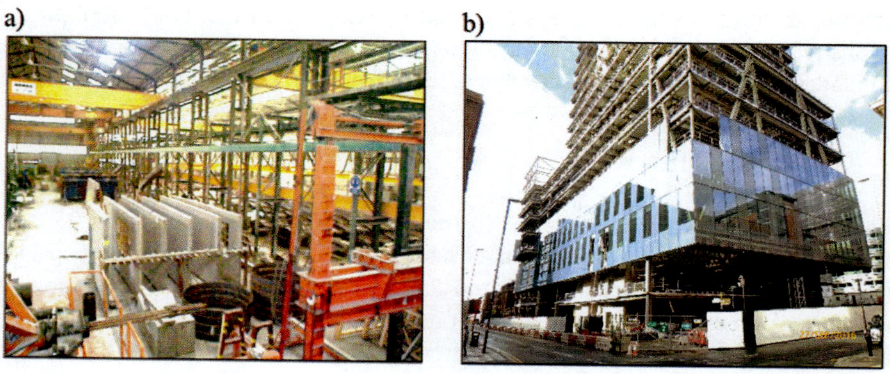

Fig. 3. (a) Precasting facility; (b) unitised glazed facade.

Table 2. Summary of research sample of 35 elements of P&M components.

P&M element	Critical path	Programme shortening	Programme saving
Package pumping stations	No	No	0%
Precast concrete retaining wall	No	No	0%
Precast concrete frame	Yes	Yes	23%
Precast concrete frame	Yes	Yes	23%
Precast concrete frame	Yes	Yes	23%
Precast concrete frame	Yes	Yes	23%
Precast concrete frame	Yes	Yes	23%
Precast concrete frame	Yes	Yes	23%
Precast concrete retaining wall	No	No	0%
Modular offices[a]	yes	Yes	57%
Precast concrete piles	Yes	Yes	5%
Precast concrete frame	Yes	Yes	23%
Precast concrete frame	Yes	Yes	23%
Precast concrete frame	Yes	Yes	23%
Precast concrete frame	Yes	Yes	23%
M&E modules (corridors)	No	No	0%
Precast concrete lift shafts	No	No	0%
Steel roof truss	No	No	0%
Steel sheet piles	No	No	0%
Structural insulated panel system	No	No	0%
Bathroom pods	Yes	No	0%
Precast concrete retaining wall	No	No	0%
Precast concrete terrace units	No	No	0%
Steel footbridge	No	No	0%
Plastic service chambers	No	No	0%
Facing brickwork cladding panels	No	No	0%
Precast concrete stairs	No	No	0%
Steel cantilever	No	No	0%
Unitised glazing	No	No	0%

(continued)

Table 2. (*continued*)

P&M element	Critical path	Programme shortening	Programme saving
Steel staircase	No	No	0%
Rooftop cradle access system	No	No	0%
M&E modules (risers)	No	No	0%
Structural timber frame	Yes	Yes	5%
Attenuation tank	No	No	0%
Precast concrete stairs	No	No	0%

[a]The Modular Offices, while shown in this table, are not a typical P&M component and are therefore not considered in the results, discussion, and conclusion sections.

6 Results and Findings

The 35 individual elements reviewed, comprising 23 variants of P&M, yielded three P&M solutions with overall programme shortening (Table 3). This related to a P&M element being on the project critical path. Precast piles provided a saving of 2.5 weeks over bored piles, equating to 5% of the overall programme. The precast frame provided a saving of 5 weeks over a load-bearing brick and block frame, equating to 23% of the overall programme. The structural timber frame provided a saving of 2 weeks over steel, decking, and slabs, equating to 5% of the overall programme.

Table 3. Comparison of programme saving with P&M versus traditional.

Element	Duration	Saving	Project saving
TRAD: CFA Bored Piles	3.5w	-	-
P&M: Precast concrete piles	1w	2.5w	2.5w off 55w (5%)
TRAD: Brick, Block, Plaster	6w	-	-
P&M: Precast concrete frame	1w	5w	5w off 22w (23%)
TRAD: Steel, Decking, Slabs	17w	-	-
P&M: Structural timber frame	15w	2w	2w off 38w (5%)

Figure 4(a) shows precast concrete piles at Ravensbury Primary School in east Manchester in 2001. Figure 4(b) is of precast concrete frame panels being erected to Beaver Road Primary School in south Manchester in 2000. Figure 4(c) is the structural timber frame being erected to The Pavilion at Spinningfields in Manchester in 2017.

Fig. 4. (a) Precast piles; (b) precast frame; (c) timber frame.

7 Optimising P&M Selection in Construction

Shortening of the overall construction programme appears wholly contingent on P&M elements being deployed on critical path. A process map is offered here to support the selection and decision-making process, acknowledging this single-most important factor in achieving programme compression. This approach to P&M optimisation and selection has positive implications not only for programme, but for cost via reduced project preliminaries. The stage process is illustrated in Table 4.

The optimising of P&M commences with a benchmark construction programme with non-P&M elements (Stage 1). The review of options for conversion to P&M must be a collaborative process (steering group) with the contractor, subcontractors, consultants (architect, structural engineer, services engineer) and client-side cost adviser (Stage 2). An unambiguous agreement of whether the elements are critical path or non-critical path must be reached (Stage 3). Affordability of P&M options is crucial particularly if a P&M element is to be designated as non-critical path (Stage 4). A direct comparison programme with the P&M elements against the original benchmark non-P&M programme is to be evaluated (Stage 5). Understanding implications of P&M selection regarding additional costs relating to lead-in, storage, transport, and lifting (Stage 6). An informed decision by the steering group on the best value P&M option (Stage 7). A formal implement-or-reject gateway will avoid existing UK practices of P&M selection dominated by special interest, e.g. subcontractor (Stage 8).

- Stage 1:- "Traditional" programme developed.
- Stage 2:- P&M options identified e.g. substructure.
- Stage 3:- Evaluation of whether P&M is critical path.
- Stage 4:- Cost and affordability assessment of P&M.
- Stage 5:- "Traditional" programme re-drawn with P&M.
- Stage 6:- Appraisal of P&M benefits, knowing constraints.
- Stage 7:- Decision on programme and cost implications.
- Stage 8:- Implement or reject P&M elements.

Table 4. A route to optimisation for prefabrication and modularisation.

PROCESS	GUIDANCE
STAGE 1 Traditional Programme Development	Develop construction programme indicating traditional solutions.
STAGE 2 Identify P&M Solution Options	e.g. 1. Precast Frame. e.g. 2. Brickwork Panels. e.g. 3. M&E Modules.
STAGE 3 P&M Solution Selection and Evaluation	Critical Path - YES. Critical Path - NO. Critical Path - NO.
STAGE 4 P&M Solution Affordability Assessments	1. Precast Frame Costs. 2. Brickwork Panels Costs. 3. M&E Modules Costs
STAGE 5 Re-Programme for P&M Exercise	Redevelop traditional construction programme with P&M solutions
STAGE 6 Formal Report on P&M Benefits	Programme preliminaries, quality, lead-in, storage, transport, lifting, hoisting.
STAGE 7 Decision on P&M Report	Teams informed decision on P&M solution.
STAGE 8 P&M Implement or Reject	1. Implement P&M. 2. Reject P&M retain trad. 3. Hybrid P&M + trad.

8 Conclusions

The current routes to specification, selection, and allocation of prefabrication and modularisation in UK construction supports programme underperformance. To obtain greater value through overall programme shortening it was observed that this is almost exclusively achieved by utilising P&M elements on the project critical path. These programme gains were observed by specifying P&M solutions to precast piling, precast concrete frame, and structural timber frame. Allocating P&M elements to activities that have project float may serve to quicken localised sections of the programme but not shorten the overall programme duration. A combination of target-setting by Government and industry bodies, poor collaboration between contractors and consultants, and a lack of accurate, as-built feedback from projects allows an ignorance to flourish around P&M in the construction industry.

Government, chartered built environment institutes (CIOB, RICS, RIBA, ICE, CIBSE), and industry groups such as Constructing Excellence and UK Research and Innovation Council very much driven and dominated by self-published reports and recommendations that establish P&M aspirations for industry. There appears continuous target-setting with little contribution or awareness of the P&M selection process. A lack of expertise, knowledge, and learning around offsite construction by these bodies in general then provides some poor compensation with their attempts at referencing outdated P&M case studies.

Awareness of the need to employ a decision-support tool when determining a project's suitability for offsite construction is essential. With offsite and modular increasing in engineering construction, and current decision-support tools currently limited or outdated, addressing this lack of process in construction is crucial to drive value improvements.

Of 23 P&M variants observed over 25 years, and reviewed in this study, only three (13%) examples were found successful in shortening the overall construction programme on the projects where they were deployed. Out of these three, two yielded a 5% overall programme saving and the third yielded a 23% overall programme saving. These figures are far from exceptional and fall short of Government requests to reduce construction programme durations by 50%. While P&M benefits reach beyond programme, compression of durations is expected as a standard. With clients, contractors, and consultants espousing "P&M always good, all of the time", and encouraged by industry bodies, this incurring of premium cost with little, or no, programme benefit is poorly controlled. Clients don't want to believe their bathroom pods, service modules, and cladding panels are not compressing the overall programme.

The study aim was to harvest a 25-year archive of P&M as-built data to critically evaluate the effectiveness of P&M in construction programmes. An unintended consequence of this has been the exposing of the P&M mantra "...always good, all of the time" as flawed. That belief and narrative is persistent in the industry, echoed and enabled by groups, bodies, and organisations that have lobbying influence.

A key recommendation from this study is that this narrative undermines projects as empirical evidence herein shows that possibly as little as 13% of projects are successfully utilising P&M for overall programme shortening purposes. Follow-up case studies are recommended to provide clarity around costs incurred by employers, against the level of programme value improvements received.

References

1. Egan, J.: Rethinking construction. Department of the Environment. Transport and the Regions (1998)
2. Farmer, M.: Modernise or Die. Construction Leadership Council, London (2016)
3. Gibb, A.G.F.: Off-site Fabrication: Prefabrication, Pre-Assembly and Modularisation. Wiley, New York (1999)
4. Goodier, C., Gibb, A.: Future opportunities for offsite in the UK. Constr. Manag. Econ. **25**(6), 585–595 (2007)
5. Goodier, C., Gibb, A., Mancini, M., Turck, C., Gjepali, O., Daniels, E.: Modularisation and offsite in engineering construction: an early decision-support tool. Proc. Inst. Civil Eng. Civil Eng. **172**(6), 3–14 (2019)
6. Green, B.: Productivity in Construction: Creating a Framework for the Industry to Thrive. Chartered Institute of Building, Bracknell (2016)
7. HM Government: Industrial strategy: government and industry in partnership. Construction 2025 (2013)
8. HM Government: Industrial strategy: building a Britain fit for the future. Government White Paper (2017)
9. HM Government. Industrial strategy: Construction sector deal. HM Government, UK (2018)
10. Housing (Temporary Accommodation) Act 1944
11. Latham, S M.: Constructing the team (1994)
12. Barbosa, F., et al.: Reinventing Construction through a Productivity Revolution. McKinsey Global Institute, Washington, D.C. (2017)
13. National Audit Office. Modernising construction (2001)
14. Siebert, M., Rodrigues, L., Gillott, M., Hines, E., Rich, D.: Identifying the barriers to change in the UK housebuilding industry. Fut. Cities Environ. **4**(1), 13 (2018)
15. Smith, RE.: Prefab Architecture: A Guide to Modular Design and Construction. Wiley, New York (2010)
16. Steinhardt, D.A., Manley, K.: Adoption of prefabricated housing–the role of country context. Sustain. Cities Soc. **22**, 126–135 (2016)
17. Trinder, B.: The first Iron bridges. Ind. Archaeol. Rev. **3**(2), 112–121 (1979)
18. Trinder, B.: Ditherington Flax Mill, Shrewsbury—a re-evaluation. Text. Hist. **23**(2), 189–223 (1992)
19. Wolstenholme, A., et al.: Never Waste a Good Crisis: A Review of Progress since Rethinking Construction and Thoughts for Our Future, Constructing Excellence, London (2009)
20. Woudhuysen, J., Abley, I.: Why Is Construction So Backward? Wiley Academy, Hoboken (2004)

A Fuzzy Logic Base for Selecting Optimal Clearance of Die-Cutting Process

Arkarapon Sontamino(✉) and Chiwapon Nitnara

Department of Mechanical Engineering Technology, The College of Industrial Technology, King Mongkut's University of Technology North Bangkok, 1518 Pracharaj 1, Bangsue 10800, Bangkok, Thailand
arkarapon.s@cit.kmutnb.ac.th

Abstract. Currently, the sheet-metal parts, a both of thin sheet-metal and thick sheet-metal are increasingly fabricated in many industrial fields. To produce the sheet-metal parts in thin sheet-metal and thick sheet-metal, the conventional die-cutting theory is usually applied. However, with the same die-cutting process parameters, the comparison of cut-surface features between thin and thick sheet-metal parts has not been investigated yet. Therefore, in the present research, the comparison of the cut surface feature for die-cutting process in various materials thicknesses between thin and thick sheet-metal parts are investigated. The FEM simulation was used as a tool to investigate these cut surface features, and the laboratory experiments were also performed to validate the FEM simulation results and fuzzy logic model has been created as a tool to assist select the optimal clearance value which used data reference from experimental and FEM method. The experimental results agreed well with the FEM simulation results in terms of cut-surface features. The results that, to achieve a good cut-surface features, the die-cutting clearance should be selected to suit the material thickness and fuzzy logic model can achieve to select optimal the die-cutting clearance effectively.

Keywords: Thin sheet-metal · Thick sheet-metal · FEM simulation · Cut-surface feature · Fuzzy logic

1 Introduction

In recent years, the sheet-metal parts, fabricated by a die-cutting process are applied in almost all industrial fields such as Aircraft industry, electronic industry, and automotive industry. A principle of die-cutting process, the sheet-metal was separated by a punch is moved downward with the great enough force to generate the crack formation on that sheet-metal as shown in Fig. 1(a-1). After die cutting process, the four components of die roll, smooth cut, crack, and burr wear generated on cut surface which is commonly cut surface from die-cutting process as shown in Fig. 1(a-2).)[1–2]. In the past researches were studied in various metal die-cutting process such as a blanking process, a punching process and a shearing process and on many kinds of sheet-metal materials such as aluminum [3], carbon steel, and high strength steel [4] by using the experiment and FEM

A. Batako et al. (Eds.): GCMM 2021, LNNS 335, pp. 157–168, 2022.
https://doi.org/10.1007/978-3-030-90532-3_13

simulation to characterize the cutting mechanism, to clarify the effects of process parameters on cut-surface characteristics, For example, the parameters effects of die-cutting process on cut-surface feature in the many past researches, including clearance between cutting edge [5], cutting speed [6], clamping force [7] and tool geometry [8]. However, the studied of cutting mechanisms and cut surface features in the past researches were focused only a thin sheet metal, In the other hand, a thick sheet metal have not been investigated yet. Therefore, in the present research, the die-cutting mechanism on a thick sheet metal was studied base on the shearing process by using FEM simulation. On the basis of the stress distribution analysis, the effects of material thickness on cut-surface characteristics was investigated and clearly clarified. Laboratory experiments were also carried out to validate the FEM simulation results. The FEM simulation results showed a good agreement with the experimental results in terms of cut-surface characteristics. The FEM simulation results showed a good agreement with the experimental results in terms of cut-surface. The results elucidated that with the same die-cutting process parameters, the different cutting mechanisms and cut surface features were obtained. Specifically, the die-roll formations and crack formation were easily generated in the case of thin sheet metal. Therefore, the smooth cut surface was smallest in the case thin sheet metal. In addition, the present research was performed to determine optimal clearance values for the thin and thick sheet material. In this study, a fuzzy logic [9] model was created for selecting suitable clearance value. The results of FEM were used to reference data [10] in fuzzy logic model consists of die-roll, smooth cut and crack formation, which is a factor indicated good quality of sheet metal.

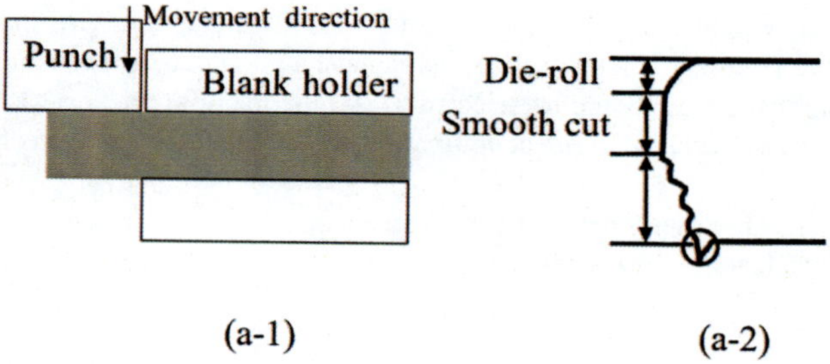

Fig. 1. Principle of shearing process and cut surface feature.

2 FEM Simulation and Experimental Procedures

FEM simulation models used in this present research, a commercial analytical code, The FEM simulation program, based on the two-dimensional, implicit quasi-static finite element method was used for the FEM simulation. In this FEM code, the solution algorithm utilized is based on the Newton-Raphson iteration. In the present research, the two-dimensional plane strain for model of shearing process as shown in Fig. 2 The punch, die, counter punch and blank holder were set as the rigid type. The work piece material (Al 1100-O) was set as an elasto-plastic type with the rectangular element of approximately 10000 elements and its properties used in this present research was referred from the author's past research [3]. The press machine 80 Tons (Accurl Y41-80) was used for experimental testing. The cut surface features obtained by experiments were observed by using profile projector. The details of FEM simulation and experimental conditions were listed in Table 1.

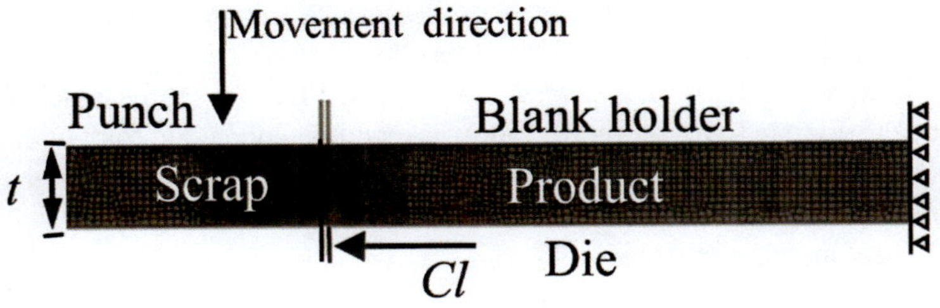

Fig. 2. Shearing process model.

Table 1. FEM simulation and experimental conditions.

Simulation model		Plane strain model
Object type		Workpiece: Elasto-plastic Punch, Die, Blank holder: Rigid
Workpiece material	Al1100-O	Tensile strength (σ_u): 92.50 MPa Elongation (δ): 43.5% Young's modulus (E): 68900 MPa Poisson's ratio (ν): 0.33
Flow curve equation	Al1100-O	$\bar{\sigma} = 153.50\bar{\varepsilon}^{0.20} + 88$
Fracture criterion equation		Normalized Cockroft & Latham
Critical fracture value	Al1100-O	1.28
Friction coefficient (μ)		0.1
Workpiece geometry		Thickness (t): 3, 7, 15 mm
		Length (WP_L): 80 mm
Shearing clearance (Cl)		3%t, 5%t, 10%t

3 Stress Distribution Analyses

Figure 3 shows the comparison of stress distribution analyses during shearing process with respect to various material thickness. The FEM simulation illustrated the same manner of stress distribution analyses of a both cases. Specifically, as the depth of cut increased, the tensile stress at cutting zone and die-roll area were increased. These manners of stress distribution analyses agree well with the shearing theory and literature [1, 2, 11]. However, at the 10% depth of cut as shown in Fig. 3(a), it was observed that the generated the tensile stress on the die-roll in the case of thin sheet metal was earliest formed as shown in Fig. 4(a-1) following by thick sheet metal. Next, at the depth of cut increased as shown in Fig. 3(b), the generated tensile stress on cutting zone was increased. It was also observed that the generated tensile stress on the cutting zone in the case of thin sheet metal was larger than thick sheet metal as shown in Fig. 3(b-1). Finally, at the 50% depth of cut as shown in Fig. 3(c), the results showed the crack formations which were easily generated in the case of thin sheet metal, as shown in Fig. 3(c-1), because of the large generated tensile stress on cutting zone. On the other hand, in the case of thick sheet metal as shown in Fig. 3(c-2), the crack formation had not been generated yet. As these results, it could be concluded that, with the same shearing process parameters except for material thickness, the die-roll formations and crack formation were easily generated in the case of thin sheet metal, following by the thickness sheet metal, respectively.

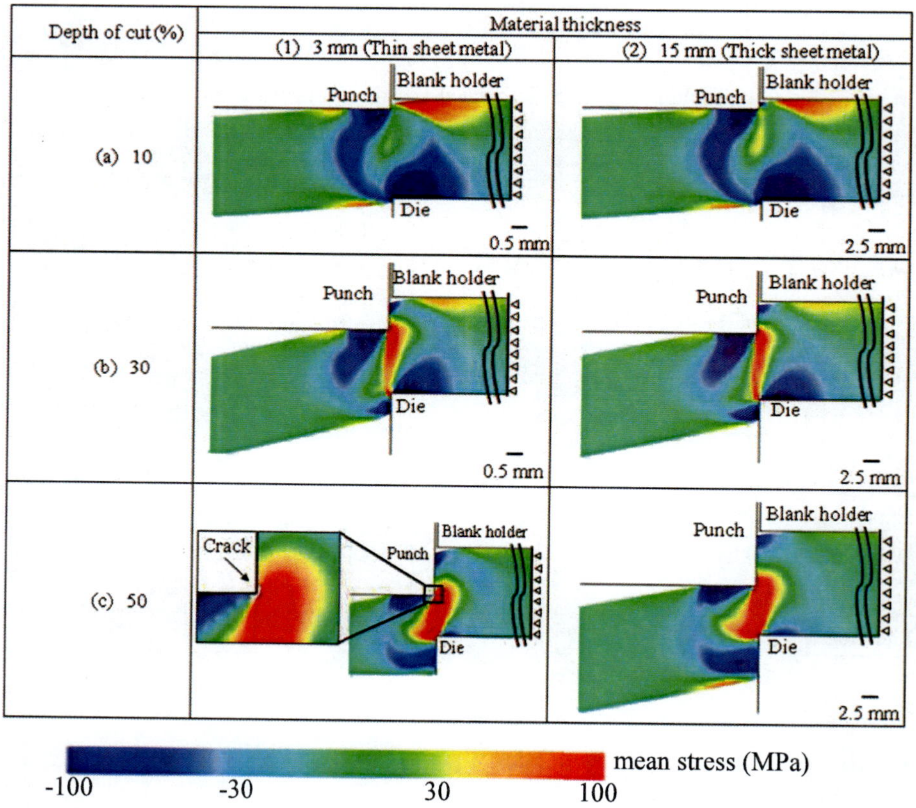

Fig. 3. Comparison of stress distribution with respect to various material thickness. (*Cl*: 3%*t*)

4 Cut Surface Features Related with Various Die-Cutting Processes

As aforementioned, on the basis of stress distribution analyses, the die-roll formation and crack formations were easily generated in the case of thin sheet metal, following by the thick sheet metal, respectively. These die-roll formation and crack formations directly affected on the smooth cut feature on cut surface feature. Specifically, as the die-roll formation and crack formations on cut surface feature were increased, the smooth cut feature decreased. Therefore, in the present research, the cut surface features related with various material thickness were examined. Figure 4 shows the cut surface features related with various material thickness. The results illustrated that the smooth cut surface was decreased as the die-roll formation and crack formations were increased. This manner of cut surface feature was generally in agreement with that reported in the shearing theory and literature [1, 2, 11]. Furthermore, the results showed that the die-roll formation and crack formations were largest and the smooth cut surface was smallest in the case of thin sheet metal, as shown in Fig. 4. As well as, the Fig. 4 showed the smallest die-roll formation and crack formations and largest smooth cut feature in the case of thick sheet metal. As these results, with the same shearing process parameters, it could be concluded that the thin sheet metal provided the smallest smooth cut surface, as well as the thick sheet metal provided the largest smooth cut surface. Therefore, the shearing clearance in case of thick sheet metal should be larger than the shearing clearance in case of thin sheet metal to obtain a easily setting because the large shearing clearance in case of thick sheet metal provided the ratio of smooth cut surface and material thickness same manner small shearing clearance in case of thin sheet metal as shown in Fig. 5.

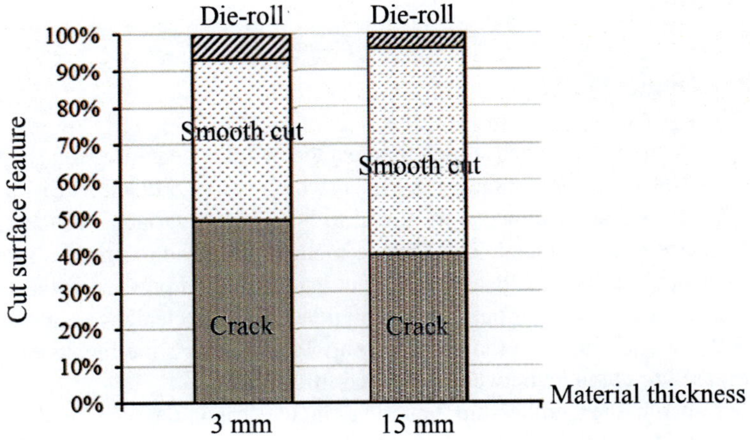

Fig. 4. Cut surface features related with various material thickness (*Cl*: 3%*t*)

5 Validation of FEM Simulation Results

To validate the accuracy of FEM simulation results, laboratory experiments were carried out. In the present research, two types of sheet metal forming including thin sheet metal

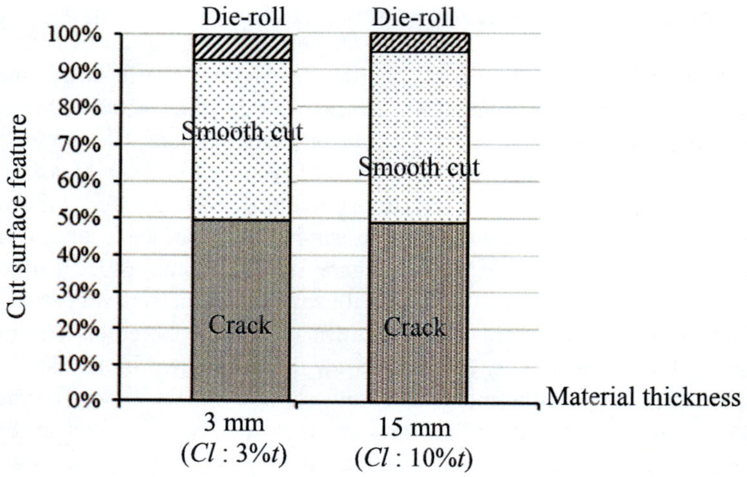

Fig. 5. Cut surface features related with various material thickness and shearing clearance

and thick sheet metal were performed as shown in Fig. 6. The FEM simulation results showed a good agreement with the experiment results, in which the errors in the amount of die-roll, smooth cut surface and crack formations compared with those obtained from the experimental results was approximately 2%. In addition, the experimental results also confirmed that the ratio of smooth cut surface and material thickness was generated on cut surface in case of thick sheet metal larger than the smooth cut surface in case of thin sheet metal.

6 Fuzzy Logic Model

The fuzzy logic model approach is used to select the optimal clearance value and created by MATLAB software [12] as shown in Fig. 7, which has significant factors consist of material thickness and clearance were found to be affected variables to the results of the good surface of sheet metal. The Sugeno method was used to create a fuzzy logic model and would be 9 rules. It has two inputs and three outputs. Material thickness and clearance are two fuzzy inputs, the optimal value of die clearance consists of die-roll, smooth cut and crack are the fuzzy output. Table 2 shows the results of the FEM simulation used to compare between FEM and fuzzy logic.

In this research, the membership function of input is triangular membership function as shown in Fig. 8, which is commonly used to create fuzzy logic. The linguistic variable of the material thickness and clearance consists of low, medium and high. A low of linguistic variable in material thickness is 3 mm, medium is 7 mm and high is 15 mm. In the clearance, a low of linguistic variable is 3%t, medium is 5%t and high is 10%t.

Figure 9 shows the outputs factors of the fuzzy logic model has been divided into 9 membership functions. In the sugeno method, the membership function of fuzzy logic model is constant value. Therefore, the output in this model consist of die-roll, smooth cut and crack are the constant value.

	Thin sheet metal (t : 3mm, Cl : 3%t)		Thick sheet metal (t : 15mm, Cl : 10%t)	
FEM simulation results	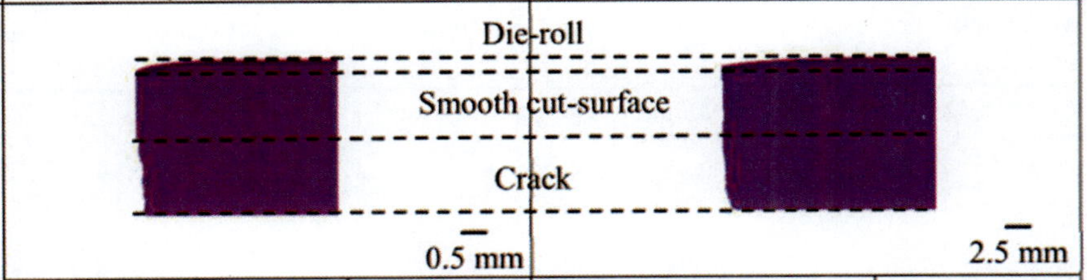			
	Die-roll /Thickness	0.21/3 mm	Die-roll /Thickness	0.78/15 mm
	Smooth cut /Thickness	1.30/3 mm	Smooth cut /Thickness	6.93/15 mm
	Crack/Thickness	1.49/3 mm	Crack/Thickness	7.29/15 mm
Experimental results	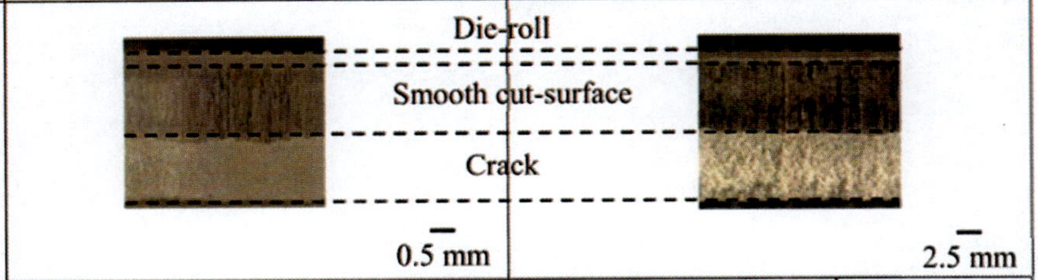			
	Die-roll/Thickness	0.20/3 mm	Die-roll/Thickness	0.80/15 mm
	Smooth cut /Thickness	1.34/3 mm	Smooth cut /Thickness	7.00/15 mm
	Crack/Thickness	1.46/3 mm	Crack/Thickness	7.20/15 mm

Fig. 6. Comparison of the cut surface features between FEM simulation and the experiental results

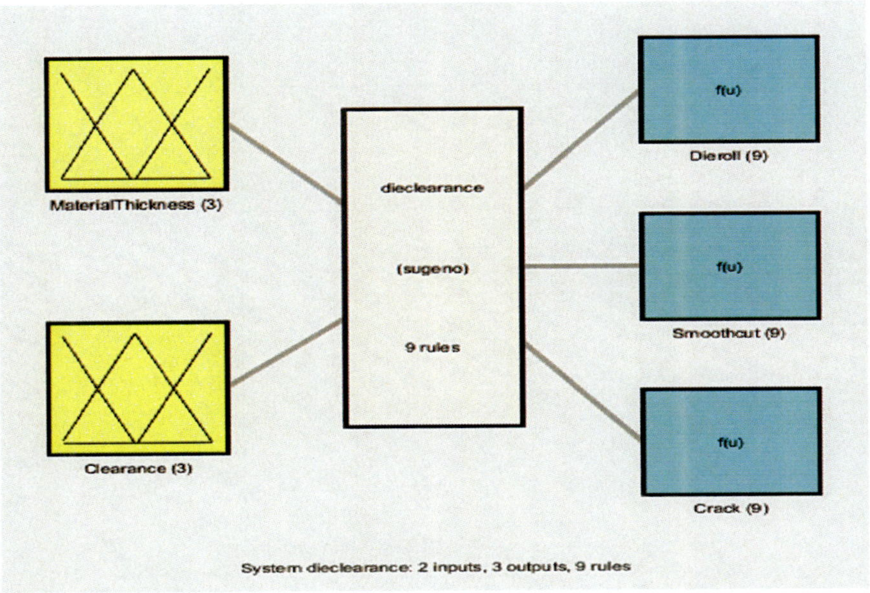

Fig. 7. Fuzzy logic model for selecting the optimal of die clearance

Table 2. Data for comparing between FEM and Fuzzy Logic

Material thickness (mm)	Clearance (%t)	Die-roll (mm)	Smooth cut (mm)	Crack (mm)
3	3	0.213	1.300	1.487
3	5	0.238	1.240	1.522
3	10	0.294	1.100	1.606
7	3	0.474	2.450	4.076
7	5	0.501	2.390	4.109
7	10	0.604	2.260	4.136
15	3	0.600	8.390	6.010
15	5	0.710	8.090	6.200
15	10	0.780	6.930	7.290

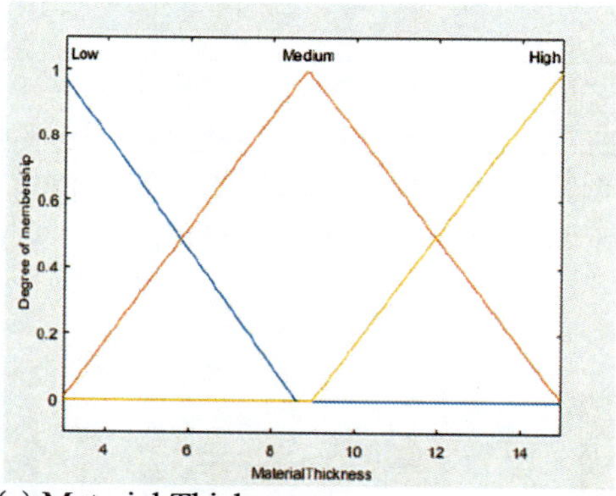

(a) Material Thickness

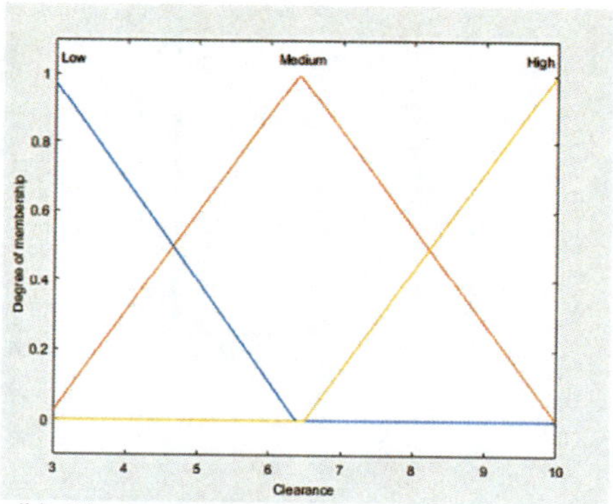

(b) Clearance

Fig. 8. Membership Function of Input

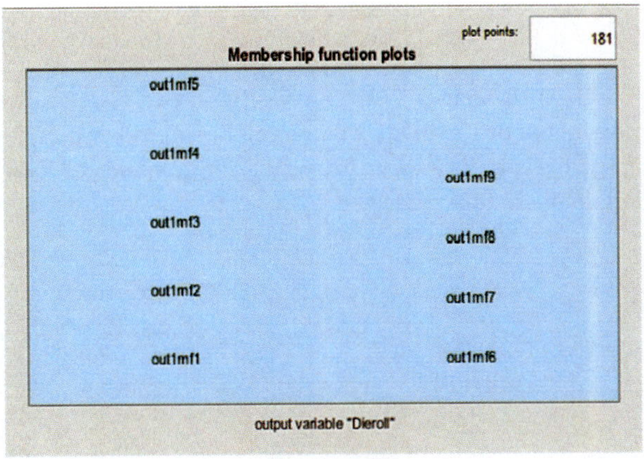

(a) die-roll

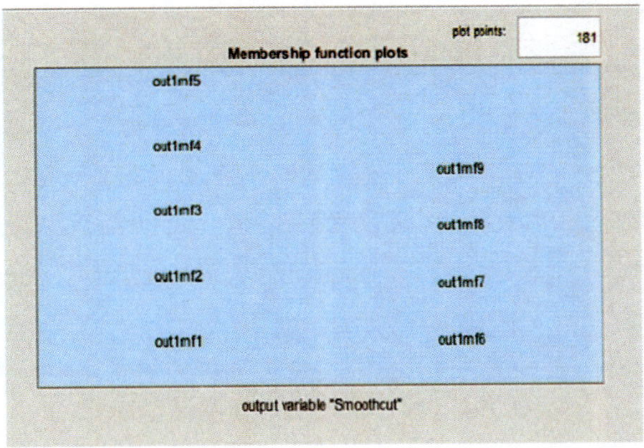

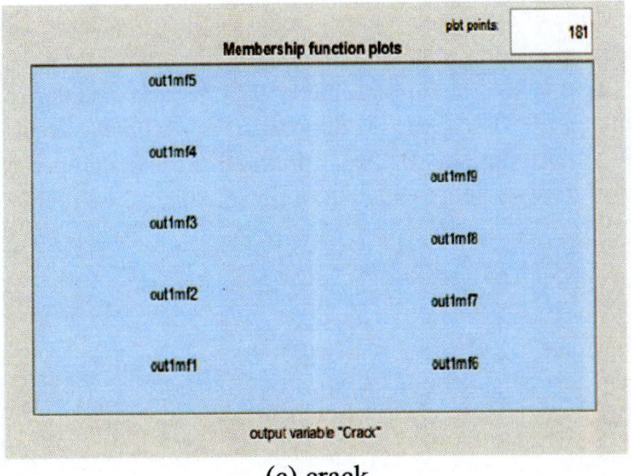

(c) crack

Fig. 9. Membership Function of Output

7 Fuzzy Logic Results

The rule viewer is used to approximate the results in MATLAB software. Figure 10 represents the rule viewer of the fuzzy logic model of the shearing process for finding the optimal value of the die clearance. When the input values changed and the fuzzy logic would calculate the output value.

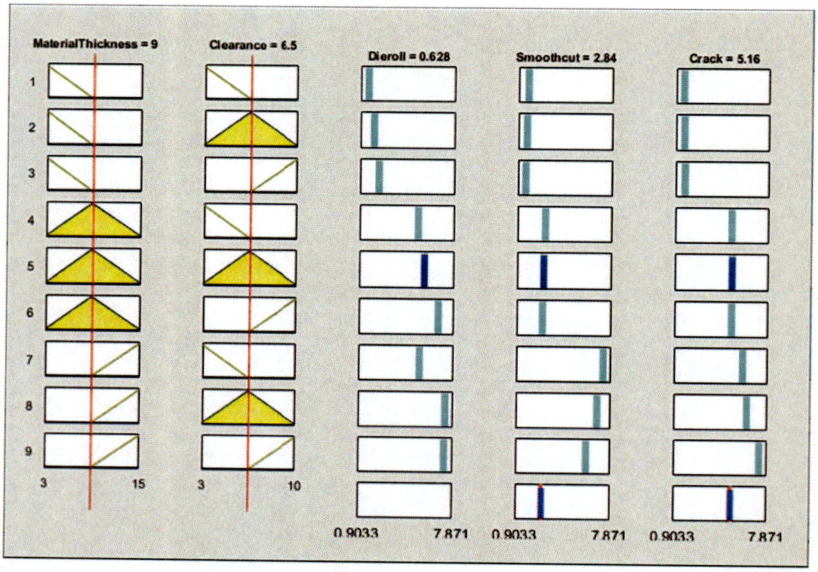

Fig. 10. Rule viewer of fuzzy logic model

Table 3 shows a comparison between the results of the FEM simulation value and the optimal selection value of the die clearance results by the fuzzy logic model. The average error value of case 1 is approximately 0.61%, case 2 is approximately 0.71%, case 3 is approximately 0.68%, case 4 is approximately 0.34%, case 5 is approximately 0.29%, case 6 is approximately 0.31%, case 7 is approximately 0.08%, case 8 is approximately 0.18% and case 9 is approximately 0.35%. It found that the developed fuzzy logic model can be used as an effectively approximate tool for selecting the suitable of die clearance.

Table 3. Comparison between the results of FEM simulation and the result of fuzzy logic

Factor		FEM			Fuzzy logic			Error
Material thickness (mm)	Clearance (%*t*)	Die-roll (mm)	Smooth cut (mm)	Crack (mm)	Die-roll (mm)	Smooth cut (mm)	Crack (mm)	Average (%)
3	3	0.213	1.3	1.487	0.213	1.298	1.512	0.61%
3	5	0.238	1.24	1.522	0.238	1.234	1.548	0.71%
3	10	0.294	1.1	1.606	0.294	1.094	1.630	0.68%
7	3	0.474	2.45	4.076	0.474	2.467	4.064	0.34%
7	5	0.501	2.39	4.109	0.501	2.403	4.097	0.29%
7	10	0.604	2.26	4.136	0.604	2.274	4.123	0.31%
15	3	0.6	8.39	6.01	0.600	8.377	6.016	0.08%
15	5	0.71	8.09	6.2	0.710	8.058	6.208	0.18%
15	10	0.78	6.93	7.29	0.780	6.857	7.290	0.35%

8 Conclusions

In this research, the FEM simulation results, as validated by laboratory experiments, show that the error in the amount of cut-surface feature compared with those from the experimental results were approximately 2%. As these results, with the same shearing process parameters except for material thickness, the cut-surface features are different, as these results, they showed that the smooth cut on cut-surface feature was increased when the material thickness increased. In addition, the Sugeno fuzzy logic model has been developed to select the optimal die clearance in the shearing process. The model was trained with input and output of FEM simulation data by using Fuzzy Logic in MATLAB software. The model is verified by comparing the fuzzy output with the FEM simulation data. The results of the FEM simulation are in a similarity to those approximated using the fuzzy logic model. It shows that the fuzzy logic model confirmed with the FEM simulation and can be effectively used to select the optimal die clearance in the die-cutting process.

Acknowledgments. This research was funded by College of Industrial Technology, King Mongkut's University of Technology North Bangkok, (Grant No. Res-CIT0271/2021).

References

1. Lange, K.: Handbook of Metal Forming. McGraw-Hill, New York (1985)
2. Schuler, M.: Forming Handbook. Springer, Berlin (1998)
3. Sontamino, A., Thipprakmas, S.: Development of a shaving die design for reducing rollover. Int. J. Adv. Manuf. Technol. **103**(5–8), 1831–1845 (2019). https://doi.org/10.1007/s00170-019-03639-9

4. Thipprakmas, S., Sontamino, A.: Fabrication of clean cut surface on high strength steel using a new shaving die design. J. Mech. Sci. Technol. **34**(1), 301–317 (2020). https://doi.org/10.1007/s12206-019-1231-z
5. Gustafsson, E., Oldenburg, M., Jansson, A.: Experimental study on the effects of clearance and clamping in steel sheet metal shearing. J. Mater. Process. Technol. **229**, 172–180 (2016)
6. Slavic, J., Bolka, S., Bratus, V., Boltezar, M.: A novel laboratory blanking apparatus for the experimental identification of blanking parameters. J. Mater. Process. Technol. **214**, 507–513 (2014)
7. Gustafsson, E., Oldenburg, M., Jansson, A.: Design and validation of a sheet metal shearing experimental procedure. J. Mater. Process. Technol. **214**, 2468–2477 (2014)
8. Thipprakmas, S.: Application of Taguchi technique to investigation of geometry and position of V-ring indenter in fine-blanking process. Mater. Des. **31**(5), 2496–2500 (2010)
9. Zadeh, L.A.: Fuzzy sets. Inf. Control **8**(3), 338–353 (1965)
10. Rao, A.V.S., Pratihar, D.K.: Fuzzy logic-based expert system to predict the results of finite element analysis. Knowl.-Based Syst. **20**(1), 37–50 (2007)
11. Sontamino, A., Thipprakmas, S.: Cut surface features in various die-cutting process. Key Eng. Mater. **719**, 127–131 (2016)
12. Syn, C.Z., Mokhtar, M., Feng, C.J., Manurung, Y.H.: Approach to prediction of laser cutting quality by employing fuzzy expert system. Expert Syst. Appl. **38**(6), 7558–7568 (2011)

Hydrophobic Coating of Plywood Using Natural Oils and Waxes

Ronald Wu Chong Mook[1], Seri Rahayu Ya'akub[1], Fethma M. Nor[2,3], and Denni Kurniawan[1,4](✉)

[1] Mechanical Engineering Program Area, Universiti Teknologi Brunei, Gadong 1410, Brunei Darussalam
denni.kurniawan@utb.edu.bn
[2] Department of Mechanical Engineering, Curtin University, 98009 Miri, Malaysia
[3] School of Design, Universiti Teknologi Brunei, Gadong 1410, Brunei Darussalam
[4] Center for Research on Agri-Food Science and Technology, Universiti Teknologi Brunei, Gadong 1410, Brunei Darussalam

Abstract. Protection against moisture on wood can be done by inducing hydrophobicity through coatings. Plant based oils and natural waxes have the potential for this purpose. This study intends to identify the feasibility of using plant-based oils and natural waxes to induce hydrophobicity on a plywood sample. The plant-based oils are tung oil and linseed oil and the natural waxes are beeswax and carnauba wax which were applied on the plywood samples by dip coating. Combinations of the natural oils and waxes were also applied. Contact angle, surface roughness, and sliding angle are the measured properties. By taking into account the criteria for contact angle, surface roughness, and sliding angle, all the coatings induced hydrophobicity to the plywood except linseed oil. The highest contact angle of 110° was attributed to tung oil and carnauba wax combination. The combinations of oil and wax appears to be the most effective in inducing hydrophobicity compared to stand-alone oil and wax.

Keywords: Coating · Wood · Hydrophobic · Plant oil · Wax

1 Introduction

Wood has wide range of applications in the construction industry. Wood is mechanically robust, lightweight, has good coefficient of thermal conductivity, dimensionally stable relative to temperature, and noise absorbing [1]. From sustainability point of view, wood is renewable and CO_2 neutral [2]. However, continuous exposure to water can restrict wood use by undesirably changing its properties and appearance, increasing its maintenance costs and reducing its usable lifetime. Rooting from its hydrophilicity, wood has severe interaction with water, which is linked to a number of processes, including fungi growth, wood surface erosion, and unwanted dimensional changes as it gains and loses moisture [3].

With respect to dimensional stability of wood, the most effective way is to reduce the rate of water uptake of wood is by rendering them hydrophobic, i.e., water repelling. This

A. Batako et al. (Eds.): GCMM 2021, LNNS 335, pp. 169–179, 2022.
https://doi.org/10.1007/978-3-030-90532-3_14

way, water cannot penetrate the wood since its capillaries are partly blocked, limiting the rate of water absorption [4]. Coatings are applied towards making wood hydrophobic. Silane based coatings [5, 6], acrylic, polyurethane, and alkyd-based coatings are some of chemical-based coatings for this purpose [7]. Recent studies even introduced nanoparticles (e.g., carbon and alumina based) to create superhydrophobic coatings on wood through the sol-gel process, direct growth, layer by layer deposition, or as coating with nanoparticle suspensions [8–10]. For sustainability, natural coatings are of interest in this study.

Plant based oils and natural waxes have been long known as hydrophobic coatings of wood [11]. Plant based oils form an elastic protective film on wood surface through oxidation and polymerisation. These oils have triglycerides of saturated and unsaturated fatty acids. Once they are applied on wood surface and exposed to the air, the unsaturated parts of the fatty acid will be oxidised by the oxygen containing atmosphere, regardless of the temperature. The oxidation process is preceding a radical polymerisation which causes cross-linking reactions and the elastic film is resulted [12]. Examples of these plant-based oils are tung and linseed oils. Tung oil is extracted by pressing the nut's seed of a tung tree (*Vernicia fordii* and *Vernicia montana*). Tung oil contains predominantly α-eleostearic acid in excess of 80% [13]. Tung oil dries more rapidly compared to other plant-based oils. Even without the additional drying agents, tung oil can form touch-dry films within few days [14]. It provides water, alkali, and chemical resistance, making it valuable in modern varnish manufacture [15]. Linseed oil is also widely used in the paint and varnish. The oil is obtained from the seed of flax plants. Linseed oil is among of the most unsaturated plant-based oils. During application, linseed oil works by blocking the lumen of the wood and upon entering the interior of the wood, it forms a film on pore surfaces [16].

Natural waxes are also commonly used as hydrophobic coating for wood or as an additive for other coatings to preserve the glossy appearance of wood [17]. Waxes consist of a complex blend of long-chained lipophilic compounds, are soluble in water and in organic solvents, and are in low viscosity upon melting [18]. Two waxes of interest in this study are beeswax and carnauba wax. Beeswax is produced by *Apis mellifera*, the western honeybee. It contains compounds including long-chain alkanes, esters, polyesters, hydroxy esters, and acids [19, 20]. Carnauba wax is extracted from the leaves of carnauba (*Copernicia prunifera*). It is among the hardest natural waxes. Carnauba wax composition includes esters, hydroxy fatty acids, straight-chain primary alcohols [21].

These bio-based hydrophobic coatings can be used individually as well as combined with one another, i.e., oil mixtures, wax mixtures, or plant oil and wax mixtures. Mixing of tung oil and linseed oil was done on beech and oak woods was carried out and the mixture shows better hydrophobicity to the woods compared to neat tung oil and linseed oil [22]. Mixture of tung oil and beeswax was applied on spruce wood [23]. It was found that the wood becomes more hydrophobic when using the mixture compared to neat oil or wax [23].

For tung oil, linseed oil, beeswax, and carnauba wax, their mixtures have not been completely studied previously. Summary on the natural coating materials and the studies which address the contact angle, sliding angle, and surface roughness when they are applied on wood surface is mentioned in Table 1. Hence, this study aims at identifying the hydrophobicity effect when wood sample is applied by mixtures of those plant-based oils and natural waxes.

Table 1. Summary of studies on natural coating materials Measured contact angle, sliding angle, and surface roughness of plywood samples coated with. Positive (+) sign means a positive impact is made for that particular measure, negative (–) sign means a negative impact is made for that particular measure, N/A means no available data is present for that particular measure.

Natural coating materials	Contact angle	Sliding angle	Surface roughness	References
Tung oil	+	+	N/A	[16, 18, 22]
Linseed oil	–	–	N/A	[18, 24]
Tung oil + Linseed oil	+	N/A	N/A	[22]
Beeswax	+	–	+	[23]
Carnauba wax	+	–	+	[21, 24]
Tung oil + Beeswax	+	–	+	[23]

2 Methodology

The wood sample used in this study is locally sourced plywood available from hardware store. The plywood samples were received as is (no thickness adjustment was done), and cut in dimensions of 50 mm × 40 mm. The coating materials used were tung oil (Craftiviti, Malaysia), linseed oil (Arto's, Malaysia), beeswax (Cangzhou Bee Industry, China), and carnauba wax (Kosten Keunen, Brazil).

Tung oil (TO) and linseed oil (LO) were used as is since they are already at liquid state albeit being viscous (Fig. 1). A mixture of tung oil and linseed oil (TO/LO) was processed by placing a 1:1 ratio of tung oil to linseed oil in a beaker and manually stirring to ensure proper mixing. The waxes were first melted inside individual beaker using a heating mantle at 65 °C for beeswax (BW) and 85 °C for carnauba wax (CW). Image of the waxes is shown in Fig. 2. As for their mixture, a 1:1 ratio of beeswax to carnauba wax (BW/CW) were placed in a beaker and heated at 85 °C. The mixture was then manually stirred to ensure all components are properly mixed. For the combinations of plant oil and natural wax, the wax was first placed in a separate beaker to be melted. After the wax is molten, the plant oil of equal ratio was added and manually stirred.

To apply the coatings, the plywood sample was dipped into each prepared coating solutions for 10 s and then air dried at room temperature for up to five days until dry to touch. It was noted that the wax dried within minutes as it cools down to room temperature.

The hydrophobicity of the coated wood samples was tested through measurement of contact angle, sliding angle, and surface roughness. The contact angle (CA) was measured using an in-house system which consists of a smartphone with a macro lens

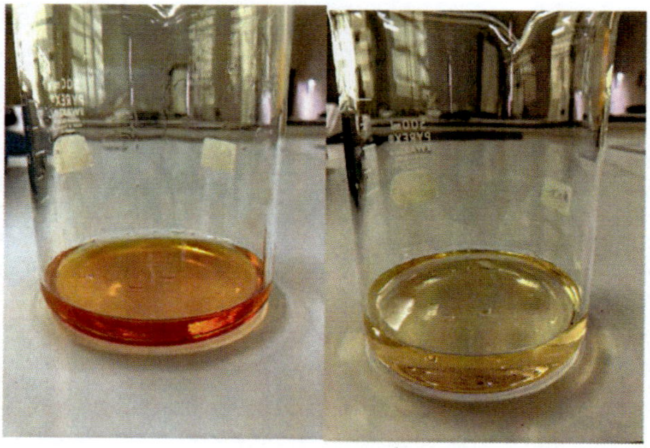

a b

Fig. 1. Images of the (a) tung oil and (b) linseed oil

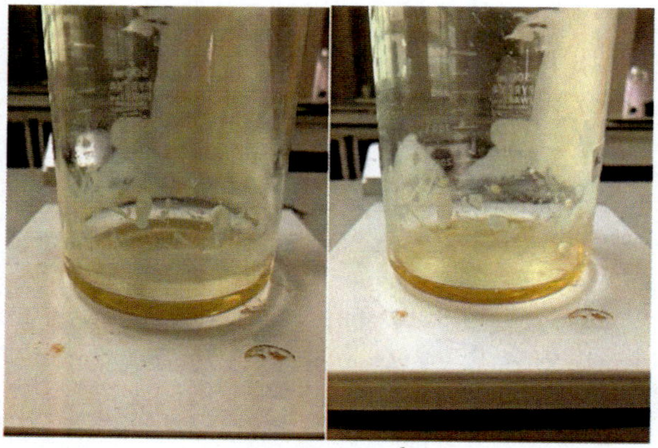

a b

Fig. 2. Images of the (a) beeswax and (b) carnauba wax

attachment (Fig. 3) to capture the image of water droplet on the sample and an open-source image editing software (ImageJ) to calculate the contact angle from the captured images.

The functionality of the in-house system was tested against an established contact angle system over a range of angles from 24° to 135°. The in-house system gets results within ±2° of those acquired from the same droplet of distilled water on a glass surface using the established system [24]. To measure the contact angle, a water droplet was dispensed onto each wood sample with a pipette. The process was video captured using the smartphone at a distance of 1–2 cm away. The frame in which the water droplet was most prominent on the wood surface was then captured. This image was then uploaded into the image analysis software (ImageJ) to measure the water droplet's static water contact angle. Each sample had three static contact angle measurements taken at random positions along the wood surface, with the average taken.

a b

Fig. 3. In house system for contact angle and sliding angle measurement which consists of (a) smartphone and (b) macro lens attachment

For sliding angle (SA) measurements, droplets of equal volume were dispensed onto either ends of the coated wood surface with a pipette. The sample is then slowly tilted upwards manually by hand until it starts to slide off towards the opposite end of the wood surface. A video was taken for the entire process. Sliding angle was determined by the angle that causes the water droplet to start sliding off the surface. An average of three measurements were then taken for each sample.

Surface roughness measurement was performed using a stationary surface measuring station (Marsurf GD20, Mahr Inc., Providence, RI, USA). Cut-off length of 17.5 mm was used to generate the surface profile and calculated the arithmetic roughness, Ra. The results were the average of three Ra measurements for each sample.

3 Results and Discussion

Coating using plant-based oils and natural waxes as well as their mixtures has been applied on the plywood samples. There are a total of ten designations for different coatings on the samples and an untreated plywood sample as comparison. Table 2 shows the overall results.

For validation, contact angle measurement for tung oil and linseed oil was compared with previous work on the same oils. For tung oil (TO) and linseed oil (LO) coated plywood, an average contact angle of 91.3° and 81.1° was calculated, respectively. Image of the latter is shown in Fig. 4. These values are in line with the readings previously obtained [22], in which the contact angle for tung oil was shown to be more than 90° and for linseed oil it was less than 90°. As for their combination, i.e., tung oil and linseed oil (TO/LO) coated plywood, the contact angle readings were approximately 97° [22] whereas ours was 94.3°. Our plywood coated with carnauba wax (CW) has been calculated to have a contact angle of 99.1°, while previous experiment [25] on wood with carnauba wax has also shown contact angles of approximately 100°. Using these comparisons, it is safe to consider our measured contact angle values are valid.

The surface roughness of the uncoated plywood was measured to be Ra = 20.53 μm. The surface roughness of plywood subjected to coatings decreased to Ra of below 15 μm.

Table 2. Measured contact angle, sliding angle, and surface roughness of plywood samples coated with.

Coating designation	Surface roughness, Ra (μm)	Contact angle (◦)	Sliding angle (◦)
Uncoated	20.53	54.6 ± 2.4	41.3 ± 0.9
TO	6.66	91.3 ± 0.9	36.2 ± 1.8
LO	5.61	81.1 ± 1.5	38.2 ± 1.4
TO/LO	6.86	94.3 ± 1.0	37.2 ± 1.4
BW	14.67	96.3 ± 1.3	33.0 ± 0.9
CW	11.83	99.1 ± 0.9	39.0 ± 1.7
BW/CW	7.86	100.8 ± 0.9	36.9 ± 1.6
TO/BW	1.97	104.7 ± 1.5	34.2 ± 1.6
TO/CW	9.28	109.8 ± 1.3	33.6 ± 1.2
LO/BW	1.66	102.2 ± 1.4	35.1 ± 1.3
LO/CW	11.00	103.2 ± 1.6	24.8 ± 1.8

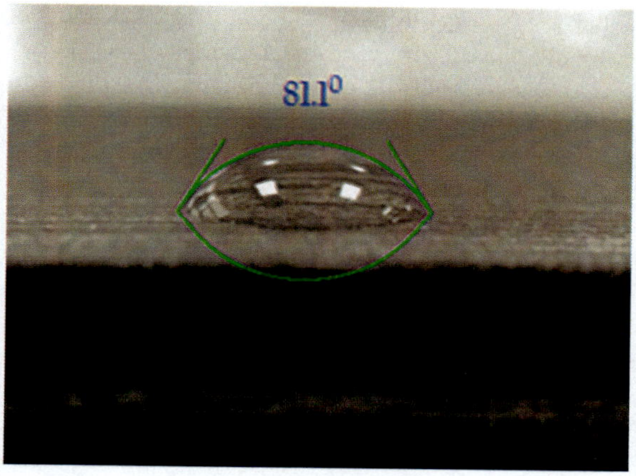

Fig. 4. One of the contact angle measurements of the linseed oil coated on plywood surface.

This points towards a decrease in initial roughness of the wood surface due to the application of coating. This indicates that the coating fills some peaks and valleys on the plywood surface, making the finer surface. This is preferred since this means the coatings were evenly distributed on the sample surface.

Surface roughness can influence hydrophobicity as a secondary factor. A surface with exceptional hydrophobicity could be created by combining hierarchical surface roughness with a coating material with low surface energy [25]. Surface roughness does not significantly transform the nature of the surface, in which changing surface roughness itself does not convert a hydrophilic surface into a hydrophobic surface, or vice versa. Surface roughness, however, has a significant impact on the degree of hydrophobicity or hydrophilicity of a surface [26]. Like in this case, the uncoated wood has surface chemical structure which is hydrophilic in nature. Changing its surface roughness alone would not cause it to be hydrophobic.

It could also be observed that the oils mixed with carnauba wax have a higher surface roughness Ra compared to oils mixed with beeswax, which hints at the apparent rougher surface provided by carnauba wax. This could be contributed to carnauba wax having better emulsification properties and oil-binding capacity, which causes carnauba wax particles within the wax film to increase surface roughness [27].

The plywood sample in its uncoated form is hydrophilic according to its measured contact angle of 54.6°. From Table 1, amongst the plant oils, tung oil (TO) is seen to have a contact angle greater than 90°, hinting at its hydrophobicity. Whereas linseed oil (LO) seems to be hydrophilic (CA < 90°). This difference in contact angle could be attributed to the slightly greater surface roughness Ra of tung oil compared to linseed oil. Overall, tung oil showed acceptable contact angles shortly after its drying. Meanwhile, linseed oil remained hydrophilic even after drying, hinting at incomplete oil polymerisation [22]. For natural waxes, the contact angle of both beeswax (BW) and carnauba wax (CW) are almost within range of each other.

The combination of plant oils, i.e., tung oil and linseed oil (TO/LO) has a contact angle of 94.3°, which fulfils the contact angle criterion for hydrophobicity (CA > 90°). This is also true for combination of natural waxes, i.e., beeswax and carnauba wax (BW/CW) which has a contact angle of 100.8°. For combination between oil and wax, the coating that has the highest contact angle is tung oil and carnauba wax combination (TO/CW) at 109.8° (Fig. 5). The combination of tung oil and beeswax (TO/BW) has a moderate contact angle at 104.7°, with linseed oil and beeswax combination (LO/BW), and linseed oil and carnauba wax combination (LO/CW) following closely at 102.2° and 103.2°, respectively.

Mobility of a droplet on a surface is measured through sliding angle measurement. The sliding angle, is the inclination when water droplets completely roll off the surface without external force. A low sliding angle is crucial in preventing water from adhering to the surface.

The sliding angle of uncoated plywood was measured to be 41.3°, one of the images is shown in Fig. 6, which was higher than those of coated plywood. Since a low sliding angle is desired, the benchmark for sliding angles of coated plywood should be less than that of the hydrophilic uncoated plywood, i.e., SA < 41.3°.

Neat beeswax has the lowest sliding angle when compared to tung oil, linseed oil, and carnauba wax. This means that water droplets are able to slide off more easily on beeswax, which could be attributed to beeswax having the higher surface roughness, signifying the link between high Ra and low sliding angle. For coating mixtures, all coated plywood samples have values that are well below than that of uncoated plywood which fulfills the criterion stated for a low sliding angle. LO/CW has the smallest value of 24.8° followed by TO/CW at 33.6°. Whereas TO/BW, and LO/BW has sliding angles of 34.2° and 35.1°, respectively. Image of sliding angle measurement on LO/CW is shown in Fig. 7.

Since a low sliding angle is crucial in preventing water from adhering to the surface, LO/CW is the most promising. Since all coated samples present a sliding angle that is

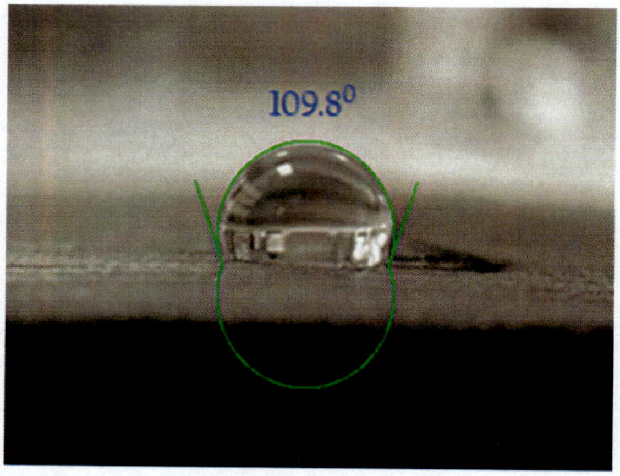

Fig. 5. One of the contact angle measurements of the combination between tung oil and carnauba wax (TO/CW) coated on plywood surface.

Fig. 6. One of the sliding angle measurements of the uncoated plywood surface.

less than the stated criterion (SA < 41.3°), it is safe to say the coating was successful in preventing adhesion of water droplet on the surface. That said, when considering both contact angle and sliding angle criteria, the combination of plant oils and wax showed more appealing values compared to the rest of the coating designations. In addition, the surface roughness Ra of natural oil and wax combination coated plywood were also well below that of uncoated plywood. This shows that the rougher, uncoated hydrophilic wood surface was successfully layered with a hydrophobic film which was made by the natural-based, coating materials.

Slight improvement in hydrophobicity of the plywood surface was achieved when using mixtures of both oil and wax. Out of all the coating designations used in this study, tung oil and carnauba wax combination (TO/CW) seems to be the most effective

Fig. 7. One of the sliding angle measurements of the combination between linseed oil and carnauba wax (LO/CW) coated on plywood surface.

across all three variables, especially in terms of contact angle and surface roughness. This is suggested to be due to the synergistic actions by the oil which can better penetrate the wood surface to a certain degree and by the wax which can better provide surface hydrophobicity on the wood surface [23].

Hierarchical surface roughness paired with a low surface energy of the coating material would result in a surface with hydrophobicity. The combination of naturally rough wood and a low-surface-energy coating was found to affect a wooden surface's overall hydrophobicity, rendering it resistant to liquid water. By taking into account the criteria for contact angle, surface roughness, and sliding angle, all coating designations were effective in providing hydrophobicity to the surface of plywood, except linseed oil. Therefore, using natural based coating materials, i.e., plant oils and natural waxes, as well as their combinations to induce hydrophobicity on wood surface is definitely feasible.

4 Conclusion

In the present work, the feasibility of using natural materials of plant-based oils (tung oil and linseed oil), natural waxes (carnauba wax and beeswax) and their combinations (oil-oil, wax-wax, and oil-wax mixtures) as hydrophobic coating on plywood surface was studied. The application of said materials as coating on samples of commercially available plywood was done using the dip-coating method.

Contact angle analysis on the surface of the wood samples indicated that all plant-based coating materials, except linseed oil were able to successfully provide an acceptable contact angle of more than 90°, showing their capability in the hydrophobisation of wood surface. The coated samples were also shown to have appropriate surface roughness that is well below 20.53 μm (surface roughness of the uncoated samples). This shows that the rougher uncoated hydrophilic wood surface was successful layered with a hydrophobic film made by the coating materials, thereby reducing its roughness. Sliding angle measured through the tilting method has also shown promising values below that of the uncoated wood samples, preventing the surface adhesion of water droplets.

Of all the materials used in this study, tung oil and carnauba wax combination (TO/CW) seems to be the most effective across all three variables, especially in terms of contact angle and surface roughness. This is suggested to be due to the synergy between penetrating action by the oil and surface hydrophobisation by the wax on the plywood surface.

Acknowledgement. The authors gratefully acknowledge the funding by Universiti Teknologi Brunei, Brunei Darussalam. The presentation of this work is made possible with the sponsorship by Sarawak Convention Bureau, Malaysia, to which the authors really appreciate.

References

1. Pajchrowski, G., Noskowiak, A., Lewandowska, A., Strykowski, W.: Wood as a building material in the light of environmental assessment of full life cycle of four buildings. Constr. Build. Mater. **52**, 428–436 (2014)
2. Goverse, T., Hekkert, M.P., Groenewegen, P., Worrell, E., Smiths, R.E.H.: Wood innovation in the residential construction sector; opportunities and constraints. Resour. Conserv. Recy. **34**, 53–74 (2001)
3. Tiemann, H.: Effect of Moisture upon the Strength and Stiffness of Wood. U.S Department of Agriculture (1906)
4. Banks, W.B., Voulgaridis, E.: The performance of water repellents in the control of moisture absorption by wood exposed to the water. In: Records of Annual Convention, pp. 43–45. British Wood Preservation Association (1980)
5. Shang, H.M., Wang, Y., Limmer, S.J., Chou, T.P., Takahashi, K., Cao, G.Z.: Optically transparent superhydrophobic silica-based films. Thin Solid Films **472**, 37–43 (2005)
6. Mahltig, B., Bottcher, H.: Modified silica sol coatings for water-repellent textiles. J. Sol-Gel. Sci. Technol. **27**, 43–52 (2003)
7. Checco, A., Rahman, A., Black, C.T.: Robust superhydrophobicity in large-area nanostructured surfaces defined by block-copolymer self-assembly. Adv. Mater. **26**, 886–891 (2014)
8. Razavi, S.M.R., et al.: Superhydrophobic surfaces made from naturally derived hydrophobic materials. ACS Sustain. Chem. Eng. **5**, 11362–11370 (2017)
9. Schellenberger, S., et al.: Facing the rain after the phase out: performance evaluation of alternative fluorinated and non-fluorinated durable water repellents for outdoor fabrics. Chemosphere **193**, 675–684 (2018)
10. Zhao, X., Hu, T., Zhang, J.: Superhydrophobic coatings with high repellency to daily consumed liquid foods based on food grade waxes. J. Colloid Interface Sci. **515**, 255–263 (2018)
11. Teaca, C.A., et al.: Natural bio-based produtcs for wood coating and protection against degradation: a review. BioResources **14**, 4873–4901 (2019)
12. Olsson, S. Enhancing UV-Protection of Clear Coated Wood by Utilizing Reactive UV-Absorber and Epoxy Functionalized Soybean Oil. Licentiate thesis, Kungliga Tekniska Hogskolan, Stockholm (2014)
13. Van der Vossen, H.A.M., Mkamilo, G.S. (eds.): Vegetable Oils, Backhuys, Wageningen (2007)

14. Honzicek, J.: Curing of air-drying paints: a critical review. Ind. Eng. Chem. Res. **58**, 12485–12505 (2019)
15. Karak, N.: Vegetable oils and their derivative. In: Karak, N. (ed.) Vegetable Oil Based Polymers Properties Processing Applications, pp. 54–95. Woodhead Publishing, Sawston (2012)
16. Ulvcrona, T., Lindberg, H., Bergsten, U.: Impregnation of Norway spruce (*Picea abies L. Karst.*) wood by hydrophobic oil and dispersion patterns in different tissues. Forest. Int. J. Forest Res. **79,** 123–134 (2005)
17. Liu, C., Wang, S., Shi, J., Wang, C.: Fabrication of superhydrophobic wood surfaces via a solution-immersion process. Appl. Surf. Sci. **258**, 761–765 (2011)
18. Bulian, F., Graystone, J.A.: Wood Coatings: Theory and Practice. Elsevier, Amsterdam (2020)
19. Naderizadeh, S., et al.: Superhydrophobic coatings from beeswax-in-water emulsions with latent heat storage capability. Adv. Mater. Interfaces **6**, 1801782 (2019)
20. Nemeth, R., Tsalagkas, D., Bak, M.: Effect of soil contact on the modulus of elasticity of beeswax-impregnated wood. BioResources **1**, 1574–1586 (2015)
21. De Freitas, C.A.S., De Sousa, P.H.M., Soares, D.J., Da Silva, J.Y.G., Benjamin, S.R., Guides, M.I.F.: Carnauba wax uses in food - a review. Food Chem. **291**, 38–48 (2019)
22. Arminger, B., Jaxel, J., Bacher, M., Gindl-Altmutter, W., Hansmann, C.: On the drying behavior of natural oils used for solid wood finishing. Prog. Organ. Coat. **148,** 105831 (2020)
23. Janesch, J., et al.: Superhydrophobic coatings on wood made of plant oil and natural wax. Prog. Org. Coat. **148**, 105891 (2020)
24. Functionalsurfaces. Composer. Simple Contact Angle Measurement Using a Smartphone. [Video Recording]. Durham University (2015)
25. Gindl, M., Reiterer, A., Sinn, G., Stanzl-Tschegg, S.E.: Effects of surface ageing on wettability, surface chemistry, and adhesion of wood. Holz Als Roh-und Werkst **62**, 273–280 (2004)
26. Ahmed, A.O.M.: Design and fabrication of different textured surfaces with improved hydrophobicity for photovoltaic cells' applications. M.Sc. thesis, King Fahd University of Petroleum and Minerals, Saudi Arabia (2014)
27. Lozhechnikova, A., Vahtikari, K., Hughes, M., Osterberg, M.: Toward energy efficiency through an optimized use of wood: the development of natural hydrophobic coatings that retain moisture-buffering ability. Energy Build **105**, 37–42 (2015)

ANN-FPA Based Modelling and Optimization of Drilling Burrs Using RSM and GA

Nripen Mondal[1], Sudip Mandal[2], Madhab Chandra Mandal[1], Santanu Das[3](✉), and Barun Haldar[4]

[1] Mechanical Engineering Department, Jalpaiguri Government Engineering College, Jalpaiguri 735102, India
[2] Electronics & Communication Engineering Department, Jalpaiguri Government Engineering College, Jalpaiguri 735102, India
[3] Mechanical Engineering Department, Kalyani Government Engineering College, Kalyani 741235, India
sdas.me@gmail.com
[4] Mechanical & Industrial Engineering Department, College of Engineering, Al Imam Mohammad Ibn Saud Islamic University, Riyadh 11432, Saudi Arabia

Abstract. In this work, drilling experiments are performed on an aluminum alloy following the L_{27} orthogonal array of response surface methodology (RSM) to find out suitable drilling parameters to minimize burr height and thickness. Drilling is done using backup support. Two types of predictive models namely, artificial neural networks (ANN), and a combination of ANN with flower pollination algorithm (ANN-FPA) are constructed using the experimental data to predict burr height and burr thickness. The developed ANN-FPA predictive model is found to be more accurate than the only ANN model. Mathematical model of burr height and burr thickness are developed using this experimental input data and ANN-FPA based output data and it is optimized using Genetic Algorithm (GA) and RSM desirability function optimization techniques. Burr height observed is found minimum at a cutting velocity of 10 m/min, feed of 0.08 mm/rev, and depth of hole within backup support of 0.2 mm, whereas burr thickness is the lowest in this work at a cutting velocity of 3.9 m/min, a feed rate of 0.095 mm/rev and depth of hole inside backup support of 0.56 mm. The validation tests are performed at these optimized process conditions and the prediction error is found to be 6.53% using GA and 6.68% using RSM desirability function for measurement of burr height, and 4.70% using GA a well as RSM desirability function in burr thickness measurement. Finally, developed multi-objective functions of burr height and thickness are optimized through GA. These results are useful to implement in manufacturing practice.

Keywords: Drilling · Burr · Burr height · Burr thickness · ANN-FPA · GA

1 Introduction

Burr is a protruding edge or small portion of workpiece material that remains attached to the workpiece after machining processes. Burr formation drastically hampers production

A. Batako et al. (Eds.): GCMM 2021, LNNS 335, pp. 180–195, 2022.
https://doi.org/10.1007/978-3-030-90532-3_15

rate, and incurs extra deburring cost as well as it affects product quality, and causes many more problems [1, 2]. Removal of burr after machining, also called deburring, increases overall processing time as well as manufacturing labour cost. It may increase by 25% of the cost of finished parts [3].

Drilling experiment has been conducted [4] on Ti-6Al-4V plate using 0.4mm twist drill according to L_{27} method by some researchers; they have varied drilling parameters like feed rate, speed, and air pressure, and have observed spindle speed and air pressure to be the most significant factor to produce dimensional accuracy of drill holes. In orthopedic surgery, bone drilling has been performed [5, 6] by some researchers and during bone drilling, huge heat and force generation have been experienced to cause damage to bone cells. They have developed a multi-response objective function that has been optimized using ANOVA and a grey-based fuzzy algorithm. The experiment study of drilling has been conducted under the minimum quantity of lubrication (MQL) conditions. Optimization of process parameters has been performed based on the response surface method (RSM) [7, 8].

To minimize burr height, a drilling experiment has been conducted [9–11] according to the L_{27} orthogonal arrayusing an HSS twist drill on alloy steel. In their work, an artificial neural network (ANN) model has been developed and compared with experimental burr height. The response surface methodology (RSM) based second-order mathematical model has been developed using the experimental input drilling parameters, like feed rate, speed, point angle, drill diameter, and clearance angle to optimize using a genetic algorithm (GA) [12–14]. The RSM based mathematical model has been developed using input parameters during drilling cobalt alloy. Finally, the multi-objective function has been optimized using ANOVA [15–17].

In the micro-milling process,the experiment has been conducted [18–20] under a minimum quantity of lubrication conditions (MQL) on titanium alloy (Ti6Al4V). Speed and feed per teeth are the most significant factor to achieve surface integrity. Experimental investigation on drilling has been conducted on AISI 4140 material using four types of novel tool geometry. They have found that drill geometry type 4 has been superior concerning other tool geometry [21]. Milling operation has been conducted according to the L_{27}orthogonal array. The optimum process parameters have been found out using a genetic algorithm (GA) and particle swarm optimization [22, 23].

Drilling experiments have been carried under MQL condition and process parameters have been optimized using fuzzy logic [24]. To understand the mechanism of the burr formation, some researchers have performed finite element (FE) simulation of drilling using 2-D and 3-D finite element (FE) analysis and the results have been validated experimentally [25–27]. A drilling experimental study has been conducted [28] on AISI 316 L stainless steel according to the Box-Behnken design method. Fiveinput parameters and their three levels have been considered to make a total of 46 experimental runs. Finally, burr height and burr thickness have been optimized using a genetic algorithm (GA). Surface roughness and cutting force model have been developed in milling experimental study. Milling process parameters have been optimized using an adaptive neuro-fuzzy inference system (ANFIS) coupled with the GA and particle swarm optimization (PSO) technique [29, 30]. To minimize burr height and burr thickness, a drilling experiment has been conducted on aluminum alloy by varying speed, feed rate, and point angle. Finally, optimum drilling process parameters have been found out using the flower pollination algorithm (FPA) [31].

Many researchers have worked on this research area, but very few of them have used the flower pollination algorithm in this. In this paper, two types of burr height and burr thickness predictive models utilizing ANN and ANN-FPA has been developed. In the ANN modeling, an error backpropagation training algorithm (EBPTA) is normally used which works on the iterative gradient method to update weight and bias gradually to minimize the error. But, it can stick to the local minima, giving the problem to its convergence. For this reason, authors have used a metaheuristic optimization technique which is used to identify the optimal set of weights and bias values of ANN, so that the ANN model can predict the output with more accuracy, and overcome the problem of sticking at local minima. A flower pollination algorithm has also been used for this purpose. The rest of the paper is structured as follows: Sect. 2 describes the materials and methodology that has been used in this research work. Next, results are described followed by discussion and conclusion.

2 Materials and Methodology

Drilling experiment is performed on aluminum alloy as per Taguchi's L_{27} orthogonal array design method. Through the analytical approach, burr height and burr thickness prediction models are developed; the first model uses an artificial neural network (ANN) and the other one utilizes an artificial neural network (ANN) combined with a flower pollination algorithm (ANN-FPA) based on the experimental data. The obtained results are validated experimentally. Next, individual models of burr height and burr thickness are optimized through RSM optimizer and genetic algorithm (GA) to obtain minimum burr height and burr thickness and the corresponding optimum drilling process parameters within this experimental domain. Finally, the multi-objective function of burr height and thickness has been optimized through RSM optimizer and genetic algorithm (GA) approach. The methodology adopted in this work is shown in the Fig. 1.

2.1 Experimental Setup and Materials Details

The experimental work is carried out on a Radial drilling Machine (Manufacturer: Pathak Industries Pvt. Ltd., Kolkata) and the chemical composition of the workpiece is found to be 0.88% magnesium, 0.61% silicon, 0.25% copper, 0.25% zinc, 0.38% manganese and remaining aluminum. Aluminum alloy of size 160 mm × 60 mm × 12 mm is chosen as workpiece material. The experimental work has been performed in an HSS uncoated twist drill of 10 mm diameter. The burr height and burr thickness have been measured using a digital Vernier caliper. The schematic of the experimental setup is shown in Fig. 2.

2.2 Experimental Design

Taguchi's L_{27} orthogonal array design has been used for this experimental work to determine the optimal drilling process parameters to minimize burr height and burr thickness. Different combinations of input process parameters and their corresponding height and thickness of the burr are shown in Table 1.

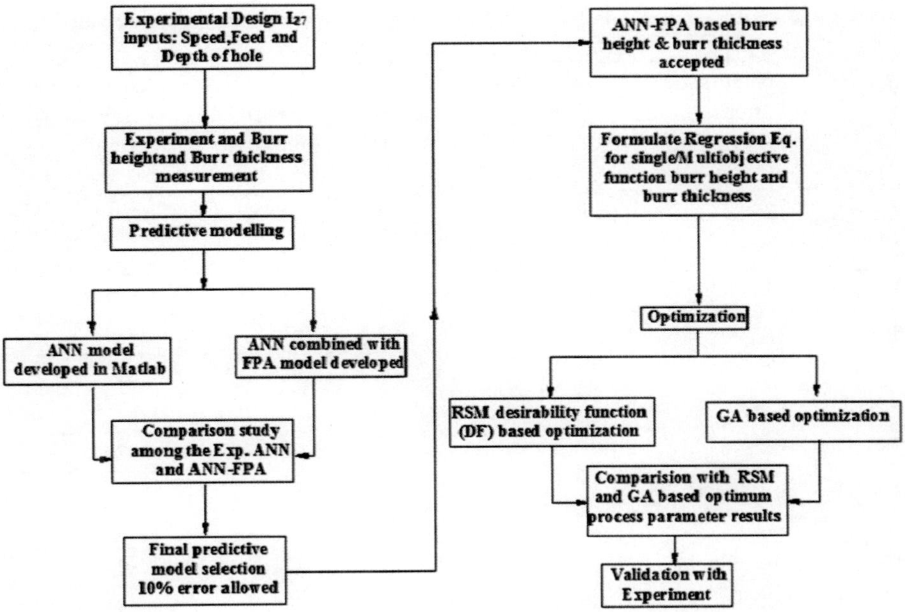

Fig. 1. Proposed methodology for this research work

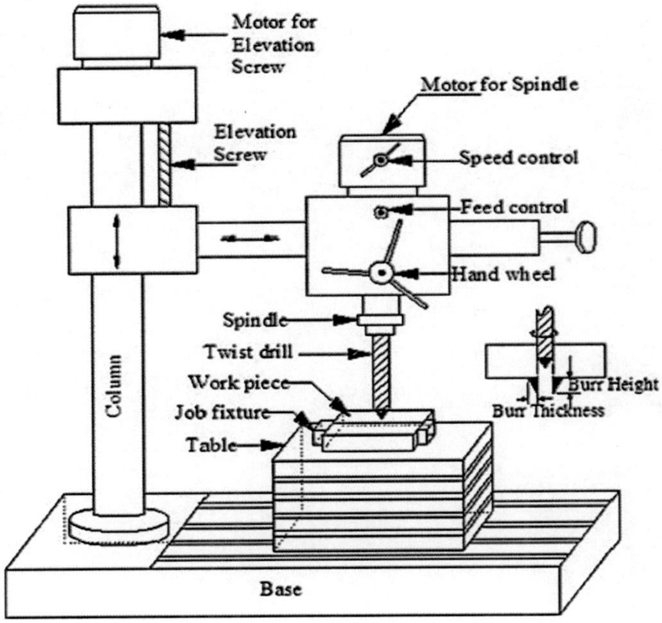

Fig. 2. Schematic of the experimental setup

Table 1. Experimental design using L_{27} orthogonal array and their response value

Exp. run	Spindle speed, S (rpm)	Feed rate, F (mm/min)	Depth inside the backup support, D (mm)	Experimental burr height, H (mm)	Experimental burr thickness, T (mm)
1	125	0.08	0.2	1.71	0.94
2	125	0.08	0.25	2.57	0.91
3	125	0.08	0.56	2.11	0.69
4	125	0.4	0.2	2.61	0.88
5	125	0.4	0.25	2.98	0.87
6	125	0.4	0.56	2.77	0.62
7	125	1.2	0.2	1.91	0.82
8	125	1.2	0.25	2.93	0.71
9	125	1.2	0.56	3.01	0.58
10	200	0.08	0.2	2.11	0.89
11	200	0.08	0.25	2.77	0.95
12	200	0.08	0.56	3.38	0.76
13	200	0.4	0.2	2.23	0.81
14	200	0.4	0.25	3.12	0.72
15	200	0.4	0.56	3.67	0.74
16	200	1.2	0.2	2.13	0.78
17	200	1.2	0.25	3.13	0.82
18	200	1.2	0.56	3.98	0.78
19	320	0.08	0.2	0.79	0.72
20	320	0.08	0.25	1.53	0.79
21	320	0.08	0.56	2.49	0.89
22	320	0.4	0.2	1.56	0.69
23	320	0.4	0.25	2.82	0.77
24	320	0.4	0.56	4.01	0.89
25	320	1.2	0.2	1.34	0.78
26	320	1.2	0.25	2.91	0.85
27	320	1.2	0.56	4.81	0.93

2.3 Artificial Neural Network (ANN) Modelling

The behavior of an artificial neural network (ANN) is similar to the human brain which is able to learn from the natural environment. Similarly, ANN is highly capable of modeling non-linear and bounded algebraic function whose value depends upon the weight and bias factor. For the present work, the Levenberg-Marquardt [32] error backpropagation training algorithm (EBPTA) is used for the training purpose which is widely used for training. The ANN structure for the present case is shown in Fig. 3.

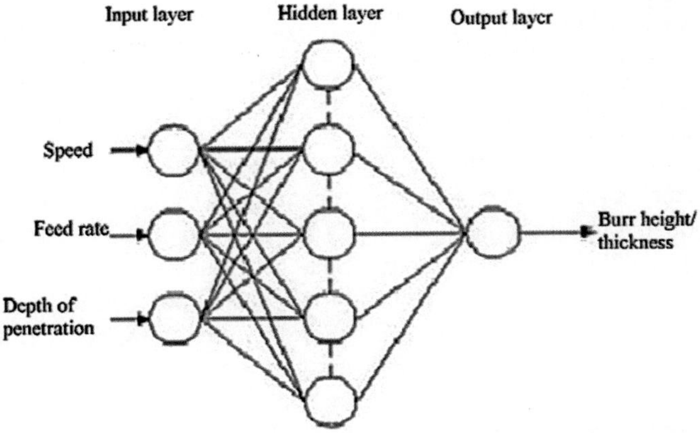

Fig. 3. Multilayer feed-forward neural network structure for the present study

In the present investigation, the ANN model for burr height and thickness consists of three connected layers, each layer consisting of one or more number of the neuron(s). According to More [32], one hidden layer is sufficient for the optimum structure of an ANN model for this problem. However, the numbers of neurons in the hidden layer are chosen as 10 which are sufficient for satisfactory accuracy for the present problem of research. All input and output data are normalized within a range of 0 to 1. The 80% data have been used for training purposes, and 10% data for testing and validation of 27 experimental datasets respectively. ANN regression plots for burr height and burr thickness are shown in Fig. 4.

2.4 Artificial Neural Network-Flower Pollination Algorithm (ANN-FPA) Hybrid

The conventional algorithm such as the EBPTA algorithm work based on the iterative gradient method to update the values of weights and bias gradually along with iteration by minimizing error. However, the conventional Back Propagation Algorithm has a major disadvantage that it can stick to local minima. Hence, in spite of using a conventional training algorithm, authors have used a metaheuristic optimization technique which is used to identify the optimal set of weights and bias values of ANN so that the ANN model is able to predict the output with more accuracy and overcomes the problem of sticking at local minima. In this problem, authors have used Flower Pollination Algorithm (FPA) [33, 34] for the optimization of ANN model weights and bias. Among the popular

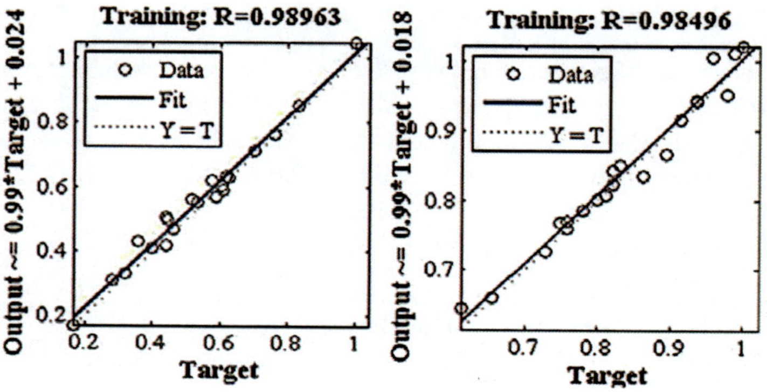

Fig. 4. Training performance with developed ANN predicted burr height and burr thickness

metaheuristic techniques, FPA provides better convergence and accuracy [34, 35]. FPA is inspired by pollen transfer for the flowering of plants or reproduction. Insects, birds, and bats which act as pollinators are mainly responsible for such transfer. This metaheuristic optimization method is based on some simplified rules of pollination.

Therefore, in this proposed ANN-FPA hybrid methodology, ANN is trained using FPA. For an ANN structure as shown in Fig. 3, there are $3 \times 10 = 30$ weight parameters between input and hidden layers, $10 \times 1 = 10$ bias for the neuron of the hidden layer, $10 \times 1 = 10$ weights between the hidden layer and output node, and $1 \times 1 = 1$ bias for the output neuron. So, a total of 51 numbers of weights and bias values are needed to be identified or optimized. The regression plot corresponding to ANN-FPA predicted burr height and burr thickness training performance is shown in Fig. 5.

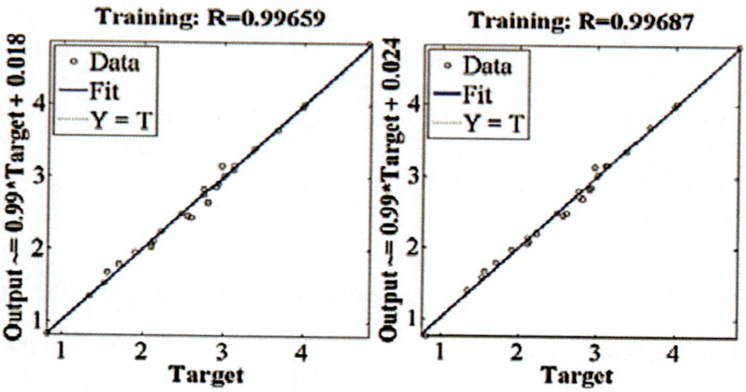

Fig. 5. Training performance of ANN-FPA predicted burr height (left) and burr thickness (right)

2.5 Response Surface Methodology Based Model

The response surface methodology (RSM) is the statistical and mathematical technique which helps to develop the mathematical model as well as analyzing and optimizing various problems. The RSM involved relationship between the output and the input variables is mapped to attain the objective of maximizing or minimizing the response parameters [36]. The second-order regression equations of height and thickness of burr are given in Eq. (1) and Eq. (2) respectively.

$$\begin{aligned}\text{Burr height} &= -4.04916 + 0.0106345 * \text{S} + 1.43289 * \text{F} + 31.6541 * \text{D} - 4.98955\text{e}^{-5} * \text{S}^2 \\ &+ 0.004054 * S * F + 0.0271067 * S * D - 1.87878 * F^2 + 2.2783 * F * D - 45.3978 * D^2\end{aligned} \quad (1)$$

$$\begin{aligned}\text{Burr thickness} &= 1.22229 - 0.00151693 * \text{S} - 0.49585 * \text{F} - 0.681238 * \text{D} - 1.65812\text{e}^{-6} * \text{S}^2 \\ &+ 0.00089423 * S * F + 0.0057058 * S * D + 0.168341 * F^2 + 0.114072 * F * D - 0.9789 * D^2\end{aligned} \quad (2)$$

where S: Spindle speed, F: Feed rate, D: Depth inside the backup support.

2.6 Optimization Using GA

A genetic algorithm is a population-based metaheuristic technique to solve the optimization (maximization or minimization) problems in engineering and many more fields of research. The RSM based mathematical model of burr height and burr thickness, i.e. Eqs. 1 and 2, are optimized using GA algorithm to achieve minimum burr height and burr thickness, and corresponding appropriate drilling process parameters are evaluated.

Objective Function: RSM based mathematical model of burr height and thickness is optimized to find out Speed (S), Feed rate (F), and Depth inside the backup support (D), lower and upper bounds of input parameters subject to 125 rpm $\leq$ S $\leq$ 320 rpm; 0.08 mm/min $\leq$ F $\leq$ 1.2 mm/min; 0.2 mm $\leq$ D $\leq$ 0.56 rpm.

2.7 Multi-objective Functions Development

In the above analysis, individual burr height and burr thickness models are optimized through RSM and GA analysis, and experimental validation of the obtained optimum process parameters is done. But, in the manufacturing aspect, it is necessary to minimize both the height and thickness of burr at the time and to find out the corresponding optimal drilling process parameters. For this purpose, individual burr height and burr thickness objective functions are combined to form a fitness function according to Taguchi's principle [37], which is defined as a multi-response signal-to-noise ratio. The fitness function is given below.

$$\text{Fitness} = -10\ \log_{10}(\text{H}_\text{f}^2) - 10\ \log_{10}(\text{T}_\text{f}^2) \quad (3)$$

where, H_f^2 and T_f^2 are the burr height and burr thickness predicted by ANN-FPA modeling using Table 2. In this fitness function model, equal importance is given to height as well asthe thickness of the burrs.

3 Results and Discussion

In the present research work, different results of two types of burr height and burr thickness predictive model namely ANN and ANN-FPA comparison have been presented. Among these two predictive models, the ANN-FPA model has more accuracy. Next using L_{27} experimental input data according to Table 1 and the ANN-FPA predicted burr height and burr thickness, individual burr height and burr thickness function have been optimized through RSM optimizer and GA. Then the RSM and GA obtain individual burr height and burr thickness drilling process parameters have been validated in the new experimental results. Finally, individual burr height and burr thickness function combined to form a multi-objective function which is optimized in RSM and GA. Now, the comprehension study of the experimental value, ANN, and ANN-FPA model obtain burr height and burr thickness value, also percentages of error ANN and ANN-FPA models are given in Table 2.

The graphical representation of burr height with the comparison among the experimental, ANN, and ANN-FPA predicted ones are shown in Fig. 6. From Fig. 6, it is detected that the maximum percentage of errors in the ANN and ANN-FPA model-based burr height prediction are 21.46 and 8.01 respectively. For this reason, the ANN-FPA model is accepted for further analysis.

Similarly, burr thickness comparison is shown in Fig. 7 from where it can be observed thatthe maximum percentage of errors in ANN and ANN-FPA model-based burr height prediction are 18.76 and 6.38 respectively.

Now, interaction effects among input variables are discussed one by one. Figure 8(a–c) is the surface plot depicting the effect of various factors on burr height and Fig. 9(a–c)is the surface plot that shows the parametric effect on burr thickness. From Fig. 8(a), it is observed that the minimum value of depth inside backup support corresponds to the lowest burr formed in this work, while Fig. 8(b) shows minimum burr formation under minimum depth inside backup support and maximum speed. From Fig. 8(c), it is seen that at minimum feed and corresponding to the maximum speed, a minimum burr is formed.

Regarding burr thickness analysis, from Fig. 9(a), it is clear that at a lower value of depth inside backup support and middle level of feed rate, minimum burr thickness formed, when from Fig. 9(b), it is observed that higher value of depth inside backup support and corresponding lower value of speed result in minimum burr thickness formed. Fig. 9(c), however, depicts that middle level of feed rate and low value of speed give low burr thickness.

Next, the individual response of burr height and burr thickness function is optimized through RSM and GA approach. After minimization of burr height and thickness, obtained optimal process parameters for burr height and burr thickness are shown in Fig. 10 and Fig. 11 respectively. In the RSM and GA optimization technique, the lower [125 0.08 0.2] and upper [320 1.2 0.56] boundary conditions are chosen. In GA, parameters are taken as population size: 100, cross over fraction: 0.8, mutation rate: 0.1, initial penalty: 10, penalty factor: 100, and total number of generations: 100. From Fig. 10, it has been observed that for both the optimization techniques of RSM and GA, minimum

Table 2. Experiment study, ANN and ANN-FPA model predicted burr height and thickness values and percentage of error

Exp. run	Exp. Burr height H_e (mm)	ANN-predicted H_a (mm)	ANN-FPA predicted H_f (mm)	Exp. Burr thickness T_e (mm)	ANN-predicted T_a (mm)	ANN-FPA predicted T_f (mm)	% of error in-ANN H_a value	% of in-ANN-FPA H_f value	% of error in-ANN T_a value	% of in-ANN-FPA T_f value
1	1.71	2.077	1.798	0.94	0.962	0.935	−21.46	−5.14	−2.34	0.53
2	2.57	2.651	2.462	0.91	0.956	0.904	−3.15	4.20	−5.05	0.65
3	2.11	2.436	2.081	0.69	0.691	0.695	−15.45	1.37	−0.14	−0.72
4	2.61	2.39	2.448	0.88	0.798	0.895	8.43	6.20	9.31	−1.70
5	2.98	3.037	3.161	0.87	0.871	0.862	−1.91	−6.07	−0.11	0.91
6	2.77	2.982	2.767	0.62	0.625	0.617	−7.65	0.10	−0.80	0.48
7	1.91	1.968	1.956	0.82	0.794	0.819	−3.04	−2.40	3.17	0.12
8	2.93	2.837	2.913	0.71	0.73	0.712	3.17	0.58	−2.81	−0.28
9	3.01	3.013	3.015	0.58	0.611	0.566	−0.09	−0.16	−5.34	2.41
10	2.11	2.005	2.041	0.89	0.897	0.918	4.97	3.27	−0.78	−3.14
11	2.77	2.527	2.838	0.95	0.972	0.914	8.77	−2.45	−2.31	3.78
12	3.38	3.42	3.396	0.76	0.761	0.756	−1.18	−0.47	−0.13	0.52
13	2.23	2.258	2.244	0.81	0.658	0.773	−1.25	−0.62	18.76	4.56
14	3.12	2.864	3.106	0.72	0.731	0.766	8.20	0.44	−1.52	−6.38
15	3.67	3.67	3.651	0.74	0.745	0.741	0	0.51	−0.67	−0.13
16	2.13	2.398	2.113	0.78	0.801	0.785	−12.58	0.79	−2.69	−0.64
17	3.13	3.023	3.164	0.82	0.763	0.815	3.41	−1.08	6.95	0.60
18	3.98	3.936	3.973	0.78	0.695	0.784	1.10	0.17	10.89	−0.51
19	0.79	0.8	0.821	0.72	0.721	0.723	−1.26	−3.92	−0.13	−0.41
20	1.53	1.576	1.547	0.79	0.809	0.791	−3.00	−1.11	−2.40	−0.12
21	2.49	2.697	2.501	0.89	0.896	0.893	−8.31	−0.44	−0.67	−0.33
22	1.56	1.609	1.685	0.69	0.749	0.693	−3.14	-8.01	−8.55	−0.43
23	2.82	2.735	2.651	0.77	0.767	0.763	3.01	5.99	0.38	0.90
24	4.01	4.095	4.021	0.89	0.949	0.888	−2.11	−0.27	−6.62	0.22
25	1.34	1.477	1.357	0.78	0.781	0.764	−10.22	−1.26	−0.12	2.05
26	2.91	2.899	2.869	0.85	0.824	0.854	0.37	1.40	3.05	−0.47
27	4.81	5.019	4.846	0.93	0.906	0.929	−4.34	−0.74	2.58	0.10

burr height is obtained to be 0.737 mm and 0.738 mm respectively, and the corresponding optimum process parameters are obtained at a speed of 320 rpm (cutting velocity of 10 m/min), a feed rate of 1.2 mm/min and depth inside the backup supportof 0.2 mm.

Similarly, for burr thickness minimization, as shown in Fig. 11, using RSM and GA technique, the minimum value of burr thickness is obtained to be 0.565 mm and corresponding process parameters obtained are speed of 125 rpm (cutting velocity of 4 m/min), a feed rate of 0.951 mm/min and depth inside the backup support of 0.56 mm.

Optimal process parameters evaluated through the RSM and GA are validated through some new experimental results. Burr height and burr thickness values observed through those new experiments and that evaluated using RSM and GA models are tabulated in Table 3.

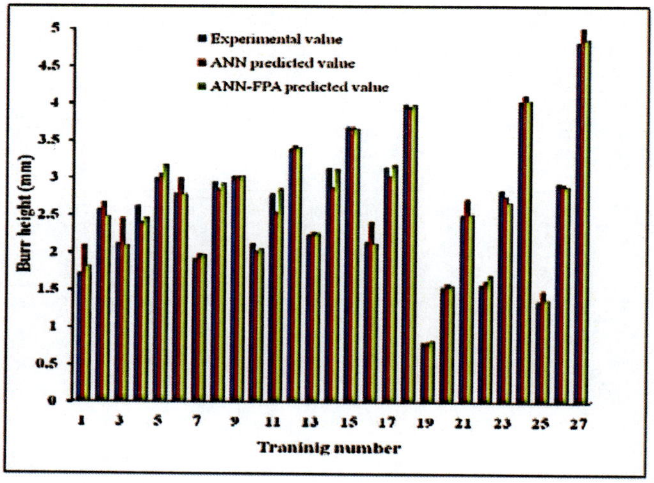

Fig. 6. Comparison among the experiment, ANN and ANN-FPA burr height

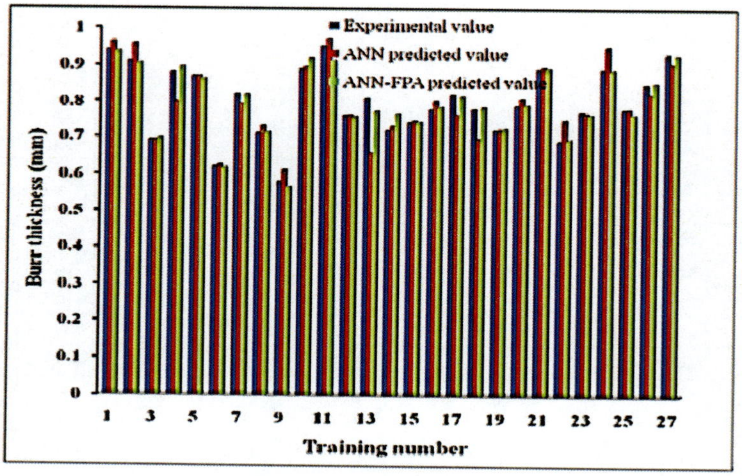

Fig. 7. Comparison among the experimental, ANN and ANN-FPA estimated burr thickness

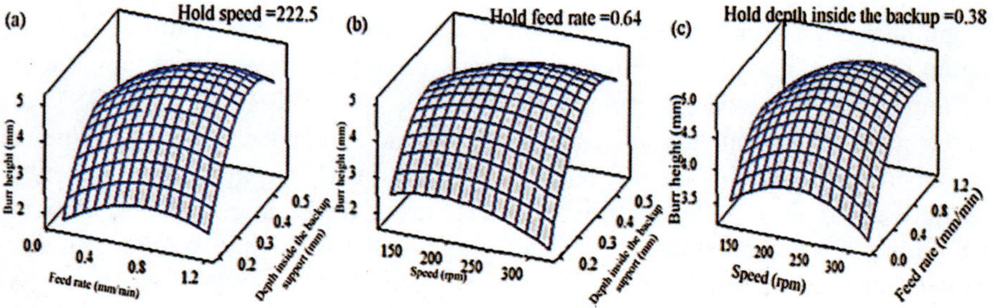

Fig. 8 Interaction effect of input parameters (a) depth inside the backup vs. feed rate (b) depth inside the backup vs. speed (c) feed rate vs. speed on the burr height formation

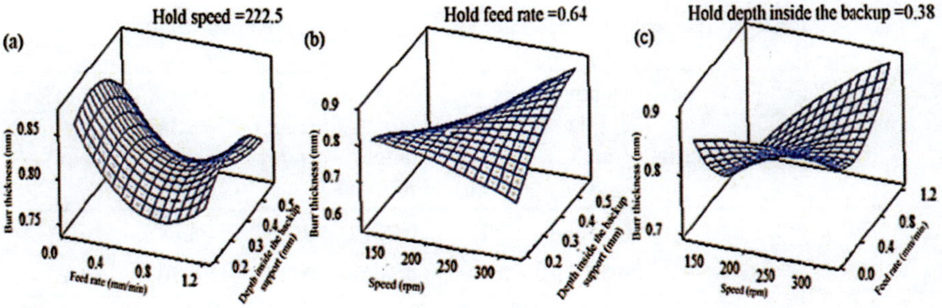

Fig. 9 Interaction effect of input parameters (a) depth inside the backup vs. feed rate (b) depth inside the backup vs. speed (c) feed rate vs. speed on the burr thickness formation

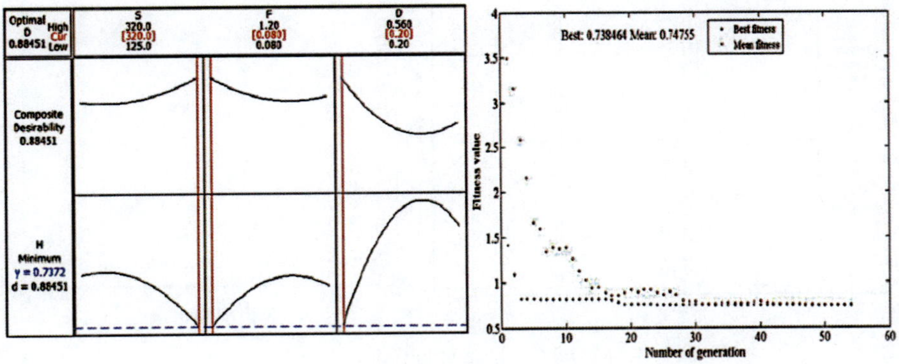

Fig. 10. Response surface and GA optimization in burr height formation

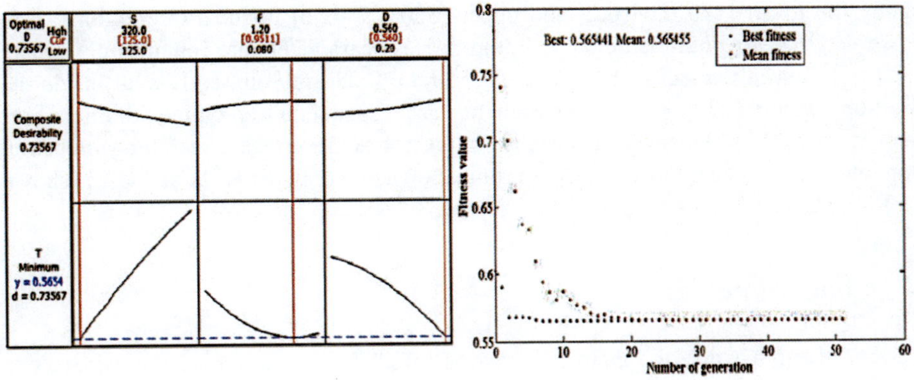

Fig. 11. Response surface and GA optimization in burr thickness formation

Table 3. Experimental validation of RSM and GA obtained results

Test no.	Spindle speed, S (rpm)	Feed rate, F (mm/min)	Depth of cut, D (mm)	Exp. Burr height (mm)	RSM burr height (mm)	GA burr height (mm)	Error in RSM (%)	Error in GA (%)
1	320	1.2	0.2	0.79	0.7372	0.7384	6.68	6.53
2	125	0.95	0.56	0.54	0.5654	0.5654	4.70	4.70

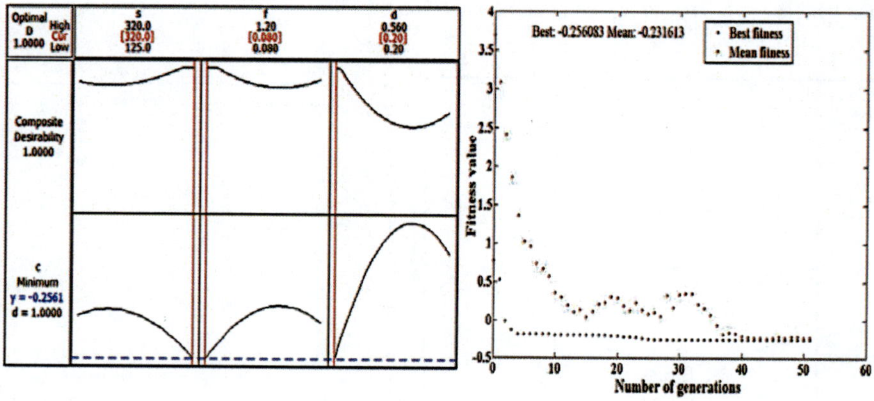

Fig. 12. RSM and GA based multi-objective optimization

The multi-objective fitness function is optimized through RSM and GA techniques using the lower [125 0.08 0.2] and upper [320 1.2 0.56] boundary conditions. It is observed that for both methods, optimum process parameters are found to be- speed: 320 rpm (cutting velocity: 10 m/min), feed rate: 0.08 mm/min, and depth inside the backup support: 0.20 mm. These parameters are responsible to produce minimum burr height as well as thickness. This is an important achievement concerning industrial manufacturing. The multi-objective optimization using both the RSM and GA is shown in Fig. 12.

4 Conclusions

In this research work, the main objective is to obtain optimal process parameters to minimize burr height and burr thickness. The following conclusion can be made from the present work:

- The comparison among the experimental, ANN, and ANN-FPA hybrid models in the burr height and burr thickness prediction is carried out. It is observed that the maximum absolute errors of ANN and ANN-FPA hybrid models are 21.46% and 8.01% respectively in the burr height estimation, while the maximum absolute error of ANN and ANN-FPA hybrid models are 18.76% and 6.38% respectively during the estimation of burr thickness.

- Using RSM optimizer and GA, minimum burr height is found out to be 0.737 mm and 0.738 mm respectively, and corresponding process parameters are obtained with a speed of 320 rpm (10 m/min cutting velocity), feed of 0.08 mm/min, and depth inside the backup support of 0.20 mm.
- Similarly applying RSM optimizer and GA method, minimum burr thickness is observedto be 0.565 mm, and corresponding speed is 125 rpm (cutting velocity: 4 m/min), feed is 0.94 mm/min and depth inside the backup support is 0.56 mm.
- Optimum process parameters obtained using RSM optimizer and GA are validated through some experimental results. It is observed that burr height estimation gives a maximum error of 6.68% and 6.53% respectively. In the burr thickness assessment, the maximum percentage of error becomes 4.7 for both the optimization methods.
- Multi-objective function optimization utilizing both RSM optimizer and GA indicates optimum process parameters to be a speed of 320 rpm (10 m/min cutting velocity), feed of 0.08 mm/min, and depth inside the backup support of 0.20 mm at which both burr height and thickness become minimized within the domain of the present experimentation and may be recommended for practice.

Acknowledgments. This research did not receive any specific grant from funding agencies in the public, commercial, or not-for-profit sectors. There is no conflict of interest related to this work.

References

1. Koelsch, J.: Divining edge quality by reading the burrs. Qual. Mag. **24–28**, 343–348 (2001)
2. Gaitonde, V.N., Karnik, S.R., Achyutha, B.T.: Methodology of Taguchi optimization for multi-objective drilling problem to minimize burr size. Int. J. Adv. Manuf. Technol. **34**, 1–8 (2006)
3. LaRoux, G.K.: Deburring precision miniature parts. Prec. Eng. **1**, 189–198 (1979)
4. Prasanna, J., Karunamoorthy, L., Raman, V.M., Prashanth, S., Chordia, R.D.: Optimization of process parameters of small hole dry drilling in Ti-6Al-4V using Taguchi and grey relational analysis. Measurement **48**, 346–354 (2014)
5. Pandey, R.K., Panda, K.K.: Multi-performance optimization of bone drilling using Taguchi method based on membership function. Measurement **59**, 9–13 (2015)
6. Pandey, R.K., Panda, S.S.: Optimization of bone drilling parameters using grey-based fuzzy algorithm. Measurement **47**, 386–392 (2014)
7. Mia, M.: Mathematical modeling and optimization of MQL assisted end milling characteristics based on RSM and Taguchi method. Measurement **121**, 249–260 (2018)
8. Saha, P.P., Das, S.: Burr minimization in face milling: an edge beveling approach. Proc. IMech Eng. Part B: J. Eng. Manuf. **225**, 1528–1534 (2011)
9. Kundu, S., Das, S., Partha, P.D.: Optimization of drilling parameters to minimize burr by providing back-up support on aluminium alloy. Procedia Eng. **97**, 230–240 (2014)
10. Mondal, N., Sardar, B.S., Halder, R.N., Das, S.: Observation of drilling burr and finding out the condition for minimum burr formation. Int. J. Manuf. Eng. **2014**, 1–12 (2014)

11. Dey, B., Mondal, N., Mondal, S.: Experimental study to minimize the burr formation in drilling process with artificial neural network analysis. IOP Conf. Ser. Mater. Sci. Eng. **377**, 1–6 (2018)
12. Gaitonde, V.N., Karnik, S.R., Achyutha, B.T., Siddeswarappa, B.: Genetic algorithm-based burr size minimization in drilling of AISI 316L stainless steel. J. Mater. Process. Technol. **197**, 225–236 (2008)
13. Mondal, N., Mandal, C.M., Dey, B., Das, S.: Genetic algorithm based drilling burr minimization using ANFIS and SVR. Proc. IMechE. Part B: J. Eng. Manuf. **221**(12), 1–13 (2019)
14. Kilickap, E., Huseyinoglu, M., Yardimeden, A.: Optimization of drilling parameters on surface roughness in drilling of AISI 1045 using response surface methodology and genetic algorithm. Int. J. Adv. Manuf. Technol. **52**(1–4), 79–88 (2011)
15. Warsi, S.S., Agha, M.H., Ahmad, R., Jaffery, S.H.I., Khan, M.: Sustainable turning using multi-objective optimization: a study of Al 6061 T6 at high cutting speeds. Int. J. Adv. Manuf. Technol. **100**(1–4), 843–855 (2018). https://doi.org/10.1007/s00170-018-2759-2
16. Saidi, R., Fathallah, B.B., Mabrouki, T., Belhadi, S., Yallese, A.M.: Modelling and optimization of the turning parameters of cobalt alloy (Stellite 6) based on RSM and desirability function. Int. J. Adv. Manuf. Technol. **100**(9–12), 2945–2968 (2019)
17. Shaomin, L., Deyuan, Z., Daxi, G., Zhenyu, S., Hui, T.: Modeling and drilling parameters optimization onburr height using harmony search algorithm in low-frequency vibration-assisted drilling. Int. J. Adv. Manuf. Technol. **101**(9–12), 2313–2325 (2018)
18. Hassanpour, H., Sadeghi, H.M., Rezaei, H., Rasti, A.: Experimental study of cutting force, microhardness, surface roughness and burr size on micromilling of Ti6AL4V in minimum quantity lubrication. Mater. Manuf. Process. **31**(13), 1654–1662 (2016)
19. Gupta, K.M., Sood, K.P., Sharma, S.V.: Machining parameters optimization of titanium alloy using response surface methodology and particle swarm optimization under minimum quantity lubrication environment. Mater. Manuf. Process. **31**(13), 1671–1682 (2015)
20. Sen, B., Mia, M., Mandal, K.U., Mondal, P.S.: GEP-and ANN-based tool wear monitoring: a virtually sensing predictive platform for MQL-assisted milling of Inconel 690. Int. J. Adv. Manuf. Technol. **105**(1–4), 395–410 (2019)
21. Meral, G., Sarikaya, M., Mia, M., Dilipak, H., Seker, U.: Optimization of hole quality produced by novel drill geometries using the Taguchi S/N approach. Int. J. Adv. Manuf. Technol. **101**(1–4), 339–355 (2019)
22. Ghosh, G., Mandal, P., Mondal, S.C.: Modeling and optimization of surface roughness in keyway milling using ANN, genetic algorithm, and particle swarm optimization. Int. J. Adv. Manuf. Technol. **100**(5–8), 1223–1242 (2017). https://doi.org/10.1007/s00170-017-1417-4
23. Kalita, K., Shivakoti, I., Ghadai, K.R.: Optimizing process parameters for laser beam micro-marking using a genetic algorithm and particle swarm optimization. Mater. Manuf. Process. **32**, 1101–1108 (2017)
24. Nandi, K.A., Davim, P.J.: A study of drilling performances with minimum quantity of lubricant using fuzzy logic rules. Mechatronics **19**, 218–232 (2009)
25. Min, S., Dornfed, D., Kim, J., Shyu, B.: Finite element modelling of burr formation in metal cutting. Mach. Sci. Technol. **5**, 307–322 (2001)
26. Tandon, P., Khan, M.R.: Three dimensional modeling and finite element simulation of a generic end mill. Comput. Aided Des. **41**, 106–114 (2009)
27. Zhang, Z., Babitsky, I.V.: Finite element modelling of a micro-drill and experiments on high speed ultrasonically assisted micro-drilling. J. Sound Vibrat. **330**, 2124–2137 (2011)
28. Gaitonde, V.N., Karnik, S.R., Siddeswarappa, B., Achyutha, B.T.: Integrating Box-Behnken design with genetic algorithm to determine the optimal parametric combination for minimizing burr size in drilling of AISI 316L stainless steel. Int. J. Adv. Manuf. Technol. **37**, 230–240 (2008)

29. Asadi, R., Yeganefar, A., Niknam, A.S.: Optimization and prediction of surface quality and cutting forces in the milling of the aluminium alloys using ANFIS and interval type 2 neuro fuzzy network couple with population-based meta-heuristic learning methods. Int. J. Adv. Manuf. Technol. **105**, 2271–2287 (2019)
30. Gaitonde, V.N., Karnik, S.R.: Minimizing burr size in drilling using artificial neural network (ANN)-particle swarm optimization (PSO) approach. Int. J. Adv. Manuf. Technol. **23**, 1783–1793 (2012)
31. Mondal, N., Mandal, S., Mandal, C.M.: FPA based optimization of drilling burr using regression analysis and ANN model. Measurement **152**, 107327 (2020)
32. More, J.J.: The Levenberg-Marquardt algorithm: implementation and theory. Num. Anal. Lect. Notes Math. **630**, 105–116 (1987)
33. Yang, X.-S.: Flower pollination algorithm for global optimization. In: Durand-Lose, J., Jonoska, N. (eds.) UCNC 2012. LNCS, vol. 7445, pp. 240–249. Springer, Heidelberg (2012). https://doi.org/10.1007/978-3-642-32894-7_27
34. Yang, X.S., Karamanoglu, M., He, X.S.: Flower pollination algorithm: a novel approach for multiobjective optimization. Eng. Optim. **46**(9), 1222–1237 (2014)
35. Ghosh, A., Mandal, S., Nandi, G., Pal, P.K.: Metaheuristic based parametric optimization of TIG Welded joint. Trans. Indian Inst. Met. **2018**, 1–11 (2018)
36. Hashmi, K.H., Zakria, G., Raza, M.B., Khalil, S.: Optimization of process parameters for high speed machining of Ti-6Al-4V using response surface methodology. Int. J. Adv. Manuf. Technol. **85**(5–8), 1847–1856 (2015). https://doi.org/10.1007/s00170-015-8057-3
37. Padke, M.S.: Quality Engineering Using Robust Design. Prentice Hall, Englewood Cliffs (1989)

Mechanical and Physical Properties of Kapok Tree Sawdust Particleboard Manufactured with Three Ecologic Binders

Soviwadan Drovou[1,2,4](✉), Ayaréma Afio[1], Kodjo Attipou[1], Komlan Assogba Kassegne[1], Antonio Pizzi[4,5], and Komlan Sanda[2,3]

[1] National Advanced School of Engineers, University of Lomé, (ENSI – UL), Lomé, Togo
pascal.drovou@gmail.com
[2] Laboratory of Research on Agricultural Resources and Environmental Health, University of Lomé, (LARASE – UL), Lomé, Togo
[3] University of Kara, (UK), Kara, Togo
[4] Laboratory for the Study and Research on Wood Material, University of Lorraine (LERMAB – UL), Épinal, France
[5] Department of Physics, King Abdulaziz University, Jeddah, Saudi Arabia

Abstract. The aim of this work is to develop carpentry residues coming from the kapok tree with tannic powders of African locust bean pod (*Parkia Biglobosa*), peel of Indian tamarind (*Pitecellobium Dulce*) and pearl bone glue, for the manufacture of particle board to solve possible environmental problems due to sawdust and the use of conventional binders. First, particleboards were prepared with the sawdust of kapok tree by varying the rate of binder. Second, the physical and mechanical properties of the particleboard prepared were tested by three – points bending, tensile strength, internal bond (IB) strength, water absorption and thickness swelling. The bending test allowed to calculate the modulus of elasticity (MOE) and the modulus of rupture (MOR). The tensile test allowed to determine Young's modulus (E) the modulus of rupture (MOT), the internal bond (IB) strength. The threshold set by the ANSI A208.1 – 2012 for MOE, MOR and IB standards, has thus been largely attained. The values of density used allow to classify the particleboards of sawdust kapok tree in the categories of medium and high density. This shows that tannic powders can replace conventional binders in the manufacture of particleboard.

Keywords: Particleboard · Sawdust of kapok tree · Tannic powder pod of African locust bean · Indian tamarind · Pearl bone glue · Mechanical and physical properties

1 Introduction

The market for particleboard and particle-based wood residues and agricultural products has been marked by strong growth in the last decade [1]. The importation of fibreboard, particleboard, laminates and veneer sheets have also increased exponentially and the

A. Batako et al. (Eds.): GCMM 2021, LNNS 335, pp. 196–214, 2022.
https://doi.org/10.1007/978-3-030-90532-3_16

demand continues to grow [2]. In developing particleboard, binders such as urea – formaldehyde, phenol formaldehyde, isocyanates and melamine resins are used [3]. Urea formaldehyde resins due to their moderate cost are much used as binders in the particleboard industry for indoor panels. The panels based on this binder don't resist in water and emit formaldehyde dangerous to man and his environment [3]. Indeed, formaldehyde is now classified as a carcinogen by the International Agency for Research on Cancer [4]. As for phenols formaldehyde resins, high cost and their possible discharge of phenol in the environment may further limit their use [5]. For this reason, today, extension of their use is avoided when possible. Polyisocyanated adhesives are also expensive and the free isocyanate group remains dangerous for public health [3].

Such problems of conventional binders have led to search for alternative solutions. One solution is to reduce formaldehyde emission levels, but this reduction has a negative effect on the mechanical properties of particleboard and also does not eliminate formaldehyde emission [6]. The second option is the chemical modification of resins based on formaldehyde: it also does not prevent the emission of formaldehyde [7]. The third option is to avoid formaldehyde emission by adding a surface barrier to the panel. However, this solution is only temporary because the life of the barrier is limited [8].

The ineffectiveness of the solutions presented above led to research for other less polluting, less toxic alternatives, to replace conventional binders. Thus, biological glues (animal and plant) attract nowadays renewed interest. These are gelatins, white of egg, casein, fats derivatives, polysaccharides, vegetable proteins, lignin and tannins. In his works Nénonéné [9] explored the vegetal tannin such as those of the African locust bean tree pods and the peel of Indian tamarin and pearl of bone glue. His research gives a base to solve the problem of the choice of a binder without emission of formaldehyde, and to achieve good mechanical strength panels that are resistant to insects and microbial agents. In the central and north region of Togo, abounds a tannic waste: the African locust bean husk and the peel of Indian tamarin.

In this study, we present the results of the tests of panels obtained using carpentry residues, namely kapok tree sawdust, bonded with the unconventional adhesives, the tannic powder of African locust bean husk (*Parkia Biglobosa*) and the peel of Indian tamarind (*Pitecellobium Dulce*).

2 Material and Methods

2.1 Material

The raw material used in this study:

The Sawdust of Kapok Tree. They are collected from carpentry. The Latin name *Ceiba Pentadra*, and common name: Kapok tree bottle, ouatier, cotton wood, or vôoka Gounga. It was put in the hot air oven Memment type at 70 °C for three days.

The Pod of African Locust Bean (*Parkia Biglobosa*). African locust bean (*P. Biglobosa*) is one of spontaneous plants in sub-Saharan Africa the most protected by local people because of its many uses (food, feed, fertilizer and soil restoration, crop protection and animal health, energy resources, habitat, tannery, toxic in water, etc. [10–12].

The wood is use, in the manufacture of paper pulp, for fermentation of seeds. The grain is use to produce a condiment sold in all local and regional markets and known under various names: Iru Nigeria, Dawa Dawa Cameroon, Sumboula Mali, Tsodi or Tsodou in Togo, [10, 13]. The pods are used to buff coloured fabrics [10] and for tanning hides, in Togo, to treat the floors and walls of mud houses against humidity. The bark of the stem is used for preparing several traditional medicines [14], and to disinfect water [14].

The Peel of Indian Tamarind (*Pitecellobium Dulce*). Indian tamarind (*P. Dulce*) is from tropical and subtropical America [15, 16], from the Fabaceae family, from the genus Pithecellobium and particularly Pithecellobium dulce. It is commonly called Indian Tamarind, Sweet Pea, etc. In Togo, it is known as false barrier. It has several synonyms: Acacia Obliquifolia M. Martens & Galeotti, Albizia dulcis (Roxb.) Kuntze, Mimosa dulcis Roxb., Inga dulcis (Roxb.) Willd., etc. [10, 16]. It is a very widespread in all tropical regions. It can reach 10–12 m of height, with a wide crown and it develops with irregular branches. The leaves are bipinnate, carrying at their base pairs of thin spines 2 to 15 mm in length. The inflorescence consists of small white flowers in flower heads. The fruit is a pod 10 to 15 cm long and wound in a spiral, dehiscent at maturity and contains 5 to 10 shiny black seeds covered with whitish arils. The grey bark and the coiled-tight pod are characteristic of this species. The root system is powerful and relatively large [10, 15, 16]. P. dulce B. is a low-demanding species for climatic and soil conditions. In Togo, P. dulce is mainly used for hedges.

Laboratory Equipment. The different equipment of laboratory used for different test are below:

- a hot air oven at maximum temperature of 140 °C Memment type,
- a brand hammer mill, RETSCH Type SK 1000, with a 0.125 mm sieve to transform the pod of African locust bean *(Parkia Biglobosa)* and Indian tamarin in powder size of less than 0.125 mm,
- a manual hydraulic thermal press CARVER type, maximum pressure of eleven (11) bars, equipped with a mold of 300 mm × 300 mm, facilitating the pressing of particleboard,
- a brand mixer Laurent Perrier, for mixing the raffinate and the binder,
- a brand mechanical test bench Instron 4467 type equipped with various accessories for the various tests namely, tensile, compression, bending three points and four points, internal bond and the Brinell hardness
- Tests are conducted in four laboratories, namely:
- Laboratory of the Research on Agricultural Resources and Environmental Health (LARASE) of the University of Lomé in Togo,
- metallurgy laboratory and testing of the National high School of Engineers (ENSI) at the University of Lomé in Togo,

Laboratory for the Study and Research on Wood Material (LERMAB) of the National School of Technology and Wood Industry (ENSTIB) of the University of Lorraine at Épinal in France.

2.2 Methods

Sample Preparation. Sawdust of kapok from carpentry, are steamed at 70 °C for 72 h. The pod of African locust bean and the peel of Indian tamarind once steamed, is cut and ground with a brand hammer mill, RETSCH Type SK 1000, with a 0.125 mm sieve. The tannic powder obtained is mixed with the sawdust in different rate: 5; 6.5; 7; 7.5; 8; 10; 12.5 and 15%. The particle, sawdust of kapok tree, binder and water were mixed for ten minutes, is pressed at a temperature of 160 °C at a pressure of 11 bars for 15 min.

The panels were made in the Laboratory of Research on Agriculture resources and Environment health (LARASE) of University of Lomé – Togo.

These particleboards are cut into different specimens at the Laboratory for Study and Research on Wood Material (LERMAB) at the National advanced School of Technology and Wood Industries (ENSTIB) of the University of Lorraine France.

The samples have the dimensions of: 150 mm × 50 mm for bending; 50 mm × 50 mm for internal bond; 150 mm × 20 mm for tensile. The Table 1 show the number of samples for the different tests.

Table 1. Numbers of samples for the different tests and the determined parameters.

Tests	Parameters	Dimensions (mm)	Number of samples
Bending	MOE, MOR	150 × 50	864
Tensile	MOT, E	150 × 20	864
Facial tensile	IB	50 × 50	864
Density	Density (ρ)	50 × 50	1440
Water immersion	TS2H, TS24H, WA2H, WA24H	50 × 50	1728

Mechanical Properties Determination. The mechanical tests are: tree points bending, tensile and internal bond are performed according to EN 312-2 2004 [17]. From these tests, the mechanical properties such as modulus of elasticity (MOE), modulus of rupture (MOR) for the bending, the Young's modulus (E), and modulus of rupture (MOT) for tensile and internal bond (IB) are calculated.

The norms: NBN EN 310: 1999 and EN 310 NF B51 - 124, 1993 [18] particleboard in terms of advocating a modulus of elasticity which strength F is taken between 10 and 40% of the breaking force.

Internal Bond (IB) of each particleboard is made according to the NF-EN 319 with ten specimens which dimensions are 50 mm × 50 mm subjected to perpendicular tensile on the faces, the internal bond is calculated.

According to ANSI A208.1–2009 [19], the reference values for mechanical and physical properties of particleboard is presented in Table 2 below. They serve as a reference for the validation of our experimental results.

Physical Properties Determination. The physical characteristics which are the density (ρ), the thickness swelling (TS) and water absorption (WA), are determined as follows:

Table 2. Characteristics required for particleboard according to the US standard ANSI A 208.1

Class	MOR (MPa)	MOE (MPa)	IB (MPa)	Hardness (N)	Maximal tension (%)
H-1	16,5	2400	0,90	2225	NS
H-2	20,5	2400	0,90	4450	NS
H-3	23,5	2750	1,00	6675	NS
M-1	11,0	1725	0,40	2225	0,35
M-S	12,5	1900	0,40	2225	0,35
M-2	14,5	2225	0,45	2225	0,35
M-3	16,5	2750	0,55	2225	0,35
LD-2	5,0	1025	0,15	NS	0,35
LD-1	3,0	550	0,10	NS	0,35

MOR: Modulus of rupture; MOE: modulus of elasticity; IB: Internal bond.
NS: not specified; H: high density panels: density > 800 kg/m^3 (H-1, 2, 3 Industrial use);
M: medium density panels: density is from 640 to 800 kg/m^3 (M-1, MS Commercial Use, M-2, M-3 Industrial use);
LD: low-density panels: density < 640 kg/m^3 (LD Used for door soul).

The density is determined according to EN 323 [20] on ten samples of 50 mm × 50 mm for each of six panels prepared. That means, we have 480 samples of each binder. So, for the three times this number for the 3 binders. The swelling test consists of immersing in water the samples of each panel whose mass and thickness are determined before. Then raises the weight and the thickness of the samples after 2 h for six samples and 24 h for the rest. It means, for this experience we have 12 samples × 48 for each binder. This determines the rate of water absorption (WA) of panels and its thickness swelling rate (TS) according to EN 317, 1993 [21].

3 Results and Discussion

3.1 Physical Properties

Density. The density of the panels varies between 613.83 and 666.39 kg.m^{-3} for the particleboard made with pearl bone glue. Those of the tannic powder of African locust bean pod (*P. Biglobosa*) are between 662.87 and 681.27 kg.m^{-3}. The tannic powder of Indian tamarind peel (P. Dulce) particleboard density is between 706.85 and 709.12 kg.m^{-3} The American norm ANSI A208.1 classifies those particleboards developed in to the medium density panels (MD–1 to MD–2) [19]. The Fig. 1 show the variation of the density according to the binder rate and its type.

The particleboards which have the high density are those made by the Indian tamarind peel, those of little density are made with pear bone glue. It can be explained by the binder density.

These results confirmed those of [9] who characterized the particleboard of kenaf at 10% of tannic powder of hush of *P. Biglobosa* and the peel of Indian tamarind. He finds

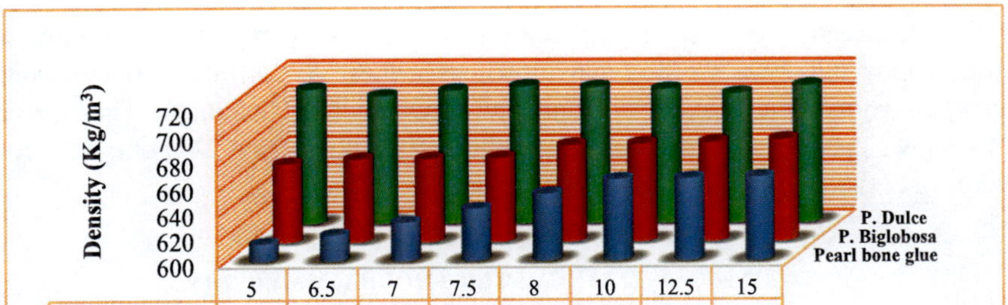

	5	6.5	7	7.5	8	10	12.5	15
Pearl bone glue	613.83	620.62	630.64	641.65	653.49	664.67	665.24	666.39
P. Biglobosa	662.87	666.59	666.71	667.26	677.35	678.26	679.3	681.27
P. Dulce	706.85	701.41	705.65	708.75	707.9	706.33	702.88	709.12

Fig. 1. Histogram of the density according to the type of binder and it's rate

the density of 473,93 kg/m^3 for the particleboard made with tannic powder of African locust bean pod and this of 492,62 kg/m^3 for the powder of the peel of Indian tamarind.

The density increases slightly with the rate of binder. The rate of binder is a parameter which significantly influences the physical properties of particleboards. According to [22], the increase in the content of binder panels leads to an improvement in the physical behaviors of particleboards [23–25] with the particleboard of kenaf at 8 and 10% of UF binder). The ANOVA analysis on R gives results which is in Table 3 below.

Table 3. Effect of factors on the density of particleboard of sawdust of kapok tree

Response density					
Factors	Df	Sum. Sq	Mean Sq	F. Value	Pr (>F)
Type of binder	1	8278	8278	67.7976	1.382 e^{-14} ***
Binder rate	7	20802	2972	11.060	<2.2 e^{-16} ***
Type of binder × Binder rate	7	295	42	0.3450	0.93238
Residuals	229	27959	122		

Signification codes: 0. '***' 0.001 '**' 0.01 '*' 0.05 '.' 0.1 '.' **: significative; ***: very significative

At the 5% threshold, the density varies according to the rate of binder and the rate of binder. According to the interactions between the different factors, there is a very significant interaction between the type of binder and the rate of binder. On the other way, there is no interaction between the type of binder and the rate of binder.

Analysis of the mean with the Tukey test in the "R software", shows that the first five identical binder rates (5; 6.5; 7; 7.5 and 8) and the last three identical (10; 12.5 and 15) are different. This means that under the value for money, we had to consider that two rates (5 and 10) instead of eight to characterize all the panels.

The Thickness Swelling (TS) of the Particleboards. All particleboards of the sawdust of kapok tree have swelled above 50% regardless of the rate of tannic powder or pearl bone glue. They can therefore only be used in a dry environment. The histogram below shows the variation of thickness swelling rate after 2- and 24- hours immersion according to the type of binder and its rate (Fig. 2 and Fig. 3).

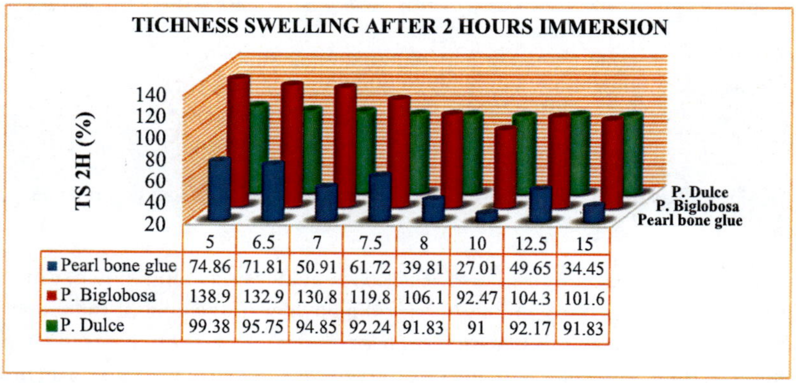

	5	6.5	7	7.5	8	10	12.5	15
Pearl bone glue	74.86	71.81	50.91	61.72	39.81	27.01	49.65	34.45
P. Biglobosa	138.9	132.9	130.8	119.8	106.1	92.47	104.3	101.6
P. Dulce	99.38	95.75	94.85	92.24	91.83	91	92.17	91.83

Fig. 2. Histogram of the thickness swelling rate after 2 h immersion according to the type of binder and its rate

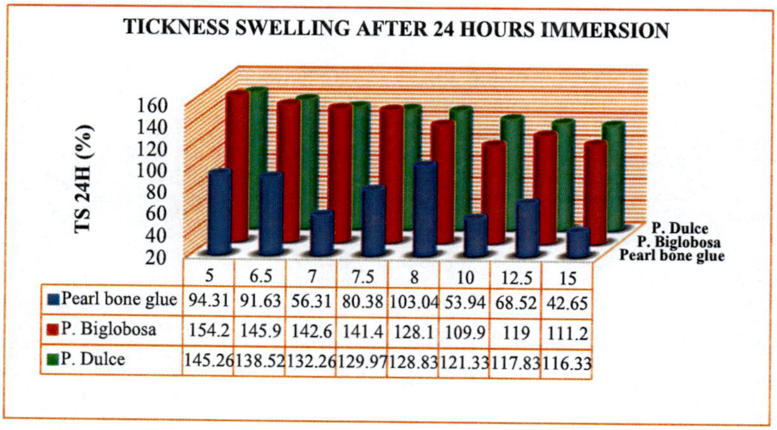

	5	6.5	7	7.5	8	10	12.5	15
Pearl bone glue	94.31	91.63	56.31	80.38	103.04	53.94	68.52	42.65
P. Biglobosa	154.2	145.9	142.6	141.4	128.1	109.9	119	111.2
P. Dulce	145.26	138.52	132.26	129.97	128.83	121.33	117.83	116.33

Fig. 3. Histogram of the thickness swelling rate after 24 h immersion according to the type of binder and its rate

It emerges from the analysis of the results that panels made with pearl bone glue have a lower swelling rate than tannin powder from African locust bean pods and those made with tannic powder from the peel of Indian tamarind. This is explained by the fact that the African locust bean pods contains more tannin (45%) than the bark of the Indian tamarind (35%) according to the work of [9].

These results confirm those of [26] who characterized the kenaf and cotton panels with pearl bone glue and found that the rate of thickness swelling decreases with the rate of binder.

The results obtained from the analysis of variance using the R software are grouped in the Tables 4 and 5.

The results reveal that all the factors (type of binder and binder rate) have a significant effect on the thickness swelling after 2 h and 24 h immersion in water.

Table 4. Effect of factors on the thickness swelling after 2 h immersion in water

Response TS_2H					
Factors	Df	Sum. Sq	Mean Sq	F. Value	Pr (>F)
Type of binder	1	7945	7945	115.0990	<2.2 e^{-16} ***
Binder rate	7	13006	1858	26.9153	<2.2 e^{-16} ***
Type of binder × Binder rate	7	242	35	0.5010	0.8333
Residuals	229	15808	69		

Table 5. Effect of factors on the thickness swelling after 24 h immersion in water

Response TS_24H					
Factors	Df	Sum. Sq	Mean Sq	F. Value	Pr (>F)
Type of binder	1	65506	65506	1167.5358	<2 e^{-16} ***
Binder rate	7	18235	2605	46.4308	<2 e^{-16} ***
Type of binder × Binder rate	7	230	33	0.5853	0.7676
Residuals	229	12848	56		

Signification codes: 0. '***' 0.001 '**' 0.01 '*' 0.05 '.' 0.1 '.' **: significative; ***: very significative

The interactions between the type of binder and the type of wood, the type of binder and the particle size, the type of wood and the particle size, and between the type of wood and the level of binder are slightly significant [27, 28].

Analysis of the average with the Tukey test in the R software shows that the tree binders are different in relation to the thickness swelling rate. As for the binder rates, the first six rates (5; 6.5; 7; 7.5; 8 and 10) and the last two (12.5 and 15) are different.

This means that, in terms of price/quality, only the two rates of binder (5 and 12.5) instead of the eight are to be considered to characterize all the panels.

The Water Absorption (WA) of the Particleboards. The panels produced with the African locust bean pods absorbed between 117.66 and 152.07% after two (2) hours immersion and between 137.73 and 177.98% after twenty - four (24) hours immersion. Those made with the tamarind bark absorbed between 108.31 and 151.16% after two (2) hours immersion and between 135.15 and 183.21% after twenty-four (24) hours immersion. Also, those made with the pearl bone glue absorbed between 64.81 and 145.59% after two (2) hours immersion and between 80.38 and 164.19% after twenty-four (24) hours immersion. It appears that the kapok tree panels have little resistance to water. The increasing of binder improves the behavior in water of the panels produced regardless of the type of the binder.

These panels must be used in a dry environment and their protection against humidity is necessary.

The histogram below shows the variation of water absorption rate after 2- and 24-hours immersion according to the type of binder and its rate (Fig. 4 and Fig. 5).

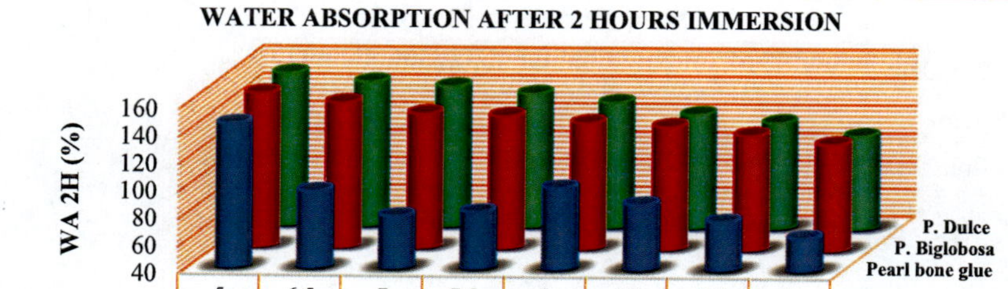

	5	6.5	7	7.5	8	10	12.5	15
Pearl bone glue	145.59	96.58	78.72	82.16	100.11	88.29	76.19	64.81
P. Biglobosa	152.07	145.68	138.12	136.64	132.23	129.79	123.96	117.66
P. Dulce	151.16	145.25	142.15	137.34	131.14	122.81	117.59	108.31

Fig. 4. Histogram of the water absorption rate after 2 h immersion according to the type of binder and its rate

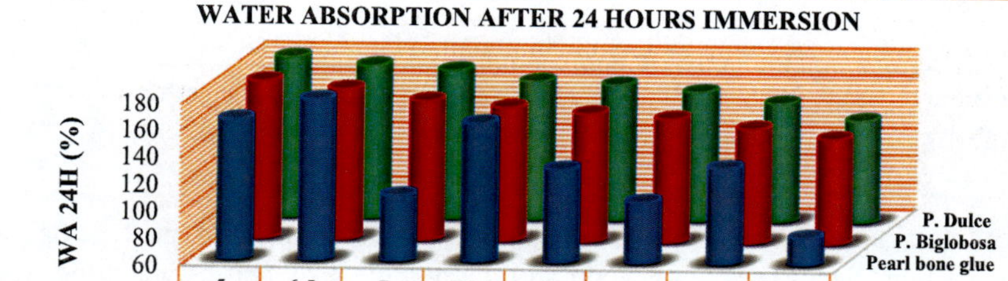

	5	6.5	7	7.5	8	10	12.5	15
Pearl bone glue	164.19	180.22	109.14	162.82	129.39	105.91	131.19	80.38
P. Biglobosa	177.98	172.37	163.85	159.06	154.84	151.37	144.56	137.73
P. Dulce	183.21	174.22	171.12	163.23	161.25	156.16	147.96	135.15

Fig. 5. Histogram of the water absorption rate after 24 h immersion according to the type of binder and its rate

Analysis of the results shows that:

- the panels made with the tannic powder of African locust bean pods absorbed less water than those made with the tannic powder from the bark of P. dulce. This could be explained by the fact that the African locust bean pods contains more tannin (45%) than the bark of P. dulce. (35%) according to the work of [9].

– the rate of water absorption followed the shape of the rate of swelling in thickness as a function of the rate of binder thus confirming those of [26] which characterized the panels of kenaf and cotton with pearls bone glue and [29].

The analysis of the variance in water absorption generated by the R software is grouped in Table 6 for 2 h and Table 7 for 24 h immersion.

Table 6. Effect of factor on the water absorption after 2 h immersion in water

Response TA_2H					
Factors	Df	Sum. Sq	Mean Sq	F. Value	Pr (>F)
Type of binder	1	1	0.6	0.0054	0 941397
Binder rate	7	48990	6998.6	66.6945	<2.2 e^{-16} ***
Type of binder × Binder rate	7	270	38.6	0.3682	0.920119
Residuals	229	24030	104.9		

Signification codes: 0. '***' 0.001 '**' 0.01 '*' 0.05 '.' 0.1 '.' **: significative; ***: very significative

Table 7. Effect of factor on the water absorption after 24 h immersion in water

Response TA_24H					
Factors	Df	Sum. Sq	Mean Sq	F. Value	Pr (>F)
Type of binder	1	323	323.3	2.0876	0.1498706
Binder rate	7	52989	7569.9	48.8862	<2.2 e^{-16} ***
Type of binder × Binder rate	7	176	25.2	0.1624	0.9920995
Residuals	229	35460	154.8		

Signification codes: 0. '***' 0.001 '**' 0.01 '*' 0.05 '.' 0.1 '.' **: significative; ***: very significative

The rate of water absorption of kapok tree sawdust particleboards varies considerably depending on the rate of binder. However, the type of binder has no effect on the variation of the absorption rate. The analysis of the interactions of the factors shows that the type of binder and binder rate have no significant effect on the water absorption rate in 2 h and 24 h.

Analysis of the average with the Tukey test in the R software shows that the tree binders are different with respect to the rate of water absorption. As for the binder rates, the first five rates (5; 6.5; 7; 7.5 and 8) and the last three (10; 12.5 and 15) are different.

This leads us to say that in terms of value for money, we had to consider the two binder rates (5 and 10) instead of the eight to characterize all the panels.

3.2 Mechanical Properties

Six (06) samples per panel were made. Six particleboards were prepared for each rate of binder for each mechanical test. The results are the means from the various mechanical tests according to American norm [19].

Modulus of Elasticity (MOE). The MOE grow with the binder rate. For panels made with tannin powder from India, the growth is slight between 8 and 15% of the powder [29]. Those of pearl bone glue grow with the rate of glue. Figure 6 shows the variation of the modulus of elasticity as a function of the binder and its rate.

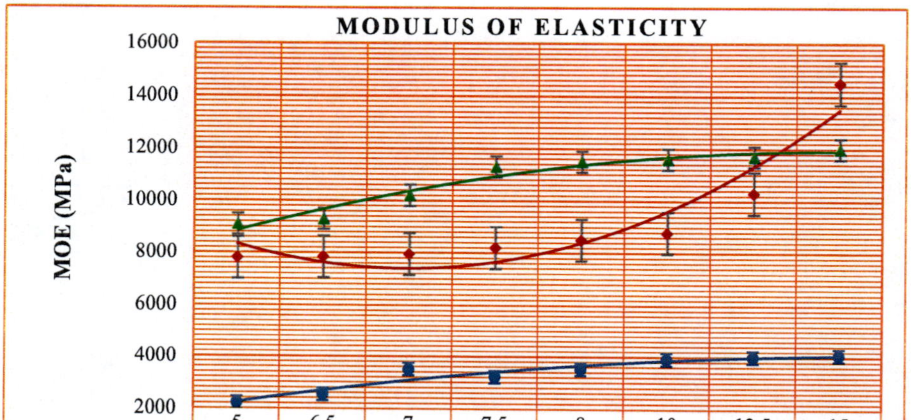

	5	6.5	7	7.5	8	10	12.5	15
● Pearl bone glue	2221	2551	3510	3225	3530	3910	4004	4081
◆ P. Biglobosa	7804	7840	7950	8200	8500	8780	10300	14500
▲ P. Dulce	9100	9300	10200	11300	11500	11600	11700	12000

Fig. 6. Variation of the modulus of elasticity according to the type of binder and its rate

The modulus of elasticity is greater than 1725 MPa, that is satisfying the requirements of American standard, the ANSI A208-1, 2009 norm (MOE = 1725 MPa). These results are superior than those obtained by [30] (MOE = 125 MPa), and [31] (950 to 1,750 MPa) in the case of panels without binder, treated by injection of water vapor, under a pressure of 6 MPa, at a temperature of 190 °C. in the presence of water vapor and for durations of 7 to 25 min.

The panels are medium density class 1 (M1) according to standard ANSI A208.1, (2009) norm.

The variance and the interaction of the factors on the MOE generated by the R software are grouped in Table 8.

The MOE of kapok tree sawdust particleboards varies considerably depending on the rate of binder. Analysis of the interactions of factors shows that the type of binder and the rate of binder have a very significant effect on the MOE.

Analysis of the mean with the Tukey test in the R software shows that the three binders are different compared to the MOE. As for the binder rates, the six first rates (5; 6.5; 7; 7.5 8; and 10) and the two last (12.5 and 15) are different.

Table 8. Effect of factor on the MOE of kapok tree sawdust particleboard.

Response MOE					
Factors	Df	Sum. Sq	Mean Sq	F. Value	Pr (>F)
Type of binder	1	5236925	5236925	8.9245	0.0031199 **
Rate of binder	7	1048189	14974141	25.5183	<2.2 e^{-16} ***
Type of binder × Rate of binder	7	1289559	1842180	3.1394	0.0034706 **
Residuals	229	1343778	586801		

Signification codes: 0. '***' 0.001 '**' 0.01 '*' 0.05 '.' 0.1 '.' **: significative; ***: very significative

This means that in terms of value for money, to assess the MOE of all the panels, only the two binder rates (5 and 12.5) instead of the eight considered.

Modulus of Rupture (MOR). The Analysis of the results of the MOR shows that the panels of kapok sawdust produced with the three types of binder are classified as medium density class 1 (M – 1) according to the same standard ANSI A208.1, (2009) norm. The MOR values are compiled in the following Fig. 7.

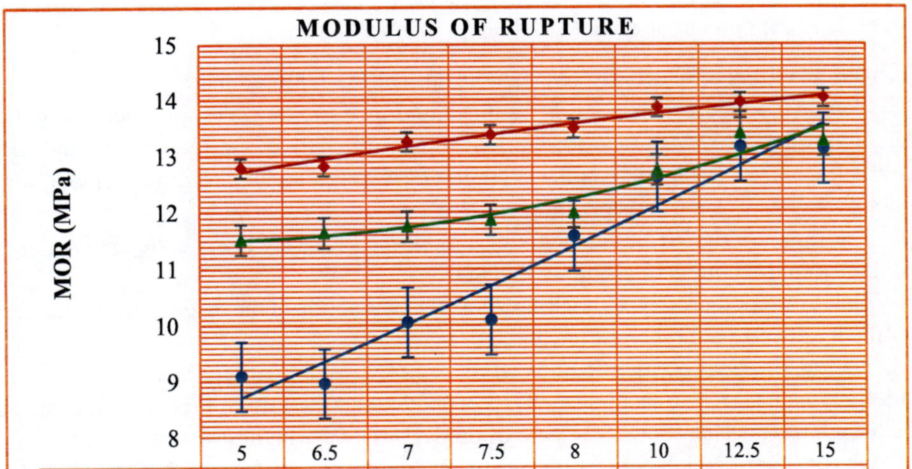

	5	6.5	7	7.5	8	10	12.5	15
● Pearl bone glue	9.09	8.96	10.05	10.09	11.57	12.61	13.15	13.11
◆ P. Biglobosa	12.79	12.82	13.25	13.37	13.48	13.85	13.94	14.02
▲ P. Dulce	11.52	11.64	11.75	11.86	11.98	12.74	13.39	13.25

Fig. 7. Variation of the modulus of rupture according to the type of binder and its rate

The results of the analysis of the variance of the data relating to the resistance to bending rupture (MOR) generated by the R software are recorded in Table 9.

The MOR of kapok tree sawdust particleboards variation considerably depending on the type of binder and the rate of binder. The analysis of the interactions of factors shows that the type of binder and the level of binder have no effect on the MOR.

Table 9. Effect of factor on the MOR of kapok tree sawdust particleboard.

Response MOR					
Factors	Df	Sum. Sq	Mean Sq	F. Value	Pr (>F)
Type of binder	1	101.5	101.47	60.7375	2 268 e^{-13} ***
Rate of binder	7	567.4	81.06	48.5211	<2.2 e^{-16} ***
Type of binder × Rate of binder	7	15.8	2.26	1.3528	0.2265
Residuals	229	382.6	1.67		

Signification codes: 0. '***' 0.001 '**' 0.01 '*' 0.05 '.' 0.1 '.' **: significative; ***: very significative

The analysis of the mean with the Tukey test in the R software shows that the first five rates (5; 6.5; 7; 7.5 and 8) while having identical effects between them are distinguished from the last three (10; 12.5 and 15) have similarly important effects.

This means that in terms of price quality, we will consider only the three types of binder but only with two rates (5 and 10) instead of the eight.

We notified that the modulus of rupture (MOR) conform to those obtained by [22] with panels containing red oak sawdust and [32] with panels based on bark of poplar, trembling aspen. They notified that the mechanical properties are improved with the increase in the content of binder.

Tensile Modulus of Elasticity (E). The results of the longitudinal elasticity modules (E) resulting from the tensile test are grouped in Fig. 8 for the panels produced with the tannic powder of African locust bean pods, that of the bark of the Indian tamarind and pearls bone glue. Each value is the average of the longitudinal elastic modulus of test on six samples of six panels per binder rate.

The values of the longitudinal modulus of elasticity (E) are lower compared to those of the flexural modulus of elasticity (MOE).

The results of the analysis of variance and the interaction of the factors on E generated by the R software are reported in Table 10.

The longitudinal elastic modulus (E) of kapok sawdust particle boards varies considerably depending on factors (the type of binder and the rate of binder). Analysis of factor interactions shows that the type of binder has a very significant effect on E.

On the other hand, the interaction between the type of binder and the rate of binder have no effect on the longitudinal elastic modulus.

Analysis of the means with the Tukey test in the R software shows that the three binders are different on the E.

Tensile Modulus of Rupture (MOT). The variation of the MOT is recorded in Fig. 9 below.

The results obtained show that the modulus of rupture increases with the rate of binder. The standard (ANSI A208.1., 2009) fixed the threshold of the tensile rupture module at 11 MPa.

The results relating to the breaking stress (MOT) respect the shape of the breaking stress in bending. It reaches the maximum at 15% of binder.

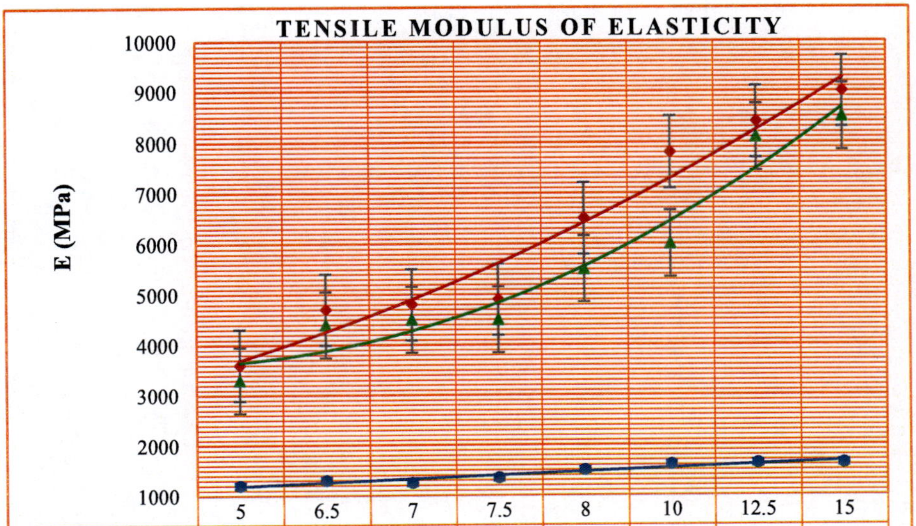

	5	6.5	7	7.5	8	10	12.5	15
● Pearl bone glue	1200	1300	1250	1350	1500	1610	1630	1634
◆ P. Biglobosa	3600	4700	4800	4900	6500	7800	8400	9000
▲ P. Dulce	3300	4400	4500	4500	5500	6000	8100	8500

Fig. 8. Variation of the tensile modulus of elasticity according to the type of binder and its rate

Table 10. Effect of factor on the tensile modulus of elasticity of kapok tree sawdust particleboard.

Response E					
Factors	Df	Sum. Sq	Mean Sq	F. Value	Pr (>F)
Type of binder	1	2824424	282442	30.965	7.311 e^{-08} ***
Rate of binder	7	5114034	730576	8.0096	1.059 e^{-08} ***
Type of binder × Rate of binder	7	240755	34394	0.3771	0.915149
Residuals	229	2088776	91213		

Signification codes: 0. '***' 0.001 '**' 0.01 '*' 0.05 '.' 0.1 '.' **: significative; ***: very significative

These results are in perfect agreement with results obtained by [26] on kenaf particle boards using bone glue in pearls at contents of 5%, 7.5%, 10%, 12.5% and 15%.

Table 11 group together the results of the analysis of the variance and the interaction of the different factors on the MOT generated by the "R software".

The variation in tensile rupture modulus (MOT) as a function of the factors (type of binder and rate of binder) of kapok sawdust particleboards is important. The Analysis of the interactions of these factors shows that the type of binder and the rate of binder have no effect on the modulus of rupture. The Analysis of the means with the Tukey test in the R software shows that the three binders represent distinct effects. As for the binder rates, the first seven rates (5; 6.5; 7; 7.5; 8; 10 and 12.5) and the last (15) are different.

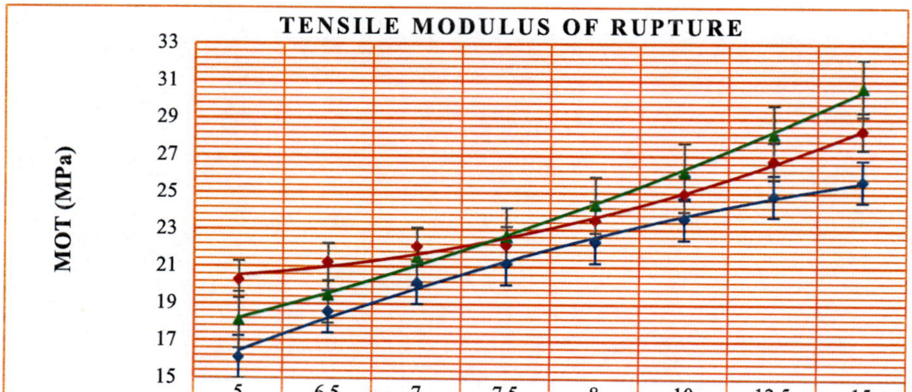

	5	6.5	7	7.5	8	10	12.5	15
◆ Pearl bone glue	16.11	18.58	20.15	21.19	22.35	23.61	24.85	25.66
◆ P. Biglobosa	20.29	21.24	22.09	22.16	23.54	24.99	26.75	28.36
▲ P. Dulce	18.11	19.48	21.53	22.66	24.35	26.16	28.18	30.66

Fig. 9. Variation of the tensile modulus of rupture according to the type of binder and its rate

Table 11. Effect of factor on the tensile modulus of rupture of kapok tree sawdust particleboard.

Response MOT					
Factors	Df	Sum. Sq	Mean Sq	F. Value	Pr (>F)
Type of binder	1	45.80	45.80	17.604	3.891 e^{-05} ***
Rate of binder	7	185.69	26.53	10.197	4.16 e^{-11} ***
Type of binder × Rate of binder	7	2.24	0.32	0.123	0.99669
Residuals	229	595.72	2.60		

Signification codes: 0. '***' 0.001 '**' 0.01 '*' 0.05 '.' 0.1 '.' **: significative; ***: very significative

This means that in terms of value for money we will only have the panels, the three types of binder, but only with two rates (5 and 15) instead of eight.

Internal Bond (IB). The panels made with tannic powder from the pod of the husk of African locust bean are more consistent than those made with tannic powder of the peel of P. Dulce. The best internal bond is of those from pearls bone glue. The internal bond strength increases with the binder rate (Fig. 10). It is over the 0.4 MPa threshold set by ANSI A208.1–2009.

The results confirm those of [23] who characterized the anegre tannin as an adhesive resin; [29] with the kapok tree panels with the pod of the husk of African locust bean. The results obtained from the analysis of variance using the R software are collated in Table 12 below.

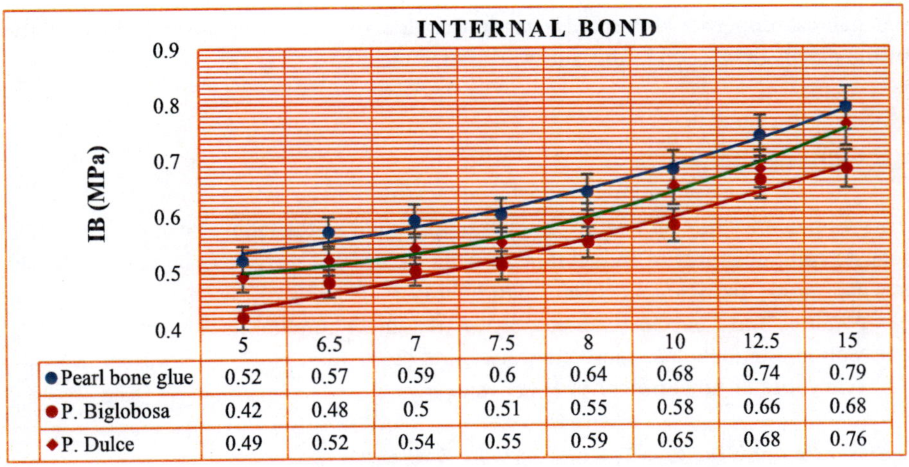

Fig. 10. Variation of the internal bond to the type of binder and its rate

Table 12. Effect of the factors on the modulus of elasticity (E) of mahogany sawdust particleboards

Response IB					
Factors	Df	Sum. Sq	Mean Sq	F. Value	Pr (>F)
Type of binder	1	0.6242	0.6242	32.8076	3.817 e^{-06} ***
Binder rate	7	1.1936	0.1705	8.9616	8.995 e^{-06} ***
Type of binder × Binder rate	7	0.1648	0.0235	1.2377	0.31609
Residuals	229	0.5328	0.0190		

Signification codes: 0. '***' 0.001 '**' 0.01 '*' 0.05 '.' 0.1 '.' **: significative; ***: very significative

These results reveal that the factors (type of binder and binder rate) have a significant effect on internal bond. The interactions between the type of binder and the binder rate are not significant.

The mean analysis with the Tukey test in the R software shows that the two binders are different compared to the IB. As for the binder rate, they are divided into three homogeneous groups (5; 6.5; 7 and 7.5); (8; 10) and (12.5 and 15) which are different from each other [11, 12, 23, 24].

The mechanical characteristics, in bending (MOE, MOR), tensile (E, MOT and IB) of the particleboards of mahogany sawdust, determined, respected the requirements of standard ANSI A208.1, 2009. The panels are medium density (MD–1) for those made with tannic powder of the pod of the husk of African locust bean.

The mechanical properties are improved with the increase of the binder content. [6] reaches the same conclusion as [25] with regard to the proportional increase in the mechanical properties of the panels with the content of binder. The mechanical properties of the panels developed in this study are satisfactory.

4 Conclusion

This work shows that an unconventional binder such as the tannic powders, of African locust bean pods, peel of Indian tamarind and pearls bone glue can be used to develop sawdust wood particleboards, and particularly that of kapok tree sawdust. The panels obtained have mechanical properties that meet international standards as with conventional binders that cause environmental problems.

The value of mechanical properties are:

- ***Bending modulus of elasticity (MOE) is between:***
- 2221 to 4081 MPa for panel made with pearls bone glue
- 7804 to 14500 MPa for panel made with tannic powder of African locust bean pods
- 9100 to 12000 MPa for panel made with tannic powder of the peel of Indian tamarind.
- ***Bending modulus of rupture (MOR) is between:***
- 8.96 to 15.15 MPa for panel made with pearls bone glue
- 12.79 to 14.02 MPa for panel made with tannic powder of African locust bean pods
- 11.52 to 13.39 MPa for panel made with tannic powder of the peel of Indian tamarind
- ***Tensile modulus of elasticity (E) is between:***
- 1200 to 1634 MPa for panel made with pearls bone glue
- 3600 to 9000 MPa for panel made with tannic powder of African locust bean pods
- 3300 to 8500 MPa for panel made with tannic powder of the peel of Indian tamarind
- ***Tensile modulus of rupture (MOT) is between:***
- 16.11 to 25.66 MPa for panel made with pearls bone glue
- 20.29 to 28.36 MPa for panel made with tannic powder of African locust bean pods
- 18.11 to 30.66 MPa for panel made with tannic powder of the peel of Indian tamarind
- ***Internal bond strength (IB) is between:***
- 0.52 to 0.79 MPa for panel made with pearls bone glue
- 0.42 to 0.68 MPa for panel made with tannic powder of African locust bean pods
- 0.49 to 0.76 MPa for panel made with tannic powder of the peel of Indian tamarind

These particleboards have the density from 613 to 800 kg.m^{-3}. They are classified among medium density particleboard (MD–1).

The very good mechanical properties values that respects ANSI A208.1 – (2009) norm is obtained. The particleboard made with kapok tree sawdust particles with the three unconventional binder (tannic powder of African locust bean pods, peel of Indian tamarind and pearls bone glue) can replace non-green panels.

The value of physical properties are:

- ***The thickness swelling rate (TS) rate:***
- over 34% after 2 h and 42% after 24 h for panel made with pearls bone glue;
- over 101% after 2 h and 111% after 24 h for panel made with tannic powder of African locust bean pods;

- over 91% after 2 h and 116% after 24 h for panel made with tannic powder of the peel of Indian tamarind;
- ***The water absorption rate (WA) rate:***
- over 64% after 2 h and 80% after 24 h for panel made with pearls bone glue;
- over 117% after 2 h and 137% after 24 h for panel made with tannic powder of African locust bean pods;
- over 108% after 2 h and 135% after 24 h for panel made with tannic powder of the peel of Indian tamarind;

The panels developed indicate that their main use will be under dry conditions.

It will be very good to determinate the thermal conductivity and phonic absorption to know if these particleboards can be used for insulating.

Acknowledgements. The authors thank the Ministry of Agriculture for the financial support of this project by WAAPP–TOGO (PPAAO–TOGO). They are also grateful to the officials of LERMAB, Prof. André MERLIN in particular. The director of the LERMAB of the University of Lorraine, Prof. Philippe GERARDIN, and the responsible of FEMTO–ST, namely Prof. Emmanuel FOLTETE, of the University of Franche Comté are thanked for hosting laboratories for experimentation.

References

1. EPF: EU research framework programs; cost development of particleboard industry in Europe (2006)
2. DCN TOGO: Direction de la comptabilité Nationale. Togo (2010)
3. Nghoue Yemele, M.C.: Développement de panneaux de particules à base d'écorce d'épinette noire et de peuplier faux tremble. PhD University LAVAL (2008)
4. IARC: IARC monographs on the evaluation of carcinogenic risks to humans, vol. 62. Wood dust and formaldehyde, Lyon France (1995)
5. Huang, M.T., Ferraro, T.: Phenolic – Compounds in food and cancer prevention, ACS Symposium, vol. 507 (1992)
6. Beyer, R.A.: Method of scavenging formaldehyde using a low mole ratio melamine – urea – formaldehyde resin. United State Patent 5684118
7. Lavisci, P., Berti, S., Pizzo, B., Zanuttini, R.: A determination test for structural wood adhesives used in thick joints. Eur. J. Wood Prod. **1–2**, 153–154 (2001). https://doi.org/10.1007/s001070050487
8. ORICA LIMITED: Urea formaldehyde resin production process (1999)
9. Nénonéné, A.Y.: Elaboration et caractérisation des panneaux de particules des tiges de kénaf à base des bioadhésifs, de la colle d'os, des tannins ou de mucilage. These INP de Toulouse (2009)
10. Berhaut, J.: Flore Illustrée du Sénégal. Dakar, Ministère du Développement Rural. Direction Des Eaux et Forêts (1975)
11. Tchiégang-Megueni, C., Mapongmetsem, P.M., Akagou, Z.H.C., Kapseu, C.: An ethnobotanical study of indigenous fruit trees in northern Cameroon. Forests Trees Livelihoods **11**, 149–158 (2001)
12. Ndir, B., Lognay, G., Wathelet, B., et al.: Composition chimique du nététu, condiment alimentaire produit par fermentaire des graines du caroubier Africain Parkia Biglobosa (Jacq.) Benth. Biotechnol. Agron. Soc. Environ. **4**(2), 101–105 (2000)

13. Mapongmetsem, P.M., Nkongmeneck, B.A., Denedoumba, D.: Germination et croissance initiale d'une Mimosaceae alimentaire locale: Parkia biglobosa (Jacq.) Benth. Procédés Biologiques et Alimentaires. (2) Revue N°2 (2004). http://www.fao.org/ag/aga/agap/frg/afris/fr/Data/137. HTM
14. Fafioye, O.O., Adebisi, A.A., Fagade, S.O.: Toxicity of Parkia biglobosa and Raphia vinifera extracts on Clariasgariepinus juveniles. Afr. J. Biotech. **3**, 627–630 (2004)
15. Duke, J.A.: Phytochemical and Ethnobotanical Databases (1983). http://www.arsgrin.gov/cgi-bin/duke/ethno-actlist.pl
16. Houérou, L.: Pithecellobium dulce (Roxb.) Benth (2000). http://www.fao/agP/AGPC/Gbase/DATA/Pf000410.HTM
17. EN 312 – 2: Panneaux de particules. Type P2, Stand P3 – MH, P4 CTBH (2004)
18. EN 310 NF B51 – 124: Panneaux à base de bois – Détermination du module d'élasticité et de la résistance à la flexion AFNOR (1993)
19. ANSI A208 1: Medium density fibreboard, Published by National Particleboard Association, Gaithersburg, MD, 11P (2009)
20. EN 323 B51-122: Panneaux à base de bois. Détermination de la masse volumique (1995)
21. EN 317: Particleboards and fiberboards, Determination of Swelling in Thickness after Immersion European Standardization Committee, Brussels (1993)
22. Chow, P.: Phenol adhesive bonded medium – density fibreboard from Quercusrubra L. bark and sawdust. Wood Fibre Sci. 11(2), 92–98 (1978)
23. Singh, S.P., Singh, J.P., Rawat, S.S.: Utilization of bark from populus deltoids for particleboard, Forest Products Division, Research Forest Institute, Dehradun Journal-of-the-Indian-Academy-of-wood-Science - Dehradun: recd, 15–19, 26–27, 1–2 (1996)
24. Place, T.A., Maloney, T.M.: Internal bond and moisture response properties of three-layer Forest prod. J. Wood Bark Boards **27**(3), 50–54 (1977)
25. Kalaycioglu, H., Nemli, G.: Producing composite particleboard from kenaf (Hibiscus cannabinus L.) stalks Industrial Crops and Products, vol. 24, pp. 177–180 (2006). https://doi.org/10.1016/j.indcrop.2006.03.011
26. Kadja, K.: Elaboration et Caractérisation mécanique et thermique des panneaux de kénaf (Hibiscus cannabinus *L.*) et de cotonnier (Gossypium hirsutum *L.*). Thèse de Doctorat, Université de Lomé (2012)
27. Geng, X., Deng, J., Zhang, S.Y.: Characteristics of pulp and paper sludge and its utilization for the manufacture of medium density fiberboard. Wood Fiber Sci. **2**, 345–351 (2007)
28. Xing, S., Riedl, B., Koubaa, A., Deng, J.: Mechanical and physical properties of particleboard made from two pulp and paper mill secondary sludges. World J. Eng. 9(1), 31–36 (2012)
29. Drovou, S., Kassegne, K., A.; Sanda, K.: Élaboration et caractérisation mécanique et physique des panneaux de Particules de sciure de kapokier avec la poudre tanifère de la cosse de Gousse de néré Lomé (Togo). Eur. Sci. J. Edn. **11**(6) ISSN 1857–7881 (Print) e - ISSN 1857-7431 (2015)
30. Xiao Z., et al.: Interaction between the retinoblastoma protein and the oncoprotein MD M2 (1995)
31. Xu, J., Han, G., Wong, E.: Development of binderless particleboard from kenaf core using steam-injection pressing. J. Wopod Sci. **49**(4), 327–332. ISSN 1435-0211 (2003)
32. Villeneuve, E.: Utilisation de l'écorce de peuplier faux – tremble pour la fabrication de panneaux de particules. Maîtrise en science du bois. Faculté de foresterie Université LAVAL (2005)

Vibration-Assisted Grinding of Mild and Hardened Steel: Effect on Power Consumption and Material Removal Rates

Vaios Tsiakoumis(✉)

General Engineering Research Institute, Liverpool John Moores University, Byrom Street, Liverpool L3 3AF, UK
vaiostsiakoumis@gmail.com

Abstract. A major part in manufacturing processes is occupied by grinding which is mainly used as finishing process. However, several processes use grinding as a rough process. The fact that grinding belongs to the latest stages of a part's manufacturing chain it makes the need for higher precision quality components more and more demanding. The need for process improvement has led to the development of alternative non-conventional machining methods. In the present work a low-frequency vibration assisted surface grinding of mild and hardened steel using different types of aluminium oxide (Al_2O_3) wheels is examined. The vibration is applied from a piezoelectric actuator to a jig which is used to hold and oscillate the workpiece at a certain frequency during dry surface grinding process. The aim is to achieve lower power consumption during the process as well as higher material removal rates. Here, three types of mild and hardened steel and three types of aluminium oxide wheels with different grade and grain sizes were employed during conventional and vibration-assisted grinding processes. The benefits of this novel method over conventional grinding are clear with 21.8% reduction in power consumption and 66.6% increase in material removal rate.

Keywords: Grinding · Vibration · Machining · Piezoelectric

1 Introduction

Over the last 50 years the vibration-assisted machining subject has attracted a number of researchers and engineers who tried to improve the process performance in terms of tool forces, material removal rates, tool life and product quality. Two of the most well-known works are these of Bespalova in 1957 [1] who studied the theory of vibro-impact mechanisms and Blekhman in 1979 [2] who examined the theory of vibration of non-linear mechanical systems. The term 'Vibration-Assisted Machining' has been used extensively by researchers to describe mainly the application of low amplitude, high frequency (in the region of ultrasonic vibrations) displacement of the tool during a machining process. A remarkable study was the one of Brehl and Dow in 2007 [3] where they presented a review on the effects of tool ultrasonic vibration during main machining processes. However, their work has been focused on the tool life and the effect on

A. Batako et al. (Eds.): GCMM 2021, LNNS 335, pp. 215–226, 2022.
https://doi.org/10.1007/978-3-030-90532-3_17

vibration on cutting forces during turning some of the results showed 40–95% reduction on work-piece surface roughness. Ultrasonic-assisted machining has also been applied to grinding processes. One of the most recent works is this of Wdowik in 2018 [4] where they surface ground ceramic and carbide samples using a diamond grinding pin that vibrated at 20 kHz and 10 μm vibration amplitude. After a number of experiments they did not found any significant difference on surface texture between conventional and ultrasonic-assisted grinding. Similar process and parameters has been used for grinding advanced materials such as ceramics and super-alloys used in medical and aerospace applications with very promising results in terms of tool-life, wheel loading, grinding forces and residual stresses [5]. Also, in another study [6] a Titanium alloy work-piece was vibrating ultrasonically at 20.4 kHz and 5, 10 and 15 μm amplitude during grinding process as opposed to tool vibration. A reduction of 13.5% in grinding forces and improvement of 10% in surface roughness compared to conventional grinding process has been observed. Regarding the reduction in grinding forces Zhang et al. [7] designed a piezo-table that vibrated the workpiece across the spindle axis during the grinding process. Even though most of the studies focus on oscillations of the tool or workpiece in high frequencies of 20 kHz or above, there are works who apply low frequencies vibrations such as Zhong and Yang [8] who developed a device that vibrated the workpiece in three directions. Their results showed a reduction in cutting force and improved quality of machined surface of silicon wafers and ceramics and in another study when low frequency vibrations were applied, lower values of surface roughness were observed [9]. One of the most recent and comprehensive studies was conducted in 2018 where a model for vibration-assisted surface grinding was developed. The frequency range was 20 Hz to 100 kHz with and amplitude of up to 50 μm. Simulation results along with experimental results proved that the superimposed vibration to the grinding process secured a superior performance of the machining in terms of cutting forces, material removed and actual depth of cuts. The application of vibration provided low cutting forces, with high removal rates along with an extended wheel life (G-ratio) [10].

The present work looks into a simplified mechanism for a new generation of stacked and self-contained actuators that could be used in machining of small and medium parts. The mechanisms are designed to operate at resonant frequencies which are below 1 kHz. Moreover, this study was based on theory of vibro-impact systems [11]. In a vibro-impact system, the impact pair can be set with negative, positive or zero clearance. If in the system, the impact pair is set with interference (preload), the clearance Δ is considered as negative (see Fig. 1).

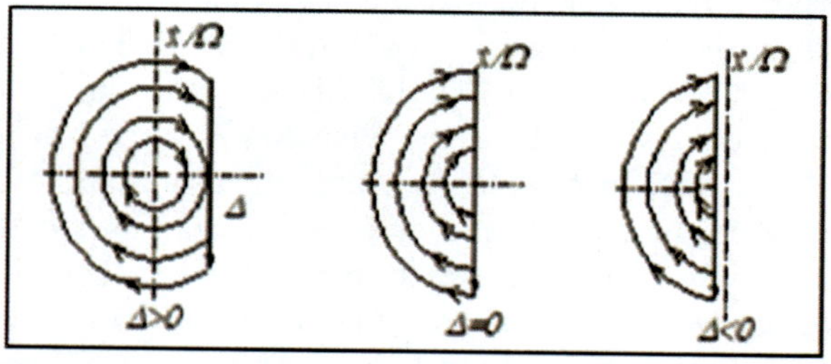

Fig. 1. Phase portrait of system motion with impact [11]

If the clearance Δ between the tool (wheel) and the workpiece were greater than zero, this would lead to a hammering process against the friable wheel with subsequent rapid wear of the wheel. Therefore, in this study it is assumed that between the impact (wheel/workpiece) there is no clearance ($\Delta = 0$) or the clearance is a negative i.e. $\Delta \leq 0$. However, depending on the table feed rate, the clearance Δ has been negative for most of the experiments undertaken in this study. When $\Delta < 0$, a vibro-impact regime can occur only when an additional hard excitation is supplied. In the present study this excitation was provided by the piezoelectric actuator. In this case the vibration velocity was lower than the work-speed ($2\pi af < Vw$, where a is the vibration amplitude and f is the vibration frequency). Even though the theory of vibro-impact systems has been applied to the present study, the superimposed vibration frequencies are not in the region of ultrasound. Moreover, in the present study, the workpiece is vibrated and not the cutting tool.

2 Machine Tool and Vibrating Jig Frequency Response

Before commencing any experimental work, the identification of frequency response of the surface grinding machine tool is necessary. The reason is to identify the resonance frequencies of the spindle unit of the machine during dynamic loading conditions. The knowledge of the machine tool dynamic characteristics is necessary and similar works can be found in the literature [12–14]. In the present study a piezo-electric actuator was used to excite the spindle unit of the machine tool at rising frequencies and a displacement sensor was recording the values of displacement at each frequency (sweep-sine test). The data recorded using a National Instrument data acquisition system (DAQ) and all the devices were calibrated before use. Figure 2 depicts the frequency response of the spindle unit of the grinding machine used for this work [15].

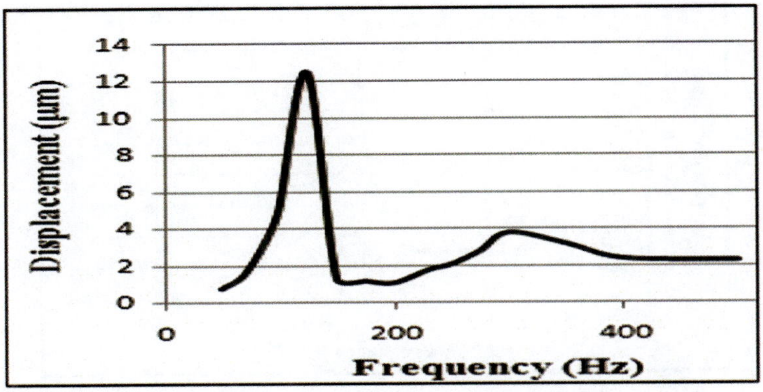

Fig. 2. Frequency-amplitude response of the spindle unit [15].

The natural frequency of the spindle unit was found to be around 125 Hz with a second peak at 300 Hz. The reason for this test was to give us an indication of what frequencies to avoid while designing the vibrating jig for the experimental work. The natural frequency and thus the excitation frequency of the vibrating jig must not coincide with the natural frequency of the spindle unit of the surface grinding machine tool.

The same method (sweep-sine test) has been used to identify the frequency response of the vibrating jig. The vibrating jig consisted of two plates connected by two flat springs. A work-piece holder was mounted on the top plate. A piezoelectric actuator transmitted the oscillations to the work-piece through a connecting rod (see Fig. 3).

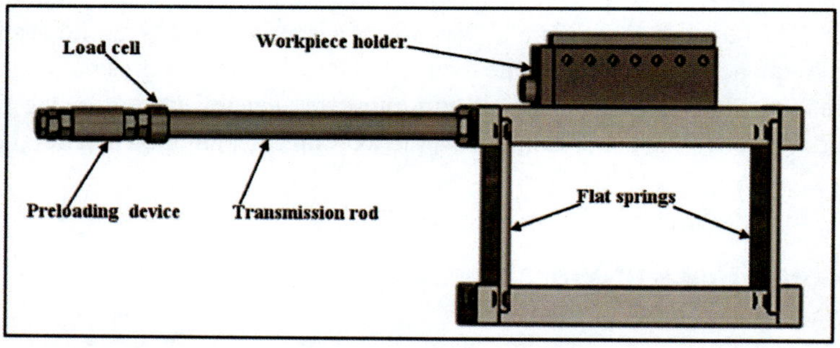

Fig. 3. Vibrating jig [16]

The actuator was placed outside the working environment in order to avoid any interaction with grinding swarf and debris. All parts of the jig were ground flat in order to secure absolute flatness of the workpiece. Also, an accelerometer was attached to the opposite side of the top plate in order to obtain the results during the sweep-sine test. The natural frequency of the jig has been found to be 275 Hz (see Fig. 4) and the results were verified by a modal analysis using finite element methods (see Fig. 5) [16].

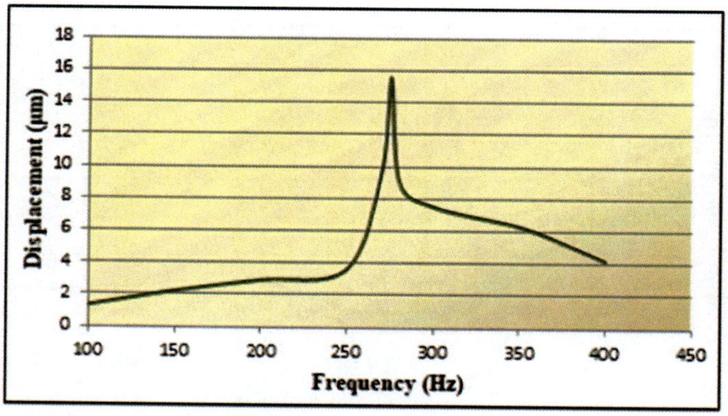

Fig. 4. Sweep sine test of the vibrating jig [16]

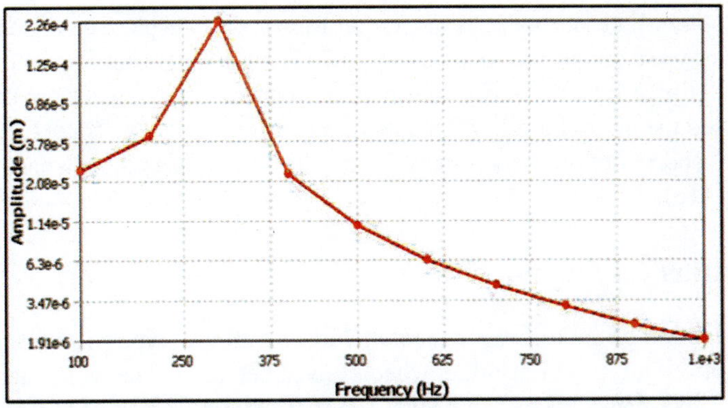

Fig. 5. Modal analysis of the vibrating jig [16]

3 Experimental Set-up

The next picture depicts the actual configuration of the vibrating jig used throughout all the experimental work (see Fig. 6).

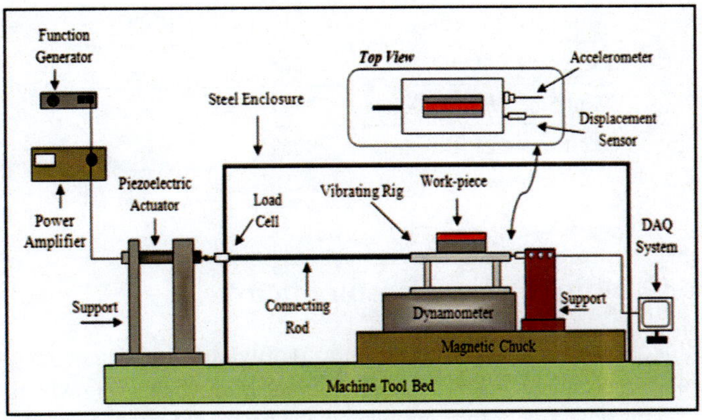

Fig. 6. Experimental configuration of the vibrating jig

All grinding tests were conducted using the Abwood Series 5025 conventional surface grinding machine. A piezoelectric actuator was used to excite the rig at the desired frequency. The actuator was located on the machine tool moving axis but outside the enclosure for safety measures; thus a steel rod was used to transmit the vibration to the jig. The length of the rod was dictated by machine tool configuration rather than a design parameter where it should have been a function of the wave length. A high resolution, water resistant displacement sensor - which was also calibrated - was attached to the other end of the jig and it was used to take real-time measurements during the test. The acceleration was recorded through an accelerometer in order to obtain the frequency domain characteristics of the system. The jig was mounted on a dynamometer in order to obtain both normal and tangential forces. The displacement sensor and the accelerometer were connected to a Data Acquisition (DAQ) system in order to observe and record all the results in real time during the experiment. The data were collected and post-processed using the Labview 8.5 software package [17]. The forces results during conventional and vibration-assisted surface grinding are presented on a past paper of the same author [16].

4 Experimental Parameters

For the conduction of the full experimental work, 3 different types of work-piece materials and 3 different types of grinding wheels have been used. The work-piece materials are: a) En31 (BS 534A99) b) Mild Steel (BS970 080440) and c) M2 Tool Steel (BS M2). The grinding wheels are: a) Al_2O_3 (454 601 L 7G V3) medium grain, medium grade, semi-friable, dense to open structure, b) Al_2O_3 (89A 60 K 5A V 217) medium grain, soft grade, friable, dense structure and c) Al_2O_3 porous wheel, long grain, Altos. Table 1 below depicts the different workpiece materials and their hardness values in Rockwell C scale.

Table 1. Workpiece materials and their hardness (HRC)

Workpiece materials	Hardness (HRC)
En31 (BS 534A99)	64.2
Mild Steel (BS970 080440	11
M2 Tool Steel (BS M2)	62

5 Results-Grinding Power Consumption

In this experiment (grinding power consumption) only one work-piece material is used, En31 (BS 534A99) with 64.2 HRC hardness and one type of grinding wheel Al_2O_3 (454 601 L 7G V3). This objective is to examine how the grinding parameters affect the power consumption during conventional and vibration assisted grinding. The fixed parameters are described on the table next (Table 2).

5.1 Effect of Depth of Cut

The applied depth of cut was 5, 10, 15, 20, 25 μm and was set by the dial of the handle on the machine tool and does not represent the actual depth after machining. The vibration frequency was chosen to be the resonant frequency of the vibrating rig in order to achieve high displacements.

The decrease of tangential forces observed in past work [16] during vibration-assisted grinding led to proportional decrease in power consumption of the process, which is seen in Fig. 7. It is observed in this graph that with the increase of depth of cut the power drawn for the process decreases.

Table 2. Fixed grinding parameters-power consumption test

Grinding parameter	Value
Grinding wheel type	Al_2O_3 (454A 601 L 7G V3)
Workpiece material	EN 31 (BS 534A99) HRC 64.2
Wheel feed	Traverse
Grinding condition	Dry
Vibration frequency	275 Hz
Vibration amplitude	15 μm (Peak)

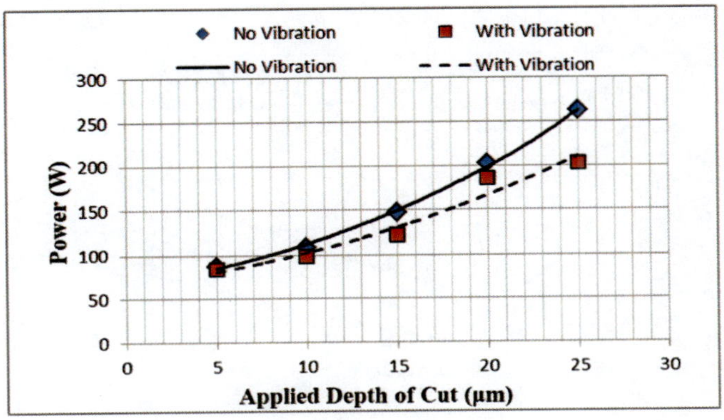

Fig. 7. Effect of depth of cut

5.2 Effect of Work Speed

Here, (see Fig. 8) five work-speeds were used namely 10, 20, 50, 100 and 200 mm/s. A depth of cut of 20 μm was chosen because at higher depths of cut the vibration effect was more obvious. It has to be mentioned that these values are the applied depth of cut and not the actual.

It is shown here that the application of vibration reduced the power consumption though the gain is not significant. For this particular machine and experimental configuration this result shows that at low work speed the vibratory grinding performed at the same level as conventional grinding in terms of power requirements.

5.3 Effect of Wheel Speed

Here, the peripheral speed of the grinding wheel was changed up to 35 m/s, whereas the other parameters remained constant (see Fig. 9).

It is observed here that the vibration-assisted grinding required less power than conventional. The reduction was more than 21.8% in some cases. The increase of the wheel speed led to increase of grinding power as power and wheel speed are proportional.

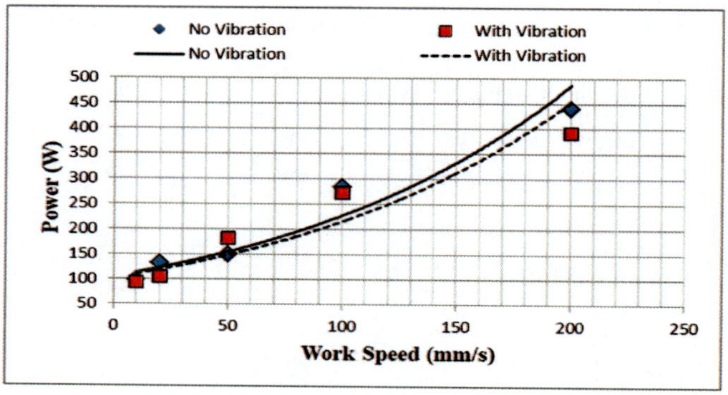

Fig. 8. Effect of work-speed

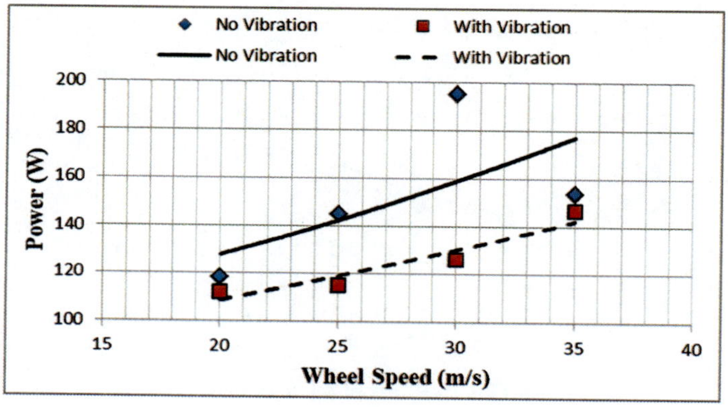

Fig. 9. Effect of wheel-speed

6 Results-Material Removal Rate

One of the most important factors in grinding process is the material removal rate Q as it gives a better indication of the performance of a wheel or a whole process. The material removal rate Q is given from the following formula [18]:

$$Q = b_w\, a_e\, V_w \tag{1}$$

Where a_e is the real depth of cut, b_w is the contact width and Vw the work-speed.

The specific material removal rate Q' is defined as the removal rate per unit grinding contact width and allows the results to be presented in a generic way. The units are in mm^3/mm/s [18].

$$Q' = a_e\, V_w \tag{2}$$

The grinding/experimental parameters for the specific test are presented below (Table 3).

The wheel speeds of 30 m/s was chosen due to the lowest amount of vibration produced to the spindle unit in past experiments. The work-speed of 50 mm/s was

Table 3. Fixed grinding parameters-specific material removal rate test

Grinding Parameter	Value
Wheel speed (Vs)	30 m/s
Work speed (Vw)	50 mm/s
Grinding condition	Dry
Wheel feed	Traverse
Vibration frequency	275 Hz
Vibration amplitude	15 μm (Peak)
Depth of cut	15 μm

chosen in a manner that vibration produced negative clearance according to vibro-impact theory. Moreover, at that value the effect of vibration was not high and it was interesting to examine the performance of different wheels and workpieces at that work-speed.

6.1 Mild Steel (BS970 080440, HRC 11)

The following graph presents the material removal rate of the three wheels used to grind mild steel.

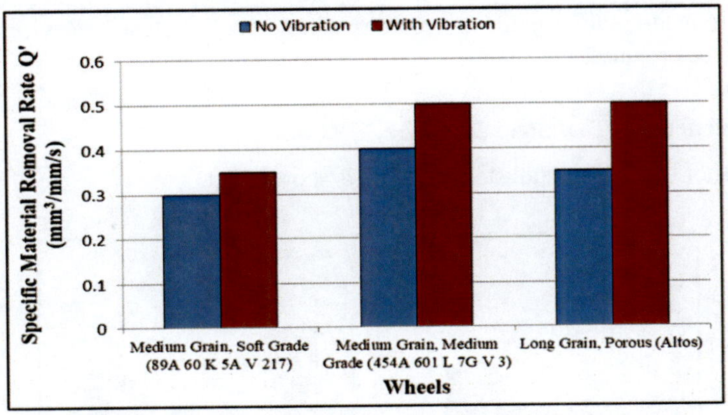

Fig. 10. Specific material removal rate results – Mild steel

According to Fig. 10 the superimposed vibration contributed to higher material removal rates for all three wheels. In this set of results, when grinding mild steel the medium grade and porous wheels reached up to 42.9% percentage improvement over conventional. Also, the nature of the wheel along with vibration increased the material removal rate as friable abrasives have large contact area. Large contact areas accomplish rapid stock removal. These wheels have the ability to expose new sharp edges more easily. The sharper the grain the better it cuts and it consequently produce higher material removal rates.

6.2 Mild Steel (BS970 080440, HRC 11)

The next set of results concern the hardened En31 workpiece material.

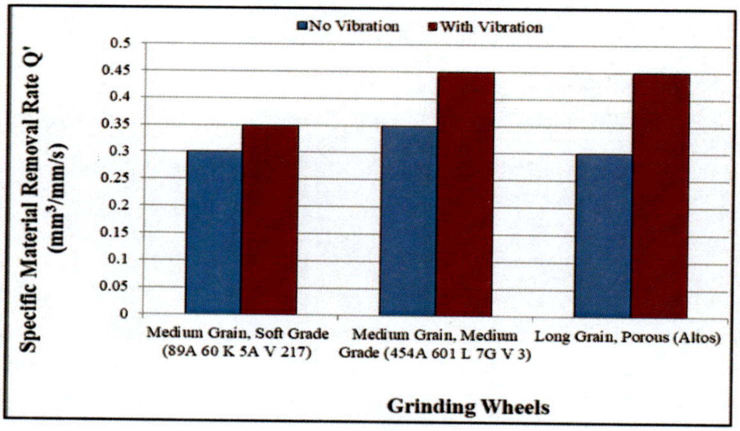

Fig. 11. Specific material removal rate results – En31

In Fig. 11 where the hardened En31 steel is machined, similar performance is observed. The superimposed vibration secured better cutting efficiency with the porous Altos wheel outperforming conventional by 50%. These results indicated that the medium grained-medium grade and porous wheels could be used for vibro-grinding for better performance.

6.3 Hardened M2 Tool Steel (BS BM2, HRC62)

The results of the hardest tool steel material is shown next.

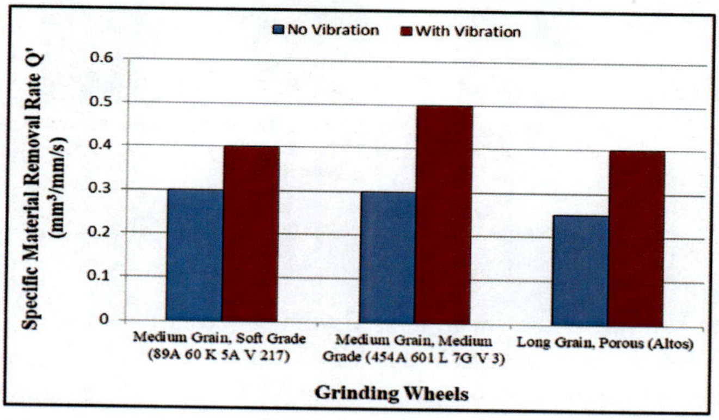

Fig. 12. Specific material removal rate results – M2

During vibration assisted grinding much higher values of material removal rate were achieved regardless of the type of grinding wheel (Fig. 12). In the present case the highest percentage of material removal increase was 66.6% and was detected during grinding with the medium grade wheel. The medium grade wheel is classified as semi-friable. Hence, its grains are not fractured very easily. The superimposed vibration allowed the grains to cut with two edges without breaking easy and achieve high material removal rate.

7 Conclusions

In the present work the effect of superimposed vibration on the work-piece during surface grinding process has been examined. A special jig that can accommodate and vibrate the work-piece during the process has been developed and its natural frequency has been established. Moreover, the natural frequencies of the machine tool spindle unit have been identified. A piezoelectric actuator used to excite the vibrating jig at its natural frequency in order to achieve high values of amplitude. The driving frequency of the system was chosen to not match the natural frequency of the wheel spindle unit in order to avoid resonance. The effect of superimposed vibration on power consumption and material removal rates during the process has been examined. It has been found that vibration improved the process performance in by reducing the power consumption of the system during grinding En31 (HRC 64.2) material. Specifically, it was observed that for small depths of cut the application of vibration performed as conventional grinding but as the depths of cut increased the reduction of power was obvious. However, it has been found that at low work-speeds the applied vibration did not contribute to power consumption decrease. Regarding the material removal rate, three different types of work-piece materials and three different types of aluminium oxide (Al_2O_3) grinding wheels have been utilised. One of the key findings is that the induced vibration in almost all cases improved the performance of the process by contributing to higher material removal rates. In some cases the increase of the removal rates reached more than 65%. Also, the nature and properties of each abrasive wheel play a vital role in specific applications. Grain size, hardness and structure of each wheel should be taken into account when targeting stock removal. Each wheel is manufactured to work under certain circumstances and grinding parameters (wheel speed, work speed) and with specific workpiece materials. Here, it was found that even when the combination wheel-material-parameters were not the ideal the application of vibration still secured an advantage over conventional processes.

References

1. Bespalova, L.V.: The theory of vibro-impact mechanisms. Izvestiya ANSSR, OTN, Moscow (1957). (in Russian)
2. Blekhman, I.I.: Vibration of nonlinear mechanical systems. In: Vibration in Engineering, 2nd edn. Moscow (1979). (in Russian)
3. Brehl, D.E., Dow, T.A.: Review of vibration-assisted machining. Precis. Eng. Cent. **32**(3), 153–172 (2007). https://doi.org/10.1016/j.precisioneng.2007.08.003
4. Wdowik, R.: Measurements of surface texture parameters after ultrasonic assisted and conventional grinding of carbide and ceramic samples in selected machining conditions. Procedia CIRP **78**, 329–334 (2018)

5. Dambatta, Y.S., Sarhan, A.A.D., Sayuti, M., Hamdi, M.: Ultrasonic assisted grinding of advanced materials for biomedical and aerospace applications - a review. Int. J. Adv. Manuf. Technol. **92**, 3825–3858 (2017)
6. Nik, M.G., Movahhedy, M.R., Akbari, J.: Ultrasonic-assisted grinding of Ti_6Al_4V alloy. Procedia CIRP **1**, 353–358 (2012)
7. Zhang, B., Hu, Z., Luo, H., Deng, Z.: Vibration-assisted grinding – piezotable design and fabrication. Nanotechnol. Precis. Eng. **4**(4), 282–290 (2006)
8. Zhong, Z.W., Yang, H.B.: Development of a vibration device for grinding with microvibration. Mater. Manuf. Processes **19**(6), 1121–1132 (2004)
9. Zhong, Z.W., Rui, Z.Y.: Grinding of single-crystal silicon using a microvibration device. Mater. Manuf. Processes **20**, 687–696 (2005)
10. Bechcinski, G., et al.: A model and application of vibratory surface grinding. J. Manuf. Sci. Eng. Am. Soc. Mech. Eng. (ASME) MANU-17-1727 **140**(10), 725 (2018). https://doi.org/10.1115/1.4040725
11. Babitsky, V.I.: Theory of Vibro-Impact Systems and Applications (Russian revised translation, Nauka, Moscow, 1978) (1998)
12. Cheng, C.H., Tony, L., Schmitz, G., Scott, D.: Rotating tool point frequency response prediction using RCSA. Mach. Sci. Technol. **11**(3), 433–446 (2007)
13. Filiz, S., Cheng, C.-H., Powell, K.B., Schmitz, T.L., Ozdoganlar, O.B.: An improved tool-holder model for RCSA tool-point frequency response prediction. Precis. Eng. **33**(1), 26–36 (2009). https://doi.org/10.1016/j.precisioneng.2008.03.003
14. Erturk, A., Ozguven, H.N., Budak, E.: Analytical modelling of spindle-tool dynamics on machine tools using Timoshenko beam model and receptance coupling for the prediction of tool point FRF. Int. J. Mach. Tool Manuf. **46**, 1901–1912 (2006)
15. Tsiakoumis, V., Batako, A.: Vibration assisted grinding of mild and hardened steel: a novel design vibrating jig and process performance. Trans. Am. Soc. Mech. Eng. ASME **4**, 141–146. ISBN 978-0-7918-4487-8 (2012)
16. Batako, A.D., Tsiakoumis, V.: An experimental investigation into resonance dry grinding of hardened steel and nickel alloys with element of MQL. Int. J. Adv. Manuf. Technol. **77**(1–4), 27–41. ISSN 0268-3768 (2014)
17. Tsiakoumis, V.: An Investigation into Vibration Assisted Machining – Application to Surface Grinding Processes, Ph.D. thesis (2011). https://doi.org/10.24377/LJMU.t.00006113
18. Rowe, W.B.: Principles of Modern Grinding Technology Chapter 2 (2009). https://doi.org/10.1016/B978-0-8155-2018-4.50008-2

Sustainable Flexible Job Shop Scheduling: A Systematic Literature Review

Jatoth Mohan(✉), Krishnanand Lanka, Neelakanteswara A. Rao, and Vijaya Kumar Manupati

Department of Mechanical Engineering, National Institute of Technology Warangal, Warangal 506004, India
mohanjatoth@gmail.com

Abstract. Job shop rescheduling has been classified according to applied methodologies or application context. With the beginning of the fourth industrial revolution, new perceptions and challenges for production rescheduling arise, mainly regarding sustainable and energy efficient rescheduling in stochastic environment. By means of a systematic literature review, this paper surveyed and analyzed various articles systematically related to flexible job shop scheduling problem in the context of sustainability, stochasticity and flexibility management. To identify the most appropriate papers, the search has been conducted with search strings framed using keywords in Web of Science, Scopus and Science Direct databases. Consequently, after final scrutiny 30 most related papers were selected and analyzed that further evaluated by identifying the issues and gaps in the existing empirical knowledge. Finally, this paper discusses studies that show how rescheduling affects the performance of production system considering sustainability and stochasticity for future research directions.

Keywords: Rescheduling · Sustainability · Stochasticity · Flexibility management · Dynamic scheduling

1 Introduction

Sustainability in manufacturing industries is gradually becoming an urgent requisite for today's manufacturing firms due to numerous well-known and evolving causes such as environmental alarms, shrinking non-renewable resources, firmer legislation and inflated energy costs, increasing customer preference for environmentally friendly products, etc. In actual production systems, scheduling is one of the significant factors at the operational phase that influence production efficiency, quality and cost.

Flexible Job shop Scheduling problem (FJSP) is extensively discussed in the literature Nouiri et al. [1]. The objective of this kind of problems is to properly allocate available resources to processes in order to optimize an objective function usually related to time, like makespan, tardiness, maximum lateness, etc. Most of the existing solution methods are proposed to emphasis on deterministic problems assuming that there is no source of uncertainties and that the problems data parameters are static and

A. Batako et al. (Eds.): GCMM 2021, LNNS 335, pp. 227–246, 2022.
https://doi.org/10.1007/978-3-030-90532-3_18

known in advance Jensen [2]. For example, the job processing time is generally considered constant, all the jobs are available at the release date and no disruptions occur on the job shop. Many new efficient meta-heuristics methods have been developed to get near-optimal solutions for deterministic FJSP. Among these methods, one can find the simulated annealing optimization algorithm Kaplanolu [3] and Particle Swarm Optimization algorithm Nouiri et al., Singh and Mahapatra [1, 4]. These methods aim to solve FJSP problem while considering mainly the minimization of the makespan value as the primary objective function of the optimization model. Currently, energy efficiency is taken into consideration in these problems. Energy costs are an increasingly important issue in real-world scheduling, for both economic and environmental reason Trentesaux and Giret [5] and some methods for energy consumption optimization in flexible job shop scheduling problem have been reported in the literature He et al. [6]. Nowadays, industrial sustainability plays a fundamental role within manufacturing systems, and accordingly, energy efficiency (EE) interventions are increasingly gaining practical interest as a component of sustainability strategy Tonelli et al. [7].

The structure of the paper is as follows. Section 2 describes the research methodology applied for this paper, which follows a SLR five-step approach. Section 3 presents the findings of the SLR organized according to the conceptualization described in Fig. 1. Section 4 discusses the obtained results, presenting meaningful research suggestions and directions for future work. Section 5 provides conclusions of the paper.

2 Methodology

This literature review paper adopted a systematic literature review (SLR) approach [8], which is a reproducible, explicit method, and different from the conventional literature survey. SLR methodologies begin with defining the review protocol that states the research questions to be addressed and the techniques which are required to conduct the review. Systematic reviews are based on a distinct search strategy that aims to identify as much of the related literature as possible.

As proposed by Denyer et al. [9], a background study of the research area was carried out before the SLR to (i) simplify the existing methods for the work to be formulated; (ii) state the adaptability of the planned SLR into the present literature; (iii) define appropriate concepts; (iv) frame the research queries to search research articles. Therefore, this paper systematically analyses related articles following a five-step approach, as mentioned by Denyer et al. [9] and Wong et al. [10]:

- Formulation of research questions
- Finding the studies
- Study selection and evaluation
- Data analysis and synthesis
- Reporting and using results

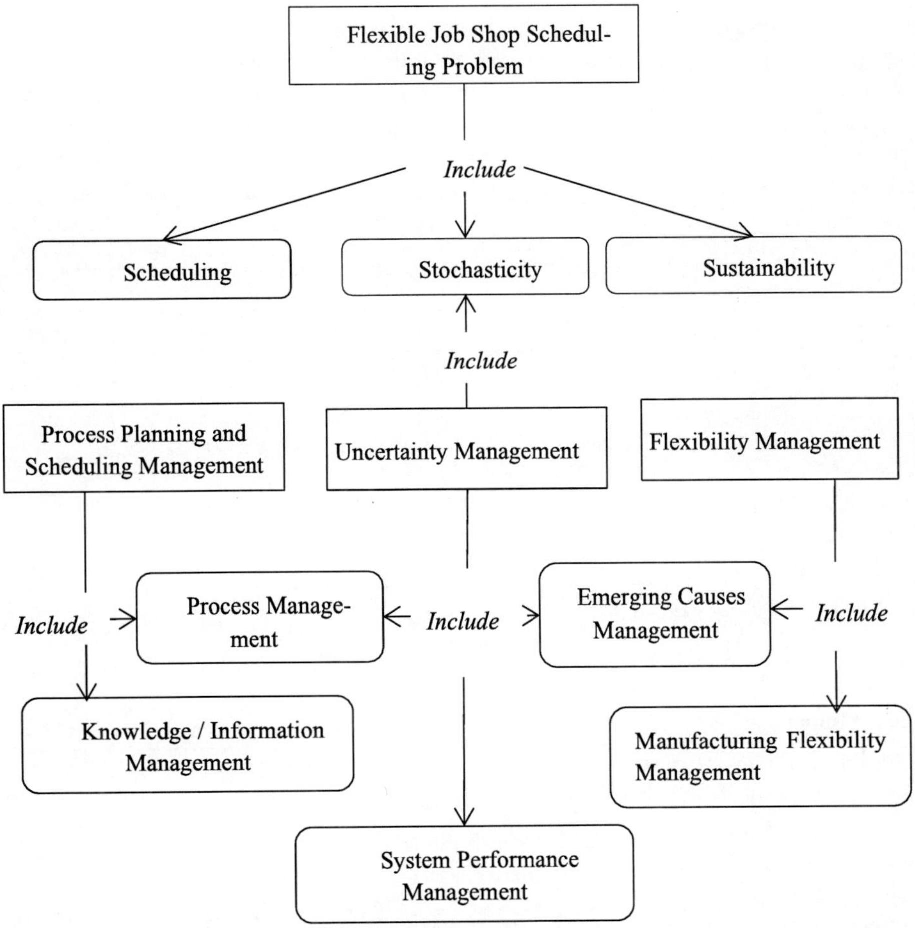

Fig. 1. Framework job shop scheduling considering stochasticity, sustainability and flexibility management.

***Step 1**: The following research queries were formulated for the present study*

RQ1: How sustainability and rescheduling are applied in flexible job shop scheduling problems?
RQ2: How to develop a research framework considering rescheduling and sustainable parameters in flexible job shop scheduling?

***Step 2:** Identifying the studies*

The search strings to find the papers were constructed through the combination of the Boolean operators "or, and" among the following five keywords: "flexible job shop scheduling" "rescheduling", "sustainability or sustainable", "simulation" and "optimization". The search strings were tightly aligned with the research questions. Following similar literature reviews [11, 12] the papers were collected from three reputed bibliographic databases, which cover the leading journals in engineering and other science areas: Scopus (www.scopus.com), Science Direct (www.sciencedirect.com), and Web of Science (www.webofknowledge.com). These databases cover a significant percentage of the published articles on internationalization, including the most relevant peer-reviewed journals on the area. The number of search results was shown in Tables 1, 2 and 3.

Table 1. Search strings and displayed corresponding results from Web of Sciences

Search string	Search field	Date of search	No. of articles
"Flexible job shop scheduling" and "Rescheduling"	Topic	26-11-2019	42
"Flexible job shop scheduling" and "Sustainability" (or) "Sustainable"	Topic	26-11-2019	17
"Flexible job shop scheduling" and "Simulation and Optimization" (or) "Simulation or Optimization"	Topic	26-11-2019	94

Table 2. Search strings and the displayed corresponding results from Scopus

Search string	Search field	Date of search	No. of articles
"Flexible job shop scheduling" and "Rescheduling"	Article Title, Keywords	26-11-2019	41
"Flexible job shop scheduling" and "Sustainability" (or) "Sustainable"	Article Title, Keywords	26-11-2019	40
"Flexible job shop scheduling" and "Simulation and Optimization" (or) "Simulation or Optimization"	Article Title, Keywords	26-11-2019	26

The initial search strings using the three bibliographic databases resulted in the identification of 1719 articles, i.e. 655 for Web of Science, 306 for Scopus and 758 for Science Direct.

Step 3*: Study selection and interpretation*

In the study selection and interpretation section, some inclusion and exclusion conditions were defined to select only the more significant studies to include in the review, i.e. the studies which address the research questions. A 5-year time horizon was first established (2015–2019) to focus the review on recent studies and recent methodologies and technologies. Further, only studies published in peer-reviewed journals in English were considered. Colicchia and Strozzi [13] claim that by restricting the search only to peer-reviewed journals, the quality control of search results can be improved prior to publication.

Table 3. Search strings and the displayed corresponding results from Science Direct

Search string	Search field	Date of search	No. of articles
"Flexible job shop scheduling" and "Rescheduling"	Title, Keywords	26-11-2019	64
"Flexible job shop scheduling" and "Sustainability" (or) "Sustainable"	Title, Keywords	26-11-2019	202
"Flexible job shop scheduling" and "Simulation and Optimization" (or) "Simulation or Optimization"	Title, Keywords	26-11-2019	180

The above procedure decreased the number of articles that has to be reviewed to 706. After checking duplicates, by using Mendeley and EndNote (first in each search string and after, considering the whole set), titles and abstracts of the selected studies were analyzed for significance. After, exhaustive review only papers that fulfil the below mentioned three criteria are chosen for further evaluation.

I. Articles specially related to rescheduling in flexible job shop scheduling
II. Articles related to sustainable, flexible job shop scheduling and rescheduling
III. Articles related to simulation and optimization of flexible job shop scheduling and rescheduling. In this step, finally reduction of articles done up to 72.

In order to classify critical issues, the content of each paper deeply investigated by the reviewers with their expertise. In this process, some of the articles did not align with the subject area were filtered out, and finally, 49 papers were listed out in Table 4. The comprehensive investigation of journals has been carried out by the authors, using two reviewers when their inquiries and findings are useful to reduce the errors, to sort out any differences. Detailed information from these 49 articles was classified based on methodologies and needed contributions. The later parts of the paper describe the outcome of the current work in a systematic procedure.

3 Findings

This section presents the following papers reviewed by the authors and highlights the problem description, the methodology followed, and tools used to solve the objectives. The studies for rescheduling, sustainability, simulation and optimization for FJSSP are listed year-wise in Table 5, Table 6 and Table 7 below respectively.

Table 4. Summary of the systematic review articles selection and evaluation

Bibliographic database analysis	Search 1	Search 2	Search 3	Total
Web of sciences	42	17	94	153
Scopus	41	40	26	107
Science Direct	64	202	180	446
Inclusive / Exclusive procedure for Web of sciences				
Time period	42	17	94	153
Document type	38	15	87	140
Research area	36	15	82	133
Language	36	15	82	133
Inclusive / Exclusive procedure for Scopus				
Date range	41	40	26	107
Document type	35	38	23	96
Research area	33	37	20	90
Language	33	37	20	90
Inclusive / Exclusive procedure for Science Direct				
Date range	64	202	180	446
Document type	53	178	145	376
Research area	46	135	114	353
Language	46	135	114	295
After checking duplicates (in each search)	48	92	79	219
After checking duplicates (in all search)	32	55	61	148
Title and abstract evaluation	72			
Finally detailed article analysis	49			

Table 5. List of studies for the Rescheduling-FJSSP relationship.

SI. No.	Reference	Findings
1	Bojun Liu et al. (2015)	Focused on the dynamic, flexible job-shop scheduling problem having fuzzy processing time. Estimation of distribution algorithm (EDA) is developed to solve the post-transforming problems
2	Kai Zhou Gao et al. (2015)	Models of combined processing constraint and virtual operation are analyzed to simplify and transform FJSP with combined processing constraints into FJSP
3	Kai Zhou Gao et al. (2016)	Improved Artificial Bee Colony algorithm is proposed for FJSSP with fuzzy processing time

(*continued*)

Table 5. (*continued*)

SI. No.	Reference	Findings
4	Tao Ning et al. (2016)	A simulation model was developed for solving the dynamic scheduling of flexible job-shop problems, and an improved hybrid multi-stage particle swarm algorithm is proposed
5	Kai Zhou Gao et al. (2016)	Two-stage ABC algorithms and their variants have been proposed for solving the FJSP, considering two constraints, namely new job insertion and fuzzy processing time
6	Kai Zhou Gao et al. (2016)	Investigated flexible job shop rescheduling problem after each new job insertion to minimize the makespan, the average of earliness and tardiness, maximum machine workload, and total machine workload
7	Sicheng Zhang (2016)	A hybrid approach for flexible job-shop rescheduling problems under a dynamic environment is presented by combining a multiagent system (MAS) and ant colony optimization (ACO)
8	Kai Zhou Gao et al. (2016)	Presented a flexible job shop rescheduling problem with new job insertion and a discrete Jaya (DJaya) algorithm
9	Ehsan Ahmadi et al. (2016)	Scheduling of multi-objective problems in flexible job shop scheduling with random machine breakdown is addressed. Two evolutionary algorithms, NSGA-II and NRGA, are applied to combine the improvement of makespan and stability simultaneously
10	Alper Hamzadayi et al. (2016)	In this paper, simulated annealing and seven dispatching rulebased complete rescheduling approaches are proposed for the dynamic m identical parallel machines scheduling problem
11	Xixing Li et al. (2017)	A hybrid artificial bee colony algorithm based on the ABC algorithm and Tabu Search algorithm has been developed to solve the FJSSP, and three rescheduling strategies have been introduced to address dynamic events
12	Sajad Karimi et al. (2017)	This paper incorporates the transportation times between the machines into the flexible job-shop scheduling problem

(*continued*)

Table 5. (*continued*)

SI. No.	Reference	Findings
13	Zigao Wu et al. (2018)	The job shop scheduling problem (JSP) with random machine breakdowns (RMBs) is studied
14	Lei Wang et al. (2018)	Proposed a dynamic rescheduling method based on a variable interval rescheduling strategy (VIRS) to deal with the DFJSSP considering machine failure, urgent job arrival, and job damage as disruptions. It also proposed an improved genetic algorithm (GA) for minimizing makespan with a mix of random initialization population by combining initialization machine and initialization operation with
15	Melissa Shahgholi Zadeh et al. (2018)	Variable Time FJSP model is proposed based on ABC algorithm that minimizes makespan considering the variation in the processing time of operations
16	Zhengcai Cao et al. (2019)	The adaptive scheduling algorithm is proposed to address the makespan minimization in the DJSSP
17	Alejandro Rossit et al. (2019)	Proposed a data-driven architecture for scheduling, in which the system has real-time access to data. Then, scheduling decisions are made ahead of time, on the basis of more information. This approach is based on the architecture of cyber-physical systems, with a data-driven engine that uses, Big Data techniques to extract vital information for Industry 4.0 systems
18	Zhen Wang et al. (2019)	A mixed-integer programming model is established to address the DJSSP when new jobs randomly arrived. The event-driven rescheduling policy and the match-up strategy are combined with rescheduling strategies
19	Shokraneh et al. (2019)	Proposed a new method of rescheduling for new order arrival, based on the concept of dynamic pegging in multi-level production

Table 6. List of studies for the Sustainability-FJSSP relationship.

SI. No.	References	Findings
1	Ying Liua et al. (2015)	An optimization approach developed based on NSGA-II is proposed for reducing energy consumption for carbon fiber reinforced polymer (CFRP)
2	Xu Gong et al. (2015)	Proposed a generic mixed-integer linear programming model to enable the job scheduling on a single machine to minimize the energy cost without exceeding the due date

(*continued*)

Table 6. (*continued*)

SI. No.	References	Findings
3	Hamed Piroozfard et al. (2016)	Minimized total carbon footprint and total late work criterion simultaneously, proposing a multi-objective genetic algorithm to obtain high-quality non-dominated schedules
4	Hadi Mokhtari et al. (2017)	Energy-efficient scheduling in a shop floor industrial environment is designed
5	Lvjiang Yin et al. (2017)	A Multi-objective Genetic Algorithm (MOGA) based on the design of a simplex lattice is used to solve the proposed low-carbon FJSP model
6	Yingfeng Zhang et al. (2017)	A dynamic game theory-based two-layer scheduling method was developed to reduce makespan, the total workload of machines, and energy consumption to achieve real-time multi-objective flexible job shop scheduling
7	Silviu Raileanu et al. (2017)	Proposed an agent-based approach for measuring in real-time energy consumption of resources in job-shop manufacturing processes
8	Qiong Liu et al. (2018)	A product carbon footprint-oriented multi-objective flexible jobshop scheduling optimization model is proposed. NSGA-II is adopted to solve the proposed model
9	Jin Wang et al. (2018)	A multi-period production planning based real-time scheduling (MPPRS) approach for the IoT-enabled low-carbon flexible job shop (LFJS) is presented
10	Xu Gong et al. (2019)	Proposed an energy and labor aware flexible job shop scheduling problem (EL-FJSSP) under dynamic electricity pricing and developed an NSGA-III for many-objective optimization of this problem
11	Pablo Vallejos Cifuentes et al. (2019)	Proposed two production-scheduling problems for flow shop and job shop environments considering the processing speed as the primary energy-related decision variable. An elitist MOGA was used to solve them
12	Yang Kou et al. (2019)	Investigated a dynamic rescheduling problem for a flexible machining system with random machine breakdown and urgent job arrivals
13	Shu Luo et al. (2019)	The multi-objective flexible job-shop scheduling problem (MOFJSP) with variable processing speeds is Investigated. A multiobjective grey wolf optimization (MOGWO) algorithm is proposed

(*continued*)

Table 6. (*continued*)

SI. No.	References	Findings
14	Oussama Masmoudi et al. (2019)	Job-shop scheduling problem with energetic aspects is considered to minimize production costs in terms of energy consumption
15	Xiuli Wua et al. (2019)	A multi-objective optimization model is formulated for the flexible job-shop scheduling problem considering deterioration effect and energy consumption simultaneously
16	Maroua Nouiri et al. (2019)	An energy-efficient scheduling and rescheduling method named Green Rescheduling Method (GRM) was proposed considering routing flexibility as the mean to handle perturbations

Table 7. List of studies for simulation and Optimization-FJSSP relationship.

SI. No.	References	Findings
1	Ye Xu et al. (2015)	An effective teaching learning-based optimization algorithm (TLBO) is proposed to solve the flexible job shop problem with fuzzy processing time
2	Hadi Mokhtari et al. (2015)	A joint production and maintenance scheduling problem was investigated where a flexible job-shop scheduling system is working, and operating machines should be maintained preventively
3	Deming Lei et al. (2016)	In this study, FJSP with the minimization of workload balance and the total energy consumption is considered, and the conflicting between two objectives is analyzed
4	Nilsen Kundakci et al. (2016)	Introduces an efficient hybrid Genetic Algorithm (GA) methodologies for minimizing makespan considering dynamic events such as random job arrivals, machine breakdowns, and changes in processing time
5	Wenjun Xua et al. (2016)	Proposed a sustainable scheduling optimization of manufacturing equipment services based on joint energy and production model
6	Maroua Nouiri et al. (2017)	A two-stage particle swarm optimization algorithm has been proposed in order to find a robust and stable predictive schedule considering machine breakdowns
7	Raviteja Buddala et al. (2017)	A two-stage teaching-learning-based optimization (2S-TLBO) method is used to solve the flexible job shop scheduling problem under machine breakdown

(*continued*)

Table 7. (*continued*)

SI. No.	References	Findings
8	E.R.R. Kato et al. (2018)	An integrated Particle Swarm Optimization algorithm with the auxiliary metaheuristics, Random-Restart Hill Climbing (PSO+RRHC), was presented for the resolution of the Flexible Job Shop Scheduling Problem (FJSP)
9	Chao Chen et al. (2018)	Focused on multi-objective dynamic, flexible job shop scheduling problem (MODFJSP) with machine breakdown
10	Yang et al. (2018)	Established a multi-objective optimization scheduling model for FJSP and developed a parallel hybrid optimization algorithm for production rescheduling under interference
11	Tianhua Jiang et al. (2018)	Bi-population based discrete cat swarm optimization algorithm (BDCSO) was presented to solve the low-carbon flexible job shop scheduling problem
12	Huaping Mu (2019)	A method for measuring the disturbance of different behavior subjects in the workshop production system and NSGA-II algorithm is proposed to meet the need for real-time solutions for production scheduling interference
13	Farbod Amiri et al. (2019)	Investigates uncertain resource assignment and job sequence problem using a simulation-optimization approach based on multiresponse efficiency in an automated, flexible job shop
14	Dai Min et al. (2019)	A multi-objective optimization model with the objective of minimizing energy consumption and makespan is formulated for a flexible job shop scheduling problem with transportation constraints

3.1 Scheduling

The Scheduling is the process of generating a production schedule for a given set of jobs and resources. Rescheduling is the process of updating an existing production schedule in response to disruptions or other changes. This includes the arrival of new jobs, machine failures, and machine repairs. The rescheduling strategy identifies the list of jobs that need to include in the new schedule. It also specifies when and how rescheduling is done. The events that trigger rescheduling may be predictable or unpredictable. Different rescheduling methods may be used in different situations. In the literature reported, there are mainly three categories of dynamic scheduling technologies, which are completely-reactive, predictive–reactive, and pro-active scheduling. Among them, predictive–reactive scheduling is the most commonly used. It has a scheduling/rescheduling process where previous schedules are revised to adapt to the new environment caused by dynamic events. In job shop scheduling problem, each operation of a job should be processed on a predefined machine only once in a fixed operation sequence. However, the wide employment of multi-purpose machines in the real-world job shop makes it more general that an operation can be managed by several machines, i.e., there are alternative routings, which is the called as flexible job shop scheduling problem (FJSSP).

3.2 Stochasticity

Stochasticity in scheduling concerns scheduling problems with random attributes, such as random processing times, random due dates, random weights, and stochastic machine breakdowns. Major applications arise in manufacturing systems, computer systems, communication systems, logistics and transportation, machine learning, etc. The objective of the stochastic scheduling problems can be regular objectives such as minimizing the total flowtime, the makespan or the total tardiness cost of missing the due dates, etc. The performance of such systems, as evaluated by a regular performance measure or an irregular performance measure, can be significantly affected by the scheduling policy adopted to prioritize over time the access of jobs to resources. The goal of stochastic scheduling is to identify scheduling policies that can optimize the objective.

Stochastic scheduling problems can be classified into two broad types: problems concerning the scheduling of a batch of stochastic jobs, and problems concerning the scheduling of queueing systems. These two types are usually under the assumption that complete information is available in the sense that the probability distributions of the random variables involved are known in advance. When such distributions are not fully specified, and there are multiple competing distributions to model the random variables of interest, the problem is referred to as incomplete information. The Bayesian method has been applied to treat stochastic scheduling problems with incomplete information.

3.3 Sustainability

Recently, increasing energy consumption has become a major concern in our world. Total demand for energy consumption has become double within the last 40 years and it is forecasted to become double once again till 2030. The industrial sectors are facing the huge challenges in terms of both environmental and economic issues. It is, therefore, crucial to reducing the energy consumption in the industrial sector via energy efficient planning and optimization [14]. Moreover, a major capacity of enhancing energy efficiency, there exists in industrial plants [15]. For this reason, an important policy of industrial plants is to design efficient mechanisms to reduce energy consumption for saving the cost and developing environment consistent operations [16]. One of the possible solutions, which is a relatively new paradigm in literature, is to cope with the shop floor scheduling of industrial operations. Although many scheduling problems have been developed in the literature, the field of energy efficient scheduling is relatively new [17]. Although there are numerous studies in the fields of sustainable supply chain management, sustainable operations management, and energy saving in the industry, the scheduling is rarely considered as a tool for improving sustainability in terms of energy efficiency. Most of the traditional scheduling models just considered the economic objectives to plan the industrial operations. There is, therefore, increasing need to devise new energy-based objectives in order to optimize the trade-off between economic and energy performance objectives. This orientation becomes more important, particularly when these objectives have conflict behaviors with each other. Hence, increasing efforts have been attracted to design scheduling problems with the objective of energy efficiency [18]. Different studies discussed several approaches to design energy efficient operations in various levels of industry [19–22]. Among industrial operations, the production and maintenance are most effective functions [23]. From the operational perspective, optimization of the production and maintenance scheduling is an appropriate and efficient mechanism for industrial plants to decrease energy consumption without any redesign in structural settings. By the appropriate planning of these operations, it is possible to gain tangible improvements in energy consumption and cost. The scheduling addresses to the problem of achieving the suitable assignment and sequence of production and maintenance operations with the aim of optimizing some special objectives [24]. As a common approach in literature, both of these operations have been studied separately. The maintenance optimization is usually carried out in distinct maintenance systems without any detailed scheduling of production operations [25]. However, as we know, this cannot be a reasonable setting in real-world industries. There are usually strong interactions between production and maintenance operations in the shop floor industrial plant. For this reason, joint scheduling of the production and the maintenance operations has become a common concern in many industrial plants. Therefore, it is necessary to collaborate on production and maintenance scheduling in order to optimize system performance and energy efficiency.

3.4 Process Planning and Scheduling Management

Detailed process planning is part of work preparation and defines feasible process parameters of process control variables (e.g. feed or cutting speed). Despite recent developments in the field of machine data acquisition and analysis, detailed process planning, especially for turning processes, is often carried out manually based on the experience of the machine operator. This reduces the flexibility of the production due to the dependence on experienced personnel. In addition, the overall process knowledge of the company is not fully exploited due to different machine operators as well as limited documentation of the processes. A knowledge-based planning of process parameters could help to overcome these problems and ensure the selection of suitable process parameters even when production orders are rescheduled on short notice.

Process Management: Detailed process planning is the last step in work preparation before the start of production and followed by production control [26]. Since the choice of process parameters depends strongly on machine tool characteristics, re-scheduling requires a revision of the selected process parameters [27]. For the subsequent data analysis, the use of machine learning methods is investigated in addition to existing planning methods. An approach considering data mining can be found at Grobe Bockmann et al. [28].

Information Management: The ontology is intended to support transversal process-related management allowing interaction between the different functionalities. As a result, databases can be well located and their data are readily available and can be transformed into useful information. One key aspect of the decision-making process is information extraction, which should ultimately result in information quality. Information quality refers to the precision of the information in terms of the time, content and clarity required for a specific task. Specifically, optimization techniques require accurate information for reaching meaningful results for decision-making. Indeed, the interoperability and exchange of data from different systems and hierarchical levels are one of the most critical aspects in the daily operation of many organizations. This activity can be drastically improved by combining knowledge and expertise management approaches.

3.5 Flexibility Management

Scheduling plays an important role in energy efficient machining [29], and has attracted much research attention to address scheduling under system characteristics, e.g., single machine, job shop and flexible job shop. For instance, Wang et al. [29] investigated a bi-objective single machine batch scheduling problem with the time-of-use electricity prices and different energy consumption rates of the machines. May et al. [30] focused on the effects of production scheduling policies to improve productive and environmental performance in a job shop system. Zhang et al. [31] proposed an integrated model based on nonlinear process planning to implement flexible job-shop scheduling for energy savings. Similar work can be found in [32], where a two-stage energy saving optimization method is proposed to select machine tool for each task first followed by scheduling of all tasks on a given machine tool.

In a practical machining system, the presence of a variety of unexpected disruption to the original schedule, such as machine breakdown, urgent jobs, job cancelation and due date changing, etc., makes the scheduling problem much challenging [33]. In fact, a new type of scheduling by using reactive strategies, sometimes called rescheduling, has to be applied in dealing with such disruption to minimize any negative impact on production. With the increasing concern on energy consumption of machining, a group of researchers started to investigate the energy efficient rescheduling problem. For instance, Nouiri et al. [34] focused on reducing energy consumption and makespan when resolving dynamic flexible job-shop scheduling problem under machine breakdowns only. Tang et al. [35] considered both machine breakdowns and urgent job arrivals in a flexible flow shop, and proposed an approach to address the energy and production efficiency of the dynamic scheduling problem. A similar work can be found in [36] with only single objective of the minimum energy consumption. Note that all of these works assumed that each job has a predetermined sequence of operations. Process routing and machine selections have significant impact on scheduling. Such impact was analyzed in Zhang et al. [37] to develop an energy saving approach for a flexible manufacturing system.

Manufacturing Flexibility Management: Today's business is set in a global environment which demands improved manufacturing and services. The competition in the global market is forcing the companies to develop strategies for manufacturing varieties of the parts to meet the ever-growing and ever-changing demands of the customer. Technological advancement and emerging markets have opened up new ways of recognizing, adapting and optimizing the flow of materials, information and resources. To achieve these requirements manufacturing companies not only need to be flexible and responsive to the changes, but also be able to produce a variety of products in a short span of time and with minimum cost.

The development in technology, material and customer preferences has resulted in products with shorter life span. New products and designs require changes in production facilities. Moreover changing customer preferences and intense competition have resulted in uncertainty in the demand patterns. Recent breakthroughs in technology and management technologies have made manufacturing sector cope up with the changing environment. There is an ever increasing pressure from suppliers, existing competitors, and new competitors. The need of the hour for manufacturing enterprise is to change from hard automation to flexible automation. Flexible Manufacturing System (FMS) helps to fulfil this requirement of the manufacturing enterprise. Browne et al. [38] has defined FMS as an integrated, computer-controlled complex system of automated material handling devices and numerically controlled tools that can simultaneously process medium-size volumes of medium variety parts. They have also proposed two basic manufacturing flexibility types: machine flexibility and routing flexibility. Based on these basic flexibility types other types of flexibility like product flexibility, process flexibility, operation flexibility, volume flexibility, expansion flexibility and production flexibility can be derived. Chen and Chung [39] have proposed three alternative sequencing models aimed at minimizing the operation assignments and demonstrated the importance of utilizing an alternate routing policy in FMS environment. Caprihan and Wadhwa [40] investigated the impact of routing flexibility on an FMS with makespan as the performance measure. They concluded that oversupply of flexibility could not guarantee a better performance, but in contrast, it increases the makespan due to the tool set-up compensation.

Emerging Cause's Management: There are many uncertainties in job-shop Scheduling Problem (JSP) process, and these uncertainties mainly come from disturbance factors that cannot be accurately described in the real world. Including the temporary departure of processing personnel or sudden failure of the processing machine and other unexpected events, as well as the floating data field of the workpiece with uncertain installation time, uncertain completion time and uncertain delivery date. The interference management is mainly based on the degree and nature of the disturbance of the problem, establishing a corresponding optimization model and designing an effective strategy to locally adjust the initial scheme to generate a minimized disturbance update scheme. The key to interference management is to control the adjustment of the initial scheme within the smallest possible range to achieve the least effect of the disturbance.

Researchers mainly studied the problem of workshop scheduling disturbance under uncertain conditions from the aspects of fuzzy Researchers processing time and fuzzy delivery time. For example Ehmke and Campbell [41] proposed a multi-objective flexible batch scheduling algorithm based on multi-population genetic algorithm. Gevaers et al. [42] proposed a hybrid particle swarm optimization algorithm that combines Baldwinian learning strategy with simulated annealing. Yap et al. [43] uses an improved adaptive particle swarm algorithm based on hormone regulation to solve JSP problems.

3.6 Uncertainty Management

Most of the researchers assume that the scheduling environment is a static environment in which no unexpected event occurs during the execution of the schedule. However, the real production environments are dynamic, and numerous random or unforeseen events such as machine breakdowns, uncertain processing times, the arrival of a new job, and cancellation of jobs may occur. Therefore, scheduling under uncertainty has been given significant focus in recent years. In fact, the occurrence of disruptions and unexpected events in scheduling problems makes the obtaining of stable solutions more valuable than the finding of optimal solutions that ignore these disruptions. In these environments, the best schedule is the one with high systematic performance and low deviation from the initial schedule after disruptions. To overcome the disruptions, three fundamental approaches have been employed, which are the reactive, proactive and proactive-reactive approaches [44, 45].

System Performance Management: Manufacturing industries generate and update production schedules, which are plans that state when certain controllable activities like processing of jobs by resources should take place. In dynamic, stochastic manufacturing environments, managers, production planners, and supervisors must not only generate high-quality schedules but also react quickly to unexpected events and revise schedules in a cost-effective manner. These events, generally difficult to take into consideration while generating a schedule, disturb the system, generating considerable differences between the predetermined schedule and its actual realization on the shop floor.

Rescheduling is required in order to minimize the effect of disturbances in the performance of the system. The various types of disruptions in the production schedule are machine failures, change in processing time, rush orders, quality problems, and unavailable material. As Bean et al. [46] state that, rescheduling is a dynamic approach that responds to disruptions considering future events. In practice, rescheduling is done periodically to plan activities for the subsequent time period based on the state of the system. It is also done occasionally in response to significant disruptions. Rescheduling is the process of updating a present production schedule in response to disruptions or other changes. This includes the arrival of new jobs, machine failures, and machine repairs Disruptions can change system status and affect performance. The system performance will deteriorate drastically and will lead to rescheduling to reduce the impact. The frequent unexpected events that trigger rescheduling are machine failure, urgent job arrival, job cancellation, change in the due date, shortage or defect of raw materials, change in job priority, change in process time and etc.

4 Discussion and Research Agenda

Some prospects can be drawn from the review of the literature and the projects. First of all, despite the increasing efforts of researchers, and given the future evolutions, scheduling taking sustainability into consideration should still be urgently addressed, but taking means other than energy into consideration (including output means), to design globally effective sustainable scheduling systems. Multiple objective methods take on their full meaning in this context. Second, it will also be compulsory to propose hybrid scheduling methods since manufacturing systems are dynamic by nature. Unforeseen situations such as machine failures, new jobs entering the system, dynamic available energy profile or cost will be very common during the normal execution of production lines in the near future. As a consequence, the link with the tactic level, and especially planning activities, must be carefully studied. This level can provide a broader view of the manufacturing system along with a larger sliding time window, which would help the manager to react in short term according to a more global view, avoiding, in this way, local inefficient or ineffective (thus, unsustainable) reactive decisions. For example, closing the loop by considering output means as possible input means (e.g., energy losses re-used to heat processes, etc.) has rarely been addressed in manufacturing and the linking between operational and tactic levels may ease the integration of such a mechanism.

These first two needs, leading to more global attention being focused on sustainable scheduling, are clearly highlighted in the prospective paper [93], in which it is proposed that managing input means such as energy consumption and output means such as carbon foot print should be achieved both predictively and reactively in order to make decisions according to real-time events. Typically, the definition of opportunistic behavior in scheduling will help to save means or to cope with the increasing volatility of input mean availability. The main issue faced by researchers is that designing classical, time-based optimization scheduling methods is already a hard task, mainly due to the well-known NP-hard features of the optimization problems, as introduced earlier in this paper. Thus, adding new constraints, new conflicting and multi-dimensional objectives, the need to define new precise multi-physics behavioral modeling, new variables, and considering all this in a volatile environment, may be perceived as unsolvable for researchers. Consequently, the authors think that changes in paradigms must occur to override this major difficulty. A possible improvement would be to couple classical "top-down" scheduling approaches, centralized or hierarchized, with emerging "bottom-up" approaches, where decisions can also be made close to the events that generated the need to make them. In this context innovative approaches and concepts such as cyber-physical systems, intelligent products, and product-service systems to name a few, can provide some insights to contribute to the development of effective and efficient sustainable scheduling systems. Other scheduling techniques such as bio-inspired ones may also provide good insights.

5 Conclusions

Manufacturing industries have a large number of scheduling problems. Since most of these problems are dynamic and there are many disruption, recovery techniques are needed to re-establish the original scheduling as soon as possible. Moreover, industries are facing increasing requirements for sustainability, so energy consumption processes should be minimized. New features of future JSP for Industry 4.0 are highlighted as decentralized scheduling, in contrast to the traditional centralized scheduling. Therefore the future trend of the JSP scheduling problem should lie in the development and utilization of the smart agent in the Industry 4.0. When the traditional centralized job shop scheduling is turned into the smart distributed scheduling, the computational workload can be greatly reduced and the system will become more flexible and agile.

The actual job shop scheduling system is becoming more and more complex, dynamic and flexible. According to the development direction of the system, we not only need to consider JSP, but multi-machines, multi-resources, even a number of factories and logistics system. So there are four types of job shop scheduling, which are JSP, Flexible JSP, Multi-resource FJSP and Multi-plants-based FJSP. However overwhelming majority of researchers still focus on the JSP and FJSP model although it is different from the actual complex production dramatically.

With the development of large data, high performance computing, cloud computing, deep learning and simulation technologies, different smart agents can predict results by themselves when scheduling is changed. With the prediction outputs, agents can construct their own multi-objective optimization models and solve them to achieve the best scheduling with current algorithms.

References

1. Nouiri, M., Bekrar, A., Jemai, A., Niar, S., Ammari, A.C.: An effective and distributed particle swarm optimization algorithm for flexible job-shop scheduling problem. J. Intell. Manuf. **29**(3), 603–615 (2015). https://doi.org/10.1007/s10845-015-1039-3
2. Jensen, M.T.: Generating robust and flexible job shop schedules using genetic algorithms. IEEE Trans. Evol. Comput. **7**(3), 275–288 (2003)
3. Kaplanoğlu, V.: An object-oriented approach for multi-objective flexible job-shop scheduling problem. Expert Syst. Appl. **1**(45), 71–84 (2016)
4. Singh, M.R., Mahapatra, S.S.: A quantum behaved particle swarm optimization for flexible job shop scheduling. Comput. Ind. Eng. **1**(93), 36–44 (2016)
5. Trentesaux, D., Giret, A.: Go-green manufacturing holons: a step towards sustainable manufacturing operations control. Manuf. Lett. **1**(5), 29–33 (2015)
6. He, Y., Li, Y., Wu, T., Sutherland, J.W.: An energy-responsive optimization method for machine tool selection and operation sequence in flexible machining job shops. J. Clean. Prod. **15**(87), 245–254 (2015)
7. Tonelli, F., et al.: Assessment of mathematical programming and agent-based modelling for off-line scheduling: application to energy aware manufacturing. CIRP Ann. **65**(1), 405–408 (2016)
8. Tranfield, D., Denyer, D., Smart, P.: Towards a methodology for developing evidence-informed management knowledge by means of systematic review. Br. J. Manag. **14**(3), 207–222 (2003)
9. Denyer, D., Tranfield, D.: Producing a systematic review. In: The Sage Handbook of Organizational Research Method, pp. 671–689 (2009)
10. Wong, C., Skipworth, H., Godsell, J., Achimugu, N.: Towards a theory of supply chain alignment enablers: a systematic literature review. Supply Chain Manage. Int. J. **17**(4), 419–437 (2012)
11. Camargo Pérez, J., Carrillo, M.H., Montoya-Torres, J.R.: Multi-criteria approaches for urban passenger transport systems: a literature review. Ann. Oper. Res. **226**(1), 69–87 (2014). https://doi.org/10.1007/s10479-014-1681-8
12. Hassini, E., Surti, C., Searcy, C.: A literature review and a case study of sustainable supply chains with a focus on metrics. Int. J. Prod. Econ. **140**(1), 69–82 (2012)
13. Colicchia, C., Strozzi, F.: Supply chain risk management: a new methodology for a systematic literature review. Supply Chain Manage. Int. J. (2012)

14. May, G., Taisch, M., Stahl, B., Sadr, V.: Toward energy efficient manufacturing: a study on practices and viewpoint of the industry. In: Emmanouilidis, C., Taisch, M., Kiritsis, D. (eds.) APMS 2012. IAICT, vol. 397, pp. 1–8. Springer, Heidelberg (2013). https://doi.org/10.1007/978-3-642-40352-1_1
15. Weinert, N., Chiotellis, S., Seliger, G.: Methodology for planning and operating energy-efficient production systems. CIRP Ann. **60**(1), 41–44 (2011)
16. Apostolos, F., Alexios, P., Georgios, P., Panagiotis, S., George, C.: Energy efficiency of manufacturing processes: a critical review. Procedia CIRP **1**(7), 628–633 (2013)
17. Mansouri, S.A., Aktas, E., Besikci, U.: Green scheduling of a two-machine flowshop: trade-off between makespan and energy consumption. Eur. J. Oper. Res. **248**(3), 772–788 (2016)
18. Ding, J.Y., Song, S., Wu, C.: Carbon-efficient scheduling of flow shops by multi-objective optimization. Eur. J. Oper. Res. **248**(3), 758–771 (2016)
19. Devoldere, T., Dewulf, W., Deprez, W., Willems, B., Duflou, J.R.: Improvement potential for energy consumption in discrete part production machines. In: Takata, S., Umeda, Y. (eds.) Advances in life cycle engineering for sustainable manufacturing businesses, pp. 311–316. Springer London, London (2007). https://doi.org/10.1007/978-1-84628-935-4_54
20. Jin, K., Zhang, H.C., Balasubramaniam, P., Nage, S.: A multiple objective optimization model for environmental benign process planning. In: 2009 16th International Conference on Industrial Engineering and Engineering Management, pp. 869–873. IEEE, 21 October 2009
21. Sun, Z., Biller, S., Gu, F., Li, L.: Energy consumption reduction for sustainable manufacturing systems considering machines with multiple-power states. In: International Manufacturing Science and Engineering Conference, vol. 44311, pp. 99–103, 1 January 2011
22. Hu, S., Liu, F., He, Y., Hu, T.: An on-line approach for energy efficiency monitoring of machine tools. J. Clean. Prod. **1**(27), 133–140 (2012)
23. Yu, A.J., Seif, J.: Minimizing tardiness and maintenance costs in flow shop scheduling by a lower-bound-based GA. Comput. Ind. Eng. **1**(97), 26–40 (2016)
24. Puzis, R., Shirtz, D., Elovici, Y.: A particle swarm model for estimating reliability and scheduling system maintenance. Enterp. Inf. Syst. **10**(4), 349–377 (2016)
25. Zhu, H., Li, M., Zhou, Z.: Machine scheduling with deteriorating and resource-dependent maintenance activity. Comput. Ind. Eng. **1**(88), 479–486 (2015)
26. Henning, H.: Modellbasierte Prozess-Adaption in der Feinplanung fertigungstechnischer Prozessketten. PZH, Produktionstechn. Zentrum (2012)
27. Schmidt, J.: Integrated work planning and production control based on status information. PZH Verlag, TEWISS-Technik und Wissen GmbH (2015)
28. Große Böckmann, M., Krappig, R., Stolorz, M., Schmitt, R.: Data-Mining in der Produktion*. Neue Methoden für eine robuste Prozessentwicklung. Werkstattstechnik Online **103**(11/12), 921–925 (2013)
29. Wang, S., Liu, M., Chu, F., Chu, C.: Bi-objective optimization of a single machine batch scheduling problem with energy cost consideration. J. Clean. Prod. **20**(137), 1205–1215 (2016)
30. May, G., Stahl, B., Taisch, M., et al.: Multi-objective genetic algorithm for energy-efficient job shop scheduling. Int. J. Prod. Res. **53**, 7071–7089 (2015)
31. Zhang, Z.W., Tang, R.Z., Peng, T., et al.: A method for minimizing the energy consumption of machining system: integration of process planning and scheduling. J. Clean. Prod. **137**, 1647–1662 (2016)
32. Wang, H., Jiang, Z.G., Wang, Y., et al.: A two-stage optimization method for energy-saving flexible job-shop scheduling based on energy dynamic characterization. J. Clean. Prod. **188**, 575–588 (2018)
33. Shen, X.N., Yao, X.: Mathematical modeling and multi-objective evolutionary algorithms applied to dynamic flexible job shop scheduling problems. Inf. Sci. **298**, 198–224 (2015)

34. Nouiri, M., Bekrar, A., Trentesaux, D.: Toward energy efficient scheduling and rescheduling for dynamic flexible job shop problem. IFAC PapersOnLine **51**(11), 1275–1280 (2018)
35. Tang, D.B., Dai, M., Salido, M.A., et al.: Energy-efficient dynamic scheduling for a flexible flow shop using an improved particle swarm optimization. Comput. Ind. **81**, 82–95 (2016)
36. Zhang, L., Li, X., Gao, L., Zhang, G.: Dynamic rescheduling in FMS that is simultaneously considering energy consumption and schedule efficiency. Int. J. Adv. Manuf. Technol. **87**(5–8), 1387–1399 (2013). https://doi.org/10.1007/s00170-013-4867-3
37. Zhang, C.Y., Jiang, P.Y., Zhang, L., et al.: Energy-aware integration of process planning and scheduling of advanced machining workshop. J. Eng. Manuf. **80**, 1–16 (2016)
38. Browne, J., Dubois, D., Rathmill, K., Sethi, S.P., Stecke, K.E.: Classification of flexible manufacturing systems. FMS Mag. **2**(2), 114–117 (1984)
39. Chen, I.J., Chung, C.H.: Effects of loading and routeing decisions on performance of flexible manufacturing systems. Int. J. Prod. Res. **29**(11), 2209–2225 (1991)
40. Caprihan, R., Wadhwa, S.: Impact of routing flexibility on the performance of an FMS—a simulation study. Int. J. Flex. Manuf. Syst. **9**(3), 273–298 (1997)
41. Ehmke, J.F., Campbell, A.M.: Customer acceptance mechanisms for home deliveries in metropolitan areas. Eur. J. Oper. Res. **233**(1), 193–207 (2014)
42. Gevaers, R., Van de Voorde, E., Vanelslander, T.: Cost modelling and simulation of last-mile characteristics in an innovative B2C supply chain environment with implications on urban areas and cities. Procedia Soc. Behav. Sci. **125**, 398–411 (2014)
43. Yap, M.D., Correia, G., Van Arem, B.: Preferences of travellers for using automated vehicles as last mile public transport of multimodal train trips. Transp. Res. Part A Policy Pract. **1**(94), 1–6 (2016)
44. Aytug, H., Lawley, M.A., McKay, K., Mohan, S., Uzsoy, R.: Executing production schedules in the face of uncertainties: a review and some future directions. Eur. J. Oper. Res. **161**(1), 86–110 (2005)
45. Vidal, T.: The many ways of facing temporal uncertainty in planning and scheduling. In: Proceedings. 11th International Symposium on Temporal Representation and Reasoning. TIME 2004, pp. 9–10. IEEE, 3 July 2004
46. Bean, J.C., Birge, J.R., Mittenthal, J., Noon, C.E.: Matchup scheduling with multiple resources, release dates and disruptions. Oper. Res. **39**(3), 470–483 (1991)

Aircraft Brake Pad Reconstruction and Fabrication

P. Jeyapandiarajan, P. Aswath(✉), and M. Anthony Xavior

School of Mechanical Engineering, VIT Vellore, Vellore 632014, India
ashwath.p@vit.ac.in

Abstract. In recent times, copper matrix based composite material for aircraft brake pads were commonly in use. Copper extracted from brake pads can have lethal effects on water quality in cities, the marine environment and the environment. A need for a new brake pad material was hence needed to be explored. For improvement of braking performance as well as to reduce the adverse effect to environment. In this research project aluminium powder composite type brake pads. Evaluating the various combinations in composites manufactured of matrix material AA 2900 and LM6 with reinforcements such as SiC, Al_2O_3 and graphene novel findings on the material for replacing the material for manufacturing the brake pad material for aerospace application has been arrived. Considering the brake pad material strength – microstructural relationship and tribological properties are very important aspects are to be achieved. Based on this AA 2900 and LM6 are evaluated for property testing in three stages. Microwave processing of composites in as sintered and furnace cooled condition, sintered and solution treated condition and tribological studies on the as sintered and solution treated condition. AA 2900 – 6 wt. % Al_2O_3 exhibited good hardness properties and compressive strength value and frictional coefficient values in both as sintered and solution treated condition. For an aerospace brake pad application, AA 2900 – 6 wt. % Al_2O_3 is strongly recommended as a replacing composite material where the necessary properties like strength – microstructural relationship, tribological behavior and aging kinetics are satisfied without any compromise.

Keywords: Copper matrix · Brake pads · Marine life · Braking performance · Aluminium powder composite

1 Introduction

Since past asbestos, semi metallic, non-asbestos organics, low steel and copper etc. have been used as brake pad materials, along with copper as a major braking material. Copper is used in higher weight percentage due to various reasons. Being a very good conductor and easily allows heat transfer and dissipation of heat from friction of the brake pads. It is used as a binder in the matrix of the friction composite and prevents shuddering and squeaking of brake pads. It easily held up in the cold conditions and kept the surface of the rotor clean due to its abrasive nature [1].

A. Batako et al. (Eds.): GCMM 2021, LNNS 335, pp. 247–256, 2022.
https://doi.org/10.1007/978-3-030-90532-3_19

It was found that an aircraft brake pads needs to be replaced once the lining material gets less than 0.098 inches [2]. This wear off of the lining material is achieved in just 120 landings by an aircraft. Jeffrey A. Nason et al. (2012) mentioned in his research that spills may be important sources of metals such as Cu, Zn, Cd, Pb and Ni, for example among these metals, copper is particularly worrying because it affects very small Catfish species. Recent studies have shown that dissolved copper concentrations above 0.18 to 2.12 μg/l (≤3 μg/l) can affect the species of carp. The optimum copper content is (0.5–1 μg/l) any concentration above this is found to be affecting marine life [3]. S. Y. Aku and co-workers (2012) mentioned in their study that asbestos fibers are carcinogenic and, after inhalation or ingestion of asbestos fibers, can cause cancer and chronic respiratory disease, thereby avoiding the use of asbestos fibers [4].

Rathod Abhik et al. (2014) have developed SiC-reinforced lightweight alloy brake pads to increase strength and abrasion resistance and provide the benefits of a low-density matrix. In this metal matrix composite, aluminum is a metallic matrix in 2014 and SiC is a reinforcing material [5]. There are two examples. In the first sample, aluminum was 80% in 2014 and the SiC improvement rate was 20%. In the second sample, aluminum in 2014 was 90% and the SiC content was 10%. It was concluded that the abrasion rate was higher (20%) than the second sample (10% increase) due to the lower density of the first sample. At high density, the hardness increases. Therefore, Sample 1 (20% addition) has a higher hardness than Sample 2 (10% increase). As the porosity of the steel decreases, the porosity increases by 7%. As the BHN (Brinell hardness value) increases, the overall wear rate increases by 48%.

RK Uyyuru et al. (2006) used pin-on-disk (asymmetric, compliant) testing techniques to study the tribological behavior of Al-Si/SiCp compounds that are hybridized in vehicle brake pads in their research data [6]. The sheet material is a metal mesh composite material (Al-MMC), although the material of the brake pad is made as a bolt. The material used for the plate is Al-MMC of various formulations. The pen is made of cushioning material and the polymer matrix composite (PMC) is scratch resistant and scratch resistant, both the wear rate and the wear factor depend on the normal load and the sliding speed. However, the wear rate or friction coefficient is different with respect to changes in the normal load and the slip rate. The normal load is the most important parameter for the wear performance. Then the effect of the speed and concentration of the abrasive seems to depend again on the bond. When the gain of SiCp in the matrix is large, a higher wear rate and an erosion coefficient (friction) are observed compared to a matrix containing an improvement in size.

The layer of friction is a thin layer of adhesive which serves primarily as a mixture of grinding tests and as a protective layer on the matrix. As a result, the formation of the contact layer plays an important role in the behavior of the Al-Si/SiCp metal-based composite material and the wear characteristics of the brake linings during use. This paper examines the chemical and microstructural changes that occur during the brake pad material surface brake simulation test. The optimal concentration of copper and other heavy materials in water is found to be 0.5–1 μg/L. Carbon brake pads are also one of the best alternatives for copper and its use is getting popular due to its increased life and ability to provide effective breaking in higher temperatures (aircraft brake temperature due to friction--) but due to its high cost its use is limited.

Aluminium is used as an alternative to copper, as a major component in the friction material. Aluminium has a good specific heat capacity, conduction, lighter and also very inexpensive compared to other alternatives used such as steel and carbon brake pads. Aluminium powder matrix seems to be a convincing material which can obtain an equivalent braking effect, durability and ability of withstanding the enormous heat released during the braking of an aircraft.

This project aims to reconstruct brake pad material constituting of aluminium powder as a major braking component with minimum percentage of copper (0.1–0.5% by weight). Testing various composition and surface treatment of the prepared brake material to achieve the conventional number of braking cycles and to provide a cheap and environment friendly braking material. That can attain the functional requirements of the aircraft brake pad.

2 Methodological Study

Powder metallurgical path have been taken for brake pad manufacturing and the various stages involved in the manufacturing the sample pillets are shown in Fig. 1. Precations are taken at all the stages of manufacturing the samples. Samples manufactured in the forms of pillet are shown in Fig. 2. Samples were also prepared through squeeze casting process since it's associated with high pressure die casting process.

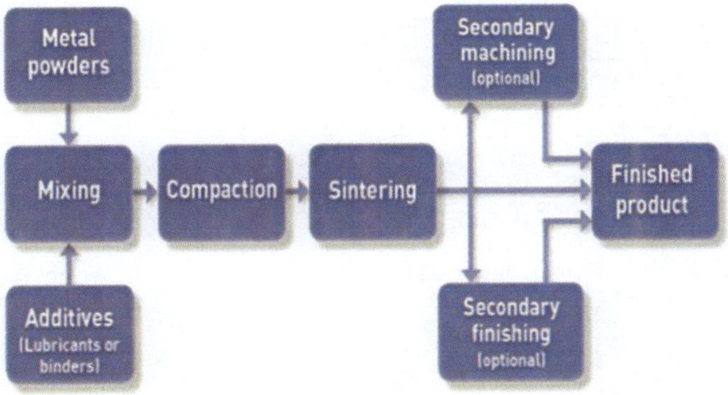

Fig. 1. Process of powder metallurgy

3 Experimental Procedure

Powders taken for the study are AA 2900 and LM6 along with graphene of 10 nm is used as one of the reinforcement material for manufacturing the MMC samples. The chemical composition of the aluminium alloys 2900 and LM 6 are shown in Table 1. Powders were mixed with 0.2 weight % of graphene with 6 weight percent of aluminium oxide mixed properly, then compacted and sintered to get the samples. Prepared samples were initially tested for density analysis as shown in Table 2. Figure 3 shows the density level

Fig. 2. Fabricated samples

for the selected process. Prepared samples were taken to analysis the various mechanical and metallurgical properties. As the method of testing various and the sample required for testing also varies. Samples were prepared and tested as per their standards.

Table 1. Chemical composition of the aluminium alloy 2900 and LM6

Elements	Cu	Mg	Si	Fe	Sn	Al
AA2900 Weight %	3.25	1.47	0.31	0.1	0.67	Remaining
LM6 Weight %	0.95	0.1	11.3	0.4	0.05	Remaining

Table 2. Samples of AA2900 and LM 6 with varying reinforcement and processes

Denisty – Sintered and casted					
S. No	Metal matrix	Method of fabrication	Reinforcement	Weight %	Density
1	LM6	PM	Graphene + Al_2O_3	0.2 + 6	
2	LM6	SC	Graphene + Al_2O_3	0.2 + 6	
3	AA 2900	PM	Graphene + Al_2O_3	0.2 + 6	
4	AA 2900	SC	Graphene + Al_2O_3	0.2 + 6	

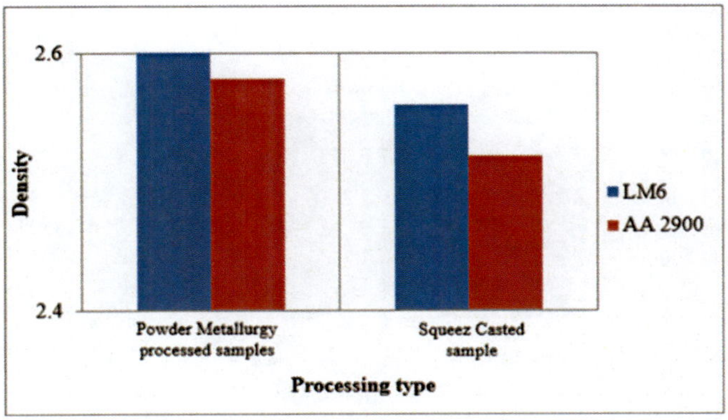

Fig. 3. Density level for selected process

4 Results and Discussion

Sample 1, 2, 3, 4 are prepared as per the Table 1. Prepared samples as shown in Fig. 2 are taken for the hardness testing with Rockwell machine. The scale used remain as B and with testing force 30 kgf with dwell time 10 s. The value shown by the indenter remains as shown in Table 3 and Fig. 4.

Table 3. Experimental values for the various mechanical testing

Sample	Hardness (HV)	Compressive strength (MPa)	Shear strength (MPa)	Tensile strength (MPa)
1	75	292.5	60	262.5
2	71	276.9	56.8	248.5
3	93	325.5	74.4	325.5
4	87	304.5	69.6	304.5

AA2900 shows higher hardness value as 93 and 87 through PM and SC process respectively. LM6 shows lower value as 75 and 71 for PM and SC process respectively.

For the compressive strength analysis the sample taken as total weight of the sample remain 57 g, diameter of the sample remain 20 mm and the height of the sample remain as 40 mm. Compressive strength has been measured and the results are shown in Table 3. Figure 5 shows the graphical representation for the same. AA2900 shows higher compressive strength compare to the LM6. Powder metallurgical sample shows higher compressive strength then the squeezed casted sample. Shear strength are measured at shear testing equipment and the values are tabulated in Table 3. Graphical representation for the same has been shown in Fig. 6. Tensile test for the same has been shown in the Table 3 and the graph referring to the tensile strength are shown in Fig. 7.

Wear test remains the key measurement frictional studies, here the frictional co-efficient for the samples are tabulated in Table 4. Wear test was carried out through

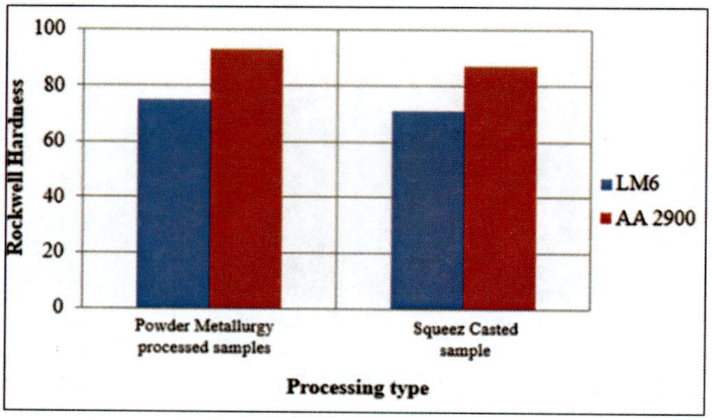

Fig. 4. Rockwell hardness vs processing type

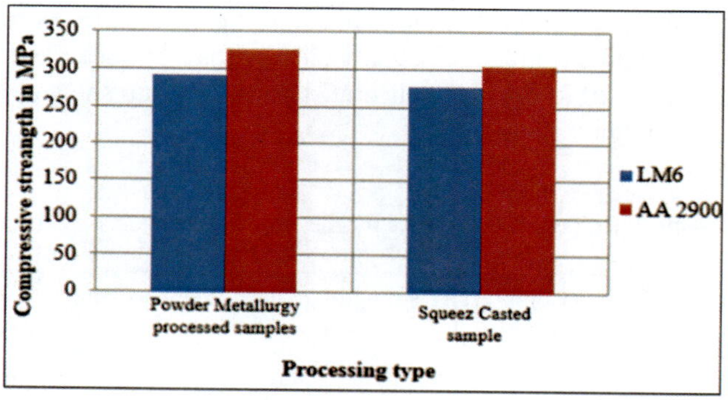

Fig. 5. Compressive strength vs process type

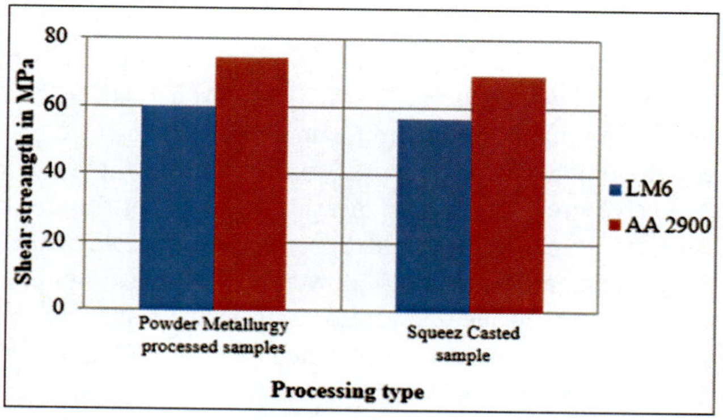

Fig. 6. Shear strength vs process type

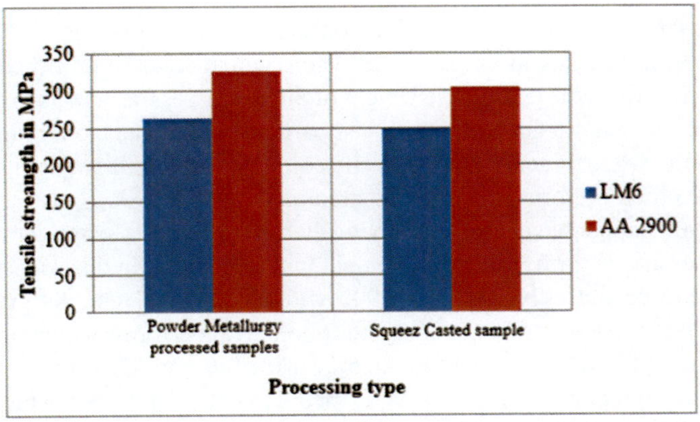

Fig. 7. Tensile strength vs process type

Table 4. Wear test for the sample

Samples	Wear (mm)	COF	No. of landings completed
1	2.1	0.412	90
2	2.3	0.433	97
3	1.6	0.376	102
4	1.68	0.391	109

pin-on disc apparatus. Since the brake pad of the aircraft undergoes wear and tear at its operation. Wear loss expected was 1.5 mm and CoF as 0.32 in the cases of LM6 wear has crossed the target limit by 50% but considering the AA2900 at the PM and SC conditions wear falls shot by only 10%. Considering the landing trails expected was 120 AA2900 brake pad only falls shot by the 11 trails which can be improved further in the heat treatment cycle. Looking into the results Al_2O_3 and graphene as reinforcing materials, it was found that the Al_2O_3 improvers increase the hardness of the matrix, and the addition of graphene resulted in lower hardness values compared to Al_2O_3 and SiC. Although the difference is not large, Al_2O_3 works better than silicon carbide for better bond. It was found that the compressive strength was increased to 6% by weight of the reinforcing agent, and when the reinforcing material was added, a tendency to reduce the compressive strength was observed. Therefore, the threshold increases in percentage to achieve maximum compressive strength. More than 6% by weight of reinforcing agent is added. The values for density and compressive strength were reduced, but a higher hardness was observed. Therefore, regardless of the purpose of increasing hardness and abrasion resistance, a higher reinforcing ratio can be used.

Among the aluminum alloy matrixes AA 2900, and LM6, the copper alloyed matrix exhibited improved strength to microstructure relationship, tribological characteristics compared to LM6 with SiC, Al_2O_3 and graphene. Si doped with the matrix material improves wetting and allows uniform distribution of the reinforcing particles, particularly in compounds reinforced with 6 wt.% Al_2O_3 during microwave irradiation. Among them, LM7 with 12.7% Si leads to excessive diffusion, which leads to reinforcement in the matrix.The increase in the percentage weight of SiC increases the defects in the structure of the matrix, such as vacuum and improved particle which leads to a reduction in the specific gravity, strength and hardness of the compound. Overall strength, tribological and microstructural performance was observed better in AA 2900.

In AA 2900 copper have a major alloying element, there is a scope for heat treating studies on the sample for further improvement in the property. In this aspect the further studies are carried out only with AA 2900 with SiC and Al_2O_3. As graphene as a reinforcement exhibiting lesser strength and tribological performance compared to SiC and Al_2O_3 the aging studies are carried out only using SiC and AL_2O_3 as reinforcement. Reinforcement weight percentage is also restricted to 3 to 9 wt. % as the higher ceramic content till 20 wt. % which leads to a reduction of the mechanical properties of the compound.

4.1 Tribological Behavior of the Samples

A dry consumption test was conducted using a plate wear tester using a 60 HRC steel cup. These pins are aluminum-metal matrix composites reinforced with various reinforcing materials (SiC, Al_2O_3, and graphene particles having an average particle size of 10 microns and 10 nm. The important imformation are drawn in this study:

1. The improvement in MMC wear depends to a large extent on the type of reinforcement and its total weight.
2. Increasing the reinforcement of the metal matrix composite body is expected to increase wear resistance and reduce the coefficient of friction. For AA 2900 and LM6, a copper alloy with an aluminum matrix has better wear properties than LM6.
3. MMC have better abrasion resistance, especially if the samples are reinforced with Al_2O_3 instead of silicon carbide and graphene due to their good load-bearing capacity and good adhesion.
4. When the sample was quenched and aged, a coefficient of friction was observed and the MMC sample with Al_2O_3 matured for 2 h showed better wear factor and weight loss than other compositions.
5. When the sample is sintered with a microwave oven, a 6 wt% Al_2O_3 enhancer has an improved coefficient of friction and weight loss, regardless of the matrix material, charge, and effects of aging.
6. At low pressure, a through groove and some damaged areas formed on the worn surface of the sample, so that silicon carbide was used as the reinforcing material regardless of the matrix material. Observe the chain of evenly inclined and longitudinally damaged areas at higher pressures in the wear direction.
7. The friction coefficient decreases as the load increases to reach the minimum, and then the friction coefficient increases again when it is determined that the sample reaches the resistance limit for additional wear.
8. When the sample is sintered with a microwave oven, Al_2O_3 has better tensile strength and friction coefficient when extracting the particles than the SiC-reinforced MMC.

Aging samples play an important role in improving the erosion properties of the sample, with an aging of 6 wt% Al_2O_3 and AA 2900 for microwave sintering of 2 h showing less weight loss and coefficient of friction.

5 Conclusion

Evaluating the various combinations in composites manufactured of matrix material AA 2900 and LM6 with reinforcements such as SiC, Al_2O_3 and graphene novel findings on the material for replacing the material for manufacturing the brake pad material for aerospace application has been arrived. Considering the brake pad material strength – microstructural relationship and tribological properties are very important aspects are to be achieved. Based on this AA 2900 and LM6 are evaluated for property testing in three stages. Microwave processing of composites in as sintered and furnace cooled condition, sintered and solution treated condition and tribological studies on the as sintered and solution treated condition. AA 2900 – 6 wt. % Al_2O_3 exhibited good hardness properties and compressive strength value and frictional coefficient values in both as sintered and solution treated condition. For an aerospace brake pad application, AA 2900 – 6 wt. % Al_2O_3 is strongly recommended as a replacing composite material where the necessary properties like strength – microstructural relationship, tribological behavior and aging kinetics are satisfied without any compromise.

References

1. Cahill, T.A., Barnes, D.E., Spada, N.J., Lawton, J.A., Cahill, T.M.: Very fine and ultrafine metals and ischemic heart disease in the California central valley 1: 2003–2007. Aerosol Sci. Technol. **45**(9), 1123–1134 (2011)
2. AMT school (a.n.d.) Brake Inspection and Service (Part One). Flight Mechanic
3. Nason, J.A., Bloomquist, D.J., Sprick, M.S.: Factors influencing dissolved copper concentrations in oregon highway storm water runoff. J. Environ. Eng. **138**(7), 734–742 (2012)
4. Aku, S.Y., Yawas, D.S., Madakson, P.B., Amaren, S.G.: Characterization of periwinkle shell as asbestos-free brake pad materials. Pacific J. Sci. Technol. **13**(2), 57–63 (2012)
5. Abhik, R., Umasankar, V., Xavior, M.A.: Evaluation of properties for Al-SiC reinforced metal matrix composite for brake pads. Procedia Eng. **97**, 941–950 (2014)
6. Uyyuru, R.K., Surappa, M.K., Brusethaug, S.: Tribological behavior of Al–Si–SiCp composites/automobile brake pad system under dry sliding conditions. Tribol. Int. **40**(2), 365–373 (2007)
7. Eriksson, M., Bergman, F., Jacobson, S.: On the nature of tribological contact in automotive brakes. Wear **252**, 1–11 (2002)
8. Österle, W., Griepentrog, M., Gross, T.I.: Chemical and microstructural changes induced by friction and wear of brakes. Wear **252**(1–12), 1–8 (2001)
9. Gurunath, P.V., Bijwe, J.: Friction and wear studies on brake-pad materials based on newly developed resin. Wear **263**, 1–8 (2007)
10. Compression Machines, C.T. (n.d.). https://www.testresources.net/test-machines/by-test-type/compression-test-machines/
11. Wear Measurement, W.T. (n.d.). http://emrtk.unimiskolc.hu/projektek/adveng/home/kurzus/korsz_anyagtech/1_konzultacio_elemei/wear_testing_measurement.htmlm

12. Hardness calibrations, A.S. (n.d.). https://www.hardnesstesters.com/test-types/rockwell-hardness-testing
13. Squeeze casting, S. (n.d.). https://www.efunda.com/processes/metal_processing/squeeze_casting.cfm
14. Powder metallurgy process, T.P. (n.d.). https://www.pm-review.com/introduction-to-powder-metallurgy/the-powder-metallurgy-process/
15. Shear strength Testing, S.S. (n.d.). https://www.ametektest.com/learningzone/testtypes/shear-strength-testing
16. Abarghouie, S.M., Reihani, S.S.: Aging behavior of a 2024 Al alloy-SiCp composite. Mater. Des. **31**(5), 2368–2374 (2010)
17. Asif, M., Chandra, K., Misra, P.S.: Development of aluminium based hybrid metal matrix composites for heavy duty applications. J. Miner. Mater. Charact. Eng. **10**(14), 1337 (2011)
18. Bayraktar, E., Katundi, D.: Development of a new aluminium matrix composite reinforced with iron oxide (Fe_3O_4). J. Achievements Mater. Manuf. Eng. **38**(1), 7–14 (2010)
19. Bekheet, N.E., Gadelrab, R.M., Salah, M.F., El-Azim, A.A.: The effects of aging on the hardness and fatigue behavior of 2024 Al alloy/SiC composites. Mater. Des. **23**(2), 153–159 (2002)

Robotic Grinding for Surface Repair

Mohammed Sufian and Xun Chen(✉)

General Engineering Research Institute, Liverpool John Moores University, Byrom Street, Liverpool L33AF, UK
x.chen@ljmu.ac.uk

Abstract. Robotic grinding is considered as an alternative machining towards an efficient and intelligent machining of components due to its flexibility, intelligence and cost efficiency, particularly in comparison with the current mainstream manufacturing modes such as CNC machines. The advances in robotic grinding during the past years aims to solve problems of precision machining in small scale surfaces and other emphasizes on the efficient machining of large-scale surfaces. In this work, a method was investigated to improve surface repair accuracy by eliminating the workpiece datum error by directly engaging the grinding wheel. In fact, the proposed method uses acoustic emission sensing technique to detect grinding contact so as to estimate correct reference datum. Process variables based on machining parameters such as depth of cut, wheel speed, feed speed, dressing condition and system time constant is used to as a key parameter for controlling the robot to conduct the grinding process. The geometrical relationship and machining precision level developed has reached an accuracy level of 30 μm and error is been controlled by considering the process variables such as depth of cut, wheel speed, feed speed, dressing condition and system time constant which is the key for controlling the robot to conduct grinding process. The recorded data provide a significant evidence to support the viability of implementing a 6-axis robotic system for various grinding applications, combining more quality and critical surface finishing practices, and an increased focus on the size and form of generated components.

Keywords: Robot grinding · Datum identification · Error compensation

1 Introduction

Robotic grinding produces a significant challenge due to its flexibility and accuracy particularly in comparison with the current mainstream manufacturing methods such as the use of CNC machines. For example, the precision concerns related to measurement area associated of the workpiece is crucial due to the geometric error of the surface of the component. It is an efficient machining process due its flexibility, cost efficiency and intelligence. They have the ability to create an enormous amount of cutting cycles which makes it flexible to machine or repair parts.

The challenge in measuring the accuracy control is that it is difficult to accurately measure the actual position of the robot when locating the workpiece which tends to

A. Batako et al. (Eds.): GCMM 2021, LNNS 335, pp. 257–272, 2022.
https://doi.org/10.1007/978-3-030-90532-3_20

provide failure to the algorithm matching the machining purposes causing incorrect values, profile errors and distribution problems [1].

In general, the geometrical accuracy of the machined product mainly depends on the kinematics of the machine, location of workpiece and tool location [2]. The geometric errors influence the location and orientation of the location of the workpiece and lead to misalignments of the workpiece. Researchers have proposed different methods based on error models, these models focus on the position and posture in relation to the joints of the robot. For example, Xiong et al. discussed the inner force distribution and load capacity of fixtures as well as the contact forces [3]. Marin and Ferreria discussed the impact of error on the location of geometry and tolerance of the workpiece [4]. Both concluded that the designed models need to have a direct relationship to achieve the required accuracy and the precision using a robot.

However, when repairing a surface on a component using a robot, it is difficult to obtain information of the without building a relationship between the workpiece and the robot. Establishing such a relation mode involves multi-sensor data in combination with the robot motion, which provide the necessary action guide to avoid any system errors caused by improper sensor resolution and installation [5]. Therefore, it is necessary to put forward robot motion control strategies which are suitable for collaborative machining and develop collaborative control software to realize the task allocation and interference avoidance.

This paper presents a method for damaged surface repair by combining welding and grinding operations. Welding is used to recover lost the material and grinding is used to reconstruct the functional surface with an adaptive strategy. Such a strategy will allow grinding cycle to achieve the required surface finish in a minimum machining time for the best economic efficiency.

2 Contact Registration for Workpiece Datum

A contact registration is a procedure used to position and tolerate an object in the robot work envelop to create a reference system for geometrical position measurement. The traditional method used in many robotic applications to determine the workpiece datum is mainly using CMM (Coordinate measuring machine) touch prob. The inspection procedure is dependent on the dimension of the workpiece. However, the type of tools for probing may also be different from one to another too. For example, sensors or dial gauge method can be used to detect the surface of the workpiece to obtain the tool offset. Serval researchers has approached different methodologies to define a datum. For example, Jin and Jiyong developed a mathematical algorithm using three different coordinate systems to find a reference point to the workpiece to try to eliminate the influence error based on 3D measurements [6]. Chaiprapat and Rujikietgumjorn developed a model to predict the geometrical variation of the workpiece surface and datum features at a given workpiece [7]. Khondaygan analyzed errors in the workpiece–fixture–cutting tool system, he proposed to a relationship between the locating errors and their sources [8]. Lizarralde et al. presented a simulation software tool to facilitate grinding machine to achieve stable conditions [9].

Given the fact that the force control may cause machining errors. Pan and Zhang highlighted a control method based on compensations of deflections and adaptive material removal rate [10]. Bisu et al. used a frequency-based method to measure the dynamic response of the robot when cutting at designated points, his method was not directly involved to machining path [11]. Zaghbani et al. have collected the cutting forces signals and vibrations in order to find a reliable dynamic stability machining with respect to spindle speed [12]. Zhao et al. investigated the effects of grain sizes, contact force, linear velocity and feed rate on the surface roughness in abrasive belt grinding of aviation blades by analyzing the response of surface [13]. Dumas et al. evaluated joint stiffness based on consideration of translational and rotational displacement of the robot end effector at a given force and torque, they concluded that joint stiffness values can be used for motion planning to optimize robot machining process but results were not validated making it un-reliable to use [14].

However, by establishing the geometrical relationship between reference datum and probe point of measurement, the error level of datum accuracy could be assessed based on the repeatability where the repeatable points are detected on the surface of the workpiece to assist the robot to define the reference point on the workpiece. In this work the acoustic emission sensor is used to provide real-time feedback to the robot system to monitor and control the detection process using the robot tool tip to define the reference point by eliminating the errors existed between the grinding wheel and conventional measuring prober.

3 Theoretical Modeling

A theoretical model based on repeatability measurement is established to define the workpiece datum. The actual datum position of the workpiece can be estimated based on multiple points collected from the surface of the workpiece through a vector model to minimise the error measurement between the real and nominal datum. Through this model, the actual position of the workpiece is estimated and error could be compensated by applying suitable machining strategy.

A common formula of a plane in space can be presented as:

$$\mathbf{aX_i + bY_i + cZ_i + d = \varepsilon_i \; for \; i, \ldots n} \tag{1}$$

Considering the measurement errors $\boldsymbol{\varepsilon_i}$ the measured points on plane should satisfy:

$$\mathbf{aX + bY + cZ + d = 0} \tag{2}$$

As for variables **X**, **Y** and **Z** are coordinates taken from the different point of the surface of the plane on the block, **a**, **b**, **c** and **d** are constants that defines the plane position, and $\boldsymbol{\varepsilon_i}$ is the geometrical error.

$$\mathbf{Z_i = b_0 + b_1X_i + b_2Y_i + \varepsilon_i \; for \; i, \ldots, n} \tag{3}$$

where $b_0 = -d/c$, $b_1 = -a/c$ and $b_2 = -b/c$. By defining $\mathbf{Z} = (Z_1, Z_2, \ldots, Z_n)'$, $\mathbf{B} = (b_1, b_2, \ldots, b_n)'$, $\boldsymbol{\varepsilon} = (\varepsilon_1, \varepsilon_2, \ldots, \varepsilon_n)'$.

The geometrical error function becomes

$$\mathbf{Z = BX + \boldsymbol{\varepsilon}} \tag{4}$$

Having established the error in each plane, the plane equation can be derived based on random points selected on the surface of each plane. For example, Eq. 5 below reflects the surface equation for the XY plane.

$$\mathbf{Z} = \mathbf{d} + (\mathbf{aX}) + (\mathbf{bY}) \tag{5}$$

Note that the plane value, XY in the above case, is now reflective of any given point within the total XY plane. Consequently, transposition of the surface equation above, to find the constant 'd', yields the axis equation for the square. As shown in Eq. 6 below.

$$\begin{aligned} d_1 &= (a_1X) + (b_1Y) - Z \\ d_2 &= (a_2X) + (c_2Z) - Y \\ d_3 &= (b_3Y) + (c_3Z) - X \end{aligned} \tag{6}$$

In three-dimensional space and integration of the variables and constants specified for each plane, the datum point X_0, Y_0, Z_0 of the robot work envelope can be established by solving Eq. 6 and be arranged accordingly to determine error values of each plane.

$$[\mathbf{E}] = [\mathbf{A_0}] + [\mathbf{A}][\boldsymbol{\alpha}] \tag{7}$$

Where $\mathbf{A_0}$ denotes the origin points, **A** represents the points developed from the axial equations and $\boldsymbol{\alpha}$ is the results generated in the repeatability testing for each plane.

The transition angle can be corrected through vector models to compensate the transitional error of the datum point as shown in Fig. 1. The existence error between the real datum and nominal datum of the workpiece can be estimated and implemented into the robot accordingly. The datum is corrected through vector models to compensate the error and define the actual position of the workpiece to start grinding.

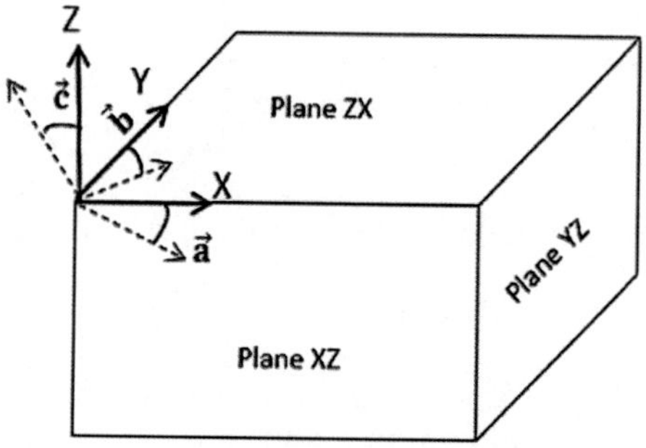

Fig. 1. Transition angle correction for constructed geometry

$$\begin{aligned} \textit{Transition Angle } \vec{a} &= \pm\left(\tan^{-1}\left(\tfrac{\Delta Y}{\Delta L_x}\right)\right) + \left(\tan^{-1}\left(\tfrac{\Delta Z}{\Delta L_x}\right)\right) \\ \textit{Transition Angle } \vec{b} &= \pm\left(\tan^{-1}\left(\tfrac{\Delta X}{\Delta L_Y}\right)\right) + \left(\tan^{-1}\left(\tfrac{\Delta Z}{\Delta L_Y}\right)\right) \\ \textit{Transition Angle } \vec{c} &= \pm\left(\tan^{-1}\left(\tfrac{\Delta Y}{\Delta L_Z}\right)\right) + \left(\tan^{-1}\left(\tfrac{\Delta X}{\Delta L_Z}\right)\right) \end{aligned} \tag{8}$$

Where:
ΔX, ΔY, ΔZ = Repeatability Values.
L_x, L_y, L_x = Distance between each detection point.

4 Accuracy Improvement with Robotic Grinding

The contact state between the work and the cutting tool presents a challenge to the robot compliance due to the difficulties in accurate modelling of force control. When machining such components, it is difficult to obtain information without building a relationship between the workpiece and the robot. This machining mode involve multi sensor data combination to allow the robot motion to provide the necessary outputs to avoid any error development. Therefore, is necessary to put forward robot motion control strategies which are suitable for collaborative machining and develop collaborative control software to realize the task allocation and interference avoidance.

However, the existing force and position control in robotic grinding aims to reduce the surface roughness of parts and pays less attention to the accuracy of form and position. The selection of the optimum grinding cycle parameters depends on the knowledge of deflections performance of the cycle. Therefore, to compensate the effect of deflections during grinding, it is essential to observe the robot compliance performance in relation to robot grinding infeed and sparkling within the cycle. As the magnitude of the grinding force changes with material removal rate and grinding wheel surface condition, it is often necessary to set a conservative operation conditions to perform grinding [15]. This means that most grinding cycles are not optimized for minimum cycle time and take longer time than required.

To improve grinding performance, the system time constant is a good measure of compliance of the grinding system. The time constant is the combined effect of the system compliance and the grinding forces during deflection between the machine and workpiece. The compliance represents the rate of deflection per unit force which depends on the geometrical factors of the workpiece as well as the grinding wheel and material properties. In this experiment, the time constant is used to allow for a more consistent control of spark out time during grinding as shown in the equation below:

$$T = T_0.e^{-\frac{t}{\tau}} \tag{9}$$

where **T** being the remaining stock, $\mathbf{T_0}$ being the initial stock, **t** is the contact time and $\boldsymbol{\tau}$ is the time constant estimated of the grinding system. It is important to determine the number of infeed and spark-outs for each grinding pass. The selection of the optimum grinding cycle parameters depends on the knowledge of deflections performance of the cycle. Therefore, to compensate the effect of deflections during grinding, it is essential to include a spark-out period to give us an idea of the infeed rates employed within the cycle to give more flexibility of the part, grinding wheel and the machine. In this experiment, the time constant is used to allow for a more consistent control of spark out time during grinding removal.

The grinding cycle model is started by taking into consideration the infeed stage as the first step, after that, the spark-out cycle is implemented to remove as the residual stock. As shown in Fig. 2, the amount of deformation is proportional to the normal forces and the real depth of cut during the grinding pass. Where:

$$F_n \propto \delta' \tag{10}$$

$$F_n \propto \varepsilon = F_n \propto \delta \tag{11}$$

$$\varepsilon \propto \delta \tag{12}$$

From hooks law relationship the stiffness (k) is introduced;

$$k\varepsilon \propto \delta \tag{13}$$

$$\varepsilon = \frac{1}{k}\delta \tag{14}$$

Therefore,

$$\varepsilon = \alpha\,\delta \tag{15}$$

Where $\boldsymbol{\alpha}$ is the coefficient of deformation and 1/k is the stiffness. In the spark out stage, the residual stock continues to be removed until the wheel is retracted as shown in Figs. 5, 6 and 7 were the grinding wheel is in contact with the workpiece. If the real depth of cut is $\boldsymbol{\delta}$ it should theoretically be cut to that position, but due to elastic deformation of the process system, it is elastically deformed due to the tip of the tool and the workpiece and only cuts to δ'. However, the infeed cycle can now be designed by taking into account the number of grinding strokes needed to reach the real depth of cut as shown in Fig. 2.

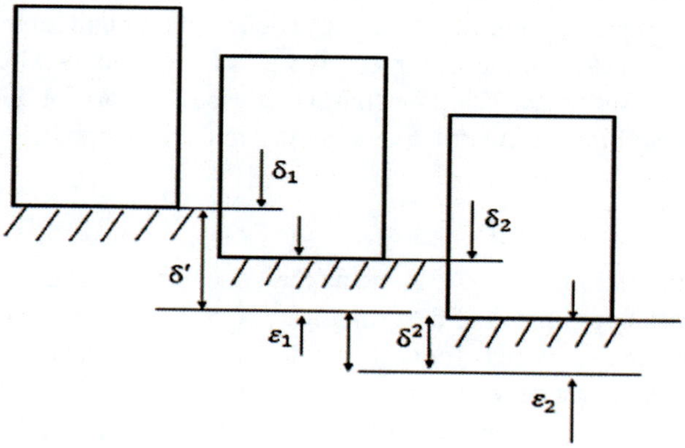

Fig. 2. Infeed process

If the theoretical depth of cut is δ', the elastic defamation amount per stoke is $\varepsilon_1, \varepsilon_2 \ldots$. the real depth of cut is as follows:

$$\delta_1 = \delta' - \varepsilon_1 \tag{16}$$

Second stroke

$$\delta_2 = 2\delta' - \delta_1 - \varepsilon_2 \tag{17}$$

Where $= \delta_1 = \delta' - \varepsilon_1$

$$\delta_2 = 2\delta' - (\delta' - \varepsilon_1) - \varepsilon_2 = \delta' + \varepsilon_1 - \varepsilon_2 \tag{18}$$

$$\delta_i = \delta' + \varepsilon_{i-1} - \varepsilon_i \tag{19}$$

Where, **i** is more number of infeed in the cycle.
Therefore,

$$\varepsilon_1 = \alpha\,\delta_1 \ldots\ldots\ldots \varepsilon_i = \alpha\delta_i \tag{20}$$

$$\delta_1 = \delta' - \alpha\,\delta_1 \tag{21}$$

$$\delta' = \delta_1(1 + \alpha) \tag{22}$$

$$\delta_1 = \frac{1}{1+\alpha}\delta'$$

Carrying over from previous feed, Therefore,

$$\delta_i = \left[1 - \left(\frac{\alpha}{1+\alpha}\right)^i\right]\delta \tag{23}$$

Where:
$\frac{\alpha}{1+\alpha}$ = Infeed Ratio.
However, the overall actual depth of cut is normally smaller than the total theoretical depth of cut, so the required dimensional accuracy is not achieved. Therefore, it is necessary to carry out the spark-out cleaning and gradually eliminate the residual deformation to achieve the required finish: For spark-out process shown in Fig. 3 below.

$$\delta_{1i} = \varepsilon_1 - \varepsilon_{1i} \tag{24}$$

$$\delta_{ni} = \varepsilon_n - \varepsilon_{in(n-1)} \tag{25}$$

Where:
ε_{1i} = first spark out deformation
δ_{1i} = first spark out depth of cut

$$\varepsilon_i = \alpha\delta_i \ldots\ldots \varepsilon_{in} = \alpha\delta_{in} \tag{26}$$

Substituting into Eq. 5–12:

$$\delta_{i1} = \frac{\alpha}{1+\alpha}\delta_1 \tag{27}$$

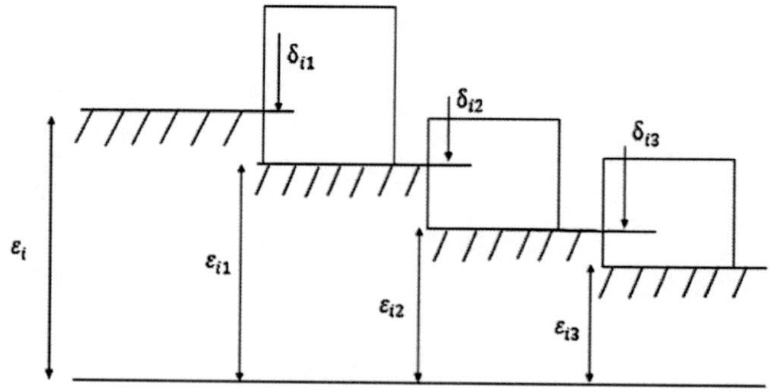

Fig. 3. Spark-out process

$$\delta_{in} = \left(\frac{\alpha}{1+\alpha}\right)^n \delta_i \tag{28}$$

Bringing the infeed stroke

$$\delta_i = \left(\frac{\alpha}{1+\alpha}\right)^n \left[1 - \left(\frac{\alpha}{1+\alpha}\right)^i\right] \delta' \tag{29}$$

As the focus is on improving the accuracy of the robot to perform repair grinding. The level of error relative to its response is defined by measuring various locations on workpiece surface and implementing a mathematical model to generate a relationship between the workpiece and the cutting tool to predict the datum reference point and associated error. The datum is then corrected according to the robot position to provide a reliable and accurate grinding movement. By establishing the geometrical relationship between reference datum and the robot tool, a grinding theoretical model is developed to perform grinding for reconstruction of the surface to achieve a smooth surface finish. The optimum grinding cycle behavior depends on the knowledge of deflections between the grinding wheel and the workpiece resulting from the grinding force and system stiffness. Therefore, to compensate the effect of deflections during grinding, it is essential to observe the number of in-feeds and spark-outs within the cycle.

Finally, the theoretical model developed in this section provides a clear step to obtain the experimental process. The empirical model provides a strategy for repair that can achieve the maximum workpiece quality, minimum machining time and economic efficiency by making a selective adaptation strategy and chosen parameter selection.

5 Experimental Setup

A KUKA KR16 robot was used in this investigation. Throughout the planning phase of the experiments, some control variables for the robot is determined to the robot. The control variables are presented in Table 1, which lists related information including robot moving speed, payload, work envelope location, motion type.

Table 1. Robot control variables.

Control variables	Normal range	Setting
Working speed	0–2 m/s	0.005 m/s
Spindle speed	0–2500 rpm	2500 rpm
Robot payload	0–16 kg	16 kg
Grinding wheel load	0–200 g	160 g
Workspace location	0–1610 mm	4 points within workspace
Motion type	Discrete	Point to point, Linear

The acoustic emission (AE) sensor used to assist the user to directly engage the grinding wheel to the workpiece to minimize the effect of error between the tool and the workpiece. The AE sensor is capable of generating feedback by sensing the position of the workpiece in relation to the tip of the tool on the robot which makes it one of the most promising processes for detection and monitoring methods. Average AE signal characteristics are shown in Table 2. The effect of signal characteristics was not investigated in this work, as the aim of using the AE sensors is to detect the contact of workpiece and provide a real-time feedback to the robot controller.

Table 2. A.E average signal detection

Detection time (s)	Amplitude (V)	RMS	Energy (J)	Peak frequency (Hz)
0.65124	34.2475	0.0042	93.1427	58.1562

At the point of contact between the workpiece and the robot tool tip, an analog signal provides a visual reference indicating the tool is contacted with the workpiece. At this stage, the interrupt command is declared in the robot software and the points are saved accordingly in the system. The sphere grinding tool is set at a 45° angle due to avoid zero cutting speed during operations. Figure 4 below shows a general overview of the experimental set up.

5.1 Datum Measurement Setup

One of the challenges in robotic machining is the setup of workpiece datum in relation to the robot origin datum. The traditional methods used in robotic applications is mainly using a touch probe as used on a CMM (Coordinate measuring machine). The proposed work aims to define the datum by using a mathematical algorithm which aims to eliminate the influence of geometrical error which is the key for controlling the robot to conduct grinding process. The error level could be assessed based on the repeatability measurements and defined measuring points for full plane inspection. For correct implementation of the mathematical model, a full inspection of the plane is carried out

(a)

(b)

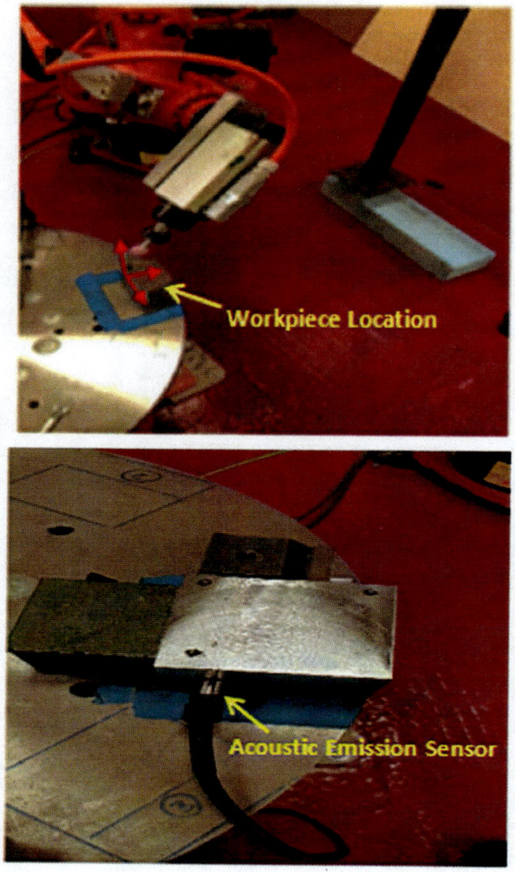

Fig. 4. (a) Workpiece location, (b) Acoustic emission sensor experimental setup

by detecting four points on each plane surface to obtain the accuracy error level. The accuracy level is assessed based on the repeatability measurements of the relative datum position between workpiece and the cutting tool. After that, the mathematical model is used to predict the real datum point and then modified accordingly to the robot tool position. The identified workpiece datum will act as the reference point to perform the grinding operation. Figure 5 illustrates a 3D model of the three surfaces of the block used for detection.

5.2 Repair Grinding Setup

Weld at random positions is created so that the model can be scanned to generate the required a tool path to be implemented with the robot to perform the grinding operation. This is a general way to perform repair engineering were small welding zones is created on the workpiece and restored to its original shape by grinding as shown in Fig. 6.

The operating parameters such as spindle speed, feed rate were kept constant and depth of cut is set to 0.3 mm for roughing. Within the finishing process the depth of

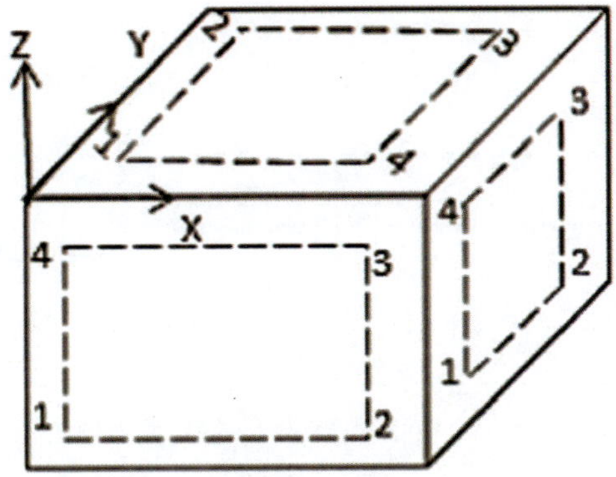

Fig. 5. Measurements points of detection

cut of is set to 0.1 mm that is the minimum setting of the robot. During these stages, the material continues to be removed from the workpiece and monitored by the acoustic emission sensor to give feedback to the operator of material removal. The theoretical model developed based on system time constant is used to predict the material removal in roughing, finishing and sparking process to achieve high accuracy level. The cycle has been divided into three mains stages 1) roughing 2) finishing and 3) Sparking as shown in Fig. 7 schematic diagram below.

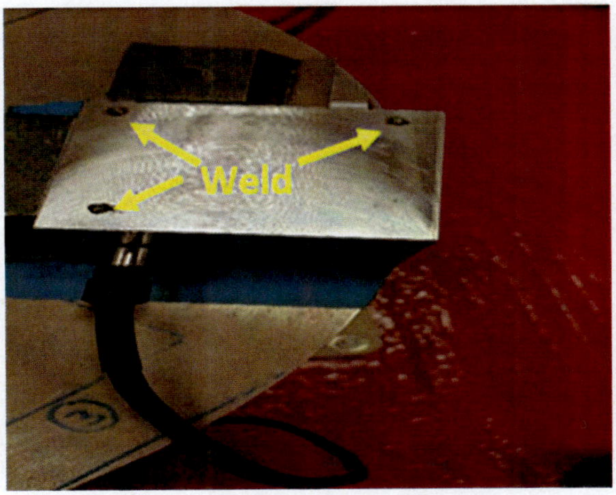

Fig. 6. Weld locations at ZX plane

6 Results and Discussions

An AE sensor used to monitor and measure the points by directly engaging grinding wheel to the workpiece to eliminate the effect of error between the tool and the workpiece. This is done by registering multiple points in each surface of the block under an

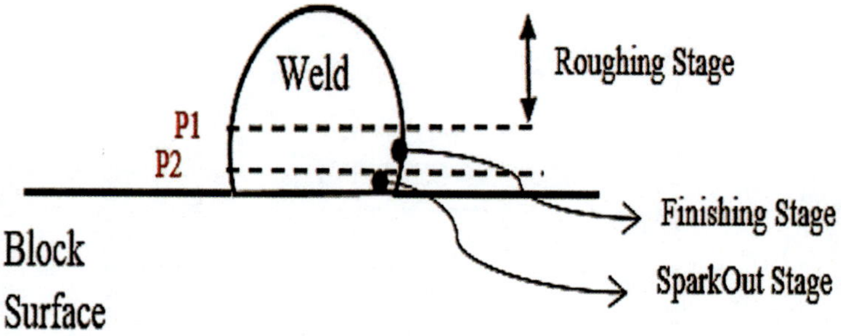

Fig. 7. Grinding cycle schematic view

operational speed of 0.005 m/s which is the minimum infeed the robot can achieve. The result in Table 3 show all points within number of repeatable times which shows that the use of acoustic emission sensor was useful for accuracy detection purposes.

The procedure is assessed based on repeatability were points are detected on each surface of the block to identify the geometric error and datum reference. Hence, the error model is developed towards the collected data to closely observe the datum point. This is done by visualizing the performance of robot repeatability and deriving a mathematical model for error compensation to locate the geometry datum points which proves the reliability and compatibility of using the acoustic emission sensor by engaging the tool into the workpiece.

Table 3. Variation in robot repeatability tests.

No. of detection times	Measurement of coordinates under 0.01 m/s		
	Plane XY	Plane YZ	Plane ZX
1	−10.84361	−13.3375	−12.5153
2	−10.85271	−13.35154	−12.5264
3	−10.8525	−13.35482	−12.5157
4	−10.85291	−13.35384	−12.5157
5	−10.85113	−13.34399	−12.5161
6	−10.88352	−13.347548	−12.5143
7	−10.85952	−13.347558	−12.5145
8	−10.85952	−13.357548	−12.5148
9	−10.88552	−13.357948	−12.5151
10	−10.87589	−13.35848	−12.5149

After that, the datum is aligned directly to the robot which aims to increase the accuracy of the grinding processes. According to results, the mathematical error accuracy achieved is 30 μm. Based on repeatability measurements of the relative positions between workpiece and robot datum the mathematical model predicted datum points as.

Workpiece Datum Coordinates:

$X_0 = 12.7302$
$Y_0 = 13.2228$
$Z_0 = 13.6362$

Datum Correction Trajectories

$\vec{a} = i + 0.01432\,j + 0.00397\,k$
$\vec{b} = -0.01230\,i + j \mp 0.00201\,k$
$\vec{c} = -0.01054\,i + 0.008401\,j + k$
Transition angle $\vec{\mathbf{a}} = \pm 0.066°$
Transition angle $\vec{\mathbf{b}} = \pm 0.059°$
Transition angle $\vec{\mathbf{c}} = \pm 0.053°$

6.1 Grinding Cycle Control

During the roughing stage, the influence of the depth of cut has been used to calculate the coefficient of deformation α in the system which is proportional to the normal forces and the real depth of cut. The theoretical model suggests that the roughing stage requires a seven number of roughing cuts, two finishing cuts and two spark-out passes to achieve final surface with minimum residual error. These cuts need to be controlled in a way where the weld is completely removed from the surface of the block without making any damage to the workpiece. Figure 8 below demonstrates the achieved experimental and theoretical results.

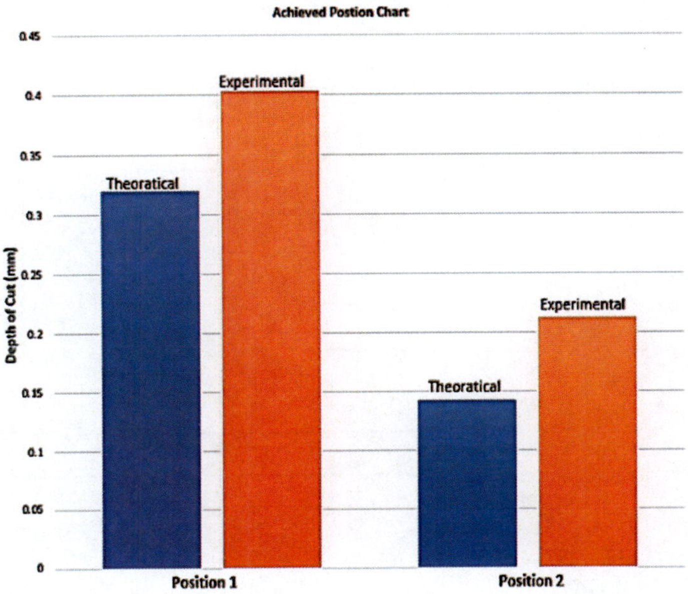

Fig. 8. Achieved positions

In each stage, the robot tool cuts the material at a controlled depth of cut to ensure the material removal achieved. As can be seen from the results above, the theoretical and experimental stage at position 1 is achieved at 0.3199 mm and 0.4030 mm. Position 2 in finishing stage achieves 0.1421 mm and 0.2130 mm. The influence in the difference is related to the abrasive grains on the wheel in the contact area on the wheel due to the high forces affecting the wheel from stiffness of the joint causing residual stock on the workpiece. Due to the robot rigidity at a small area, the coefficient of deformation α value had an effect of the results because it was only calculated through the first cut. Also, the abrasive grains and the contact area were not taken into account in the calculations which also affects the contact stiffness of grinding wheel which is normally supported by the stiffness of single abrasive grain. It is known that the higher the surface roughness the lower the residual of workpiece and the higher the contact stiffness of grinding wheel. From such a viewpoint, this project aims to investigate the total grinding accuracy in order to perform grinding and not the effect of stiffness during grinding. Therefore, grinding operation was carried using a single grinding wheel and the residual stock removal of the workpiece was measured based on the depth of cut and calculations performed.

However, at the position at which the finishing stage ends, spark outs have to be carried out to ensure material is completely removed to smoothen up the surface. The advantage of spark-outs is to provide closer tolerances by removing the remaining stock, therefore a number of two spark outs have been made according to the theoretical model. A profile measurement is conducted to measure the weld area to observe how much residual material is left. The machine allows to capture the contour profiles of the boundary using a single probe to observe how much material is been removed from the surface (Fig. 9).

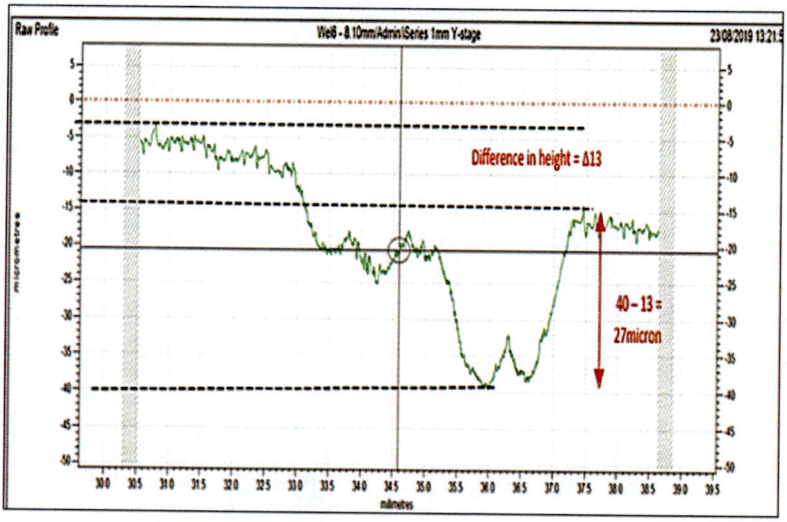

Fig. 9. Surface profile measurement area at ZX

The residual is approximate 30 μ below the surface level, this could be due to large forces generated in the wheel caused from the stiffness of the robot. Also, a small lip noticed due to the tool being lift of the surface of the block which is due to an error related to the dynamic behavior of the robot arm which mainly occurs from structural deformations, stiffness and robot compliance. The effect of these conditions is difficult to control because the maximum of depth of cut is limited to 0.1 mm and the angle between the grinding spindle axis and the surface tangent (usually 45°) overt the grinding process. This eventually causes errors between the tool and geometry resulting to unexpected material on the surface of the block. Preventing unwanted motions possess more challenging design problem, which can significantly affect the performance. However, the depth of cut achieved is approximately less than 30 μ which validates the accuracy level achieved form method developed which proves it is efficient and could be used for repair engineering.

7 Conclusions

Robotic grinding is an effective technique that could revolutionize the repair works in manufacturing industry. This work establishes a repair method to perform grinding using an articulated 6-axis robot. Through this method a geometrical relationship between workpiece reference datum and grinding wheel can be established and the error level is assessed based on the repeatability and defined measuring points on the surfaces of the workpiece. An optimum grinding cycle is designed by taking advantages of abrasive machining to support and guide the selection of infeed speed and number of passes required to verify the final grinding repair operation. This provides a suitable solution for precision material removal for repairing components in manufacturing and maintenance operations. The work concludes the following points:

- Based on repeatability measurements of the relative positions between workpiece and robot datum, the mathematical model developed is used to predict the workpiece datum. The geometrical datum error achieved is less than 30 μm which supports the process monitoring and control strategy to provide a reliable and accurate grinding movement using the robot.
- The developed grinding cycle has improved the machining repair accuracy to a level of 30 μm. The investigation considers the process variables such as depth of cut, wheel speed, feed speed, dressing condition and system time constant as a key variable in the design of the grinding cycle.

Finally, this work provides a suitable solution for precision measurement to repair components in manufacturing and maintenance operation using a robot in many industrial sectors. The main novelty of this work is defining the error accuracy by using the cutting tool as a probe in the robot system using acoustic emission monitoring technology that modifies robot commands accordingly. A mathematical model is developed for compensating machining errors in relation to its geometrical position by utilizing system relaxing technique that satisfies the need of eliminating the residual error during robot grinding.

References

1. Tao, B., Zhao, X.W., Ding, H.: Mobile-robotic machining for large complex components: a review study. Sci. China Technol. Sci. **62**(8), 1388–1400 (2019). https://doi.org/10.1007/s11431-019-9510-1
2. Quin, G., Zhang, W., Wan, M.: A mathematical approach to analysis and optimal design of a fixture locating scheme. Int. J. Adv. Manuf. Technol. **1**(29), 349–359 (2006)
3. Xiong, Y., Ding, H., Wang, M.: Quantitative analysis of inner force distribution and load capacity of grasps and fixtures. J. Manuf. **2**(124), 444–455 (2002)
4. Marin, R., Ferreria, P.: Analysis of the influence of fixture locator errors on the compliance of work part features to geometric tolerance specifications. J. Manuf. **3**(125), 609–616 (2003)
5. Kubla, T., Singule, V.: Investigation of position accuracy of industrial robots and online methods for accuracy improvement in machining processes. In: International Conference on Electrical Drives and Power Electronics, Brno (2015)
6. Jin, T., Jiyong, K.: A 3-D point sets registration method in reverse engineering. Comput. Ind. Eng. **53**(2), 270–276 (2007)
7. Chaiprapat, S., Rujikietgumjorn, S.: Modelling of positional variability of fixture workpiece due to locating error. J. Manuf. Sci. Technol. **2**(121), 273–281 (2008)
8. Khodaygan, S.: Manufacturing error compensation based on cutting tool location correction in machining processes. Int. J. Comput. Integr. Manuf. **27**(11), 969–978 (2014). https://doi.org/10.1080/0951192X.2013.874580
9. Lizarralde, R., Barrenetxea, D., Gallego, I., Marquinez, J.: Practical application of new simulation method for the elimination of geometric instabilities in centerless grinding. CIRP Ann. Manuf. Technol. **54**(1), 273–276 (2005)
10. Pan, Z., Zhang, H., Zhu, Z., Wang, J.: Chatter analysis of robotic machining process. J. Mater. Process. Technol. **173**(3), 301–309 (2006). https://doi.org/10.1016/j.jmatprotec.2005.11.033
11. Bisu, C., Cherif, M., Gerard, A., Nevez, J.: Dynamic behavior analysis for a six axis industrial machining robot. In: Proceedings ICASAAM, Bucharest (2011)
12. Imed Zaghbani, M., Lamraoui, V., Songmene, M.: Robotic high speed machining of aluminum alloys. Adv. Mater. Res. **188**, 584–589 (2011). https://doi.org/10.4028/www.scientific.net/AMR.188.584
13. Zhao, T., Shi, Y., Lin, X., Duan, J., Sun, P., Zhang, J.: Surface roughness prediction and parameters optimization in grinding and polishing process for IBR of aero-engine. Int. J. Adv. Manuf. Technol. **74**(5–8), 653–663 (2014). https://doi.org/10.1007/s00170-014-6020-3
14. Dumas, C., Caro, S., Garnier, S., Furet, B.: Joint stiffness identification of six-revolute industrial serial robots. Robot. Comput. Integr. Manuf. **27**(4), 881–888 (2011). https://doi.org/10.1016/j.rcim.2011.02.003
15. Allanson, D.R., Rowe, W.B., Chen, X., Boyle, A.: Automatic dwell control in computer numerical control plunge grinding. Proc. Inst. Mech. Eng. Part B J. Eng. Manuf. **211**(7), 565–575 (1997). https://doi.org/10.1243/0954405971516518

Neuro Fuzzy Based Runner Selection for Plastic Injection Moulding Design

Chiwapon Nitnara[1] (✉) and Suthep Butdee[2]

[1] Department of Mechanical Engineering Technology, Faculty of College of Industrial Technology, King Mongkut's University of Technology North Bangkok, 1518 Pracharaj I, Bangsue, Bangkok 10800, Thailand
chiwapon.n@cit.kmutmn.ac.th

[2] Department of Production Engineering, Faculty of Engineering, King Mongkut's University of Technology North Bangkok, 1518 Pracharaj I, Bangsue, Bangkok 10800, Thailand

Abstract. Plastic injection moulding is an effective and widely used for mass production. The major factors which influence product quality are machine, mould, and material. The mould design is one of the most difficulties and complicated particularly for a multiple cavity mould. It needs to stand for high pressure of hot plastic processing via a gate and a runner. Therefore, the gate and the runner design are the most critical aspect for obtaining a high-quality plastic part. Traditionally, they are selected based on an experienced injection mould designer. However, it is time consuming and expensive. This paper proposes the novel concept of selection the gate and runner types to suite with mould cavity shapes and dimensions using Neuro-Fuzzy method. The research method is focuses on an injection hot runner design for plastic injection moulding. The case study is illustrated and discussed in details.

Keywords: Runner system · Plastic injection moulding design · FEM · Fuzzy Fuzzy

1 Introduction

Plastic injection moulding is one of the important industrial processes; due to the plastic injection process has a high production rate, short production time and easily creates complex part [1]. Therefore, plastics are widely used to substitute for metals such as automotive parts, electrical appliance components, household goods and many other common plastic items. In the production process, molten polymer is injected under high pressure and with high speed into the mould cavity [2]. This is followed by the cooling of the resulting from melt polymer and, in the final stage, the ejection of the product from the mould. From this process, the injection moulding process will require factors such as injection pressure, filling time, packing time and cooling [3], resulting in good surface and less warpage. However, in order to make good quality part requires good design which the runner and gate. It is important to consider.

A. Batako et al. (Eds.): GCMM 2021, LNNS 335, pp. 273–287, 2022.
https://doi.org/10.1007/978-3-030-90532-3_21

In the design process of plastic injection mould, runner is responsible for conveying the melt polymer into the gate so that the melt polymer can flow into the mould cavity at a reasonable rate. If the designed part is come out with good quality, the runner and gate must be suitable designed the right size in order to reduce the warpage of the part to obtain perfect finish goods. For finding the right size of the runner and the gate can reduce time and cost [4]. Traditionally, Finite Element (FEM) simulation is a widely used method [5]. It can simulate the size of runner and gate in many ways to show the warpage effect of the part. The simulation tool can save production cost and time.

In this research, the suitable sizes of the runner and gate sizes are determined. The case study is illustrated the simple box shape of plastic cavities. The dimension is 50 × 50 × 10 mm at the based design. Then, the study is designed by different thickness. They are 2 mm, 2.5 mm, and 3 mm. The material property is on Polypropylene (PP). FEM simulation method is mostly used to find the optimum size of the warpage [6, 7]. Neuro Fuzzy is used to learn the simulation results warpage of the part, in order to be able to select the runner and gate that the appropriate part and obtains minimal warpage.

2 Injection Mould Theory

2.1 Injection Process Analysis

Filling process is to fill the mould cavity with the melt polymer at high pressure which is a process to transform the melt polymer into a shape as in the Eq. (1).

$$\frac{\partial \rho}{\partial t} + \vec{\nabla} \cdot (\vec{\nabla} \vec{V}) \tag{1}$$

ρ is the plastic density and V is the vector velocity.

Newton's second law comes from momentum, also known as the force generated by plastic flow, as in the equation [2].

$$\rho \cdot [\frac{\partial \vec{V}}{\partial T} + (\vec{V} \cdot \vec{\nabla})\vec{V}] = -\vec{\nabla} + P + \vec{\nabla}. \rightarrow \tau + P\vec{f} \tag{2}$$

P is the flow pressure; f, the body force; τ is the stress tensor.

In the cooling process, the plastic part is completely in contact with the mould cavity without thermal resistance. Plastic flow distribution and heat transmission are relationships with the cooling process. The equation can be shown as (3).

$$\frac{\partial^2 \mathrm{y}}{\partial \mathrm{a}^2} + \frac{\partial^2 \mathrm{y}}{\partial \mathrm{b}^2} + \frac{\partial^2 \mathrm{y}}{\partial \mathrm{c}^2} = 0 \tag{3}$$

The Eqs. (1) , (2) and (3) are only preliminary for plastic injection analysis. In a simulation with FEM, multiple factors are considered for accurate of analysis results.

2.2 Hot Runner

The hot runner system developed from conventional runner system has been added with the Manifold, Nozzle, Heater, Thermocouple, and Temperature Control Unit by installing the Manifold and Nozzle, resulting in heating of the runner and area [8] as shown in Fig. 1. In the process of injecting melt polymer, the liquid plastic flows into the gate of the mould and then flows into mould cavity. After that, the liquid plastic in the cavity is changes the status to be a solid hardened plastic part. However, the liquid plastic in the runway that is heated does not harden because the temperature is higher than the melting point of the plastic used, so there is no spur & runner attached to the part. This plastic injection mould for hot runner system can save the plastic wasted on the runner system [9] and save production cost.

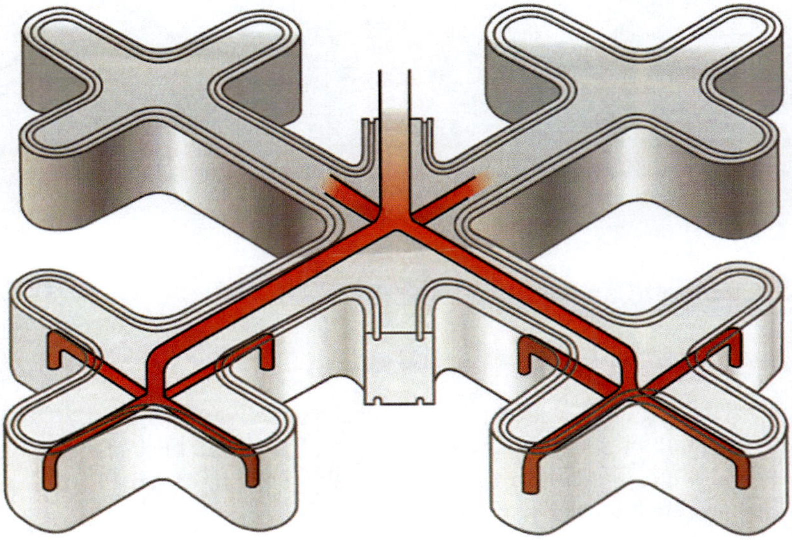

Fig. 1. Manifold of hot runner system [10].

2.3 Hot Runner Gate

In this research, the gate of hot runner system has been specified as the valve gate. It is a favour for using and developing gate of hot runner system. It consists of a spring valve and hydraulic or pneumatic control valve. The valve gate is responsible for opening the plastic flow smoothly. When the pressure drops, the valve gate is closed, the liquid plastic does not flow into the mould cavity [11–13]. It helps eliminate of weld line and reduce nozzle pressure. The simple valve gate is design as shown in the Fig. 2.

Fig. 2. Valve gate [10].

3 Experimental and Simulation

3.1 Experimental Parts

The dimensions of the box-shaped part used for simulation study are 50 × 50 × 10 mm. with different wall thickness. They are three cases; 2 mm wall thickness, 2.5 mm wall thickness and 3 mm wall thickness. The material is Polypropylene (PP). The general view of the part are presented in Fig. 3.

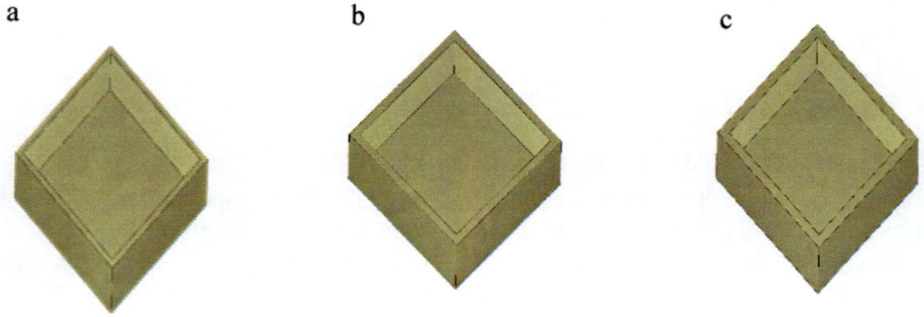

Fig. 3. (a) thickness 2 mm; (b) thickness 2.5 mm; (c) thickness 3 mm.

The hot runner system is called melt channel, the size and length of the melt channel are defined to flow length not longer than 80 mm. In this experiment, the flow length is 78.5 mm and there are 2 types of design which are type 1 and type 2 as shown in Fig. 4. In addition, the gate of hot runner system has been specified as the valve gate.

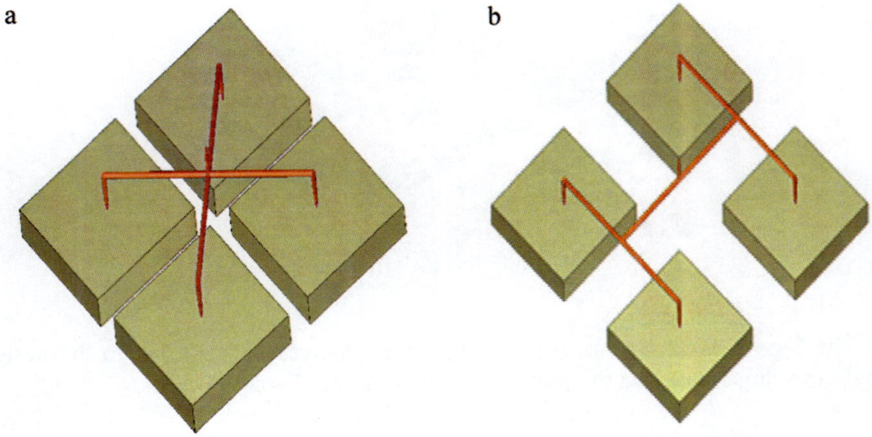

Fig. 4. (a) type 1; (b) type 2.

The objective of the previous research is to determine the optimum the control parameters for minimizing the warpage [14, 15]. Several parameters were selected for FEM simulation by Moldex3d software, such as: (1) gate diameter, (2) gate length, (3) part dimension and (4) melt channel design type. Table 1 shows the factors and case study designed in the experimental. There are 3 cases are studied.

Table 1. The factors and case study designed in experimental.

Factors	Case 1	Case 2	Case 3
Gate diameter (mm)	0.8	0.9	1
Gate length (mm)	2.5	3	3.5
Part dimension (mm)	50 × 50 ×10 × 2	50 × 50 ×10 × 2.5	50 × 50 × 10 × 3
Melt channel design type	1	-	2

The Fig. 5 shows the filling melt front time with four cavities with hot runner designed. The filling process is started in the middle area of the part which is shown by the red color.

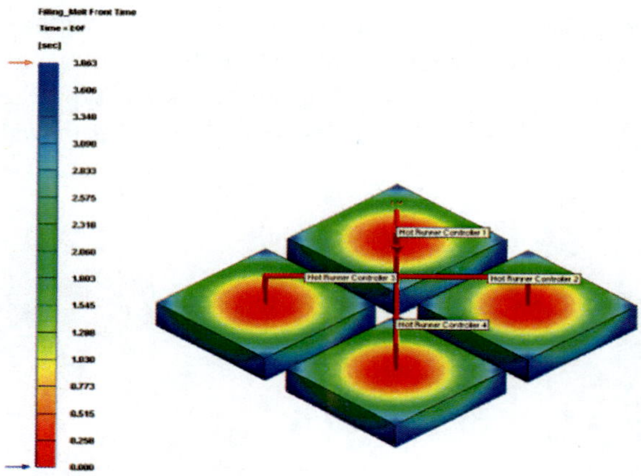

Fig. 5. Filling melt front time.

The Fig. 6 shows the warpage of the simulation result according to the design conditions. It is found that the warpage occurred is very less.

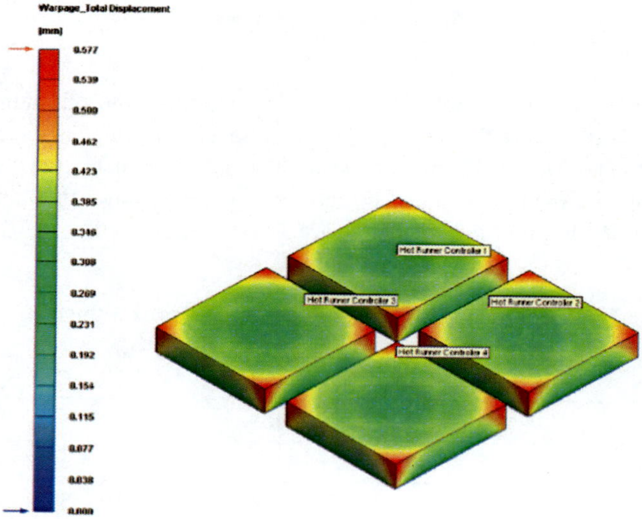

Fig. 6. Warpage total displacement.

3.2 Finite Element

Finite element (FEM) simulation is used to simulate with melt channel in order to find the minimum and maximum warpage [16]. In our research study, as shown in the Fig. 5, the filling process of experimental material (PP) is used for filling time at 4 s and packing time at 5.36 s. Figure 6 shows the final results of the FEM analysis to the warp total displacement (mm) simulation. In the example simulation figure of experiment No. 3 in

Table 2, with the gate diameter 0.8 mm, gate length 2.5 mm, dimension 50 × 50 × 10 × 2.5 mm and melt channel design type 1.

Table 2 shows the experimental study and results using FEM. There is a total of 54 cases and the round gate shape is selected for the runner system. The diameters are varied from 0.8 mm to 1.0 mm. The gate length is varied from 2.5 mm to 3.5 mm. It includes varied part dimensions and melt channel type. The simulation result carried out with warpage sizes. It is 0.371 mm in the case 1, 0.577 mm in the case 3 and 0.878 mm in the case 54.

Table 2. The experimental and results of simulation.

No.	Gate diameter (mm)	Gate length (mm)	Part dimension (mm)	Melt channel design type	Warpage (mm)
1	0.8	2.5	50 × 50 × 10 × 2	1	0.371
2	0.8	2.5	50 × 50 × 10 × 2	2	0.387
3	0.8	2.5	50 × 50 × 10 × 2.5	1	0.577
.	.	.	.	.	.
.	.	.	.	.	.
.	.	.	.	.	.
52	1	3.5	50 × 50 × 10 × 2.5	2	0.568
53	1	3.5	50 × 50 × 10 × 3	1	0.869
54	1	3.5	50 × 50 × 10 × 3	2	0.878

Figure 7 shows the mould base and cooling channel of simulation. There are 8 channels of this study.

In second step, the result of the simulation by FEM in each of the gate diameter and the part dimension that has the least warpage are selected. In order to simulate the appropriate melt channel diameter this causes the warpage to be less [17]. The melt channel diameter of the experiment consists of 1.5 mm, 2 mm and 2.5 mm. The simulation results are shown in Table 3.

Figure 8, shows an example of warpage simulation results. In the example, the factors of the experiment consists of gate diameter 0.8 mm, gate length 2.5 mm, part dimension 50 × 50 × 10 × 2.5 mm, melt channel design type 2 and melt channel diameter 1.5 mm. The warpage total displacement of example is 0.522 mm. In the next step, the melt channels of various diameters that cause the least warpage were selected for the creation of the Neuro Fuzzy model.

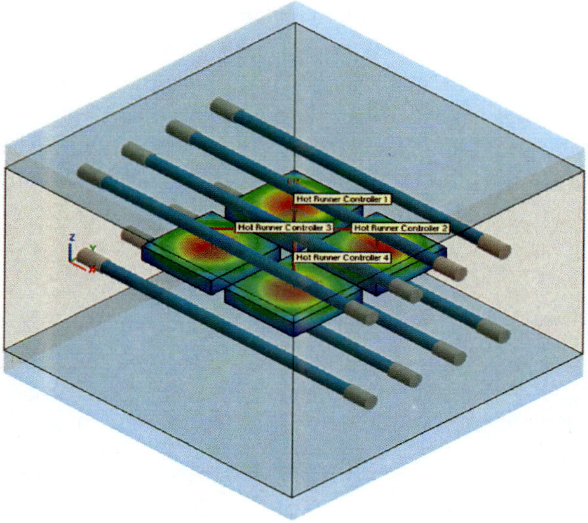

Fig. 7. Mould base and cooling channel.

Table 3. The results of FEM simulation in difference melt channel diameter.

Gate diameter (mm)	Gate length (mm)	Part dimension (mm)	Design type	Warpage		
				Diameter 1.5 mm	Diameter 2 mm	Diameter 2.5 mm
0.8	2.5	50 × 50 × 10 × 2	1	0.366	0.371	0.376
0.8	2.5	50 × 50 × 10 × 2.5	2	0.552	0.520	0.567
0.8	2.5	50 × 50 × 10 × 3	2	0.884	0.743	0.896
0.9	3.5	50 × 50 × 10 × 2	1	0.363	0.371	0.376
0.9	2.5	50 × 50 × 10 × 2.5	2	0.571	0.520	0.592
0.9	3.5	50 × 50 × 10 × 3	1	0.870	0.869	0.892
1	2.5	50 × 50 × 10 × 2	1	0.372	0.371	0.376
1	2.5	50 × 50 × 10 × 2.5	2	0.553	0.520	0.566
1	2.5	50 × 50 × 10 × 3	2	0.880	0.743	0.894

3.3 Neuro Fuzzy Model

Due to the FEM model is not suitable for a large number of analyses, because of the high computational time. However, it can be collocated to other tools to improve its efficiency. This research developed the FEM Method collaborative with Neuro Fuzzy. The ability of Neuro Fuzzy is effectively training linear and nonlinear data characteristics [18, 19]. Consequently, the new method can save time for designing the plastic injection mould. The FEM method was used to simulate the warpage of experiment. There are 9 data of the experiment from the results of the FEM simulation are undertaken for leaning in

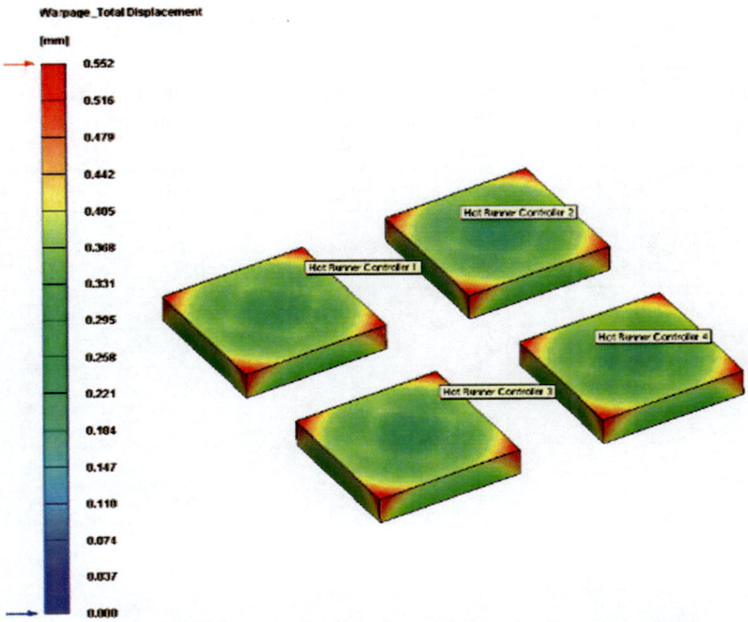

Fig. 8. Warpage simulation results of melt channel diameter 1.5 mm.

Neuro Fuzzy Model by using MATLAB software. To develop the Neuro Fuzzy intelligent system, there are two main processes:

1. Creating input and output data for training in the Neuro Fuzzy system.
2. Defining a method of generating Fuzzy Logic model and training data.

In the first process, the results of simulation are selected for leaning in Neuro Fuzzy system as shown in Table 4. After that data from simulation are created to input and output parameter as shown in Fig. 9.

Table 4 shows the data training for Neuro Fuzzy system including gate diameter, gate length, part dimension, melt channel type and warpage.

In the final process, the factors of the Neuro Fuzzy model consist of gate diameter, gate length, and part dimension, melt channel design type and melt channel diameter. The sub clustering method is used to generate the Fuzzy Logic system. Due to the data is non-matrix, it cannot find nearby results. The hybrid method is used to training data for optimizing Fuzzy Logic model which has 100 epoch to Neuro Fuzzy system leaning as shown in Fig. 10 and creating neural network structure as shown in Fig. 11.

Table 4. Data training for Neuro Fuzzy system.

Gate diameter (mm)	Gate length (mm)	Part dimension (mm)	Melt channel design type	Melt channel diameter	Warpage
0.8	2.5	50	1	1.5	0.366
0.8	2.5	62.5	2	2	0.520
0.8	2.5	75	2	2	0.743
0.9	3.5	50	1	1.5	0.363
0.9	2.5	62.5	2	2	0.520
0.9	3.5	75	1	2	0.869
1	2.5	50	1	2	0.371
1	2.5	62.5	2	2	0.520
1	2.5	75	2	2	0.743

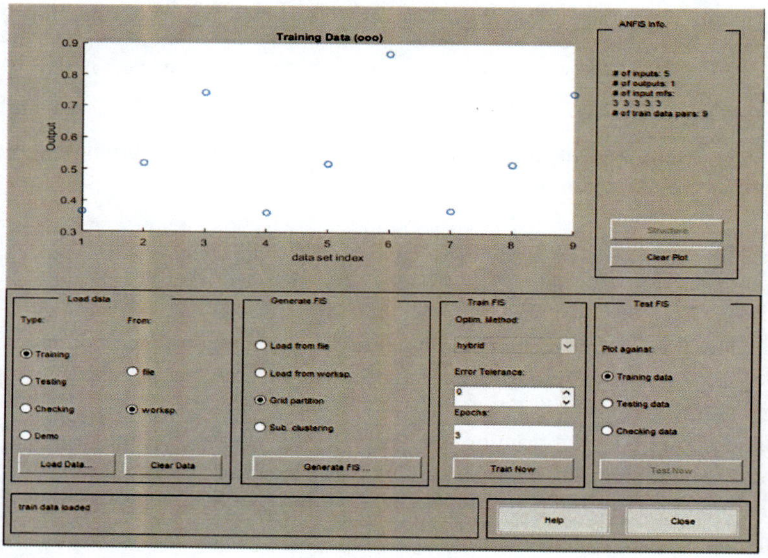

Fig. 9. Input and output data for training in the Neuro Fuzzy system.

Figure 10 shows the data of the Neuro Fuzzy model, found that it has good training characteristics. Notice from the red dot which drop into the data set and the average testing error value is 0.000000422046%. From this value, it can be concluded that Neuro Fuzzy is effectively approximate results data. Figure 11 shows the neural network structure consist of number of nodes 116 nodes, number of linear 54 parameters, number of nonlinear 90 parameters and number of training data pairs 9 data and number of fuzzy rules 9 rules.

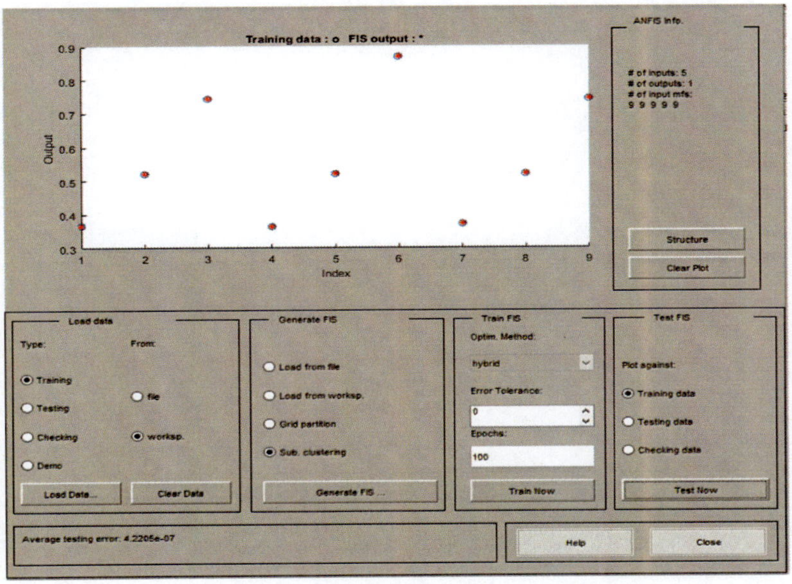

Fig. 10. Data of the Neuro Fuzzy model.

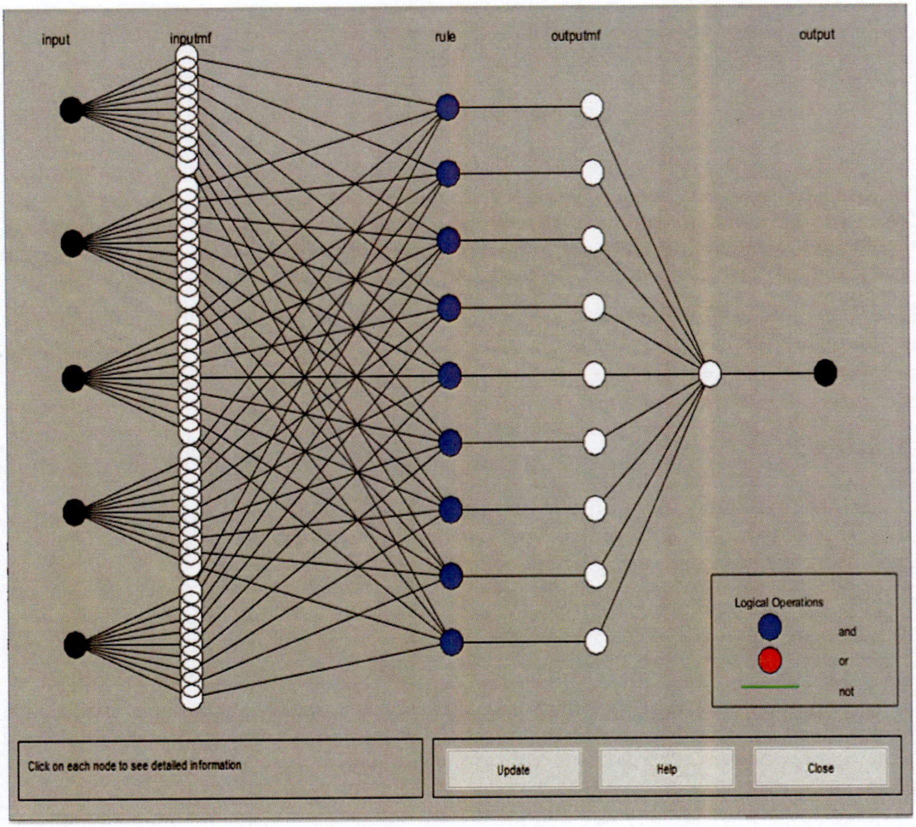

Fig. 11. Neural network structure.

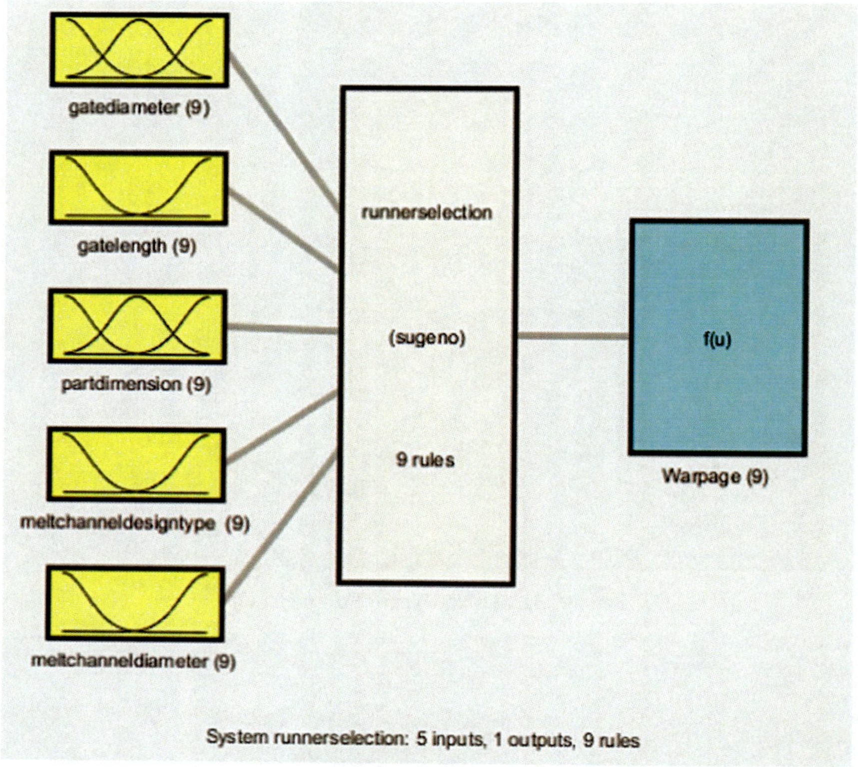

Fig. 12. Fuzzy Logic model.

According to Neuro Fuzzy training, the Fuzzy Logic model is divided into 5 inputs and 1 output [20] according to as shown in Fig. 12, which includes gate diameter, gate length, part dimension, melt channel design type and melt channel diameter. The membership function of inputs is Gaussian functions and there are have 3 levels of the membership function. In the output, there are 9 constant values.

4 Result

In order to find the results of the Neuro Fuzzy model that use rule viewer function of MATLAB software as shown in Fig. 13. The main purpose is to select the gate diameter, gate length, melt channel design type and melt channel diameter affected to minimum warpage of box shape parts. For example, the part dimension is $50 \times 50 \times 10 \times 2$ mm, the minimum warpage is 0.366 mm, with the selection of the gate diameter = 0.8 mm, gate length = 2.5 mm, melt channel design type = 1 and melt channel diameter = 1.5 mm. Table 5 and Fig. 14 shows a comparison between the simulation value and the optimal selection value of the approximated results by the Neuro Fuzzy model. The approximate error value of warpage is 0.000000422046%. Therefore, it has been shown that the process parameter for suitable selection hot runner-system design performance can be effectively obtained by Neuro Fuzzy method.

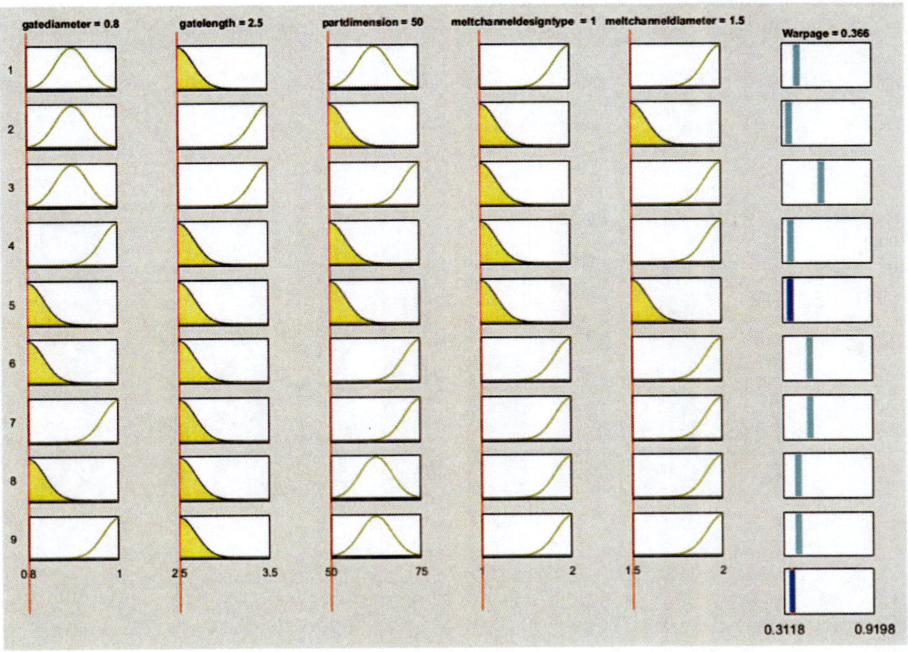

Fig. 13. Rule viewer function.

Table 5. Comparison between the Neuro Fuzzy and FEM simulation

Warpage	Simulation method		Error
	FEM	Neuro Fuzzy	
Case 1	0.366	0.366	0%
Case 2	0.520	0.520	0%
Case 3	0.743	0.743	0%
Case 4	0.363	0.363	0%
Case 5	0.520	0.520	0%
Case 6	0.869	0.869	0%
Case 7	0.371	0.371	0%
Case 8	0.520	0.520	0%
Case 9	0.743	0.743	0%

The comparison between Neuro Fuzzy and FEM can be concluded that all cases have an error value of 0%. The Neuro Fuzzy approximation is similar to the simulation results of FEM. The developed Neuro Fuzzy model can be used as an effectively approximate tool for selecting suitable hot runner design.

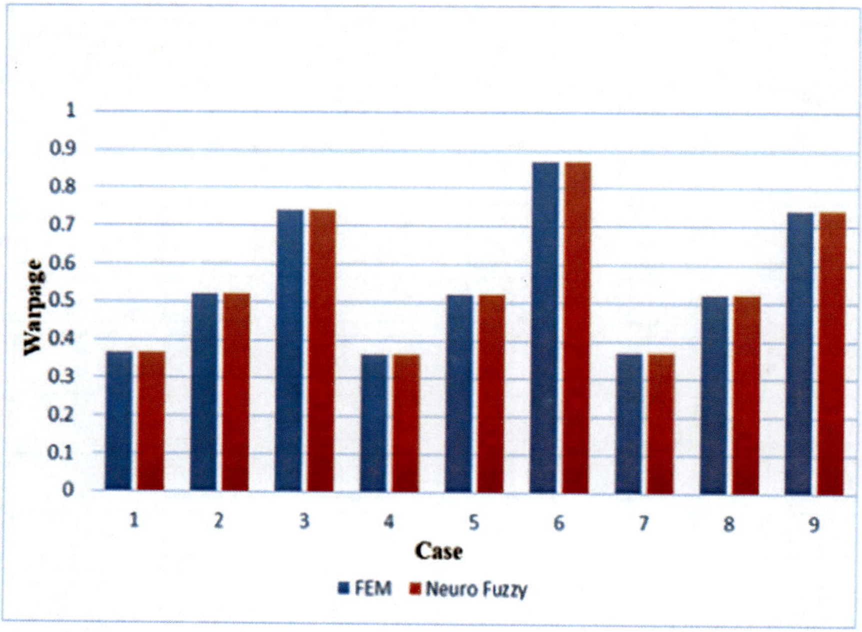

Fig. 14. Comparison between Neuro Fuzzy and FEM.

5 Conclusion

In this research, the FEM method and Neuro Fuzzy are collaborated for minimizing warpage to simulate a sample of box shape parts manufactured by injection moulding with a hot runner mould type. To achieve the minimum warpage, the appropriate factors of hot runner design are determined in various conditions consist of gate diameter, gate length, part dimension, melt channel design type and melt channel diameter. FEM simulation results were used to training data of the Neuro Fuzzy model which can approximate the warp of the injection process. In the selection hot runner design, the Neuro Fuzzy model is used to approximate warpage from various designs. It is found that the Neuro Fuzzy model and the FEM simulation can be used to select hot runner in different design effectively. Therefore, the result developed in this research can be used to select suitable hot runner design. In addition, it is a tool to assist decisions making for designing plastic injection mould efficiently.

References

1. Oktem, H., Erzurumlu, T., Uzman, I.: Application of Taguchi optimization technique in determining plastic injection molding process parameters for a thin-shell part. Mater. Des. **28**(4), 1271–1278 (2007)
2. Moayyedian, M., Abhary, K., Marian, R.: Gate design and filling process analysis of the cavity in injection molding process. Adv. Manuf. **4**(2), 1–11 (2016). https://doi.org/10.1007/s40436-016-0138-5

3. Shi, F., Lou, Z.L., Zhang, Y.Q., Lu, J.G.: Optimisation of plastic injection moulding process with soft computing. Int. J. Adv. Manuf. Technol. **21**(9), 656–661 (2003)
4. Li, C.S., Shen, Y.K.: Optimum design of runner system balancing in injection molding. Int. Commun. Heat Mass Transfer **22**(2), 179–188 (1995)
5. Zhou, H.: Computer Modeling for Injection Molding: Simulation, Optimization, and Control. John Wiley & Sons, Hoboken (2013)
6. Sadeghi, B.H.M.: A BP-neural network predictor model for plastic injection molding process. J. Mater. Process. Technol. **103**(3), 411–416 (2000)
7. Yen, C., Lin, J.C., Li, W., Huang, M.F.: An abductive neural network approach to the design of runner dimensions for the minimization of warpage in injection mouldings. J. Mater. Process. Technol. **174**(1–3), 22–28 (2006)
8. Spina, R.: Injection moulding of automotive components: comparison between hot runner systems for a case study. J. Mater. Process. Technol. **155**, 1497–1504 (2004)
9. Demirer, A., Soydan, Y., Kapti, A.O.: An experimental investigation of the effects of hot runner system on injection moulding process in comparison with conventional runner system. Mater. Des. **28**(5), 1467–1476 (2007)
10. Wang, M.L., Chang, R.Y., Hsu, C.H.D.: Molding Simulation: Theory and Practice. Carl Hanser Verlag, Munich (2018)
11. Park, C.H., Pyo, B.G., Choi, D.H.: Design automation of valve gate locations and open timing for injection molding of an automotive instrument panel. In: 2012 IEEE International Conference on Automation Science and Engineering (CASE), IEEE, p. 843–845 (2012)
12. Chen, S.C., Chien, R.D., Tseng, H.H., Huang, J.S.: Response of a sequential-valve-gate system used for thin-wall injection molding. J. Appl. Polym. Sci. **98**(5), 1969–1977 (2005)
13. Javierre, C., Fernández, Á., Aísa, J., Clavería, I.: Criteria on feeding system design: conventional and sequential injection moulding. J. Mater. Process. Technol. **171**(3), 373–384 (2006)
14. Kovács, J.G., Siklo, B.: Test method development for deformation analysis of injection moulded plastic parts. Polym. Testing **30**(5), 543–547 (2011)
15. Ozcelik, B., Erzurumlu, T.: Comparison of the warpage optimization in the plastic injection molding using ANOVA, neural network model and genetic algorithm. J. Mater. Process. Technol. **171**(3), 437–445 (2006)
16. Chen, Y., Zhu, J.: Warpage analysis and optimization of thin-walled injection molding parts based on numerical simulation and orthogonal experiment. IOP Conf. Ser. Mater. Sci. Eng. **688**(3), 033027 (2019). https://doi.org/10.1088/1757-899X/688/3/033027
17. Moayyedian, M., Abhary, K., Marian, R.: Optimization of injection molding process based on fuzzy quality evaluation and Taguchi experimental design. CIRP J. Manuf. Sci. Technol. **21**, 150–160 (2018)
18. He, W., Zhang, Y.F., Lee, K.S., Liu, T.I.: Development of a fuzzy-neuro system for parameter resetting of injection molding. J. Manuf. Sci. Eng. **123**(1), 110–118 (2001)
19. Nauck, D., Klawonn, F., Kruse, R.: Foundations of Neuro-Fuzzy Systems. John Wiley & Sons, Hoboken (1997)
20. Butdee, S., Nitnara, C.: A fuzzy logic combined with LP model for performance evaluation to distribute purchase orders in cluster manufacturing. Proc. Manuf. **30**, 19–25 (2019)

Improving the Efficiency of Mechatronic Systems for a Rapid Adoption and Integration of Industry 4.0

Raul Turmanidze[1], Giorgi Popkhadze[1], and Andre D. L. Batako[2](✉)

[1] Department of Mechanical Engineering Manufacturing Technologies, Georgian Technical University, Tbilisi, Georgia
[2] Liverpool John Moores University, Byrom Street, Liverpool L3 3AF, UK
a.d.batako@ljmu.ac.uk

Abstract. The presented work here has a direct relevance in solving issues related to the adoption of I4.0 in mechatronics. The potential problems, which may occur in the implementation of the industry - 4.0 reform have been explored. It was found that the pace and level of development of this reform will be determined to a large extent by the effectiveness of the individual part used and the entire mechatronic system. It has also been established that is a result of systematic miniaturization of the parts/subassembly of radio-electronic equipment and microelectronic equipment and microelectronic technology. The main problem of these reforms and the implementation of complex technological processes is instrumentation and technological support, especially with micro-cutting tools and equipment. This work presents an example of these investigations, with practical methods for improving the performance of micro-drilling process in printed circuit boards. High sensitivity measuring devices were used to accurately gauge drilling forces and torques acting on the tool.

Keywords: Industry - 4.0 · Mechatronic systems · Microdrills · Drilling PCB · Design optimization manufacturing

1 Introduction

Today it has already been unambiguously recognized that at the beginning of the 21st century the whole world is on the verge of the fourth scientific and technological revolution, which should fundamentally change the style and level of thinking, i.e. the rules of life for every person and especially the young generation in all countries of the world. This is due to the fact that, according to many scientists and authors of wide-profile studies to study the state of the necessary conditions for ensuring a worthy meeting of such major reforms, the upcoming change is assessed as the most comprehensive and large-scale in the history of all mankind. It will be held under the abbreviated name "Industry - 4.0" [1].

During the first industrial revolution, which lasted longer than two centuries, water and steam were used to mechanize certain industrial operations. As a result of the second

A. Batako et al. (Eds.): GCMM 2021, LNNS 335, pp. 288–303, 2022.
https://doi.org/10.1007/978-3-030-90532-3_22

revolution, on the basis of the use of electricity, mass production of many products was created in various areas of the national economy. During the third revolution, through the use of electronic and information technology, production processes became automated. And now, based on the results of the third revolution, the fourth revolution is developing, which relies on digital technologies, the development of which was started in the second half of the last century. It implies the fusion of several modern technologies and the disappearance of any boundaries between the physical, digital and biological spheres, that is, the creation of cyber - physical systems [2–8].

In other words, the ultimate goal of the Industry 4.0 reform is: full automation and remote control of complex technological processes and administrative and financial operations through the use of ultra-modern mechatronic systems [1, 8–10].

The results of the first three revolutions were general and applied for all countries, for every enterprise and, practically, for every person. However, the process of development of the 4th revolution and the degree of consistent use of the results of its individual stages in practice will have a peculiar character for different industries. Of course, the basic principles will be general, but since each specific industry has its own special modern, multi-operational and multi-parameter technologies, special knowledge and an individual approach will be needed for their design and management.

To create the aforementioned mechatronic systems that determine the level and pace of development of the Industry 4.0 reform, high-precision technological equipment and special micro-tools for various purposes are required.

Back in the 90s of the last century, we made a classification of all the main micro-tools used at enterprises of microelectronics and microelectronic technology, which are divided into three main groups: cutting, assembly and assembly. Each group includes subgroups with different types of tools and specific areas of their application.

To significantly increase the reliability of microcircuits and, accordingly, the final product, it is imperative to note another group of microtools, control tools - devices that enable us, before landing on a printed circuit board, to check all the operational characteristics of a ready-made and inserted closed microcircuit and on special stands to investigate the stability of these characteristics in a wide range of changes in electrical, mechanical and thermal loads. This work has designed, manufactured, tested and patented several variants of such a device for different standard sizes of microcircuits, their packaging along with dies and molds have been created for their production.

After testing and selection of microcircuits, such methods can practically exclude early failures of complex microcircuits and expensive equipment and devices during operation, which provides a great economic effect. The designs and operation of the above all types of microtools will be similarly reported during the presentation.

All these tools are used in a fairly large number, since a significant part of modern technology, from household equipment to space equipment, is a combination of mechanical assemblies, hydro and pneumatic equipment and microelectronic units or entire control systems, that is, it is a complex mechatronic system. Their manufacture requires the implementation of many technological operations of various profiles.

The presented report shows the classification of the micro-tools used for the production of the main units of modern mechatronic systems, as well as ways to systematically improve the efficiency of using individual groups of tools, taking into account the process of intensive miniaturization of electronic devices and the increase in the level of requirements for their operational characteristics.

However, it should be noted that among them, cutting micro-tools, especially twist drills, are repeatedly used in large quantities. This is due to the fact that in the process of manufacturing units of mechatronic systems, it is often necessary to process holes of small diameters (about 1 mm or less), especially on the details of hydraulic and pneumatic equipment. As for the production of microelectronic assemblies, in the technological processes of manufacturing basic parts - printed circuit boards, a significant part of the work comes in the operations of drilling a huge number of small-diameter holes.

Various methods are used to make holes in printed circuit boards, but practice has shown that the most acceptable method, especially when processing multilayer printed circuit boards with subsequent metallization of the hole surfaces, is still the drilling method.

Drilling is carried out with carbide microdrills, the optimal values of the geometric parameters of which have been established by repeated experiments and experiments of the respective industries. In particular: the optimal rake angle and, accordingly, the angle of inclination of the spiral groove is approximately 300, and the clearance angle is 180. They are re-sharpened through every 1000 holes and are designed for 3–4 regrinding.

The production of printed circuit boards is mass production, where, in order to increase productivity, a package composed of several plates is drilled, that is, deep drilling occurs, where the drilling depth exceeds $8 \div 10$ times to the diameter.

Downtime of expensive technological equipment, especially in mass production, is associated with significant economic losses. In the production of printed circuit boards, downtime is associated not only with the envisaged replacement of the tool for the purpose of regrinding, but also unforeseen, caused by brittle fractures even before the first regrinding. The likelihood of brittle fracture increases significantly with deep drilling of a package of printed circuit boards. In this case, the fracture zone is located in the vicinity of the end of the spiral groove.

2 Research Methodology

Providing for the mass production of printed circuit boards, a slight increase in durability, including the fragile durability of microdrills and, accordingly, an increase in the productivity of drilling processes, can give a significant economic effect. Research work aimed at increasing the durability of carbide microdrills and the productivity of the deep hole drilling process for a package of printed circuit boards was carried out in the laboratory of precision microinstruments of the Department of Production Technologies of Engineering Mechanics of the Georgian Technical University in close cooperation with specialists from the Magdeburg Technical University. Otto von Guricke (Germany).

The studies were started by studying the nature of the change in power indicators - torque and axial reinforcement, depending on the depth and drilling modes of a package of printed circuit boards made of fiberglass.

To measure the axial force, a device was manufactured on the basis of known methods and existing analogues, the measuring element of which is a system of highly sensitive Swiss made (mod. SW-2L) strain gauges mounted on an elastic body (Fig. 1).

As for the measurement of torque, in our case the existing indirect method, in which the measurement is carried out by measuring the power of the cutting process, is unsuitable, since when drilling with microdrills, the power indicators are so low that it is practically impossible to fix its changes. Therefore, it is necessary to use a method that will make it possible to measure the torque directly, while with high accuracy.

Fig. 1. Instrument for measuring axial efforts

Fig. 2. The device for measurement of torque

For this purpose, we have designed and manufactured a special device (Fig. 2), in which the table of workpieces for drilling is installed on an axis that has the ability to rotate angularly. For its part, the axis of the table is equipped with a face, the plane of which runs on the axis of symmetry and acts on the elastic measuring element with a certain shoulder. As a measuring element, a system of Swiss made (mod. TEDEA-1004) strain gauges with high measurement accuracy (with a total error of $\pm 0.02\%$) mounted on an elastic body is also used here.

The experiments were carried out with drills made of WC6OM grade hard alloy with a diameter of 0.9 mm and a long working part e = 10 mm. The rake angle and, accordingly, the angle of inclination of the helical groove, the clearance angle was 18°.

Drilling of a package of printed-circuit boards made of fiberglass with a thickness of 1.6 mm, composed of 5 plates with a total thickness of 8 mm (Fig. 3).

3 Results and Discussion

The workpieces were drilled at different cutting modes and the readings of the measuring instruments were recorded through each 1mm depth. The experimental results are shown in Figs. 4 and 5.

It is clearly seen from these results that, as expected, force loads progressively increase with depth. In this case, if the axial force increases by approximately 1.5 times, then the magnitude of the torque increases by 3–4 times.

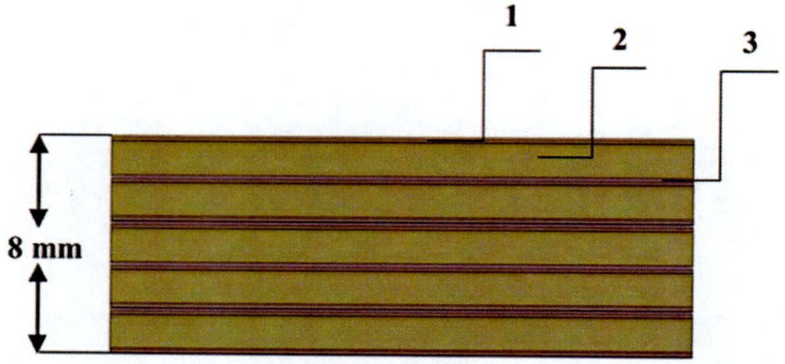

Fig. 3. Package diagram printed circuit boards of 5 plates. 1 -copper layer, 2-fiber, 3-double layer of copper foil.

The question is, what caused such an increase in strength indicators? The cutting conditions at the cutting edge of the drill do not change with increasing depth. The only reason for this may be an increase in the contact area of abrasive chips with the surface of the machined hole and the friction forces arising from it [11–19].

It should be noted that the sharp drop in the axial force in the sections with depths 2–3 and 5–6 (see Fig. 5) is due to the absence of a copper layer in these sections of the workpiece being processed (Fig. 3).

Of these two strength indicators, the main factor in increasing the probability of brittle fracture of the cutting tool may not be an increase in axial force, but a more progressive increase in torque, since the compressive strength indicators of carbide materials significantly exceed torsional forces.

Acceleration of the process of chip removal from the cutting zone would help to reduce friction forces and, accordingly, increase the reliability of the drilling process. The problem of removing chips when drilling deep holes in different cases is solved in different ways. For example, when drilling with large drills, this is carried out by flushing with the help of a cutting fluid, which is fed into the hole through channels made in the drill body itself. In other cases, when the dimensions of the drills do not allow the supply of coolant by the above method, in order to remove the chips, the intermittent drilling method is used, in which, after drilling a certain depth, the drill is periodically withdrawn from the hole at a rapid pace.

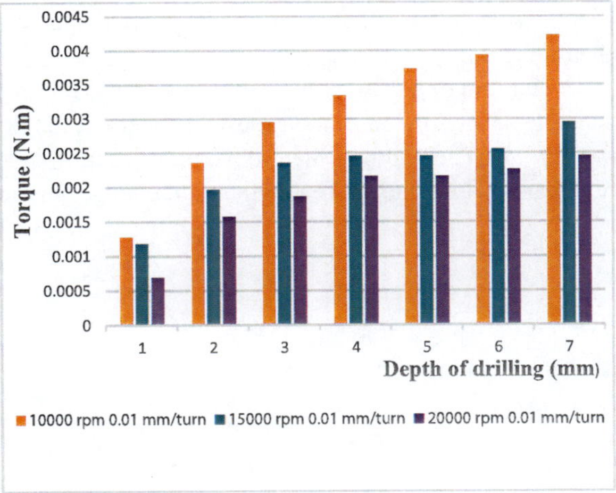

Fig. 4. The chart for standard drills $\omega = 30°$

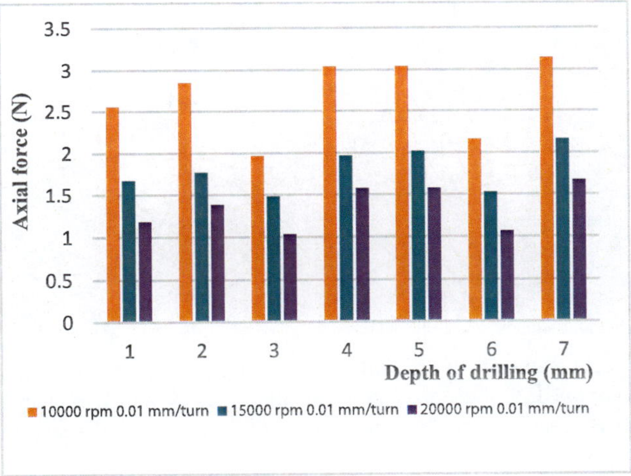

Fig. 5. The chart of changes of axial efforts for standard drills $\omega = 30°$

The use of these methods in our case is unacceptable because firstly, we are dealing with micro drills, and at the same time, the use of cutting fluid in the production of printed circuit boards is not permissible. The intermittent drilling method is also inappropriate, as this would lead to a strong drop in productivity. When machining deep holes with micro drills, an increase in the pitch of the spiral groove, that is, a decrease in the angle of inclination, can become an accelerating factor in chip removal, but this would lead to a deterioration in cutting conditions due to a decrease in the rake angle of the drill.

If the design of the drill is made in such a way that at the tip the most effective rake angle and, accordingly, the angle ω for the cutting process, and gradually decrease its value in the direction of the end of the spiral groove, that is, cut the spiral groove with a variable angle of inclination and gradually increase its pitch, this is would lead to faster chip removal and easier drilling conditions.

In Fig. 6 shows a diagram of a drill with a variable angle of inclination of the spiral groove, where the angle of inclination of the groove at the tip of the drill ω_0, and at the end of the working part ω_1. The width of the groove in the section of the perpendicular groove B_n does not change along the entire length, but it changes in the end section and at the apex is:

$$B_{T0} = \frac{B_n}{\cos \omega_0}, \tag{1}$$

and at the end of the working parts

$$B_{T1} = \frac{B_n}{\cos \omega_1}, \tag{2}$$

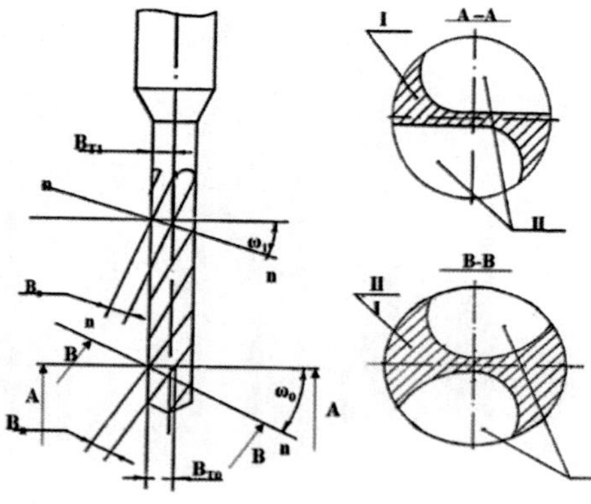

Fig. 6. Geometrical parameters of drill with variable angle of inclination of the spiral grooves, AA is the section of the drill of the plane perpendicular to its axis: I - Useful area of the drill, II - Groove profile. B-B - section of the drill in the plane of the perpendicular spiral

The useful end section of the drill also changes, the value of which is at the top:

$$S_0 = \frac{\pi d^2}{4} - \frac{2S_n}{\cos \omega_0}, \tag{3}$$

and at the end of the working part:

$$S_1 = \frac{\pi d^2}{4} - \frac{2S_n}{\cos \omega_1}, \tag{4}$$

where S_n is the area of the groove in the perpendicular section, d is the diameter of the working part of the drill.

Taking into account that $\omega_0 > \omega_1$, it turns out that in the direction of the end of the working part, the useful section of the drill increases. That is, such drills, in comparison with standard ones, must withstand increased loads.

The manufacture of such drills is associated with certain difficulties. In the manufacture of standard drills with a constant angle of inclination of the spiral groove, a special device carries out mutual coordination of two movements - rotation of the drill workpiece and its movement in the axial direction in such a way that for each revolution the workpiece is evenly moved by one spiral step. In the case of a variable pitch, the axial feed is carried out with a certain acceleration, so that a uniform decrease in the angle of inclination of the spiral groove along the entire length of the cut is ensured. Because of this, we had to modernize the equipment and, in the cam feed mechanism, replace the cam with an Archimedean spiral with a cam with a profile made according to our calculations, based on the law of changing the angle of inclination and, accordingly, the pitch of the spiral groove of the drill. In addition, due to the variable angle of inclination of the spiral groove, when grinding such grooves, it is imperative during processing to change the orientation of the grinding wheel relative to the axis of the drill workpiece, respectively, to the change in the angle of inclination of the groove. This can be done in two ways: keep stationary the axis of grinding spindle, then rotate the workpiece spindle axis around the point of intersection of these axes by the angle value (see Fig. 7), or leave the workpiece spindle axis motionless and rotate the grinding spindle axis (see Fig. 8).

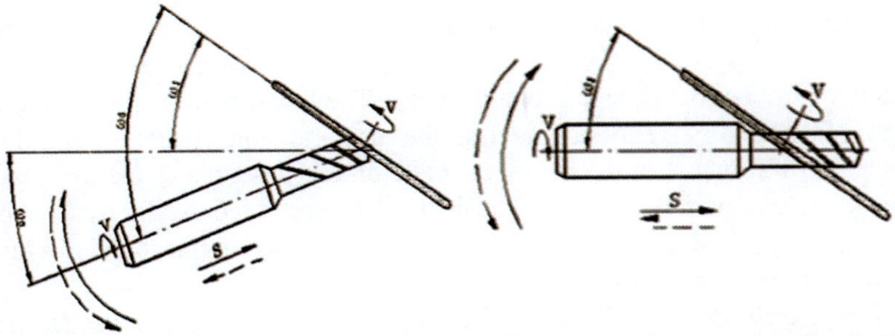

Fig. 7. Schemes of change of orientation of axes of a spindle of preparation of a drill and grinding spindle. Method of turn of an axis of a spindle of preparation.

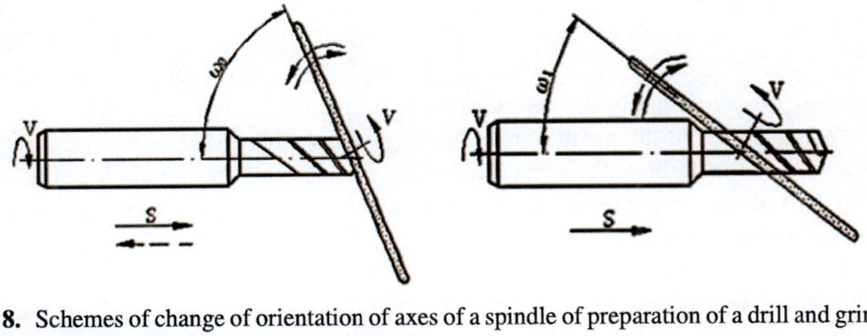

Fig. 8. Schemes of change of orientation of axes of a spindle of preparation of a drill and grinding spindle. Method of turn of a grinding spindle.

Proceeding from constructive reasons the preference was given by us to the first option (Fig. 7) and in the course of modernization of the equipment it was equipped with the additional mechanism of turn providing when cutting spiral flutes, turn of an axis of preparation of a drill relatively to an axis of a grinding spindle at a corner size $\omega = \omega_0 - \omega_1$ thus depending on are long the cutting part of a drill turn is carried out by the linear law:

$$\omega_x = l_x \cdot K_\omega, \tag{5}$$

Where ω_x - the current size of an angle of rotation of an axis of preparation, l_x - the current coordinate of length of the cutting part of a drill, K_ω - the size of change of a tilt angle of a spiral flute per unit length the cutting part of a drill.

A schematic diagram of such a modernized device with a lever swing mechanism is shown in Fig. 9. It should be noted that in order to achieve higher accuracy, instead of the lever mechanism, a rack and pinion pair can be used.

The device for grinding spiral grooves towards the drill workpiece 1 is installed on the rotation axis 2, mounted on the grinding machine table - 3D642. In this case, the axis of rotation of the device must necessarily pass to the point "O" of intersection of the axis of the drill workpiece and the vertical axis of symmetry of the grinding wheel 3, tucked in accordance with the profile of the spiral groove of the drill. The other end of the device rests on a circular guide 4. The rotation is carried out by a lever mechanism, the support axis 5 of which is fixedly mounted on the base 6 of the device. With the axial movement of the upper slide 7 and, accordingly, the drill workpiece, the pin 8 rigidly connected to it rotates the right arm of the lever 9. The left arm of the lever is connected to the pin 10, which is fixedly mounted on the machine table. Thus, turning the lever causes the entire structure of the device to rotate. In this case, the elements of the lever mechanism are designed so that when the drill workpiece is fed axially by the length of the screw thread, the device rotates by the amount of change in the angle of inclination of the spiral groove: $\Delta_\omega = \omega_0 - \omega_1$.

For this, the distance between the left fixed finger and the axis of rotation of the device is calculated using the formula:

$$R = l \cdot \frac{L_\pi}{L_n} \cdot \frac{360}{2\pi \Delta\omega}, \tag{4}$$

Where l - is the length of the spiral cut of the drill, mm; L_π - the length of the left arm of the lever; L_n - the length of the right arm of the lever; $\Delta\omega$ - the amount of change in the angle of inclination of the spiral grooves of the drill.

In the structures of the positioning/clamping device, there is also the possibility of adjusting the value of the axial feed of the workpiece and the starting angle at the tip of the drill ω_0 within a certain range. With this approach, prototypes of drills were designed and manufactured with different ranges of variation in the angle of inclination of the spiral groove: $\omega = 30 - 70°$, $\omega = 35 - 20°$, $\omega = 40 - 22°$ and $\omega = 43 - 23°$.

The same experiments were carried out on all these drills, as on standard ones and the experimental results for drills with $\omega = 30 - 70°$ and $\omega = 35 - 20°$ are shown in Figs. 10, 11, 12 and 13.

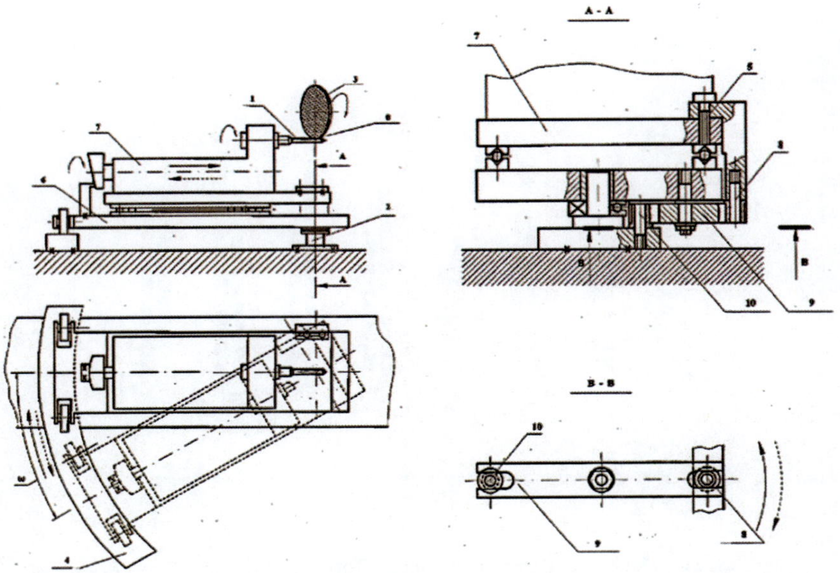

Fig. 9. Schematic diagram of the rotation axis of the drill mechanism workpiece relative to the plane of the grinding wheel at Grinding of helical grooves with a variable tilt angle.

The analysis of these graphs shows the following:

For the drills $\omega = 30 - 70°$, the axial force values are almost the same as for the standard drill $\omega = 30$, which is as expected, since these drills have the same rake angle and, accordingly, the cutting conditions at the cutting edge. As for the torque values, they are underestimated by 12–16% on the drill $\omega = 30 - 70°$, depending on the cutting conditions.

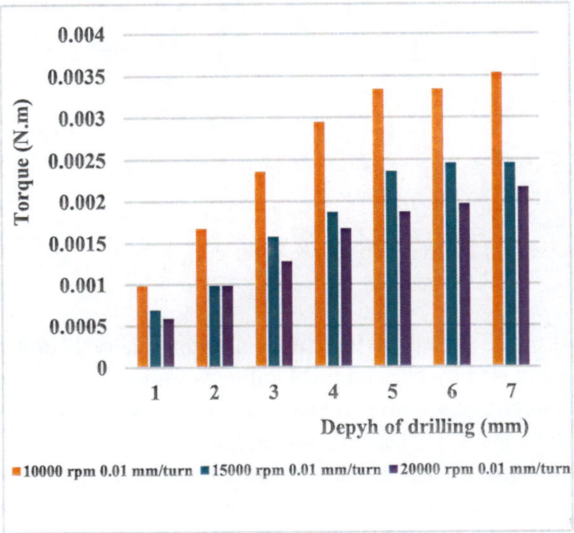

Fig. 10. The chart of change of a torque for drills $\omega = 30 - 17°$

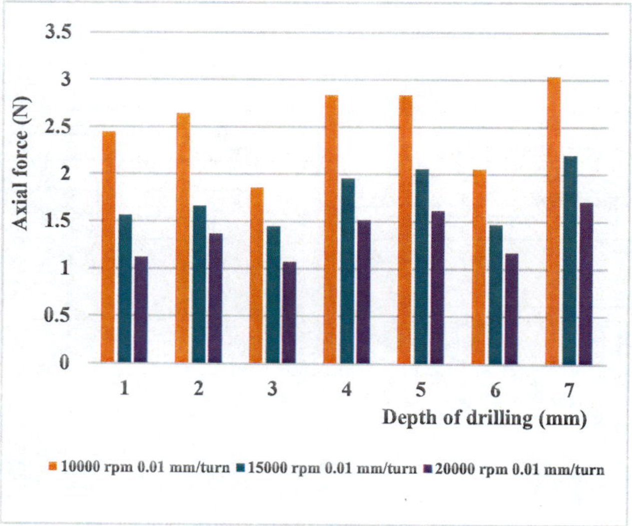

Fig. 11. The chart of changes of axial efforts for drills $\omega = 30 - 17°$

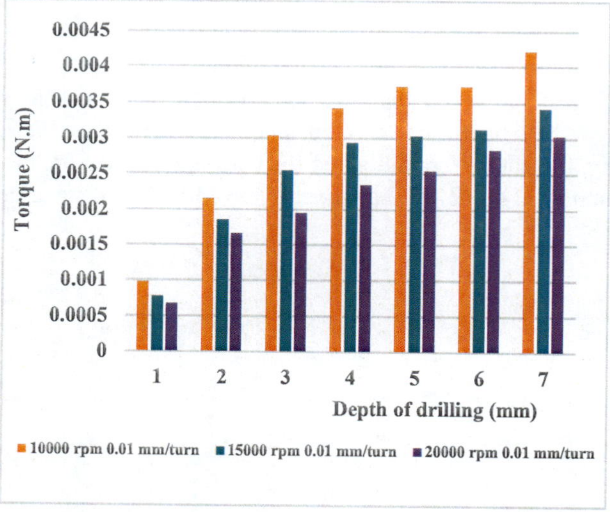

Fig. 12. The chart of torque for drills $\omega = 35 - 20°$

For drills $\omega = 35 - 20°$, the axial force indicators are relatively underestimated, since the rake angle is increased by 5% and this facilitates the cutting process, but the torque value grows more intensively with increasing drilling depth and exceeds the indicators of both previous samples. It is clear that the underestimation of the torque on the drills $\omega = 30 - 17°$ compared to standard drills $\omega = 20°$ is due to a gradual increase in the pitch of the spiral flute and, accordingly, a decrease in the contact area of the formed chips with the walls of the hole being machined. An increase in the angle of inclination

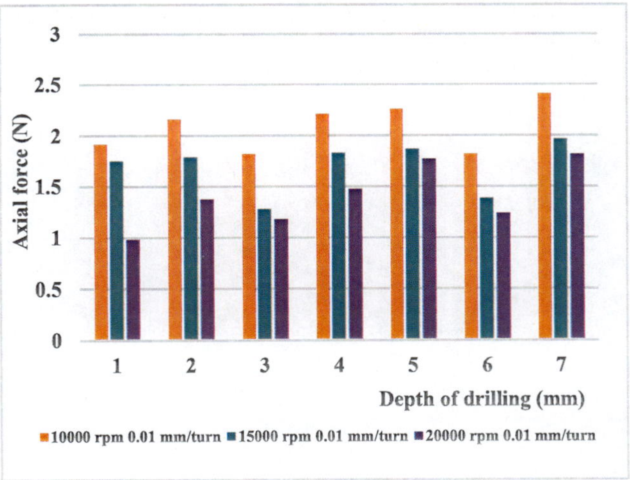

Fig. 13. The chart of changes of axial efforts for drills $\omega = 35 - 20^0$

and, accordingly, a decrease in the helix pitch on drills $\omega = 35 - 20°$ again causes the opposite process - the torque indicators increase intensively.

As illustrated in Figs. 14, 15, 16 and17, when drilling with drills of $\omega = 40 - 22°$ and $\omega = 43 - 23°$, the power indicators are relatively underestimated, which is explained by the improvement in cutting conditions due to a significant increase by 5–8° in the rake angle at the cutting edge. However, a decrease in the taper angle during further operation causes an increase in the intensity of wear, and they reach only 600–800 holes without regrinding.

To get a clearer picture, it was necessary to continue the comparative experiments between standard drills $\omega = 30°$ and drills with a variable angle spiral flute $\omega = 30 - 17°$ Experiments were continued until the drills broke. At the same time, force indicators were checked through every 200 holes, which gradually increase with increasing wear of the cutting edge.

After 1000 holes, the check was carried out every 100 holes, as the probability of drill breakage increases. Throughout the series of experiments for both types of drills, the indicators of wear of the cutting edge and axial force remained almost the same. In terms of torque, standard drills $\omega = 30°$ have always exceeded those of drills with a variable angle helical flute $\omega = 30 - 17°$.

The statistical analysis has shown that the breakage of standard drills $\omega = 30°$ occurs from 1200 to 1300 holes, whereas the drills with variable angle $\omega = 30 - 17°$ had a better performance and failed from –1400 to –1500 holes.

The relevance of the research work carried out and the need to improve the methods of conducting experiments over time is becoming acute and necessary. This is due to the fact that the constant miniaturization of units and parts of radio equipment and computer technology is forcing specialists in the design and implementation of technological processes for their processing, respectively, to reduce the size of the main actuating surfaces. This is evidenced by the fact that 15–20 years ago, drills with a minimum

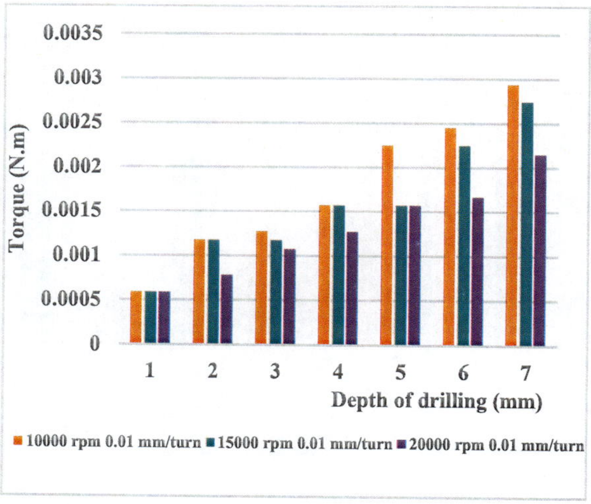

Fig. 14. The chart of change of a torque for a drill $\omega = 40 - 22°$

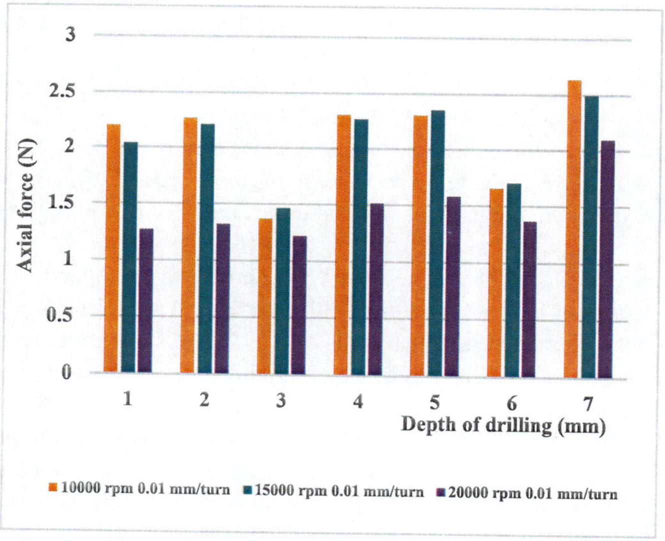

Fig. 15. The chart of change of axial effort for drills $\omega = 40 - 22°$

working diameter of 0.8–0.9 mm were used to drill printed circuit boards, at present, according to the latest data, these dimensions have decreased to 0.1–0.2 mm, and even less for single special boards. The process of miniaturization is still going on and it is not difficult to imagine an approximate situation, which will be, for example, in 20–30 years. At this time, the process of optimizing the geometric parameters of the drills will become even more important because it will be necessary to use every micron of the drill's working diameter as efficiently as possible. The same situation will be for

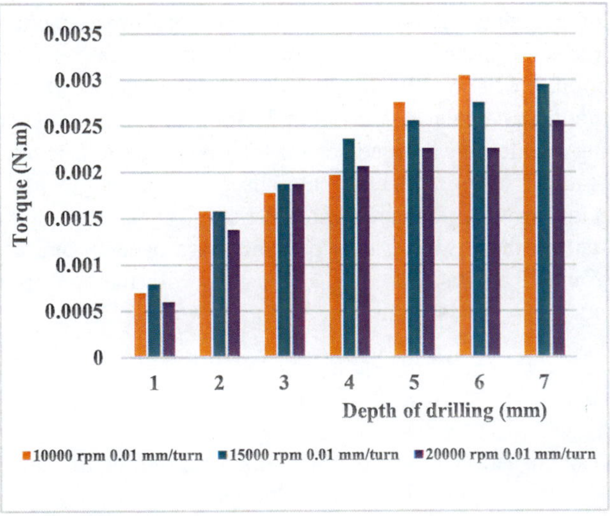

Fig. 16. The chart of change of a torque for drills $\omega = 43 - 23°$

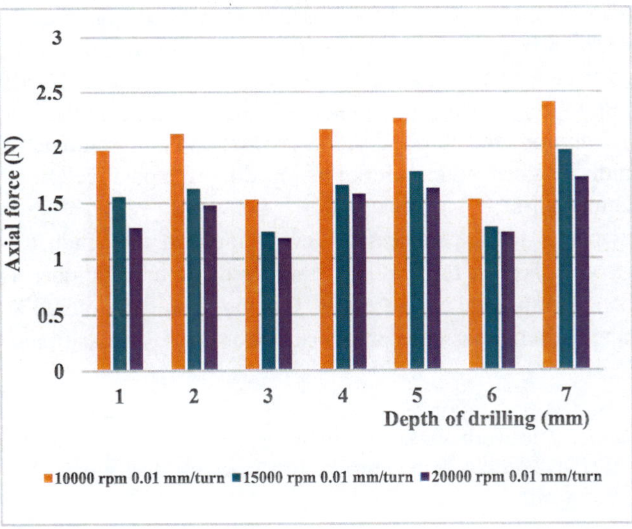

Fig. 17. The chart of change of axial effort for drills $\omega = 43 - 23°$

increasing the efficiency of operation of other instruments used in the production of electronic equipment.

Currently, similar experiments to determine the effectiveness of drills with a variable angle of inclination of the spiral groove are carried out in the processing of metal materials widely used in industry, such as titanium and aluminium alloys and non-ferrous metals with different physical and mechanical properties.

As a result of the intensive process of miniaturization of manufactured products and other related processes in the microelectronic industry, at every certain time in many areas of production, it will periodically arise the need to create more accurate and more acceptable methods for calculating the optimal parameters of critical parts and tools, as well as designing technological processes for their production. It is undeniable that all these transition periods will require new materials, new technologies and new generation equipment with improved performance characteristics. The most important thing is that in order to carry out these works, new personnel are needed, and especially a new generation of young specialists, who will be armed with the appropriate knowledge, which is what our Universities should do.

4 Conclusion

1. It has been established that the rate of development of the "Industry- 4.0" reform is strongly dependent on the efficiency of the operational characteristics of the used mechatronic systems.
2. It has been proved that the output characteristics of individual units of the used mechatronic systems can be significantly improved by optimizing the design and geometric parameters and the processes of their manufacture. For example, the durability of carbide precision twist drills with a working diameter of less than 1 mm and the drilling performance of printed circuit boards can be increased by at least 25–30%.
3. By optimizing the geometric parameters of carbide micro drills, it was possible to significantly improve the chip formation process and chip removal from the cutting zone when drilling printed circuit boards, leading to a positive effect on the quality of the machined surface.
4. It was shown that for the successful development of the "Industry- 4.0" reform, along with other issues, the level of training of young engineering personnel is of fundamental importance. Therefore, for this should be included into teaching curriculum to form the basic necessary skills for future applications.

Acknowledgements. This work was supported by the National Science Foundation. Shota Rustaveli (SRNSF) [PHDF-19-2224, Improving the Efficiency of Mechatronic Systems to Support the Industry 4.0 Reforms].

References

1. What is Industry 4.0. https://www.epicor.com/en/resource-center/articles/what-is-industry-4-0/
2. The Evolution of Industry 1.0 to 4.0. https://www.seekmomentum.com/blog/manufacturing/the-evolution-of-industry-from-to-4
3. The birth of the electric machines. https://royalsocietypublishing.org/doi/10.1098/rsta.2014.0208
4. The invention of the electric motor. https://www.eti.kit.edu/english/1376.php

5. History of mobile phones and the first mobile phone. https://www.uswitch.com/mobiles/guides/history-of-mobile-phones/
6. What is the internet? History and stages of development. http://www.ecommerce-digest.com/history-of-the-internet.html
7. Zujevs, A., Osadcuks, V., Ahrendt, P.: Trends in robotic sensor technologies for fruit harvesting: 2010-2015. Proc. Comput. Sci. **77**, 227–233 (2015). https://doi.org/10.1016/j.procs.2015.12.378
8. Ivan, A.: Research regarding optimization of industrial robots for machining applications. Doctoral thesis, Universiti "Politehnica" of Bucharest (2011)
9. Jang, B.: Nanotube fiber reinforced composite materials and method of producing fiber reinforced composites, 23 December 2003
10. Turmanidze, R., Bachanadze, V., Popkhadze, G.: Basic technological processes required to create mechatronic sistems to meet today's needs and problems of their instrumental support of "Industry 4.0" challenges. Int. Sci. J. "Inov." **5**(3), 123–128 (2017). (Sofia, Bulgaria, ISSN Print 1314-8907, ISSN WEB 2534-8469123-128)
11. Karpushewski, B., Turmanidze, R., Dübner, L., Kushnarenko, O.: Erhohung der standzeit und prozesssicherheit von mikrobohrern durch die entwicklung neuer werkzeuggeometrien. In: Collection of Scientific Works "Modern Technologies in Mechanical Engineering", Publishing House of the Kharkov National Technical University "KPI". Release 2. Kharkov, pp. 27–32 (2008)
12. Adamia, D., Gviniashvili, Z., Bachanadze, V.: Peculiarities of formation of shavings grooves of spiral drill of hard alloy with variable inclination. Sci. Tech. J. **3**(15), 133–140 (2009). Transport and Machinebuilding
13. Turmanidze, R., Kushnarenko, O., Adamia, D., Gviniashvili, Z.: Small-sized hard metal spiral drills with variable angle of inclination chip grooves. In: Collection Scientific Works "Modern Technologies in Mechanical Engineering", Publishing House of the Kharkov National Technical University "KPI". Kharkov, Release 5, pp. 318–327 (2010)
14. Turmanidze, R., Gviniashvili, Z.: Peculiarities of manufacture of fine-sized spiral drills with variable inclination angle of chip grooves and results of their tests. In: The 66th International Congress of Precision Machining ICPM 2011, 13–15 September 2011, pp. 253–258. Liverpool John Moores University (2011). ISSN: 1013-9826. ISBN-13: 978-3-03785-297-2
15. Lei, X., Shen, B., Cheng, L., Sun, F., Chen, M.: Influence of pretreatment and deposition parameters on the properties and cutting performance of NCD coated PCB micro drills. Int. J. Refract. Metal Hard Mater. **43**, 30–41 (2014). https://doi.org/10.1016/j.ijrmhm.2013.10.016
16. Kang, M.C., Je, S.K., Kim, K.H., Shin, B.S., Kwon, D.H., Kim, J.S.: Cutting performance of CRN-based coatings tool deposited by hybrid coating method for micro drilling applications. Surf. Coat. Technol. **202**(22–23), 5629–5632 (2008). https://doi.org/10.1016/j.surfcoat.2008.06.130
17. Rajan, K., Prabhushankar, N., Nagarajan, N.: A review of current micro drilling processes. Int. J. Innov. Sci. Eng. Technol. (IJISET) **3**(1), 60–64 (2016)
18. Zheng, L.J., Wang, C.Y., Fu, L.Y., Yang, L.P., Qu, Y.P., Song, Y.X.: Wear mechanisms of micro-drills during dry high speed drilling of PCB. J. Mater. Process. Technol. **212**(10), 1989–1997 (2012). https://doi.org/10.1016/j.jmatprotec.2012.05.004
19. Karabegović, I., Kovačević, A., Banjanović-Mehmedović, L., Dašić, P.: Handbook of Research on Integrating Industry 4.0 in Business and Manufacturing. IGI Global, Hershey, 661p. (2020). ISBN 978-1-7998-2725-2. https://doi.org/10.4018/978-1-7998-2725-2

Asymmetry in the Tension and Compression Flow Stress and the Effect of Sub-cell Size on the Hardness of a Selective Laser Melt 316L Stainless Steel

Mustafa Güden[1](✉), Samed Enser[1], Burcu Arslan Hamat[2], Alptuğ Tanrıkulu[2], and Hakan Yavaş[2,3]

[1] Department of Mechanical Engineering, İzmir Institute of Technology, Urla, Izmir, Turkey
mustafaguden@iyte.edu.tr

[2] Turkish Aerospace Industries Inc., R&D and Prototype Operations, Kahramankazan, Ankara, Turkey

[3] Turkish Aerospace Industries Inc., Aerostructures, VP, Operations Directorate, Additive Manufacturing Division, Kahramankazan, Ankara, Turkey

Abstract. An asymmetry between tension and compression tests was determined experimentally in the Selective Laser Melt (SLM) stainless steel 316L alloy in the building direction. The asymmetry was ascribed to the used biaxial scanning strategy which resulted in a strong alignment of ⟨110⟩ along the build direction (fiber texture) and a random distribution of ⟨100⟩, ⟨110⟩ and ⟨111⟩ directions normal to the building direction. The strong fiber texture in the building direction induced lower twinning stress in tension than in compression, while the tension and compression twining stresses were found similar in the normal to building direction. The favored twinning in the specimens tested in the building direction resulted in a higher tensile true fracture strain; hence, a higher ductility. Lastly, the hardness measurements made on the specimens having similar gain sizes, but different sub-cell sizes processed using higher and lower laser powers tended to support that the sub-cell boundaries in SLM-316L alloy acted as imperfect barriers to the dislocation motion.

Keywords: Additive · SLM · 316L · Asymmetry · Flow stress · Sub-cell

1 Introduction

Extremely high cooling rates (103 to 108 K s^{-1}) [1, 2] associated with high thermal gradients in the Selective Laser Melt processing of stainless steel 316L alloy (SLM-316L) results in a distinct microstructure development. The microstructure consists of columnar grains in the building direction and a sub-micron size sub-cell structure in which the cell boundaries are decorated by a dislocation network (high dislocation density) and higher solute atom concentrations of heavy atoms like Mo and Cr [3]. The yield strength of SLM-316L was found much higher than that of commercial alloy, which was ascribed

A. Batako et al. (Eds.): GCMM 2021, LNNS 335, pp. 304–312, 2022.
https://doi.org/10.1007/978-3-030-90532-3_23

to a relatively higher dislocation density of SLM-316L, ~$1.14 \times 1015\ m^{-2}$ [4]. Besides, a crystallographic fiber texture of <100> in the building direction was reported [5–10]. Applying a rotation strategy between neighboring islands (island scanning) however altered the texture development; a strong fiber texture of ⟨110⟩ along the build direction (z-axis) and a random distribution of ⟨100⟩, ⟨110⟩ and ⟨111⟩ directions along the normal to the building direction (x- and y-axis, ±45° to the scanning directions) are formed [11]. A twin-induced-plasticity (TWIP) proceeding with twinning and slip was commonly reported for SLM-316L [4, 9, 12, 13]. The flow stress is usually higher in the normal to the building direction and the ductility is vice-verse [4, 9]. It is noted however that there has been so far no report on the asymmetry in the tension and compression of SLM-316L alloy. This study was therefore conducted to investigate the asymmetry in the tension and compression of an SLM-316L alloy fabricated by a biaxial scanning strategy in two different directions: in the building and normal to the building direction. The observed asymmetry was explained based on the texture in the studied structure. The effect of sub-cell size on the hardness was further investigated on the alloys having similar sizes of grains but different sizes of sub-cell.

2 Materials and Methods

The compression and tension test specimens were fabricated in a Laser-powder bed fusion AM Concept Laser M2 Cusing device (400 W continuous-wave fiber laser) using gas atomized 316L powder with an average particle size of 40 μm under nitrogen atmosphere. The power, scanning rate, hatching space, and spot size of the incident beam were sequentially 370 W, 900 mm s^{-1}, 115 μm, and 160 μm. A multidirectional scanning (biaxial scanning) laser-pattern rotated 90° between adjacent layers was used to construct the samples with a constant layer thickness of 30 μm. In few samples, the process parameters were adjusted to give a smaller size of sub-cell by keeping a constant grain size. In this case, the power, scanning rate, hatching space, and spot size of the incident beam were sequentially 180 W, 700 mm s^{-1}, 95 μm, and 140 μm. This alloy is referred to as a lower- laser-power alloy and was merely used to determine the effect of sub-cell size on the hardness.

The compression test specimens were machined from 6 × 13 × 130 mm rectangular bars; the long axis was fabricated in the building (z-direction) and normal to the building direction (x- and y-direction) depicted in Figs. 1(a) and (b), respectively. After removing the supports and machining the bar 0.5 mm from each side, the cylindrical compression test samples, 5 mm in diameter and 5 mm in length were prepared using an electro-discharge machine. The compression loading axes of the specimens are in the y-direction (Fig. 1(a)) and z-direction (Fig. 1(b)). As is seen in Fig. 1(c), the microstructure of the biaxially scanned SLM-316L consists of biaxial melt pools (±45°) in the x-y plane and the cross-sections of the biaxial melt pools in the x-z and y-z planes. The tensile test specimens were machined from the cylindrical SLM-316L bar samples, 6 mm in diameter and 80 mm in length, fabricated in the z- and y-direction. Tensile test specimens were machined from these fabricated bars in accord with the ASTM E8/E8M−13a standard "Standard Test Methods for Tension Testing of Metallic Materials" with a gage-diameter of 2.5 mm and a gauge length of 16 mm.

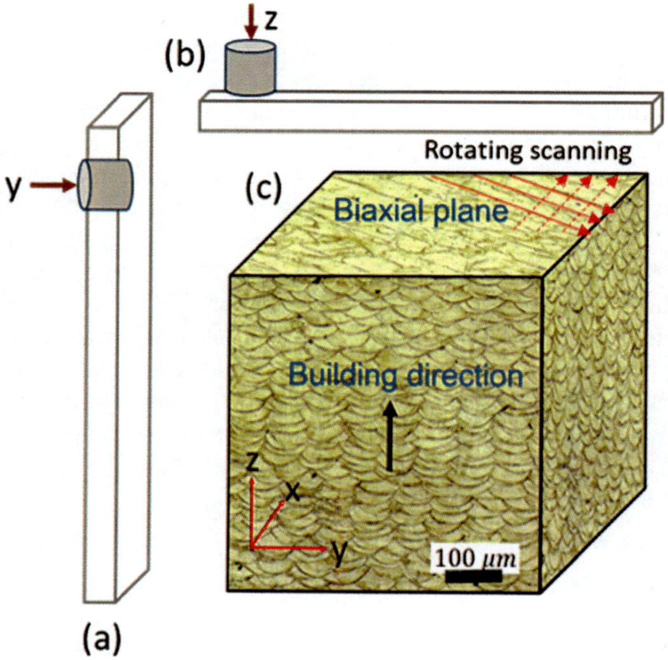

Fig. 1. SLM-316L bars and compression specimen extraction (a) in y- and (b) z-direction and (c) the schematic of microstructure and directions.

Quasi-static compression and tension tests were performed at the strain rate of ~$1 \times 10^{-3}\ s^{-1}$ in a Shimadzu AG-X universal test machine. At least three tests were performed for each group of specimens and the specimen displacements were recorded using a video extensometer. The metallographic samples prepared by the standard metallography in a mirror finish were etched using a solution of 10% HNO3, 20% HCl and 20% Glycerol. The microscopic analyses were performed in a Meiji IM7 100 optical microscope and an FEI Quanta 205 FEG and Philips XL 30SFEG scanning electron microscope (SEM). The average sub-cell and grain sizes were determined by the linear intercept method. Hardness Vickers (HV) tests were performed on the polished and polished-etched surfaces of mounted samples in a Shimadzu Micro Vickers Hardness Tester by applying 0.49 and 19.61 N loads for 10 ms. The crystallographic structure of as-built specimens was determined using a Philips X'Pert Pro X-Ray Diffraction (XRD) device (Cu-Kα radiation, $\lambda = 1.54$ A°, and 40 kV).

3 Result and Discussion

Figure 2 shows the compression and tension true stress-strain curves of three SLM-316L specimens in the y-direction (45° to the melt pools in the biaxial plane) and z-direction (building direction). The flow stresses of compression and tension tests at 6% strain and fracture stresses and strains of the tension tests are further tabulated in Table 1. The compression true stress-strain curves of the specimens tested in the x- direction, not shown here, were found to be very similar to those of the specimens tested in the y-direction, proving isotropic mechanical properties in the biaxial plane. As is noted in Fig. 2, the compression stress-strain curves in the y- and z-direction are comparable with each other, except the specimens tested in y-direction show slightly higher compressive stresses at low strains, while they become very much similar at increasing strain. Also, the compression tested specimens did not show any indication of shear banding and fracture until about large strains, 35–55%. It is further noted in Fig. 2 that the tension flow stresses are slightly higher than those of compression in the y-direction, while the difference vanishes at increasing strains and they become very much comparable with each other. On the other side, a distinct difference between the tension and compression stresses of the specimens tested in the building direction is seen in Fig. 2. In the building direction, the compression stresses are higher than those of tension at low strains and the difference between them increases as the strain increases. Not only the flow stresses but also the tension and compression strain hardening behavior of the specimens tested in the z-direction are also very different from each other. A much lower strain hardening is seen in the tension specimens in the building direction, while the compression and tension specimens in the y-direction and the compression specimens in the z-direction show very much similar work hardening, depicted in Fig. 3. These confirmed an asymmetry in the compression and tension flow stresses in the z-direction. An abrupt change in the work hardening rate in Fig. 3 indicates the initiation of the twinning mode of deformation. The twinning mode of deformation is noted to start at the strains lower than 10% in both compression and tension tests as seen in the same figure. The microscopic observations on the specimens deformed until about various final strains also confirmed the twinning deformation starting at about 10% strain.

It was previously shown that the rotation of laser scans between neighboring layers in SLM resulted in a mismatch of the positions of melt pools in each layer, breaking the epitaxy of columnar growth and resulting in a fiber texture in the building direction [11]. A similar observation was also made in the present study. Figure 4(a) shows the microstructure of studied SLM-316L in the y-direction (normal to melt pools). The columnar grains are seen to extent within only few melt pools, proving a break of development of columnar grains. The EBSD micrograph of melt pools seen Fig. 4(b) also confirms this, the breaking of the epitaxial growth through building direction. The fiber texture of ⟨110⟩ along the build direction is evident with the highest XRD peak in (111) plane as reported in a recent study [11]. As seen in the inset of Fig. 4(b), the XRD pattern of the as-built sample exhibits the highest peak in the (111) plane.

The microscopic observations have indicated the TWIP deformation of the specimens tested both in tension and compression in both directions. All compression and tensile tested specimens exhibited localized twinning in certain regions on a very fine scale as previously proposed [42]. Figures 5(a) and (b) show the optical micrographs of twinned regions in compression and tension tested specimens (z-direction). The twins in the same micrographs are inclined to the loading axis and the localized twinning is also evident by the formation of the steps at the melt pool boundaries. The thickness of the localized twinned regions was found higher in the tension specimens tested in the z-direction (3.8 μm) than the tension specimens tested in the y-direction (~2 μm). The fraction of twinning was also found higher in the tension specimens tested in the z-direction.

Table 1. Compression and tension flow stresses and tensile fracture stresses and strains.

Test	True stress at 6% strain, MPa	True fracture stress, MPa	True fracture strain
Com.y-axis	674 ± 4.16	-	
Com.z-axis	645 ± 0.88	-	
Ten.y-axis	711 ± 2.9	1192	1.32
Ten.z-axis	604 ± 4.44	1215	1.48

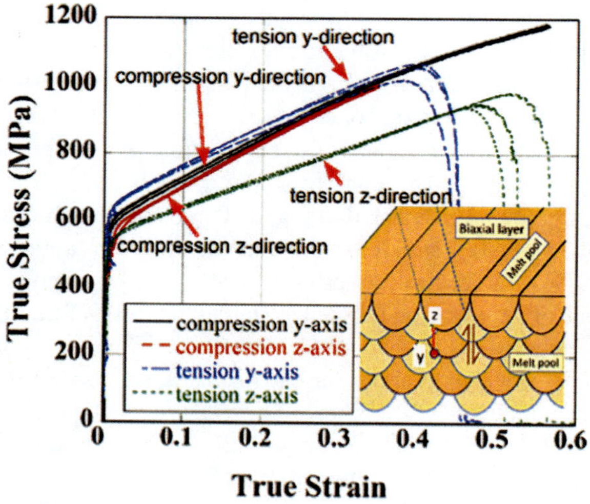

Fig. 2. Tension and compression true stress-strain curves in the y- and z-direction.

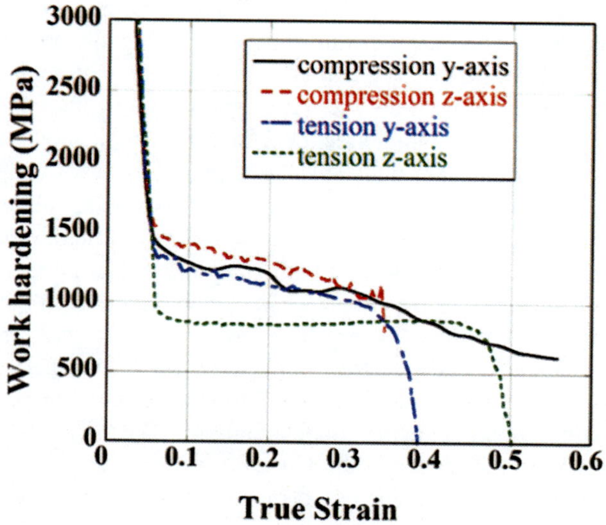

Fig. 3. Work hardening vs. true strain.

In the twinned regions, the distance between individual twins was measured less than 100 nm, smaller than the sub-cell size, 1.06 μm, (see the inset of Fig. 5(a)). Furthermore, bent-twins seen in Figs. 5(a) and (b) confirm the concurrent plastic deformation via slip by dislocation motion.

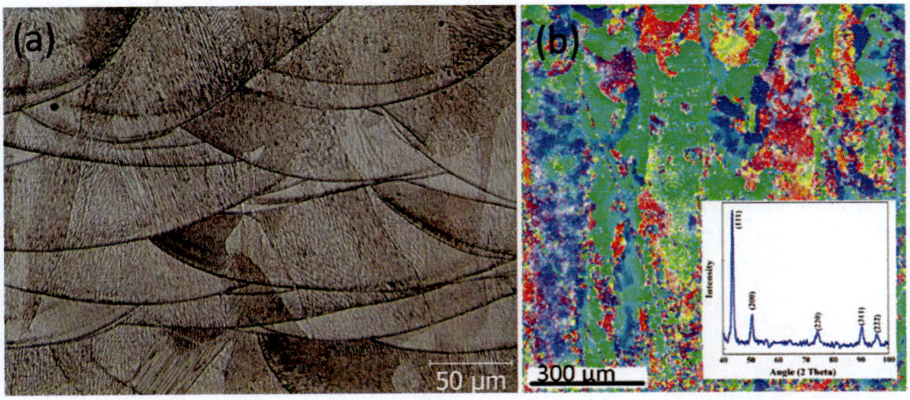

Fig. 4. Optical micrographs in the z-y plane showing melt pools and the XRD pattern and (b) EBSD micrograph showing disrupted columnar grains in the building direction.

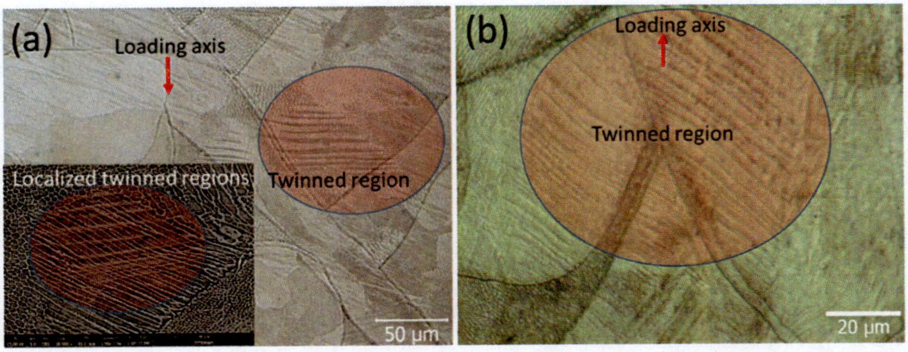

Fig. 5. Optical micrographs showing the twinned regions in (a) compression specimen (0.2 strain) and (b) tension specimen (0.4 strain) tested in the z-direction.

The formation of mechanical twinning depend on the applied stress and the stacking fault energy (SFE) [14]. The stacking faults were proposed to serve for the nucleation of twinning [15]. For stainless steels, the typical range of SFEs for slip was reported as >45 mJ m^{-2}, for twinning as 20–45 mJ m^{-2} and/or phase transformation as <20 mJ m^{-2} [12]. The lower values of SFE induce a larger separation between the Shockley partials which prevent the cross-slip. Twinning occurs when the resolved shear stress in the twining plane reaches as critical resolved shear stress in the twinning direction as similar with slip. Different from slip, twinning is polarized; tension and compression loading may initiate different twinning modes at different stresses. For example, the hcp materials with a strong fiber texture of basal plane through the loading axis favor the twinning in compression but not in tension [16]. In the materials without a fiber texture, the twinning stresses for tension and compression are similar, while in textured materials an asymmetry in compression and tension stresses is found [16]. Above may also explain the observed asymmetry in the applied compression and tension stresses in the building direction. The fiber texture in the building direction induces twinning at lower stress in tension than in compression, while the twining stress is very much similar in tension and compression in the y-direction in which the directions are randomly oriented due to the rotating strategy used in the SLM. The favored twinning in the building direction also results in higher true fracture strains than the normal to the building direction as tabulated in Table 1. The true fracture strains in Table 1 were calculated from the diameter of the necked-fractured sections. The higher ductility also results in a higher true fracture strength in the building direction (Table 1). The asymmetry may also be raised by the residual stresses developed in the building direction. A tensile residual stress in the building direction will induce a lower tensile yield stress but a higher compressive yield stress. But the difference in the work hardening behavior between the specimens tested in tension and compression tended to conclude that texture played a dominant in the detected asymmetry.

Figures 6(a) and (b) show the SEM pictures of the sub-cell structure of SLM-316L samples processed using higher and lower laser power to obtain two different sizes of sub-cell. The grain sizes were measured 30.5 and 29.2 μm, and the sub-cell sizes were measured 1.06 and 0.66 μm for higher and lower laser power processed alloys, respectively. These two alloys have similar grain but different sub-cell sizes in the biaxial plane. Figures 7(a) and (b) show the hardness measurements inside a grain (0.49 N) and over the melt-pool grains (19.6 N), respectively. The average hardness values are found 286 ± 6.53 and 282 ± 17 HV at 0.49 N and 245 ± 9.35 and 243 ± 7.9 HV at 19.61 N for higher and lower laser power processed specimens, respectively. These results show a slightly higher harnesses of the larger sub-cell size specimens. This also agrees with the previous studies reported that the dislocation cell walls (sub-cell) behaved as imperfect barriers to the dislocation motion [3, 17].

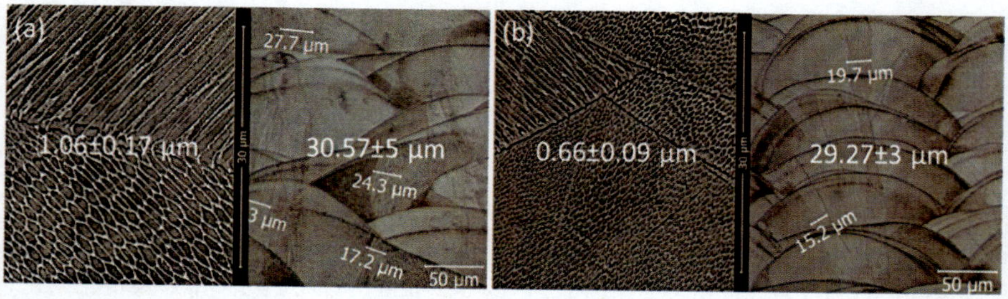

Fig. 6. The sub-cell and grain micrographs and average grain size of SLM-316L processed with (a) higher laser power and (b) lower laser power.

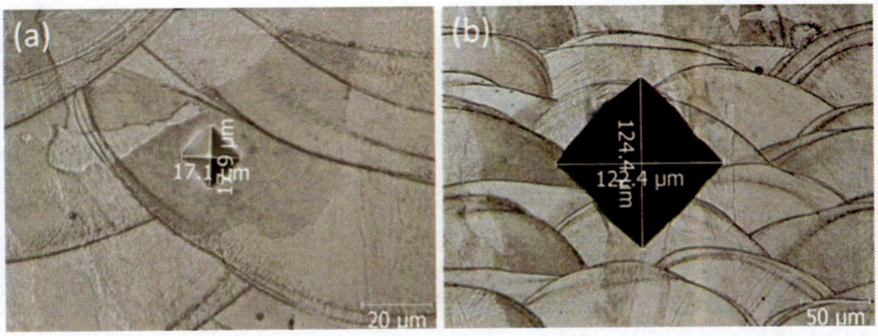

Fig. 7. Hardness test (a) inside a grain and (b) over the grains of a lower laser power processed SLM-316L

4 Conclusions

The asymmetry in the tension and compression of an SLM-316L alloy fabricated by a biaxial scanning strategy in two different directions (in the building and normal to the building direction) was experimentally investigated. The effect of sub-cell size on the hardness was further investigated on the alloys having similar grain but different sub-cell sizes. The asymmetry between tension and compression tests in the building direction was ascribed to the used biaxial scanning strategy which resulted in a strong alignment of ⟨110⟩ along the build direction (fiber texture) and a random distribution of ⟨100⟩, ⟨110⟩ and ⟨111⟩ directions normal to the building direction. The strong fiber texture in the building direction induced lower twinning stress in tension than in compression, while the tension and compression twining stresses were found similar in the normal to building direction. The favored twinning in the specimens tested in the building direction resulted in a higher tensile true fracture strain; hence, a higher ductility. Lastly, the hardness measurements made on the specimens having similar gain sizes, but different sub-cell sizes processed using higher and lower laser powers tended to support that the sub-cell boundaries in SLM-316L alloy acted as imperfect barriers to the dislocation motion.

References

1. Song, B., Zhao, X., Li, S., et al.: Front. Mech. Eng. **10**(2), 111–125 (2015). https://doi.org/10.1007/s11465-015-0341-2
2. Loh, L.-E., Chua, C.-K., Yeong, W.-Y., et al.: Int. J. Heat Mass Transf. **80**, 288–300 (2015). https://doi.org/10.1016/j.ijheatmasstransfer.2014.09.014
3. Liu, L.F., Ding, Q.Q., Zhong, Y., et al.: Materi. Today **21**(4), 354–361 (2018). https://doi.org/10.1016/j.mattod.2017.11.004
4. Yin, Y.J., Sun, J.Q., Guo, J., Kan, X.F., Yang, D.C.: Mater. Sci. Eng. A **744**, 773–777 (2019). https://doi.org/10.1016/j.msea.2018.12.092
5. Sun, Z., Tan, X., Tor, S.B., Chua, C.K.: NPG Asia Mater. **10**(4), 127–136 (2018). https://doi.org/10.1038/s41427-018-0018-5
6. Sun, S.-H., Ishimoto, T., Hagihara, K., Tsutsumi, Y., Hanawa, T., Nakano, T.: Scripta Mater. **159**, 89–93 (2019). https://doi.org/10.1016/j.scriptamat.2018.09.017

7. Niendorf, T., Leuders, S., Riemer, A., et al.: Adv. Eng. Mater. **16**(7), 857–861 (2014). https://doi.org/10.1002/adem.201300579
8. Andreau, O., Koutiri, I., Peyre, P., et al.: J. Mater. Process. Technol. **264**, 21–31 (2019). https://doi.org/10.1016/j.jmatprotec.2018.08.049
9. Wang, X.L., Muniz-Lerma, J.A., Shandiz, M.A., Sanchez-Mata, O., Brochu, M.: Mater. Sci. Eng. A-Struct. Mater. Prop. Microstruct. Process. **766**, 16 (2019). https://doi.org/10.1016/j.msea.2019.138395
10. Niendorf, T., Leuders, S., Riemer, A., Richard, H.A., Tröster, T., Schwarze, D.: Metallur. Mater. Trans. B **44**(4), 794–796 (2013). https://doi.org/10.1007/s11663-013-9875-z
11. Marattukalam, J.J., Karlsson, D., Pacheco, V., et al.: Mater. Des. **193**, 11 (2020). https://doi.org/10.1016/j.matdes.2020.108852
12. Woo, W., Jeong, J.S., Kim, D.K., et al.: Sci. Rep. **10**(1), 15 (2020). https://doi.org/10.1038/s41598-019-55606-9
13. Pham, M.S., Dovgyy, B., Hooper, P.A.: Mater. Sci. Eng. A **704**, 102–111 (2017). https://doi.org/10.1016/j.msea.2017.07.082
14. Rémy, L., Pineau, A.: Mater. Sci. Eng. **26**(1), 123–132 (1976). https://doi.org/10.1016/0025-5416(76)90234-2
15. Idrissi, H., Renard, K., Ryelandt, L., Schryvers, D., Jacques, P.J.: Acta Mater. **58**(7), 2464–2476 (2010). https://doi.org/10.1016/j.actamat.2009.12.032
16. Christian, J.W., Mahajan, S.: Progr. Mater. Sci. **39**(1–2), 1–157 (1995). https://doi.org/10.1016/0079-6425(94)00007-7
17. Birnbaum, A.J., Ryou, H., Steuben, J.C., Iliopoulos, A.P., Wahl, K.J., Michopoulos, J.G.: Mater. Lett. **280**, 128570 (2020). https://doi.org/10.1016/j.matlet.2020.128570

Investigation of the Risk Factors Causing Safety and Delay Issues in Oil and Gas Pipeline Construction Projects

Raj Shah(✉), Layth Kraidi, Wilfred Matipa, and Fiona Borthwick

School of Civil Engineering and Built Environment, Faculty of Engineering and Technology, Liverpool John Moores University, Liverpool, England, UK
r.shah@ljmu.ac.uk

Abstract. Building the new Oil and Gas Pipelines (OGPs) without analyzing the potential Risk Factors (RFs) that influence the safety of these pipelines causes time and cost overrun in these projects. Quantifying the impact of the RFs on the duration of the projects is essential to provide an accurate estimate about the project duration and recognize the potential RFs that causing delay and safety issues in the projects. Identifying the RFs in the projects via literature review and analyzing their impact on the delivery time of the projects via a questionnaire survey and computer-based risk analysis model are the key methods adopted in the study. This study uses the Monte Carlo Simulation algorithm, which is integrated within @Risk Simulator to quantify the delay impact of the RFs. The @Risk simulator has the flexibility to integrated different risk distribution methods for each RF, which helps in overcoming the limitations of using ASTA Powerproject risk simulator to quantify the delay the projects, as done in a prior study. The @Risk simulator is useful in analyzing the delay in each task, working stage and overall project duration. The results provided by risk simulator is useful in understanding the sensitivity and the criticality of each RF that might cause delay issues in the projects. The outcomes of this study will help to the stakeholders, the decision-makers and the policy-makers of OGP projects to make sound decisions and enable them to take preventive actions of risk management while starting a new OGP project, which helps in minimizing the delay in the projects.

Keywords: Oil and Gas Pipelines · Risk Factors · Monte Carlo Simulation · @Risk · Delay · Time overrun · Construction projects · Stakeholders judgements · Computer model

1 Introduction

Project management involves making schedules for project activities in order to monitor the project's time progress [1]. In the field of construction industry, delay is one of the most common problems in the majority of the projects in both developed and developing countries [2]. Delay may happen in every project during the construction stage, but it varies between the different projects and the different countries [3]. Understanding the

A. Batako et al. (Eds.): GCMM 2021, LNNS 335, pp. 313–326, 2022.
https://doi.org/10.1007/978-3-030-90532-3_24

delay factors and their level of impact on a project may help to avoid or minimize the project delay [4]. To do so, providing good knowledge about the Risk Factors (RFs) RFs and using analytical or simulation techniques are the most effective methods of risk assessment [5].

It is essential, therefore, to make an accurate check to quantify the impact of the RFs on the duration of new pipeline projects. This is because of the current results of risk assessment help in making the correct reactions and strategies towards the RFs during the planning and construction stages of projects, which will help in avoiding and/or minimizing the construction delay in these projects. Otherwise, these projects will be subject to delay problems, which have a significant impact on a country's economy.

Analyzing the impact of the RFs on the duration of the projects at the planning and design stage could help the stakeholders to make sound decisions in response to risk management to keep the delay interruption in the projects to a minimum, as much as possible. However, there is a lack of studies analyzing and quantifying the impact of the RFs on the duration of the projects in the developing countries, such as Iraq. Oil and Gas Pipelines (OGPs) in Iraq are some of the projects that significantly affected by the security situation of the country and suffering delay, which make the projects do not deliver at the targeted/planned time. Construction delay in OGP projects in Iraq obstructs the government's plans of increasing the oil and gas expert rate after 2003, which has a direct impact on the economic development and the economic situations of the country.

The aim of this paper, therefore, is to develop a Risk Management Framework (RMF) which will be used to identify, assess, and quantify the impact of the RFs associated with OGPs during the construction stage of these projects. The RMF will be used to quantify the delay impact caused by the delay and RFs and estimate whether or not the projects could be delivered on time.

2 Literature Review

The literature review revealed that several techniques could be used to analyze the risks in construction projects. For instance, checklists, interviews with the stakeholders, brainstorming, surveys and the Delphi technique. Some of these methods of risk and delay analysis are discussed below.

For example, C. Kog [6] identified and ranked the delay factors in construction projects in Portugal, the UK, and the US via examining 13 studies about the problem of construction delay in these countries. Kadry et al. [7] used qualitative documents analyses to analyze the delay factors in construction projects in 16 countries with a high geopolitical risk. These two studies were limited to analyzing the delay factors in construction projects in the mentioned countries only. As well as, these studies did not make any assessment about the delay factors or quantify their impact on the projects. For example, they did not use any kind of survey, computer modelling or simulation methods to analyze the delay factors and quantify their impact on project duration.

R. Shah [8] identified the comparative delay factors in construction projects in countries such as Australia, Ghana and Malaysia via a questionnaire survey and recommended the potential measures to reduce their impact on the projects. This study has analyzed the possible minimum, the mean and the maximum duration of construction projects and the sensitivity of the work activities in these projects in the mentioned countries. Prasad et al. [9] used a questionnaire survey to identify and analyze the delay factors in transportation, power and water projects in India. Another questionnaire survey was carried out by Chiu and Lai [10] to analyze the frequency and the severity levels of the delay factors in the construction of electrical projects in Hong Kong. Mpofu [11] analyzed the delay factors in construction projects in the United Arab Emirates (UAE) via exploring the perceptions of the clients, the contractors and the consultants about the delay problem in their projects. Shebob et al. [4] analyzed the possible minimum, the mean and the maximum duration of a contraction project in Libya and the UK using Monte Carlo Simulation (MCS). However, the risk assessment methods used in these studies are limited to their regions of study, which means they cannot be effectively applied to analyze the impact of the delay factors in oil and gas projects and improve the level of safety of these projects elsewhere. Moreover, the oil and gas industry is complex by nature [12]. Oil and gas projects are recognized as highly technical projects by nature which are complex, have a high level of uncertainty, demand careful risk assessment and require appropriate risk management strategies [13]. Therefore, managing the RFs in such complex projects, like OGP projects, is difficult [14] and requires a high level of experience in risk management [15].

M. Fallahnejad [16] used document analysis and a questionnaire survey to identify the main delay factors and analyze their impact on pipeline projects in Iran. Similarly, Sweis et al. [17] used a questionnaire survey to identify the root causes of the delay factors in gas pipeline projects in Iran. Ruqaishi and Bashir [18] investigated the delay factors in the construction of oil and gas projects in Oman. Rui et al. [19] carried out a comprehensive study to identify the RFs that affect the schedule of oil and gas projects in Nigeria. However, the risk assessment methods used in these studies are limited to their regions of study, which means they cannot be effectively applied to analyze the impact of the delay factors in oil and gas projects and improve the level of safety of these projects elsewhere particularly, where the projects are subject to different types of RFs such as security related risks.

In summary, the methods of delay assessments methods as discussed above (e.g. using the records, checklists, interviews, and surveys) are inactive to make an accurate estimation of the delay impact caused by the RFs in OGP project elsewhere. In other words, these methods cannot be used to analyze the delay in a developing and insecure country like Iraq because of a number of the following gaps in knowledge. For example, the records about delay and risk factors are either not available or not accessible in OGP projects in Iraq and the security level in Iraq is low, which makes the projects subject to different types of RFs particularly, the risks related to the security situations.

3 Methodology

The aim of this paper is to analyze the potential delay in a case study project caused by the associated RFs. The pipeline is going to be built in the south of Iraq. The length of the pipe is 164 km. It links Badra oil and gas field with the export point on the Gulf in Basra via Gharraf–An Nassiriyah [20]. This project has been under planning since May 21, 2019 and the targeted delivery date is January 13, 2023. This means the overall duration of the project is estimated as 3 years and 238 days (1334 days) [20] (Fig. 1).

Fig. 1. Iraq oil fields and pipelines [33]

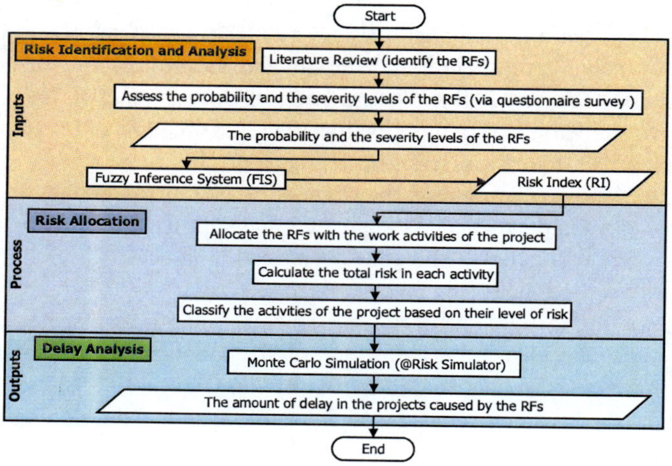

Fig. 2. The information flow chart of the Risk Management Framework (RMF) adopted in this paper.

As shown in the figure above, the RMF is structured under three main components: inputs (which is about risk identification and analysis), process (which is about calculate the risk level in the activities of the project) and outputs (which is about the amount of delay in the project), as described below (Fig. 2).

1. **Inputs**, the RFs were identified via an extensive literature review about the risks in OGP projects worldwide [21, 22]. The Risk Probability (RP) and Risk Severity (RS) levels of the RFs were assessed via conducting a questionnaire survey of the

stakeholders in OGP projects in Iraq [23–25]. The results of the survey (i.e. the RP and RS levels of the RFs) were used as inputs for a computer-based risk assessment model, which used the fuzzy theory to calculate the Risk Index (RI) of the RFs [25, 26]. Table 1 Explains the identified RFs and their degree of impact on OGP projects in Iraq.

The results of the survey and the computer-based risk assessment model indicated that terrorism and sabotage, corruption, low public legal and moral awareness, insecure areas and thieves are the most critical RFs that affect the safety of OGP projects in Iraq. On other hands it was found that the RFs that have the less impact on the projects are geological risks, natural disasters and weather conditions, hacker attacks on the operating or control systems, vehicular accidents and animal accidents.

2. **Process**, the process part of the RMF focuses on calculating the risk levels of the project activities as follows.

2.1 *Allocating the RFs to the Project Activities.*
The RFs were allocated to the work activities depending on the type of RF and the nature of the activity. Professional knowledge was used to achieve this task. The subjective and objective analysis of a technical report [27] was used to justify the process of risk allocation because it explained what was required in each activity, the nature of each activity and the potential RFs that could affect that activity based on vast experience and a review of the construction of OGP projects worldwide.

2.2 *Calculate the Total Activity Risk (TAR) Each Activity Using* Eq. 1.
The TAR calculates the summation of the RI values of the RFs allocated to the project activities.

$$\boldsymbol{TAR\ of\ an\ activity} = \sum \boldsymbol{RI\ of\ the\ relevant\ RFs\ to\ the\ activity} \tag{1}$$

2.3 *Calculate the Total Activity Risk Ratio (TARR) of the Activity from 100% Using* Eq. 2.

$$.TARR\,(100\%) = \frac{The\ total\ risk\ of\ an\ activity}{\sum Total\ risk\ of\ the\ project\ \times 100\%} \tag{2}$$

2.4 *Classify the Project Activities Based on Their Level of Risk as Follows.*

2.5 The activities with [0–1] total risk were considered as Very Low (VL) risk activities; the activities with [1, 2] total risk have a Low (L) risk; those with [2, 3] total risk have a Moderate (M) risk; those with [3, 4] total risk have a High (H) risk; and those with [4, 5] total risk have a Very High (VH) risk. Based on the risk level of the activity, this paper has set up five different levels of risk variation on the duration of the project as shown in Table 2.

3. **Outputs**, this section explains how Monte Carlo Simulation (MCS) works to simulate the impact RFs on the duration of the project. After allocating the RFs to the work activities of the project, MCS will calculate the duration of

Table 1. The identified RFs and their degree of impact on OGP projects in Iraq (the inputs of the RMF).

The findings of the literature review	The results of the survey		The result of the FIS
RFs	RP*	RS^	RI
Terrorism and sabotage	3.995	4.490	3.99
Corruption	3.980	4.192	3.87
Low public legal and moral awareness	3.712	3.859	3.80
Insecure areas	3.717	4.106	3.76
Thieves	3.692	4.081	3.75
Corrosion and lack of protection against it	3.687	3.990	3.72
Lack of proper training	3.646	3.773	3.71
Improper safety regulations	3.687	3.949	3.70
Exposed pipelines	3.667	3.682	3.70
Improper inspection and maintenance	3.227	3.924	3.69
Conflicts over land ownership	3.495	3.611	3.68
Shortage of IT services and modern equipment	3.667	3.652	3.68
Weak ability to identify and monitor the risks	3.631	3.899	3.67
Design, construction and material defects	3.333	3.848	3.64
Lack of risk registration	3.566	3.697	3.60
Easy access to pipeline	3.631	3.646	3.57
Limited warning signs	3.626	3.571	3.56
Little research on this topic	3.621	3.697	3.55
Lawlessness	3.606	3.682	3.54
Stakeholders not paying proper attention	3.530	3.143	3.51
Public poverty and education level	3.449	3.409	3.49
Inadequate risk management	3.227	3.505	3.48
Leakage of sensitive information	2.980	3.505	3.38
Threats to staff	3.323	3.399	3.35
Operational errors	3.101	3.611	3.30
Geological risks	2.747	3.182	3.17
Natural disasters and weather conditions	2.652	3.066	3.10
Hacker attacks on the operating or control systems	2.237	2.970	3.03
Vehicular accidents	2.465	2.712	2.80
Animal accidents	1.894	2.020	1.95

*In the survey, the probability level of the RFs were calculated using Five-points Likert Scale, as follows (Almost certain, Likely, Possible, Unlikely and Rare) (23).

^In the survey, the severity level of the RFs were calculated using Five-points Likert Scale, as follows (Catastrophic, Major, Moderate, Minor and Negligible) (23).

Table 2. The total risk and risk levels of the project's main working activities.

Activities	Equation 1	Equation 2	Risk level	The impact level on the duration of the project
Concept and definitions	18.11	0.86	VL	95%–105%
Life-cycle plan	71.8	3.41	H	80%–120%
Choosing the route	76.65	3.64	H	80%–120%
Route approval	73.14	3.47	H	80%–120%
Design and development	43.44	2.06	M	85% –115%
Installation procedure	29.28	1.39	L	90%–110%
Risk assessment	49.67	2.36	M	85% –115%
Time schedule	22.08	1.05	L	90%–110%
Cost estimation	22.08	1.05	L	90%–110%
Communications	25.43	1.21	L	90%–110%
Materials order	18.41	0.87	VL	95%–105%
Survey, staking and setting out	75.77	3.60	H	80%–120%
Clearing and grading the right-of-way	73.46	3.49	H	80%–120%
Topsoil stripping	57.88	2.75	M	85%–115%
Buildings, roads and river crossings	76.63	3.64	H	80%–120%
Pipe transportation to site	59.02	2.80	M	85%–115%
Temporary fencing and signage	51.09	2.43	M	85%–115%
Trenching	54.05	2.57	M	85%–115%
Temporary erosion control and side support	57.48	2.73	M	85%–115%
Pipe set-up	43.84	2.08	M	85%–115%
NDT tests	32.77	1.56	L	90%–110%
Welding, fabrication and installing	36.28	1.72	L	90%–110%
Sandblast	32.82	1.56	L	90%–110%
Painting	32.81	1.56	L	90%–110%
Coating	54.69	2.60	M	85%–115%
Lowering pipe and backfilling	46.71	2.22	M	85%–115%

(*continued*)

Table 2. (*continued*)

Activities	Equation 1	Equation 2	Risk level	The impact level on the duration of the project
Cathodic protection of the pipe	68.64	3.26	H	80%–120%
Final fitting	32.61	1.55	L	90%–110%
As-built survey	32.48	1.54	L	90%–110%
Hydro, pressure test	29.1	1.38	L	90%–110%
Backfilling	36.16	1.72	L	90%–110%
Fencing and signage	61.49	2.92	M	85%–115%
Final clean-up	40.11	1.90	L	90%–110%
Right-of-way reclamation	54.03	2.57	M	85%–115%
Safety barriers	55.53	2.64	M	85%–115%
Operation within design limits	97.54	4.63	VH	75%–125%
Commissioning operation value	97.54	4.63	VH	75%–125%
Measure the performance and efficiency	29.26	1.39	L	90%–110%
Enhanced performance and efficiency	97.54	4.63	VH	75%–125%
Monitoring and inspection	42.57	2.02	M	85%–115%
Maintenance	59.54	2.83	H	80%–120%
Risk control	36.31	1.72	L	90%–110%

each activity by applying the iterations between the minimum and maximum duration of the using @Risk Simulator. The initial planned duration of the project was 3 years and 238 days (1334 days). After analyzing the potential RFs that affect the work activities of the project, it was found that the average delay in the project is 1374.94 days, which means the delay in the project is 40.94 days (Fig. 3).

Moreover, @Risk could be used to analyze the delay in the duration of the individual activities of the project after considering the impact of the associated RFs. The results revealed that the activities with the highest delay in the project are the hydro and pressure test, the as-built survey, the trenching activity, the cathodic protection of the pipe, the temporary fencing and signage and of the sand blast activities, with a potential delay of 142.07 days, 137.04 days, 80.17 days, 57.62 days, 53.71 days and 47.88, respectively (see Table 3).

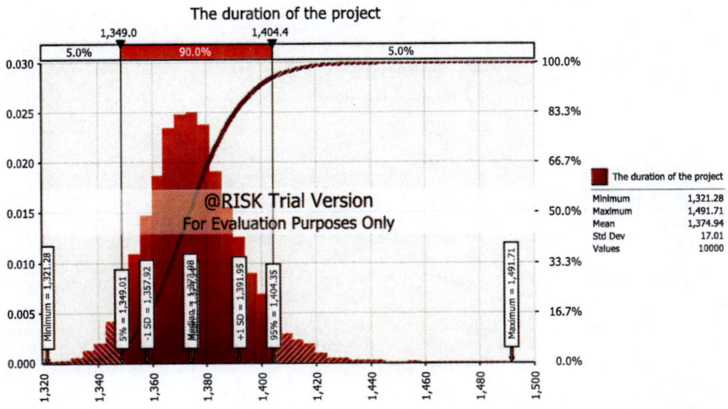

Fig. 3. The results of simulating the duration of the project

Table 3. The duration and the potential delay of the activities of the project.

Activity	Planned duration (day)	Median (day)	Delay (day) = median – planned duration
Concept and definitions	60	81.82	21.82
Life-cycle plan	60	82.56	22.56
Choosing the route(s)	90	136.40	46.40
Route(s) approval	90	128.53	46.4
Design and development activity	90	123.65	38.53
Manufacturing and installation (procedure/plan)	30	54.10	33.65
Risk assessment and management plans	90	128.56	24.1
Time schedule	45	61.24	38.56
Staking for construction and communications	30	61.24	31.24
Survey, staking and setting out	5	42.35	37.35
Materials order activity	30	6.01	−23.99
Clearing and grading the Right-Of-Way (ROW)	30	41.71	11.71
Topsoil stripping and front-end grading	30	43.12	13.12
Buildings, roads and rivers crossings	45	42.95	−2.05
Temporary fencing and signage	7	60.71	53.71

(*continued*)

Table 3. *(continued)*

Activity	Planned duration (day)	Median (day)	Delay (day) = median – planned duration
Pipe transporting to sit	90	22.43	−67.57
Trenching	60	140.17	80.17
Temporary erosion control and side support	60	83.07	23.07
Pipe set-up	99	91.73	−7.27
Welding, fabrication and installing pipe	100	143.91	43.91
NDT tests	100	147.66	47.66
Sand blast	100	147.88	47.88
Painting	100	147.72	47.72
Cathodic protecting the pipe	90	147.62	57.62
Coating	100	131.84	31.84
Lowering pipe in and backfilling	100	147.79	47.79
As-built survey	10	147.04	137.04
Final fitting	100	14.28	−114.28
Hydro pressure test	5	147.07	142.07
Backfilling	30	6.04	−23.96
Fencing and signage	14	41.65	27.65
Final clean-up	21	17.19	−3.81
Right-of-way reclamation	29	28.31	−0.69
Safety barriers	29	38.42	9.42
Fencing and signage	41	70.79	29.79

4 Findings and Discussion

In conditions of data scarcity, the RFs will be mainly identified based on the literature review. Therefore, the RFs in OGP projects were identified based on an extensive and worldwide literature review about the risks that affect the safety of the pipelines. Based on the investigations of literature review, thirty RFs were identified in the projects and. The findings of the literature review in order to overcome the problem of data scarcity about the RFs in OGP projects in Iraq, which represent the first finding and contribution of this paper. The second finding and contribution of this paper come from analyzing the RP and RS levels of the RFs based on a questionnaire survey that reflect the reality of the problems after engaging with the stakeholders in the projects. However, there is a potential problem associated with assessing the RFs depending on the stakeholders' judgement only, as they may not always yield to a consistent and accurate ranking of the RFs [28]. This is because the stakeholders have different views of the impact levels of the RFs [29]. Moreover, assessing and ranking the RFs in the projects based only on the results of the survey could cause inaccurate ranking of the RFs. For example, an RF with a high value of RS could still be considered as a critical RF that needs to be dealt with a matter of urgency. However, the same RF could not come at the top of the ranking if it had a low RP. This is similar if the RP of the RF is high and the RS is low, which is one of the limitations of the traditional risk assessment methods. Therefore, in order to reduce the uncertainty associated with analyzing the RFs, the calculation of the degree of impact of the RFs on the OGP projects will be performed using the fuzz theory [30]. The fuzzy uses approximate ranges rather than exact values of processing and controls. In other words, the use of ranges of very low, low, moderate, high and very high risk, the interpolation between these ranges, and the aggregate results of RP and RS of the RFs have helped in reducing the uncertainty of calculating the RI values of the RFs, which represent the third finding and contribution of this paper [31].

After identifying the RFs that might affect the safety of the pipelines and analyzing their degree of impact on the projects, this paper has allocated the RFs to activities of the projects based on the nature of the RFs and the activities. Then, Eqs. 1 and 2 were used to calculate the amount of risk and level of risk in each activity of the project. This represent the fourth finding and contribution of this paper, which help the stakeholders in the projects to identify the RFs and activities that have the highest impact on the duration of the project.

MCS integrated with @Risk simulator was used to quantify the delay impact of the RFs on the duration of the project. The results of MCS explained that the minimum and maximum duration of the project are 1329.30 days and 1441.84 days, respectively. The project has a 5% chance of being completed between 1329.30 and 1349.1 days or between 1404.5 days and 1441.84 days. The project has a 90% probability of being finished between 1349.1 days and 1404.5 days. The median duration of the project is 1374.94 days, which means that the project has a 50% probability of being completed in this duration, which means the delay in the project is 40.94 days, which is the main finding of this paper.

@Risk simulator helps to apply different distribution methods for each RF and activity at the same time, which will enhance the risk simulation results and add more confidence regarding the project completion probability. Applying different risk simulation methods for risk simulation make the results of risk analysis different. For example, [32] used MCS integrated with ASTA Powerproject software to analyze the RFs that cause delay in the same project. ASTA Powerproject has only four methods of risk distribution (which are uniform, normal, skewed normal and skewed triangular distribution), but only one distribution method could be applied on time during the process of risk simulation, which does not reflect the reality of the RFs on the duration of the projects. As different RFs have different impact of the activities of the projects. This is one of the limitations of using ASTA Powerproject quantifies the impact of the RFs on the duration of the projects. As well as, ASTA Powerproject software could not be used to analyze the delay in the different stages of the projects. In summary, compared to the ASTA Powerproject, @Risk simulator is a more useful and powerful tool to analyze the RFs and the project delay. This is because @Risk can use more and different risk distribution methods than ASTA. Moreover, it could be used to analyze the delay in the duration of the individual activities, by the stages of the projects and by the overall duration of the projects. Meanwhile, the ASTA risk simulator is useful to analyze the delay that affects the overall duration of the projects only. Table 4 summarizes the results of the ASTA Powerproject and the @Risk simulator.

Table 4. The difference in using the ASTA risk simulator and the @Risk simulator to analyse the delay in the project.

Program	Results (delay)	Reference
ASTA risk simulator	14–15 days (using four different distribution methods)	[32]
@Risk simulator	41 days	This study

The difference in the results of ASTA and @Risk is because the ASTA Powerproject applies one risk distribution method for all RFs and project activities at a time, which makes the RFs and the project activities give the same impact regarding the duration of the project, which is not accurate. On the other side, the @Risk Simulator applies different risk distribution methods for the RFs and the project activities, rather than one distribution method at a time, with a degree of impact on the duration of the project. For example, RiskTriang (0,0.7,1) distribution was assigned to the stealing the products and the materials RFs, which is different from assigning Uniform, Normal, Triangular or Skewed Triangular with no degree of impact on the duration of the project, as done in ASTA [32]. Assigning different risk distribution methods for the RFs and the project activities with a degree of impact on the duration of the project was the reason behind the difference in the results of ASTA and @Risk.

5 Conclusion

After considering the impact of the RFs on the duration of the project, it was found that the potential delay in the project is 41 days. The activities that have the highest impact on the duration of the project are hydro and pressure test, the as-built survey, the trenching activity, the cathodic protection of the pipe, the temporary fencing and signage and of the sand blast activities.

The advantage of using the @Risk simulator rather than the ASTA risk simulator is that the @Risk simulator has more flexibility in applying different risk distribution methods for the same RFs and work activities at the same time. Additionally, the @Risk simulator could help the researchers to analyze the delay by the activities of the project, which could not been done using the ASTA risk simulator.

This research has developed a systematic and integrated RMF, which was useful to quantify the delay impact in the OGP projects. The RMF designed in this study was used to provide a wide range of knowledge about identifying the RFs and analyzing their impact on OGP projects is a systematic and accurate way. The RMF that designed in this paper is a useful tool that could be used to analyze the construction delay in OGP during the planning and design stage of these projects.

Acknowledgements. The authors are extremely grateful for the financial support from the Ministry of Higher Education and Scientific Research (MOHERS), Iraq and AL-Muthanna University.

References

1. Shah, R.K., Dawood, N.: 24th CIB W78 Conference Bringing ICT Knowledge to Work., Maribor, Slovenia, pp. 499–504 (2007)
2. Shebob, A., Dawood, N., Shah, R.K.: Development of a methodology for analysing and quantifying the impact of delay factors affecting construction projects. J. Constr. Eng. Proj. Manage. **2**, 17–29 (2012)
3. Alaghbari, W., Kadir, M.R.A., Salim, A., Ernawati: The significant factors causing delay of building construction projects in Malaysia. Eng. Constr. Architect. Manage. **14**(2), 192–206 (2007). https://doi.org/10.1108/09699980710731308
4. Shebob, A., Dawood, N., Shah, R.K., Xu, Q.: Comparative study of delay factors in Libyan and the UK construction industry. Eng. Constr. Architect. Manage. **19**(6), 688–712 (2012). https://doi.org/10.1108/09699981211277577
5. Ruwanpura, J., Ariaratnam, S.T., El-assaly, A.: Prediction models for sewer infrastructure utilizing rule-based simulation. Civil Eng. Environ. Syst. **21**(3), 169–185 (2004). https://doi.org/10.1080/10286600410001694192
6. Kog, Y.C.: Major construction delay factors in Portugal, the UK, and the US. Pract. Period. Struct. Des. Constr. **23**(4), 04018024 (2018). https://doi.org/10.1061/(ASCE)SC.1943-5576.0000389
7. Kadry, M., Osman, H., Georgy, M.: Causes of construction delays in countries with high geopolitical risks. J. Construct. Eng. Manage. **143**(2), 04016095 (2017). https://doi.org/10.1061/(ASCE)CO.1943-7862.0001222
8. Shah, R.K.: An exploration of causes for delay and cost overruns in construction projects: case study of Australia, Malaysia and Ghana. J. Adv. Coll. Eng. Manag. **2**, 41 (2016)
9. Prasad, K.V., Vasugi, V., Venkatesan, R., Nikhil, B.: Analysis of causes of delay in Indian construction projects and mitigation measures. J. Financ. Manag. Prop. Constr. **24**, 58–78 (2019)
10. Betty, W.Y.C.H.I.U., Joseph, H.K.L.A.I.: Project delay: key electrical construction factors in Hong Kong. J. Civil Eng. Manage. **23**(7), 847–857 (2017)
11. Mpofu, B., Ochieng, E.G., Moobela, C., Pretorius, A.: Profiling causative factors leading to construction project delays in the United Arab Emirates. Eng. Constr. Architect. Manage. **24**(2), 346–376 (2017). https://doi.org/10.1108/ECAM-05-2015-0072
12. Lameda, R.L., Van den Berg, F.G.: SPE Digit. Energy Conference on Exhibition, Society of Petroleum Engineers (2009)
13. Almadhlouh, A.A.: Risk Management Approaches in Oil and Gas Industry Projects: A Qualitative Study. Colorado Technical University (2019)
14. Laufer, A., Shapira, A., Telem, D.: Communicating in dynamic conditions: how do on-site construction project managers do it? J. Manage. Eng. **24**(2), 75–86 (2008). https://doi.org/10.1061/(ASCE)0742-597X(2008)24:2(75)
15. Aarseth, W., Rolstadås, A., Andersen, B.: Managing organizational challenges in global projects. Int. J. Manag. Projects Bus. **7**(1), 103–132 (2013). https://doi.org/10.1108/IJMPB-02-2011-0008
16. Fallahnejad, M.H.: Delay causes in Iran gas pipeline projects. Int. J. Project Manage. **31**(1), 136–146 (2013). https://doi.org/10.1016/j.ijproman.2012.06.003

17. Sweis, R., Moarefi, A., Amiri, M., Moarefi, S., Saleh, R.: Causes of delay in Iranian oil and gas projects: a root cause analysis. Int. J. Energy Sect. Manage. **13**(3), 630–650 (2019). https://doi.org/10.1108/IJESM-04-2018-0014
18. Ruqaishi, M., Bashir, H.A.: Causes of delay in construction projects in the oil and gas industry in the gulf cooperation council countries: a case study. J. Manage. Eng. **31**(3), 05014017 (2015). https://doi.org/10.1061/(ASCE)ME.1943-5479.0000248
19. Rui, Z., et al.: A comprehensive investigation on performance of oil and gas development in Nigeria: technical and non-technical analyses. Energy **158**, 666–680 (2018). https://doi.org/10.1016/j.energy.2018.06.027
20. Gulf Oil & Gas, Gulf Oil Gas (2020)
21. Kraidi, L., Shah, R., Matipa, W., Borthwick, F.: In: Dubai (Ed.), Facebook Restaurant Week, Proceedings, 22nd – 26th May, 2017, Faculty of Engineering and Technology James Parsons Building, Byrom Street, Liverpool, L3 3AF, Liverpool, UK, 2017, pp. 133–148 (2017)
22. Kraidi, L., Shah, R., Matipa, W., Borthwick, F.: Creating Construction Conference 2018 – Proceedings of Budapest University of Technology and Economics, Ljubljana, Slovenia, 2018, pp. 663–670 (2018)
23. Kraidi, L., Shah, R., Matipa, W., Borthwick, F.: Periodica Polytechnica Architecture (2019)
24. Kraidi, L., Shah, R., Matipa, W., Borthwick, F.: Creating Construction Conference 2018 – Proceedings of Budapest University of Technology and Economics, Ljubljana, Slovenia, 2018, pp. 304–311 (2018)
25. Kraidi, L., Shah, R., Matipa, W., Borthwick, F.: As a Working Paper ARCOM 34th Conference 3–5 September, Belfast, UK (2018)
26. Kraidi, L., Shah, R., Matipa, W., Borthwick, F.: Using stakeholders' judgement and fuzzy logic theory to analyze the risk influencing factors in oil and gas pipeline projects: case study in Iraq, stage II. Int. J. Critic. Infrastruct. Protect. **28**, 100337 (2020). https://doi.org/10.1016/j.ijcip.2020.100337
27. (FracTracker Alliance) F.T.A.: Pipeline Construstion Step by Step (2019)
28. Tang, K.H.D., Dawal, S.Z. M., Olugu, E.U.: Proceedings of One Curtin International Postgraduating Conference, Miri, Sarawak, Malaysia, 2018, pp. 325–331 (2018)
29. Lavasani, S.M., Ramzali, N., Sabzalipour, F., Akyuz, E.: Utilisation of fuzzy fault tree analysis (FFTA) for quantified risk analysis of leakage in abandoned oil and natural-gas wells. Ocean Eng. **108**, 729–737 (2015). https://doi.org/10.1016/j.oceaneng.2015.09.008
30. Urbina, A.G., Aoyama, A.: Measuring the benefit of investing in pipeline safety using fuzzy risk assessment. J. Loss Prevent. Process Indus. **45**, 116–132 (2017). https://doi.org/10.1016/j.jlp.2016.11.018
31. Kraidi, L., Shah, S., Matipa, W., Borthwick, F.: ASC 2019 International Conference 10–13 April, Denver, Colorado, USA (2019)
32. Kraidi, L., Shah, R., Matipa, W.O.F.: In: T. Leathem (Ed.), International Conference on 56th Annual Associated Schools of Construction International Conference, EasyChair, EPiC Series in Built Environment, Liverpool, UK, 2020, pp. 356–363 (2020)
33. World Map (2014)

Why and How to Educate Selected Industry 4.0 Stakeholders in Technical and Non-technical Issues

Jan Kazmierczak(✉)

Faculty of Organization and Management, Silesian University of Technology, Roosevelt Street 26-28, 41-800 Zabrze, Poland
jan.kazmierczak@polsl.pl

Abstract. In this paper, the author tries to present and analyse their thoughts, as well as introduce and discuss the results of their earlier research focused on surveying the needs and possibilities of educating stakeholders of processes related to the implementation of the Industry 4.0 (I4.0) paradigm into practice. Special attention was – and still is – paid to Micro, Small and Medium-Sized Enterprises (MSME) as a specific group participating in I4.0 processes. The first part of the paper focuses on presenting and classifying the I4.0 stakeholders, who are taking part in the mentioned processes. The second part briefly presents the results of investigating the level of knowledge about I4.0 in selected groups of individuals potentially engaged in the fourth industrial revolution in Poland. In the next part of the paper, selected results of a wider survey of Polish MSMEs are presented. Based on the results of this research, the author formulates their opinions on the educational needs of the specified group of I4.0 stakeholders and presents some proposals of methods and tools in order to increase the level of their readiness to participate in the mentioned processes. As a particular problem of educating, the possibility and/or necessity of assessing the social impacts of new (innovative) products and technologies appearing in the modern world (Technology Assessment-TA) are treated. Finally, a few personal comments from the author are presented on the Covid-19 pandemic impact on modern activities in industry.

Keywords: Industry 4.0 · SME · Education · Technology Assessment

1 Introduction

Referring to the first sentence from the previous publication of the author of the article [14], undoubtedly the present is a time of revolutionary changes in the history of mankind. Of course, various revolutions, understood as profound changes in the ways of thinking and acting, system solutions, means, and methods used in a finite time, occurred in the past, occur in the present, and are likely to occur in virtually all areas of human life and activity. However, as a starting point for this study, it was assumed that the present time is characterised by a peculiar accumulation of revolutionary changes.

To avoid overly general considerations, this study focuses on revolutionary phenomena that occurred, are occurring, and are likely to occur in such a specific area of human

A. Batako et al. (Eds.): GCMM 2021, LNNS 335, pp. 327–343, 2022.
https://doi.org/10.1007/978-3-030-90532-3_25

activity as the production of various goods and services. These types of breakthroughs in the field of technology development are most often called technological revolutions. It is generally accepted that three such revolutions have occurred so far (the Industrial Revolution (1780–1840), the Technological Revolution or the Second Industrial Revolution (1870–1920) [19], and the Information and Telecommunications Revolution, also known as the Digital Revolution or the Third Industrial Revolution) (1975–2010) [23, 24]).

Nowadays, there are also fundamental changes occurring, especially in the industrial and economic sphere. This is the fourth industrial revolution known as Industry 4.0. The term Industry 4.0 (originally 'Industrie 4.0') derives from the strategy of the German government, generally focused on the development and use of highly advanced technologies in industrial practice, especially informatization and digitalization (computerisation) of production processes. The term was first used at the Hanover fair in 2011. The new opportunities offered by Information and Communication Technologies (ICT) solutions [11] are first considered as the revolutionary changes of technological origin that are the core of this revolution. It can be concluded that the fourth industrial revolution is the implementation of the latest discoveries in industry and economy, especially in areas such as obtaining, collecting, processing, and using data, information, and knowledge.

As in the case of earlier industrial revolutions, revolutionary technological changes not only entail very profound changes in the workplaces of active people, but also in the places of life of the whole of society (not only professionally active people). The same 'pillars' [6] support the creation of Industry 4.0 (Internet of Things, IT systems integration, the use of Cloud Computing technology, etc.) and are the basis for the creation and implementation of modern solutions transforming the living space (e.g. the concept of 'Smart Cities' [3, 12, 22] modern systems and means of public transport, health care systems and finally solutions for the natural environment in which the modern person lives.

It can be observed that revolutionary processes in industry are embedded in an appropriate social background. For example, the first industrial revolution developed as a result of profound social and economic changes in the late 18th and early 19th centuries. The Second Revolution was clearly 'correlated' with the First World War [19, 29]. The third industrial revolution brought such concepts in the social dimension as the new energy programme of the 21st century [24, 26], the next phase of the European integration process [23, 24], and finally the development of an alternative production ecosystem [21]. For the above reasons, research on the relationship 'human-technology' (or 'technology - human') is also being developed in the ethical dimension [2].

Finding a place to reflect on the relationship between technology and the human factor (not only understood in terms of anthropotechnical systems) is perhaps one of the key elements of the current I4.0 revolution. The processes that make up Industry 4.0 needed, need, and probably will need an appropriate social 'foundation'. It is no coincidence that, according to many researchers, the emergence of a 'new industry' is associated with social phenomena and the processes of transforming society (the concept of Society 4.0 [18] or even Society 5.0 [27]. In the I4.0 paradigm, according to the author of this study, not only scientific research on the connections between the functioning of society and human technological activity, but also practical means and methods enabling the optimisation of the 'human-technology' relationship - the effect of 'rising' revolutionary changes observed in the three previous industrial revolutions in the industry of a properly defined and recognised social background. Perhaps a good illustration of this is the emergence in the mid-20th century, not only in the field of interest of scientists but also practitioners of the problem of assessing the impact of new technologies and products on society, on the social groups or individuals that make up modern society. A new interdisciplinary branch of science has emerged, actually it is emerging, known as 'Technology Assessment–TA' [1, 25]. Both identifying the need for such considerations and making efforts to encapsulate the idea of TA by an adequate set of tools and methods require - according to the author of this study - both the appreciation of these issues by scientists and the effective persuasion of all stakeholders of the 'New Industry' (perhaps 'New Economy'?) about the need to expand knowledge and qualifications to the extent of enabling participation in such assessments. Thus, it can be stated in a very simplified way, that the novelties concerning the functioning of society imply the need for learning, also seen in an interdisciplinary perspective [13].

Therefore, one might ask, what the stakeholders of the fourth industrial revolution should learn? And ask, what are the consequences of not trying to acquire knowledge and qualifications to function in the (rapidly) changing reality? The problem of nontechnological aspects of the fourth industrial revolution is willingly taken up in the literature (e.g., [7, 20]).

If a revolution occurs, one must expect its effects, which are and will be observed in the reality affected ('burdened') by revolutionary changes. The changes of revolutionary origin can affect our reality or some parts of this reality, either positively or negatively. Figuratively speaking, a revolution can be both a blessing and a curse. From the experience of humanity, it can observe that with each revolution there are both beneficiaries and victims. French revolutionist Pierre Vergniaud is credited for saying that "It is to be feared that a revolution, like Saturn, can devour all of its children one by one". He was probably right, although–as stated above–usually only some part of the revolution child is at risk of being devoured.

Limiting once again the range of considerations presented in this paper, one can state that Industry 4.0 (probably?) brings a lot of positive effects for the family of their beneficiaries (or most of them?). However, if Vergniaud was right, it can also bring (more or less significant) dangers for particular members of this family. Considering the I4.0 revolution, the question appears: who and how–if at all–can be "devoured" by Industry 4.0? And, if such a danger exists, how to protect all I4.0 stakeholders or a selected group of them against threats deriving from the Industry 4.0 paradigm?

The above questions were the basis for the assumption that the group potentially threatened by the effects of the I4.0 revolution may be - for various reasons [28] - micro, small and medium-sized enterprises (MSME). Moreover, it was assumed that by means of "getting out of the danger zone" this group of I4.0 stakeholders can and should be properly prepared (based on the identified needs) and educational activities should be conducted.

Considerations leading to the author's intention to obtain the response to the above questions, but limited to the predetermined group of stakeholders, which are Micro,

Small and Medium-sized Enterprises (MSMEs), are presented in the next part of this article.

2 About the Relationship Between Technology and the Human Factor in Industry 4.0

If one tries to think about the idea of Industry 4.0 not only as a purely technical set of problems to be solved, but also to consider the presence of a human factor in these problems, a necessity appears–firstly–to identify and classify humans involved in the industrial processes into predefined categories (groups).

It seems justified to carry out such identification and classification with reference to the 'classical' industrial system (the term 'classic' is understood here to mean the industrial system from the pre-Industry 4.0 period). The 'classical' system understood in this way is a natural and obvious reference to the search for changes leading to the system compliant with I4.0. Taking into account the initial assumptions adopted for the preparation of this article, the next step should be to analyse the needs and possibilities of preparing various groups of stakeholders for the implementation of new tasks, in this case related to the I4.0 paradigm [9].

As a consequence of using the 'classical' model of the industrial system as a reference level, the classification (grouping) of stakeholders can be carried out, e.g., according to such criteria as:

1. The type (size) of the entity performing production tasks and/or providing services (especially logistics services in business-to-business (B2B) relations ([4, 10]),
2. Roles (functions) performed by individuals in the structure and business activities, and many others.

For the purposes of this article, it was decided to limit the research on the needs of stakeholders in terms of their preparation for various challenges of the Industry 4.0 revolution to the following groups:

- groups of micro, small and medium-sized enterprises (criterion 1),
- groups of owners and top management of enterprises (criterion 2).

Such a choice was dictated by the conviction of the planners of the discussed research task that it is the leaders who have the most important roles to play in guiding social groups, organisations, or, finally, economic entities (enterprises) through revolutionary phenomena. To play these roles properly, they should be properly prepared.

One might show a wide range of possibilities to enlarge the knowledge and practical competences, both for researchers and for 'involved in the practice of' I4.0 stakeholders. To put such possibilities into practice, carefully planned and carried out research projects are needed, focussed on solving precisely identified tasks. The main research assumption of this paper is that the education of selected stakeholders has to be included in the considerations on Industry 4.0 as an essential factor, that up to now was not adequately noticed in the research dedicated to revolutionary changes in the global industry, as well as in activities aimed at applying the results of such research in industrial practice.

In successive parts of this paper, some research activities are introduced, which were dedicated to recognising the needs and expectations of the fixed, above-mentioned groups of I4.0 stakeholders.

An additional element of the research project was an attempt to test the effectiveness of the applied research tools and methods, taking into account the specificity of selected I4.0 stakeholder groups. This factor was primarily taken into account in the initial research described in Sect. 3 of this study. It was assumed that the group of owners and managers of MSMEs should have the greatest amount of knowledge about the Industry 4.0 model. Gaining information about the unconscious needs of the studied groups was another 'hidden goal' of the discussed research. Therefore, in this part of the research, the respondents were given the opportunity to supplement their responses with their own comments. The basic assumptions of these studies and their results are presented in the next two parts of the article.

3 Brief Presentation of the Results of the Initial Research

The thesis 'Business entities belonging to the micro, small and medium-sized enterprise sector require special attention and appropriate support in view of being participants of Industry 4.0 processes' was the basis for a survey, which was carried out in the region of Upper Silesia in Poland between October 2018 and January 2019. Some results of this survey, which were presented in detail in [14], are briefly introduced below. The Silesian Voivodeship, especially the central part of the Voivodeship (GZM Metropolis) area, is heavily urbanised and strongly saturated with micro, small and medium-sized production and service enterprises (MSMEs). Therefore, it was assumed that the results of the initial survey, dedicated to this class of units, may be treated as partially representative on the initial stage of the research. The basic assumptions of the survey are presented in Table 1.

The substantive content of the survey for the needs of the first round of research was prepared on the basis of the results presented in available publications (e.g., [5, 28]) and as a result of discussing the suggested questions with selected representatives of the MSME sector as well as with CEOs of regional and local Chambers of Commerce.

A list of questions is introduced in Table 2. The questionnaire form was preceded by a short (cca. 1 page) introduction, which was aimed at supporting the volunteers in carrying out the study, as well as for respondents to get to know and understand the key points of the Industry 4.0 paradigm.

A list of questions is introduced in Table 2. The survey form was preceded by a short (1 page) introduction, aimed at supporting the volunteers in carrying out the study, as well as respondents to get to know and understand the key points of the Industry 4.0 paradigm.

In the described study, the volunteers conducting the survey were students of the Faculty of Organization and Management of the Silesian University of Technology who, after conducting an appropriate briefing on the rules for selecting respondents and explaining any doubts of these respondents reported during the research, were issued 200 copies of the survey form.

Table 1 Basic assumptions of the initial survey of MSME class enterprises

1.	The questionnaire was addressed exclusively to micro-, small- and medium-sized enterprises
2	The group of respondents was limited to owners and/or managers of the examined units
3	The area of activity (range of activities/services, branch of industry, etc.) was not considered
4	The questionnaire was designed to be as simple as possible, to be filled in by respondents
5	Each of the questions gave the respondent the opportunity to choose a single answer
6	Every answer could be provided with a comment; however, it was not obligatory
7	All respondents could provide information in the survey about their company or they could remain anonymous
8	The questionnaire was distributed by volunteers who were helping the respondents fill in the questionnaire
9	The (subjective) choice of respondents was exclusively made by the volunteers

Table 2 Questions asked to respondents and variants of responses (VoR)

Question	VoR 1	VoR 2	VoR 3	VoR 4	VoR 5
Have you already heard about the idea of the 'fourth industrial revolution INDUSTRY 4.0 (I4.0)'?	Yes	No	-	-	-
Do you think that I4.0 is already affecting/will affect the functioning of your company in the nearest future?	Yes	No	-	-	-
If the answer to question 2 is yes: the impact of I 4.0 on the functioning of your company is/will be:	Significant	Negligible	-	-	-
If the impact of I4.0 on the functioning of your company is/will be significant, then this applies to:	Internal organization and functioning of the company	Co-operation of the company with external partners	Both mentioned factors	-	-
Does the impact of I4.0 on the functioning of your company require/will require taking by you (by your company) any measures or adjustments?	Yes	No	-	-	-

(*continued*)

Table 2 (*continued*)

Question	VoR 1	VoR 2	VoR 3	VoR 4	VoR 5
If the response to question 5 is yes: Will the necessary measures/adjustments to be undertaken primarily include – in your opinion – reinforcement (development, extension):	Knowledge and competences of the company management	Knowledge and competencies of all employees	Technical and/or ICT equipment	Two of the mentioned factors (which ones and why)	All mentioned factors
Is it necessary for your company – in your opinion – to initiate and carry out some measures focused on the activities/adjustments related to I4.0:	By the company in your own range (utilizing your "own forces")	With the use of external support to the full extent of previously identified needs	With the use of external support in a selected scope of needs	Other solution	-
Will the detailed recognition (inventory) of your company's needs in the implementation of solutions related to the I4.0 model – in your opinion – be carried out:	By the company in your own range (utilizing your "own forces")	With the use of external support	By a combined team of company staff and external experts or advisors	Other solution	-

Finally, seventy-four responses were obtained (37%). An analysis of the survey results has revealed interesting statements. For instance, it can already be noticed that the responses given to question 1 were somewhat unexpected. A relatively high percentage of respondents (46%) declared their unfamiliarity with the Industry 4.0 concept. Some of them declared that they first learnt about the concept of I4.0 after reading the short informational part of the questionnaire form. This result, however, became less puzzling in comparison with the responses received to question 2. This question, regarding the significance of the impact of Industry 4.0 on the functioning of the surveyed entities, got over 70% of positive responses, stating that such an impact will be significant. Analysing, subsequently, the comments to the responses to questions 1 and 2, we can see that many respondents understand modern challenges in the economy and industry, but have not yet encountered nor understood the terms 'the fourth industrial revolution' and/or 'Industry 4.0'.

These results also indicate that the attention devoted to educating the 'community' of MSME owners and managers by educational sector units was and still is inadequate to their needs. The situation, in which only big economic players and scientific circles are familiar with (or declare they are) and discuss Industry 4.0, but for a significant group of smaller players, not possessing the proper knowledge, this discussion is not fully understood, is not acceptable.

Negative effects of this lack of knowledge about the concept of 'Industry 4.0' are confirmed in further results of the presented survey. About 10% of respondents declaring knowledge of the concept of Industry 4.0 and 64% of those who declared their unfamiliarity of this concept believe that its impact on the functioning of the examined entity will be negligible or none. There are various arguments, ranging from "my business activity is of 'niche nature', therefore I do not intend to be implied by the Industry 4.0 processes" to "I have my order book full for the next 10 years, so I will not be forced into engaging myself in these new concepts". There were also responses like "my company is too small to notice the effects of the industrial revolution" or "I have already signed contracts with external partners who guarantee a stable operation without having to make any changes in my company". It is surprising to see the lack of responses to question 4 in 19% of the questionnaires. The comments (or the lack thereof) do not indicate that, according to the respondent, both the 'internal' and 'external' situation of the company does not require changes in the Industry 4.0 perspective.

The research ('first round'), briefly presented above and described in more detail in [14], was repeated after one year [15] with the same territorial area and the same group of respondents, as well as the same research tools. The aim of this 'second round' was to verify the results of the first round of research and to obtain knowledge (taking into account the effects of limiting the territorial area and a statistically insignificant number of respondents) whether and what changes in the opinions of respondents belonging to the selected group can be observed after one year? To sum up, it was stated that the research described in this part of the paper is worth continuing and extending. Thus, a short (7 items) list of questions concerning the problem of educating MSME people in the problems of Industry 4.0 was added to the broadly developed survey of Polish entities within this category. The survey as well as its selected results are presented in the next part of this paper.

4 Surveys Dedicated to Problems that Selected Polish SME's Face with the Industry 4.0 Paradigm

The research, the results of which are presented below (in the selected part which refers to the educational problems), was conducted by the scientific research group dedicated to the problems of MSME in Industry 4.0 and formed in the Faculty of Organization and Management, Silesian University of Technology (Poland). In particular, the questionnaires began to be planned in October 2019, to be carried out from December 2019 to March 2020. The survey was based on an extended questionnaire form, dedicated to SMEs throughout Poland, developed by members of the research group. Comparing the initial assumptions of the previous survey with its results, as presented in the previous chapter, it has been decided not to consider 'micro-size' enterprises in this extended survey.

The original plan of the described survey was to cover 900 entities (selected according to the principles of sample representativeness). However, due to the situation caused by the Covid-19 pandemic, the study was eventually suspended at the end of March 2020 after obtaining 611 (six hundred and eleven) responses to the questionnaire. It is planned to continue this research at a more favourable time in the future. The authors of the survey expect that, in the second part of the survey, the results obtained will make it possible to identify and analyse, to some extent, the impact of the pandemic on the functioning of MSMEs and to plan further research, adequate to the results obtained earlier. The results of currently observed reactions of students and academic teachers of the author's home University on challenges arising from pandemics, in particular the ability of the mentioned groups of the academic society to adapt themselves and function effectively in the system of distance learning, are offering some possibilities for extending the plan of research introduced here. Because of earlier interests in educational problems, the author of this paper has decided to propose a questionnaire with a group of questions focused on the needs and expectations of university participants regarding the education of SME employees in the field of preparing the personnel of an enterprise to participate in I4.0 processes. These questions are listed in Table 3.

Table 3 Questions asked to SME'S owners and CEO's

Question	Content of the question
1	Adaptation of the company's operations to the requirements of INDUSTRY 4.0 (I4.0) requires adaptation of the knowledge and competence of the company's employees to these requirements
2	The company is able to carry out information and training activities related to the requirements of I4.0 on its own
3	The company is interested in consulting in the implementation of I4.0
4	The company is interested in external training services in the field of preparing employees for the implementation of I4.0
5	Training services related to the implementation of I4.0 should only cover certain employees of the company (indicated by the management)
6	Training services related to the implementation of I4.0 should cover all employees of the company, including top management
7	Requirements for learning outcomes focused on I4.0 should be specified in cooperation between the training provider and the company management

The survey questionnaire gave respondents the possibility to answer by use of a slightly modified 5-level Likert's scale [17] that contained the following possibilities to answer: (1) Yes, (2) Probably Yes, (3) Difficult to Say, (4) Probably Not and (5) No.

This and no other method of surveying clearly influenced the form of the obtained data and, consequently, the possibility of using specific methods and tools to evaluate this data. The author of this study assumed that the simplest methods would be used, adequate both to the structure of the available data and to the expected results of the analysis. The structure of the data is clearly determined by the adopted rating scale (Likert scale). The expectations as to the results of the analysis of the obtained data are influenced by both the MSME sample selected for the study (relatively numerous, but only quantitatively oriented, and not distinguishing such important features of the entities covered by the survey, such as the industry and type of activity, location and territorial scope, legal form and many others). These expectations are also determined by the material scope of the questions asked. Both of these key factors will be, according to the author of this study, the subject of further research. For the above reasons, it was decided to use, at the present stage of the research, possibly simple tools for data visualization and analysis.

In the first step, the obtained results were shown by means of pie charts, which refer to the absolute number of responses according to one of the five available levels. Examples of the results of such visualisations, for the sample given by 611 responses, are shown in Figs. 1 and 2.

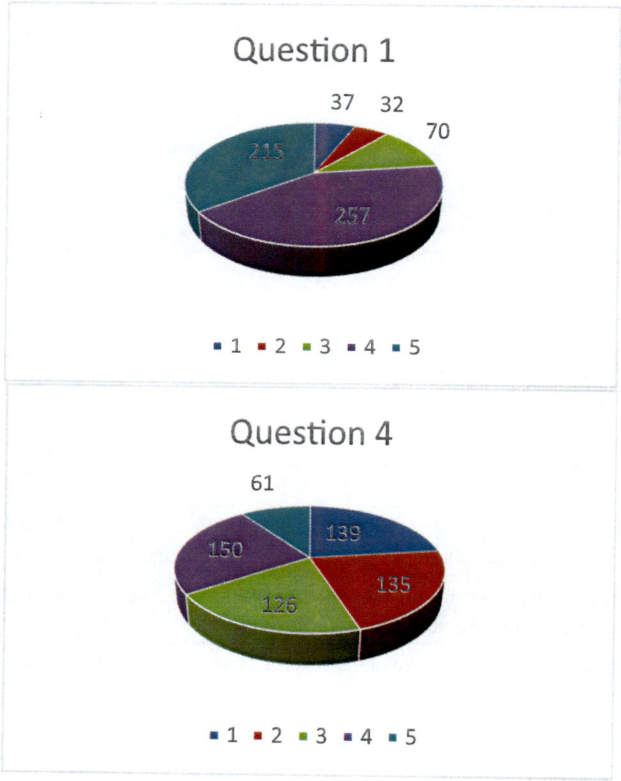

Fig. 1. Distribution of responses to questions 1 and 4

Further in-depth analyses of the data obtained in the presented survey seem interesting. In the next part of this paper, the used methods of these analyses are introduced and the results of applying these methods in the reported case are briefly discussed.

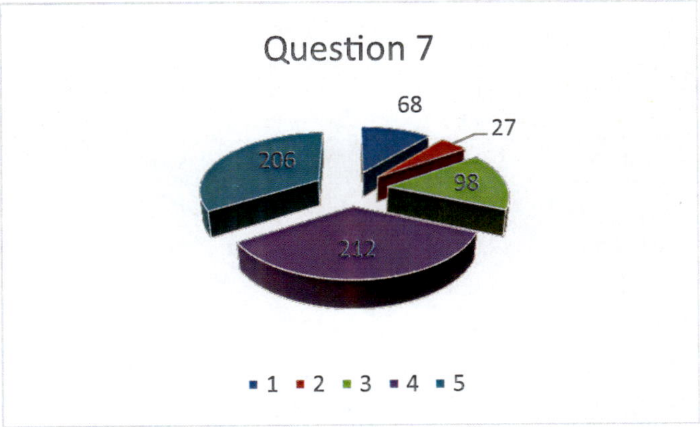

Fig. 2. Distribution of responses to question 7

5 Brief Analysis of the Results of the Presented Survey

As mentioned above, it was decided to use, at the present stage of the research, what possibly are simple tools for data analysis and visualization.

Even if the visualisation as shown in Figs. 1 and 2 allows for the interpretation of data practically only in the qualitative aspect and concerns 'raw data', it allows for an interesting analysis of this data, for example an analysis of the share of individual responses in the range of possibilities of providing such a response to individual questions. In particular, at this stage of data analysis, attention was drawn to the possible dominance of one choice of response to a given question.

This type of analyses of the data collected in the presented survey seems to give some interesting responses but can also create a background for many new questions. For instance, it seems to be worth explaining why the responses to questions 5 and 6 are so equally distributed throughout the Likert's scale. Perhaps further surveys and tools used for analysing data ought to be completed with some additional descriptors of the investigated enterprises, like the kind of activities or type of production. In Polish circumstances, the geographical location of the enterprise in the country or the geographical range of activities have to be considered too. It also seems worth considering, e.g., the responses to question 3, which are almost equally distributed, also when we take into account the responses from "yes" to "probably not", but the response "no" is clearly less frequent than the other ones.

As mentioned above, it was decided to use, at the present stage of the research, what possibly are simple tools for data analysis. In particular, a simple data aggregation was performed (the total number of respondents' responses to each of the seven questions presented in Table 3 in absolute terms and in percentage relative to the total number of responses). The results of such aggregation are shown in Table 4 (note: each percentage is rounded to one decimal number, so the sum of these values in the rows of Table 4 may not be equal to 100).

Even a cursory look at the results shown in Table 4 indicates some lack of consistency in the obtained responses. For instance, respondents are overly optimistic about the level

Table 4 Responses for the questions listed in Table 3 (aggregated)

Question number	Yes + (%)	Probably Yes + (%)	Difficult to Say + (%)	Probably Not + (%)	No + (%)
1	37 (6.1%)	32 (5.2%)	70 (1.5%)	257 (42.1%)	215 (35.2%)
2	86 (14.1%)	139 (22.7%)	127 (20.8%)	175 (28.6%)	84 (13.8%)
3	120 (19.6%)	124 (20.3%)	138 (22.6%)	161 (26.4%)	68 (11.1%)
4	139 (22.7%)	135 (22.1%)	126 (20.6%)	150 (24.6%)	61 (10%)
5	112 (18.3%)	98 (16%)	117 (19.1%)	194 (31.8%)	90 (14.7%)
6	118 (19.3%)	126 (20.6%)	126 (20.6%)	142 (23.2%)	99 (16.2%)
7	68 (11.1%)	27 (4.4%)	98 (16%)	212 (37.7%)	206 (33.7%)

of competences and knowledge of their employees on the requirements of Industry 4.0. According to the responses to question 1 (77.3% of positive responses – see Fig. 1), these competences and knowledge do not need to be adapted, since they seem appropriate at this moment. On the other hand, responding companies are clearly interested in external training services to prepare the employees for the implementation of 'INDUSTRY 4.0' (responses to question 4 indicate a clear "yes" and "probably yes" advantage: 48% of positive responses with 20.6% of responses being "neutral").

Therefore, it seems clear that the responses to questions 1 and 4 are somewhat 'incompatible' and it is worth thinking – if and how to enrich employees' knowledge and competences. It also seems interesting as to why the respondents are so negative (71.4% of aggregated choices of "probably not" and "no") about specifying requirements for learning outcomes focused on I4.0 in cooperation between the training provider and the company management (question 7 – see Fig. 2).

It was also decided to evaluate the sets of responses to individual questions, determining the median value for each of these sets. It is intended to enable the identification of those questions to which the respondents were more likely to answer negative (1 or 2 on a 5-point Likert scale) or positive 4 or 5 on a 5-point Likert scale). The results of this analysis are shown in Table 5.

Table 5 Median of answers to questions.

No of the question	Median value from all responses
Question 1	**4** (Probably Yes)
Question 2	**3** (Difficult to Say)
Question 3	**3** (Difficult to Say)
Question 4	**2** (Probably Not)
Question 5	**3** (Difficult to Say)
Question 6	**4** (Probably Yes)
Question 7	**4** (Probably Yes)

These results, in the opinion of the author of this study, justify the conclusion that for questions 1, 6, and 7 there was a predominance of 'positive choices' (4 or 5 in the scale used), whereas the responses to question 4 were mostly 'negative choices' (1 or 2 on the scale used). In the remaining questions (questions 3 and 5, the distribution of responses was relatively balanced).

Moreover, as a possible basis for deducting the mutual equivalence (the possibility of a similar interpretation by the respondents) of individual questions, it was decided to determine the correlation coefficients for the responses for each of the possible pairs of questions. The results of this assessment are shown in Table 5.

Table 6 The correlation coefficient estimated for responses to questions in possible pairs.

	Q1	Q2	Q3	Q4	Q5	Q6	Q7
Q1	1						
Q2	0,1434	1					
Q3	0,2746	0,0131	1				
Q4	0,2842	- 0,023	**0,8081**	1			
Q5	0,1593	0,066	0,3749	0,3596	1		
Q6	0,2827	0,1464	0,4409	0,3972	0,0175	1	
Q7	0,295	0,0799	0,5070	0,5322	0,4080	0,4767	1

The analysis of the values of the correlation coefficients shown in Table 6 allows to conclude, taking into account the limitations resulting from the interpretation of the indicator used, that the responses to individual questions do not show any significant correlation. A fairly significant level of correlation of responses to questions 3 and 4 seems to be intuitively justified: respondents who have a lack of knowledge about I4.0 processes and believe that they need expert support in this area, are also interested in the possibility of training in this area of issues (Table 6).

As can be seen from the considerations presented in this chapter, they generate more questions than they provide adequate responses for the research. Such questions concern in particular the needs of owners and managers of SMEs in the scope of supplementing the knowledge about the Industry 4.0 concept, but also the degree of their awareness of such needs.

According to the author of this paper, there are at least three possible sources to these questions:

1. Members of the management team in the enterprise are often not sure of their own competences,
2. Members of management appreciate the knowledge and competences of the training service provider and do not want to "disturb" them
3. It is more convenient to leave all activities covered by the external order, including proposing training results, training service providers, which ultimately relieves the enterprise managers from responsibility for the results of the training.

Even if the above variants 2 and 3 are – in their base – similar, the interpretation of the answers for question 7 clearly gives us some additional paths for planning further research. On the other hand, the results presented above indicate the reasons why the findings of the research obtained so far are not fully satisfactory regarding the assumed goal of conducting this research.

The author of this article suspects that the above-mentioned factors, identified during the analysis of the survey results, may have two basic sources:

1. They result from the wording of questions, not fully understood by the respondents (this may apply to the situation of an 'even distribution of responses').
2. They are the result of substantive deficiencies in the questions asked.
3. Deficiencies of the second kind, appearing in the discussed studies, were probably caused by the need to achieve a compromise between the number of questions asked (too many would be 'indigestible' for respondents) and their content (too complex would make it difficult to respond to).

6 Conclusions

According to the author of this paper, the above considerations allow to partially answer the questions formulated in the title of this article. Certainly, the introduced questionnaire, in a sense due to its pilot nature, as indicated also in the previous part of the article, does not allow for providing full responses to all possible research questions concerned with a preparation of SME class stakeholders to fulfil the needs of the Industry 4.0 revolution. The response, which can be formulated, is clearly conditioned by the questions asked. In addition, the obtained results generate further questions, which is probably natural in the search for nontrivial responses.

Considering the above objections, the results obtained, first of all, give us a basis for the statement that for the group of small and medium-sized enterprises, as well as for 'micro' enterprises (considering the results of the initial research), knowledge and competences in the area related to the Industry 4.0 paradigm of both the management and other employees in such enterprises until now leave much to be desired.

Moreover, in relation to the research works described in the available publications, it seems justified to state that it is, first and foremost, a necessity to broaden the self-awareness of respondents of this type of research in the area, as well as the awareness of those conducting such research, referring both to the real needs of these respondents, as well as to the abilities to meet such needs on the side of external providers of services, especially those of a consultative and educational nature.

To this day, there is a widespread situation in which the offer of training entities includes issues, in which such entities feel 'strong' (competent). These entities assume that potential recipients of their services will choose what they need from their offer. On the other hand, recipients' awareness of their needs for educational services is burdened with the conviction that providers of consulting and training services know best what to teach. And so, the 'vicious circle' closes. Breaking this cycle is probably the most adequate response to the question: "why educate?".

Going back to the questions included in the title, if the considered group of Industry 4.0 stakeholders in this paper is limited to SMEs, one can state that:

- It is relatively easier to tell why stakeholders belonging to this group should be educated: because of an identified lack of knowledge and an attachment to old habits, e.g., in terms of thinking about relationships with consultancy and educational service providers. In broader terms, most, if not all, old habits relate to a significant proportion of SMEs' contacts with the science and research sector.
- The research results discussed here give less possibilities to answer the question of how to conduct training in the discussed area. The statement that they should be carried out effectively is undoubtedly trivial, but also contains an element of rationality. One can identify specific feedback here: the better the needs are identified, the better the trainers are able to reach the trainees with a competent message, resulting in more effective training.

To sum up, the problem of identifying the needs in the scope of supplementing knowledge and qualifications of people employed in SMEs, also in the Industry 4.0 problem area, as well as the problems of the selection of adequate knowledge transfer methods, remains open and should undoubtedly be the subject of further research.

However, if we start talking about further research, it is necessary to mention the most important problem, not considered up until now. When the questionnaires reported in this paper were initiated, nobody had heard about the Covid-19 pandemic. Thus, nobody was able to foresee that the paradigms describing various processes that occur in the modern world–among others, the paradigm of the fourth industrial revolution (Industry 4.0)–does not contain yet an element of unimaginable value: the problem of the personal health and safety of humans, who are involved in the subsequent processes.

Undoubtedly, the Covid-19 pandemic has already influenced, is influencing, and will significantly influence the functioning of the economy as an important element of the functioning of society. The specific mechanism of such an influence is also noticeable in the research intention described in this study. Both the conditions in which the research plan was created and the basic conditions of such a plan can be clearly divided into 'before' and 'after' times. March of 2020 creates a time turning point for both the way of thinking and the ways of functioning of practically all social groups. There have been profound changes, among others, ways of unprecedentedly developed remote work or ways of running a business.

The effects of a pandemic can also lead to a reassessment of the basis of research. For example, in research on the impact of technology on society (Technology Assessment - TA), the set of criteria used so far for such an assessment will undoubtedly require the rebuilding of priorities and significant supplements. It can be assumed that the issue of health safety will take a prominent place in the ranking of such criteria.

All stakeholders need to find the right means and ways to use experiences, skills, and qualifications (e.g., in the field of 'remote working life') to restore certain areas of human activity to a state that is not a simple reconstruction of the 'pre-pandemic' state, but allows every member of society to function effectively in the new reality, including industrial reality.

The author of this study is convinced that the research tasks undertaken so far require a reassessment. Undoubtedly, the challenge of the highest priority 'for now' is to find effective measures and ways to combat the pandemic. The challenges for the 'day after the pandemic' can be seen in the example illustrated by the research described in this paper. The problem of perception of specific challenges as well as understanding and accepting the need to acquire knowledge and qualifications, enabling them to cope with the challenges in the specific relationship of the community of micro, small, and medium-sized enterprises versus Industry 4.0 still needs to be examined, but the solutions prepared so far in this area not only can, but should be modified. Paradoxically, this is also a unique opportunity for scientists: they have to re-plan and carry out their research plans, also already undertaken and partially implemented, in accordance with new plans.

References

1. Banta, D.: What is technology assessment? Int. J. Technol. Assess. Health Care **25**, 7–9 (2009)
2. Bendel, O.: Die Industrie 4.0 aus ethischer Sicht. HMD Praxis der Wirtschaftsinformatik **52**(5), 739–748 (2015). https://doi.org/10.1365/s40702-015-0163-z
3. Caragliu, A., del Bo, C., Nijkamp, P.: Smart cities in Europe. J. Urban Technol. **18**, 65–82 (2011)
4. Confetto, M.G., Conte, F., Vollero, A., Covucci, C.: From dual marketing to marketing 4.0: the role played by digital technology and the internet. In: Palazzo, M., Foroudi, P., Siano, A. (eds.) Beyond Multi-channel Marketing: Critical Issues in Dual Marketing, pp. 141–161. Emerald Publishing Limited, Bingely (2020). https://doi.org/10.1108/978-1-83867-685-820201012
5. Coşkun, S., Kayıkcı, Y., Gençay, E.: Adapting engineering education to industry 4.0 vision. Technologies **7**, 10 (2019)
6. Erboz, G.: How to Define Industry 4.0: The Main Pillars of Industry 4.0. Managerial Trends in the Development of Enterprises in Globalization Era, pp. 761–767 (2017)
7. Erol, S., Jäger, A., Hold, P., Ott, K., Sihn, W.: Tangible industry 4.0: a scenario-based approach to learning for the future of production. Procedia CIRP **54**, 13–18 (2016). https://doi.org/10.1016/j.procir.2016.03.162
8. Faller, C., Feldmüller, D.: Industry 4.0 learning factory for regional SMEs. Proc. CIRP **32**, 88–91 (2015). https://doi.org/10.1016/j.procir.2015.02.117
9. Hizam-Hanafiah, M., Soomro, M.A., Abdullah, N.L.: Industry 4.0 readiness models: a systematic literature review of model dimensions. Information **11**(7), 364 (2020). https://doi.org/10.3390/info11070364
10. Immonen, M., Hallikas, J., Pynnönen, M.: Antecedents of system purchasing in B2B services. J. Purch. Supply Manag. **22**, 205–213 (2016)
11. Jorgenson, D.W., Vu, K.M.: The ICT revolution, world economic growth, and policy issues. Telecommun. Policy **40**, 383–397 (2016)
12. Karwot, J., Kaźmierczak, J., ... Przystałka, P.: Smart water in smart city: a case study. In: Proceedings of the 16th International Multidisciplinary Scientific Conference SGEM2016, Book 3, 1(SGEM2016 Conference Proceedings, ISBN 978–619–7105–61–2/ISSN 1314–2704), pp. 391–398 (2016)
13. Kaźmierczak J.: About utilizing the concept of the Engineering of Needs (EoN) in the Technology Assessment (TA) approach. Int. J. Eng. Sci. Invent. **6**(10), 2319–6726 (2017). www.ijesi.org. (ISSN (Online): 2319–6734, ISSN (Print): 2319–6726)
14. Kaźmierczak, J.: SME's versus "Industry 4.0": extended case study based on results of initial research in Polish conditions. Multidisc. Aspects Prod. Eng. **2**, 305–314 (2019)
15. Kaźmierczak, J.: Do the owners and managers of polish MSME's recognize and understand properly the challenges of "industry 4.0"? Multidisc. Aspects Prod. Eng. **3**, 405–421 (2020)
16. Kitchin, R.: The ethics of smart cities and urban science. Philos. Trans. R. Soc. A Math. Phys. Eng. Sci. **374**(2083), 20160115 (2016). https://doi.org/10.1098/rsta.2016.0115

17. Likert, R.: A technique for the measurement of attitudes. Arch. Psychol. **140**, 44–53 (1932)
18. Mazali, T.: From industry 4.0 to society 4.0, there and back. AI & Soc. **33**(3), 405–411 (2017). https://doi.org/10.1007/s00146-017-0792-6
19. Mokyr, J.: The second industrial revolution, 1870–1914. Wave Motion **51**, 758–767 (1988)
20. Müller, J.M., Buliga, O., Voigt, K.I.: Fortune favors the prepared: how SMEs approach business model innovations in Industry 4.0. Technol. Forecast. Soc. Chang. **132**, 2–17 (2018)
21. Naboni, R., Paoletti, I.: The third industrial revolution. In: Advanced Customization in Architectural Design and Construction. SAST, pp. 7–27. Springer, Cham (2015). https://doi.org/10.1007/978-3-319-04423-1_2
22. Nam, T., Pardo, T.A.: Conceptualizing smart city with dimensions of technology, people, and institutions. In: ACM International Conference Proceeding Series, pp. 282–291 (2011)
23. Rifkin, J.: The third industrial revolution. Eng. Technol. **3**, 26–27 (2008)
24. Rifkin, J.: The third industrial revolution: how the internet, green electricity, and 3-D printing are ushering in a sustainable era of distributed capitalism. World Finan. Rev. 1–8 (2012)
25. Rip, A.: Technology assessment. In: International Encyclopedia of the Social and Behavioral Sciences: Second Edition, pp. 125–128. Elsevier Inc. (2015)
26. Robinson, D.J.: Building a green economy. In: The Energy Economy, pp. 173–191. Palgrave Macmillan US, New York (2015). https://doi.org/10.1057/9781137469274_9
27. Salgues, B.: Society 5.0, Its Logic and Its Construction in: Society 5.0, pp. 1–21. John Wiley & Sons Inc., Hoboken (2018)
28. Sommer, L.: Industrial revolution - Industry 4.0: are German manufacturing SMEs the first victims of this revolution? J. Indus. Eng. Manage. **8**, 1512–1532 (2015)
29. Walt, S.M.: Revolution and War. World Politics **44**, 321–368 (1992)

Evaluation of the Effectiveness of Standard Scheduling Rules – An Educational Approach

Ryszard Perłowski[1], Arkadiusz Gola[2](✉), and Katarzyna Antosz[1]

[1] Faculty of Mechanical Engineering and Aeronautics, Rzeszow University of Technology, Powstańców Warszawy 8, 35-959 Rzeszów, Poland
{rpztmiop,katarzyna.antosz}@prz.edu.pl

[2] Faculty of Mechanical Engineering, Lublin University of Technology, Nadbystrzycka 38, 20-618 Lublin, Poland
a.gola@pollub.pl

Abstract. Proper sequencing of tasks on technological machines remains a crucial element in each manufacturing company. Designing an optimal schedule may seem to be a fairly simple problem. Nevertheless, scheduling is a subject to numerous constraints which limit and prevent the use of effective scheduling methods in a real production environment. Although there is a considerable amount of academic literature in the field of scheduling production jobs, the proposed solutions have a limited educational value. Because of NP-hard character of schedule's optimization problem, it is usually necessary to use specific and often very expensive software to teach students or process engineers the idea of scheduling, and how different scheduling rules presented in the literature influence production effectiveness. Therefore, we decided to develop a universal and simple IT tool that supports an education process in the area of operation management and allows to present a general idea of schedule design, to compare different scheduling algorithms and to find an optimal production schedule by means of complete search algorithm. The paper presents an original solution dedicated to educational purposes that allows to teach students or future process engineers how to design manufacturing schedules, what the priority rules are and how the schedules can be optimized. In order to develop the presented solution, the MS Excel spreadsheet was used. Moreover, the integrated optimizer and VBA language were used to make the necessary calculations and optimization process. The mode of action of the developed solution was illustrated with an example including 10 production tasks with 10 different operations.

Keywords: Scheduling · Sequencing · Optimization · Manufacturing · MS Excel · VBA · Operations management · Education

1 Introduction

Planning, scheduling and the division of resources are the basic decision areas within the production process management. They occur independently from the forms of production organisation, production type or the kind of production system [1]. The correct

A. Batako et al. (Eds.): GCMM 2021, LNNS 335, pp. 344–357, 2022.
https://doi.org/10.1007/978-3-030-90532-3_26

scheduling (task sequencing) and the proper schedule implementation are the key elements here because they decide about the timely realization of production orders as well as the optimal use of available production capacity [2, 3].

In general, the notion of production scheduling describes the sequence of the realized tasks and operations on the particular production stands so as to obtain the best use of production resources in regard to determined aim criterion [4]. The developed schedule most often takes the form of a graph or a description of particular operations that takes into account the scheduling units in which these operations will be realized [5]. As far as developing a schedule itself is not a demanding task, its optimization creates a significant problem [6]. The basic difficulty related to production scheduling is the problem of a great computational complexity. It happens because the problem of task scheduling belongs to NP-hard problem class. As a consequence, it causes a great difficulty in finding a precise solution [7]. This complexity rises with the increase of the task number as well as with the number of machines realizing production processes. Then, it is only possible to find an acceptable solution [8].

The issue of production scheduling was significantly influenced by the computer technologies development. The application of a computer in this area aimed at creating a system helpful in the management and supervision of a production process. The systems were supposed to take their own decisions in relation to the production task scheduling. The development of these solutions resulted in the creation of MRP class systems (Material Requirements Planning), which evolved later into MRP II systems (Manufacturing Resource Planning) and now to the commonly used ERP class systems [9, 10]. In order to increase the effectiveness of a scheduling process, companies often apply complex and very expensive tools which support a decision process. The example of such systems are APS systems (Advanced Planning System) which use advanced optimization algorithms based on linear programming (e.g.by means of a branch and bound method) and genetic algorithms [11, 12].

Despite the fact that the problem of production scheduling has been the subject of extensive research for many years and enterprises possess more and more advanced technologies which support processing engineers work, the education process with regard to the rules and optimization of production tasks scheduling remains a challenge. Contrary to the systems used in industry, the systems dedicated to the needs of education have to be characterized by a certain degree of universalism and simplicity which enable the cognition of the scheduling idea (the aim of an education process in not to obtain an optimal solution but familiarizing students with the essence of scheduling and the influence of particular factors, e.g. of the priority rules used, on the possibility of quick designing of a schedule that is close to a suboptimal solution. Therefore, we decided to develop an original solution dedicated to educational purposes that allows to teach students or future process engineers how to design manufacturing schedules, what the priority rules are and how the schedules can be optimized. In order to develop the presented solution, we used a well-known and commonly used MS Excel spreadsheet.

The further part of this article comprises of four parts. Point 2 presents the main issues of the production scheduling problem as well as the idea and the use of priority rules. Sect. 3 describes a developed solution that enables scheduling and that is dedicated to educational purposes. Sect. 4 shows the example of the possibilities of effectiveness evaluation of the chosen priority rules in a scheduling process with the use of the developed tool. The last point 5 summarizes and concludes the performed research.

2 Research Problem and Methodology

Formal methods of production scheduling began to develop at the end of 19th century. The process of creating production planning offices started at that time. The aim of the created offices was plan designing, resource management and operation monitoring. Moreover, some ideas of developing the lists which would consider a sequence of tasks performance during the day appeared [9, 13]. The predecessor of the design of production task scheduling methodology was Henry Gantt who designed a series of charts which allow to observe and analyse the realised production processes [14]. The graphic representation of a production progress was presented as a chart and it is used in production scheduling nowadays. It is known as a Gantt chart (schedule) [5].

The problem with production scheduling is the right allocation of production tasks (comprising of a certain number of operations) to the resources, which are available machines and stands. The issue may be considered in different production environments. The literature distinguishes three basic environments, such as: flow-shops, job-shops and open-shops [15]. In industry practice task scheduling is most often considered in a job-shop system that assumes a full order of tasks resulting from technological limitations [16, 17].

In the area of production scheduling diverse methods which allow to create schedules with the adopted aim criteria are used. General methods are divided into [18, 19]:

- Precise methods (also known as optimization) that are intended to find a precise (from the point of view of the adopted aim function) solution of the considered problem,
- Approximate methods (also called heuristic), which do not guarantee the finding of an optimized solution, are characterized by a shorter time of searching for a solution.

In spite of the existence of a number of production scheduling methods and proposed algorithms, their application in the education of future production processing engineers and managers is not an easy issue. That is because, in most cases, the enterprises with a variable and multi-assortment unit and small-batch production do not possess professional software which allows them to optimize the developed production schedules. The proper application of the existing priority rules becomes the key significance in a task scheduling process [20]. The use of priority rules is limited to the choice of the tasks which are supposed to be performed on a particular machine under technological limitations, and the production schedule is created sequentially on successive machines without the necessity of optimizing the developed schedule. In the literature on the subject there are several priority rules dedicated to the needs of production task sequencing. For instance, Kaban, Othman and Rohman in [21] present 44 rules, Conway – 92 [22], and Wojakowski [23] as many as 170. However, while reviewing the literature in the field of task scheduling in the real production systems, it should be stated that most common rules are as follows [24]:

- *First Come First Service* (FCFS), according to this rule the first to be chosen during scheduling from among these awaiting in a queue is the operation which arrived first to the queue. It can be represented with the expression:

$$P_{IJ}(t) = min\{z_{ij}(t) = R_{ij}\}(i,j) \in A(t), \tag{1}$$

where R_{ij} is the date of arrival of j-nth operation of i-nth task to the queue of the considered stand.

It is the simplest priority rule, also known as "natural regulations". The results of the studies obtained with this method are usually used for comparing the results with other rules. This rule is used in arguable situations when as a result of the functioning of any other of the rules some arguable situations are created, e.g. in the situation when the rule based on the time of operation encounters the operation of the same processing times.

- *Last Come First Service* (LCFS) - this rule is opposite to FCFS – the operation which arrived the last is chosen from among the operations awaiting in a queue:

$$P_{IJ}(t) = min\{z_{ij}(t) = -R_{ij}\}, (i,j) \in A(t), \tag{2}$$

This rule causes the performance of operations in the order reverse to the order of arrival. The functioning of this rule usually generates a huge dispersion in the times of production cycles.

- *Shortest Processing Time* (SPT) - the aim of this rule is to select from among the operations awaiting to be executed, the one which is characterized by the shortest processing time. Thus:

$$P_{IJ}(t) = min\{z_{ij}(t) = to_{ij}\}(i,j) \in A(t), \tag{3}$$

where to_{tj} is the time of operation j of a task i.

This rule is most often applied in practice. It is characterized by simplicity – it orders operations in a queue according to the increasing values of priority indicators (operation times). It is the example of the rule relating to the operation processing times.

- Longest Processing Time (LPT) – this rule chooses these operations from a queue which are characterized by the longest processing times:

$$P_{IJ}(t) = min\{z_{ij}(t) = to_{ij}\}(i,j) \in A(t), \tag{4}$$

This rule maximizes all the values which SPT rule minimizes. That's why, many authors assess it negatively. However, this rule works out in the situations when the fast load decrease of the chosen production stands is essential. This rule is good for the operations of a high stand-consumption [25]. The rules mentioned above are most commonly used priorities in the area of production scheduling in a real production system. Furthermore, they represent the most important groups of scheduling rules. The main argument that decides about their great popularity is their simplicity and the low cost of implementation as well as its rate of action what enables an efficient development of the designed production schedules.

3 Configuration of MS Excel Spreadsheet for Production Scheduling

Spreadsheet is a popular and widely used tool for data processing. Supported by the VBA programming language (Visual Basic for Applications) it gives huge possibilities of solving engineering problems. Applying this tool in solving the problems of production task sequencing/scheduling allows to design own tools which conduct the analyses found in commercial applications. The following assumptions were taken while designing the application:

- The maximum number of tasks – 10,
- The maximum number of operations in a task – 10,
- The planning horizon – 1500 units of time,
- The possibility of setting any route (any random order of operations in a task with the possibility of repeating a stand),
- Each operation is performed on one stand,
- Sequencing takes place on the level of two production tasks.

The designed application allows to sequence tasks by means of standard priority rules such as:

- FCFS – task sequencing according to the order of task arrival,
- LCFS – sequencing according to the reverse order of task arrival,
- SPT – sequencing according to the shortest total processing time of a task,
- LPT – sequencing according to the longest total processing time of a task.

The application also allows to sequence tasks with a complete search method regarded as a method searching for a precise solution. The choice of a task sequence with this method gives the possibility of obtaining an optimized solution from the point of view of a minimal length of all tasks sequencing with the optimal use of resources. It is an active schedule in which all the operations are moved as far as possible to the left side of a time axis. The application of the presented methods requires the use of nonstandard formulas written in a VBA code.

The application contains two spreadsheets: RULES in which the tasks are sequenced according to standard priority rules, and PERMUTATIONS with a method of complete search. Each spreadsheet has three basic elements: input data (a route and task times), a Gantt chart and chosen measures.

order of operations											processing time										
1	2	3	4	5	6	7	8	9	10	order	1	2	3	4	5	6	7	8	9	10	total processing time
1	2	3	4	5	6	7	8	9	10	1	29	78	9	36	49	11	62	56	44	21	395
1	3	5	10	4	2	7	6	8	9	2	43	90	75	11	69	28	46	46	72	30	510
2	1	4	3	9	6	8	7	10	5	3	91	85	39	74	90	10	12	89	45	33	568
2	3	1	5	7	9	8	4	10	6	4	81	95	71	99	9	52	85	98	22	43	655
3	1	2	6	4	5	9	8	10	7	5	14	6	22	61	26	69	21	49	72	53	393
3	2	6	4	9	10	1	7	5	8	6	84	2	52	95	48	72	47	65	6	25	496
2	1	4	3	7	6	10	9	8	5	7	46	37	61	13	32	21	32	89	30	55	416
3	1	2	6	5	7	9	10	8	4	8	31	86	46	74	32	88	19	48	36	79	539
1	2	4	6	3	10	7	8	5	9	9	76	69	76	51	85	11	40	89	26	74	597
2	1	3	7	9	10	6	4	5	8	10	85	13	61	7	64	76	47	52	90	45	540

Fig. 1. Input data area.

Input data are the consecutively numbered tasks - column L, Columns B – K are the numbers of consecutively performed operations in a task, whereas column M – V are the processing times of consecutive operations. The last column contains the total time of a task.

The sequencing results (Gantt chart) are displayed in columns that correspond to the particular agreed time units (from 1 to 1500). The rows indicate particular stands in which, in the right time units, the numbers of assigned tasks are displayed. Additionally, they are marked with colours according to the colour palette in the input data (Fig. 2).

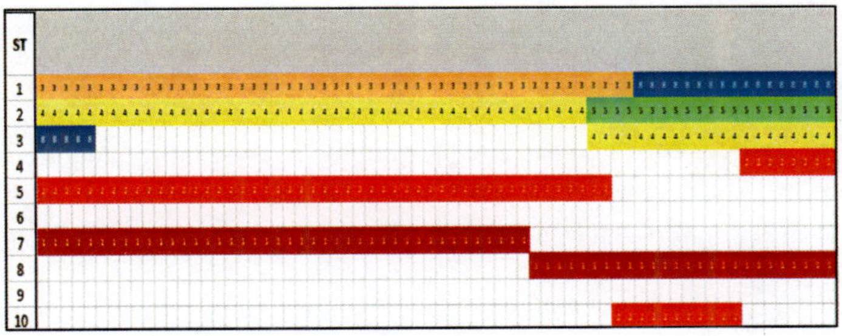

Fig. 2. Gantt chart excerpt.

The result of sequencing is also presented in the form of numbers: the start and finish of a task, the real processing time of a task and breaks. Certainly, they are the chosen, relatively simple measures that can be easily developed on the basis of the created Gantt chart (Fig. 3).

The basic calculations of a task order are conducted with a VBA code. In the presented rules the basic step is the right (in a descending and ascending order) classification of a task list in input data according to the order of task arrival column L) for FCFS and LCFS rules, or to the total processing time (column W) for the SPT and LPT rules. The example of a code sorting tasks for the FCFS rule was presented in Fig. 4.

The tasks sorted in a right way are then planned on the load matrix (mapa), specially defined for this purpose, of a size corresponding to the maximum planning horizon (1500 units of time)and the number of stands (10 tasks). Planning takes place according to standard Finish to Start rule, meaning that the next operation in a task can be started

	start	end	real processing time	break	%
1	1	395	395		
2	30	583	554	44	9%
3	108	862	755	187	33%
4	199	973	775	120	18%
5	1	1055	1055	662	168%
6	15	1151	1137	641	129%
7	304	968	665	249	60%
8	207	1165	959	420	78%
9	79	1349	1271	674	113%
10	505	1410	906	366	68%
			total	3363	

Fig. 3. Input data – sequencing counters.

```
Range("A3:W12").Select
ActiveWorkbook.Worksheets("Reguly").sort.SortFields.Clear
ActiveWorkbook.Worksheets("Reguly").sort.SortFields.Add Key:=Range( _
    "L3:L12"), SortOn:=xlSortOnValues, Order:=xlAscending, DataOption:= _
    xlSortNormal
With ActiveWorkbook.Worksheets("Reguly").sort
    .SetRange Range("A3:W12")
    .Header = xlGuess
    .MatchCase = False
    .Orientation = xlTopToBottom
    .SortMethod = xlPinYin
    .Apply
End With
Range("A1").Select
```

Fig. 4. Task sorting code.

after finishing of the preceding one. The programme analyses if the required, in a given time, stand is available for placing the operation, alternatively it moves it forward. This action is repeated for all the tasks in a sequencing order according to the selected rule (Fig. 5).

The method of complete search requires generating of all possible combinations of task order, and then checking a main measure for each combination, i.e. the sequencing length. The mechanism of stand loading is identical as in case of rules. Generating of particular combinations takes place in 10 For-Next loops with an additional condition that checks if the given combination is not repeated (Fig. 6).

The programme calculates the sequencing length every time, and if the result is more favourable than the one determined previously, it corrects the load matrix and Gantt chart. This way, while checking all the combinations it is possible to determine an optimal solution. In case of a maximum size of a task possible to be solved in an application (10 orders), the number of combinations is 3 628 800 solutions. The time needed for solving is relatively long and equals to about 140 min. However, in this case, according to the authors, the educational aspect has a greater significance than the rate of the obtained solution.

```
Sub mapa_obciazenia(czas, kolejna_oper, start, zlecenie)

   For v = start To 1500 - czas
      If mapa(kolejna_oper, v) = 0 Then
                For w = 1 To czas - 1
                     If mapa(kolejna_oper, v + w) <> 0 Then
                        start = v + w
                        v = v + w - 1
                        GoTo 1
                     End If
                Next w
                For a = v To czas + v - 1
                     mapa(kolejna_oper, a) = zlecenie
                Next a
                start = v + czas
                Exit Sub
      End If
1:
  Next v
End Sub
```

Description of the variables used:

czas – processing time

kolejna_oper – selected operation

start – end time of the preceding operation

zlecenie – task number

Fig. 5. Code of task scheduling on stands.

```
For a = lo_tablicy To up_tablicy
 For b = lo_tablicy To up_tablicy
  If tablica(b) <> tablica(a) Then
   For c = lo_tablicy To up_tablicy
    If tablica(c) <> tablica(a) And tablica(c) <> tablica(b) Then
     For d = lo_tablicy To up_tablicy
      If tablica(d) <> tablica(a) And tablica(d) <> tablica(b) And tablica(d) <> tablica(c) Then
       For e = lo_tablicy To up_tablicy
        If tablica(e) <> tablica(a) And tablica(e) <> tablica(b) And tablica(e) <> tablica(c) And tablic
         For f = lo_tablicy To up_tablicy
          If tablica(f) <> tablica(a) And tablica(f) <> tablica(b) And tablica(f) <> tablica(c) And tabl
           For g = lo_tablicy To up_tablicy
            If tablica(g) <> tablica(a) And tablica(g) <> tablica(b) And tablica(g) <> tablica(c) And ta
             For h = lo_tablicy To up_tablicy
              If tablica(h) <> tablica(a) And tablica(h) <> tablica(b) And tablica(h) <> tablica(c) And
               For i = lo_tablicy To up_tablicy
                If tablica(i) <> tablica(a) And tablica(i) <> tablica(b) And tablica(i) <> tablica(c) An
                 For j = lo_tablicy To up_tablicy
                  If tablica(j) <> tablica(a) And tablica(j) <> tablica(b) And tablica(j) <> tablica(c)
                     z(1) = tablica(a) 'zlecenie jako pierwsze
                     z(2) = tablica(b) 'zlecenie jako drugie
                     z(3) = tablica(c) 'zlecenie jako trzecie
                     z(4) = tablica(d) 'zlecenie jako czwarte
                     z(5) = tablica(e) 'zlecenie jako piąte
                     z(6) = tablica(f) 'zlecenie jako szóste
                     z(7) = tablica(g) 'zlecenie jako siódme
                     z(8) = tablica(h) 'zlecenie jako ósme
                     z(9) = tablica(i) 'zlecenie jako dziewiąte
                     z(10) = tablica(j) 'zlecenie jako dziesiąte
```

Description of the variables used:

tablica() – a variable that stores the iteratively assigned task number i.e tablica(a)' zlecenie jako pierwsze' – task numbered as first

z() – a variable storing a randomly selected, unique combination of tasks

Fig. 6. Part of the code creating task order combinations.

4 The Evaluation of the Effectiveness of the Chosen Priority Rules in the Process of Scheduling from Spreadsheet – An Illustrative Example

Spreadsheet the evaluation of the possibilities of the designed spreadsheet application was conducted by choosing a test set of 10 tasks that included 10 operations (stands) (Fig. 1). The results of task scheduling were presented in Figs. 7, 8, 9, 10 and 11. The minimal sequencing length (makespan) was the main measure of the solution quality. Additionally, the length of breaks in the particular tasks realization, which result from waiting time for the stand availability, was evaluated (Fig. 13 and 14).

order of operations											processing time										total processing time
1	2	3	4	5	6	7	8	9	10	order	1	2	3	4	5	6	7	8	9	10	
1	2	3	4	5	6	7	8	9	10	1	29	78	9	36	49	11	62	56	44	21	395
1	3	5	10	4	2	7	6	8	9	2	43	90	75	11	69	28	46	46	72	30	510
2	1	4	3	9	6	8	7	10	5	3	91	85	39	74	90	10	12	89	45	33	568
2	3	1	5	7	9	8	4	10	6	4	81	95	71	99	9	52	85	98	22	43	655
3	1	2	6	4	5	9	8	10	7	5	14	6	22	61	26	69	21	49	72	53	393
3	2	6	4	9	10	1	7	5	8	6	84	2	52	95	48	72	47	65	6	25	496
2	1	4	3	7	6	10	9	8	5	7	46	37	61	13	32	21	32	89	30	55	416
3	1	2	6	5	7	9	10	8	4	8	31	86	46	74	32	88	19	48	36	79	539
1	2	4	6	3	10	7	8	5	9	9	76	69	76	51	85	11	40	89	26	74	597
2	1	3	7	9	10	6	4	5	8	10	85	13	61	7	64	76	47	52	90	45	540
																		Makespan		FCFS	1410

(a)

	start	end	real processing time	break	%
1	1	395	395		
2	30	583	554	44	9%
3	108	862	755	187	33%
4	199	973	775	120	18%
5	1	1055	1055	662	168%
6	15	1151	1137	641	129%
7	304	968	665	249	60%
8	207	1165	959	420	78%
9	79	1349	1271	674	113%
10	505	1410	906	366	68%
			total	3363	

(b)

Fig. 7. a. FCFS rule - order of the realised tasks. b. FCFS rule - sequencing results.

While analysing the sequencing results obtained with the rules, LPT rule with the result of 1272 time units occurred to be the most optimal. The remaining analysed rules gave the results of respectively: FCFS – 1410, LCFS – 1483, SPT – 1487 (Fig. 12). The delay results are interesting. The best result was obtained here for SPT rule. The total time of delays is about 50% shorter (1818 time units) than in the remaining rules, from 3216 to 3929 time units (Fig. 15).

order of operations											processing time										total processing time
1	2	3	4	5	6	7	8	9	10	order	1	2	3	4	5	6	7	8	9	10	
2	1	3	7	9	10	6	4	5	8	10	85	13	61	7	64	76	47	52	90	45	540
1	2	4	6	3	10	7	8	5	9	9	76	69	76	51	85	11	40	89	26	74	597
3	1	2	6	5	7	9	10	8	4	8	31	86	46	74	32	88	19	48	36	79	539
2	1	4	3	7	6	10	9	8	5	7	46	37	61	13	32	21	32	89	30	55	416
3	2	6	4	9	10	1	7	5	8	6	84	2	52	95	48	72	47	65	6	25	496
3	1	2	6	4	5	9	8	10	7	5	14	6	22	61	26	69	21	49	72	53	393
2	3	1	5	7	9	8	4	10	6	4	81	95	71	99	9	52	85	98	22	43	655
2	1	4	3	9	6	8	7	10	5	3	91	85	39	74	90	10	12	89	45	33	568
1	3	5	10	4	2	7	6	8	9	2	43	90	75	11	69	28	46	46	72	30	510
1	2	3	4	5	6	7	8	9	10	1	29	78	9	36	49	11	62	56	44	21	395
																		Makespan		LCFS	1483

(a)

	start	end	real processing time	break	%
1	228	1483	1256	861	218%
2	185	1413	1229	719	141%
3	360	1270	911	343	60%
4	279	1187	909	254	39%
5	32	1103	1072	679	173%
6	160	1081	922	426	86%
7	231	903	673	257	62%
8	1	797	797	258	48%
9	1	729	729	132	22%
10	1	540	540		
			total	3929	

(b)

Fig. 8. a. LCFS rule - order of the realised tasks. b. LCFS rule - sequencing results.

order of operations											processing time										total processing time
1	2	3	4	5	6	7	8	9	10	order	1	2	3	4	5	6	7	8	9	10	
3	1	2	6	4	5	9	8	10	7	5	14	6	22	61	26	69	21	49	72	53	393
1	2	3	4	5	6	7	8	9	10	1	29	78	9	36	49	11	62	56	44	21	395
2	1	4	3	7	6	10	9	8	5	7	46	37	61	13	32	21	32	89	30	55	416
3	2	6	4	9	10	1	7	5	8	6	84	2	52	95	48	72	47	65	6	25	496
1	3	5	10	4	2	7	6	8	9	2	43	90	75	11	69	28	46	46	72	30	510
3	1	2	6	5	7	9	10	8	4	8	31	86	46	74	32	88	19	48	36	79	539
2	1	3	7	9	10	6	4	5	8	10	85	13	61	7	64	76	47	52	90	45	540
2	1	4	3	9	6	8	7	10	5	3	91	85	39	74	90	10	12	89	45	33	568
1	2	4	6	3	10	7	8	5	9	9	76	69	76	51	85	11	40	89	26	74	597
2	3	1	5	7	9	8	4	10	6	4	81	95	71	99	9	52	85	98	22	43	655
																		Makespan		SPT	1487

Fig. 9. a. SPT rule - order of the realised tasks.

	start	end	real processing time	break	%
1	21	441	421	26	7%
2	50	795	746	236	46%
3	464	1122	659	91	16%
4	555	1487	933	278	42%
5	1	393	393		
6	15	693	679	183	37%
7	128	652	525	109	26%
8	227	977	751	212	39%
9	93	1144	1052	455	76%
10	176	943	768	228	42%
			total	1818	

Fig. 10. b. SPT rule - sequencing results.

order of operations											processing time										total processing time
1	2	3	4	5	6	7	8	9	10	order	1	2	3	4	5	6	7	8	9	10	
2	3	1	5	7	9	8	4	10	6	4	81	95	71	99	9	52	85	98	22	43	655
1	2	4	6	3	10	7	8	5	9	9	76	69	76	51	85	11	40	89	26	74	597
2	1	4	3	9	6	8	7	10	5	3	91	85	39	74	90	10	12	89	45	33	568
2	1	3	7	9	10	6	4	5	8	10	85	13	61	7	64	76	47	52	90	45	540
3	1	2	6	5	7	9	10	8	4	8	31	86	46	74	32	88	19	48	36	79	539
1	3	5	10	4	2	7	6	8	9	2	43	90	75	11	69	28	46	46	72	30	510
3	2	6	4	9	10	1	7	5	8	6	84	2	52	95	48	72	47	65	6	25	496
2	1	4	3	7	6	10	9	8	5	7	46	37	61	13	32	21	32	89	30	55	416
1	2	3	4	5	6	7	8	9	10	1	29	78	9	36	49	11	62	56	44	21	395
3	1	2	6	4	5	9	8	10	7	5	14	6	22	61	26	69	21	49	72	53	393
																	Makespan	LPT			1272

Fig. 11. a. LPT rule - order of the realised tasks.

	start	end	real processing time	break	%
1	389	1228	840	445	113%
2	346	1272	927	417	82%
3	151	760	610	42	7%
4	1	655	655		
5	32	1177	1146	753	192%
6	177	1077	901	405	82%
7	375	1162	788	372	89%
8	1	981	981	442	82%
9	1	681	681	84	14%
10	242	1037	796	256	47%
			total	3216	

Fig. 12. b. LPT rule - sequencing results.

order of operations											order of operations										
1	2	3	4	5	6	7	8	9	10	order	1	2	3	4	5	6	7	8	9	10	total processing time
1	3	5	10	4	2	7	6	8	9	2	43	90	75	11	69	28	46	46	72	30	510
3	1	2	6	4	5	9	8	10	7	5	14	6	22	61	26	69	21	49	72	53	393
3	2	6	4	9	10	1	7	5	8	6	84	2	52	95	48	72	47	65	6	25	496
2	1	4	3	7	6	10	9	8	5	7	46	37	61	13	32	21	32	89	30	55	416
2	3	1	5	7	9	8	4	10	6	4	81	95	71	99	9	52	85	98	22	43	655
1	2	4	6	3	10	7	8	5	9	9	76	69	76	51	85	11	40	89	26	74	597
2	1	3	7	9	10	6	4	5	8	10	85	13	61	7	64	76	47	52	90	45	540
2	1	4	3	9	6	8	7	10	5	3	91	85	39	74	90	10	12	89	45	33	568
3	1	2	6	5	7	9	10	8	4	8	31	86	46	74	32	88	19	48	36	79	539
1	2	3	4	5	6	7	8	9	10	1	29	78	9	36	49	11	62	56	44	21	395
																				Makespan	1046

Fig. 13. a. Complete search method - order of the realised tasks.

	start	end	real processing time	break	%
1	163	1017	855	460	116%
2	1	510	510		
3	402	1034	633	65	11%
4	72	924	853	198	30%
5	1	472	472	79	20%
6	134	646	513	17	3%
7	1	731	731	315	76%
8	328	1044	717	178	33%
9	87	950	864	267	45%
10	317	1046	730	190	35%
			total	1769	

Fig. 14. b. Complete search method - sequencing results.

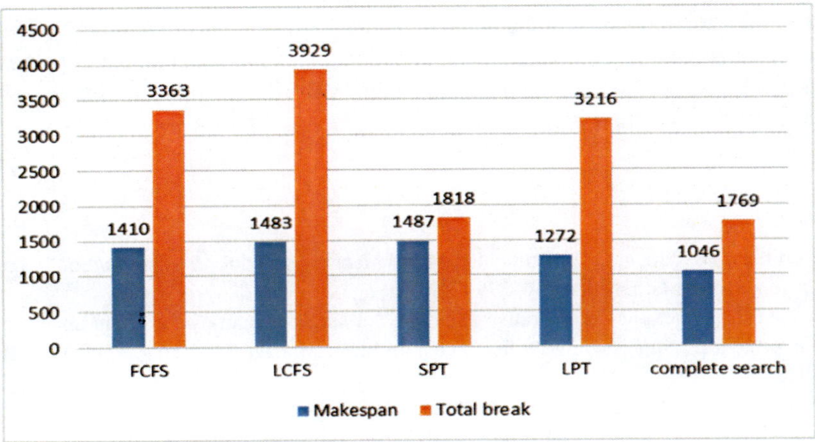

Fig. 15. Summary of the obtained results.

The complete search method shows, as it might be expected, the best result of 1046 time units and the shortest break time.

5 Summary and Conclusions

Education of production schedulers and future managers requires the use of the tools similar in their functionality to those used in real industrial conditions. The education in the area of production task sequencing methods as well as production schedules optimization is particularly challenging in this regard. Because of the computational complexity of the analysed problem, it needs to employ, on one side, universal tools (which allow for a multi-aspect presentation of the analysed issue), and, on the other side, the tools which possess complex functionality that allows, among others, to obtain an optimal solution and task sequencing with the chosen priority methods. Unfortunately, despite the fact that there are many systems supporting production planning and scheduling tasks on the market, their value is limited from the point of view of the possibilities of using them in an educational process. That is why, the education at universities or training courses on the subject often takes place in a theoretical way or with a limited use of one or a few commercial systems dedicated to industrial implementation.

In response to the existing needs, within the realized research, a universal and simple IT tool that allows to present a general idea of schedules design, to compare different scheduling algorithms and to find an optimal production schedule with the complete search algorithm was developed. In particular, the designed application allows to sequence tasks with the standard methods such as FCFS, LCFS, SPT and LPT as well as with the complete search method (included in the methods searching for a precise solution), which makes it possible to obtain an optimal solution from the point of view of a minimal sequencing length of all the tasks with the minimal use of resources. In order to evaluate the possibilities and functionality of the designed application of spreadsheet, numerical analyses were conducted by choosing a test set of 10 tasks with 10 operations (stands) each.

The developed solution can be useful in an educational process and it allows to teach how to design manufacturing schedules, what the priority rules are and how schedules can be optimized. In order to make the application universal, the MS Excel spreadsheet was used. Moreover, the integrated optimizer and VBA language were used to perform any necessary calculations and an optimization process.

References

1. Abedinnia, H., Glock, C.H., Schneider, M.D.: Machine scheduling in production: a content analysis. Appl. Math. Model. **50**, 279–299 (2017)
2. Varela, M.L.R., et al.: Collaborative paradigm for single-machine scheduling under just-in-time principles: total holding-tardiness cost problem. Manage Prod. Eng. Rev. **9**(1), 90–103 (2018)
3. Perłowski, R., Antosz, K., Zielecki, W.: Optimization of the medium-term produciton planning in the company – case study. Lecture Notes Electr. Eng. **505**, 369–379 (2018)
4. Proth, J.M.: Scheduling: new trends in industrial environment. Annu. Rev. Control. **31**(1), 157–166 (2007)

5. Sobaszek, Ł, Gola, A., Świć, A.: Predictive scheduling as a part of intelligent job scheduling system. Adv. Intell. Syst. Comput. **637**, 358–367 (2018)
6. Nielsen, I., Dang, Q.V., Nielsen, P., Pawlewski, P.: Scheduling of mobile robots with preemptive tasks. Adv. Intell. Syst. Comput. **290**, 19–27 (2014)
7. Sitek, P., Wikarek, J.: A multi-level approach to ubiquitous modeling and solving constraints in combinatorial optimization problems in production and distribution. Appl. Intell. **48**(5), 1344–1367 (2017). https://doi.org/10.1007/s10489-017-1107-9
8. Sotskov, Y., Shakhlevich, N.V.: NP-hardness of shop-scheduling problems with three jobs. Discret. Appl. Math. **59**, 237–266 (1995)
9. Hermann, J.W.: A history of production scheduling. Int. Ser. Oper. Res. Manage. Sci. **89**, 1–22 (2006)
10. Brzozowska, J., Gola, A.: Computer aided assembly planning using MS Excel Software – a Case Study. Appl. Comput. Sci. **15**(1), 18–35 (2021)
11. Chen, K., Ji, P.: A genetic algorithm for dynamic advanced planning and scheduling (DAPS) with frozen interval. Exp. Syst. Appl. **33**, 1004–1010 (2007)
12. Chen, K., Ji, P.: A mixed integer programming model for advanced planning and scheduling (APS). Eur. J. Oper. Res. **81**, 515–522 (2007)
13. Weaver, P.: A Brief History of Scheduling – Back to the Future. In: myPrimavera Conference, Hyatt, Canberra (2006)
14. Błażewicz, J., Domschke, W., Pesch, E.: The job shop scheduling problem: conventional and new solution techniques. Eur. J. Oper. Res. **33**, 1–33 (1996)
15. Burduk, A., Musial, K., Kochanska, J.: Tabu search and genetic algorithm for production process scheduling problem. LogForum **15**(2), 181–189 (2019)
16. Sobaszek, Ł., Gola, A., Kozłowski, E.: Application of survival function in robust scheduling of production jobs. In: Ganzha, M., Maciaszek, M., Paprzycki, M. (eds.) Proceedings of the 2017 Federated Conference on Computer Science and Information Systems (FEDCSIS), IEEE, New York, pp. 575–578 (2017)
17. Vilcot, G., Billaut, J.: A tabu search and a genetic algorithm for solving a bicriteria general job-shop scheduling problem. Eur. J. Oper. Res. **190**, 398–411 (2008)
18. Gola, A.: Economic aspects of manufacturing systems design. Actual Prob. Econ. **156**(6), 205–212 (2014)
19. Bocewicz, G., Nielsen, I., Banaszak, Z.: Production flows scheduling subject to fuzzy processing time constraints. Int. J. Comput. Integr. Manuf. **19**(10), 1105–1127 (2016)
20. Klimek, M.: Techniques of generating schedules for the problem of financial optimization of multi-stage project. Appl. Comput. Sci. **15**(1), 18–35 (2019)
21. Kaban, A.K., Othman, Z., Rohmah, D.: Comparison of dispatching rules in job-shop scheduling problem using simulation: a case study. Int. J. Simulat. Model. **11**(3), 129–140 (2012)
22. Conway, R.W.: An Experimental Investigation of Priority Assignment in a Job Shop. Santa Monica, California (1964)
23. Wojakowski, P.: Metoda projektowania przepływu produkcji w warunkach zmiennego zapotrzebowania, praca doktorska, Kraków (2012)
24. Pergher, I., Frej, E.A., Roselli, L.R.P., de Almeida, A.T.: Integrating simulation and FITradeof method for scheduling rules selection in job-shop production systems. Int. J. Prod. Econ. **227**, 107669 (2020)
25. Gola, A., Pastuszak, Z., Relich, M., Sobaszek, Ł, Szwarc, E.: Scalability analysis of selected structures of a Reconfigurable manufacturing system taking into account a reduction in machine tools reliability. Eksploatacja i Niezawodnosc **23**(2), 242–252 (2021)

Efficiency Analysis of Hybrid Forecasting Models Supporting Manufacturing Companies in Production Planning, Maintenance and Quality Management

Maria Rosienkiewicz(✉)

Department of Laser Technologies, Automation and Production Organization, Wroclaw University of Science and Technology, Wyb. Wyspiańskiego 27, 50-370 Wrocław, Poland
maria.rosienkiewicz@pwr.edu.pl

Abstract. The paper addresses the problem of forecasting in manufacturing systems. The main aim of the paper is to present new hybrid models combining traditional forecasting techniques based on time series (autoregressive-integrated moving average, simple exponential smoothing, Holt's model, trigonometric exponential smoothing, simple moving average, exponential moving aver-age, weighted moving average, zero-lag exponential moving average, Syntetos-Boylan method) with artificial intelligence-based methods. Results of the literature review and experiences gained from cooperation with industrial companies lead to a conclusion that hybrid models seem to provide more accurate forecasts in comparison to traditional forecasting methods (both parametric and nonparametric). Research methodology presented in the paper includes an assessment of the proposed hybrid models by a comparison with traditional methods and is based on relative forecast error ex post and coefficient of determination. The hybrid models are applied in the three areas of the production system, namely: production planning, maintenance and quality management. Empirical verification of the proposed models is built upon real data from different manufacturing companies. The models aim at obtaining more effective forecasts in comparison to application of individual forecasting methods and at solving a problem of insufficient data or the lack of available data, which is still a common issue in industrial practice. The proposed models can support the process of implementation of Industry 4.0 technologies in manufacturing companies and their transformation into smart factories.

Keywords: Hybrid forecasting · Production planning · Maintenance · Quality management

Nomenclature

AI	Artificial intelligence
ANN	Artificial neural network model
ARIMA	Autoregressive-integrated moving average
BIC	Bayesian information criterion

A. Batako et al. (Eds.): GCMM 2021, LNNS 335, pp. 358–369, 2022.
https://doi.org/10.1007/978-3-030-90532-3_27

ECO	Econometric model
EMA	Exponential moving average
Holt	Holt's model
hybrid_ANN	Hybrid forecasting artificial neural network model
hybrid_ECO	Hybrid forecasting econometric model
hybrid_SVM	Hybrid forecasting regression support vector machine model
I	Relative forecast error *ex* post
R^2	Coefficient of determination
SBA	Syntetos-Boylan method
SES	Simple exponential smoothing
SMA	Simple moving average
SVM	Support vector machine
TES	Trigonometric exponential smoothing
WMA	Weighted moving average
ZLEMA	Zero-lag exponential moving average

1 Introduction

According to Duffuaa et al., the major elements of a production system include maintenance, quality and production [1]. In order to meet customer demand the production has to be properly scheduled, which may result in utilizing machines to their full capacity. In consequence, although the productivity may be increased, simultaneously machine availability can be decreased because of machine failure. What is more, product quality may also be worsened because of potential machine degradation. At the same time another relation can be indicated – "preventive maintenance minimizes the deterioration of equipment (by minimizing wear and breakdowns) but interrupts the production schedule resulting in a delay for customer orders" [1]. The three listed elements of the production system influence each other both directly and indirectly. Each of them should be therefore efficiently managed in order to ensure the fulfilling of customer's needs. The successful management is based on the right decision making process which often is dependent on accurate forecasting. In terms of production planning, "a number of sensitive decisions like capacity planning, order release, lot sizing, and scheduling have to be made carefully" [2]. Manufacturing companies do not know their future demands and therefore are forced to rely on demand forecasts in order to make production planning decisions [3].

Nowadays, in the era of Industry 4.0, the increasing complexity of production planning and control can be observed due to the high dynamics of markets [4]. In order to answer this challenge industrial companies use different types of IT systems as a support in decision-making processes related to accurate production planning. Nevertheless, it has to be underlined that "IT systems often do not provide a reliable forecast of delivery dates, because the planning models are implemented uniquely and have never been adapted due to changes in the production system" [4]. Therefore there is a need to develop new and efficient forecasting algorithms which can be easily implemented in the IT systems used in manufacturing companies.

2 Literature Review on Hybrid Demand Forecasting in Production Systems

Literature review on forecasting leads to a conclusion that two major trends in research papers can be identified – application of artificial intelligence (AI) techniques [5] and increasing number of hybrid forecasting approaches [6].

A detailed and up-to-date review of research on the development of hybrid forecasting methods can be found i.a. in [6–9]. Analysis of these publications shows that although numerous research papers on development of new hybrid models dedicated to forecasting challenges can be observed, the vast majority of them are not applied to issues crucial from the production systems point of view. As mentioned in the Introduction section, in order to ensure efficient management of the production, accurate forecasting methods should be applied to production planning, maintenance and quality control.

When analyzing – from this perspective – different case studies described in the literature it can be noticed that the production-related problems addressed in terms of hybrid forecasting concern mainly:

- In production planning area: product demand and/or sales forecasting [10–13], price forecasting [14, 15], production forecasting [16, 17].
- In maintenance area: long-term machine state forecasting [18], spare parts demand forecasting [19–22], forecasting number of inspections [23],
- In quality control area: quality forecasting [24, 25].

In comparison to other research topics, where hybrid forecasting is applied, it can be stated that relatively little has been done in terms of development of hybrid forecasting dedicated to production planning, maintenance and quality control and there still exists a gap which should be filled in. The hybrid solutions described in the literature in general are composed of: (1) AI-based method and a traditional quantitative forecasting technique; (2) AI-based method, a traditional quantitative forecasting technique and a traditional qualitative forecasting technique; (3) a few traditional quantitative forecasting techniques or (4) data mining technique and other classical mathematical or statistical method. Moreover, an analysis of the proposed hybrid approaches presented in above-listed research papers leads to a conclusion that the majority of them is combining only a few forecasting models – usually two or three. Hence it can be concluded that as – "combining different models can enhance forecasting accuracy compared with individual models used separately" [6] – it is expected that combining larger numbers of forecasting methods should lead to a higher accuracy of obtained forecasts. Bearing in mind this assumption and above-mentioned conclusions the main aim of this paper is to propose new artificial intelligence-based hybrid forecasting models combining a significant number of forecasting methods and to assess their accuracy in comparison to traditional techniques. The accuracy assessment will be based on the real data from three production system case studies addressing production planning, maintenance and quality control.

3 Research Methodology

In order to answer the challenges identified within the literature review and conclusions coming from the projects realized with manufacturing companies, three forecasting hybrid models are prosed: (1) a hybrid forecasting econometric model (hybrid_ECO), (2) a hybrid forecasting artificial neural network model (hybrid_ANN), (3) a hybrid forecasting regression support vector machine model (hybrid_SVM). They aim at obtaining more accurate forecasts than traditional forecasting methods. The models also try to answer the challenge of insufficient data to create a set of potential explanatory variables. The proposed hybrid algorithms combine the existing methods in a way that – according to the in-depth literature analysis – has not been done so far. A scheme of a research methodology is presented in Fig. 1. It is composed of two parts – the first part presents an algorithm for the forecasting accuracy assessment and optimal method selection, which was developed in R language (3.6.3 version). The second part presents schematically production system areas in which the algorithm will be applied.

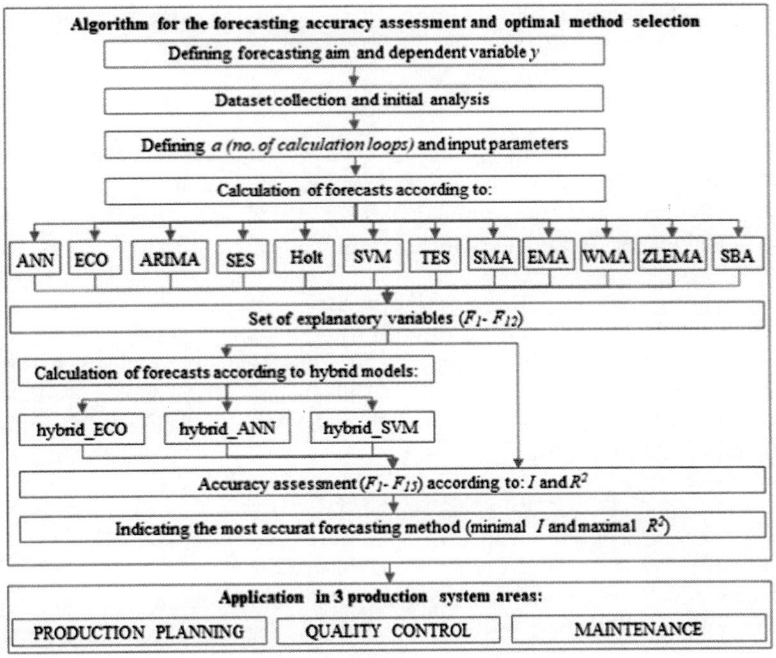

Fig. 1. Research methodology diagram.

According to the algorithm for the forecasting accuracy assessment presented in Fig. 1 at first a prediction aim should be defined and a dependent variable y indicated. Next, the appropriate data should be gathered and initially analyzed. Afterwards the input parameters should be set, as well as the number of computation loops a in the algorithm. The variable a indicates also from how many periods forecasts from average-based prediction methods should be calculated. Subsequently, the forecasts are computed – 12 methods are applied, namely: F_1 artificial neural network model (ANN) [26], F_2 econometric model (ECO), F_3 autoregressive-integrated moving average (ARIMA) [27], F_4 simple exponential smoothing (SES) [28], F_5 Holt's model (Holt) [28], F_6 support vector machine (SVM) [29], F_7 trigonometric exponential smoothing (TES) [30], F_8 simple moving average (SMA) [27], F_9 exponential moving average (EMA) [31], F_{10} weighted moving average (WMA) [27], F_{11} zero-lag exponential moving average (ZLEMA) [32], and F_{12} Syntetos-Boylan method (SBA) [33]. Calculated forecasts from the 12 listed methods are then composing a set of explanatory variables (F_1-F_{12}). This set is an input to the proposed hybrid models (hybrid_ECO, hybrid_ANN and hybrid_SVM). Each of the models is developed to predict the $\hat{y}$, which is a forecasted value of the variable y. Hybrid econometric model is given by the following formula:

$$\hat{y} = \widehat{\alpha}_0 + \widehat{\alpha}_1 F_1 + \widehat{\alpha}_2 F_2 + \cdots + \widehat{\alpha}_m F_m, \tag{1}$$

where: $\widehat{\alpha}_i$ – parameters, F_i – explanatory variable composed of the forecasts. In order to select the right subset of explanatory variables, the Bayesian information criterion (BIC) is applied [34]. Hybrid ANN model is developed on the basis of neurons. The basic model of the neuron can be described by the following expression [26]:

$$\hat{y}_t = f\left(\sum\nolimits_{p=1}^{P} w_p u_p + u_0\right), \tag{2}$$

where u_p, $p = 1,2,\ldots, P$ – inputs (F_1-F_{12} for hybrid model), u_0 - bias, w_p – weights, $f(\cdot)$ – activation function. A mathematical function converts the weighted sum of the signals to form the output of the neuron. In case of the hybrid forecasting support vector machine model for regression the following formula is used [35]:

$$\hat{y} = w^t \varphi(x) + b, \tag{3}$$

where w and b represent the weights and the bias, respectively $\varphi(x)$ is a nanolinear function which transfers x_k into a high-dimensional space, and x represents input variables (in this case $x = F_1, F_2, \ldots, F_{12}$). In hybrid_SVM the radial basis function (RBF) kernel function is applied [29]:

$$K(X_i, X_j) = \exp(-\gamma |X_i - X_j|^2), \tag{4}$$

where $K(X_i, X_j) = \phi(X_i) \cdot \phi(X_j)$, ϕ – transformation.

The main aim of the algorithm is to indicate the best forecasting method for a given forecasting problem. The best in this context means the most accurate. The primary criterion used for the accuracy assessment is the lowest value of the relative forecast error *ex post* I and the secondary criterion is the highest value of the coefficient of determination R^2. They are given by the following formulas:

$$I = \sqrt{\frac{\sum_{t=1}^{m} (y_t - \hat{y}_t)^2}{\sum_{t=1}^{m} y_t^2}}, R^2 = \frac{\sum_{i=1}^{n} (\hat{y}_t - \overline{y})^2}{\sum_{i=1}^{n} (y_t - \overline{y})^2}. \tag{5}$$

The coefficient of determination R^2 value represents the goodness of fit between the observed values and the actual values [36].

According to the presented methodology, the discussed algorithm will be next applied to the three elements of the production system: production planning, quality control and maintenance.

4 Accuracy Analysis of the Hybrid Models

In order to empirically verify and assess the proposed hybrid models according to the methodology presented in the previous chapter, real datasets were applied. The Author's intention was to implement the hybrid models to datasets coming not only from different areas of the production system, but also from different manufacturing sectors, therefore to verify the proposed models 3 different companies were selected and addressed. The idea behind this approach was to check if the developed algorithms are versatile and comprehensive enough to meet forecasting challenges from various industries.

In production planning data from a furniture factory were used (to investigate the case study see [13]), in quality control data from a company manufacturing ceramic insulators were used (to investigate the case study see [37]), and in maintenance data from an underground copper mine were used (to investigate the case study see [38, 39]) – in all three case studies monthly values were taken into account. In the first case study the forecasted value of $\hat{y}$ was monthly demand for a product – calculations were made for 7 products (furniture types: PP1-PP7). In terms of quality control the forecasted $\hat{y}$ was monthly level of defects – 2 cases were considered forested level of defects based on Al_2O_3 type A (QC1) and B (QC2). In the maintenance case study $\hat{y}$ was monthly demand for spare parts or consumable materials, namely air filter element (MN1), bearing grease (MN2), brake fluid (MN3), engine oil (MN4), hydraulic oil (MN5), inner tube (MN6), and tyre (MN7).

For each case and each investigated method/ model relative forecast error *ex post* I and coefficient of determination R^2 were computed. Obtained results are presented in Table 1 and Table 2. The most accurate outcomes are marked in bold – the lowest values of the relative forecast error *ex post* I is presented in Table 1, whereas the highest values of the coefficient of determination R^2 in Table 2.

The computed I and R^2 show that the proposed hybrid models are definitely more accurate than other investigated traditional forecasting methods. For all 16 investigated cases (PP1-PP7, QC1-QC2, MN1-MN2) the most accurate forecasting methods (according to both I and R^2 criteria) were the hybrid models – 10 out of 16 hybrid_ECO and 6 out of 16 hybrid_ANN. In order to analyze the efficiency of all 15 researched methods in the 3 analyzed production system areas, an average I and an average R^2 were computed and presented in Fig. 2 and Fig. 3.

An analysis of the average I and R^2 shows that the obtained accuracy of forecasts from the proposed hybrid models is satisfactory in each of the investigated production system areas. For production planning the most efficient turned out to be hybrid_ECO ($I = 7\%$, $R^2 = 96\%$). The other proposed hybrid models also gave acceptable results – respectively hybrid_ANN: $I = 19\%$, $R^2 = 80\%$, and hybrid_SVM: $I = 23\%$, $R^2 = 81\%$. The other traditional forecasting methods were significantly less accurate ($26\% \leq I \leq 61\%$, $24\% \leq R^2 \leq 71\%$). In terms of quality control all three hybrid models were characterized by low errors and good fit – hybrid_ECO: $I = 11\%$, $R^2 = 93\%$,

Table 1. Relative forecast errors *ex post*.

Case	PP1	PP2	PP3	PP4	PP5	PP6	PP7	QC1
Method	a = 5	a = 7	a = 5	a = 2	a = 6	a = 7	a = 7	a = 5
ANN	0,257	0,292	0,216	0,186	0,329	0,286	0,230	0,117
ECO	0,302	0,285	0,301	0,330	0,296	0,373	0,364	0,131
ARIMA	0,400	0,470	0,399	0,369	0,423	0,505	0,392	0,302
SES	0,546	0,446	0,445	0,396	0,458	0,547	0,490	0,317
Holt	0,464	0,467	0,445	0,378	0,428	0,512	0,438	0,311
SVM	0,299	0,270	0,262	0,287	0,350	0,450	0,294	0,207
TES	0,357	0,447	0,437	0,345	0,351	0,505	0,314	0,275
SMA	0,437	0,709	0,466	0,406	0,509	0,568	0,377	0,340
EMA	0,434	0,656	0,436	0,385	0,475	0,531	0,404	0,321
WMA	0,435	0,583	0,440	0,400	0,455	0,517	0,382	0,322
ZLEMA	0,506	0,445	0,443	0,529	0,418	0,492	0,443	0,327
SBA	0,658	1,021	0,489	0,466	0,599	0,623	0,421	0,428
hybrid_ECO	**0,102**	**0,043**	0,098	**0,100**	**0,091**	**0,030**	**0,055**	0,096
hybrid_ANN	0,267	0,159	**0,087**	0,176	0,258	0,145	0,208	**0,089**
hybrid_SVM	0,230	0,243	0,151	0,189	0,305	0,293	0,188	0,142
Case	**MN1**	**MN2**	**MN3**	**MN4**	**MN5**	**MN6**	**MN7**	**QC2**
Method	**a = 5**	**a = 3**	**a = 7**	**a = 7**	**a = 7**	**a = 5**	**a = 5**	**a = 7**
ANN	0,383	0,135	0,142	0,339	0,170	0,368	0,347	0,161
ECO	0,474	0,409	0,159	0,364	0,272	0,532	0,401	0,195
ARIMA	0,484	0,386	0,180	0,408	0,289	0,567	0,483	0,302
SES	0,537	0,516	0,170	0,388	0,284	0,651	0,572	0,317
Holt	0,531	0,523	0,174	0,410	0,293	0,596	0,522	0,310
SVM	0,415	0,360	0,130	0,323	0,263	0,522	0,406	0,272
TES	0,500	0,386	0,159	0,381	0,283	0,529	0,417	0,275
SMA	0,542	0,564	0,220	0,496	0,334	0,566	0,480	0,347
EMA	0,501	0,505	0,219	0,444	0,314	0,568	0,489	0,332
WMA	0,523	0,541	0,188	0,444	0,309	0,567	0,490	0,331
ZLEMA	0,598	0,695	0,185	0,416	0,295	0,653	0,590	0,326
SBA	0,503	0,504	0,172	0,519	0,493	1,216	0,603	0,430
hybrid_ECO	0,341	**0,113**	**0,066**	0,159	**0,101**	0,354	0,202	**0,131**
hybrid_ANN	**0,214**	0,121	0,128	**0,058**	0,144	**0,259**	**0,190**	0,174
hybrid_SVM	0,342	0,201	0,118	0,319	0,176	0,315	0,257	0,187

Table 2. Coefficients of determination.

Case	PP1	PP2	PP3	PP4	PP5	PP6	PP7	QC1
Method	a = 5	a = 7	a = 5	a = 2	a = 6	a = 7	a = 7	a = 5
ANN	0,579	0,717	0,814	0,770	0,720	0,770	0,604	0,929
ECO	0,400	0,730	0,479	0,211	0,747	0,604	0,128	0,909
ARIMA	0,218	0,661	0,110	0,197	0,522	0,366	0,034	0,518
SES	0,001	0,621	0,119	0,150	0,457	0,312	0,144	0,488
Holt	0,022	0,690	0,118	0,202	0,487	0,359	0,240	0,493
SVM	0,456	0,801	0,606	0,415	0,677	0,462	0,426	0,847
TES	0,200	0,656	0,106	0,276	0,672	0,366	0,197	0,613
SMA	0,084	0,688	0,045	0,122	0,444	0,286	0,000	0,463
EMA	0,071	0,674	0,075	0,173	0,513	0,353	0,020	0,529
WMA	0,074	0,703	0,086	0,135	0,508	0,361	0,043	0,503
ZLEMA	0,002	0,679	0,185	0,063	0,567	0,426	0,117	0,458
SBA	0,157	0,522	0,014	0,169	0,512	0,344	0,001	0,553
hybrid_ECO	**0,931**	**0,994**	0,942	**0,927**	**0,976**	**0,997**	**0,973**	0,951
hybrid_ANN	0,591	0,918	**0,956**	0,774	0,832	0,940	0,624	**0,958**
hybrid_SVM	0,692	0,858	0,886	0,768	0,783	0,835	0,863	0,926
Case	MN1	MN2	MN3	MN4	MN5	MN6	MN7	QC2
Method	a = 5	a = 3	a = 7	a = 7	a = 7	a = 5	a = 5	a = 7
ANN	0,455	0,930	0,457	0,436	0,892	0,731	0,647	0,860
ECO	0,143	0,315	0,306	0,348	0,703	0,420	0,443	0,790
ARIMA	0,148	0,387	0,297	0,344	0,687	0,420	0,265	0,500
SES	0,158	0,143	0,317	0,358	0,694	0,295	0,129	0,473
Holt	0,127	0,086	0,351	0,318	0,690	0,351	0,173	0,477
SVM	0,404	0,573	0,547	0,508	0,723	0,492	0,488	0,681
TES	0,099	0,392	0,322	0,359	0,693	0,447	0,396	0,597
SMA	0,055	0,013	0,149	0,073	0,592	0,420	0,264	0,457
EMA	0,120	0,085	0,200	0,158	0,634	0,413	0,249	0,513
WMA	0,099	0,050	0,220	0,171	0,639	0,405	0,228	0,482
ZLEMA	0,119	0,080	0,381	0,351	0,694	0,288	0,124	0,436
SBA	0,033	0,000	0,287	0,036	0,499	0,449	0,346	0,534
hybrid_ECO	0,556	**0,948**	**0,880**	0,875	**0,959**	0,731	0,856	**0,905**
hybrid_ANN	**0,836**	0,941	0,605	**0,985**	0,917	**0,864**	**0,873**	0,864
hybrid_SVM	0,617	0,906	0,635	0,528	0,879	0,796	0,783	0,859

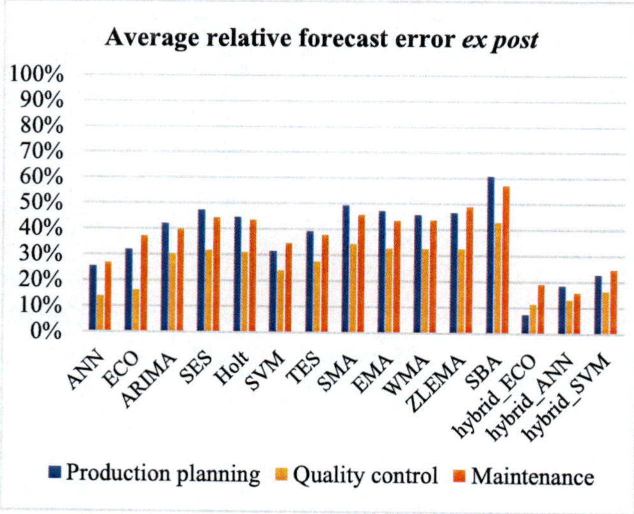

Fig. 2. Average relative forecast error *ex post*.

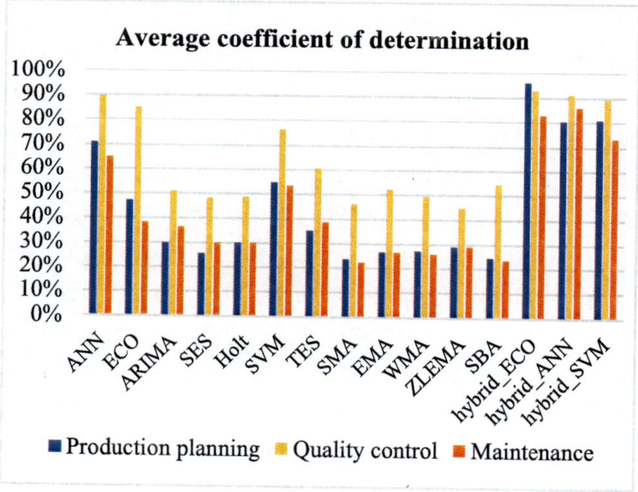

Fig. 3. Average coefficient of determination.

hybrid_ANN: $I = 13\%, R^2 = 91\%$ and hybrid_SVM: $\mathrm{I} = 16\%, R^2 = 89\%$. Also ANN and ECO proved to be satisfactory. In maintenance case, the most accurate turned out to be hybrid_ANN ($I = 16\%, R^2 = 86\%$), whereas hybrid_ECO ($I = 19\%, R^2 = 83\%$) and hybrid_SVM ($I = 25\%, R^2 = 73\%$) were slightly less efficient. Moreover, it can be noticed that all the other investigated traditional forecasting methods were characterized by rather high errors and low fit in case of maintenance ($27\% \leq I \leq 57\%, 22\% \leq R^2 \leq 65\%$) and quality control (results apart from ANN and ECO: $24\% \leq I \leq 43\%, 45\% \leq R^2 2 \leq 76\%$).

5 Conclusions

The main aim of this paper was to propose new hybrid forecasting models and to assess their accuracy in comparison to traditional techniques – the obtained results turned out to be satisfactory as the new models proved to be significantly more accurate than the other 12 investigated methods (ANN, ECO, ARIMA, SES, Holt, SVM, TES, SMA, EMA, WMA, ZLEMA and SBA). The empirical verification based on real data coming from the 3 production system areas – production planning, maintenance and quality control – showed that these hybrid models can serve as efficient forecasting tools for manufacturing companies. What is more, the proposed models help solve the common problem in industrial practice, which is related to insufficient data or the lack of available data. The introduced algorithm for the forecasting accuracy assessment and optimal method selection applied to the IT systems of manufacturing companies can support their process of implementation of Industry 4.0 technologies and thus their transformation into smart factories.

In the future research it is planned to further verify and apply the proposed hybrid forecasting models in different manufacturing sectors. What is more, as the proposed algorithm for the forecasting accuracy assessment is in general versatile, it can be applied for any forecasting problem related to the manufacturing process of a product at any production stage. The application area of the algorithm is not limited to the investigated in this paper production planning, maintenance and quality control, but can be implemented also to other prediction problems occurring in a manufacturing process e.g. in household appliance, automotive, aerospace or railway industries.

References

1. Duffuaa, S., Kolus, A., Al-Turki, U., El-Khalifa, A.: An integrated model of production scheduling, maintenance and quality for a single machine. Comput. Ind. Eng. **142**, 106239 (2020). https://doi.org/10.1016/j.cie.2019.106239
2. Gansterer, M.: Aggregate planning and forecasting in make-to-order production systems. Int. J. Prod. Econ. **170**, 521–528 (2015). https://doi.org/10.1016/j.ijpe.2015.06.001
3. Xie, J., Lee, T.S., Zhao, X.: Impact of forecasting error on the performance of capacitated multi-item production systems. Comput. Ind. Eng. **46**, 205–219 (2004). https://doi.org/10.1016/j.cie.2003.12.020
4. Schuh, G., Potente, T., Hauptvogel, A.: Methodology for the evaluation of forecast reliability of production planning systems. Proc. CIRP. **17**, 469–474 (2014). https://doi.org/10.1016/j.procir.2014.01.069
5. Hall, O.P.: Artificial Intelligence Techniques Enhance Business Forecasts. Computer-based analysis increases accuracy, Graziadio Business Review 5 (2002)
6. Hajirahimi, Z., Khashei, M.: Hybrid structures in time series modelling and forecasting: a review. Eng. Appl. Artif. Intell. **86**, 83–106 (2019). https://doi.org/10.1016/j.engappai.2019.08.018
7. Tascikaraoglu, A., Uzunoglu, M.: A review of combined approaches for prediction of short-term wind speed and power. Renew. Sustain. Energy Rev. **34**, 243–254 (2014). https://doi.org/10.1016/j.rser.2014.03.033
8. Deb, C., Zhang, F., Yang, J., Lee, S.E., Shah, K.W.: A review on time series forecasting techniques for building energy consumption. Renew. Sustain. Energy Rev. **74**, 902–924 (2017). https://doi.org/10.1016/j.rser.2017.02.085

9. Pradeepkumar, D., Ravi, V.: Soft computing hybrids for FOREX rate prediction: a comprehensive review. Comput. Oper. Res. **99**, 262–284 (2018). https://doi.org/10.1016/j.cor.2018.05.020
10. Więcek, P.: A hybrid demand forecasting method in Logistics [in Polish]. Logistyka. **4**, 9 (2015)
11. Grzeszczyk, T.A.: Integrated forecasting method in enterprise management [in Polish]. Zastosowania metod statystycznych w badaniach naukowych **II**, 8 (2003)
12. Jurczyk, K., Kutyba, A.: Forecasting sales volume with the use of artificial neural networks on the example of a flower industry company [in Polish]. Logistyka 323–333 (2015)
13. Rosienkiewicz, M., Kowalski, A., Helman, J., Zbieć, M.: Development of lean hybrid furniture production control system based on glenday sieve. Artif. Neural Netw. Simulat. Model. Drvna Industrija. **69**, 163–173 (2018). https://doi.org/10.5552/drind.2018.1747
14. Zhang, X., Wang, J., Gao, Y.: A hybrid short-term electricity price forecasting framework: cuckoo search-based feature selection with singular spectrum analysis and SVM. Energy Econ. **81**, 899–913 (2019). https://doi.org/10.1016/j.eneco.2019.05.026
15. Areekul, P., Senjyu, T., Toyama, H., Yona, A.: Notice of violation of IEEE publication principles - a hybrid arima and neural network model for short-term price forecasting in deregulated market. IEEE Trans. Power Syst. **25**, 524–530 (2010). https://doi.org/10.1109/TPWRS.2009.2036488
16. Yang, L., Li, B.: The Combination forecasting model of grain production based on stepwise regression method and RBF neural network. Adv. J. Food Sci. Technol. **7**(11), 891–895 (2015). https://doi.org/10.19026/ajfst.7.2528
17. Rathod, S., Mishra, G.C., Singh, K.N.: Hybrid time series models for forecasting banana production in Karnataka state, India. J. Indian Soc. Agri. Stat. **71**(3), 193–200 (2017)
18. Pham, H.T., Tran, V.T., Yang, B.-S.: A hybrid of nonlinear autoregressive model with exogenous input and autoregressive moving average model for long-term machine state forecasting. Exp. Syst. Appl. **37**, 3310–3317 (2010). https://doi.org/10.1016/j.eswa.2009.10.020
19. Amin-Naseri, M.R., Tabar, B.R.: Neural network approach to lumpy demand forecasting for spare parts in process industries. In: 2008 International Conference on Computer and Communication Engineering, IEEE, Kuala Lumpur, Malaysia, pp. 1378–1382 (2008). https://doi.org/10.1109/ICCCE.2008.4580831
20. Hua, Z.S., Zhang, B., Yang, J., Tan, D.S.: A new approach of forecasting intermittent demand for spare parts inventories in the process industries. J. Oper. Res. Soc. **58**, 52–61 (2007). https://doi.org/10.1057/palgrave.jors.2602119
21. Hua, Z., Zhang, B.: A hybrid support vector machines and logistic regression approach for forecasting intermittent demand of spare parts. Appl. Math. Comput. **181**, 1035–1048 (2006). https://doi.org/10.1016/j.amc.2006.01.064
22. Bounou, O., El Barkany, A., El Biyaali, A.: Parametric approaches for spare parts demand, international journal of supply chain. Management **7**, 432–439 (2018)
23. Ruiz-Aguilar, J.J., Turias, I.J., Jiménez-Come, M.J.: Hybrid approaches based on SARIMA and artificial neural networks for inspection time series forecasting. Transp. Res. Part E: Logistic. Transp. Rev. **67**, 1–13 (2014). https://doi.org/10.1016/j.tre.2014.03.009
24. Ömer, F.D.: A hybrid neural network and ARIMA model for water quality time series prediction. Eng. Appl. Artif. Intell. **23**, 586–594 (2010). https://doi.org/10.1016/j.engappai.2009.09.015
25. Yin, S., Liu, L., Hou, J.: A multivariate statistical combination forecasting method for product quality evaluation. Inf. Sci. **355–356**, 229–236 (2016). https://doi.org/10.1016/j.ins.2016.03.035
26. Korbicz, J., Kościelny, J.M., Kowalczuk, Z., Cholewa, W.: Process diagnostics. Models. Artificial intelligence methods. Usage [in Polish]. Wydawnictwa Naukowo-Techniczn (2002)

27. Hyndman, R.J., Athanasopoulos, G.: Forecasting: principles and practice ; [a comprehensive indtroduction to the latest forecasting methods using R ; learn to improve your forecast accuracy using dozenss of real data examples (2018)
28. Gruszczyński, M., Kuszewski, T., Podgórska, M.: Econometrics and operations research [in Polish]. Wydawnictwo Naukowe PWN, Warszawa (2009)
29. http://documentation.statsoft.com/STATISTICAHelp.aspx?path=Machin
30. eLearning/MachineLearning/Overviews/SupportVectorMachinesIntroductoryOverview. Accessed 9 Mar .2020
31. De Livera, A.M., Hyndman, R.J., Snyder, R.D.: Forecasting time series with complex seasonal patterns using exponential smoothing. J. Am. Stat. Assoc. **106**, 1513–1527 (2011). https://doi.org/10.1198/jasa.2011.tm09771
32. Praekhaow P., Determination of Trading Points using the Moving Average Methods. In: Bangkok, p. 6 (2010)
33. http://www.mesasoftware.com/papers/ZeroLag.pdf. Accessed 10 Feb 2020
34. Croston, J.D.: Forecasting and stock control for intermittent demands. J. Oper. Res. Soc. **23**, 289–303 (1972). https://doi.org/10.1057/jors.1972.50
35. Schwarz, G.: Estimating the dimension of a model. Ann. Stat. **6**, 461–464 (1978). https://doi.org/10.1214/aos/1176344136
36. Essa, F.A., Abd, E.M., Elsheikh, A.H.: An enhanced productivity prediction model of active solar still using artificial neural network and Harris Hawks optimizer. Appl. Therm. Eng. **170**, 115020 (2020). https://doi.org/10.1016/j.applthermaleng.2020.115020
37. Zhang, F., Murphy, M.D., Shalloo, L., Ruelle, E., Upton, J.: An automatic model configuration and optimization system for milk production forecasting. Comput. Electron. Agric. **128**, 100–111 (2016). https://doi.org/10.1016/j.compag.2016.08.016
38. Kowalski, A., Rosienkiewicz, M.: ANN-based hybrid algorithm supporting composition control of casting slip in manufacture of ceramic insulators. In: Graña, M., López-Guede, J.M., Etxaniz, O., Herrero, Á., Quintián, H., Corchado, E. (eds.) SOCO/CISIS/ICEUTE -2016. AISC, vol. 527, pp. 357–365. Springer, Cham (2017). https://doi.org/10.1007/978-3-319-47364-2_34
39. Rosienkiewicz, M., Chlebus, E., Detyna, J.: A hybrid spares demand forecasting method dedicated to mining industry. Appl. Math. Model. **49**, 87–107 (2017). https://doi.org/10.1016/j.apm.2017.04.027
40. Rosienkiewicz, M.: Accuracy assessment of artificial intelligence-based hybrid models for spare parts demand forecasting in mining industry. In: Wilimowska, Z., Borzemski, L., Świątek, J. (eds.) ISAT 2019. AISC, vol. 1052, pp. 176–187. Springer, Cham (2020). https://doi.org/10.1007/978-3-030-30443-0_16

Design and Development of a Dual Flow System for Fluid Delivery in Grinding Application

V. K. Tipparthi[1](✉), D. Novovic[2], X. Chen[1], and A. D. L. Batako[1]

[1] General Engineering Research Institute, Liverpool John Moores University, Byrom Street, Liverpool LA3 3AF, UK
vijaykumar7841@gmail.com

[2] Rolls-Royce Plc, Wilmore Road, Gate 6, PC-32, Derby DE24 9BD, UK

Abstract. The application of fluid delivery in metal cutting process remains crucial matter requiring further research. The growing demand and rising cost of cutting fluid usage and environmental aspect of fluid disposal are a key driver for research into alternative fluid delivery techniques. The minimum quantity lubrication (MQL) has emerged as an alternative means to reduce cutting fluids usage. MQL dispenses mist spray using small amount of oil in a compressed air jet and provides efficient lubrication, directed to the cutting zone. MQL is capable of maintaining low economical and production costs or even reducing temperature, minimizing the high flow rates of conventional flood coolant application.

This research work attempts to design a multi-purpose novel nozzle system for a new approach to fluid delivery in MQL technique. One important factors in effective coolant delivery is nozzle design. An emphasis is made on the design concept and an approach to the development of adequate flow rate using low supply pressure to improve nozzle exit performance. The results obtained demonstrate that the method developed achieved better results and can deliver a comparable performance to flood delivery under the conditions investigated.

Keywords: Nozzle design · MQL · Fluid delivery · Coherent length

1 Introduction

The coolant has important roles in grinding processes such as cooling, lubrication, chip transportation, grinding wheel cleaning and corrosion reduction, as well as reducing the heat transferred to the workpiece and tool. One of the most important and often overlooked factors in effective coolant delivery is nozzle design. An emphasis is made on the design concept and an approach to the development of adequate flow rate using low supply pressure to improve nozzle exit performance.

The application of a nozzle is a very broad topic in machining technology and it is used to achieve the best possible cooling and lubrication effect on the tool and workpiece. A coherent jet type nozzle developed by Webster [1] helped to reduce the air barrier and to achieve a coolant jet velocity matching with grinding wheel speed. Poor nozzle design and plumbing problems lead to losses in the pipes and dispersed jet. Changes in pipe

A. Batako et al. (Eds.): GCMM 2021, LNNS 335, pp. 370–380, 2022.
https://doi.org/10.1007/978-3-030-90532-3_28

diameter, elbow and edges also cause turbulence preventing the full development of the fluid to a coherent jet.

A concave internal profile of the jet nozzle provides a better and longer coherent jet reaching the grinding contact zone from a relative distance. Coherency is better achieved with large contraction ratios and a smooth internal surface [1]. Jet coherency can be improved by incorporating flow conditioning plates that straighten the flow before or within the nozzle [2].

The cooling effects of cutting fluids can be maximized by good nozzle design. The lengths of pipe within a flow system must be adequately selected to attenuate the effect of elbows and transitions along the fluid path. The shape and size of the aperture affects the profile of the jet and its coherency. The following guidelines must be observed when designing an efficient nozzle for grinding application. Matching the speed of the fluid jet with the wheel peripheral speed is desired but recent finding suggested that this is not always the best. An optimum jet speed appears to be in the region of 0.6 ~ 0.8 ($V_j = 0.6\ V_s$) of wheel speed, [3, 4]. The work of [5, 6] and [7, 8] on nozzle position noticed that the coolant delivery via a tangential jet, tend to increase the side leakage. Raising the nozzle slightly (12 mm from the horizontal), reduces the side leakage. A key parameter for establishing the efficiency of fluid delivery in the contact region is useful flow, which is the flow that passes through the grinding contact zone [9–13], along with the depth of fluid penetration into the grinding wheel and flow rate through the grinding contact zone [14, 15]. Gviniashivili, [16] used wheel speed, jet velocity and power required for fluid acceleration to determine a suitable nozzle outlet required to achieve a required fluid film thickness in the grinding zone. Babic, [17] used an air jet injection system to create a mist of air and water as a coolant fluid, thus removing ecological hazards. Barczak, [18] Determined the process performance and effectiveness of the MQL method on fine cut plane surface grinding.

2 Investigation into Internal and External Fluid Flow Profiles

In general, plastic "click-fit" nozzles may be adequate for tool room common application. Also commercial MQL systems are equipped with standard spot nozzles which are typical not adequately configured for MQL application to grinding process. For high volume production, these nozzles are inadequate as they are turbulent thus, the jet spray is not contained nor controlled. In addition, these nozzles cannot be held in correct position for long period which prevents uniform fluid delivery. A research has been made on several existing models on internal & external fluid flow behavior. A new design is proposed to increase the pressure inside the nozzle chamber with a decreasing area of cross section which in turn increases the pressure prior exit and this allows increasing in exit velocity. Several design optimizations and modifications has been undertaken in order to select the best performing profile.

2.1 Internal Nozzle Exit Cavity Fluid Pattern

In grinding, the wheel at higher speeds entrains a thick boundary layer of air. The relatively low pressure fluid delivered via common nozzles is unable to penetrate the developed layer and the fluid delivery is highly inefficient. By considering all these issues, the nozzle internal profile and exit cavity is designed in such a way that nozzle deliver fluid in an air blade type jet to supply at high velocity speeds that penetrates and break through the boundary layer.

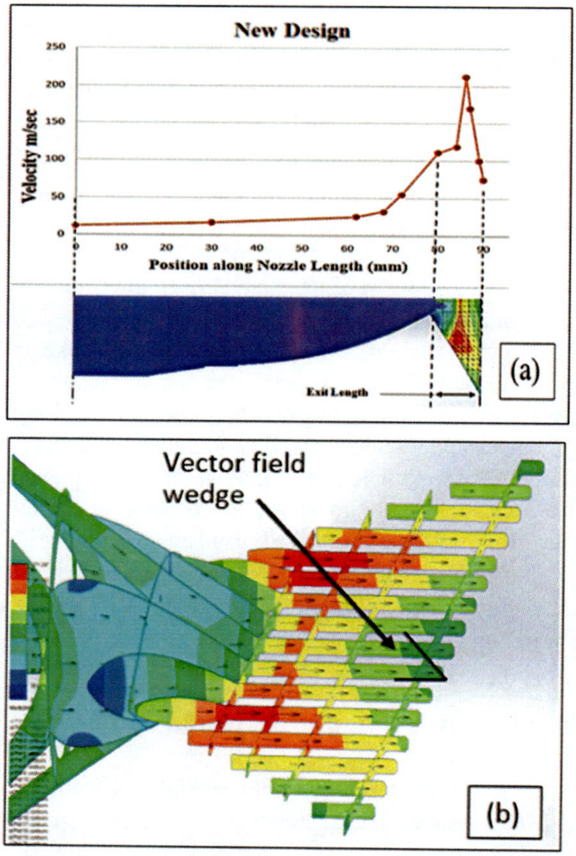

Fig. 1. (a) Velocity profile drawn centre line of new design nozzle with Dovetail exit profile; (b) Vector field in cross section of Dovetail profile.

In Fig. 1 (a), new design, the fluid velocities are explored throughout the profile this is acheieved by sectioning the entire nozzle using longitudinal & horizontal cut planes and recording the velocities along the nozzle internal cavity. Using a cut plane passing through the centre line of the nozzle 90 mm length, a graph of the velocities along the nozzle is obatined. It shows that peak velocoity of 220 m/s is reached in the area of cross firing. This graph reveals that by optimising the length of the exit cavity and its shape it is possible to get exit velocity close to 180 m/s or higher.

Figure 1 (b) depicts a 3D velocity profile of the flow in the nozzle throat and the exit cavity where the variations of velocities can be observed and a field of wedge with it acute angle located at the bottom area. However, due to fluid attachment to nozzle walls, this vector field close to the walls regain the shape of classical velocity profile of a fluid flowing in a duct.

2.2 External Nozzle Exit Jet Velocity Profiles

The simulations using ANSYS CFX have been an indispensable tool that helped visualizing internal and external fluid flow patterns of this new design concept of dual flow. The below Fig. 2 (a) presents the external fluid flow behavior pattern of the new design nozzle drawn using with 5-mm interval. It shows that jet keeps a stable average velocity of around 120 m/s across the nozzle at a distance of in between 50 to 75 mm. Here the line denoted by "0 mm" is the velocity at the immediate nozzle exit, where higher jet velocities present a turbulent profile due to the effect of exit edge shape and burst into atmospheric pressure. However, at 50 mm away from the exit the jet displays a profile similar to laminar flow. Therefore, it is preferred to position the nozzle a bit back about 50 mm for improved performance or very close to the wheel for cleaning purpose.

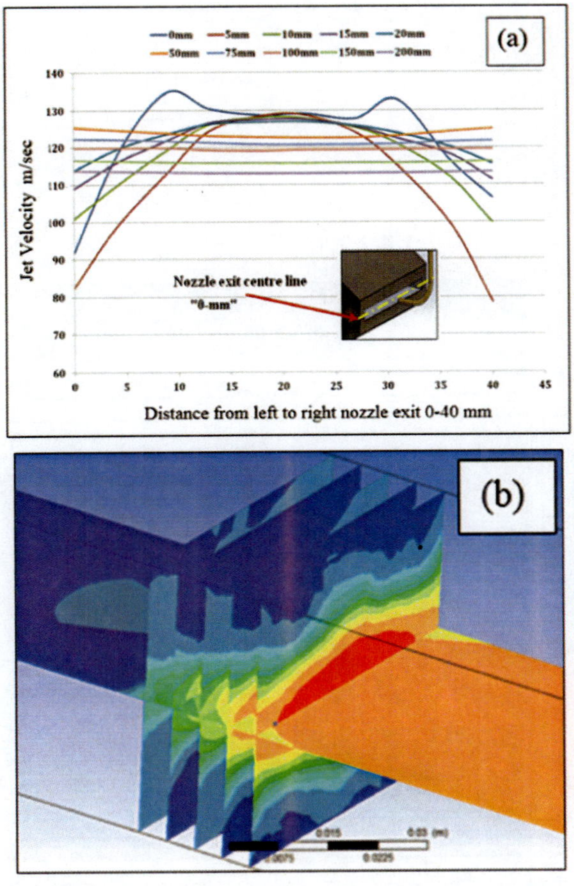

Fig. 2. (a) External velocity profiles drawn on new design nozzle with 5-mm interval; (b) Velocity cut plots with 5-mm interval.

Figure 2 (b) displays a close view of the vertical cut plots drawn to investigate the jet velocities. By using 5 mm interval the jet velocity profile across nozzle aperture at a distance of 5 to 200 mm with various intervals, with a flow rate of 50 L/min were captured. It shows that jet keeps a stable average velocity of around 120 m/s across the nozzle at a distance of in between 50 to 75 mm.

2.3 Fluid Flow in Conventional Nozzle and New Design (Minimized Flowrate)

Using 5 L/min flowrate on both conventional and new design Fig. 3 (a) depicts the velocity in a plane (velocity line) passing through the center line of the conventional nozzle along the nozzle length. This plot reveals that a low uniform flow followed up to the distance of 0.05 m and after crossing this region the fluid start accelerating up to the throat; here the fluid acceleration is further increased due to the structure of throat and exit cavity of the nozzle reached with a peak velocity of 21.7 m/s.

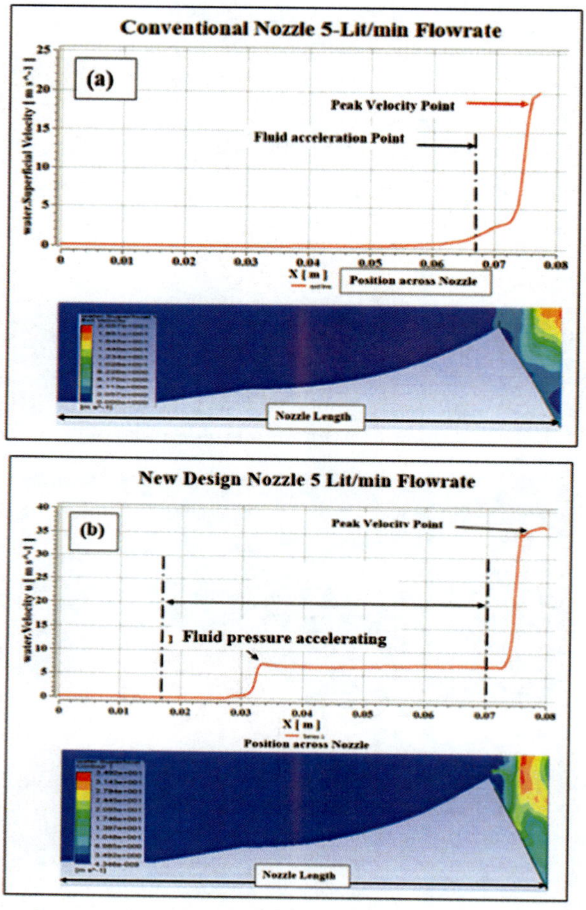

Fig. 3. (a) Velocity profile drawn across center of the conventional nozzle; (b) Velocity profile drawn on new design nozzle.

Figure 3 (b) illustrates the velocity plot across nozzle length and gives a good insight of influence of the New Design. The inlet fluid flow reached uniformly at a distance of 0.03 m where, the velocity increased due to the new design and the flow uniformly increased up to the outlet tip of the throat is indicated by arrow. After the immediate exit of the nozzle throat, the fluid velocity increased in exit cavity with jet velocity of 35 m/s. Here, the nozzle exit cavity is shortened by 10-mm to build up fully flow.

3 Nozzle Performance Testing

Experiments were carried out on a converted Jakobson surface grinding machine. The traversed Pitot tube measurement system was fitted and enclosed in an inspection chamber. The system was attached to the manually operated machines. Since the traverse Pitot tube system can move in both y and z directions, the only other x-axis movement was achieved using a sliding system measured by varying the distance of the Pitot tube from the nozzle aperture. This meant that the effective jet thickness could be measured at any distance from the nozzle exit ($x = 0$ mm) up to a physically operated sliding system allowed maximum of $x = 1$ m distance.

The external flow measurements of the nozzle is conducted by using LabVIEW tool and test rig equipment. To obtain an accurate data with pressure transducer there are a number of step by step process is required prior to collecting the nozzle exit flow data. This system needs consideration of the appropriate voltage and pressure range, zeroing of the transducer i.e., air in sensor and calibration of the entire system. Experimental work took place using nozzle with minimized different flowrates of 5 L/min & 10 L/min respectively and MQL using 100 ml/h at 6-bar air pressure, and the best performed conventional nozzle and new design were presented.

A Pitot tube system was designed and implanted to investigate the jet coherency for flows exiting of a nozzles. The traverse motion of the probe allows capturing the jet stream flow from 0 to 100 mm or up to its coherence length. The Pitot tube is positioned accurately in line to the nozzle aperture. This method is considered as an effective and economical means of measuring jet velocity and coherence length of the jet stream. Using the Pitot tube measurement system, the jet velocity profiles and coherence length was determined from actual pressure readings. The coherence length is vital to position the nozzle to the grinding contact zone. It is accepted that the fluid jet should cover the entire grinding contact zone. Any misdirected of the nozzle position and distance coverage during grinding will not be cooled effectively. The true effectiveness of jet is performed only when it is close to contact zone. For this reason, most attention was focused on the close range spectrum at values for z approaching zero. To build a complete 3D surface plots of the jet stream however, the measurements for x ranged from zero up to 100 mm from the nozzle orifice. This confirms that the total thickness and width of the jet should cover the workpiece area and contact zone.

3.1 Fluid Jet Velocity Profiles Conventional Nozzle and New Design Nozzle

The actual flow measurements are very complicate to perform and not all situations possible to measure. 'CFD' computational fluid dynamics modelling results are validated here with experimental results. The nozzle tests were conducted using the flowrates of 5 & 10 L/min using coolant (Phenol) captured on nozzle aperture middle layer, Fig. 4 (a) represents comparison between conventional & new design nozzle, the jet velocity profiles using 5 L/min achieved 21.1 m/s has captured after immediate exit of the nozzle, however, with the same flowrate used in new design the jet exit velocity achieved was 25.75 m/s, the difference was 4.45 m/s which is an increase of 22.3%. As in the case of 10-L/min for conventional nozzle, the maximum jet exit velocity achieved 25.1 m/s. using the same flowrate with new design, output jet velocity jumps to 35.3 m/s, shown in above Fig. 4 (b); this provides a step up of 10.2 m/s which is an increase of 40.6%. This plot reveals that the effect of new design performance on both the nozzles.

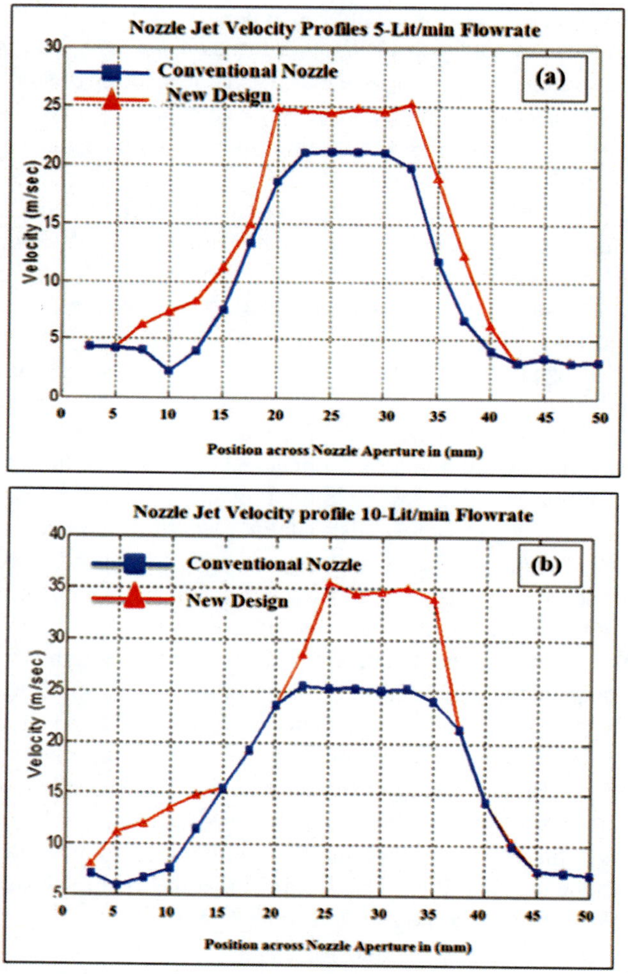

Fig. 4. (a) Jet velocity profiles for 5-L/min (b) Jet velocity profiles for 10-L/min.

3.2 Nozzle Supply Input Flowrate

To give a visual representation of the jet stream for comparison with simulation, the nozzle jet stream is schematically developed by a multi-layered 3-D surface plots represented by top, middle and bottom layers using MATLAB® showing the results from experimentation with a surface plot connecting the results to build up the multi-layered three-dimensional profile. The surface plots shown area direct comparison of the velocity at a point across the fluid stream. Each of the curves represents one measurement plane downstream of the nozzle aperture. The velocity profile of the conventional nozzle of supply flowrate of 10-L/min. Here, using phenol as a coolant, the velocity profile appeared completely different compared to the nozzle with air as a fluid. Due to the mass of the coolant, the jet profile formed like a diverged shape and the middle layer registered peak velocities at the centre region of the nozzle aperture (Table 1) (Fig. 5).

Table 1. .

Input flowrate (L/min)	Conventional nozzle exit velocity (m/sec)	New design nozzle exit velocity (m/sec)	Percentage increase (%)
Simulation results			
5 – L/min	16	19.5	21.8%
10 – L/min	21.7	35.5	63.6%
Experimentation Results			
5 – L/min	21.1	25.75	22.3%
10 – L/min	25.1	35.3	40.6%

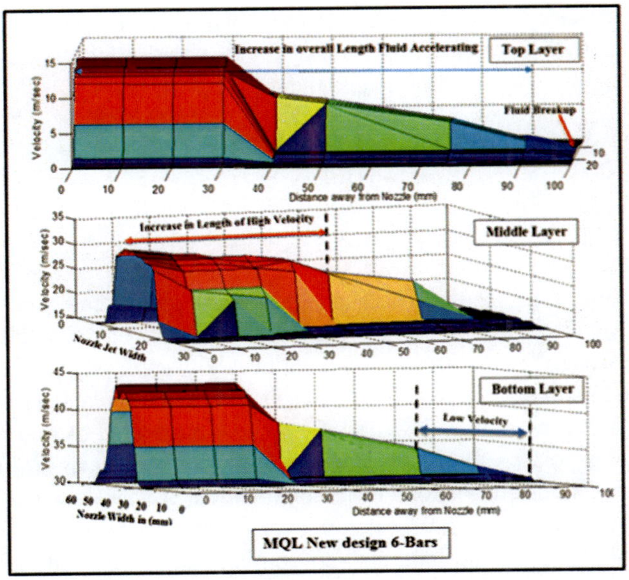

Fig. 5. MQL multi-layered 3-D surface plot.

The experimentation is continued and measured with top, middle and bottom layers to visualize the external actual fluid flow behavior pattern, in this conventional nozzle test the jet gradually increased from top to bottom layer. In the top layer it is seen that the jet is achieved velocity of 12 m/s and the peak jet velocity coherence length covered away from the nozzle is 10-mm, later 10–40 mm the fluid jet gradually decreased has a low speed up to the jet break up, in middle layer recorded as 32 m/s and in bottom layer reached peak velocity of 52 m/s. however, the fluid maintained an average velocity up to a distance of 10–40 mm beyond this the fluid velocity drops gradually down the stream, the maximum distance measured away from the nozzle aperture was 0 to 100 mm.

From the inspection of the new design, a clear trend of the jet thickness increasing with distance is apparent. The velocity profiles relax and become increasingly centered showing the peak velocity and increasing jet width. The surface plot shows after the coherent region general decreasing trend visualizing the effect of jet break up starting point downstream of the nozzle exit. The new design, top layer has velocity of about 15 m/s, the middle layer recorded 35 m/s whereas the bottom layer secures up to 55 m/s. Using new design, the length covered by high jet velocity coherence length is almost doubled (up to 75 mm) away from the nozzle exit compared to conventional nozzle test. Theoretically the jet break up length is approximately 100 mm from the nozzle exit. The triple-layer plots reveal a new finding in the jet behavior. It is observed that the three layers display different peak velocities, this interesting behavior of the jet has never been shown before neither experimentally or theoretically.

One would assume that the middle layer would have the average kinetic energy. However, it has been discovered here that the bottom layer gives the highest jet velocity. This can be explained by the effect of gravitational forces acting on the jet, thus bringing the bulk of the fluid to the bottom part of the jet. This discovery might change the perception of nozzle positioning in machining application. Here with a nozzle aperture of 0.3 mm, it is now shown that one must take in consideration that the lower layer provides high jet velocities compared to any other part of jet. Therefore the nozzle must be positioned in a way that the lower third of the jet hits the cutting zone or the acute angle (wedge) between the workpiece and the grinding wheel.

4 Conclusion

This project brought up an advanced understanding of the fundamental of research and development of a multipurpose nozzle system that could be used for both flood and MQL coolant application. The internal shape, geometry and exit cavity of the nozzle has a strong effect on the jet coherence and peak velocity. The peak velocity is individual to each nozzle design i.e., using Conventional nozzle, New Design nozzle, and MQL application gives a solution to achieve the efficient & effective coolant in grinding contact zone. For the first time a dual flow nested nozzle is presented to the manufacturing communities to exploit the advantage of this multipurpose nozzle as it offers the use of one, two or three fluids either alone or in combination. It was shown as never before that the jet velocity increases from the top layer to the bottom, From these graphs, it is concluded that for optimum fluid delivery to the cutting zone with the expected jet velocity, one should position the nozzle in such a way that the bottom third of the jet is targeted to the grinding contact zone with the upper two third contouring the wheel surface.

This work also provides an advanced understanding of MQL performance and provides evidence that MQL can be effective to cover the wheel width. The most notable result achieved from this present work is the concept of new design method improved jet velocity and it was proven that variations is observed in nozzle external jet velocity this was shown by 3-D multi-layer surface plots experimentally.

Acknowledgements. This work is partly supported by Rolls-Royce Plc and Liverpool John Moores Univeristy.

References

1. Webster, J.A., Cui, C., Mindek, Jr., R.B.: Grinding Fluid Application System Design. University of Connecticut, USA, Centre for grinding research and development (1995)
2. Brinksmeier, E., Heinzel, C., Wittmann, M.: Friction, Cooling and Lubrication in grinding, Division of Production Technology and Manufacturing Technologies. In: Bremen University, Germany (1999)
3. Batako, A.D., Rowe, W.B., Morgan, M.N.: Temperature measurement in high efficiency deep grinding. Int. J. Mach. Tools Manuf **45**, 1231–1245 (2005)
4. Jackson, A.: An Investigation of Useful Fluid Flow in Grinding (2008)
5. Ebbrell, S., Woolley, N.H., Tridimas, Y.D., Allanson, D.R., Rowe, E.W.B.: The effects of cutting fluid application methods on the grinding process. Int. J. Mach. Tools Manuf **40**, 209–223 (2000)
6. Cambell, J.D.: "Optimised Coolant Application". In: 1st International machining and Grinding Conference, Michigan, USA, 12–14 September Society of Manufacturing engineers (1995)
7. Cui, C.: "Experimental investigation of thermofluids in the grinding zone". PhD dissertation, university of Connecticut (1995)
8. Bulloni, M., Desilvestri, S., Liu, Y., Magni, V.: "High resolution measurement of jet stream thickness by optical ranging". In: Optics Communications, vol. 66, May 1998, pp.133–143 (1998)
9. Jackson, A., Batako, A.D., Morgan, M.N.: "Fluid Delivery in Grinding-A Review of the Term Useful Flow", Submitted to GARS-2005, LJMU, UK, 22 June (2005)
10. Chang, C.C.: An Analysis of Coolant Flow and heat transfer in Grinding, PhD. Thesis, University of Pittsburgh (1994)
11. Rowe, W.B., Ebbrell, S., Morgan, M.N.: Process requirements for cost effective precision grinding. CIRP Ann. Manuf. Technol. **53**(1), 255–258 (2004)
12. Gviniashvili V.K., Woolley N.H., Rowe W.B.: Useful coolant flow rate in grinding. In: International journal of machine tools manufacture, vol. 44, May 2004, pp 629–636 (2004)
13. Morgan M.N., Jackson A.R, Wu. H., Baines V.A., Batako A., Rowe W.B.: Optimisation of fluid application in grinding, AMTReL, Liverpool John Moores University, UK (2008)
14. Chong-Ching, C.: An application of lubrication theory to predict useful flow rate of coolants on grinding porous media. In: Department of Mechanical Engineering, Kung-Shan University of Technology, Taiwan, Tribology International vol. 30, no.8, pp. 5 (1997)
15. Morgan, M.N., Baines-Jones, V.: "On the coherent length of fluid nozzles in grinding" AMTReL, General Engineering Research Institute. In: Faculty of Technology and Environment, Liverpool John Moores University, Byrom Street, Liverpool, L3 3AF, UK (2009)

16. Gviniashvili, V.K., Woolley, N.H., Rowe, W.B.: Useful Coolant Flow Rate in Grinding. School of Engineering, Liverpool John Moores University, AMTTREL (2003)
17. Babic, D., Murray, D.B., Torrance. A.A.: Mist jet cooling of grinding processes, department of mechanical and manufacturing engineering. Trinity College, Dublin, Ireland (2005)
18. Barczak, Ł.: "Application of Minimum Quantity Lubrication (MQL) in Plane Surface Grinding". Liverpool John Moores University, UK (2010)

Influence of Nano-particles on the Properties of Recycled Plastics

K. Han Sheng[1], Ezutah Udoncy Olugu[1(✉)], Cik Suhana Binti Hassan[1], V. Karunamoothei[2], and Kah Hou Teng[1,3]

[1] Faculty of Engineering, Technology and Built Environment, UCSI University, Kuala Lumpur, Malaysia
olugu@ucsiuniversity.edu.my

[2] Clean Growth UK, Faculty of Engineering and Technology, Liverpool John Moores University, Liverpool L3 3AF, UK

[3] Built Environment and Sustainable Technology Institution (BEST), Faculty of Engineering and Technology, Liverpool John Moores University, Liverpool L3 3AF, UK

Abstract. The disposal of plastic wares has been one of the most difficult challenges in the past few decades. High volume of used plastics have found their way to the oceans which is a threat to aquatic life. Recycling has been identified as one of the techniques to eliminate the incessant disposal of plastics. In this study, the influence of nanoparticles on the properties of recycled plastics was examined. Nine samples were prepared which were classified into three categories such as nano-based HDPE-sugarcane bagasse powder composite, nano-based HDPE composite and HDPE without nanoparticles and sugarcane bagasse powder. Samples of nanoparticles-recycled plastic composites were prepared by adding nanoparticles (iron oxide, copper oxide and aluminum oxide) to the high-density polyethylene. These samples were subjected to various tests, such as fatigue, compressive, and water absorption test. The sample containing 4.55% aluminum nanoparticles and 95.45% HDPE obtained the best performance in fatigue tests and water absorption tests. In compressive test, the sample with copper oxide nanoparticles and sugarcane bagasse powder had the highest compressive force and compressive strength. Among the three types of nanoparticles, aluminum oxide nanoparticles had the best performance. Results indicated that nanoparticles enhanced the properties of recycled high-density polyethylene. The study concluded with some recommendations for further studies.

Keywords: High-density polyethylene · Nanoparticles · Fatigue test · Compressive test · Water absorption test

1 Introduction

Sequel to certain characteristics of plastics such as availability, durability, versatility, lightweight, inexpensive, flexibility, moldability and others, plastics have found application in almost every facet of consumer products such as furniture, cooking utensils, containers, plastic bag, food packaging and many more. Study have found that about

A. Batako et al. (Eds.): GCMM 2021, LNNS 335, pp. 381–389, 2022.
https://doi.org/10.1007/978-3-030-90532-3_29

8.3 billion tonnes of plastic have been produced between 1950 and 2015 of which 5.8 billion tonnes of plastic waste have been generated [1]. It was further stated that only 12% has been incinerated, 9% recycled, and around 60% sent to landfills or to the environment [1, 2]. In fact, it is estimated that more than 100 million tonnes of plastics are manufactured across the globe annually. The plastic wastes account for about 10% of the total global wastes generated annually [3]. The incessant use of plastics has resulted in accumulation of plastic waste which has become an environmental concern. According to the Environmental Protection Agency's (EPA) solid waste management guidelines (1989), most thermoplastic wastes are currently managed by landfills, which is the least popular method of waste management [3]. It is an established fact that these plastic wastes affect the quality of soil and cause environment pollution. In addition, plastic wastes have various negative impact on the earth's ecosystems which includes aquatic environment [4]. Since plastic have a long life before it can be completely degraded, plastic wastes form major component part of marine litter [5]. The garbage contains unwanted plastics material which are indigestible in marine lives.

It is an established fact that recycling of plastics takes 88% less resources compared to the production of new plastics [5, 6]. This is evident in the amount of plastics which have been recycled in the past decade. The amount of plastics which have been recycled in past decade is estimated to be only a fraction, less than 30% of the overall produced plastics [2]. There are various challenges facing recycling of plastics. Some of these challenges include the composition of the plastics, cost of recycling, changes in mechanical properties of the recycled plastics among others [2, 6, 7].

The use of recycled plastics can be extended for various applications from garden tiles and furniture to retaining walls, railway sleepers, pedestrian bridges or even to heavy load bridges [8–11]. Hence, evaluating the properties of the recycled plastics is considered significant in order to determine their applicability and adaptability for the intended applications. Commonly, recycled properties would show deterioration of properties as compared to virgin plastic. The study aims to investigate the usage of nanoparticles and natural fibre as the additive to reinforce the properties of recycled plastics. It is hoped that the property obtained in this research would be of benefits for further research work on its applications.

2 Materials and Experimental Work

2.1 Materials Compositions

The study prepared three main materials which are used high-density polyethylene (HDPE) plastics, nanoparticles, and natural fibre. The nanoparticles used in this study were iron oxide (Fe_3O_4), copper oxide (CuO) and aluminium oxide (Al_2O_3). The nanoparticles were chosen as the additives due to their high surface area which can significantly increase interfacial interactions between the particles and plastics. Furthermore, due to their lightweight properties, the addition of nanoparticles would not incur increment in the light transmission properties of the base plastics [12]. The natural fibre used is sugarcane bagasse powder. Sugarcane bagasse fibre was chosen due to its abundant quantity where its normally just disposed and/or burned. At present, the consumption of bagasse for any other applicable field is still very low, particularly in

Asian countries [13]. Thus, finding useful utilization of the bagasse would not only help to alleviate the disposal problems but also in creating value added materials. Recycled HDPE were prepared from waste beverage bottles which were collected from resource recycle bin. These plastics were cut into small pieces so that it can melt easily. Sugarcane bagasse fibre was collected, washed, soaked with sodium hydroxide (NaOH), left to dried and ground into powder. The chemical treatment with NaOH for the fibre was performed to remove impurities on the fibre, therefore increasing the surface area for bonding between the fibre and the plastic.

Table 1. Composition of the samples

Samples	Material		
	HDPE	Nanoparticles	Sugarcane bagasse
HDPE+Fe_3O_4+sugarcane bagasse powder	10.0 g	0.5 g	0.5 g
HDPE+CuO+sugarcane bagasse powder	10.0 g	0.5 g	0.5 g
HDPE+Al_2O_3+sugarcane bagasse powder	10.0 g	0.5 g	0.5 g
HDPE + Fe_3O_4	10.5 g	0.5 g	-
HDPE + CuO	10.5 g	0.5 g	-
HDPE + Al_2O_3	10.5 g	0.5 g	-
HDPE	11.0 g	-	-

Table 1 shows different combinations of HDPE, nanoparticles, and sugarcane powder by mass. A total of 9 different samples were prepared for the experiments and classified into three categories which were nano-based HDPE-sugarcane bagasse powder composite, nano-based HDPE composite and HDPE without nanoparticles and sugarcane bagasse powder. Samples 1, 2 and 3 were prepared in combination with HDPE, nanoparticles, and sugarcane bagasse powder. Samples 4, 5 and 6 were prepared in combination with HDPE and nanoparticles only. Samples 7, 8 and 9 were prepared with HDPE only. For fatigue testing, the sample was 11 g, and the sample for compression testing and water absorption testing was 22 g. The shape of the samples for the compressive test was cylindrical. The diameter and height of the samples were 2.5 and 5 cm respectively. The percentage composition of each sample by mass is presented in Table 2. The mass of each sample was 11 g while the samples for compression testing and water absorption testing was 22 g.

2.2 Materials Compositions

In order to evaluate the influence of nanoparticles and sugarcane bagasse on the properties of recycled plastics, the fatigue, compressive and water absorption tests were conducted. The fatigue test was performed on a rotary fatigue testing machine at the speed of 30 rpm. Compressive test was performed on automatic compression machine with the rate of 10 kN/s.

The water absorption test was conducted to study the water absorption behaviour of nano-based HDPE composites [2, 14]. The samples were immersed in various containers filled with distilled water at 25 °C. The samples remained immersed for a period of 72 h

Table 2. Percentage composition of the samples by mass

Sample	Combination	%	Mass (g)
1	HDPE	90.90	10.0
	Fe3O4 nanoparticles	4.55	0.5
	Sugarcane bagasse powder	4.55	0.5
2	HDPE	90.90	10.0
	CuO nanoparticles	4.55	0.5
	Sugarcane bagasse powder	4.55	0.5
3	HDPE	90.90	10.0
	Al2O3 nanoparticles	4.55	0.5
	Sugarcane bagasse powder	4.55	0.5
4	HDPE	95.45	10.5
	Fe3O4 nanoparticles	4.55	0.5
5	HDPE	95.45	10.5
	CuO nanoparticles	4.55	0.5
6	HDPE	95.45	10.5
	Al_2O_3 nanoparticles	4.55	0.5
7	HDPE	100	11.0

and readings were taken every 24 h. The mass of the sample before immersion was taken as m_0, and the mass of the sample after immersion was m. The percentage of the mass gain as well as the percentage change in mass relative to the initial mass was calculated using Eq. 1 [15].

$$c = \frac{m - m_0}{m_0} \times 100\% \tag{1}$$

3 Results and Discussion

3.1 Fatigue Test

Table 3 shows the comparison between fatigue test of different composition of samples. Figure 1 presents the graph of fatigue life versus the different composition of the samples. Sample 1, 2 and 3 (nano-based HDPE-sugarcane bagasse powder composite) were tested with 8 kg, sample 4, 5 and 6 (nano-based HDPE composite) were tested with 6 kg and sample 7 (HDPE plastic) were tested with 4 kg. According to the data, the number of cycles of sample 3 (HDPE + aluminium nanoparticles + bagasse powder) was 49, which was the maximum number of cycles under a load of 8 kg. However, the number of cycles for sample 6 (HDPE + aluminium nanoparticles) was 92, which was the most cycled under a load of 6 kg. Under a load of 4 kg, the sample 9 had the least number of cycles.

The composition of samples 7, 8 and 9 were the same, but only the number of cycles of samples 7 and 8 were close, while the number of cycles of samples 9 was lower than both. Among the three types of nanoparticles (iron oxide, copper oxide and aluminium oxide), the sample with aluminium oxide nanoparticles had the highest number of cycles.

Table 3. Comparison between fatigue test of different composition of samples.

Load	Samples	No. of cycles		
		1st	2nd	Average cycles
8 kg	1	15	17	16
	2	10	18	14
	3	48	50	49
6 kg	4	26	36	31
	5	58	68	63
	6	88	96	92
4 kg	7	712	770	741

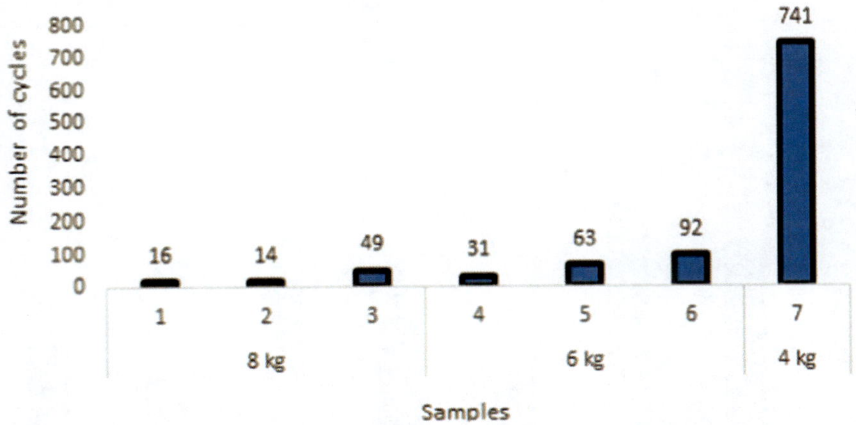

Fig. 1. Comparison between fatigue test of different composition of samples.

3.2 Compressive Test

Table 4 shows the comparison between compressive test of different composition of samples. Figure 2 shows the results of the compressive strength of different sample compositions, while Fig. 3 shows the results of the compressive strength of different sample compositions. According to the data, sample 2 (HDPE + copper oxide nanoparticles + sugarcane bagasse powder) had the highest compressive force and compressive strength. As shown in Table 4, the maximum compressive force and compressive strength of sample 2 were 28.15 kN and 1.59 MPa respectively, which contains 4.55% by mass of copper oxide nanoparticles and bagasse powder. Among the samples containing nanoparticles, the compressive force and compressive strength of sample 6 was the lowest. Of all the three types of nanoparticles, copper oxide-plastic composite had the highest compressive force and compressive strength values. This is followed by iron oxide and aluminium oxide plastic composites. The compressive force and compressive strength of the sample with sugarcane bagasse powder is greater without sugarcane bagasse powder. Therefore, it is obvious that samples containing bagasse powder can withstand greater compression and strength.

Table 4. Comparison between compressive test of different composition of samples.

Samples	Force (kN)	Compressive strength (MPa)
1	23.53	1.33
2	28.15	1.59
3	9.53	0.50
4	14.87	0.84
5	19.44	1.10
6	8.08	0.45
7	12.15	0.68

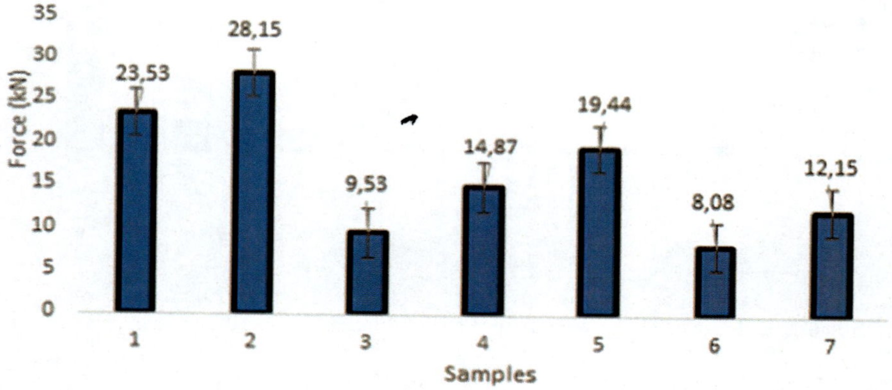

Fig. 2. Comparison between compressive force of different composition of samples.

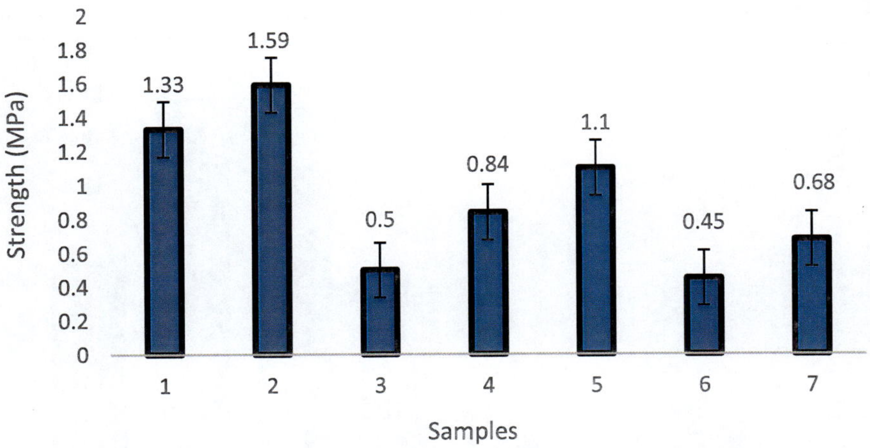

Fig. 3. Comparison between compressive strength of different composition of samples.

3.3 Water Absorption Test

Table 5 shows the comparison of water absorption of samples with different compositions in 3 days, while Fig. 4 shows the relationship between mass gain and immersion time at 25 °C for different sample compositions. According to the result obtained, samples 1, 2 and 3 (nano-based HDPE-sugarcane composite) were all very low, and there is relatively not much difference. Samples 4, 5 and 6 (nano-based HDPE composite) had higher mass gain compared to nano-based HDPE-sugarcane composite sample. The readings of samples 7 is considered medium. After 48 h, the mass of samples remained unchanged, but some samples still slightly increased. The water absorption of sample 6 (HDPE + aluminium oxide nanoparticles) had a maximum value of 0.88% after 72 h. The water absorption capacity of sample 2 (HDPE + copper oxide nanoparticles + bagasse powder) was only 0.05%. Therefore, by adding nanoparticles, the water absorption rate of the sample increased slightly, but the water absorption rate of the sample with sugarcane bagasse powder was reduced.

Table 5. Comparison between water absorption test of different composition of samples.

Sample	Initial mass (g)	Mass (g)			Mass gain (c) (%)		
		24 hrs	48 hrs	72 hrs	24 hrs	48 hrs	72 hrs
1	20.32	20.35	20.36	20.36	0.15	0.20	0.20
2	21.86	21.88	21.87	21.87	0.09	0.05	0.05
3	20.20	20.23	20.23	20.24	0.15	0.15	0.20
4	21.30	21.35	21.35	21.37	0.23	0.23	0.33
5	21.75	21.90	21.90	21.92	0.69	0.69	0.78
6	21.52	21.67	21.70	21.71	0.70	0.84	0.88
7	21.98	22.10	22.09	22.09	0.51	0.49	0.49

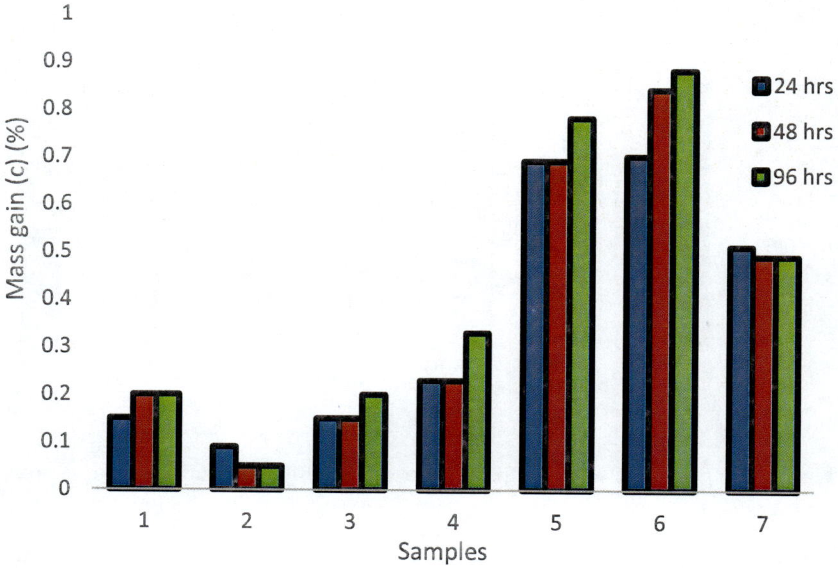

Fig. 4. Water absorption behavior of different composition of sample at 25 °C

4 Conclusion

In this study, the influence of nanoparticles on the properties of recycled plastics has been examined through various tests, such as fatigue test, compressive test, and water absorption test. The findings showed that in terms of fatigue resistance, aluminum oxide nanoparticles had better results in nano-based HDPE composites. For the composition of 4.55% by mass of aluminum oxide nanoparticles and 95.45% by mass of HDPE, the fatigue life increased to 92 cycles. Sample 2 had sufficiently high compressive force (28.15 kN) and compressive strength (1.59 MPa) which contains 4.55% by mass of copper oxide nanoparticles, 4.55% by mass of bagasse powder and 90.0% by mass of HDPE. Among the three nanoparticles, copper oxide nanoparticles had the highest compressive force and compressive strength. For the composition of 4.55% by mass of aluminum oxide nanoparticles and 95.45% by mass of HDPE, the water absorption increased to 0.88% after 72 h. Therefore, by adding nanoparticles, the water absorption rate of the sample increased slightly, but the water absorption rate of the sugarcane bagasse powder sample decreased. Samples containing aluminum nanoparticles achieved better performance in fatigue tests and water absorption tests as compared to other nanoparticles plastic composites. Further study is recommended to investigate the application of other nano-particles on the properties of recycled HDPE.

Acknowledgements. The authors gratefully acknowledge Faculty of Engineering, Technology and Built Environment, UCSI University Malaysia and Faculty of Engineering and Technology, Liverpool John Moores University for support to conduct this research work.

References

1. Geyer, R., Jambeck, R.J., Law, K.L.: Production, use, and fate of all plastics ever made. Sci. Adv. **3**(7), e1700782 (2017)
2. Tsakona, M., Rucevska, I.: UNEP Baseline report on plastic waste at Plastic Waste Partnership working group First meeting Beau Vallon, Seychelles, 2–5 March 2020 (2020)
3. Post, W., Susa, A., Blaauw, R., Molenveld, K., Knoop, R.J.I.: A review on the potential and limitations of recyclable thermosets for structural applications. Polym. Rev. **60**(2020), 359–388 (2020)
4. Prata, J.C., et al.: Solutions and integrated strategies for the control and mitigation of plastic and microplastic pollution. Int. J. Environ. Res. Public Health **16**(2019), 1–19 (2019)
5. Henderson, L., Green, C.: Making sense of microplastics? Public understandings of plastic pollution. Marine Pollut. Bullet. **152**, 110908 (2020)
6. Santos, J., Pham, A., Stasinopoulos, P., Giustozzi, F.: Recycling waste plastics in roads: a life-cycle assessment study using primary data. Sci. Total Environ. **751**, 141842 (2021)
7. Shahrajabian, H., Sadeghian, F.: The investigation of alumina nanoparticles' effects on the mechanical and thermal properties of HDPE/rPET/MAPE blends. Int. Nano Lett. **9**(3), 213–219 (2019). https://doi.org/10.1007/s40089-019-0273-7
8. Bajracharya, M., Manolo, A.C., Karunasena, W., Lau, K.T.: An overview of mechanical properties and durability of glass-fibre reinforced recycled mixed plastic waste composites. Mater. Des. **62**, 98–112 (2014)
9. Chong, W.K.O., Hermreck, C.: World's first recycled plastic bridges. In: ICSDC 2011: Integrating Sustainability Practices in the Construction Industry, pp. 585–593. ASCE, USA (2011)
10. Wood, K.: Tough I-beam bridge for tank traffic. Composite Technology **15**, 6 (2009)
11. Prochazkova, Z., Kralik, V., Nemecek, J., Sejnoha, M.: Recyled plastic material properties defined by nanoindentation. Adv. Mater. Lett. **7**(1), 78–82 (2016)
12. Zdiri, K., Elamri, A., Hamdaoui, M., Harzallah, O., Khenoussi, N., Brendle, J.: Reinforcement of recycled PP polymers by nanoparticles incorporation. Green Chem. Lett. Rev. **11**(3), 296–311 (2018)
13. Mustafa Al Bakri, A.M., Liyana, J., Norazian, M.N., Kamarudin, H., Ruzaidi, C.M.: Mechanical properties of polymer composites with sugarcane baggase filler. Adv. Mater. Res. **740**, 739–744 (2013)
14. Muñoz, E., García-Manrique, J.A.: Water absorption behaviour and its effect on the mechanical properties of flax fibre reinforced bioepoxy composites. Int. J. Polymer Sci. **2015**, 16–18 (2015)
15. Khalili, P., Liu, X., Zhao, Z., Blinzler, B.: Fully biodegradable composites: Thermal, flammability, moisture absorption and mechanical properties of Natural fibre-reinforced composites with nano-hydroxyapatite. Materials **12**(2019), 1–13 (2019)

Hygroscopic Performance of Sulphate-Based MgO Construction Boards

Ayodele Olofinjana[1](✉), Nicholas Jays[1], David Young[2], Jitendra Mata[3], and Rezwanul Haque[1]

[1] University of the Sunshine Coast, Sippy Downs, QLD 4556, Australia
aolofinj@usc.edu.au
[2] Charles Darwin University, Darwin, NT 0909, Australia
[3] Australian Centre for Neutron Scattering, Lucas Heights, NSW 2234, Australia

Abstract. Magnesium oxide (MgO) construction boards are currently being used to replace fibre cement boards in various construction applications. This include usage in external claddings where reports of hygroscopic degradation were linked to presence of residual chlorides originating from the manufacturing process. The industry has responded with new products that are being marketed as Magnesium Oxysulphate (MOS) boards that are still essentially MgO based but are manufactured from non-chloride precursor. In this paper the hygroscopic performance of the new MOS based MgO is compared with some other commercially Chloride containing samples. It is shown that MOS based samples exhibited lower water absorption comparable to fibre cement boards. No evidence of chloride in the new MOS based MgO boards was found by Energy Dispersive Spectroscopy (EDS). Small angle neutron scattering (SANS), of dry samples and hydrated samples using deuterated water and normal water showed that the better hygroscopic properties of MOS based MgO board was not related to structural morphological modification and are more likely due to the absence of residual chloride in the new generation MgO boards.

Keywords: Magnesium oxide board · Magnesium Oxysulphate (MOS) · Hygroscopic performance

1 Introduction

The ability of magnesia (MgO) based cement to bind and consolidate filler materials ranging from granite to sawdust and give acceptable compressive and tensile strength has led to the development of MgO construction materials [1–3]. They are now widely distributed and are available in various forms as wall and floorboards for both interior and exterior uses. They are generally favored over cement-based materials for their better fire resistance, strength and other environmental factors. However, there has been recent concerns over their water absorption and hygroscopic properties [4, 5]. They are known to have shown a lot of variability in performance with little understanding. There are currently no classification methods and difficult to separate performance of many types MgO board on function.

A. Batako et al. (Eds.): GCMM 2021, LNNS 335, pp. 390–401, 2022.
https://doi.org/10.1007/978-3-030-90532-3_30

Commercially available MgO construction boards generally consist of 48–50% magnesia (MgO) with hydration reaction with other salts to form a cementing matrix binder that hold the filler materials. The most common MgO based cement is based on reaction of MgO with Magnesium Chloride to form Magnesium Oxychloride (MOC). The cementation action in MOC is based on the hydration reaction to form phase 3 ($3Mg(OH)_2 \cdot MgCl_2 \cdot 8H_2O$) or phase 5 ($5Mg(OH)_2 \cdot MgCl_2 \cdot 8H_2O$).

Though phase 3 has been accepted as the key dominant phase in many MOC products, the dominant phase is often determined by mole ratio of $MgO - MgCl_2$ during production [6–8]. There is no doubt that the possible variation in phase compositions of the hydrated oxychloride adds to the variability in the performance of the MOC based MgO construction boards and many reports exist on the widespread of performance on mechanical and water absorption properties.

Recent disturbing reports [5, 7, 8] on the chloride containing MgO boards has led to development of Chloride free boards. The most prominent of these are based on the binding action from formation of magnesium oxysulphate (MOS). The hydration reaction for formation of oxysulphate are generally based on the reaction [9, 10] to form a binding product of the form: $xMg(OH)_2 \cdot yMgSO_4 \cdot zH_2O$.

The actual values of x, y, and z depends on reaction conditions iluding mole ratio of precursors, pH, and temperature. The stichometry values of x, y, and z in the ensuing oxysulphate are generally used to describe the dominant oxysulphate phase. Several phases including 318, 513, 516, 517 and many other combinations of x, y, z has been reported [3, 4, 10]. Many works [8, 11–13] on the mechanical and water absorption properties of the many possible Magnesium oxysulphate phases has shown that phase 517 ($5Mg(OH)_2 \cdot MgSO_4 \cdot 7H_2O$) has the best combination of properties to function as binder in MgO based construction boards. Processes have been developed to engineer the dominance of the 517 oxysulphate phase in MOS MgO boards and most Cl free MgO are typically based on 517 oxysulphate phase.

Marketing of MgO cement-based boards are however rarely so designated and there is still a lot of confusion on what type of MgO board with designation like low Chloride and Chloride-free products. In this work, we compared the water absorption properties of commercial samples of MOC and MOS based MgO construction boards and present distinguishing test for the two types.

2 Experimental

2.1 Board Samples

Three samples of commercial MOC and MOS based MgO construction boards were obtained and cut into various sizes for water absorption tests, and electron microscopy.

2.2 Water Absorption in Ambient and Controlled Humidity

Water absorption samples were cut to 70 mm × 70 mm in size. Each specimen was dried using a drying chamber at 90 ± 5 °C until constant weight, to remove any pre-existing moisture, and allowed to cool to room temperature in a desiccator type cabinet.

The initial mass of the specimens was recorded using a 3-point balance. The specimens were then fully submerged in clean tap water for 48 h at ambient temperature (23 ± 4 °C). The water temperature was monitored using the datalogger. A plastic rack was used to separate the specimens to ensure all specimens had an equal surface area exposed. The specimens were removed from the water tank and wiped with a damp cloth. The final saturated weight of the boards was recorded using the 3-point balance. The mass increase due to water absorption was calculated using equation:

$$\text{Water absorption, mass \%} = \left(\frac{W_s - W_d}{W_s}\right) \times 100 \quad (1)$$

Where: W_s is saturated mass and W_d is dry mass.

For the controlled humidity test, new samples were prepared, with a size of 40 mm × 40 mm. For each sample, 6 specimens were prepared, with 3 specimens used for a targeted controlled humidity. Because of variation in the thicknesses of the various board, epoxy coating was applied to the edges of the samples to eliminate the effects of water absorption through edges and allow a normalised comparison of absorptivity of the working surface. Each specimen was then dried using a drying chamber at 90 ± 5 °C until constant weight to remove any pre-existing moisture, and allowed to cool to room temperature in a desiccator type cabinet. The initial mass of the specimens was recorded using a 3-point balance.

Saturated salt solutions were used to attain a controlled humidity environment. The targeted controlled humidity are based on the published equilibria values of relative humidity (RH) for saturated salts at fixed temperatures. Since equilibrium RHs are temperature sensitive, an incubation chamber was utilized for temperature regulation. The Incubation chamber was set at 32 °C. Two salts were selected to achieve two controlled RH atmosphere. The selected saturated salts were potassium sulphate to achieve constant relative humidity environments (97.0 ± 0.4% RH @ 30 °C) and sodium chloride to achieve a relative humidity environment (75.1 ± 0.2% RH @ 30 °C). In order to speed up the time to attain RH equilibrium, the fan convection was intermittently used in the chamber.

2.3 Scanning Electron Microscopy (SEM) and Energy Dispersive Spectroscopy (EDS) Analyses

Scanning Electron Microcopy (SEM) and Energy Dispersive Spectrometry (EDS) examination of board samples was conducted to monitor compositional variations in board samples and corrosion products from exposed metallic fasteners. Small samples of MgO boards were mounted on aluminum stubs and were gold coated using an argon ion sputter coater for electrical conductivity measurements. Both the cross-sectional area and the surface of the boards were examined at an accelerating voltage of 30 kV.

2.4 Thermal Analyses

Thermal analyses were conducted on a Simultaneous Thermal analyzer (STA) that follow the thermal events that represent release of absorbed water and structural decomposition by thermo gravimetry (TGA) and differential scanning calorimetry (DSC). It is thought that results from this test could provide additional information to explain observed differences in hygrothermal performance. For each of the board, a small quantity of MgO binder was ground into powder and about 50 mg powder loaded into an alumina pan. Thermal scan was conducted on the STA at the scanning rate of 10°/min from room temperature to 800 °C. The STA was programmed to collect data for differential scanning calorimetry (DSC) and Thermo-gravimetry (TGA). The TGA data were used to follow mass loss that can distinguish between absorbed water and volatiles released from structural decomposition of the binder phase.

2.5 Small Angle Nuclear Scattering (SANS)

Three The water absorption in MOS based MgO boards was further investigated by Small angle nuclear scattering (SANS) and Ultra Small Angle Nuclear Scattering (USANS) using cold neutrons with fixed wavelengths. The aim was to compare the scattering variable (Q) of dry, water (H_2O) and deuterium (D_20) saturated samples to gain information about water absorption with morphology of meso-structures.

3 Results and Discussions

3.1 Hygroscopic Behavior of Construction Boards

The effect of humidity on water absorption of MOC and MOS based board are shown in Fig. 1. It is clear that water absorption is very low (less than 10%) at low RH of 75% for all MgO products and comparable to traditional fibre cement. However, MgO products showed much higher water absorption at relative humidity of 97%. More than 20% water absorption was observed for MOC-based boards while MOS based board only indicate about half (around 10%) absorption. This humidity-based behavior is a cause for concern on the effect on aiding corrosion of metal inserts and in culturing development of mould and mildew.

Tests were conducted with galvanized steel screws inserted into the boards and exposed to 97% RH. The results indicated that screws inserted into the MOC boards were more susceptible to corrosion. Samples of screws taking from MOC and MOS boards are shown in Fig. 2. The screw inserted in the MOS board did not show any visible sign of corrosion. On the other hand, the rust evidencing electrochemical corrosion is clearly visible on the screw harvested from the MOC board (Fig. 2).

3.2 Water Immersion and Drying Tests

Simple water immersion tests were performed to follow water absorption to explain the hygroscopic behavior. The results of water immersion and drying tests are summarized in Fig. 3. In the as-received form, both MOC and MOS type MgO boards showed a mass loss about 10% when oven dried to constant weight. This indicated that at normal usage and in normal atmospheric conditions, there is little difference in the water absorption properties of these two families of MgO boards. This is consistent with the hygroscopic

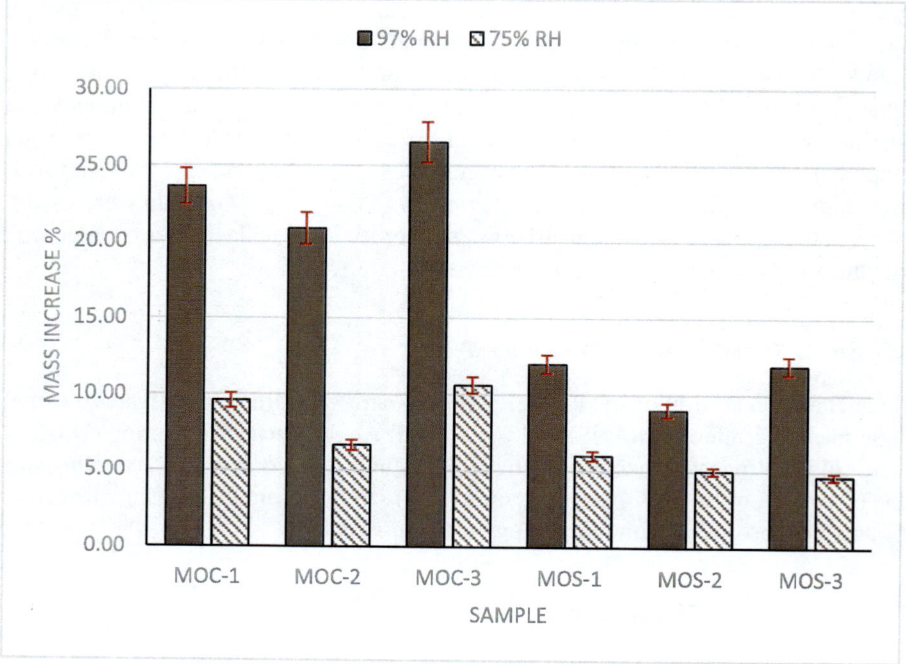

Fig. 1. Comparing the mass gain in samples of MOC and MOS based MgO board exposed to low humidity (75% RH) and exposed to high humidity (97% RH)

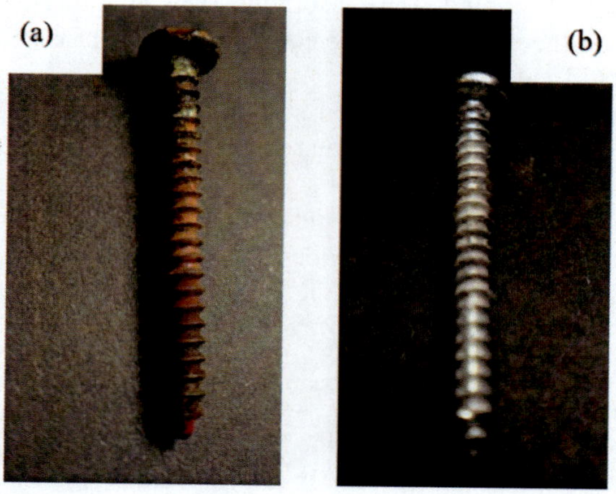

Fig. 2. Appearance of metal inserts after 55-day exposures of MgO boards to 97% RH (a) MOC-MgO board (b) MOS-MgO board

results of water loss (Fig. 1) at 75% RH. The indication of this is that, when the RH is not high, water absorptions in both MOC and MOS type MgO boards are at acceptable levels and quite comparable to fibre cement-based boards.

The results of water absorption indicated that there is a vast difference in water absorption from water immersion tests. The MOC types indicated much higher weight gained after 48 h immersion in water. MOC based boards exhibited weight gains of 25–38% while MOS based boards showed much less weight gain 12–18%. This is an indication that the water absorption of MgO construction board is very much affected by the chemical nature of the hydrating phase. This water immersion test is consistent with results of hygroscopic tests at high humidity of 97% RH presented in Fig. 1.

Figure 3 also shows the summary of overall weight loss on heating to a high temperature in a DSC. The water loss indicated by overall DSC mass loss would include decomposition of hydrating structural water. There was no clear difference in DSC weight loss which was about 40–45% for both MOS and MOC types. This is mostly due that the fact that at high temperatures up to 800 °C, the hydrating phase decomposes, and the weight loss would not only represent absorbed water but other volatiles that are part of the chemical structures. Further analysis of DSC results is presented in later section.

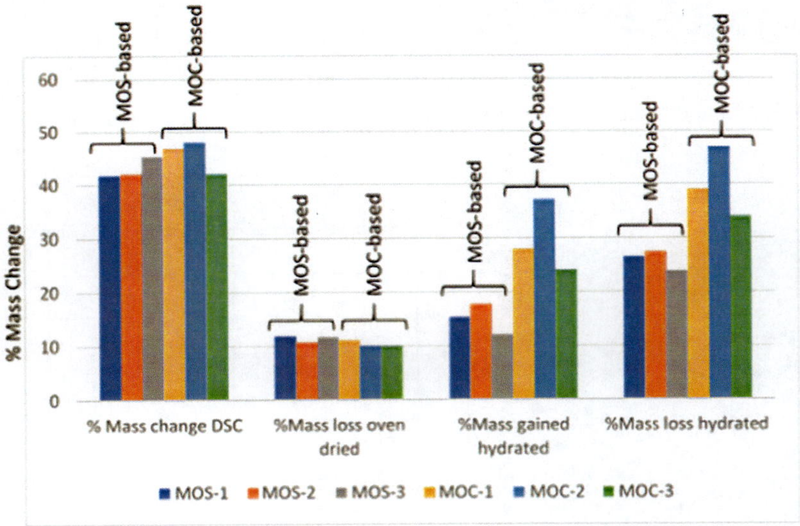

Fig. 3. Comparing water absorption in MOS and MOC based MgO construction boards.

3.3 SEM/EDS Analysis

The SEM and EDS analyses of samples from MOC and MOS boards are shown in Figs. 4 and 5. Since MgO boards consist of many filler materials, the focus was on Mg, O and components of oxysalts. It is hard to make any distinguishing differences in the morphology of phases, but the qualitative analysis of the EDS provided a marker to distinguish the binding phase. Figure 4 shows the SEM and EDS analysis for the MOC binding phase. It shows clearly, a Cl peak identifying the binder as an oxychloride phase. Figure 5 also showed the SEM and the accompanying EDS spectrum. Clear in Fig. 5 is the S peak that distinguish the binding phase as MOS. This result provided a means of distinguishing the binding phase and the type of MgO cemented construction materials. Though EDS examination could function as a quick identifier for determining water absorption or hygroscopic properties, the extensive preparation required, and the cost would not be ideal for routine tests.

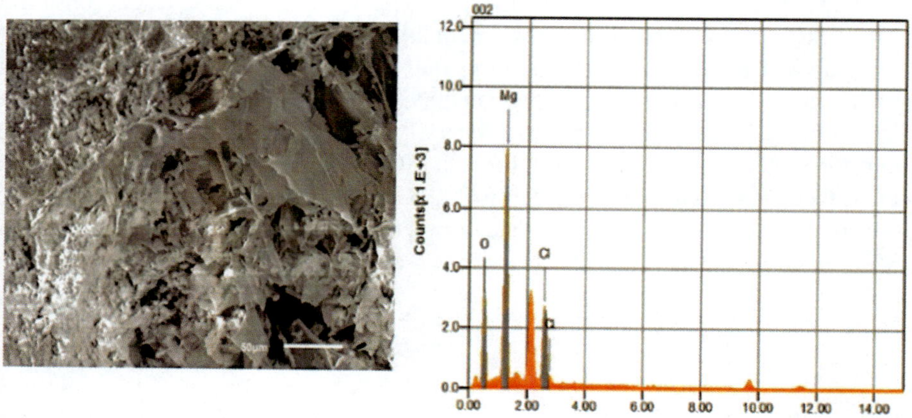

Fig. 4. SEM and EDS of MOC-based MgO board, Cl peak in EDS spectrum identifies the MOC phase

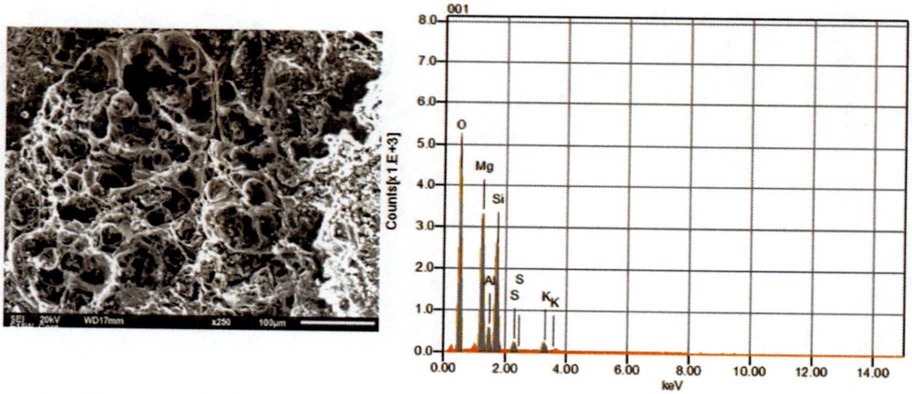

Fig. 5. SEM and EDS of MOS-based MgO board, S peak in EDS spectrum identifies the MOS phase

3.4 DSC and TGA Thermal Analysis

The thermal characteristics of MOS and MOC binders were followed in a simultaneous thermal analyzer (STA). Both DSC and TGA signals were recorded to follow water absorption and decomposition reactions at higher temperatures. Results of thermal analyses are presented in Figs. 6 and 7. MOC samples in Fig. 6 showed water loss of 20% from room temperature to about 300 °C. A rapid weight loss ~20% occurred between 300 °C and 400 °C. Juxtaposing the DSC and TGA signal indicated that the reaction between 300 and 400 °C must be a phase change and this could be associated with the decomposition and due to structural water loss of the MOC phase $xMg(OH)_2 \cdot MgCl_2 \cdot 8H_2O$. The DSC indicate a small peak around 200 °C and this is thought to be related with phase change in proportions of phase of 318 (phase 3) or 518 (phase 5). This however is not reflected in the TGA because it does not affect the stoichiometry of water of crystallisation since the mole proportion of water remain the same in both MOC phases 3 and 5. This therefore suggest that water loss indicated in the TGA trace ('20%) are mostly absorbed water. This is consistent with high water absorption recorded in the immersion tests.

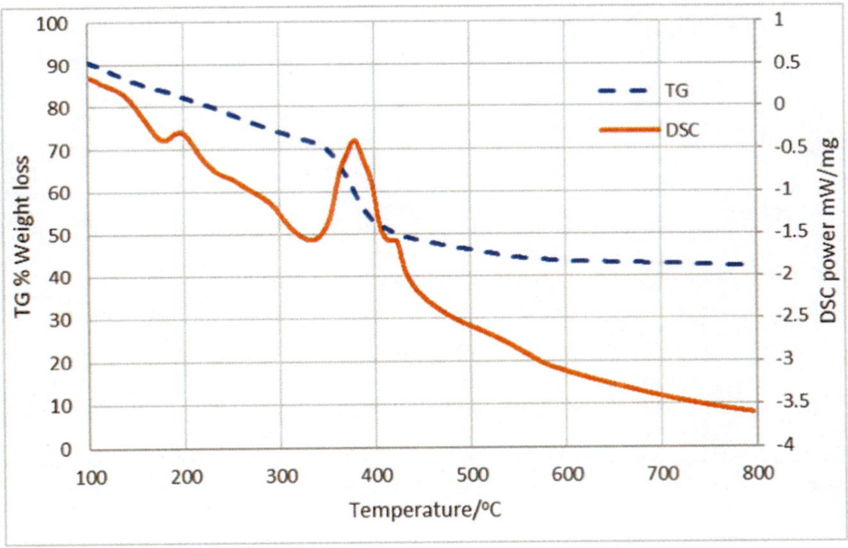

Fig. 6. Characteristic thermal analysis signal for an MOC-based MgO board, DSC single peak 100–200 °C. Indicative of single-phase change.

Figure 7 shows the thermal analysis results for the MOS binder. The TGA indicated a steady wight loss about 10% between 100 °C and 200 °C. Juxtaposing both DSC and TGA throws some light on the thermal events responsible for the weight loss. No doubt, we expect some absorbed water about 5% from immersion tests but the clear DSC double peak is an indication that thermophysical phase changes are involved. It is known that the MOS 517 phase was stabilized by pH control [9, 13] and the present results indicated transition of two phases prior to complete structural decomposition. It is being shown that [9] phase 516 in the temperature range 100–120 °C, and phase 514 above 120 °C. The two DSC peaks can therefore be identified as the following phase changes:

Step 1 (DSC peak 1): Transition from phase 517 to 516 at 120 °C with theoretical weight loss of 3.3%

$$5Mg(OH)_2 \cdot MgSO_4 \cdot 7H_2O \rightarrow 5Mg(OH)_2 \cdot MgSO_4 \cdot 6H_2O \quad (2)$$

Step 2 (DSC peak 2): Transition from phase 516 to 514 at 174 °C theoretical 6.7% weight loss

$$5Mg(OH)_2 \cdot MgSO_4 \cdot 6H_2O \rightarrow 5Mg(OH)_2 \cdot MgSO_4 \cdot 4H_2O + 2H_2O \quad (3)$$

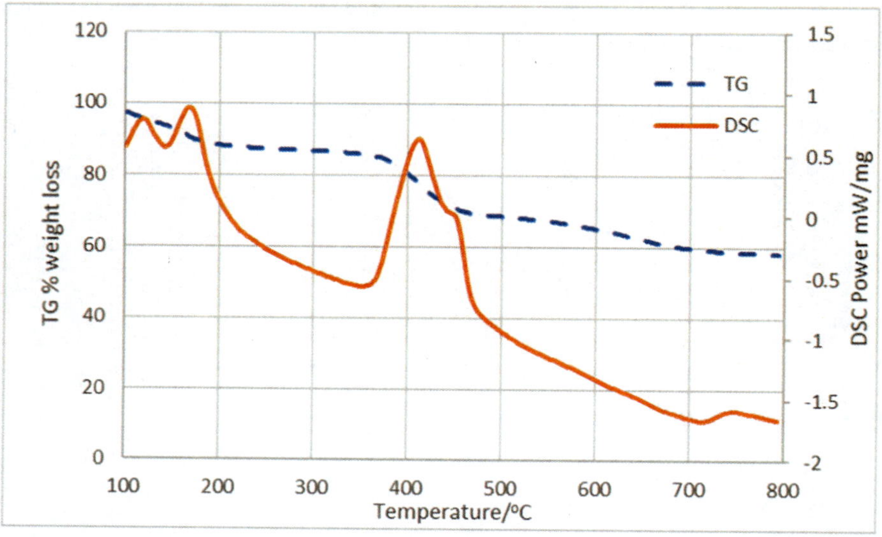

Fig. 7. Characteristic thermal analysis signal for an MOS-based MgO board, DSC double peak between 100 °C–200 °C. Indicative of double phase change.

The total weight loss of about 12% recorded in the TGA between 100 °C and 200 °C are therefore the combination of phase changes in two steps: $517 \xrightarrow{(120°C)} 516 \xrightarrow{(174°C)} 514$ in the MOS binder. The double peak of the DSC analysis can be a easy and useful method of distinguishing MOS based MgO products without the need for detailed chemical analysis.

Both MOC and MOS binders showed complete decomposition between 336 °C–445 °C and 350 °C–490 °C respectively. It is noted that the decomposition of MOC phase occurred at slightly lower temperature but weight change for both are around 20%. Total weight change from room temperature to decomposition in both types are similar and not enough to distinguish them.

The difference in hygroscopic and water absorption calls for a simple distinguishable method short of complete chemical analyses. The double peak in the MOS binder that signify phase transitions not evident in MOC binders therefore provided an effective method of sorting MgO boards as to their hygroscopic behavior.

3.5 Small and Ultra Small Angle Nuclear Scattering

Samples of MOS construction board were examine with neutron scattering methods to gain understanding of the water absorption. Scattering information were gathered from two separate instruments for Small angle neutron scattering (SANS) and Ultra small angle neutron scattering (USANS). The scattering vector Q should reflect any changes in structure that would accompany saturated sample. Figure 8 showed the wide range of Q scattering factor obtained by combining data from USANS and SANS. No observable difference in intensity between Q range of 10^{-3}–10^{-0} suggest that there was no difference in structure in dry and saturated samples in that Q range. We focused on the Q range 10^{-4}–10^{-3} to examine the SANS data for any change. The scattering factor from SANS are presented in Fig. 9 that did not show any noticeable difference for both H_2O and D_2O saturation of MOS based MgO board. Since the scattering factor did not show that saturation affected the mesostructure, it can therefore be conjectured that the water saturation and by extension, water absorption of MOS is not related to the physical structures and must be a chemical factor. This would explain that water absorption is related to chemical potential and that oxychloride in MOC binders has more affinity for water than oxysulphate in MOS binder. This would seem to explain the superior hygroscopic properties of the MOS based construction boards.

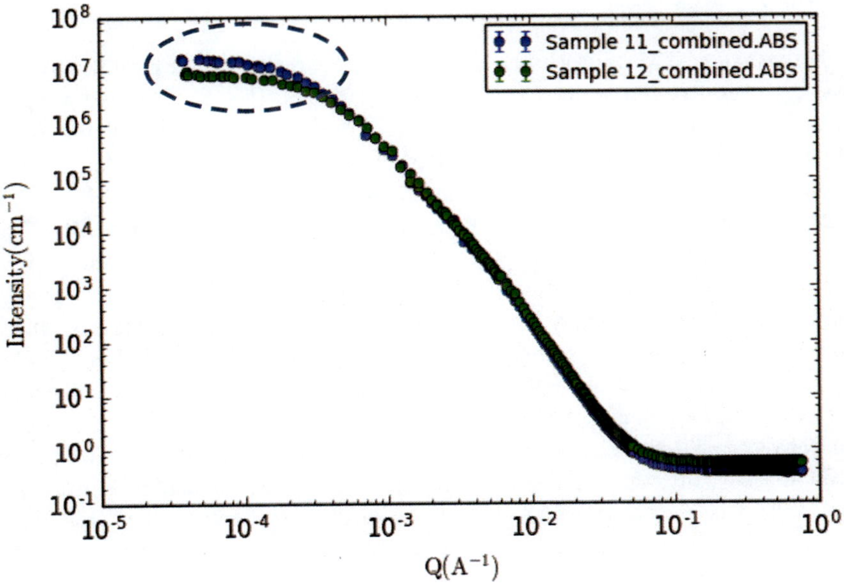

Fig. 8. Combined USANS and SANS data to show wide range of scattering factor Q.

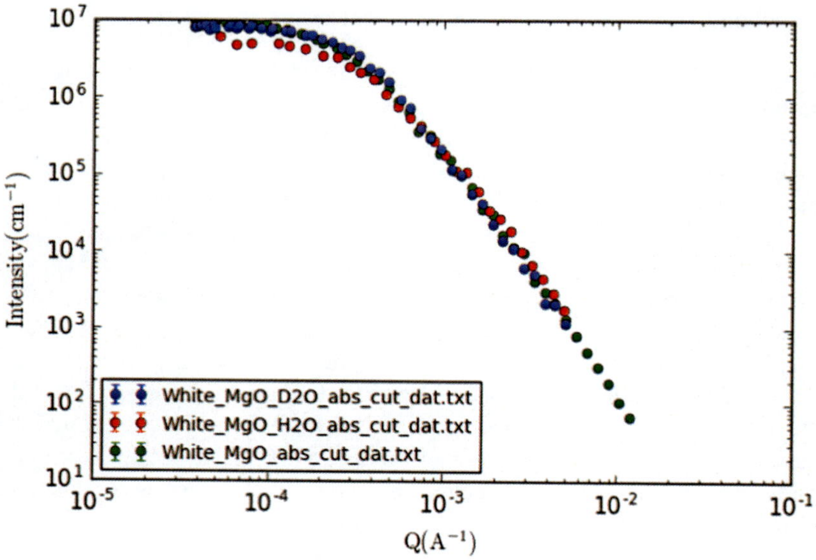

Fig. 9. SANS data to cover lower range of scattering factor Q.

4 Conclusions

It is shown that most Magnesia based cement construction boards are based on hydration of MgO to form oxychlorides (MOC) and oxysulphate (MOS). The hydrating salts provided the consolidation and are responsible for the hygroscopic performance of these MgO-based materials. They are easily differentiated by EDS of the binder that shows a clear Cl peak for MOC or a clear S peak for MOS. Both MOS and MOC based boards performed similarly in low relative humidity of 75% RH. However, MOC-based products are shown to exhibit high water absorption at relatively high humidity of 95%. Water immersion tests also showed that MOS based boards absorbs much less water compared to MOC based board. The high absorption in MOC based boards are thought to be related to the water absorbing characteristics of the oxychloride binding phase. The combination of water absorption and the chemical potential of chloride dissolution provided electrolytic action that aided the corrosion of metal inserts in MOC boards. Such corrosion actions were not observed in MOS boards. Though market products of MgO-based construction materials are generally branded as Chloride free or low chloride there is very little reference made to distinguish between these two generic types. Thermal analysis is an effective way of distinguishing the MOC and MOS bound MgO board. MOS based boards, displays a two peak DSC characteristic peak between 100 °C and 160 °C that showed phase transition from phase 517 to phase 516 and then 516 to 514 phase, while MOC binder only showed a single peak in this temperature range. Both MOC and MOS indicated complete loss of structural water between 300 °C to 400 °C where they lost their binding action. The double peak identified in the MOS is proposed as a standard test to identifying MOS based boards. Small angle Neutron Scattering (SANS) studies indicated that water absorption are not related differences in mesostructured porosity but more likely to chemical structure of the oxysalts. MOS based MgO construction boards are best for hygroscopic performance and the DSC thermal characterisation method proposed here is a quick distinguishing test to separate these products at the marketplace.

Acknowledgements. The authors acknowledge AINSE for awarding grant No. 8285 to carry out neutron diffraction studies at ANSTO.

References

1. Martins, J.A., Gomes, C.M., Fontanini, P., Dornelles, K.: Comparative analysis on thermal performance of MgO and fiber cement boards applied to light steel frame building systems. J. Build. Eng. **21**, 312–316 (2019)
2. Li, P.X., Froese, T.M., Cavka, B.T.: Life cycle assessment of magnesium oxide structural insulated panels for a smart home in Vancouver. Energy Build. **175**, 78–86 (2018)
3. Walling, S.A., Provis, J.L.: Magnesia-based cements: a journey of 150 years, and cements for the future? Chem. Rev. **116**(7), 4170–4204 (2016)
4. Wang, D.X., Di, S.J., Gao, X.Y., Wang, R.H., Chen, Z.G.: Strength properties and associated mechanisms of magnesium oxychloride cement-solidified urban river sludge. Constr. Build. Mater. **250**, 118933 (2020)
5. Jays, N., Olofinjana, A., Young, D.J.: Assessing variability in the hygrothermal performance of magnesium oxide (MgO) cladding products of the Australian market. Constr. Build. Mater. **203**, 491–500 (2019)
6. Luo, X.R., et al.: Effect of hydroxyacetic acid on the water resistance of magnesium oxychloride cement. Constr. Build. Mater. **246**, 118428 (2020)
7. Jiříčková, A., et al.: Synthesis, structure, and thermal stability of magnesium oxychloride $5Mg(OH)_2 \bullet MgCl_2 \bullet 8H_2O$. Appl. Sci. **10**(5), 1683 (2020)
8. He, P.P., Poon, C.S., Richardson, I.G., Tsang, D.C.W.: The mechanism of supplementary cementitious materials enhancing the water resistance of magnesium oxychloride cement (MOC): a comparison between pulverized fuel ash and incinerated sewage sludge ash. Cem. Concr. Compos. **109**, 103562 (2020)
9. Li, X., Qiu, R.F., Xue, F.B., Fang, L., Cheng, F.Q.: Effects of unreactive MgO and impurities in light burned MgO on the hydration process and performance of base magnesium sulfate cement. Constr. Build. Mater. **240**, 117854 (2020)
10. Guo, T., Wang, H.F., Yang, H.J., Cai, X.S., Ma, Q., Yang, S.M.: The mechanical properties of magnesium oxysulfate cement enhanced with 517 phase magnesium oxysulfate whiskers. Constr. Build. Mater. **150**, 844–850 (2017)
11. Tang, S.W., Wei, C.R., Cai, R.J., Huang, J.S., Chen, E., Yuan, J.H.: In situ monitoring of pore structure of magnesium oxysulfate cement paste: effect of $MgSO_4/H_2O$ ratio. J. Ind. Eng. Chem. **83**, 387–400 (2020)
12. Zeng, X., Yu, H., Wu, C.: An overview of study on basic magnesium sulfate cement and concrete in China (2012–2019). KSCE J. Civ. Eng. **23**(10), 4445–4453 (2019). https://doi.org/10.1007/s12205-019-0199-7
13. Qin, L., Gao, X.J., Li, W.G., Ye, H.: Modification of magnesium oxysulfate cement by incorporating weak acids. J. Mater. Civ. Eng. **30**(9), 04018209 (2018)

Energy-Saving of Conveyor Belts in the Strategy and Reporting of Corporate Social Responsibility Initiatives of Producers

Mirosław Bajda(✉), Leszek Jurdziak, Katarzyna Pactwa, and Justyna Woźniak

Department of Mining and Geodesy, Faculty of Geoengineering, Mining and Geology, Wroclaw University of Science and Technology, Na Grobli 15, 50-421 Wroclaw, Poland
miroslaw.bajda@pwr.edu.pl

Abstract. Increasing energy efficiency is objectively the right thing from an economic point of view. Besides, due to political and social pressure, it acquires additional significance due to slowing down temperature rise and climate change. Energy-saving becomes part of the Corporate Social Responsibility (CSR) strategy of many producers and a marketing tool due to the growing climate awareness of consumers. This applies to many products, including for industries with significant environmental impact. The example is conveyor belts, which are used to transport bulk materials in ports, smelters, chemical industry, and mining. To maintain raw material prices at an acceptable level, it is necessary to reduce transport costs, and conveyor belts constitute an important part of them. The reduction of belt production costs may affect their quality and durability. Therefore, attention was paid to reducing the resistance of conveyor movement and here a significant part (up to 60%) is the conveyor belt indentation resistance. Many belt manufacturers started intensive research on improving the properties of belts and achieving a significant reduction in energy consumption during belt operation. Manufacturers who have succeeded use this competitive advantage in marketing to convince customers of new solutions. They warm up their image, indicating a reduction in the carbon footprint of their products, which is one of the elements of their CSR strategy. Confirmation of energy savings requires not only laboratory tests but also long-term industrial research. The paper discusses the results of four-year comparative tests of standard and energy-saving belts operated in one of the mines in Poland.

Keywords: Energy efficiency · Conveyor belts · Belt indentation resistance · Corporate Social Responsibility (CSR) · Carbon footprint

1 Introduction

The article presents a combined perspective on the results obtained in conveyor belt laboratory tests and on the issue of social responsibility in Polish industry sector. Belt conveyors are presently the most common transportation machines used in surface and underground mining. Apart from mining industry, belt conveyors are also used in cement

A. Batako et al. (Eds.): GCMM 2021, LNNS 335, pp. 402–414, 2022.
https://doi.org/10.1007/978-3-030-90532-3_31

and chemical industries, as well as in power plants and ports. Belt transportation technology used in global mining industry has undergone constant development. The advances are most clearly visible in long-distance belt conveyor designs. The two most important belt conveyor development areas include: improving belt conveyor efficiency and single conveyor length, which entails power increase of installed drive mechanisms. High-power conveyors need considerable amounts of electric energy, and consequently, due to global increase in energy prices, they generate increased transportation costs.

Polish open-cast mines mostly use steel-cord belts, as they find application in long conveyors of highest capacities. Such conveyor systems consume large amounts of electricity. Therefore, the need to reduce their energy consumption has become an important problem, and this can be solved directly by using energy-efficient belts with appropriate parameters of the bottom cover. The savings generated by using a belt with the bottom cover of appropriate parameters are so significant because belt indentation rolling resistance in long conveyors accounts for over 60% of the total belt conveyor resistances to motion. Belts installed on conveyors having the greatest lengths and the highest capacities are of the steel-cord type, and therefore this type of belts has become the object of the tests here described.

The manufacturer of the tested conveyor belts is a part of a large international group with extensive business operations. The authors address one of the aspects related to the environmental and social responsibility of the company with respect to its energy consumption. The article also presents a dual-aspect approach to energy saving, in which the focus is placed on the one hand on the operation of belt conveyors and on the other hand on the manufacturing of belts and on the related corporate responsibility.

2 Corporate Social Responsibility Strategy and Global Reporting Initiative

The concept of Corporate Social Responsibility (CSR) dates back to the turn of the 20th century, when according to Carnegie (1868) it was viewed as a concern shown by the wealthy for the poor members of the society. Current understanding of CSR involves the need to take account of the problems found in the company's environment (1953, "Social Responsibilities of the Businessman" H. Bowen). According to the European Commission's Green Paper (2001), CSR is the responsibility of a company for its influence on the society (taking account of problems related to the society, environment, ethics, human rights and consumers). The communication of CSR initiatives is effected among other things by non-financial reporting. Reporting CSR initiatives is one of the steps aimed at repairing or changing company image, referred to as "branding" in management sciences [9].

Global Reporting Initiative (GRI) is a non-government international non-profit organization based on a network of cooperating stakeholders. It offers standards which may be used by companies to measure and report their social, environmental and economic achievements regardless of the type or profile of their business, size of the company or the location of its operations. The GRI Standards include requirements, recommendations and guidelines which have evolved over the last 20 years. The initial version of the standards (G1) was issued in 2000, and the current set of the GRI Standards comprises 3 reporting areas:

- basic information (Foundation, GRI 101), divided into three sections: 1. presents the reporting principles for defining report content and report quality, 2. defines the process for using the GRI Standards for reporting, 3. presents the methods for using the GRI Standards,
- profile indicators – for reporting contextual information related to the organization, i.e. information on its profile, strategy, ethics and integrity, governance, stakeholder engagement practices, and reporting process (General Disclosures, GRI 102 in the basic (Core) version or in the extended (Comprehensive) version); new signature GRI 102 – 1, GRI 102 – 56, management approach for each significant topic, which allows an organization to provide narrative information on why a topic is material, where the impacts occur and how the organization manages the impact (Management Approach, GRI 103 – 1, 103 – 2, 103 – 3).

Moreover, the latest version of the GRI Standards includes 33 Topic-specific Standards, arranged in a series:

- Economic, GRI 200,
- Environmental, GRI 300,
- Social, GRI 400, [19].

The Environmental set includes detailed standards related to energy, listed as follows:

- GRI 302-1 Energy consumption within the organization,
- GRI 302-2 Energy consumption outside of the organization,
- GRI 302-3 Energy intensity,
- GRI 302-4 Reduction of energy consumption,
- GRI 302-5 Reductions in energy requirements of products and services [10].

The international GRI communication standards enjoy a growing interest in Poland. Research [7] indicates that in the years 2015–2017 the GRI communication tools were used to report CSR initiatives in 76% of all reports. The provisions of a EU directive of 2014 (Directive 2014/95/UE [13]) link these areas thematically. The advantages resulting from introducing the arrangements of the EU directive, as mentioned in [6, 11, 17–19] are as follows:

- increased company transparency
- prestige (the company is viewed as a reliable partner and a good employer)
- increased competitiveness in the global scale, building a competitive advantage
- increased and improved relationships with stakeholders; means for communication
- emphasis on innovativeness and on learning
- increased confidence and general attractiveness for the clients and the investors
- risk management tool – conscious reputation risk management
- improved image, ethical issues
- building company reputation, brand
- possible comparisons within the industry branches.

Implementation of CSR initiatives remains in agreement with the principles of sustained development (SD). In the context of sustained development goals and of the energy-related issues addressed in this article, the implementation of actions consistent with the idea of SD, and thus also with the idea of CSR, should be ensured inter alia by: improving energy efficiency (including energy audits, maintenance of energy infrastructure, limited energy demand, etc.), limiting emissions and reporting their measured values, as well as using renewable energy [1, 14]. Importantly, the very fact of publicly reporting the data by companies is not equivalent to taking actions which support both the environment and the society. Nevertheless, reporting within the scope presented in this paper is a tool which allows not only the presentation, but also the acquisition of information on the current condition and on the results of changes introduced by companies. This information may be further interpreted in the context of climate changes and the binding legal regulations.

3 Energy-Efficient Conveyor Belt

In recent years, very intensive research has been performed into lowering energy consumption of belt conveyor drive mechanisms. Large-scale theoretical and experimental research showed potential for saving energy in belt conveyor individual components [8, 16]. One of the research approaches was to identify conveyor resistances [8]. The research proved that the feasibility of limiting electric energy consumption is linked to belt properties. Estimations suggest that implementing improved energy-saving belt with adequate parameters of rubber cover will allow a significant decrease of conveyor primary resistance, which will result in decreased electric energy consumption by conveyor drive mechanisms.

Analysis of belt transportation systems in Polish brown coal mines shows the scale of the problem. The KWB "Belchatow" lignite mine, which extracts above 4×10^7 t of coal and more than 1×10^8 m^3 of overlay per year, may be a good example here. An observation that material transportation at the KWB "Belchatow" mine requires belt conveyors of total length exceeding 160 km and is responsible for approximately 50% of electric energy usage, makes apparent the importance of the technological optimization of belt conveyors for lowering the energy consumption of drive mechanisms, in both economic and ecological sense [3–5].

Extensive theoretical and experimental research demonstrated the potential for saving energy in individual components of belt conveyors. One of the research approaches was to identify conveyor resistances, as analyzed among others by Spaans [15] and Gladysiewicz [8]. This research demonstrated a relationship between belt conveyor resistance to motion and belt properties. The findings imply that a proper selection of belt properties, and more specifically of the bottom cover properties, allows designing an energy-efficient belt, which may significantly reduce conveyor primary resistance, and the reduced resistance will result in reduced electric energy consumption by conveyor drive mechanisms.

The method for analytical identification of belt indentation rolling resistance developed by Gladysiewicz [8] takes into account belt elastic properties and belt damping, by introducing the relevant calculation coefficients:

- elastic modulus of belt bottom cover E_e
- damping factor of belt bottom cover ψ_e.

The values of the two coefficients are identified in laboratory tests. The tests consist in cyclically compressing rubber samples with a certain force and in measuring the deformations which occur in the rubber due to the applied load [2]. The measured load and the corresponding rubber deformation serve to prepare its hysteresis. The shape of the hysteresis and its angle are used to calculate coefficients ψe and Ee. The load range and the rubber compression frequency is adjusted so as to reflect the conditions in which the actual belt would be operated on a conveyor with consideration to belt speed and the distances between successive idler sets. Tests of belt rubber covers should be performed in a wide range of temperatures, both above and below zero degrees Celsius, as temperature was proven to have an influence on the values of both the damping factor and the elastic modulus. By means of example, Fig. 1 shows changes of the damping factor and the elastic modulus versus temperature for four rubber bottom covers of a conveyor belt.

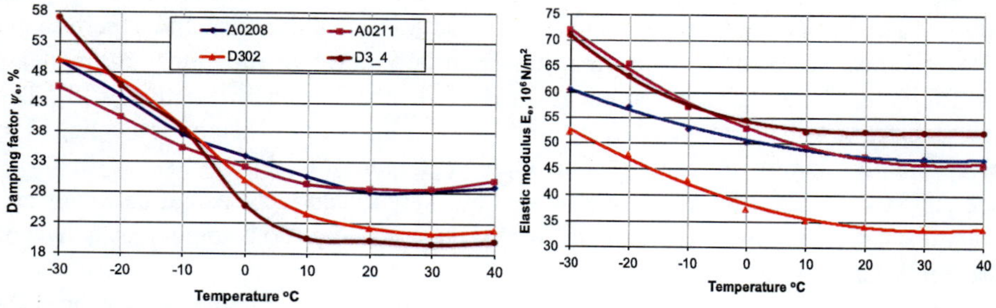

Fig. 1. Changes of the damping factor (left) and of the elastic modulus (right) for the bottom cover rubber in relationship to temperature.

The data presented in Fig. 1 indicate that as the temperature decreases, the values of both the damping factor (Fig. 1 on the left) and the elastic modulus (Fig. 1 on the right) increase. The values of the damping factor and of the elastic modulus of the belt bottom cover can now serve to calculate belt indentation rolling resistance with the use of professional belt conveyor design software (QNK-TT). The calculation method in the software uses the identification of the components of belt conveyor unit resistances based on analyses of physical phenomena accompanying the movement of the belt carrying the material along the conveyor route.

Figure 2 shows the values of belt indentation rolling resistances W_e, as calculated in the QNK-TT software for two belts designated with symbols P and K, and also the values of idler rotational resistances W_k. The idler rotational resistances were plotted in order to demonstrate that the values of resistances which are not influenced by the belt properties are several times smaller. The calculations were based on the data from one of the overburden conveyors operated in the KWB "Belchatow" lignite mine. They were performed with consideration to a number of the conveyor's technical and service data, as well as the route-related information. The object for the calculations was a 1200 m long belt conveyor, whose route is at 0° inclination to the ground. The belt installed on the conveyor is an ST type belt with steel cords 7.6 mm in diameter, with belt width of 2.250 m and with nominal tensile strength of 3150 kN/m. The top cover thickness is 14 mm and the bottom cover thickness is 7 mm. The belt moves on idlers 0.19 m in diameter, arranged in the three-roll sets. The sets are located at a distance of 1.2 m from each other and the belt moves with the speed of 5.98 m/s. The belt designated with symbol P is a regular belt, operated on a daily basis in the mine, and differs from the K1 belt only with respect to the rubber bottom cover. The bottom cover of belt K1 is made of a rubber compound, which in comparison to belt P has better dynamic parameters – lower value of damping factor ψ_e and higher value of elastic modulus E_e.

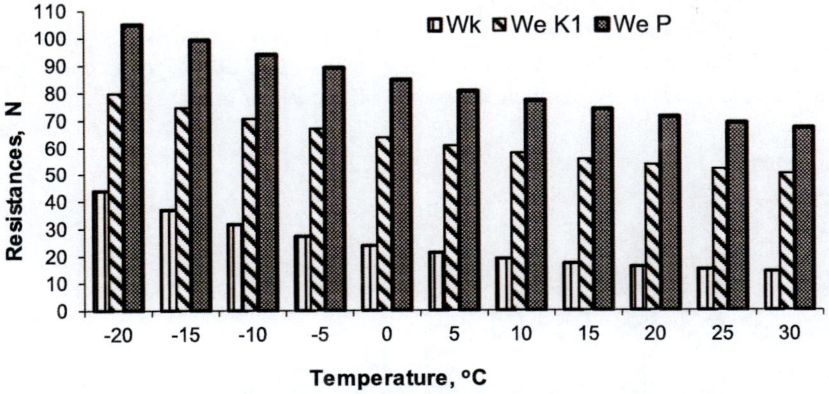

Fig. 2. Relationship between indentation rolling resistances W_k belt indentation rolling resistances W_e for the energy-efficient belt K1 and for the regular belt P.

The data presented in Fig. 2 indicate that the indentation rolling resistances of belt K1 are lower by approximately 32% than the indentation rolling resistances of belt P. Belt indentation rolling resistance is also the greatest component of the main resistances in the belt conveyor. Therefore, the analysis of this component should be given priority in the search for optimal belt conveyor design. Belt indentation rolling resistances were tested on a specially designed test rig having an inclined plane. It is used to simulate the conditions in which the idler works with the belt [2]. The test rig serves to perform research into the influence of belt parameters and various idler design solutions on conveyor resistances to motion.

The practical advantages of using conveyor belts with rubber bottom covers having reduced indentation rolling resistance were demonstrated by variant calculations of the drive power were performed for the same conveyor. The calculations are with allowance for the values of ψe and Ee resulting from ambient temperature changes. Figure 3 shows power demand of a belt conveyor drive mechanism depending on the installed conveyor belt: P – regular belt commonly used in the mine and K1 – belt showing approximately 32% lower indentation rolling resistance in comparison to the regular belt P.

The results indicate that lower temperatures negatively influence the values of belt indentation rolling resistances (Fig. 2). In the temperature range of 30 °C to –20 °C, temperature decrease is accompanied by an increase of the elastic modulus at compression, which should improve the belt rolling conditions. However, at the same time the bottom cover damping factor increases significantly. As the indentation rolling resistance linearly depends on the damping factor and exponentially, the exponent being 1/3, decreases with an increase of the belt elastic modulus [2], the result is that belt indentation rolling resistance increases with decreasing temperature. This fact has a decisive influence on the increased electric energy consumption by the conveyor drive mechanism (Fig. 3).

By installing belt K1 in place of the previously used regular belt P, the power consumption of the drive mechanism could be reduced by approx. 11% within the entire range of the analyzed temperatures. For this reason, conveyor belt K1 can be referred to as an energy-efficient belt which not only generates lower belt indentation rolling resistances, but also is less sensitive to ambient temperature changes in comparison to the regular belt P. Within the temperature range of −20 °C ÷ 30 °C, power consumption differences are at 479.1 kW for belt K1 and at 548.9 kW for belt P.

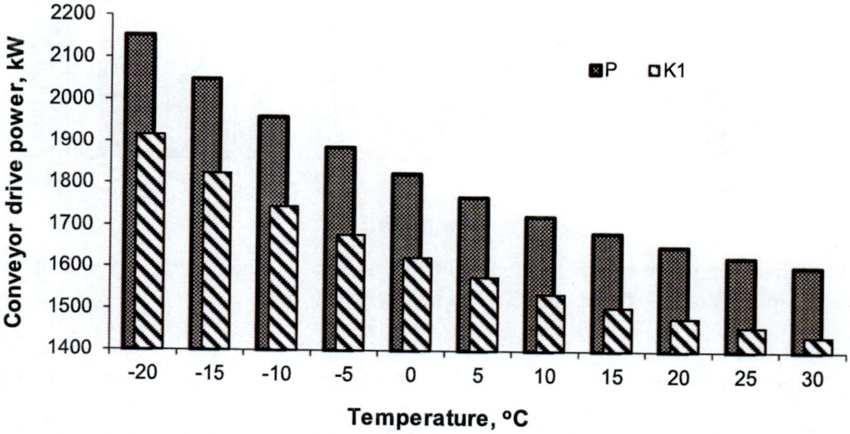

Fig. 3. Relationship between the power demand of the belt conveyor drive mechanism and temperature with the use of regular rubber P and energy-efficient rubber K1.

The most extensive information on the electric power consumption of conveyor drive mechanisms can be obtained in direct measurements performed on the conveyor. Over a period of four years, the following parameters were recorded for three selected conveyors transporting lignite to the power plant in the KWB "Belchatow" mine: operating time per month, transported mass and consumed power. The measurements were performed simultaneously and therefore the potential differences in power consumption can be assumed not to result from different operating conditions. The conveyors were of identical design, with steel-cord belts installed: two of them worked with the regular belts and one of them – with the energy-efficient belt. The analyzed conveyors were located in parallel lignite transport routes. This fact hinders the comparison, because significant differences occurred in the mass of the transported lignite and in the operating time of the conveyors, with respect to both individual months and the total time. As a result, the actual mean conveyor capacities (the quotient of the mass transported in a particular month and the operating time) varied in individual months. Therefore, a methodology was needed to compare the power consumption data, so that the comparison was reliable despite the differences [3–5]. The calculations included unit consumption of the energy needed to transport 1 Mg of lignite over a distance of 1 km. The changes of the unit energy consumption indicator were investigated from the perspective of the influence of conveyor capacities and weather conditions.

In the Bełchatów mine one of the largest lignite mines in Europe and the biggest in Poland managers of the conveyor belt department conducted four-year lasting observations of the energy consumption of 3 conveyors. The almost identical conveyors operated in parallel coal lines. At the end of each months managers recorded a large diversity of operating times, transported lignite mass and energy consumed. The data were analysed by researchers at the Faculty of Geoengineering, Mining and Geology at the Wroclaw University of Science and Technology [3–5]. Based on recorded times, mass and energy used real average capacity and the unit indicator of energy consumed for transfer 1 Mg of coal to the distance of 1 km (the length of all conveyors) were calculated for the following months of the analyzed period. The statistically significant differences in the energy consumption between conveyors were noticed. The cause of differences may be variable level of real average capacity of analysed conveyors. It was found that the capacity of conveyors is negatively correlated with the analyzed index of energy consumption. It is worth to underlain that the differences in the consumption of the energy on the same conveyor are greater than the differences between the conveyors.

The reason for statistically significant differences in the energy consumption index WskZE between the tested conveyors is the statistically significant difference in the average capacity of both conveyors Q (Fig. 4).

Analysis of selected regression curves selected for both conveyors separately and data from all conveyors together showed that changes in the energy consumption index of about 75% could be explained by differences in average capacity, as the R2 indexes for the selected reciprocal models reach this level. Such a significant impact of the conveyor load on energy consumption has vast potential for savings. An increase in average capacity from 2000 Mg/h to 2700 Mg/h can be associated with a decrease in the energy consumption index even by 80 W/Mg/km. This reduction may be achieved without any investments by controlling loads of conveyors from the Traffic Control Center (COKR) [4].

Theoretical analysis based on the power required to cover belt conveyor resistance to motion [16, 20, 21] leads to the formula on unit energy consumption indicator WskZE for horizontal conveyors in the form:

$$WskZE(v, Q_r) = \frac{Cfg}{\eta}(v\frac{m_k + 2m_t}{Q_r} + \frac{1}{3,6}) \tag{1}$$

Where:

Q_r - the real capacity of the belt conveyor,

v – the speed of the belt (constant in all analyzed conveyors 5,24 m/s),

Cf – artificial friction factor,

g - gravitational acceleration in m/s^2,

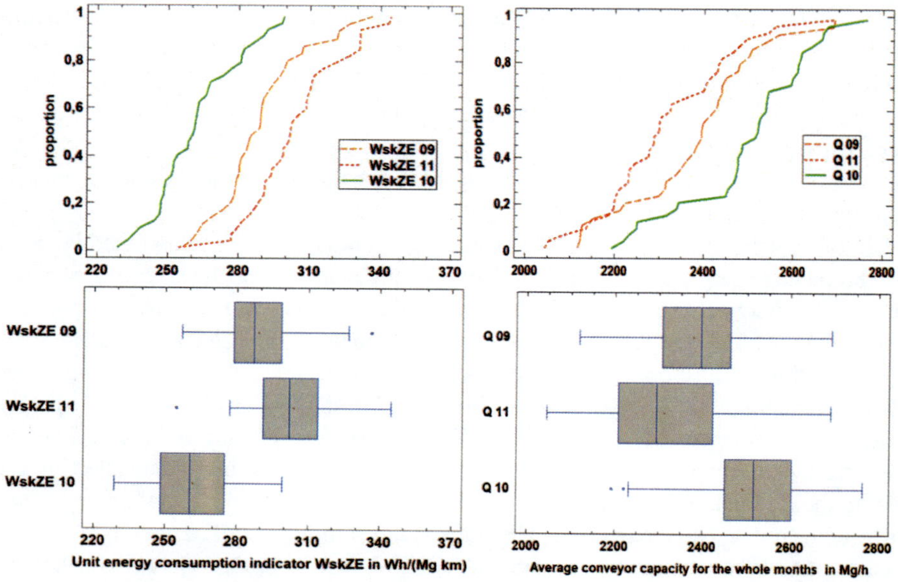

Fig. 4. Quantile plots of WskZE and Q for belt conveyors with standard belt (WskZE 09 and 11) and energy-saving belt (WskZE 10).

h - power drive efficiency in %,
m_k – the unit mass of the rotating parts of carrying and returning idler rollers (kg/m),
m_b – the unit mass of the belt (kg/m).

The output shows the results of fitting a reciprocal-X model to describe the relationship between $WskZ_{ENB}$ (indicator for conveyors with the normal/standard belt) and Q_{NB}. The formula of the fitted model is:

$$W_{sk}ZE_{NB}(Q_{NB}) = 6,45931 + 677878/Q_{NB} \tag{2}$$

Since the P-value in the ANOVA table is less than 0,05, there is a statistically significant relationship between $WskZE_{NB}$ and Q_{NB} at the 95,0% confidence level.

The R-Squared statistic indicates that the model as fitted explains 80,6961% of the variability in $WskZE_{NB}$. The correlation coefficient equals 0,89831, indicating a moderately strong relationship between the variables. The standard error of the estimate shows the standard deviation of the residuals equals 9,22522. This value is used to construct prediction limits for new observations (Fig. 5).

The same analysis was done for the conveyor with the energy-saving belt $WskZE_{ESB}$ (Conv.10). The formula describing the fitted model is:

$$W_{sk}ZE_{ESB}(Q_{ESB}) = 0,345643 + 648767/Q_{ESB} \tag{3}$$

The R-Squared statistic indicates that the model as fitted explains 72,4931% of the variability in $WskZE_{ESB}$. The correlation coefficient equals 0,851429, indicating a moderately strong relationship between the variables. A bit less than for standard (normal belt). The standard error of the estimate shows the standard deviation of the residuals equals 9,70694. Very similar to the earlier case (Fig. 5).

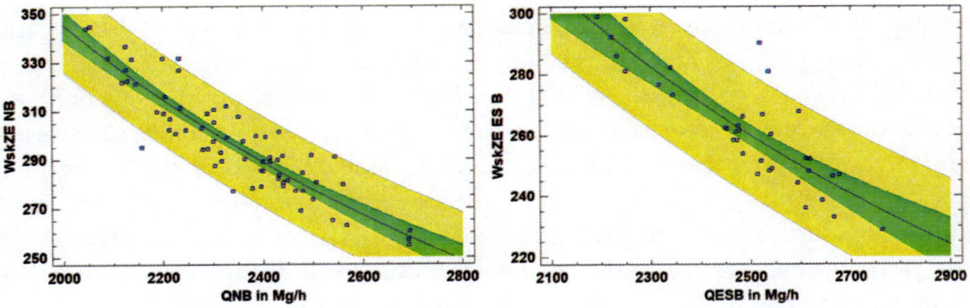

Fig. 5. Fitted *WskZE(Q)* models for the data for 2 conveyors with normal belt (2, left side) and energy-saving belt (3, right side).

After putting all curves on one chart (Fig. 6) we can see visually statistically significant differences between unit energy consumption indicators WskZE(Q) for normal and energy-saving belts. The differences are not too high from 5,98% to 6,64%, average about 18,37 Wh/(Mg km). We should take into account that it gives about 170 MWh/km per year (for 40 million Mg of lignite excavated and transported to power plant). In the Bełchatow mine, there is about 160 km of conveyors and about 1/3 transporting coal. Annual energy savings from coal conveyors can reach 9 GWh per year. Additional savings may be obtained by adjusting the speed of conveyors to reduced load as was achieved in the Konin mine [12].

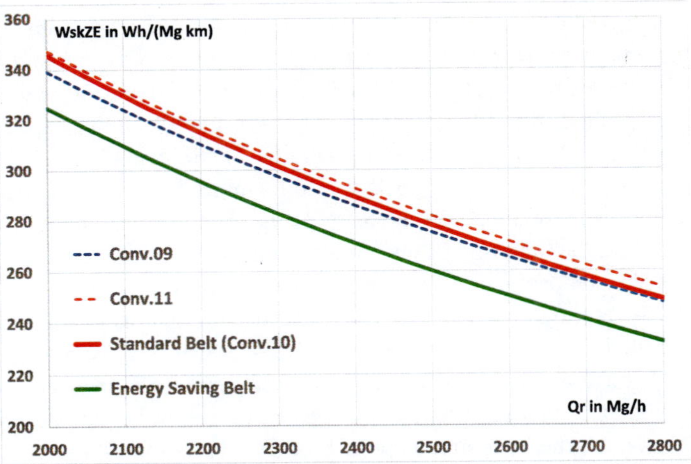

Fig. 6. Lines for fitted models for *WskZE(Q)* showing real energy savings of application of energy-saving belts on coal conveyors in the Bełchatów lignite mine.

4 Implementation of CSR in the Company Strategy

The article focuses on the concept of corporate social responsibility. Its authors address the problem of incorporating the idea of "energy-efficiency" into the strategy of "X" – a company which until present has not included such aspects in their reports. The website of the entire group contains information on non-financial reporting being performed since 2017. The most recent annual report of 2019 states that the organization promotes innovations and acts in accordance with the rules of collective responsibility, and also that it is increasingly often faced with ecological problems and social challenges. The group focuses on six primary topic areas, which include energy (consumption), raw materials and their consumption, sustained development and innovations, occupational health and safety, social standards and work conditions, and sustained development of the supply chain. Sustained development has been proven to be an important factor contributing to the long-term success of a company. The group expects their partners to meet strict requirements, especially in the industry sector.

The issues related to the area of energy consumption include the promotion of an efficient and responsible energy usage in production, with the chosen indicator being the energy intensity understood as energy consumption (kWh) per unit produced (kg or piece), and with the goal being to achieve improvement over a period of a year. The company reports that in 2019 the goal was partially achieved, and the initiative will be continued in the following years. Plans for the 2020 horizon include further optimization of energy consumption through the efficiency-related projects as well as the planning and piloting an energy concept.

The current strategy of the company "X" outlines goals for the group members. Polish industry sector, which includes both the production and sales, has guidelines e.g. for obtaining energy intensity at 2% per year until 2021 (development of energy intensity), or development of waste & scrap rate 5% p.a.

An interesting solution would be to include the results of the research on belt properties and belt energy consumption in the strategy of a conveyor belt manufacturer. The strategy of the company necessitates defining the strategic goals and identifying their meters, i.e. the methods for measuring their achievement [19], in this case the achievement of the energy-related CSR goals. In the case of the manufacturer of the tested conveyor belts, one of the strategic goals would be to minimize environmental impact by reducing the operational energy intensity of the manufactured belts in the supply chain. In such case, the reporting of this information and data would be consistent with the strategy of the companies forming the structure of the entire group.

Currently, the group's non-financial reports use the international GRI Standards, and mention only one indicator related to GRI 302: 302-1: Energy consumption within the organization. It informs that the company does not purchase renewable fuels. At the same time, it indicates that limiting energy consumption is of key importance, and the stress is placed on the energy required for the production processes, as well as for energy conversion and distribution. The recent report of the holding group provides information about the total energy usage over the years. The lowest consumption level was observed in 2019. The laboratory tests focused on energy savings in the belt transportation sector may become part of the strategy used by the company in its non-financial reports. Supplementing the reports with detailed guidelines regarding energy, such as unit energy consumption, based on international GRI 302-1-5 Standards, will become part of a complex approach to sustained reporting.

It will also allow including greenhouse gas emission indicators GRI 305: Emissions 2016 3305-1: Direct (Scope 1) GHG emissions (the emission level is directly related to the energy consumption in production) and 305-2: Energy indirect (Scope 2) GHG emissions.

5 Conclusion

The GRI Standards have been created on the basis of international standards, requirements and guidelines which have been developed over years in order to aid the implementation of responsible business systems and the reporting on sustained corporate development.

The results of both laboratory tests and computer simulations of belt indentation rolling resistance, as well as measurements of electric energy consumption in conveyor drive mechanisms confirm that installing energy-efficient conveyor belts, which generate lower indentation rolling resistances, leads to significant economic savings (reduced power demand from conveyor drive mechanisms and as a result lower electricity costs for the mine), to environmental advantages (lower CO_2 emissions to the atmosphere, and most importantly to health gains, as the quality of air improves. This is one aspect of energy-efficiency, viewed from the perspective of operating the conveyor. However, attention should be also paid to the issues related to sustained energy consumption by the belt manufacturer, which in this case is implemented and reported in non-financial reports. The authors believe that such a complex approach is recommended and should gain popularity in the future. As a result, it will contribute to the growing importance and broader application of the idea of corporate social responsibility.

Acknowledgments. The research work co-founded with the research subsidy of the Polish Ministry of Science and Higher Education granted for 2020.

References

1. Agenda 2030: Transforming our world: the 2030 Agenda for Sustainable Development; A/RES/70/1 United Nations Resolution adopted by the General Assembly on 25 September 2015 (2015). https://sustainabledevelopment.un.org
2. Bajda, M.: Influence of the rubber cover on the conveyor belt's rolling resistance over idlers, Dissertation. Wroclaw University of Technology (2009)
3. Bajda, M., Jurdziak, L., Konieczka, Z.: Comparison of electricity consumption by belt conveyors in a brown coal mine. Part 1. Study of statistical significance of differences and correlations. Surf. Min. **5**, 4–14 (2018) (in Polish)
4. Bajda, M., Jurdziak, L., Konieczka, Z.: Comparison of electricity consumption by belt conveyors in a brown coal mine. Part 2. Study of the belt conveyors capacity influence. Surf. Min. **3**, 26–38 (2019) (in Polish)
5. Bajda, M., Jurdziak, L., Konieczka, Z.: Comparison of electricity consumption by belt conveyors in a brown coal mine. Part 3. Correction of the forecasts of the unit energy consumption by temperature influence Surf. Min. **4**, 12–23 (2019) (in Polish)

6. Chen, R., Hung, S., Lee, C.: Corporate social responsibility and firm idiosyncratic risk in different market states. Corp. Soc. Respons. Environ. Mgmt. **25**, 642–658 (2018). https://doi.org/10.1002/csr.1483
7. Fijałkowska, J., Macuda, M.: CSR reporting practices in Poland. In: Długopolska-Mikonowicz, A., et al. (eds.) Corporate Social Responsibility in Poland CSR Sustainability, Ethics and Governance, pp. 195–212. Springer, Cham (2019). https://doi.org/10.1007/978-3-030-00440-8_13
8. Gladysiewicz, L.: A Method of Determining the Primary Resistance of a Belt Conveyor With a Particular Consideration of Belt Properties. Wroclaw University of Technology Press (1990)
9. Golob, U., Podnar, K.: Researching CSR and brands in the here and now: an integrative perspective. J. Brand Manag. **26**, 1–8 (2018). https://doi.org/10.1057/s41262-018-0112-6
10. GRI Standards. https://www.globalreporting.org/standards/media/1009/gri-302-energy-2016.pdf
11. Krasodomska, J.: Non-financial reporting in the light of consultative processes. Research Papers of Wroclaw University of Economics, vol. 479, pp. 99–107 (2017) (in Polish)
12. Kasztelewicz, Z., Szymanski, J.: Energy saving control method of electrical drives in conveyor belts with regulated speed in surface brown coal mine. Arch. Min. Sci. **53**, 319–329 (2008)
13. Official Journal of the European Union 2014 Official Journal of the European Union 15.11.2014 L 330/1 DIRECTIVE 2014/95/EU Of the European Parliament and of the Council of 22 October 2014 amending Directive2013/34/UE as regards disclosure of non-financial and diversity information by certain large undertakings and groups EUR-Lex Access to European Union law. http://eur-lex.europa.eu/. Accessed on 28 March 2019
14. Pactwa, K.: Scope of Implementation of Sustainable Development Goals by the Mining Sector in Poland. Faculty of Geoengineering, Mining and Geology Wroclaw University of Science and Technology (2019) (in Polish)
15. Spaans, C.: The calculation of the mine motion resistance of belt conveyors. Bulk Solid Hand. 11 **4**, 809–826 (1991)
16. Suchorab, N.: Specific energy consumption -the comparison of belt conveyors. Min. Sci. **26**, 263–274 (2019). https://doi.org/10.5277/msc192619
17. Szadziewska, A.: The role of accounting in measuring the effectiveness of CSR activities. J. Manage. Finan. Pap. Rep. Fac. Manage. Gdansk Univ. **4**, 257–277 (2013). (in Polish)
18. Wozniak, J., Pactwa, K.: Environmental activity of mining industry leaders in Poland in line with the principles of sustainable development. Sustain. **9**(11), 1903 (2017). https://doi.org/10.3390/su9101903
19. Wozniak, J.: The Role and Implementation of the Concept of Social Responsibility in the Functioning of Mining and Energy Industry. Faculty of Geoengineering, Mining and Geology Wroclaw University of Science and Technology (2019) (in Polish)
20. Zhang, S., Xia, X.: A new energy calculation model of belt conveyor. AFRICON 2009 Nairobi, pp. 1–6 (2009). https://doi.org/10.1109/AFRCON.2009.5308257
21. Zhang, S., Xia, X.: Modeling and energy efficiency optimization of belt conveyors. Appl. Energy **88**(2011), 3061–3071 (2011)

Improvement of the Properties of Polymer Composites by Mineral Basalt Modified with Tetraethoxysilane

J. Aneli[1(✉)], L. Shamanuri[1], and N. Bakradze[2]

[1] R. Dvali Institute of Machine Mechanics, 10 Mindeli Street, 0186 Tbilisi, Georgia
jimaneli@yahoo.com

[2] Georgian Technical University, 77 Kostava Street, 0177 Tbilisi, Georgia

Abstract. Ultimate strength, softening temperature, and water absorption of the polymer composites based on dean epoxy resin (type ED-20) and low-density polyethylene (PE) with unmodified and modified by tetraethoxisilane (TEOS) mineral basalt are described. Comparison of experimental results obtained for investigated composites shows that they containing modified filler have more high values of mentioned above parameters than composites with unmodified filler at definite loading. The maximum of ultimate strength for composites with modified filler is higher than that for composites with unmodified filler. First composites have also higher softening temperature and lower water absorption. Dependence of the properties of composites has an extreme character- maximums of the technical parameters appear the composites with definite concentrations of filler. Experimental results are explained in terms of structural peculiarities of polymer composites.

Keywords: Epoxy resin · Polyethylene · Basalt · Polymer composite · Mechanical and thermal properties · Water absorption

1 Introduction

Today the polymer composites attract a great attention of the scientists and engineers. High monitoring of the exploitation properties in wide interval, high durability, stability to aggressive media, lightness, easy technology of obtaining and finally low cost are the fundamental characteristics of these materials. The noted factors led to the high competitiveness to such tradition materials as metals, ceramics, wood and skin. Currently the polymer (synthetic or natural) composites with different mineral fillers are very spread [1–7]. Thanks to these fillers many properties of the composites are improved -increases the durability and rigidity, decrease the shrinkage during hardening process and water absorption, improves thermal stability, fireproof and dielectric properties and finally the price of composites becomes cheaper [3–5]. At the same time, it must be noted that the mineral fillers at high content led to some impair of different physical properties of composites. Therefore, the attention of the scientists is attracted to substances, which would be remove mentioned leaks. It is known that silicon organic substances (both low

A. Batako et al. (Eds.): GCMM 2021, LNNS 335, pp. 415–422, 2022.
https://doi.org/10.1007/978-3-030-90532-3_32

and high molecular) as "soft materials" reveal hydrophobic properties, high elasticity and durability in wide range of filling and temperatures and consequently introduction of these materials to the polymer blends, in general, can increase compatibility of ingredients and respectively increase of the mineral filler concentration in the composites [6, 7]. Since the price of mineral fillers is commonly very low than that for polymer (binder), the main investigation in the field of composite technology is directed to creation of high filled composite materials. If we foresee that the composites may be obtained on the basis of different natural or artificial remains, the application of these materials will be connected with great economical effect.

Application of method of so-called polymerization filling (polymerization on the filler particles) as a rule leads to obtaining of composites with high mechanical properties due to formation of the chemical bonds on the filler surfaces between polymer and active groups on the particle's surfaces [3, 4]. Although this method consists several stages of processes and is relatively expensive.

At this time the polymer layered silicate nanocomposites are rather popular [5–7]. There are three main methods of their synthesis: in situ polymerization, solution intercalation, and melt processing (or melt blending). A fourth method, using sol–gel techniques, can in principle be applied but is not utilized often, because of the higher temperatures needed and the tendency of the inorganic component to aggregate. Each of the three main approaches may yield exfoliated, intercalated, or a mixed of exfoliated and intercalated structures. The degree of exfoliation versus intercalation depends on a host of experimental conditions, such as monomer type, solvents, temperatures, etc.

The purpose of presented work is the investigation of effect of modify by tetraetoxisilane of the mineral basalt on mechanical strengthening, thermal stability and water adsorption.

2 Experimental

The initial TEOS were supplied from Sigma–Aldrich. Mineral -bentonite as a filler was used. The organic solvents were purified by drying and distillation. FTIR spectra were recorded on a Jasco FT/IR-4200 device.

The physical-mechanical and thermal properties of obtained composite polymer materials based on activated basalt were testified by standard methods.

Reaction of silanization of basalt with TEOS was provided by following manipulations: in a three-necked flask supplied with a mechanical mixer, thermometer and dropping funnel, to a solution of 50 g basalt in 80 ml anhydrous toluene the toluene solution of 1.5 g (0.0072 mol) in 5 ml toluene was added. The reaction mixture heated up to boiling point of used solvent during 5 h. After this the reaction mixture was filtered and dried under vacuum at 1000C up to constant weight.

High temperature condensation reaction between basalt and TEOS was carried out in anhydrous toluene solution (~35%). The masses of TEOS were 5 and 10% from the mass of filler. The reaction system was heated at the solvent boiling temperature (~110 °C) during 5–6 h by stirring. After that solvent ethyl alcohol was separate from precipitate and remained tetraethoxysilane on the rotor-evaporator and work thermal the modified basalt as at the first stage pure basalt up. The reaction goes by following scheme (the direction of reaction was improved by FTIR spectra analysis):

$$\begin{array}{lcl} \text{-OH} & & \text{-O-Si(OEt)}_3 \\ / & & / \\ \text{Si -- OH} + \text{n Si(OEt)}_4 \text{ - EtOH , °C} & \rightarrow & \text{Si --O-Si(OEt)}_3 \\ \backslash & & \backslash \\ \text{-OH} & & \text{-O-Si (OEt)}_3 \end{array}$$

On the basis of modified basalt and epoxy resin (of type ED-20) the polymer composites with different content of filler were obtained after careful wet mixing of components in mixer. After the blends with hardening agent (polyethylene- polyamine) were placed to the cylindrical forms (in accordance with standards ISO) for hardening at room temperature during 24 h. The samples hardened later were undergo to temperature treatment at 120 °C during 4 h.

It was shown that the modification goes more successfully when the masses of TEOS were near 5% from the masses of the filler. The concentration of powder basalt (average diameter up to 50 μ) was changed in the range 20–60 mass %. Filler was modified with 5 and 10 mass % of TEOS. Analogical procedures were conducted in case of composites based on PE.

Following parameters were defined for obtained composites: ultimate strength, softening temperature and water absorption.

3 Results and Discussion

On the Figs. 1 and 2 the curves of dependence of ultimate strength on the filler concentration of the composites based on dian epoxy resin (ER) and Polyethylene (PE) filled by high dispersed basalt with average diameter of particles <50 μ are presented.

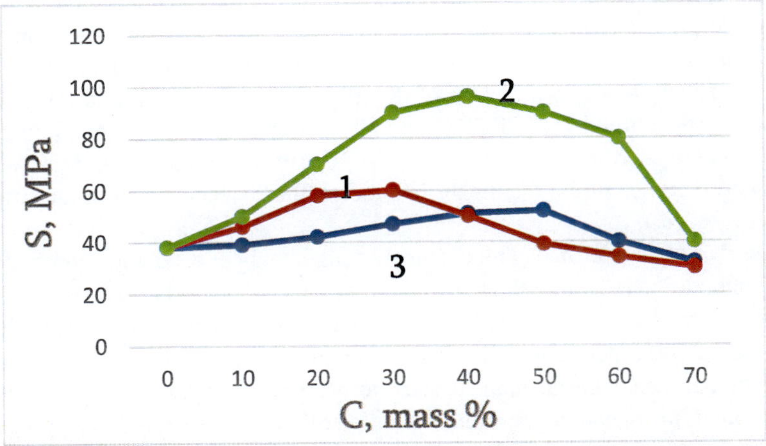

Fig. 1. Dependence of the ultimate strength on the filler concentration for composites based on epoxy resin and basalt. 1-with unmodified basalt, 2 and 3 –with modified by 5 and 10 mass % of tetraetoxisilane, respectively.

The forms of all presented on the Figs. 1 and 2 curves have an extreme character with maximums of mechanical strength. However, the place on the diagrams and the height of these maximums essentially depend on the amount of modifier TEOS.

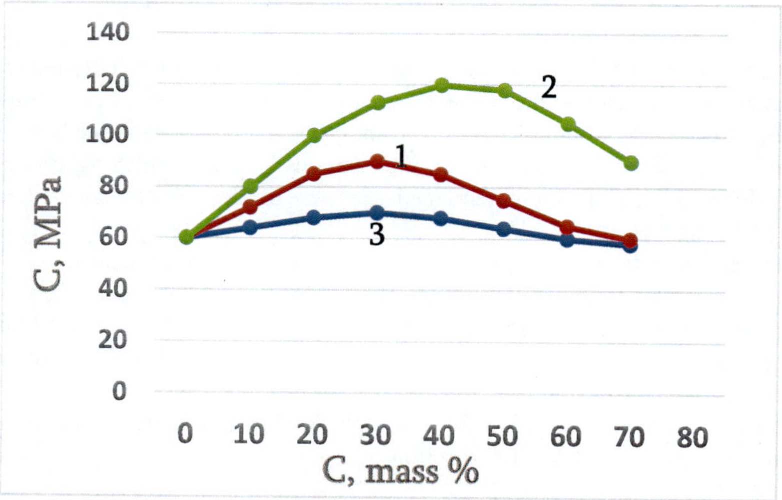

Fig. 2. Dependence of the ultimate strength on the filler concentration for composites based on polyethylene and basalt. 1-with unmodified basalt, 2 and 3 –with modified by 5 and 10 mass % of TEOS, respectively.

The curves presented on the Fig. 1 allow us to make a conception that mechanical properties of the composites increases till definite maximal significances after which decreases. The curve 1 corresponds to composite containing the filler without modifier. The shape of this curve reflects well known dependence for filled polymer composites. The mechanical strengthening of this composite with unmodified basalt has the maximum at relatively low concentration of the filler, while this maximum for analogical composite with modified with TEOS (5 and 10 wt %) basalt the noted maximum shares to more high concentrations of this filler (Fig. 1, curves 2 and 3). This result has practically important meaning – the higher is mineral filler concentration in the composite the lower is its cost.

The curves of dependence of mechanical strengthening of the composites with modified and unmodified bentonite show that modifier molecules displaced on the filler particle surface increases it activity (expressed with enhancing of composite material in result of chemical reactions between active groups of ingredients) till definite concentrations of the filler, higher which the mechanical characteristics of composite decreases. One of the main reason of this phenomenon is the formation of number structural defects (mechanical cracks, empties) in result of formation of the surfaces non reacted with organic part of the binder, amount of which increases with increasing of filler concentration. It is due to so called effect of high filling (Rebinder effect).

In the composites with modified fillers the characteristic maximums on the curve of dependence ultimate strengthening on the filler concentration modifier molecules enhance the interaction between heterogeneous phases from one side and absorb the mechanical stresses in composite body at hardening from other one. Therefore, the maximums on the curves for the composites with modified fillers share to more high concentrations of the filler. However, at following increasing of filling in the composite the modifier phase increases, which is usually soft phase and plays the role of structural defects. Increasing of this phase leads to softening or decreasing of composite mechanical strengthening.

It is interested that the mechanical properties of the composites, from the other side, are better for the composites, filler in which is modified by 5% of TEOS. The composites with fillers modified with more high concentration of the modifier (10%) are characterized with less strength. It may be proposed that a reason of this phenomenon is non full using of modifier in the reactions of filler modifications. In this case the modifiers molecules can play the role of plasticizer and therefore weaken the strength of the material. The decreasing of this parameter for composites takes place also at relatively high concentrations of the filler, which can be attributed as appearing of the regions with dry filler particles in the polymer matrix creating the defect places in it and, consequently, lead to weakening of the composites.

Analogical discuss we can provide in case of composites based on PE and basalt.

The results of investigations of thermal stability of composites obtained by method Vicat are presented as the dependences of the value of deepening of the indenter in the composite samples body at increasing temperature (Figs. 3, and 4).

In accordance with Figs. 3 the change of thermal stability of composites based on epoxy resin and different concentration reflex the character of the dependence of composite mechanical strength on the filler concentration. It can be supposed that the high mechanical properties composites are characterized with more perfect structure due to homogeneous distribution of the filler particles in the polymer matrix at relatively low concentrations of the structural defects ensuring a creating homogeneous temperature field (without gradients) in the composites.

From the curves presented in Fig. 3 it is seen that modified with TEOS basalt as filler in the composite effects to definite extent on the thermal stability of the composite. Namely the softening of the composites containing modified by TEOS basalt begins at relatively high temperatures than in case of ones containing unmodified filler. This result is in good agreement with ones obtained at investigation of the mechanical properties of corresponding materials.

Enhancing of the compatibility of the ingredients in the composites containing modified by TEOS reflected also on the characteristic of water absorption of the composites. In accordance with Figs. 4 the composites with modified filler show more high-water stability than analogues with unmodified fillers. It can be proposed that the composites containing this filler modified with more high concentration of the same modifier will be more hydrophobic than presented samples, because the named composites with high concentration of modifier as high hydrophobic component will.

The obtained experimental results may be explained in terms of composite structure peculiarities.

Silane molecules displaced on the surface of basalt particles lead to activation of them and participate in chemical reactions between active groups of ES (hydroxyl)

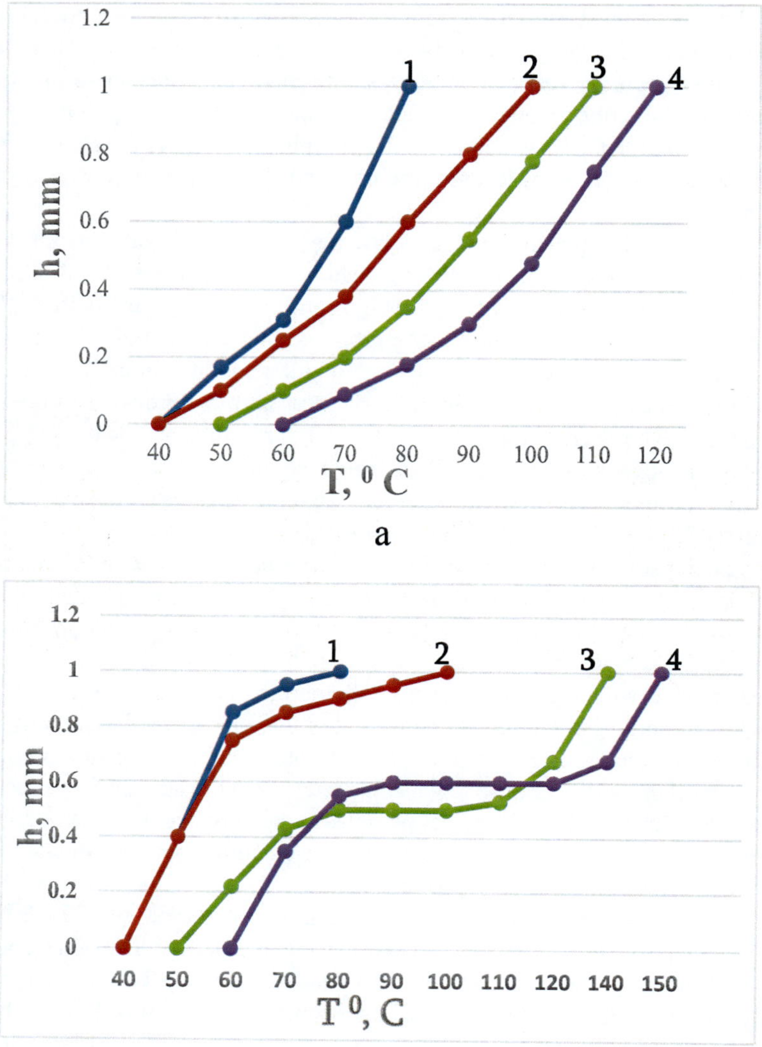

Fig. 3. Dependence of the softening of composites on temperature for composites based on epoxy resin (a) and polyethylene (b). 1,2 – with unmodified and modified by 20 mass % basalt, 3,4 –with unmodified and modified by 50 mass % basalt.

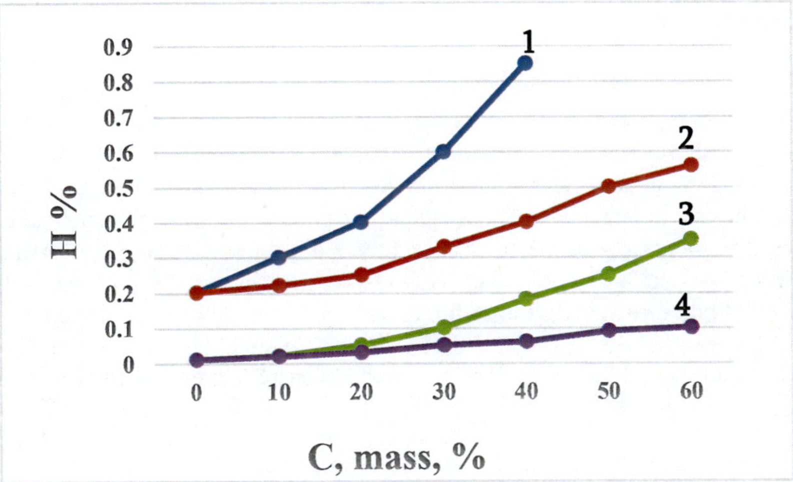

Fig. 4. Dependence of water absorption in composites on the basalt concentration. 1,2- ED-20 with unmodified and modified by 20 mass % basalt, 3,4 - PE with unmodified and modified by 20 mass % basalt

and homopolymer (epoxy group). Silane molecules create the "buffer" zones between filler and the homopolymer. This phenomenon may be one of the reasons of increasing of strengthening of composites in comparison with composites containing unmodified fillers. The composites with modified basalt display more high compatibility of the components than in case of same composites with unmodified filler. The modified filler has more strong contact with polymer matrix (due to silane modifier) than unmodified basalt. Therefore mechanical stresses formed in composites by stretching or compressing forces absorb effectively by relatively soft silane phases, i.e. the development of micro defects in carbon chain polymer matrix of composite districts and finishes in silane part of material the rigidity of which decreases.

The structural peculiarities of composites display also in thermo-mechanical properties of the materials. It is clear that softening of composites with modified by TEOS composites begins at relatively high temperatures. This phenomenon is in good correlation with corresponding composite mechanical strength. Of course, the modified filler has more strong interactions (due to modifier) with epoxy polymer molecules, than unmodified filler.

The amplified competition of the filler particles with macromolecules by TEOS displays well also on the characteristics of water absorption. In general loosening of micro-structure because of micro empty areas is due to the increasing of filler content. Formation of such defects in the microstructure of composite promotes the water absorption processes. Water absorption of composites with modified basalt is lower than that for one with unmodified filler to some extent. The decreasing of water absorption of composites containing silane compound is result of hydrophobic properties of ones.

4 Conclusions

Investigation of mechanical strengthening, softening temperature and water absorption for polymer composites based on epoxy resin and polyethylene with mineral filler basalt unmodified and modified by low amount (less 5 and 10 mass%) TEOS leads to conclusion that modify agent stipulates the formation of heterogeneous structures with higher compatibility of ingredients and consequently to enhancing of noted technical characteristics. The increasing of mechanical and thermal properties of the composites with relatively high contain of modified by TEOS basalt filler has practical aspect - saving of flexibility of polymer materials after filling and improvement of their properties.

Acknowledgments. Authors thank Shota Rustaveli National Science Foundation of Georgia for funding of the publication of this article based on the project CARYS-19–2692.

References

1. Achilias, D.S., Roupakias, C., Megalokonomos, P., Lappas, A.: J. Hazard. Mater. **149**, 536 (2007)
2. Jassim, A.K.: Procedia Manufac. **8**, 635 (2017)
3. Gorrasi, J., et al.: Polymer **44**(8), 2271 (2003)
4. Chen, H., Giannelis, E.P., Sogah, D.Y.: J. Am. Chem. Soc. **121**, 1615 (1999)
5. Shipp, D.A.: Clarkson University. Elsevier, Potsdam, NY, USA (2010)
6. Vo, L.T., Giannelis, E.P.: Macromolecules **40**, 8271 (2007)
7. Pavlidou, S., Papaspyrides, C.D.: Prog. Polym. Sci. **33**, 1119 (2008)

Influence Parameters the Surface Roughness of JIS SCMS420 Stainless Steel on Lapping Process

Somkiat Thermsuk, Suwit Thammasang(✉), Danai Sonsuphap, and Thanapol Sombat

Department Production Design Technology, Faculty of Technical Education, Rajamangala University of Technology Isan Khonkaen Campus, Khon Kaen, Thailand
suwit.tu@rmuti.ac.th

Abstract. This objective of research was to study the influence parameters into the surface roughness of JIS SCMS420 stainless steel on lapping process. Factorial experiment was applied to analyze the four-alumina powder size of 0.05, 0.30, 1.00, 3.00 μm with 9 lapping times of 30, 60, 90, 120, 150, 180, 210, 240 and 270 min respectively. In preparing of specimen surface roughness for experimental, this research used the ratio between the alumina powder of 200 g, alumina powder lubricant of 150 ml and water of 1 l in order to obtain the average surface roughness value (Ra) of 0.4–1.0 μm. The best result of statistical analysis exhibited the alumina powder size of 1.00 μm was obtained the lowest average surface roughness (Ra) of 0.0304 μm. Moreover, the statistical comparison of desirability was highest of 98.68%.

Keywords: Lapping process · Alumina powder · Factorial experiment

1 Introduction

At present, the continuously expanding industries are the automotive, sanitary, furniture and electronic industries. This requires more researches and analyses of work surfaces requiring very high surface roughness in order to increase product value for maximum return. In general, the process of polishing surface are processes which are widely used in industrial applications. In which the lapping process is a process to adjust the surface of a workpiece by polishing or removing the material. The result would be glossy skin and very high resolution value. The type of materials that will be used to fine-tune the aforementioned surface all of which affect the surface smoothness of the work surface in the final surface finishing, which is considered cleaning and surface modification of metal by a combination of mechanical and chemical processes. To eliminate burrs or spines on the surface of the material being processed (Debarring /burr Removing) including defects on the surface and make the texture of the workpiece look as desired, skin properties that are often modified by beauty, smoothness of the surface, adhesion ability, wear resistance, corrosion resistance from chemicals slowing or preventing tarnishing and electrical conductivity etc. Lapping is a method that adjusts the plane surface to increase accuracy. Abrasive machining technology [1–4] possessing extinct advantages of high

A. Batako et al. (Eds.): GCMM 2021, LNNS 335, pp. 423–430, 2022.
https://doi.org/10.1007/978-3-030-90532-3_33

efficiency, high flexibility, high accuracy and low surface damage has been extensively used in machining of flat and free form surfaces as compared with traditional machining technologies, such as cutting, milling and drilling processes with geometric tools [5–7]. The most common scrubbing method is bearing which can be found in piston in the oil pump [8]. The measurement accuracy is very important for high-precision technology relating to the work surface polishing process [9]. The influence on the process of scrubbing the surface in a thorough scrubbing consideration from the experiment depends on many factors [11]. The quality and accuracy of the surface shape can be achieved by forming the type in the lapping process which requires continuous experimentation to ensure high efficiency [10, 12, 13]. Adjust Including statistical analysis of variance (ANOVA) [14, 15]. The objective of this research is to compare the influences of 4 types of alumina polishing powder on the average surface roughness (Ra) of JIS SCMS420 stainless steel in the surface treatment process by applying the Design and Analysis of Experiment (DOE) principles to perform process parameters determination of statistical significance.

2 Experimental Procedure

Beginning with the preparation of the work piece before heat treatment by bringing the stainless flat bar steel JIS SCMS420 to cut to size 35 × 35 × 6 mm, then the workpiece is cut by milling machine to the size 30 × 30 × 5 mm. After that, storing the surface with a flat surface grinding process and using a medium aluminum oxide grinding wheel (H) No. (No) 80 on the surface of the test piece on both sides with area size of 30 × 30 mm to have the average surface roughness (Ra) of 0.4–1.0 μm. The next step is the continuous heat treatment process starting at annealing temperature of 400 °C for a period of 10 min, then increase the temperature to 820 °C for a period of 30 min. When the time limit is reached, then dip the workpiece into the plating substance of black-colored used engine oil. While oiling should have oscillations in order to thoroughly coat the workpiece in every part. Once the oil has been moistened, all of the work will be placed in a sand bucket to allow it to cool down. Last, the final step is to use sandpaper to finish the surface (Fig. 1).

Fig. 1. The process of grinding a flat surface.

Fig. 2. Lapping machine.

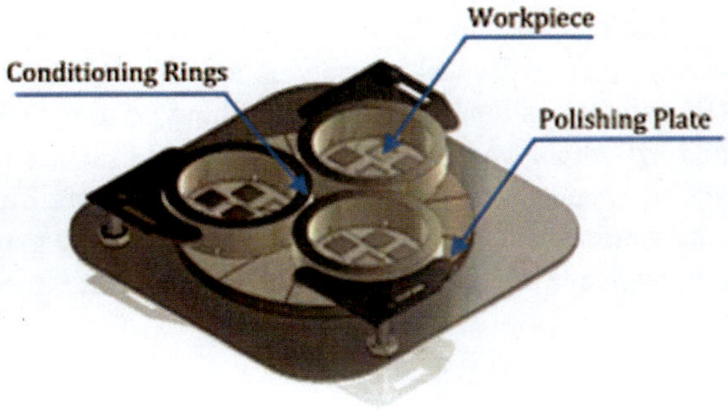

Fig. 3. Arrangement of work into the surface finishing machine.

The process of refining fine skin will start from mixing the ratio between the amount of alumina scrub powder of 200 g, lubricant, alumina scrub powder at 150 ml and clean 1 l water to get the average surface roughness (Ra) 0.4–1.0 μm. Then turn on the machine the Lapping Machine as shown in Fig. 2. and mix the ingredients first by 5 min. The next step is to take these specimens of JIS SCMS420 stainless steel to conduct the experiment by using Piece Test available on the disc's surface as shown in Fig. 3. After that, the experiment was performed by adjusting the time interval in the fine finishing process, which is 30, 60, 90, 120, 150, 180, 210, 240 and 270 min to collect the average surface roughness data: (Ra) of the specimen as shown in Fig. 4.

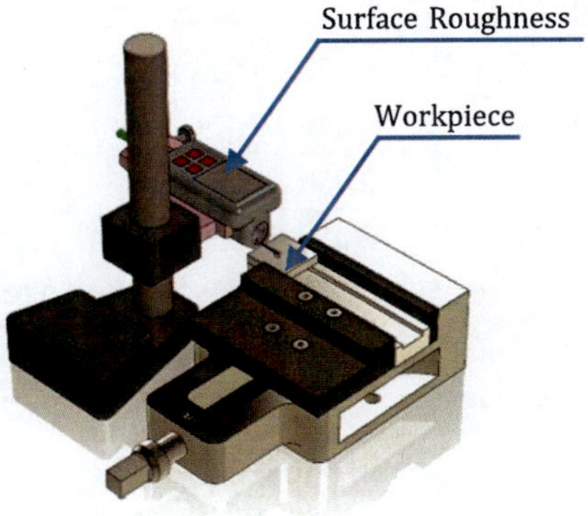

Fig. 4. Data collection of average surface roughness (Ra).

3 Results

When all the experiments were conducted according to the statistical experimental plan with the analysis of factorial experiments. With the alumina polishing powder studied in 4 types of sizes at 0.05, 0.30, 1.00 and 3.00 μm, the main parameters response for statistical analysis is the average surface roughness (Ra) of the specimen of the JIS SCMS420 stainless steel material as shown in Table 1.

The results from the factorial experiment analysis shown in Table 2 found that all 4 alumina powder (A) parameters are 0.05, 0.30, 1.00, 3.00 μm and the lapping time (B) parameters for all 9 surface finishing periods of 30, 60, 90, 120, 150, 180, 210, 240 and 270 min have influences on the average surface roughness (Ra) of the test piece for stainless steel material JIS SCMS420 in both parts. The main influence of each parameter (main effect) and the joint influence between the parameters (interaction) are at a 95% confidence level since the P-value < α value or 0.01 < 0.05.

Table 1. Average surface roughness data (Ra) for statistical analysis.

(Lapping time; min)									
(Alumina powder; μm)	30	60	90	120	150	180	210	240	270
0.05	0.510	0.118	0.143	0.103	0.098	0.050	0.098	0.042	0.050
0.05	0.246	0.290	0.144	0.103	0.115	0.050	0.087	0.038	0.049
0.05	0.169	0.126	0.100	0.098	0.104	0.112	0.075	0.034	0.057
0.05	0.211	0.378	0.151	0.181	0.095	0.075	0.080	0.188	0.033
0.05	0.270	0.076	0.142	0.103	0.103	0.132	0.070	0.043	0.037
0.30	0.415	0.249	0.100	0.063	0.081	0.086	0.067	0.041	0.058
0.30	0.085	0.102	0.117	0.054	0.107	0.071	0.057	0.065	0.071
0.30	0.253	0.327	0.137	0.114	0.074	0.097	0.111	0.070	0.038
0.30	0.284	0.081	0.111	0.084	0.121	0.087	0.053	0.048	0.035
0.30	0.167	0.278	0.117	0.124	0.053	0.093	0.053	0.053	0.056
1.00	0.146	0.077	0.051	0.049	0.056	0.055	0.051	0.045	0.030
1.00	0.054	0.058	0.071	0.071	0.051	0.036	0.035	0.030	0.033
1.00	0.031	0.075	0.050	0.044	0.039	0.053	0.052	0.024	0.032
1.00	0.045	0.035	0.056	0.056	0.050	0.048	0.047	0.038	0.026
1.00	0.076	0.059	0.065	0.065	0.049	0.047	0.045	0.026	0.031
3.00	0.090	0.088	0.079	0.087	0.056	0.065	0.062	0.061	0.056
3.00	0.091	0.107	0.082	0.077	0.072	0.078	0.067	0.044	0.049
3.00	0.085	0.083	0.081	0.069	0.082	0.076	0.072	0.054	0.050
3.00	0.093	0.070	0.078	0.075	0.085	0.066	0.049	0.047	0.051
3.00	0.083	0.078	0.072	0.077	0.077	0.067	0.071	0.052	0.048

When considering the residual analysis, the Pareto chart of the standardized effects shows the factor A is the alumina powder, B is the time to fine-tune the surface. When interaction between parameter A with parameter B is the alumina powder, lapping time is found to be beyond critical lines, so these two factors affect the workpiece roughness after fine workmanship with a significant process of skin lapping as shown in Fig. 5. When considering the results of the experiment by Minitab version 18 program for the validation of experimental forms, the examination results are shown by assuming that the form of residuals obtained from the experimental data must be in accordance with the principle of residual values and have the normal distribution from the versus fits graph which can check for the variance of the error values. The good data needs to have the variance of the error value must be consistent from the versus order graph and be free with the average close to the value of 0 in order to make the data accurate and reliable. The good data information must have characteristics of good control charts as shown in Fig. 6.

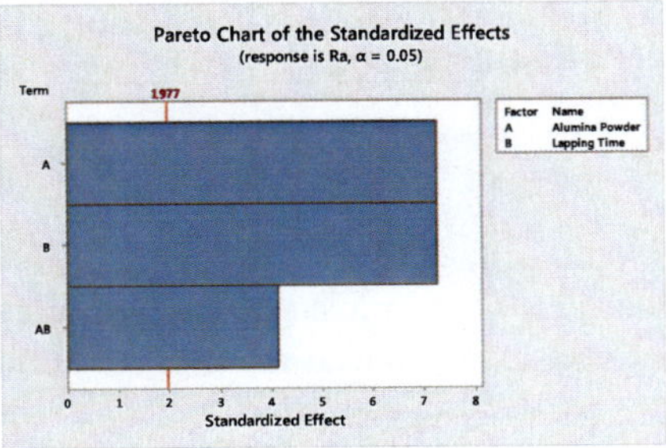

Fig. 5. Pareto chart of the standardized effect.

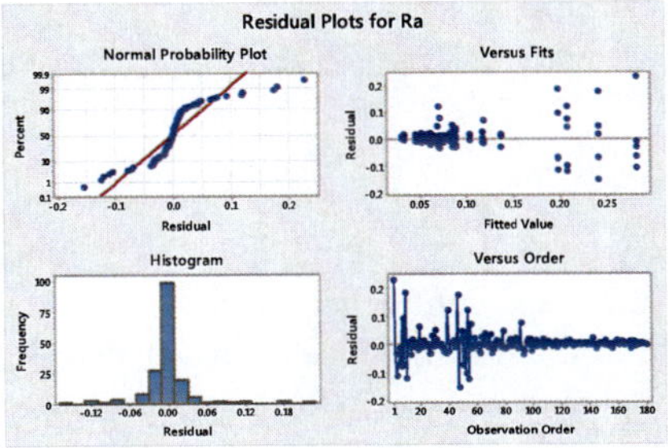

Fig. 6. (Residual plots for Ra).

When considering the main effect plot of each parameter shown in Fig. 7, it can be seen that the graph, both the alumina powder (A) parameter and the fine time adjustment lapping time (B) parameter are not parallel with the reference line in the horizontal axial. This means both of these parameters influenced the average surface roughness response (Ra) of the specimen of JIS SCMS420 stainless steel material, which gave consistent results with the results from the factorial experiment analysis as follows shown in Table 2.

Table 2. Results from factorial experiences.

Source		DF	SS	MS	F-value	P-value
Model		35	0.5714	0.016327	7.71	<0.001
Linear		11	0.4262	0.038749	18.30	<0.001
	Alumina Powder (A)	3	0.1590	0.053010	25.04	<0.001
	Lapping Time (B)	8	0.2672	0.033402	15.78	<0.001
2-Way Interactions		24	0.1452	0.006050	2.86	<0.001
	(A) * (B)	24	0.1452	0.006050	2.86	<0.001
Error		144	0.3049	0.002117		
Total		179	0.8763			

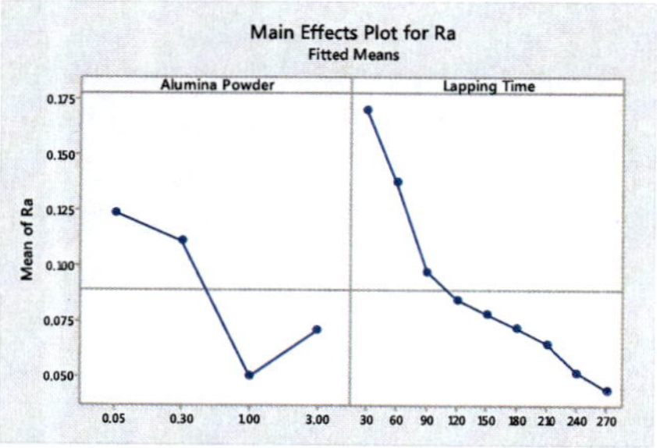

Fig. 7. (Main effects plot for Ra).

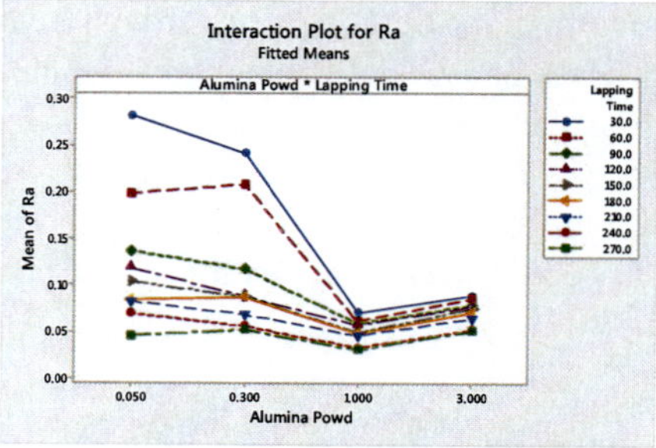

Fig. 8. (Interaction plot for Ra).

In addition, the interaction between parameters (shown in Fig. 8) can be seen that only a small part of the graph has a tendency not to be the same as most graphs. But at a statistical confidence level of 95%, it still considered a significant result. This means the common influence between these two parameters influenced the average surface roughness response (Ra) of the specimen JIS SCMS420 stainless steel material, which gave consistent results with the results from the factorial experiment analysis as follows shown in Table 2. In determining the most suitable parameters of both parameters, the alumina powder (A) and the lapping time (B), they will result in the quality of the work surface. The stainless steel material JIS SCMS420 is good, which indicates that the average roughness (Ra) is the smallest (minimum), which is the most suitable condition is the size of alumina polishing powder is 1.00 μm and the fineness adjustment time of 270 min will result in a Ra value of 0.0304 μm. In addition, when considering the satisfaction value (desirability: D) regarding statistical processing found to be very high as 98.68% as shown in Fig. 9.

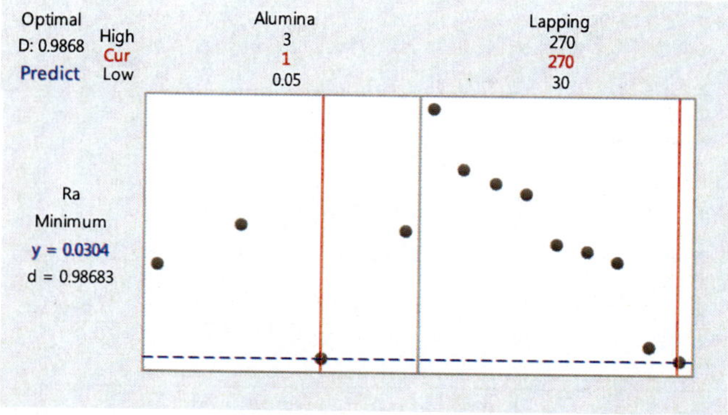

Fig. 9. (Response optimization).

4 Conclusions

By conducting a study on the influence of parameters on the surface roughness of JIS SCMS420 stainless steel in fine surface processing using the factorial design techniques, the findings are the optimum condition of this process was the size of the polishing powder alumina at 1.00 μm and the fineness is 270 min which will result in an average surface roughness (Ra) of 0.0304 μm. Moreover, these moments of fine grained increase would result in a surface roughness average (Ra) lower as well. Due to the accumulation of metal debris that replaces the furrow of the material surface until it becomes homogeneous from heat generated in the fine surface treatment process.

Acknowledgements. The author would like to thank the laboratory, Faculty of Engineering Industrial. Engineering Rajamangala University of Technology Krungthep. This facility offers the use of a lapping machine for conducting research, as well as the Minitab Version 18. software program to analyze the results. Validation of experimental forms.

References

1. Jin, M., et al.: A novel functionally graded lapping and polishing method for the improvement of material removal uniformity. J. Manuf. Process. **50**, 102–110 (2020)
2. Aurich, J.C.: Abrasive processes for micro parts and structures. CIRP Ann. – Manuf. Technol. **68**, 653–676 (2019)
3. Zhao, B., Ding, W., Chen, Z., Yang, C.: Pore structure design and grinding performance of porous metal-bonded CBN abrasive wheels fabricated by vacuum sintering. J. Manuf. Process. **44**, 125–132 (2019)
4. Qi, H., Wen, D., Yuan, Q., Zhang, L., Chen, Z.: Numerical investigation on particle impact erosion in ultrasonic-assisted abrasive slurry jet micro-machining of glasses. Powder Technol. **314**, 627–634 (2017)
5. Wang, Y., Zou, B., Wang, J., Wu, Y., Huang, C.: Effect of the progressive tool wear on surface topography and chip formation in micro-milling of Ti–6Al– 4V using Ti(C7N3)-based cermet micro-mill. Tribol. Int. **141**, 105900 (2020)
6. Yang, H., Ding, W., Chen, Y., Laporte, S., Xu, J., Fu, Y.: Drilling force model for forced low frequency vibration assisted drilling of Ti-6Al-4V titanium alloy. Int. J. Mach. Tools Manuf. **146**, 103438 (2019)
7. Wang, Y., Zou, B., Huang, C.: Tool wear mechanisms and micro-channels quality in micro-machining of Ti-6Al-4V alloy using the Ti(C7N3)-based cermet micro- mills. Tribol. Int. **134**, 60–76 (2019)
8. Babichev, A.P., Mategorin, N.V., Getmanskii, D.V., Motrenko, P.D., Nelidin, V.V.: Vibrational lapping of cylindrical parts, Russ. Eng. Res. **29**, 99–101 (2009)
9. Chang, K.-Y., Song, Y.-H., Lin, T.-R.: Analysis of lapping and polishing of a gauge block. Int. J. Adv. Manuf. Technol. **20**, 414–419 (2002)
10. Enomoto, T., Tani, Y., Orii, K.: Development of a Lapping Film Utilizing Agglomerative Superfine Silica Abrasives for Edge Finishing of a Silicon Wafer. Initiatives of Precision Engineering at the Beginning of a Millennium, pp. 391–395 (2001)
11. Makarov, V.F., Muratov, K.R., Gashev, E.A.: Optimal cutting time and speed in abrasive lapping of ceramic. Russ. Eng. Res. **37**, 916–918 (2017)
12. Qiu, Z., Fang, F.Z., Ding, L., Zhao, Q.: Investigation of diamond cutting tool lapping system based on on-machine image measurement. Int. J. Adv. Manuf. Technol. **56**, 79–86 (2011)

13. Schlecht, B., Rudolph, F., Schumann, S.: Experimental studies and simulation of hypoid gear lapping. Forsch. Ingenieurwes. **81**(2–3), 95–100 (2017). https://doi.org/10.1007/s10010-017-0214-4
14. Zhu, N., Zheng, F., Zhu, Y., Xu, S., Zuo, D.: Research of abrasive embedment-free lapping on soft-brittle lithium niobate wafer. Int. J. Adv. Manuf. Technol. **87**(5–8), 1951–1956 (2016). https://doi.org/10.1007/s00170-016-8582-8
15. Choopani, Y., Razfar, M.R., Saraeian, P., Farahnakian, M.: Experimental investigation of external surface finishing of AISI 440 C stainless steel cylinders using the magnetic abrasive finishing process. Int. J. Adv. Manuf. Technol. **83**, 1811–1821 (2016)

Knowledge Management as a Sustainable Development Supporting Method in Manufacturing Organizations – A Systematic Literature Review

Łuksz Szczekala[1] and Dorota Stadnicka[2(✉)]

[1] Doctoral School of Engineering and Technical Sciences at the Rzeszow University of Technology, Al. Powstancow Warszawy 12, 35-959 Rzeszow, Poland
[2] Rzeszow University of Technology, Al. Powstancow Warszawy 12, 35-959 Rzeszow, Poland
dorota.stadnicka@prz.edu.pl

Abstract. The importance of knowledge management (KM) is rising in every mature company as well as it is widely discussed in the literature. Many papers were published in order to discuss KM in different contexts. The aim of this work is to review and analyze literature which is connected to KM. The authors treat KM as a supporting method for sustainable development (SD) of a manufacturing system in terms of economic, environmental and social aspects. A systematic literature review was performed according to the developed research methodology, which covers research questions and keywords identification, selection and exclusion criteria and results analysis. The search was performed with the use of keywords such as knowledge management, manufacturing organization as well as sustainable development and their modifications. Afterwards, quantitative and qualitative analyses were performed. In the quantitative analysis the criteria such as a number of papers published in different years, in different journals and conference materials were used to illustrate the published works. A deeper analysis was performed on the basis of the author keywords, abstracts and full texts where available. The main goal of this paper was to analyze how deeply the aspects which concern KM and SD are discussed in the publications connected with manufacturing companies. The presented work is a part of wider research connected with the effective implementation of KM in manufacturing areas. Social aspects were identified as the least discussed in the literature and they need future studies. Additionally, the IT system to support KM and application of such emerging technologies as data mining, Augmented Reality, Internet of Things and Blockchain and Edge Computing can be further investigated in the context of SD.

Keywords: Knowledge management · Manufacturing organization · Sustainable development · Systematic literature review

1 Introduction

Knowledge management (KM) is perceived as a competitive advantage for modern high-tech companies and it is widely discussed in publications. Many literature review papers

A. Batako et al. (Eds.): GCMM 2021, LNNS 335, pp. 431–459, 2022.
https://doi.org/10.1007/978-3-030-90532-3_34

were published to discuss KM in different contexts. In the work [1] the authors discuss KM in the context of sustainability. They reviewed 45 papers directly connected with sustainable development (SD) and they discovered that there is a gap in practical aspects of KM in the context of sustainability.

This paper focuses on manufacturing companies and it reviews the publications connected with KM and SD of manufacturing organizations. Since the manufacturing companies use many resources, also non-renewable sources of materials, energy, and still engage a lot of human resources into manufacturing processes, the SD of such companies is especially important.

The second section presents an introduction to Knowledge Management (KM) and Sustainable Development (SD). Section 3 describes the research goal and methodology. Section 4 presents the review results analysis and discussion. The last section concerns conclusions and future research.

2 Knowledge Management (KM) and Sustainable Development (SD)

SD has a long history [2] although, until now many companies focused only on the chosen SD aspects. According to Porter and van der Linde [3], the concept of sustainability meets the needs of society (social aspects), is environmentally friendly (environmental aspects) and supports companies economy (economic aspects). The work [4] shows that KM powerfully mediates the associations between the entrepreneurial orientation of SMEs and their engagement in SD practices. Moreover, the work [5] indicates that KM significantly influences green innovations development. In the paper [6] the authors underline that knowledge motivates people to comply with the environmental, social, and economic SD requirements. Therefore, the KM implementation can support SD and it is worth to study how KM can be implemented.

In the paper [7] the authors present a systematic literature review of the publications connected with knowledge management systems (KMS). The authors reviewed 54 papers and they identified 14 issues connected with KMS, as follows: (1) KMS capabilities and features, (2) barriers for KMS implementation, (3) intention to use KMS, (4) KMS as a tool in the retrieval information, (5) Big Data, (6) loss of critical knowledge, (7) wrong functioning of KMS in supporting a decision process, (8) Security, (9) New technologies, (10) knowledge dissemination technologies, (11) social life, complex and dynamic environments, (12) costs of KMS, (13) performance measurement of KMS, and (14) KM process support by KMS. In the paper [8] the authors discuss Software Architecture Knowledge Management (SAKM). The work studies the ways of capturing, using, maintaining, sharing, and reusing architectural knowledge, by the systematic literature review of the published papers. They reviewed 115 works in which they identified 56 different approaches to SAKM. They also discovered that most of the approaches focus on using the already-captured knowledge. The problems of capturing, reusing, sharing and maintaining are marginally discussed.

In the work [9] the authors reviewed the published works concerning the KM strategies and processes in agile software development. They analyzed 32 studies. They discovered that in these publications 33 practices that implement a personalization strategy and 7 practices that implement codification are reported. In the work [10] the problem of sharing and the management of managers' knowledge are under review. In a structured literature review 57 papers were studied. The authors distinguished five perspectives in terms of respectively (1) practices, (2) processes, (3) organizational learning, (4) managers' viewpoints, and (5) the nature of the knowledge ("hard", "soft") involved.

The paper [11] presents a literature review of the transformation from IT-based knowledge management into KM supported by building information modelling (BIM). BIM is characterized by such features as parametric modelling, virtual visualization and a centralized platform. In the research the authors reviewed 115 papers on IT-based KM as well as 73 papers on BIM-supported KM. Such aspects as: knowledge capture, knowledge reuse, knowledge sharing, knowledge storage/retrieval, knowledge representation, knowledge classification and knowledge visualization were discussed.

In the work [12] KM on the shop floor, among others, was discussed. The authors analyzed the dependencies between: (1) knowledge management, (2) human values, (3) organizational performance, (4) teams performance, (5) organizational design, (6) teamwork, (7) awareness, (8) leadership, (9) teams autonomy and (10) knowledge sharing, presented in the literature. On the basis of the performed systematic literature review the authors defined 9 main areas of the research. Among them, the following are connected with KM: (1) Human Values vs. KM, (2) KM vs. Organizational Design, (3) KM vs. Performance and (4) KM vs. Teamwork.

In the work [13] the authors performed a systematic review of the literature looking for the information about the role and benefits of KM in World Class Manufacturing (WCM) system. They found 10 articles related to WCM and KM. According to the authors, WCM and KM present some similarities, as follows: (1) they are oriented towards people and (2) results, (3) they play a fundamental role in leadership and (4) promote effective communication, (5) they encourage teamwork and (6) seek to exceed customer expectations. In the work [14], in the systematic literature review, the authors identified the success factors critical for the knowledge management.

To sum up, it can be said that there are three pillars of the knowledge management: people, technology, and processes [7, 15, 16]. Furthermore, in KM the following aspects are crucial [17]: (1) Knowledge creation; (2) Knowledge storage and retrieval; (3) Knowledge transfer/sharing and (4) Knowledge application.

The presented aspects were taken into consideration in the systematic literature review performed in the frame of this work. Moreover, based on the authors' best knowledge, the context of SD supported by KM in manufacturing organizations was not analyzed before.

3 Research Goal and Methodology

In the study the following research goals were identified:

- RQ1: To what extent sustainable development (SD) and knowledge management (KM) are discussed in the literature in the context of manufacturing organizations?
- RQ2: What are the aspects mostly discussed in sustainable development (SD) and knowledge management (KM) in the context of manufacturing organizations?

The work methodology is based on the following steps:

- Keywords identification and searching rules development, databases identification and adoption of the selection and exclusion criteria.
- Data collection, analysis and discussion.

The following keywords were used in the search process: "knowledge management", "sustainable development" and "manufacturing processes". Since in the literature different terms can be used, some modifications of the keywords were also used: "Management of Knowledge", "Economic Sustainability", "Sustained Development", "Sustainable Growth", "Sustainable Economic Growth" and "Production".

The following searching rule was applied: *("Knowledge Management" OR "Management of Knowledge") AND ("Sustainable Development" OR "Economic Sustainability" OR "Sustained Development" OR "Sustainable Growth" OR "Sustainable Economic Growth") AND (Manufacturing OR Production).*

In the presented research four data bases were chosen to search the publications connected with the analyzed topic: Web of Science, Scopus, Science Direct and IEEE. The literature search was limited to the papers written in English. Only journal and CPs as well as book chapters were taken into consideration. Duplicated papers, which were registered in more than one database, were eliminated. It also happened that in Web of Science one paper appeared three times in the set of the retrieved papers. In the next step the papers were eliminated after reading the abstract if it was discovered that they are not in the scope. Finally 99 papers were left for the further analysis. For 67 papers a full text was available or delivered by authors.

The performed search in the databases resulted with the following number of papers connected with the discussed topics: Web of Science (43 papers), Scopus (207 papers), Science Direct (10 papers) and IEEE (22 papers).

In the first step of the analysis only titles, keywords and abstracts of the publications were considered. First of all, the analyses to present the numbers of papers connected with KM and SD in different databases were performed. Then, the analysis to see the dynamics of the increase in the number of publications was done. Next, the analysis of an article type and source were performed (article, CP, book chapter). In the next step, author keywords were analyzed in order to identify the keywords connected with KM and SD. Then, the papers which are connected with KM and SD aspects, based on the keywords, were indicated. Finally, the available full texts of the papers were studied and connections with KM and SD were discovered. The papers connected with certain aspects were indicated. The discussion concerns the research questions and indicates the aspects discussed the least.

4 Results Analysis and Discussion

99 papers were analyzed but only for 67 papers a full text was available. Table 1 presents the full list of the retrieved 99 papers including only abstracts of the papers.

Figures 1, 2 and 3 present the results of a quantitative analysis of the 99 papers. It can be seen that a large increase in the number of the published papers connected with the analyzed topic appeared in 2019 (Fig. 1).

Table 1. Papers and problems analyzed in the reviewed works – based on author keywords analysis; F – full text available, PT – publication type, JP – journal paper, CP – conference paper, BC – book chapter.

ID	Author (year)	Title		PT
P001	Roy et al. (2020)	Reactive and proactive pathways to sustainable apparel supply chains: Manufacturer's perspective on stakeholder salience and organizational learning toward responsible management	F	JP
P002	Zhou et al. (2020)	How fairness perceptions, embeddedness, and knowledge sharing drive green innovation in sustainable supply chains: An equity theory and network perspective to achieve sustainable development goals	–	JP
P003	Yusliza et al. (2020)	A structural model of the impact of green intellectual capital on sustainable performance	F	JP
P005	Abbas (2020)	Impact of total quality management on corporate sustainability through the mediating effect of knowledge management	F	JP
P006	Kaasinen et al. (2020)	Empowering and engaging industrial workers with Operator 4.0 solutions	F	JP
P007	Atyam (2019)	Knowledge management and effective leadership is the only solution for sustainable growth (Performance) of India's corporate sector: A mediation role of knowledge centered culture	F	JP
P009	Rossi et al. (2019)	A multi-criteria index to support ecodesign implementation in manufacturing products: benefits and limits in real case studies	F	JP
P010	Lin et al. (2019)	Production data analysis system using novel process capability indices-based circular economy	–	JP
P011	Naderi et al. (2019)	Sustainable operations management for industry 4.0 and its social return	F	JP

(continued)

Table 1. (*continued*)

ID	Author (year)	Title		PT
P012	Abbas and Sağsan (2019)	Impact of knowledge management practices on green innovation and corporate sustainable development: A structural analysis	F	JP
P013	Amaranti et al. (2019)	The Relationship between Absorptive Capacity, Knowledge Sharing Capability, and Green Dynamic Capability: A Conceptual Model	F	CP
P014	Aaldering et al. (2019)	Analysis of technological knowledge stock and prediction of its future development potential: The case of lithium-ion batteries	–	JP
P015	Asaul et al. (2019)	The Latest Information Systems in the Enterprise Management and Trends in their Development	F	CP
P017	Alamil et al. (2019)	Drivers for sharing knowledge in the kingdom of Saudi Arabia construction industry: An empirical study	F	CP
P018	Okorokov et al. (2019)	Building intellectual capital of specialists in the context of digital transformation of the Russian economy	F	CP
P019	Akram et al. (2019)	Knowledge management, sustainable business performance and empowering leadership: A firm-level approach	F	JP
P021	Birou et al. (2019)	Sustainability knowledge and training: outcomes and firm performance	–	JP
P023	Yusoff et al. (2019)	Does organizational learning capability allow improving business sustainability? A quantitative analysis in the manufacturing SME context	F	CP
P025	Tavasieva et al. (2019)	Knowledge management as a matter of vital importance for a modern organization	–	CP

(*continued*)

Table 1. (*continued*)

ID	Author (year)	Title		PT
P026	Paranitharan and Ramesh Babu (2019)	An empirical study on integrated manufacturing system practiced in Indian industries – A structural equation modelling approach	F	CP
P028	Matta et al. (2019)	Knowledge extracting from eco-design activity	–	CP
P030	Zhang et al. (2019)	A knowledge sharing framework for green supply chain management based on blockchain and edge computing	–	CP
P033	Reyes et al. (2019)	A method for choosing adapted life cycle assessment indicators as a driver of environmental learning: A French textile case study	F	JP
P034	Masuin et al. (2018)	Information System Development on Web-Based in Integrated Management System Through Improving Knowledge Management to Increase Organization Performance of Construction Company (A Conceptual Framework)	F	CP
P035	Guan (2018)	Research on performance evaluation model of papermaking enterprise innovation management based on organizational shared mental model	–	JP
P036	Shaikh Ibrahim and Vasudevan (2018)	Review and Analysis of issues related to the implementation of Knowledge Management practices in Indian Automotive SMEs	F	CP
P038	Marie et al. (2018)	Knowledge management system for risk mitigation in supply chain uncertainty: Case from automotive battery supply chain	F	CP

(*continued*)

Table 1. (*continued*)

ID	Author (year)	Title		PT
P039	de Guimarães et al. (2018)	The influence of entrepreneurial, market, knowledge management orientations on cleaner production and the sustainable competitive advantage	F	JP
P040	Mandal (2018)	Decision modelling in evaluating information management strategy in manufacturing industry	–	JP
P044	Sunardi (2017)	Linking human capital and enterprise sustainability in Indonesian medium-sized food manufacturing enterprises: The role of informal knowledge sharing practice	F	CP
P045	Peruzzini et al. (2017)	A social life cycle assessment methodology for smart manufacturing: The case of study of a kitchen sink	F	JP
P046	Paranitharan et al. (2017)	An empirical validation of integrated manufacturing business excellence model	F	JP
P052	Wittbrodt et al. (2017)	Challenges for industry 4.0 in management area	–	CP
P056	Menegotto et al. (2017)	Social innovation in the Brazilian wine chain: Co-creation of innovative ideas in processes, products and services in a multidisciplinary environment	–	CP
P057	Zgodavova et al. (2016)	Intelligent approaches for an organisation's management system change	–	JP
P058	Mandal and Bagchi (2016)	Strategic role of information, knowledge and technology in manufacturing industry performance	–	JP
P059	Bárdy et al. (2016)	An interactive game for applying the principles of lean production in automotive	–	CP

(*continued*)

Table 1. (*continued*)

ID	Author (year)	Title		PT
P060	Paszek et al. (2016)	Methods of knowledge representation in knowledge management systems for technological preparation of production	–	CP
P062	Bracke et al. (2016)	Contribution for Analysing, Saving and Prioritising of Lessons Learned Issues Regarding Product Improvement and Future Product Generations	F	CP
P063	Singh et al. (2016)	Investigation into the design and development of knowledge-based innovation processes in manufacturing companies	–	CP
P064	Azyabi (2016)	The role of information technology in enhancing SMEs capabilities through knowledge management	F	CP
P066	Petruse et al. (2016)	Augmented reality applications in the transition towards the sustainable organization	F	CP
P068	Kharabsheh et al. (2016)	A comprehensive model of knowledge sharing	F	CP
P074	Basmer et al. (2015)	Open production: Chances for social sustainability in manufacturing	F	CP
P075	Daud et al. (2015)	Knowledge management strategy, eco-innovation and corporate sustainability performance	–	CP
P079	Schurig et al. (2015)	Factors for effective learning in production networks to improve environmental performance	F	CP
P080	Gembarski and Lachmayer (2015)	Degrees of customization and sales support systems-enablers to sustainability in mass customization	F	CP
P084	Lillehagen and Petersen (2015)	Holistic design for continuous innovation and sustainable knowledge bases	F	CP

(*continued*)

Table 1. (*continued*)

ID	Author (year)	Title		PT
P086	Kalakul et al. (2014)	Integration of life cycle assessment software with tools for economic and sustainability analyses and process simulation for sustainable process design	–	JP
P087	Zhou et al. (2014)	The study of knowledge-based active learning grid operation experience	–	CP
P088	Vasudevan and Chawan (2014)	Demystifying knowledge management in Indian manufacturing SMEs	F	CP
P089	Armstrong et al. (2014)	Gate-to-gate sustainability assessment for small-scale manufacturing businesses: Caddisfly jewelry production	F	CP
P090	de Souza et al. (2014)	The Maintenance Function in the Context of Corporate Sustainability: A Theoretical-Analytical Reflexion	F	CP
P091	Mani et al. (2014)	Sustainability characterisation for manufacturing processes	–	JP
P093	Schrettle et al. (2014)	Turning sustainability into action: Explaining firms' sustainability efforts and their impact on firm performance	F	JP
P096	El Hachem et al. (2014)	Systemic assessment framework of a learning organization's competitive positioning	F	JP
P099	de Marchi and Grandinetti (2013)	Knowledge strategies for environmental innovations: The case of Italian manufacturing firms	F	JP
P102	Zavloschi-Dunăreanu (2013)	The knowledge management in the modern industrial company	–	JP
P105	Chen et al. (2013)	Analyzing the success factors on the implementation of product data management: A case study of electronic manufacturing industry	–	CP

(*continued*)

Table 1. (*continued*)

ID	Author (year)	Title		PT
P106	Machado et al. (2013)	Studies to minimize waste in a production process by using value stream map and knowledge management	–	CP
P107	Lanz et al. (2013)	Information flows in future advanced manufacturing ecosystems	F	CP
P109	Muñoz et al. (2013)	Considering environmental assessment in an ontological framework for enterprise sustainability	–	JP
P113	Goh et al. (2012)	Environmental knowledge management processes and supply chain management environmental performance of Malaysian manufacturing firms: An exploratory study	F	CP
P119	Efthymiou et al. (2012)	On a predictive maintenance platform for production systems	F	CP
P120	Seok et al. (2012)	Sustainability decision support system based on collaborative control theory	F	JP
P122	Machado et al. (2012)	Process understanding tools in PAT implementations	–	CP
P123	Widok et al. (2011)	Combining sustainability criteria with discrete event simulation	F	CP
P124	Aarabi et al. (2011)	Conceptual model for information systems of sustainable supply chain management	F	CP
P129	Yangand Zhang (2011)	Annotating semantics of multidisciplinary engineering resources to support knowledge sharing in sustainable manufacturing	–	BC
P132	Trotta (2010)	Product lifecycle management: Sustainability and knowledge management as keys in a complex system of product development	F	JP

(*continued*)

Table 1. (*continued*)

ID	Author (year)	Title		PT
P146	Van Der Meer and Sinnappan (2008)	Categorisation of knowledge management processes in the development of sustainable enterprises	F	CP
P147	Xie et al. (2008)	A study on the relationships between organizational culture, organizational learning, technological innovation and sustainable competitive advantage	F	CP
P152	Chen and Wang (2008)	The data mining technology based on CIMS and its application on automotive remanufacturing	F	CP
P154	Jianjun et al. (2008)	Research on the knowledge management architecture of LCED based ontologies and multi-agent system	F	JP
P155	Chen (2007)	An integrated model of innovation value chain for sustainable competitive advantage in high-tech industry	–	CP
P160	Lee et al. (2006)	The framework of information sharing in end-of-life for sustainable product development	F	CP
P161	Wallace et al. (2005)	Engineering knowledge management	F	BC
P163	Liu et al. (2004)	An empirical study on the correlation between knowledge management capability and competitiveness in Taiwan's industries	F	JP
P164	Hatch and Dyer (2004)	Human capital and learning as a source of sustainable competitive advantage	F	JP
P165	Chapman and Hyland (2004)	Complexity and learning behaviors in product innovation	F	JP
P168	Meyer et al. (2001)	Environmental knowledge management in the product development process	–	CP

(*continued*)

Table 1. (*continued*)

ID	Author (year)	Title		PT
P169	Boisot and Griffiths (1999)	Possession is nine tenths of the law: managing a firm's knowledge base in a regime of weak appropriability	–	JP
P170	Swanson (1996)	The new organizational knowledge and its systems foundations	F	CP
P176	Shahzad et al. (2020)	Relation of environment sustainability to CSR and green innovation: A case of Pakistani manufacturing industry	F	JP
P177	Abbas (2020)	Impact of total quality management on corporate green performance through the mediating role of corporate social responsibility	F	JP
P179	Martins et al. (2019)	Knowledge management in the context of sustainability: Literature review and opportunities for future research	F	JP
P181	Manavalan et al. (2019)	A review of Internet of Things (IoT) embedded sustainable supply chain for industry 4.0 requirements	F	JP
P182	Kashan et al. (2018)	A Discovery of Multiple Levels of Open Innovation in Understanding the Economic Sustainability. A Case Study in the Manufacturing Industry	F	JP
P184	Widya-Hasuti et al. (2018)	The Role of Process Innovation between Firm-Specific Capabilities and Sustainable Innovation in SMEs: Empirical Evidence from Indonesia	F	JP
P185	Wang et al. (2018)	An Integrated Open Approach to Capturing Systematic Knowledge for Manufacturing Process Innovation Based on Collective Intelligence	F	JP
P187	Illiashenko et al. (2017)	Knowledge management in the system of organization's innovative development	F	JP

(*continued*)

Table 1. *(continued)*

ID	Author (year)	Title		PT
P189	Guminski et al. (2014)	The model of knowledge management system supporting sustainable growth in machine-building industry enterprises	–	CP
P190	Chang (2012)	The Effect of Corporate Environmental Commitment on Green Product Innovation	F	CP
P192	Chen and Chen (2010)	Relationships among organizational innovation, learning and knowledge management in the information technology industry	–	JP
P193	Nathai-Balkissoon and Pun (2010)	A Holistic Approach To Organisational Learning In Manufacturing: A Model And A Research Agenda So Towards Epistemic Sovereignty: (Re)-Thinking Development In A Changing Global Political Economy	F	CP
P196	Soltani (2009)	The role of knowledge management paradigm in establishing of learning organization	–	CP
P198	Tang and Wang (2009)	Design Institute's Knowledge Sharing System Preliminary Study	F	CP
P199	Kimseng et al. (2016)	The influences of knowledge management on innovation within formal and non-formal R&D manufacturing firms, Thailand	F	CP
P200	Yuqing (2011)	Ideas and research on constructing the environmental structure system for the enterprise knowledge management	F	CP

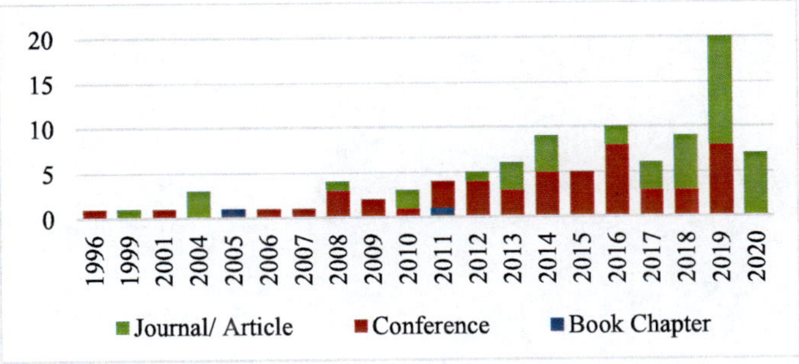

Fig. 1. Number of papers in years by type

The highest number of JPs was published in Journal of Cleaner Production (Fig. 2). This journal is highly oriented on sustainability issues, sustainable development, sustainable consumption, sustainable products and services, corporate sustainability and Corporate Social Responsibility, cleaner production and technical processes etc. Concerning the CPs, they were presented on different conferences and published in proceedings such as: "IFIP Advances in Information and Communication Technology" (4 papers), "IOP Conference Series: Materials Science and Engineering" (4 papers), Procedia CIRP (3 papers) etc. (Fig. 3).

The next analysis was based on author keywords. First of all, the identified keywords were grouped. The KM aspects were assembled into three groups including the following authors' keywords:

- *knowledge creation and acquisition* (Lesson learned; Creative thinking; Production of knowledge; Knowledge Acquisition; Knowledge management and modelling; External knowledge; Digital engineering resources; Digital storage; Semantic annotation),
- *knowledge sharing* (Semantic information; Knowledge representation; Marketing knowledge; Cooperative patent classification; Internal knowledge; Classifications; Random forest classifier; Codification; Categorisation; Collective intelligence; Information resource; Know-how; Organization of datum; Active knowledge architecture; Antecedents; Sustainable knowledge bases), and
- *knowledge storing* (Knowledge sharing; Organizational learning; Transfer of knowledge; Knowledge transfer; Expert system; Informal knowledge sharing; Knowledge sharing; Knowledge Dissemination; Information dissemination).

The aspects connected with SD were also grouped into three groups including the following authors' keywords:

- *ecological aspects* (Ecological materials; Environmental; Environmental management; Cleaner environment; Cleaner production; United Nations Environment Programme; Environmental knowledge management; Environmental toxicology),

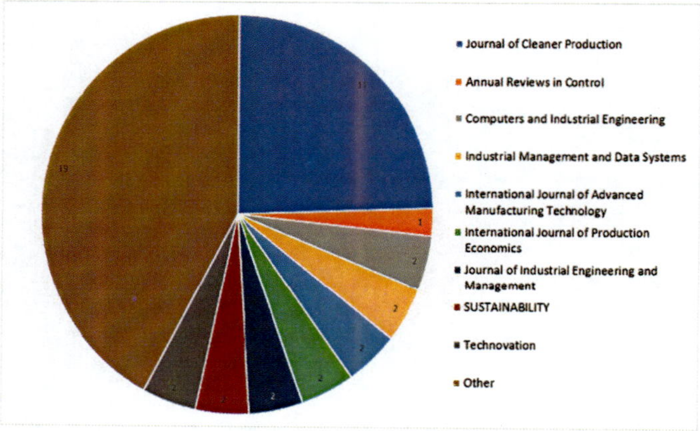

Fig. 2. Number of publications in journals

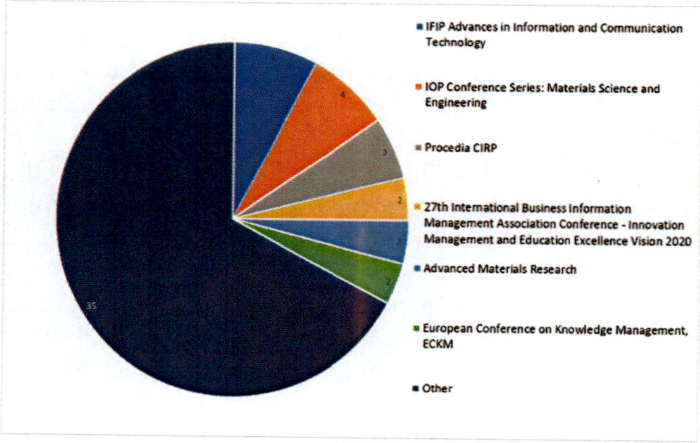

Fig. 3. Number of CPs by source

- *economic aspects* (Sustainable growth; Investments; Economic and mathematical modelling; Economic and social effects; Economic sustainability; Economical sustainability; Economics; Emerging economy; Knowledge economy; Lean production), and
- *social aspects* (Economic and social effects; Social aspects; Social capital; Social sustainability; Social Return; Social Performance; Social innovation; Social responsibility; Social Impact of Automation; Responsiveness to Knowledge).

Then, it was analyzed which papers contain the keywords from the presented groups. Based on the performed analysis, it can be concluded that only in case of 7 papers the keywords point directly to the idea that the papers are connected with the indicated KM and SD aspects. Table 2 presents the IDs of the articles which at the same time are connected with one KM and one SD aspect. The table shows that for example P059 is connected with two SD aspects (Knowledge creation and acquisition and Knowledge sharing) and with two SD aspects (Ecological aspects and Economic aspects). Moreover, it can be seen that only an abstract was available for P059. No papers presenting connections between knowledge sharing and ecological aspects were identified, based on the keywords analysis.

Table 2. Papers and problems analyzed in the reviewed works – based on author keywords analysis.

		Sustainable development		
		Ecological aspects	Economic aspects	Social aspects
Knowledge management	Knowledge creation and acquisition	P059	P059	P019
	Knowledge storing	–	P122	P107 P190
	Knowledge sharing	P059	P059 P122 P189	P019 P074

Next, the abstracts of the retrieved 99 papers or full texts if available, were analyzed to find out how they address the KM and SD aspects. It was discovered, that more KM and SD aspects were discussed than it was visible on the basis of the keywords analysis. Table 3 indicates the papers which discuss certain aspects of KM and SD.

Based on the analysis results it can be said that all the indicated SD aspects and KM elements were illustrated in 10 out of 99 analyzed articles. The main concepts discussed in these papers are presented in this section in order to show the context in which KM is discussed together with SD.

The article P005, based on the data coming from Pakistani manufacturing companies, presents how total quality management practices can facilitate the achievement of corporate sustainability objectives and the role of KM in this process. In the article P012, the authors, on the basis of the data coming from Pakistani company managers, investigate the impact of KM processes on green technologies and green innovations as well as the environment, social and economic aspects of sustainability. The authors of the article P019 discuss the role of building empowered teams by their leaders in knowledge acquisition and sharing. They investigate the impact of knowledge acquisition, knowledge dissemination and responsiveness to knowledge on sustainable business performance, taking into account not only economic but also social and environmental aspects.

The article P045 proposes a Social Life Cycle Assessment methodology to support a strategic company decision-making process, enterprise modelling and KM and to assess company sustainability. The article P093 identifies the sustainability drivers and proposes a framework showing which dimensions influence decisions concerning the sustainability and which dimensions are influenced by these decisions. The paper P096 offers a framework to assess a learning organization's competitive positioning. In the article P120 the authors reviewed the sustainability concepts focusing on the issues of supply networks. They indicate the importance of decision support systems and they suggest to use the strategies applying the Collaborative Control Theory in order to maximize the sustainability of supply networks.

In the article P146 the authors assigned KM processes to the SD practice to choose the best processes for sustainable projects. In the article P179, on the basis of the literature review, the authors present the gaps in the KM use in SD context. They identified the following research areas: implementation of KM in different industries, potentials for promoting knowledge, and KM capacity building. Finally, the article P189 presents a model of a KM system which integrates economic, social and environmental aspects in the machine-building enterprises. The authors of the article outline the role of processing and "delivering" knowledge in sustainable business growth. The IT system proposed supports decision making processes in construction, production etc.

The remaining papers were analyzed to see how they address certain KM and SD aspects.

Sustainable development – ecological aspects

Ecology aspects of SD were discussed in 55 papers. Some of them deal with the assessment of the whole system performance and some discuss chosen managerial aspects. For example, the paper P003 provides evidence that green intellectual capital has a positive association with the environmental performance in large manufacturing companies. The study was performed in Malaysian companies. Similarly, the paper P011 develops a methodology and procedures that allow to predict the system performance and its environmental impact. Furthermore, the paper P123 presents an ecological material flow analysis for production processes. The article P059 is focused on creative thinking. It can be assumed that creative thinking develops human resources and directs them towards better utilization of resources and reduction of wastes. The paper P075 presents a review of the past research connected with KM, eco-innovation as well as corporate sustainability performance.

The paper P001 presents how different sustainable supply chain management practices, connected with knowledge sharing inside a supply chain, enhanced the supply chain environmental performances of an apparel supply chains. The paper P113 assesses the influence of environmental knowledge management on the performance of green supply chain performance. The research also indicates to the need of revising the training programs and policies in the context of strengthening the environmental knowledge management. The paper P109 presents an ontological framework for information and knowledge models sharing in order to facilitate the environmental assessment in a company. The model was implemented in a supply chain to support designing and planning process.

The paper P009 describes knowledge sharing as a supporting driver in Ecological Product Development assessment. The authors propose a multi-criteria index approach called Product Impact Index. The paper P028 discusses how an eco-design approach can lead to the innovative product development. The authors investigate how designers can be guided to consider environmental criteria in the process. The authors of the paper P168 present the Environmental Knowledge Management System to support a designing process. The system can be used in data collection, technical system modelling and process optimization. The paper P086 presents a software for Life Cycle Assessment (LCA) which contains a KM tool, environmental impact estimation tool, and LCA calculation tool.

The paper P013 investigates the relationship between knowledge sharing ability, absorptive capacity, and green dynamic capability. The authors of the paper P106 apply Value Stream Mapping to identify wastes. They assume that in KM a knowledge base containing the information coming from a value stream map can be used to reduce wastes.

The paper P039 discusses the influence of the strategic drivers: Entrepreneurial Orientation, Market Orientation and Knowledge Management Orientation on SD of Small and Medium Enterprises (SME). The research shows that the integrated strategic drivers should enhance cleaner production more effectively than a separate approach to these drivers.

The paper P099 investigates the assessment and development of the knowledge needed to drive Environmental Innovations. The authors verified how KM strategies vary according to the differential importance that sustainability has for the company's innovative strategy.

Sustainable development – economic aspects

Economic aspects are discussed in 87 papers. Similarly to ecological aspects, also in relation to economic aspects, the analyzed papers assess the whole system or its part. For example, in the paper P011 the authors propose a methodology and procedures that allow predicting the system performance and its economic impact. The paper P190 discusses corporate environmental commitment and its affection on green product innovation performance. The papers P046 and describe an integrated model for manufacturing industries called Integrated Manufacturing Business Excellence System. The system is based on widely known concepts: total quality management, total productive maintenance, corporate social responsibility, knowledge management, lean manufacturing, agile manufacturing, and sustainable manufacturing. KM is a crucial element in order to achieve operational excellence. The authors propose a model which combines the management of quality, environment, safety, knowledge, productivity, lean and agile manufacturing, six sigma by using ten critical factors such as: visionary leadership and management commitment, systematic approach to management, customer focus and satisfaction, continuous improvement and innovation, organization culture, supplier quality management, workforce management, manufacturing strategy, practicing manufacturing practice audit system, and the practice of technology management information system. The paper P034 describes Management System which integrates a quality management system, environment management system, health and safety management system in the organization of a construction company in Indonesia. Moreover, the authors underline the importance of KM and an information system in order to improve the performance.

The paper P015 describes the need of using, storing and sharing information knowledge from a modern IT system in a managerial decision process in order to maximize profitability of an enterprise.

In other papers the following issues are analysed: creative thinking that can lead to wastes reduction and in consequence improves economic aspects (P059); collaboration

Table 3. Topics connected with SD aspects and KM elements discussed in the analyzed papers; F – full text available, A – abstract, np – number of papers.

Sustainable development			Knowledge management			Paper ID full text availability	NP
Ecology	Economy	Social	Creating acquiring	Storing	Sharing		
X	X	X	X	X	X	P005F, P012F, P019F, P045F, P093F, P096F, P120F, P146F, P179F, P189A	10
X	X	X	X	X		P003F	1
X	X	X	X			P011F, P123F	2
X	X	X			X	P002A, P006F, P023F, P089A, P129A	5
X	X	X				P176F	1
X	X		X	X	X	P010A, P026F, P038F, P039F, P046F, P062F, P066F, P075A, P086A, P106A, P113F, P119F, P132F, P147F, P152F, P154F, P160F, P168A, P184F	19
X	X			X	X	P084F	2
X	X				X	P001F, P009F, P021A, P035A, P090F, P109A	6
	X	X	X	X	X	P059A, P88F, P185F	2
	X	X	X	X		P177F	1
	X	X	X		X	P190F	1
	X	X	X			P196A	1
	X	X		X		P124F	1
	X	X			X	P018F, P056A, P074F	3
X			X	X	X	P079F, P087A	2
X			X		X	P091A, P099F	2
X				X		P014A	1
X					X	P013F, P028F, P030F, P033F	4
	X		X	X	X	P007F, P015F, P017F, P025A, P034F, P036F, P040A, P058A, P063A, P064F, P080F, P102A, 1P05A, P122A, P155A, P161F, P163F, P170F, P181F, P182F, P187F, P193F, P199F, P200F	24

(*continued*)

Table 3. (*continued*)

Sustainable development			Knowledge management			Paper ID full text availability	NP
Ecology	Economy	Social	Creating acquiring	Storing	Sharing		
	X		X	X		P169A	1
	X		X		X	P044F, P164F, P165F	3
	X		X			P192A	1
	X			X		P060A	1
	X				X	P152F, P198F	2
		X	X	X	X	P107F	1
		X	X		X	P057A	1
		X			X	P068F	1

which enables more value creation by utilizing knowledge (P074); knowledge creation from a data analysis which allows companies to increase competitiveness by improvements in processes (P122); green intellectual capital which has a positive association with economic performance in the large manufacturing companies (P003).

The paper P002 presents that knowledge sharing in a supply chain improves operational performance of the supply chain. The paper P056 emphasizes the need of cooperation among organizations in a value chain to exchange information and knowledge and accelerate economic growth. The paper P181 examines the potential opportunities of using Internet of Things (IoT) concept in a sustainable supply chain and Industry 4.0 transformation. The authors proposed a conceptual framework model which includes five perspectives: Business, Technology, Sustainable Development, Collaboration and Management Strategy.

The paper P010 proposes a system for the improvement of production process of bicycle parts in order to increase quality and performance. In this system knowledge can be created as improvement references, stored and used for the future analysis. The authors of the papers P058 and discovered that financial performance of an organization depends on its strategies of information and technology management. Moreover, financial performance is indirectly influenced by KM strategies. The paper P200 assesses KM processes as a competitive advantage for Chinese enterprises. The author states that KM can be used as a "strategic weapon" for long-term sustainable development.

The paper P036 investigates KM issues in automotive component manufacturing SMEs in India and focuses on the impact of each issue in an SME on the KM implementation. The authors highlight the importance of KM in building a competitive advantage in modern economy. The paper P147 studies the impact of organizational culture, learning capabilities and technological innovation on a sustainable competitive advantage. Empirical research showed that all three aspects have a significant influence on building a sustainable competitive advantage of the company. The paper P163 confirms the correlation between competitiveness and KM in Taiwan. Knowledge is described as a main part of manufacturing resources.

The paper P080 characterizes tools for translating customer's needs into technical specifications in order to improve production efficiency in Mass Customization Business Model. Such tools should positively impact the sustainability of such a model.

The paper P165 presents the association between Continuous Product Innovation with KM systems and processes. The research shows that the improvement of organizational learning can positively influence the competitiveness of an organization and enhance a product innovation. The paper also discusses different approaches that can be helpful in this improvement. In the paper P035 the authors analyzed the model of innovation management performance evaluation in a papermaking enterprise. The model is based on an organizational shared mental model. The paper P063 presents the model to support KM in the manufacturing companies' innovation processes. In the paper P182 the authors define five levels of open innovation and knowledge possession in terms of economic sustainability in Iranian manufacturing company. First, "part level" describes suppliers with no internal knowledge. They require full support for production, where the fifth "architectural level" illustrates suppliers with full knowledge to provide innovative solutions. The paper P184 investigates the connection between process innovation, firm-specific capabilities and sustainable innovation. The study shows that company-specific capabilities affect sustainable innovation through process innovation.

Sustainable development – social aspects

Social aspects are discussed in 31 papers. The article P074 addresses the social aspect of sustainability of companies underlining the role of collaboration. The article P107 outlines sociotechnical challenges needed to be solved to realize future manufacturing systems. The papers P190 and P003 discuss green human capital and its influence on green product innovations development and social performance. The authors of the paper P011 developed a methodology and procedures that allow to predict the system performance and its social impact. The authors of the paper P068 suggest that people who have strong social ties in a company are more willing to share knowledge.

The paper P018 justifies the need for the faster transfer to the intelligent model of the sustainable development of society where personal creativity is a competitive advantage of a company. The authors of the paper P129 propose an approach to multidisciplinary engineering resources semantic annotation in order to support design processes.

The paper P164 illustrates the dependencies between investments in human capital and learning performance. According to the research, the development through training significantly improves learning by doing process which leads to better performance. The paper P196 presents the issues connected with learning organization such as change of employees' mental set, operational principles and learning organization infrastructures design. The authors present the foundations of a learning organization i.e.: continuous learning, encouraging individuals to regular learning, assessment of employees' capabilities to learn, team and international learning, systemic thinking.

The paper P057 presents the environment of a web-based role-play simulation which can help in the organizational change implementation to support people to adapt more quickly to a new situation.

The paper P089 presents a comparative assessment of sustainability taking into account the analysis of a life cycle inventory, life cycle costing, and risk assessment of a worker's injury.

The paper P176 investigates the influence of Corporate Social Responsibility (CSR) on environmental sustainability and green innovation. Researchers found out that CSR activities positively influence both environmental sustainability and green innovation within manufacturing companies in Pakistan. The authors of the paper P177 examine three dimensions of CSR: social, employees, and customer and their role in the relationship between Total Quality Management (TQM) and Corporate Green Performance (CGP). The study shows that TQM positively influences CGP, and CSR plays a mediating role. Moreover, in Pakistani medium and large enterprises, TQM and CGP are equally important, whereas the CSR importance depends on the size of a company.

Knowledge Management – Knowledge Creating and Acquiring

Knowledge Creation and Acquisition is discussed in 72 papers. In the article P107 the authors emphasize the existing of tacit knowledge which employees possess and should be acquired by the company. The authors of the article P122 discuss the importance of gathering a large amount of data for an organization. It is especially important to define exactly which data are useful from the processes management point of view. The study presented in the article P190 discusses how green relationship learning can influence green product innovations performance development. In the article P059 the authors illustrate the importance of supporting knowledge creation and acquisition by developing creative thinking. The authors point out that the development and the right usage of human capital are considered as one of the biggest competitive advantages nowadays.

The paper P038 aims to describe the process of designing a KM system for the risk mitigation in a supply chain. The proposed solution should support the acquisition, dissemination and storage of knowledge.

The paper P066 discusses the role of implementing Augmented Reality (AR) technology in an organization to improve efficiency of workforce and meet sustainable development goals. It is stated that AR technology can improve a product development process, and through innovative approach to the knowledge acquisition and storing it can help workers to perform tasks. The paper P152 describes the usage of data mining technology in Computer Integrated Manufacturing System. The technology supports the information acquisition and processing in order to improve production processes in an automotive industry in China. The paper P185 presents the approach to KM system based on a fusion algorithm, collective intelligence and semantic elements. These integrated techniques implemented in Computer-Aided Process Innovation to support an innovation design through the captured and well organized knowledge base. The paper P199 analyses the impact of formal and non-formal R&D companies on KM performance in the product and process innovation. The study shows that in manufacturing companies in Thailand KM performance is better in formal R&D structures. The authors of the paper P192 study how KM is influenced by organizational innovation and learning.

The paper P193 points out Organizational Learning and KM elements such as culture, tools and instruments, learning processes and practices in order to create a holistic manufacturing organizational learning model. Such a model should support the diversification of energy-based economy, promote sustainable development and rise creativity and competences within a society. The paper P079 analyses the influence of collaborative learning through production networks. The authors state that the knowledge acquisition and sharing between different factories can improve environmental performance of companies. The paper P091 discusses classifications of manufacturing processes, indicators of sustainable manufacturing, computable metrics as well as information models and software presented in the literature. It also presents a conceptual sustainability model. The authors emphasize the significance of a measurement system to monitor manufacturing system environmental performance for comparisons.

The paper P102 is focused on the knowledge management, especially the knowledge acquisition. The author emphasizes that thanks to the knowledge capturing process time can be saved comparing to knowledge creation from scratches. Therefore, KM systems should first search and collect the existing knowledge. The paper P088 underlines an in-depth analysis of the market and knowledge acquisition through conferences, workshops, etc. as a key driving factor for the successful KM implementation.

The paper P119 presents a predictive maintenance platform based on three pillars: data acquisition, knowledge presentation and advisory indicators. The indicators particularly focus on energy and environmental performance.

The paper P132 illustrates how proper management of Product Life Cycle on early stages can support the design of sustainably designed products. The author analyses Product Lifecycle Management tools, which can formalize and then reuse knowledge in a new product development phase.

Knowledge Management – Knowledge Storing

Knowledge Storing problems are discussed in 66 papers. The paper P025 discusses the importance of KM for modern organizations. In the article P107 the authors describe the importance of an information flow within a company and outline sociotechnical challenges needed to be solved in order to realize future manufacturing systems. The authors emphasize the existing of tacit knowledge which employees possess and should share within the company. The authors of the article P122 underline that the data stored in a company should be connected to certain processes in which they are indispensable in the process management.

The paper P007 presents the impact of KM and effective leadership on the organizational performance with a mediating role of the knowledge infrastructure capability.

The paper P014 presents how knowledge stocks can influence the future development of Lithium-Ion Batteries technology.

The paper P062 proposes a method of storing and analyzing lessons learned from New Product Development process. A proper usage of lessons learned should positively influence a new generation of products.

The paper P064 analyses the influence of internal and external KM on the dynamic capabilities of the enterprises in Saudi Arabia. Moreover, the author states that the proper IT systems play a crucial role in storing and using company's knowledge in order to reduce lead times and costs in SME's.

The paper P124 describes the role of Sustainable Supply Chain Management (SSCM) in achieving sustainable development in three pillars: ecological, economic and social.

The stages of SSCM are defined as: pre-manufacture, manufacture, use and post-use. The authors propose a conceptual model of a database structure for the IT system supporting SSCM.

The paper P084 discusses the importance of collaboration and communication in diverse networks in order to sustain a competitive advantage, especially in a product design. Sustainable, active knowledge bases are key to the proper use of best practices.

The paper P161 points out main issues in engineering KM in aerospace industry. The problems of retaining knowledge and experience in the fast changing environment leads to the cost increase and may influence a competitive advantage of a company.

The paper P060 presents a method for the knowledge representation to be applied in KM system. The paper P087 presents a knowledge base with a learning ability. The authors propose to apply Bayesian classifier model which is based on weight. The paper P105 focuses on product data management. The authors of the paper P187 propose a scheme of knowledge production and management to strengthen competitive advantages by innovative development.

The authors of the paper P169 emphasize that currently a major challenge in industry is to motivate employees to contribute their knowledge to the company's know-how.

Knowledge Management – Knowledge Sharing

Knowledge Sharing problems are discussed in 88 papers. For example, in the article P059 the authors describe the effective transfer of knowledge about lean manufacturing techniques and tools through a managerial game. One of the main goals of a company should be the human resources development what improves the competitive advantages of the company. The authors of the article P074 indicate collaboration as a way to better knowledge utilization and sharing. The authors of the paper P107 describe the importance of an information flow within a company. While, the authors of the article P122 illustrate the importance of fast sharing of information and findings as an increasing necessity in knowledge-based enterprises. They emphasize the importance of graphical tools applied for sharing the information concerning process monitoring in order to control the process better. The article P190 discusses a green relationship learning process in an organization. Then, the paper P002 indicates that knowledge sharing demonstrates an impact on the green innovation in the whole sustainable supply chain.

The results of the research presented in the paper P006 show that there is a need for knowledge sharing as well as adaptive learning solutions to facilitate designing employees' work and training in the companies implementing Industry 4.0 solutions. The findings presented in the paper P023 show that the organizational learning capability has a significantly positive relationship with business sustainability. The paper P068 presents a comprehensive model of knowledge sharing inside a company.

The paper P017 explores drivers for sharing knowledge in the Kingdom of Saudi Arabia construction industry in order to increase the productivity of industry. They are: integration of different teams knowledge through interaction, setting up an environment to boost knowledge sharing, using different platforms to transfer knowledge, knowledge retaining, reflection on the self-understandings and creation of communities to share ideas. The paper P044 suggests that particular human capital traits, for example: employee's experience, educational background, and employee's skills are likely to influence the effectiveness of informal knowledge sharing which leads to the sustainable development of the company. The research was performed in food industry.

The paper P170 illustrates organizational knowledge as a competitive advantage of 21st century enterprise. Moreover, new organizational knowledge will be more system-based, core-technical, transaction-oriented and global rather than person-based, administrative, production-oriented and local. The paper P030 proposes the use of Blockchain and Edge Computing technologies in a knowledge sharing framework in order to secure sensible data. The framework is used for a green supply chain and the example of an automotive company is analyzed.

The paper P033 proposes a method of choosing correct Life Cycle Assessment indicators based on the knowledge transfer concept. Evaluation Method for Choosing Indicator was developed to facilitate the learning process and validated on French textile industry. The paper P154 describes the use of a multi-agent system and ontology in KM architecture in a life cycle engineering design. Such approach supports knowledge sharing and reusing among the whole lifecycle of a product. The paper P160 develops a framework for easier information storing and sharing in a product life cycle management based on ontology. The framework is created to support sustainable product development.

The paper P090 assesses the importance of integration between the maintenance and production information exchange in order to achieve corporate sustainability in Brazil.

In the paper P155 the author proposes an integrated model of an innovation value chain. The model integrates KM and innovation management in order to achieve a sustainable competitive advantage. Each component in the chain can add value.

The authors of the paper P021 discovered a positive impact of an organizational sustainability training and knowledge on sustainable supply chain practices as well as a positive impact of sustainable supply chain practices utilization on sustainable supply chain outcomes. This has a positive impact on the operational performance and environmental economic performance.

The paper P198 presents the issues in successful knowledge sharing implementation. Prisoner's dilemma is presented in relation to knowledge sharing. What is more, the authors perceive KM as a key factor to achieve sustainable development in enterprises.

5 Conclusions and Future Work

The paper presents a systematic literature review. The goal of the paper was to analyze and to summarize how deeply the aspects which concern KM and SD are discussed in the publications connected with manufacturing companies as well as what topics are discussed in papers in order to identify the directions of the future research and to identify emerging topics.

Based on the performed analyses the authors tried to answer the posed research questions as follows. The first research question asked to what extent sustainable development (SD) and knowledge management (KM) are discussed in the literature in the context of manufacturing organizations. It can be concluded that SD and KM are more and more discussed in the literature. The number of papers connected with the topic published in 2019 doubled in comparison to 2018, and 12 papers were published in journals in 2019. A growing trend of interest in SD and KM is visible. The most recent papers deal with sustainable supply chains, green innovations, green intellectual capital, total quality management and industrial workers engagement in building and use of knowledge to support SD in manufacturing organizations.

The second research question asked what aspects are mostly discussed in sustainable development (SD) and knowledge management (KM) in the context of manufacturing organizations. It was noticed that the highest number of papers discuss knowledge sharing (88 papers) in the context of KM, and economic aspects (87 papers) in the context of SD. This shows that the main problems are connected with effective knowledge sharing and still many works can support the process by proposing different ways and motivators for knowledge sharing. Moreover, it can be concluded that the main motivation for undertaking actions towards KM improvements is still economy. The researchers prove that KM can improve economic aspects. That is always a good motivation for the companies of which the main goal is to survive and develop and what is impossible without financial support. For example, according to the authors of the work [18], financial performance can be improved in the organizations which are able to acquire and absorb knowledge successfully.

58 papers discuss all KM aspects and 19 all SD aspects. 19 papers discuss all KM aspects with only economic and ecological aspects of SD, and 24 papers discuss all KM aspects with only economic aspects of SD. Only one paper discusses all KM aspects and only social aspects of SD. Generally, social aspects are discussed in only 31 papers. Therefore, this aspect can be indicated as the aspect in which still much work should be done. Particularly, it is because humans will have a crucial role in the Factories of the Future [19]. On the basis of the performed analysis, it can be concluded that social aspects need to be further investigated in manufacturing organizations, particularly, because knowledge creation is very closely connected with human resources. On one side social aspects may influence KM and on the other side proper KM can influence social aspects. It should be further investigated.

The knowledge storing problems are closely connected to IT systems. Some papers propose IT systems to support KM, e.g. [20–22] and [23]. However, further studies connected with safety and effective ways to store knowledge with the use of IT systems is also recommended to be conducted.

The emerging technologies identified in the literature review which appeared in the context of KM are: data mining technologies [24], Augmented Reality [25], Internet of Things [26], Blockchain and Edge Computing technologies [27]. Their use has to be further investigated.

In the future work the authors plan to analyze how IT systems as well as emerging technologies can support KM and SD.

References

1. Martins, V.W.B., Rampasso, I.S., Anholon, R., Quelhas, O.L.G., Leal Filho, W.: Knowledge management in the context of sustainability: literature review and opportunities for future research. J. Clean. Prod. **229**, 489–500 (2019). https://doi.org/10.1016/j.jclepro.2019.04.354
2. Mensah, J.: Sustainable development: meaning, history, principles, pillars, and implications for human action: literature review. Cogent Soc. Sci. **5** (2019). https://doi.org/10.1080/23311886.2019.1653531

3. Porter, M.E., Van Der Linde, C.: Toward a new conception of the environment-competitiveness relationship. In: Corporate Environmental Responsibility, pp. 61–82. Taylor and Francis (2017). https://doi.org/10.1257/jep.9.4.97
4. Iqbal, Z., Malik, M.: Entrepreneurial orientation and engagement of Pakistani small and medium enterprises in sustainable development practices: mediating role of knowledge management. Bus. Strateg. Dev. **2**, 192–203 (2019). https://doi.org/10.1002/bsd2.53
5. Abbas, J., Sağsan, M.: Impact of knowledge management practices on green innovation and corporate sustainable development: a structural analysis. J. Clean. Prod. **229**, 611–620 (2019). https://doi.org/10.1016/j.jclepro.2019.05.024
6. Mikalauskiene, A., Atkociuniene, Z.: Knowledge management impact on sustainable development. Montenegrin J. Econ. **15**, 149–160 (2019)
7. Iskandar, K., Jambak, M.I., Kosala, R., Prabowo, H.: Current issue on knowledge management system for future research: a systematic literature review. In: Procedia Computer Science, pp. 68–80. Elsevier B.V. (2017) https://doi.org/10.1016/j.procs.2017.10.011
8. Weinreich, R., Groher, I.: Software architecture knowledge management approaches and their support for knowledge management activities: a systematic literature review. Inf. Softw. Technol. (2016). https://doi.org/10.1016/j.infsof.2016.09.007
9. Ouriques, R.A.B., Wnuk, K., Gorschek, T., Svensson, R.B.: Knowledge management strategies and processes in agile software development: a systematic literature review. Int. J. Softw. Eng. Knowl. Eng. **29** (2019). https://doi.org/10.1142/S0218194019500153
10. Ouédraogo, M., Rinfret, N.: The sharing and management of managers' knowledge: a structured literature review. Knowl. Process Manag. **26**, 244–261 (2019). https://doi.org/10.1002/kpm.1606
11. Wang, H., Meng, X.: Transformation from IT-based knowledge management into BIM-supported knowledge management: a literature review. Expert Syst. Appl. (2019). https://doi.org/10.1016/j.eswa.2018.12.017
12. Bastos, F.D., Cordeiro, J.V., Drohomeretski, E.: Human values, teamwork design and knowledge management on the shop floor: a systematic literature review. Indep. J. Manag. Prod. **10**, 186 (2019). https://doi.org/10.14807/ijmp.v10i1.795
13. de Mendes, R.C., de Mattos, M.C.: Gestão do Conhecimento e World Class Manufacturing: Uma aproximação inicial a partir da revisão de literatura. Perspect. em Cienc. da Inf. (2017). https://doi.org/10.1590/1981-5344/3103
14. Sensuse, D.I., Qodarsih, N., Lusa, J.S., Prima, P.: Critical success factors of knowledge management: a systematic literature review. In: 2018 International Conference on Information Technology Systems and Innovation, ICITSI 2018 - Proceedings, pp. 26–31. Institute of Electrical and Electronics Engineers Inc. https://doi.org/10.1109/ICITSI.2018.8695926
15. Gunasekaran, A., Ngai, E.W.T.: Knowledge management in 21st century manufacturing. Int. J. Prod. Res. **45**, 2391–2418 (2007). https://doi.org/10.1080/00207540601020429
16. Baloh, P., Desouza, K.C.: Towards knowledge needs-technology FIT model for knowledge management systems. In: Proceedings of the 4th International Conference on Design Science Research in Information Systems and Technology, DESRIST '09 (2009). https://doi.org/10.1145/1555619.1555646
17. Ghobadi, S., Mathiassen, L.: Perceived barriers to effective knowledge sharing in agile software teams. Inf. Syst. J. **26**, 95–125 (2016). https://doi.org/10.1111/isj.12053
18. Shahzad, M., Ying, Q., Ur Rehman, S., Zafar, A., Ding, X., Abbas, J.: Impact of knowledge absorptive capacity on corporate sustainability with mediating role of CSR: analysis from the asian context. J. Environ. Plan. Manag. (2019). https://doi.org/ https://doi.org/10.1080/09640568.2019.1575799
19. Kaasinen, E., et al.: Empowering and engaging industrial workers with Operator 4.0 solutions. Comput. Ind. Eng. **139**, art. no. 105678 (2020). https://doi.org/10.1016/j.cie.2019.01.052

20. Guminski, A., Dohn, K, Matusek, M., Zolenski, W.: The model of knowledge management system supporting sustainable growth in machine-building industry enterprises. In: 9th International Forum on Knowledge Asset Dynamics: Knowledge and Management Models for Sustainable Growth, pp. 1856–1886 (2014)
21. Asaul, A., Voynarenko, M., Dzhulii, L., Yemchuk, L., Skorobohata, L., Mykoliuk, O.: The latest information systems in the enterprise management and trends in their development. In: 9th International Conference on Advanced Computer Information Technologies, ACIT 2019 - Proceedings, art. no. 8779874, pp. 409–412 (2019). https://doi.org/10.1109/ACITT.2019.8779874
22. Azyabi, N.G.: The role of information technology in enhancing SMEs capabilities through knowledge management. In: Proceedings of the 28th International Business Information Management Association Conference - Vision 2020: Innovation Management, Development Sustainability, and Competitive Economic Growth, pp. 1162–1175 (2016)
23. Aarabi, M., Mat Saman, M.Z., Khoei, M.R., Wong, K.Y., Beheshti, H.M., Zakuan, N.: Conceptual model for information systems of sustainable supply chain management. In: IEEE International Conference on Industrial Engineering and Engineering Management, art. no. 6117927, pp. 303–307 (2011). https://doi.org/10.1109/IEEM.2011.6117927
24. Chen, K., Wang, J.: The data mining technology based on CIMS and its application on automotive remanufacturing. In: Proceedings - 1st International Workshop on Knowledge Discovery and Data Mining, WKDD, art. no. 4470347, pp. 48–52 (2008). https://doi.org/10.1109/WKDD.2008.125
25. Petruse, R.E., Grecu, V., Chiliban, B.M.: Augmented reality applications in the transition towards the sustainable organization. In: Gervasi, O., et al. (eds.) ICCSA 2016. LNCS, vol. 9788, pp. 428–442. Springer, Cham (2016). https://doi.org/10.1007/978-3-319-42111-7_33
26. Manavalan, E., Jayakrishna, K.: A review of Internet of Things (IoT) embedded sustainable supply chain for industry 4.0 requirements. Comput. Ind. Eng. **127**, 925–953 (2019). https://doi.org/10.1016/j.cie.2018.11.030
27. Ball, P., Huaccho Huatuco, L., Howlett, R.J., Setchi, R. (eds.): KES-SDM 2019. SIST, vol. 155. Springer, Singapore (2019). https://doi.org/10.1007/978-981-13-9271-9

Productivity Gain by Combination of Lean Manufacturing and Mass Customisation Methods in Aviation Industry

Govardhana Rao Banna[1](✉), Teik-Cheng Lim[1], and Kotireddy Rama Bhupal Reddy[2]

[1] School of Science and Technology, Singapore University of Social Sciences, 463 Clementi Road, Singapore, Singapore
banna001@suss.edu.sg

[2] School of Engineering, Republic Polytechnic, 9 Woodlands Avenue 9, Singapore, Singapore

Abstract. Aviation companies are experiencing increased demand for products due to growth in air travel. Variations in product volume and variety, limited availability of skilled labour and costs are challenges for these companies. It is necessary to explore innovative ways to mediate such issues. A research study was carried out at an aviation manufacturing company to transform its assembly operations by implementing lean manufacturing and mass customisation methods. Time studies were conducted and process times of random samples were recorded before and after transformation. Hypothesis testing was done for testing statistical significance of process times reduction after transformation. Productivity gains were investigated in all four areas of its assembly operations. Regular time studies of process times continued while assembly operations ran for 5 years after transformation. Process times and productivity in each assembly area were assessed based on 5 years' data. Results show that process times reduction was statistically significant in all four areas of assembly operations. Productivity gains were reported from 25% to 27% in these areas. Trend analysis of five years' data found that productivity gains were maintained above 20% level. Literature review has identified that lean manufacturing and mass customisation practices have not been extensively investigated in the context of aviation industry. This study contributes empirical research findings to literature about the influence of combination of lean manufacturing and mass customisation methods on productivity in aviation. It contributes evidence to productivity gain for aviation companies to adopt lean manufacturing and mass customisation methods.

Keywords: Mass customization · Lean manufacturing · Productivity gain

1 Introduction

Aviation 2035 report highlights that "Towards mass customisation" and "Lean through technology adaptation" are going to be high impact and high probability megatrends [1]. Application of mass customisation and lean manufacturing methods presents research opportunity to help aviation industry, which is witnessing high growth in Asia. More

A. Batako et al. (Eds.): GCMM 2021, LNNS 335, pp. 460–474, 2022.
https://doi.org/10.1007/978-3-030-90532-3_35

than half of the new passenger traffic will be coming from this region according to IATA's 20-year air passenger forecast. Demand for variety of products and services is on raise. Satisfying customers' demand for increased product volume and variety in aviation companies has become a challenge. Adding more manufacturing facilities in these companies can be cost prohibitive and may be underutilized during demand variations. Instead, increasing productivity of existing facilities and raise capability to produce more variety would be an attractive option to aviation manufacturing companies. Avenues for achieving higher productivity levels while handling more product volume and variety should be researched. It has been reported through scholarly works that elimination of wastes and increase of value addition is possible with the application of lean manufacturing principles as proposed by Womack and Jones [2]. Mass customisation is also reported as an avenue to explore, as it brings the ability to produce high variety by using flexible and highly responsive manufacturing systems [3]. Hence, a combination of lean manufacturing and mass customisation methods seem to be the way forward for aviation companies. Productivity of manufacturing facilities can be increased by lean manufacturing methods and more product volume can be produced. Likewise, product variety increase can be by mass customisation methods. These methods together will transform the existing manufacturing facilities into highly productive to cope product volume and product variety demand.

An industry-based research study was conducted at aviation company in Singapore that manufacturers the inflight service carts. The company is in similar situation of handling more product volume and variety changes due to aviation growth. It is also facing challenges of limited availability of costly skilled labour in Singapore. Therefore, productivity improvement to produce more product volume and variety is necessary for this company. Research study was conducted first to transform company's four areas of inflight service cart assembly operations by applying selected mass customisation and lean manufacturing methods. The aim of the transformation was to reduce the non-value-added activities and enhance the capability to quickly turnaround the products. Then study went on to investigate the productivity improvement achieved in all four areas of the assembly operations due to transformation. As part of research study, this paper presents the empirical evidence and statistical validation of productivity improvement achieved during the transformation stage. It also presents the empirical evidence of continuation of productivity improvement thereafter for 4 to 5 years post transformation. It contributes an empirical research work to literature in terms of demonstrating an approach that was adopted for transformation and empirical study on how much productivity improvement possible by combination of lean manufacturing and mass customisation methods. This research also contributes to the aviation industry to find a way forward to improve productivity and to gain the ability to produce more product variety.

2 Literature Review

Lean manufacturing requires reduction of any waste that customers do not believe adds value and for which they are not willing to pay. It is an approach based on Toyota Production System [4]. It works on specific principles such as identifying value from the customer's perspective; mapping the value stream; creating flow. eliminating functional barriers and identify ways to improve lead time; establishing a pull system that starts new work only when there is demand for it; pursuing perfection with continual improvement. Implementation of lean manufacturing brings productivity and quality improvement, reduce lead times and operating costs. Similarly, the term mass customisation was coined by Davis, S. [5] as a new frontier for both manufacturing and service industries. Mass customisation aims to bring the ability to produce high variety in terms of customisations at the individual customer level and requires the use of flexible and highly responsive manufacturing systems [3]. Because, it aims for high variety to offer individual level customisations, involvement customers influence the strategies and approaches of mass customisation. Mass customisation can be developed with available information technology and flexible work processes. They permit companies to customize goods and services for individual customers in high volumes at a relatively low cost.

It is well documented in the literature that Toyota started practicing lean manufacturing methods in early 1900. The thought process of this approach was thoroughly described in the book "The Machine That Changed the World" [6]. Scholarly work was punished in the literature about principles of lean manufacturing proposed by Womack and Jones [2]. They are 1) identify what the customer values, 2) identify the value stream and challenge all wasted steps, 3) produce the product when the customer wants it and once started, keep the product flowing continuously through the value stream, 4) introduce pull between all steps where continuous flow is impossible and 5) manage toward perfection through continuous improvement. Hence, lean manufacturing aims at maximising the value addition and increase productivity within production and service systems.

Five mass customization strategies were published by Lampel and Mintzberg [7] based on the stage of customer's involvement. They are Pure standardisation, Segmented standardisation, Customised standardisation, Tailored customisation and Pure customisation. However, the success of a mass customisation strategy depends on how company makes their choice of product or service process strategy work [8]. Based on the degree of customer's involvement, Pine and Gilmore [9] advocated four categories of approaches to mass customisation. They are 1) Adaptive Customisation, 2) Transparent Customisation, 3) Cosmetic Customisation and 4) Collaborative Customisation. Mass customisation is linked to the productivity and capability of facilities in the company. Tseng and Jiao [10] defined mass customisation as "producing goods and services to meet individual customer's needs with near mass production efficiency", It is projected as company's capability to increase customisations without a corresponding increase in costs [11]. Kaplan and Haenlein [12] elaborates it as "a strategy that creates value by some form of company-customer interaction at the fabrication and assembly stage of the operations level to create customized products with production cost and monetary price similar to those of mass-produced products".

As the demand for customized products is continuously rising, the survival of manufacturing and customer service depend on lean manufacturing and mass customisation [13]. Adoption of mass customisation and lean manufacturing approaches together offer the suite of methods to harness the synergies and achieving the elimination of waste, quick turnaround and capability to produce high variety. The list of these methods includes value-stream mapping, process re-engineering, flexible and automated tooling, product modularization, postponed customisation, cellular layout, computerization and internet. Literature presents few evidences of the integration of lean practices and mass customisation methods. Efficiency improvements were reported by application of lean principles into mass customized environments of home building designs [14]. Integration of lean principles into mass customisation were explored for efficiency improvements in area of production system design in fabrication industry [15] and in the area of supply chain design in semiconductors industry [16]. Literature review highlights that lean manufacturing and mass customisation methods are widely explored individually than together. In addition, there is less exploration of such application of combined lean manufacturing and mass customisation methods in aviation manufacturing.

There are also theoretical gaps in the current literature regarding the application of lean manufacturing and mass customisation methods. First, there is less clarity on a proven approach to implement combination of lean manufacturing and mass customisation methods together. Second, there is lack of empirical evidence to validate how much productivity improvement can be achieved by these methods. This research study addresses these theoretical gaps and aims to contribute the current literature.

3 Methodology

A research study was conducted in inflight service carts assembly operations at an aviation manufacturing company in in Singapore. Inflight service carts assembly operations in the company's facility have four areas "Door Assembly", "Base Assembly", "Side Panel Assembly" and "Final Assembly" as shown in Fig. 1.

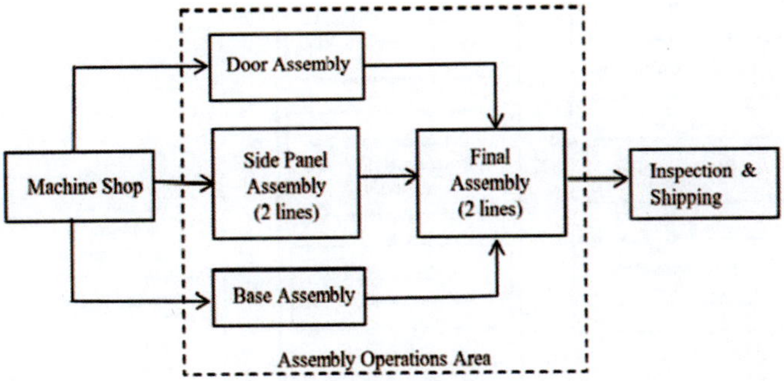

Fig. 1. Inflight service carts assembly operations

Combination of lean manufacturing and mass customisation methods were adopted as potential avenue to improve productivity in the assembly operations. This paper presents an approach for transforming assembly operations and sustaining the productivity improvement by combination of lean manufacturing and mass customisation methods.

3.1 Research Objectives

The following two research objectives are addressed, which contributes to build empirical research work on studying productivity improvement by implementing lean manufacturing and mass customisation methods.

1. To validate statistical significance of productivity improvement after the transformation of assembly operations by combination of lean manufacturing and mass customisation methods
2. To establish empirical evidence for long term sustaining of productivity improvement post transformation

3.2 Approach

This is an empirical research study and is quantitative in nature. Assembly operations were transformed for achieving improved performance. Operations performance data was collected through observations before and after the transformation and for long term period post transformation of assembly operations. The approach applied in this research study in two phases is shown in Fig. 2 below:

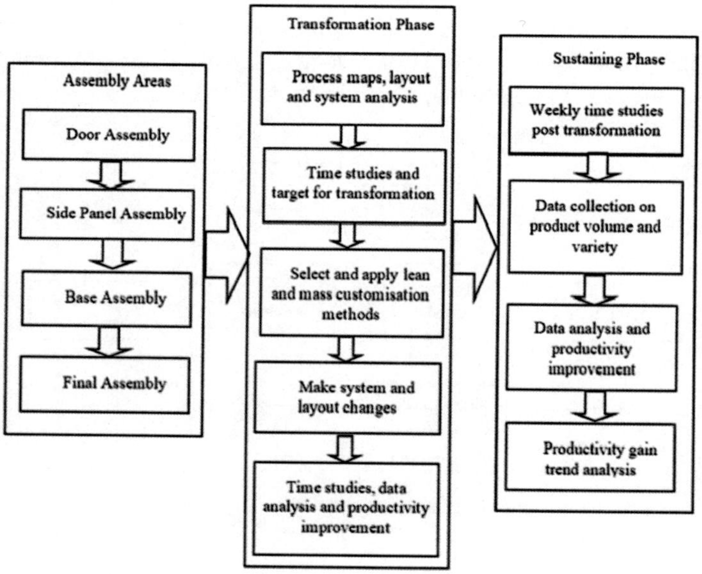

Fig. 2. Transformation stages and tasks

Quantitative research methods were used to aid the analysis of empirical evidence gathered from the performance of assembly operations. "Experimental Research", "Longitudinal Study" and "Correlational Research" [17] were deployed to study the transformation phase and the sustaining phase respectively in order to build the empirical research work on productivity improvement at the aviation manufacturing company.

Before the transformation, process mapping and value stream mapping for all areas in assembly operations was done for analysis. From these published lean manufacturing methods [18] and mass customisation methods [14–16] documented in the literature, following methods are selected for implementation in transformation as shown in Table 1 below.

Table 1. Selected lean manufacturing and mass customisation methods

Lean manufacturing	Mass customisation
i) Value stream mapping ii) Process re-engineering iii) Continuous flows iv) Organised work areas and visual controls	v) Flexible tooling and quick response automated equipment vi) Assembly cells with process modularity vii) Flexible layout configuration

i) Value stream mapping and Process reengineering

Analysis of value addition and process steps in assembly operations helped in identifying non-value-added activities for improvement. Process re-engineering was carried out to reduce the non-value-added activities and wastes. As a result, process time and waiting times in all areas of assembly operations were reduced. Value Stream Maps (VSMs) of all areas in assembly operations are shown in Fig. 3, 4, 5 and 6 below. VSMs show the costs of inflight service cart assembly in US$ values.

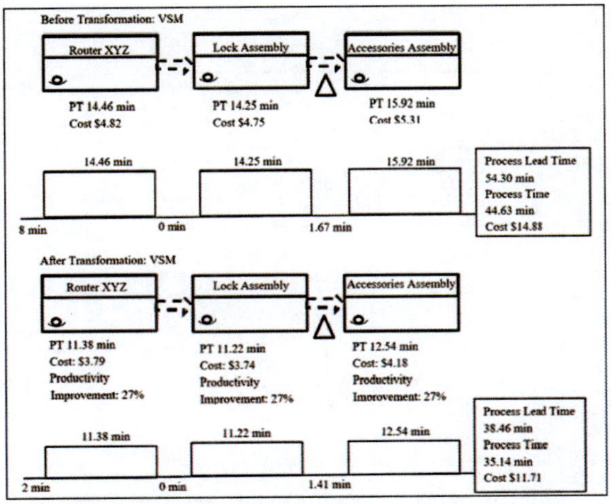

Fig. 3. Value Stream Mapping - door assembly

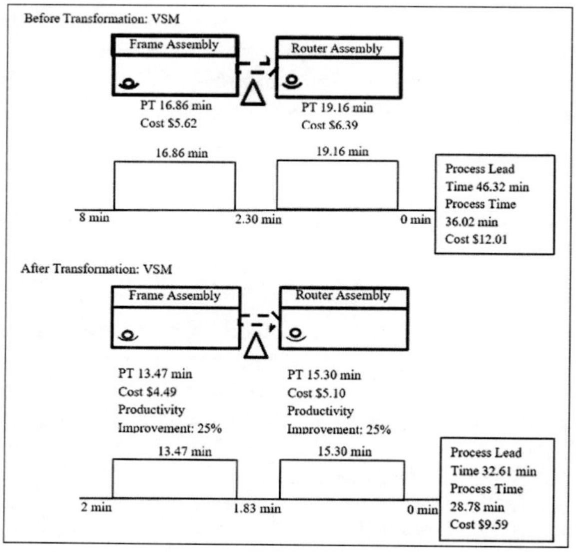

Fig. 4. Value Stream Mapping - side panel assembly

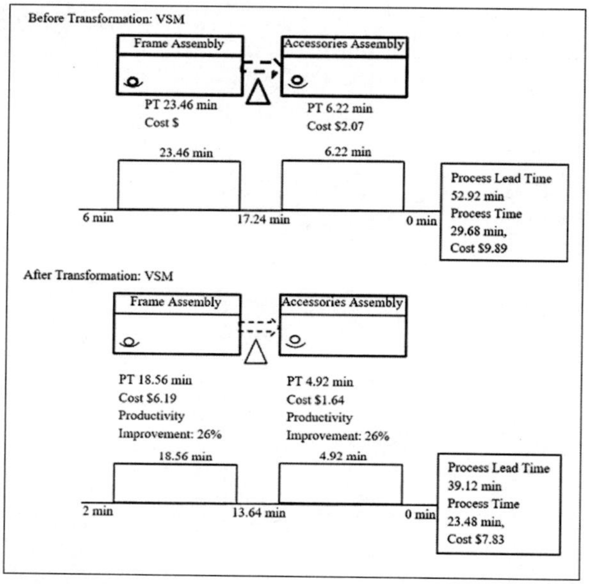

Fig. 5. Value Stream Mapping - base assembly

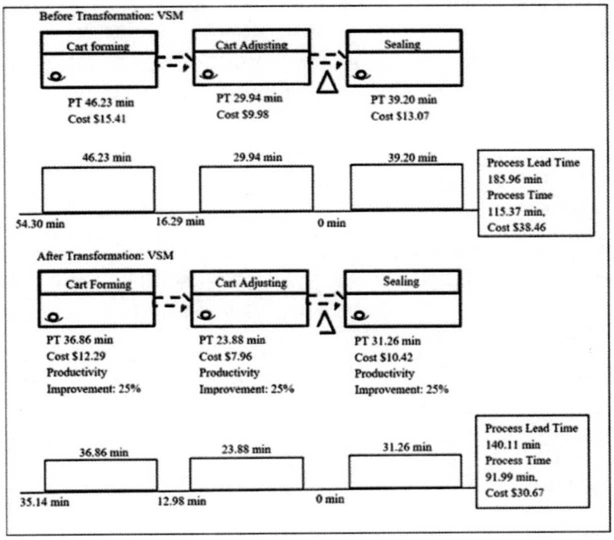

Fig. 6. Value Stream Mapping - final assembly

Based on the improved value addition due to process times reduction, annual cost savings in assembly of inflight service carts are estimated in US$ for current and projected product demand as shown in Table 2 below.

Table 2. Estimated monetary benefits of improved value addition

Parameters	Door Assembly		Side Panel Assembly		Base Assembly	
	Current	Projected	Current	Projected	Current	Projected
Product Demand	8640	11340	11520	15120	7560	5760
Cost Saving	$ 27,331	$ 35,872	$ 27,840	$ 36,540	$ 15,624	$ 44,870
Total Cost Saving	$111,946 (Current demand)			$146,929(Projected demanded)		

Existing layout in assembly operations were reconfigured as part of the transformation. Layouts were redesigned into assembly cells. Continuous flow and flexible configuration were implemented to handle product variety. Continuous flow of work enabled assembly operations to deploy multi-skilled operators and standardise cycle times. As a result, it reduced work-in-progress inventory and wait time of work items. The arrangement of machines and equipment in assembly operations were done according to the process requirements of product families to be assembled. Simulation runs were conducted to assess utilisation levels and smooth product flows in all assembly cells. As a result, assembly cells achieved improved capability of handling more product volume and flexibility to handle product variety.

iii) Organised work areas and visual controls

5S methodology was adopted in the transformation of assembly operations for organizing work areas and brining visual controls. Lineside consumable and tooling were organised to make them available to operator at proper locations in assembly operations. Floor markings created visibility of assembly work cell boundaries and locations of items. Direct access to tooling and materials reduced searching time and waiting times. As a result, it enabled quick turnaround of the work and contributed to productivity improvement.

iv) Automated quick turnaround equipment and flexible tooling
Automated equipment with jigs and fixtures were deployed for performing tasks in assembly operations. Redesigned processes and workstations were equipped flexible tooling in assembly operations. All four areas of assembly operations achieved quick-turnaround capabilities and increased customisation of assemblies. As a result, process times were reduced, productivity improved, and output capacity increased in assembly operations.

3.3 Data Sampling

Performance of assembly operations was observed, and data was collected from four assembly areas before and after the transformation phase. The performance of the transformed assembly system was also monitored post transformation and data was collected from these assembly areas during post transformation. Performance data collection was done for the main parameters of interest - process times, product volume and variety.

a) Data collection frequency
Time studies were conducted in assembly, side panel assembly, base assembly and final assembly areas of assembly operations of inflight service carts [20]. Random data samples were collected each day, for four weeks per month for 3 months before and after the transformation. A total 60 data samples each were collected before and after the transformation in the four areas of the assembly operations.

The performance of the transformed assembly system was tracked post transformation for a period of 4–5 years. Data collection included process times from time studies, product volume and product variety from customer orders. Weekly random samples, totaling 52 data samples every year were collected in the four areas of the assembly operations.

Data samples included key parameters such as process times, product volume, product variety. Average values of these parameters are chosen as variables in this research study as listed below.

PT	Average Process Time per unit
VOL	Product Volume
VAR	Product Variety
PG	Productivity Gain

They are identified by a suffix to represent the specific area or section of the assembly operations. PTd, VOLd, VARd, and PGd represent average process time per unit, product volume, product variety and productivity gain in door assembly area respectively. Suffix change in the variable represents different assembly areas.

B) Data tables
In total two sets of 60 data samples of Process time (PT) representing operations parameters before and after transformation of assembly operations was collected. The average Process time (PT) for assembly operations are shown in Table 3 below.

Table 3. Average Process Times before and after transformation

3 months Average (min/unit)	Door assembly PT_D	Side panel assembly PT_S	Base assembly PT_B	Final assembly PT_F
Process Time Before transformation	44.63	36.02	29.68	115.37
Process Time After transformation	35.14	28.78	23.48	91.99

Monthly data samples of production Volume (VOLB) and Variety (VARB) were collected from the customer orders three months before transformation in assembly operations. Likewise, monthly data samples of production Volume (VOLA) and Variety (VARA) were collected in three months after the transformation.

Transformation of door assembly is completed in 2013, whereas side panel assembly, base assembly and final assembly were completed in 2014. Performance data in assembly operations was collected to do longitudinal study for 4–5 years' period post transformation.

4 Performance Analysis

Performance in assembly operations was investigated in experimental research and longitudinal study during transformation phase and sustaining phase. The aim is to address the two research objectives, one to investigate productivity improvement level and other is to verify the continuity of productivity improvement for longer period post transformation.

4.1 Performance at Transformation Stage

Data analysis, reduction in average Process time (PT) and corresponding productivity gain (PG) are shown in Table 4 below.

It shows that average Process time (PT) is improved in all areas of assembly operations after transformation. Hence, process time and productivity have improved in all areas of assembly operations after transformation by implementing a combination of lean manufacturing and mass customisation methods.

4.2 Long-Term Performance Post Transformation

During longitudinal study, productivity gains in assembly operations are calculated based on the process times collected over 4 to 5 years post transformation. Annual average productivity gains are shown in Table 5 below.

Table 4. Productivity gain due to transformation in assembly operations

	Process Time (PT) (min/unit)			Productivity Gain (PG)
Area	Before	After	Reduction	%
Door assembly	44.63	35.14	9.49	27.00%
Side panel assembly	36.02	28.78	7.24	25.17%
Base assembly	29.68	23.48	6.20	26.39%
Final assembly	115.37	91.99	23.38	25.41%

Table 5. Productivity gain post transformation in assembly operations

	Average Productivity Gain PG			
Year	Door Assembly Operations	Side panel Assembly Operations	Base Assembly Operations	Final Assembly Operations
2014	26.62%			
2015	26.70%	24.54%	25.54%	24.90%
2016	26.53%	24.81%	25.81%	24.73%
2017	26.35%	24.46%	25.46%	24.87%
2018	26.57%	24.73%	25.73%	24.77%

5 Results and Discussion

Findings from performance in assembly operations due to transformation confirmed the reduction of process times and improved the productivity in all four areas of assembly operations. The productivity gain in door assembly is 27%, side panel assembly is 25%, base assembly is 26% and final assembly is 25%.

5.1 Transformation Stage

Data analysis of samples shows the process times are reduced and productivity improved in assembly operations due to transformation. As part of experimental research, statistically validation is carried out using 2-sample t test. Significance of process time reduction, which enabled productivity improvement is validated. Proposed hypothesis is that "transformation by implementing a combination of mass customisation and lean manufacturing methods reduce process times in assembly operations to enable productivity improvement."

Hypothesis is tested to determine whether on average, at 5% level of significance, the assembly process time in assembly improved compared to the assembly process time before the transformation. Two-Sample t-test was conducted for significance testing on difference between two population means.

Let $\mu 1$ denote population mean of process time in assembly operations before the transformation and $\mu 2$ denote population mean of process time in assembly operations after the transformation.

Null Hypothesis H0: $\mu 1 - \mu 2 = 0$ and
Proposed Hypothesis H1: $\mu 1 - \mu 2 > 0$ min, Level of significance: $\alpha = 0.05$
SE mean from the sampling distribution of the difference between two sample means is:

$$\widehat{\sigma}_{\underline{X}_1 - \underline{X}_2} = \sqrt{\frac{\widehat{\sigma}_1^2}{n1} + \frac{\widehat{\sigma}_2^2}{n2}}$$

(when samples are larger than 30).
t is calculated as below for $\mu 1 - \mu 2 = 0$

$$t = \frac{(\underline{X}_1 - \underline{X}_2) - (\mu_1 - \mu_2)}{\widehat{\sigma}_{\underline{X}_1 - \underline{X}_2}}$$

Using two set of 60 data samples of processing times from assembly operations collected before and after transformation, the statistics are calculated as shown in Table 6 below.

Table 6. Statistics from Two-sample t-tests in assembly operations

Statistics	Sample 1: Before Transformation		Sample2: After Transformation	
		Door assembly operations		
Sample size	n_1	60	n_2	60
Sample mean	$\bar{x}_1$	44.63 min	$\bar{x}_2$	35.14 min
Sample standard deviation	σ_1	2.140 min	σ_2	1.400 min
		$\hat{\sigma}_{\bar{x}_1-\bar{x}_2} = 0.33$ t = 28.74		
		Side panel assembly operations		
Sample size	n_1	60	n_2	60
Sample mean	$\bar{x}_1$	36.02 min	$\bar{x}_2$	28.78 min
Sample standard deviation	σ_1	1.676 min	σ_2	1.373 min
		$\hat{\sigma}_{\bar{x}_1-\bar{x}_2} = 0.28$ t = 25.90		
		Base assembly operations		
Sample size	n_1	60	n_2	60
Sample mean	$\bar{x}_1$	29.68 min	$\bar{x}_2$	23.48 min
Sample standard deviation	σ_1	0.857 min	σ_2	0.819 min
		$\hat{\sigma}_{\bar{x}_1-\bar{x}_2} = 0.153$ t = 40.49		
		Final assembly operations		
Sample size	n_1	60	n_2	60
Sample mean	$\bar{x}_1$	115.37 min	$\bar{x}_2$	91.99 min
Sample standard deviation	σ_1	2.140 min	σ_2	1.400 min
		$\hat{\sigma}_{\bar{x}_1-\bar{x}_2} = 0.1$ t = 233.314		

Critical value of t at $\alpha = 0.05$ and degrees of freedom = 118 (same size1 -1 + sample size2 -1, 60 + 60 – 2) from t-tables is 1.9836. From the statistics in Table 6 t > t critical value at $\alpha = 0.05$. Since the t-statistic (standardized difference between the two samples means) values in all in all areas of assembly operations are larger than value of t critical, it can be concluded that there is a significant difference between population means of process times in all areas of assembly operations. Hence it proves the proposed hypothesis that process time reduction in assembly operations is significant after transformation which enabled productivity improvement.

Based on the process time reduction, productivity is improved, and cycle time is reduced in all areas of assembly operations. The positive impact is output capacity is increased in all areas of assembly operations. Output capacity increase in all areas of assembly operations is shown in Table 7 below.

Table 7. Increased output capacity in four assembly areas

	Before Transformation		After Transformation	
Area	Demand unit/month	Output capacity unit/month	Increased Demand unit/month	Increased Output capacity unit/month
Door assembly	720	754	945	957
Side panel assembly	960	1252	1252	1568
Base assembly	480	512	630	647
Final assembly	480	518	630	652

5.2 Post Transformation

From the performance analysis during longitudinal study, there is a strong empirical evidence to confirm that productivity gain is maintained consistently well above 20% in all four areas. Over 4–5 years post transformation, it is evident from the productivity trend as shown in Fig. 7 below.

During longitudinal study, empirical evidence to output capacity increase is supported by more product volume and variety that was produced post transformation in all areas of assembly operations as shown in Table 8 below.

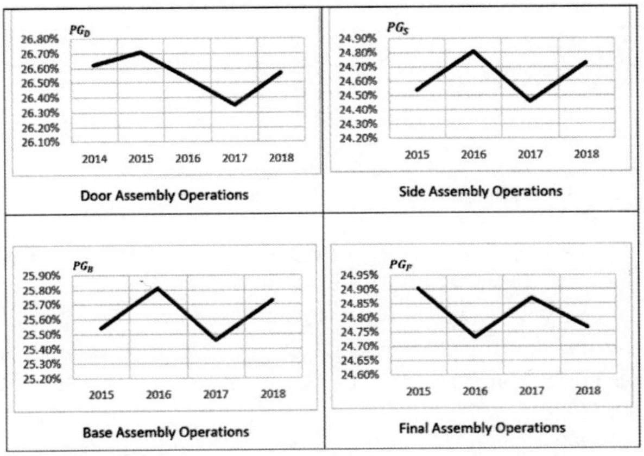

Fig. 7. Productivity gain trend – Post transformation

Table 8. Product volume and product variety increase in four assembly areas

	Before Transformation							
	Door assembly		Side panel assembly		Base assembly		Final assembly	
Year	VOL_D	VAR_D	VOL_S	VAR_S	VOL_B	VAR_B	VOL_F	VAR_F
2013	5016	16						
2014			6688	16	3344	16	3344	16
	Post Transformation							
2014	6743	29						
2015	7353	24	9500	24	4902	24	4890	24
2016	5515	27	7601	27	3889	27	3992	27
2017	6803	25	8790	25	4536	25	4525	25
2018	6251	29	8614	29	4247	29	4279	29

6 Conclusion

Lean manufacturing and mass customisation practices have not been extensively investigated in the context of aviation industry. This research study was conducted at an aviation manufacturing company to address this gap and investigate the sustaining of the productivity improvement over long period. This paper presented an approach for studying productivity improvement by lean manufacturing and mass customisation methods. Experimental research and longitudinal study methods were used in this approach to derive empirical evidence. From the data analysis, productivity gains above 25% are reported during the transformation and 2-stample t-tests confirmed their statistical significance. Productivity gains were sustained above 20% over 4 to 5 years' post-transformation, while producing more product volume and variety in the assembly operations. Hence it brought empirical evidence to support the sustaining of productivity gain by lean manufacturing and mass customisation methods.

Acknowledgements. This is an industry based research in Diethelm Keller Aviation Ltd (DKA), Singapore. It was initiated by a grant from Civil Aviation Authority of Singapore (CAAS) for the productivity improvement in the aviation industry in Singapore. The support from CAAS and DKA is appreciated.

References

1. Arthur, D.L.: Aviation 2035 -Scenarios for value chain recomposition and value sharing in aviation post the era of hyper-competition (2018)
2. Womack, J.P., Jones, D.T.: Lean Thinking. Simon & Schuster, New York (1996)
3. Pine, M.B.: Mass customizing products and services. Plan. Rev. **21**(4), 6–55 (1993)
4. Ohno, T.: Toyota Production System: Beyond Large-scale Production. New York Productivity Press, New York (1988)
5. Davis, S.M.: Future Perfect, p. 243. Addison-Wesley Publishing, Reading (1987)
6. James, P.W., Daniel, T.J., Daniel, R.: The Machine That Changed the World. Free Press (1990)
7. Lampel, J., Mintzberg, H.: Customizing customization. Sloan Management Review, pp. 91–101 (1996)
8. Kahn, B.: Dynamic relationship with customers: high variety strategies. J. Acad. Market. Sci. **26**, 45–33 (1998)
9. Pine II, B.J., Gilmore, J.H.: The Experience Economy. Harvard Business School Press 1993, Boston, Massachusetts (1999)
10. Tseng, M.M., Jiao, J.: Design for mass customisation. Ann. CIRP **45**(1), 153–156 (1996)
11. McCarthy, I.P.: Special issue editorial: the what, why and how of mass customisation. Prod. Plan. Control **15**(4), 347–351 (2004)
12. Kaplan, A.M., Haenlein, M.: Toward a Parsimonious Definition of Traditional and Electronic Mass Customisation, J. Prod. Innov. Manage. **23**(2), 168–182 (2006)
13. David, P.A.: Automation Trends. An Omron Electronics LLC Publications September 2003 issue (2003)
14. Isabelina, N.: Mass customisation strategies and their relationship to lean production in the homebuilding industry. Thesis Doctor of Philosophy in Industrial Engineering and Management Systems, The University of Central Florida, Orlando, Florida (2007)
15. Gregory, B.S.: An integrated framework for applying lean manufacturing and other strategies in mass customisation environment. Thesis-Graduate School, University of Kentucky, Kentucky (2008)
16. Sumanth, R.: Lean manufacturing in a mass customisation plant: inventory correction and shortage measurement. Thesis-Master of Engineering in Manufacturing, Massachusetts Institute of Technology, MA (2011)
17. Bhat, A.: Empirical Research: Definition, Methods, Types and Examples. QuestionPro (2018). https://www.questionpro.com/blog/empirical-research/. Accessed Dec 2019
18. Sundar, R., Balaji, A.N., Satheesh, K.R.M.: A review on lean manufacturing implementation techniques. Procedia Eng. **97**(2014), 1875–1885 (2014)
19. Banna, G.R., Lim, T.C., Reddy, K.R.B.: Process re-engineering of inflight service cart door assembly. In: 6th International Conference on Operations and Supply Chain Management. Bali, Indonesia (2014)
20. Banna, G.R., Lim, T.C., Reddy, K.R.B: Redesigning and improving the productivity of airline inflight service carts assembly system. In: 22nd ATRS World Conference, Seoul, South Korea (2018)

Analysis of Platforms Supporting Open Innovation Approach

Mariusz Cholewa(✉), Mateusz Molasy, Maria Rosienkiewicz, and Joanna Helman

Department of Laser Technologies, Automation and Production Organization, Wroclaw University of Science and Technology, Wyb. Wyspiańskiego 27, 50-370 Wrocław, Poland

mariusz.cholewa@pwr.edu.pl

Abstract. The paper discusses new phenomena and tools supporting Open Innovation development through dedicated up-to-date IT solutions to face challenges of modern industry. In particular, emphasis is placed on innovative crowdsourcing and crowdfunding models creating an effective collaborative environment, with the special focus on the Advanced Manufacturing area. A study of 31 existing platforms at international and regional level in Central European countries is presented. In-depth research based on the developed methodology is presented, which included an assessment of the functionality of platforms based on developed 6 action scenarios (pilot actions), collecting information and structuring the collected data in relation to international and local platforms. The achieved results are analyzed in order to determine the limitations of already existing solutions in the field of international and regional technology transfer and cooperation. Also, the meta-platform approach is identified and described as a promising and effective solution with the highest potential to support innovation and collaboration at various levels.

Keywords: Open Innovation · Advanced manufacturing · Crowdsourcing · Meta-platform

1 Introduction

As the fourth industrial revolution progresses, manufacturing companies face new perspectives and various challenges not only technological but also organizational. These challenges are forced by the need for continuous improvement of the products and production processes in order to meet customer's expectations [1]. A large part of the production challenges is related to the digitization of the product, process and means of production [2].

Manufacturing companies, in order to be able to keep up with market needs and maintain the competitive edge, should use various types of tools enabling new products development and processes improvement in an innovative way. One of the frequently discussed concepts of innovation in recent years is the concept of Open Innovation (OI),

A. Batako et al. (Eds.): GCMM 2021, LNNS 335, pp. 475–490, 2022.
https://doi.org/10.1007/978-3-030-90532-3_36

which enables external entities to participate in the companies' innovation process. Thus firms can benefit from innovation in exchange for sharing their knowledge and technology base with individuals. An analysis of current trends shows that increasing number of companies is implementing elements of Open Innovation into their activities [3]. Industry is gaining benefits from using i.a. crowdsourcing, crowdfunding and microworking. The usage of those new phenomena, tools and methods in industrial companies is necessary if they want to keep up with worldwide competition. Business models based on Open Innovation paradigm are still under development and therefore there is a need to continue research towards building a dedicated environment, where companies could fully benefit from the new approaches.

Platformization is another important trend affecting today's businesses including manufacturing industry. It can be defined as the transition from individual products/services to platforms as intermediaries in transactions and organizing value creation processes [4]. The goal of platform providers is to encourage the creation of ecosystems of interested parties that build their business on the platform and thus create value for users [5–7]. By their nature, platforms offer a degree of openness, and through this openness digital platforms give entrepreneurs new opportunities. Platform owners provide certain resources, such as application programming interfaces (APIs), development kits (SDKs) and other resources to enable their users to build their offer [8].

Together, OI and platforms have created a wide variety of opportunities for the companies - from serving as an input to innovation for recognized companies to participating as an equal business partner on existing platforms [9]. In the industrial context it is required to have that kind of approaches and tools to support product development process, particularly in preliminary phases, where it is highly desirable to provide original and inventive solutions. There are many platforms around the world that are designed to help users to take advantage of the new opportunities that OI provides. It is done by digitally providing tools that include idealization, innovation competitions, funding tools, innovation communities, innovation marketplaces and innovation toolkits, as well as education and process consulting for their users in various forms. Some of them have an established position and experience in the process of matchmaking however a number of new platforms that have already achieved critical project volumes is growing.

The purpose of this article is to evaluate the effectiveness of support offered by existing Open Innovation platforms for enterprises wanting to keep up with changes in the modern world in a transition towards 4th Industrial Revolution. The innovative services offered as part of the Open Innovation concept within different platforms will be assed in order to determine the platforms thanks to which industry can support their challenges.

2 Open Innovation in Industrial Application

According to Henry Chesbrough it can be stated that the innovation process is currently facing a „paradigm shift". The way how companies innovate new ideas and bring them to the market is undergoing a fundamental change – the transition from a closed innovation model to an increasingly open one [10]. Usage of the paradigm allows enterprises free flow of ideas, products and services from the outside to the inside and vice versa – mostly in order to remain competitive. One of the more frequently discussed concepts of innovation in recent years is the concept of Open Innovation. The concept is defined as “a distributed innovation process involving purposive knowledge flows across organizational boundaries for monetary or non-monetary reasons” [11]. It enables external entities to participate in the innovation process and benefit from innovation in exchange for sharing their knowledge and technology base.

Recently, not only research and publication on the topic of Open Innovation have grown exponentially, but their methodologies are spurring innovation activities in a variety of industries, being carried into the domains of theories of firms’ strategies [12]. Open innovations reflect the growing interest of industries in developing methodologies and supporting tools. Collaborative technologies facilitate the aggregation of multiple intelligences for the search of new ideas and innovative solutions within a community. Thus, the collective search of innovative solutions is the result of the aggregation of multiple intelligences [13].

The increase in digitization and the decrease in communication costs for customers has led to the exponential growth of such online platforms that were used for crowdsourcing and generating new ideas about products, services and processes by large companies [14, 15]. According to Howe “crowdsourcing is the act of taking a job traditionally performed by a designated agent (usually an employee) and outsourcing it to an undefined, generally large group of people in the form of an open call” [16].

The analysis shows that more than 80% of high-tech companies listed in the S&P 500 index have established OI platforms to benefit from customers’ inputs [14]. It can be also observed that instead of a company having its own Open Innovation platform, the platforms are framed as ecosystem level collaboration spaces or solution marketplaces. With this in mind, mainly large organizations and companies (such as Amazon, P&G, Unilever, SAP, Boeing, Roche, Electrolux, Siemens, Bayer, Rockefeller Foundation, NASA, etc.) join such platforms to increase their research and development and find solutions to their business problems. On the public sector side Open Innovation platforms are used not just for seeking technological solutions to societal challenges, but also to foster social innovation.

Currently, the innovation process on existing OI platforms is not standardized and has no common structure. For Majchrzak and Malhotra problems with architectures of existing platforms are: not enough collaboration, minimal feedback on the evolution of ideas, and individual efforts to develop new ideas [17].

What is also worth to mention is that today the concept is being adopted in wider pool of initiatives, including those that seem far away from industrial. In fighting the pandemic of Covid-19 numerous of initiatives have been introduced that corresponds well to a concept of Open Innovation. One of the reasons of adopting the concept to these issues is the fact that Open Innovation can accelerate an undertakings. Therefore, opening up speeds up innovation process and allows to take advantage of the knowledge of others [18].

The study presented in this paper will attempt to determine (1) which types of innovative services, that make up Open Innovation tools accessible for industry, are widely used on platforms, and on the other hand, (2) what can be improved in the platform architecture and (3) which services with high innovation potential are still underestimated and unnoticed by OI platform providers.

3 Introduction to Research on Open Innovation Services

Invented in 2000 and launched in 2010, Open Innovation platforms are constantly evolving, as evidenced by the double increase in the number of challenges published each year. Open Innovation platforms manage search and communication between problem givers and problem solvers, organized around a specific topic (challenges). Most often, brokering process takes place between the company and individuals (crowdsourcing of ideas) or specialists involved in scientific research or having specific specialist knowledge (searching for solutions). A platform built and working that way usually integrates knowledge bases - including users - to create mutual value and capture on platforms, thus enabling network effects [19]. Typically, OI platforms are virtual environments that serve as infrastructure for information exchange. Most of them are implemented as Internet solutions, which means that they offer an Internet interface as a mean of interaction between users. Innovation services often reflect well-organized, collaborative and Open Innovation environments. Thanks to that it can be stated that OI platforms serve as intelligent systems for creating innovation among different industrial actors.

The issue of the efficiency of OI platforms has been raised in connection with the implementation of the Central European (CE) SYNERGY project. SYNERGY, which stands for "Synergic networking for innovativeness enhancement of Central European actors focused on high-tech industry" is a project which aims at enhancement of innovativeness in European regions through strengthening linkages and beyond border cooperation to create synergy between SMEs, industry, research, intermediaries and policy makers (https://www.interreg-central.eu/Content.Node/SYNERGY.html). The project's scope is mainly oriented on advanced manufacturing with a special focus on the most promising modern industrial technologies defined as Key Project Areas (KPA): 1. additive manufacturing and 3D printing, 2. micro- and nanotechnology-related processes and materials, and 3. Industry 4.0. The SYNERGY's goal is to link functionalities of commercial and industrial oriented OI platforms with public and social oriented solutions by implementing Quadruple Helix Innovation Model.

Project partners represent entities of Triple Helix Innovation Model – 4 higher education and research institutions, 1 SME, 2 business support organizations as well as government representatives as an associated partner. Consortium supported by general public, that will use developed innovative services and pilot actions, will be able to create an ecosystem of Quadruple Helix Innovation Model [20, 21].

Within the project, different types of innovative services enhancing crowd innovation initiatives, which can be applied in manufacturing industry, have been analyzed and developed. They can be distinguished within four types: crowdsourcing (obtaining needed services, ideas, or content by soliciting contributions from a large group of people), microworking (a series of small tasks which together comprise a large unified project, and are completed by many people), crowdfunding (funding a project or venture by raising small amounts of money from a large number of people) and exchanging resources (exchanging physical resources, skills and competences of employees of one company with another entity). Each of the innovative service types is divided into several services. The addressed services are listed below:

- Crowdsourcing:
  - Social Product Development,
  - Research & Innovation projects;
- Microworking:
  - Industrial challenges,
  - Crowd innovation for companies;
- Crowdfunding:
  - Crowdfunding for research projects;
- Exchanging resources:
  - Infrastructure sharing,
  - Competences exchange.

Table 1. SYNERGY pilot actions [22]

Pilot action	Description
Simulated crowdfunding (PA1)	1. Platform user uploads research problem/innovative idea to be solved and declares budget 2. Other platform users (including companies) declare how much are they ready to spend in order to have the access to the results 3. SYNERGY project awards with the voucher the best 5 projects that reach the budget goal
Vouchers for research/innovation projects (PA2)	1. Platform user uploads project/research/innovation idea related to Key Project Areas (KPAs) 2. The same platform user is searching for a partner from different Central Europe regions in order to solve the problem together 3. New partners present their common research plan 4. SYNERGY project selects the best research plan and awards the winner with the voucher 5. Awarded partners perform research and upload the results on the platform
Rent-A-Robot (PA3)	1. User uploads offer related to robot infrastructure 2. Others declare needs for a short time robot rental/utilization 3. Tools & problems are merged, partners get in contact solving problems (sharing resources - HR & know-how) 4. Virtual marked place is turned to a real network solving problem
Crowd innovation for companies (PA4)	1. A company uploads a problem related to KPAs to be solved and defines the budget 2. Platform users solve the problems and present their results 3. A company selects the best solution and awards the winner with the declared budget

(*continued*)

Table 1. *(continued)*

Pilot action	Description
Vouchers for developed solutions of the research projects (PA5)	1. Platform users (companies, individuals, researchers) upload problems (e.g. technological, research) related to KPAs to be solved 2. SYNERGY project selects the most crucial problems (e.g. 5 of them) 3. Platform users solve selected problems and present their results 4. SYNERGY project awards the best 5 solutions with vouchers
Design and prototype model (PA6)	1. General brainstorm on SYNERGY product idea 2. Micro-working approach on product development 3. Interested idea givers submit proposed design & prototype model 4. SYNERGY project awards the best 3–5 solutions with a voucher

Based on above-presented services a set of specific 6 action scenarios (pilot actions) has been developed. Pilot projects designed in such a support implementation of Open Innovation. The possibility of carrying them out on the platform allows to define such a platform as a comprehensive OI platform. They are presented in Table 1.

Each of the actions was planned and described in detail by project team. The definition of new innovative service (in the test phase) includes: listing target groups, describing how to reach givers and how to promote pilot actions among takers to create crowd, time schedule/workflow for each pilot action, development of example of challenge definition/solution definition, description of process in simple graphics (for users), as well as proposition of criteria of assessment.

Rent-A-Robot (based on the idea of infrastructure sharing) is a good example of an innovative service and pilot action supporting sustainable manufacturing, and Open Innovation and in consequence facilitating transformation of the enterprise towards modern industry. In this service industrial companies, SMEs and research organizations were defined as a potential Infrastructure takers and providers.

The process of implementing the pilot action should be carried out based on the developed and described model, which simplified presentation is shown in the Fig. 1, where SCIP platform is a Synergy Crowd Innovation Platform developed as a tool to support project realization.

It was assumed that pilot actions and assessment of their feasibility on existing OI platforms will be used for reliable and comparable assessment of OI platforms. They will allow to evaluate their suitability for companies wanting to develop and deploy components of Industry 4.0 and other KPAs, and also indicate the most and least supported elements of the Open Innovation concept, thus discovering services with still high potential for increasing the company's efficiency but requiring greater support from OI platforms.

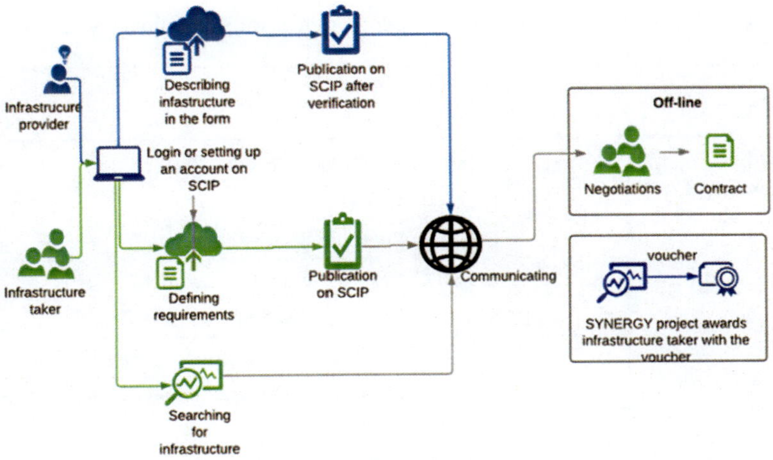

Fig. 1. Process of Rent-a-robot pilot action (own elaboration).

4 Research Methodology

To realize the abovementioned aims it was necessary to conduct research to determine state-of-the-art, tendencies of OI platforms evolution and real needs of organizations and individuals wanting to implement Open Innovation platform to support their business activities. The general plan of the research was based on the developed three-step procedure, which led to the definition of improved, more accessible and complex innovative services:

- Preliminary analysis of existing platforms to explore the research area,
- Division of platforms in terms of their main functions and offered services,
- Detailed analysis of platforms.

This study focused on research on the existing Open Innovation platforms' functionalities that can support innovation enhancement in industry. The research consisted of analyzing a number of platforms based on a previously developed methodology. Both large and well-known platforms as well as small, regional ones in several countries were analyzed.

To achieve the main goal of the analysis, the research concept was elaborated to determine which sources of information (online & offline) should be used to collect reliable data on crowd innovation approaches and how the needed data should be structured, gathered and analyzed. The research and its results were elaborated based on analysis of existing and functioning Open Innovation platforms as a tool for Open Innovation approach. One of the elements of this analysis was the assessment of possibilities of implementation of proposed/developed specific 6 action scenarios - pilot actions. The research was conducted based on primary and desk research by collecting data by telephone calls and personal meetings with platform providers (if possible) and through internet queries (including obtaining the necessary data by asking questions in e-mails).

Table 2. Breakdown of platforms by pilot activities for international platforms

Pilot action	Platform number	Platform
Simulated Crowdfunding (PA1)	1	Experiment: www.experiment.com
	2	Kickstarter: www.kickstarter.com
	3	Indiegogo: www.indiegogo.com
	4	Crowd Cube: www.crowdcube.com
	5	Startnext: www.startnext.com (regional)
	6	Crowdfundport: www.crowdfundport.neurovation.net
	7	Place2help Frankfurt RheinMain: www.place2help.org/rhein-main (regional)
	8	KITCrowd: www.kitcrowd.de (regional)
	9	ScienceShip: www.scienceship.com (regional)
Vouchers for research/innovation projects (PA2)	10	NineSigma: www.ninesigma.com
	11	InnoCentive: www.innocentive.com
	12	Chaordix: www.chaordix.com
	13	Ideascale: www.ideascale.com
	14	Neurovation: www.neurovation.net (regional)
	15	Genius Croatia: www.en.genius-croatia.com (regional)
Rent-A-Robot (PA3)	16	Hirebotics: www.hirebotics.com
	17	Envoy AI: www.envoyai.com/exchange
Crowd innovation for companies (PA4)	18	Connect + Develop: www.pgconnectdevelop.com
	19	Amazon Mechanical Turk: www.mturk.com
	20	Appjobber: www.appjobber.de (regional)
	21	Crowd Guru GmbH: www.crowdguru.de (regional)
	22	Clickworker: www.clickworker.de
	23	Mathesia: www.mathesia.com (regional)
Vouchers for developed solutions of the research projects (PA5)	24	IdeaConnection: www.ideaconnection.com
	25	Presans: www.presans.com
	26	Hypios CrowdInnovation: www.hypios-ci.com
	27	Jovoto: www.jovoto.com
Design and prototype model (PA6)	28	Quirky: www.quirky.com
	29	OpenIDEO: www.openideo.com
	30	crowdSPRING: www.crowdspring.com
	31	Invented4: www.inventedfor.com (regional)

International/global and regional/local platforms divided by pilot actions are presented in the Table 2.

Each of the platform has been described on the basis of the following criteria:

- general criteria:
  - number of users,
  - number of submitted projects (if applicable, depending on the OI service type),
  - total amount of the budget of projects (if applicable, depending on the OI service type),
  - number of supporters (if applicable, depending on the OI service type);
- specific pilot actions criteria:
  - is it possible to run the pilot action on a certain platform?
  - if so, what is the procedure for running the pilot action on a certain platform (including detailed description with needed steps, including print screens)?
  - if not, what functionalities or options are hindering the running of the pilot actions or limit it to a certain point? What functionalities are missing in this context?

For local platforms, the above criteria have been extended to technical issues to assess their level of openness and interoperability. Among other aspects investigated, these were issues such as: IT technology that has been used, openness and integration possibilities, business model open for collaboration, IPR issues.

5 Analysis of the Gathered Data and Discussion of the Outcomes

Based on the developed methodology 31 platforms, both international (21) and regional (10), were analyzed. As a result of the research a catalogue of the most used, and therefore the most effective and beneficial, also financially, for users including production companies, crowd innovation initiatives and solutions was developed. It was based on the results of research performed by all project partners.

Table 3. Example of the analysis form.

Platform name	Kickstarter
Website address	www.kickstarter.com
Number of users	16 Million
Number projects	50.590.463 of which 157.010 successful
Total amount of the projects budget	Roughly 4 billion USD
Supporters number	15.696.994
Intended SYNERGY pilot action	Crowdfunding

(continued)

Table 3. (*continued*)

Platform name	Kickstarter
Is it possible to run the pilot action on the platform?	Yes Kickstarter is open for every project. It is not specialized on research projects, but more on new products or creative works. There is the possibility to create company accounts as legal entities to secure reputable and reliable funding and implementation of the pilot actions
How to run pilot action on certain platform?	Create project page with video and description Offer rewards Keep supporters updated If successfully funded, platform fees have to be paid How to award voucher? Criteria? Comment section Rewards Not only reaching budget but also "sum" of overreaching it More information here: https://www.kickstarter.com/help/handbook?ref=global-footer The platform offers various tools to present the project and to make contact with supporters. At the tab "campaign" the project and its creator can be presented broadly through photos, text, videos, timelines and so on. You can find a FAQ and a blog to update your supporters. Moreover, there is the opportunity for the supporters to leave a comment for asking questions or giving suggestions publicly. But it is also possible to send private messages or to do a live stream To compensate the supporters, there is an extra reward section, where different rewards can be shown and described with a title, a minimum fund and the time it needs to be delivered To get an overview of the running project, Kickstarter provides numbers and facts in the dashboard, the backer report and various statistics. The data gives some indication about the supporters, where they come from, about the financial issues, the number of visits and conversion into pledges, as well as the most popular rewards All this data helps the creator to advertise more accurately, to update his project properly and to stay in contact with his community Last but not least, there is a creator's handbook where everyone can get some tips, step by step instructions and find all the information needed To start your project, you should design your project page, describe your rewards and keep your community updated with the latest on goings

(*continued*)

Table 3. (*continued*)

Platform name	Kickstarter
How to run pilot action on certain platform?	

The in-depth research covered collecting information and structuring collected data in regard to international platforms and in regards to local platforms.

The research included a developed and described set of information including possibilities and ways of realizing improved innovations services - pilot actions, that need to be verified in the test environment. All information have been described according to the developed template. In Table 3 an example of a form containing an analysis of one of the selected platforms has been presented.

All collected data has been gathered and analyzed. Table 4 presents all the analyzed platforms and some key information about their performance, functions and capabilities in the context of the possibility of using them to carry out pilot actions.

Based on the information and data gathered in this analysis, a list of observations and findings resulting from the conducted research was developed. First of all, the results have showed that 18 of existing platforms were supporting crowdsourcing, 18 microworking, 9 crowdfunding and 2 partly exchange of resources (Fig. 2b). The breakdown of innovative services that can be provided for i.e. industrial users is as follows: pilot action Simulated Crowdfunding (PA1) can be supported by 9 platforms, Vouchers for research/innovation projects (PA2) by 16, Rent-a-Robot (PA3) by 3, Crowd innovation for companies (PA4) by 12, Vouchers of developed solution of the research projects (PA5) by 17 and Design and prototype model (PA6) by 7 platforms.

Table 4. List of collected platforms with some key data and supported pilot actions (* implements partly).

Platform no.	No. of users	No. of submitted projects	PA1	PA2	PA3	PA4	PA5	PA6
1	220	858	x					
2	16 Million	50.590.463 (157.010 successful)	x					

(*continued*)

Table 4. (*continued*)

Platform no.	No. of users	No. of submitted projects	PA1	PA2	PA3	PA4	PA5	PA6
3	10 Million	>800,000	x					
4	691.000	821 successful	x					
5	1 Million	12000 of which 6800 funded	x	x		x	x	
6	Data not available	2 challenges, 24 solutions	x			x	x	
7	Data not available	54	x					
8	Data not available	Data not available	x					
9	Data not available	Data not available	x	x		x	x	
10	>2 Million	44		x			x	
11	380.000	>162.000		x			x	
12	>1.150.000	>100		x			x	x
13	4.5 Million	Data not available		x		x	x	x
14	8.300	>4.500 ideas generated		x		x	x	
15	220	500–600 submitted		x			x	
16	Data not available	Data not available			x*			
17	24	62			x*		x	
18	2000	22				x		
19	>500.000	3.8 million tasks				x		
20	10.000	>1 Million				x		
21	52.000	15 million micro jobs				x		
22	2 Million	Data not available		x		x	x	
23	>3000	42		x		x	x	
24	500 (SMEs and LE)	114		x			x	
25	Data not available	>100		x		x	x	
26	950 000 experts and 45.000 users	>160		x			x	
27	>100.000	~70.000		x			x	x
28	1.300.000	321.000						x
29	>100.000	23.079		x			x	x
30	210.000	Data not available						x
31	>3000	Data not available		x				x

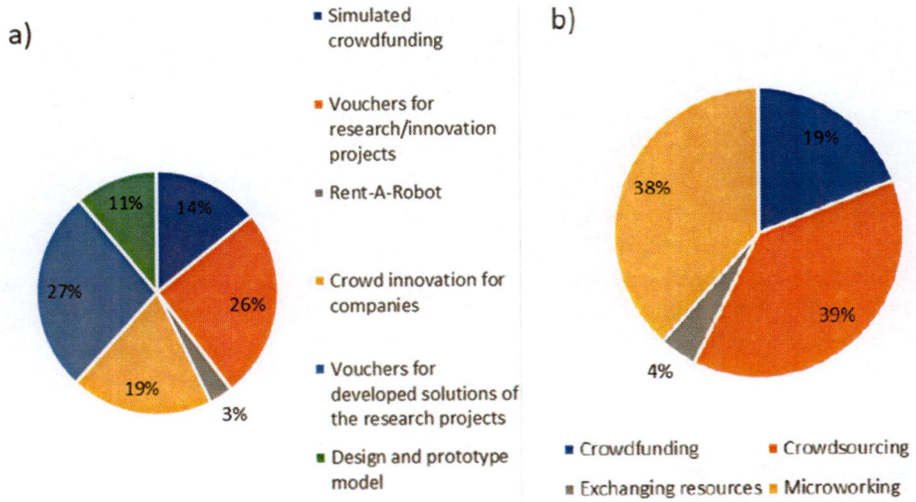

Fig. 2. Open Innovation activities supported by OI platforms: a) per pilot action, b) per innovation service.

It was also observed that there is one pilot action which is not widely supported by existing OI platforms. The action Rent-a-Robot represents exchanging resources functionality of the Open Innovation concept. Exchanging or sharing resources, which can be defined as sharing free capacity of machines and equipment, allows companies to effectively respond to the challenges of Modern Industry by simplifying access to the latest technologies or infrastructures in their neighborhood or region. It can be one of the most important elements of building market advantage and transformation of the company towards sustainable manufacturing [23]. One of the additional observations was the important role of experts/moderators, whose knowledge and experience can facilitate the process of solving problems. But also, this definitely human-dependent aspect can be effectively implemented by changing the architecture of existing platforms and using latest information technologies e.g. artificial intelligence techniques to supporting the process of finding a solution to a problem. This opens a new stage in the development of OI platforms by using intelligent algorithms and tools to support the wisdom of the crowd.

The analysis showed that regardless of the existing, well-established OI platforms there is a need to develop a new digital OI platform which will implement above findings and will respond to identified deficiencies, especially those related to exchanging resources. Therefore, within the SYNERGY project a special dedicated platform will be developed, that ensuring crowdfunding, microworking and crowdsourcing for innovative solutions for the CE society, focused on Advanced Manufacturing. The planned platform, in addition to the advantages and functions mentioned in above research, will complement these capabilities by using a multilateral, cross-border (Central Europe) approach and offering complementary services.

An important result of the study is determining the validity and significance of the concept of meta platform as a future, effective model of IT solutions supporting Open

Innovation. Meta platforms are developed to perform their tasks, using autonomous platforms (which were the subject of the study) and act as a proxy server for these platforms [24]. The research clearly showed that there is no OI platform that supports all innovative services. The use of the meta platform concept would dramatically increase their efficiency and impact area. Cooperation and interaction within metaplatform framework will allow the dissemination of the concept of Open Innovation and reaching the necessary critical mass, which will provide adequate technical and substantive resources ready to support various domains of knowledge.

Furthermore, research enabled elaboration of list containing recommendations to improve the operation and effectiveness of OI platforms. This list has been grouped into three categories, which together with their selected content are presented below:

1. Clear message:
   a. Platforms' effectiveness depends on clear and unambiguous classification of research, scientific and interest areas.
   b. Openness on modern communication channels and information formats including multimedia or video.
2. Process and quality:
   a. Reliability of the platform means quality of the content which should be assured by appropriate quality assurance processes.
   b. Efficient data propagation systems with private channels and the implemented NDA protection mechanism are of key importance for business entities.
   c. Effective management and access to the substantive content which can be described by various types of data, including legacy systems (photos, sketches, renders and prototypes, pitch video, models).
3. Platform environment:
   a. A large community is crucial for the successful operation of the platform.
   b. Keeping in touch with users during and after the campaign to inform about the latest results and final conclusions/results/reports. It should be supported by latest communication channels including forums, chats, talks or even video conferencing.

6 Conclusions

The use and implementation of Open Innovation approach through utilization of digital innovation platforms offers significant potential of improvement of companies' effectiveness. Crowd skills and intelligence can initiate or upgrade companies' inbound activities to support conceptual design of new products and processes, which are one of the main challenges the manufacturing industry faces today, due to the highly demand for creative solutions.

Presented study determined types of innovative services, that make up Open Innovation tools accessible for industry. Also underestimated OI services with high innovation potential were identified and moreover improvements in the architecture and operation of the platform were indicated. The study proved that huge potential and effectiveness of innovative services offered as part of the Open Innovation concept within different IO platforms can be implemented to support industrial companies and their challenges especially in transition towards 4th Industrial Revolution. In addition, it has been observed that the potential of the Exchanging resources service has not yet been fully exploited, which is particularly important for new, developing manufacturing companies or those which want to flexibly respond to the changing market demands.

The conducted research has proved that the idea of Open Innovation is popular and strongly supported by IT tools, usually taking the form of a platform. Almost all services supporting Open Innovations are used on different platforms, and those innovative services are often combined with each other to provide a more comprehensive set of functionalities. However, one multifunctional platform, that containing all services based on crowdsourcing, microworking, crowdfunding and resources exchange in one place is missing. What is more, it has been also observed that there is a lack of both international and regional solutions that would fully implement the assumed functionality for the pilot action: sharing the infrastructure.

In the future, it seems reasonable to undertake research on the concept of better using the potential of innovative services such as infrastructure sharing, but also on exploring the capabilities of metaplatforms.

Acknowledgements. The research leading to these results was funded by Interreg Central Europe programme priority Innovation and knowledge development under grant agreement No CE1171, project SYNERGY as well as from funds of Polish Ministry of Science and Higher Education for science in 2018–2020 for the implementation of an international project.

References

1. Kotowska, J., Markowski, M., Burduk, A.: Optimization of the supply of components for mass production with the use of the ant colony algorithm. In: Burduk, A., Mazurkiewicz, D. (eds.) ISPEM 2017. AISC, vol. 637, pp. 347–357. Springer, Cham (2018). https://doi.org/10.1007/978-3-319-64465-3_34
2. Burduk, A., Musiał, K.: Genetic algorithm adoption to transport task optimization. In: Graña, M., López-Guede, J.M., Etxaniz, O., Herrero, Á., Quintián, H., Corchado, E. (eds.) SOCO/CISIS/ICEUTE -2016. AISC, vol. 527, pp. 366–375. Springer, Cham (2017). https://doi.org/10.1007/978-3-319-47364-2_35
3. Open innovation 2.0 yearbook 2016, European Commission, Directorate-General for Communications Networks, Content and Technology (2016)
4. Bygstad, B., Hanseth, O.: Transforming digital infrastructures though platformization. In: Twenty-Sixth European Conference on Information Systems (ECIS2018), Portsmouth, UK (2018)
5. Adner, R.: Ecosystem as structure: an actionable construct for strategy. J. Manag. **43**(1), 39–58 (2017)
6. Autio, E., Thomas, L.: Innovation ecosystems. The Oxford Handbook of Innovation Management, pp. 204–288 (2014)

7. Jacobides, M., Cennamo, C., Gawer, A.: Towards a theory of ecosystems. Strateg. Manage. J. **39**(8), 2255–2276 (2018)
8. Ghazawneh, A., Henfridsson, O.: Balancing platform control and external contribution in third-party development: the boundary resources model. Inf. Syst. J. **23**(2), 173–192 (2013)
9. Nambisan, S., Siegel, D., Kenney, M.: On open innovation, platforms, and entrepreneurship, Strateg. Entrepreneurship J. **12**(3), 354–368 (2018)
10. Chesbrough, H.: Open Innovation: The New Imperative for Creating and Profiting from Technology. Harvard Business Press (2006)
11. Chesbrough, H., Bogers, M.: Explicating Open Innovation: Clarifying an Emerging Paradigm for Understanding Innovation. New Frontiers in Open Innovation, pp. 3–28. Oxford University Press, Oxford (2014)
12. Rothaermel, F.T., Hess, A.M.: Building dynamic capabilities: innovation driven by individual, firm, and network-level effects. Organ. Sci. **18**(6), 898–921 (2007)
13. Lopez, R., Belaud, J., Le Lann, J., Negny, S.: Using the collective intelligence for inventive problem solving: a contribution for open computer aided innovation. Expert Syst. Appl. **42**(23), 9340–9352 (2015)
14. Mahr, D., Lievens, A.: Virtual lead user communities: drivers of knowledge creation for innovation. Res. Policy **41**(1), 167–177 (2012)
15. West, J., Lakhani, K.R.: Getting clear about communities in open innovation. Ind. Innov. **15**(2), 223–231 (2008)
16. Howe, J.: The rise of crowdsourcing. Wired Magaz. **14**(6), 1–4 (2006)
17. Majchrzak, A., Malhotra, A.: Towards an information systems perspective and research agenda on crowdsourcing for innovation. J. Strat. Inf. Syst. **22**, 257–268 (2013)
18. Chesbrough, H.: Ind. Mark. Manage. **88**, 410–413. https://doi.org/10.1016/j.indmarman.2020.04.010
19. Adamczyk, S., Bullinger, A., Moeslein, K.: Commenting for new ideas: insights from an open innovation platform. Int. J. Technol. Intell. Plan. **7**(3), 232 (2011)
20. McAdam, M., Debackere, K.: Beyond 'triple helix' toward 'quadruple helix' models in regional innovation systems: implications for theory and practice: beyond 'triple helix' toward 'quadruple helix' models. R&D Manag. **48**, 3–6 (2018)
21. Rosienkiewicz, M., Helman, J., Cholewa, M., Molasy, M.: SYNERGY project: open innovation platform for advanced manufacturing in central Europe. In: Burduk, A., Chlebus, E., Nowakowski, T., Tubis, A. (eds.) ISPEM 2018. AISC, vol. 835, pp. 306–315. Springer, Cham (2019). https://doi.org/10.1007/978-3-319-97490-3_30
22. Application Form CE1171SYNERGY; Interreg CENTRAL EUROPE – Call 2 (2017)
23. Helman, J., Rosienkiewicz, M., Molasy, M., Cholewa, M.: Infrastructure Sharing Model as a Solution for Sustainable Manufacturing Upcoming in Smart Innovation, Systems and Technologies, Springer, Singapore (2020). https://doi.org/10.1007/978-981-15-8131-1_7
24. Daniel, F., Kucherbaev, P., Cappiello, C., Benatallah, B., Allahbakhsh, M.: Quality control in crowdsourcing: a survey of quality attributes, assessment techniques and assurance actions. ACM Comput. Surv. **51**(1), 1–40 (2018). https://doi.org/10.1145/3148148

Smart Manufacturing Platform Estimation for SME in Automotive Rubber Part Supply Chain Management

Suthep Butdee(✉)

Department of Production Engineering, Faculty of Engineering, King Mongkut's University of Technology North Bangkok, 1518 Pracharaj I, Bangsue, Bangkok 10800, Thailand
Suthep.butdee@gmail.com

Abstract. This paper proposes a smart manufacturing platform estimation to evaluation for the rubber part subcontractor environment using fuzzy logic modeling. This model can help factory manager to select a suitable design factory layout and facility as well as the components of communication. Manufacturing is currently changing to a modern and digital transformation throughout the whole life cycle based on design, planning, and control. Automotive rubber part industry is one of the biggest clusters in Thailand which is managed as a supply chain. However, it has been operated using traditional environment style which lacks of competitive advantage. Smart manufacturing platform with AI is indispensable. Fuzzy logic is developed to assist supply chain manager to estimate the platform effectively.

Keywords: Smart manufacturing platform estimation · SME · Automotive rubber part industry · Subcontractors · Fuzzy logic modeling

1 Introduction

Smart manufacturing is emerged with the edge of industry 4.0 that focuses on high value creation and using the digital communication technology such as computer integrated manufacturing, cloud manufacturing, flexible manufacturing. Design is the first stage of manufacturing that uses the digital tools for making a digital mockup to test that idea is workable, sufficiently effective via simulation and modeling. Planning can be start from a smart product planning which is much different architecture from the traditional thinking model. Process planning including production planning are quite important which is accompanied with digital tools such as Wi-Fi, internet IoT, NB-IoT, SCADA, cloud platform, ethernet routing, PLC Modbus, TCPIP protocol, mobile application, etc. In addition, manufacturing automation is extended to be more intelligent and continuous operation. Robotic, AGVs with Kankan's QR code for smart logistics are developed to link on the same protocol. Smart planning needs to be more flexible, adaptable, and rapid changeable. Control is also significant for the smart manufacturing. Instead of controlling by manual and operators, HMI is implemented to link with smart communication tools which are capable to work two ways of a feed black control. Therefore, the smart

A. Batako et al. (Eds.): GCMM 2021, LNNS 335, pp. 491–504, 2022.
https://doi.org/10.1007/978-3-030-90532-3_37

manufacturing platform is involved with flexibility and ability to respond rapidly to both design and production changes to achieve competitive advantage. With the capabilities of the internet and IoT, smart manufacturing is able to increase integration, big data storage and analysis. Cloud computing helps companies' access to highly configurable, controllable, and monitoring ability from remote area with portable server, networks, and other storage applicants to be created and released at a rapid pace. Smart manufacturing platforms support manufacturer to collect data broadcast from smart machines and data at production lines.

The automotive industry comprises of a wide range of companies and organizations involved with design, development, manufacturing, marketing. It is one of the world's largest economic sectors by revenue. There are a lot of parts which are produced before assembly. In one aspect of material used for a car, it can be divided into four categories; metal, plastic, rubber, and composite. The rubber parts are very high demands in Thailand because it is central location to produce automobiles in South East Asia. The global well-known brands are planted in Thailand such as Toyota, Honda, Nissan, Volvo, Mercedes Benzes, BMW, Hindi, and so on. The car is classified into several types: city car, saloon, luxury saloon, executive, mini cooper, minivans, sport cars, off-road cars including bus and trucks. However, most of the cars use rubber parts for examples, gasket, O-ring, grommet, seal, rubber brush, rubber floor, radiator tube, air tube, oil brake tube, and so on. Such parts are mostly produced by compression machines, injection machines, or blow mold machines. However, the machines are mostly usages of old technology which is needed to improve and retrofit to suit with the industry 4.0. The share market of the OEM is approximately 80% whereas the REM is about 60–70% of the whole share market in Thailand. For export volume, the OEM part is about 80–85% whereas REM is about 15–20%. The rubber parts are also used for bicycle, and motorcycle.

Industry 4.0 has established by German government to promote the computerization in the way of intelligent factory in aspect of adaptability, efficiency, and utilization. The infrastructure has to be changed or improved to reach the integration of new technology such as IoT and CPS (cyber-physical systems. In addition, the system includes wireless connections, sensors, HMI (Human Machine Integration), M2M (Machine to Machine communication) etc. Presently, industries are performed more and more in the forms of supply chain which links as a network between a company and its suppliers to produce and distribute a specific product to market. They might share market, customer orders, material, resources, know-how, technology transfer, people and information, as well as profits.

Automotive rubber part industry is one of the biggest clusters in Thailand. It consists of a master company which is known as the 1st tier or OEM firm and sub-contractors which are called 2nd tier. The sub-contractors are mostly SMEs. The purchasing order is responsible by the master company and distribute to the subcontractors in the supply chain. With a highly standard of rubber parts, it is therefore all of manufacturing resources need to be centralized control by the 1st tier such as machines, molds and engineering standard document. Presently, digitalization influences all industrial sectors. Automotive rubber part industry in Thailand needs to change existing management method to a new paradigm which is called smart manufacturing platform for supply chain management.

2 Literature Review

This section describes smart manufacturing environment, components, tools and technology which is particularly used in automotive industry and supply chain. Dalenogare et al. [1] presents technologies for industrial 4.0 which is considered as a new industrial stage integrated both vertical and horizontal manufacturing process in the Brazilian environment using regression analysis. The industry 4.0 integrates manufacturing operation system, information and communication technology such as IoT, CPS (Cyber Physical Systems), SCADA dashboard, embedded chip in order to reduce time to market a new product launching, reduce cost and processing time, but increase productivity and competitiveness. Jeschke et al. [2] proposes industrial internet of things and cyber manufacturing system which are accompanied with an information network of physical objects of sensors, HMI, PLC, SCADA. They allow interaction and cooperation of the objects to achieve ultimate business goals. The cyber manufacturing is a concept that involves translation of data from interconnected systems into predictive operations to achieve resilient performance. It interacts with industrial big data and smart analytics to discover and comprehend invisible issues for decision making. Frank [3] states that industry 4.0 can increase more competition within business resulting from technological innovations and changing customer requirements. It emerges and provides digital solutions including smart manufacturing, smart products, smart supply chain and smart working by the new platform of IoT, cloud services, and big data analysis. Ghadge et al. [4] proposes the framework of industry 4.0 and smart manufacturing which is divided into 3 stages. Stage 1 is employment cloud system and IoT to cooperate with ERP, SCADA, PLC and sensor. Smart supply chain is used to transform digital platforms working with remote monitoring and collaborative robots. Stage 2 applies big data analytics and accompanies with automation and visualization. Digital platform is used for suppliers in the supply chains. The stage 3 brings the big data analytics to support flexibility of the flexible production lines and links with customers. The impact of industry 4.0 on supply chains by simulation analysis offers benefit challenges to join in a new way of the network. It is found that the industry 4.0 can predict a new challenges and opportunities for future supply chains. The key drivers and barriers of industry 4.0 consist of agility which enable the supply chains to work with real-time planning and control, to predict future events of customer behavior, to delivery and route tracking in logistics, to customize to serve variety of product needed, to determine accuracy data to support the efficiency of making data on time. However, there are some constraints such as financial management, expertise, and so on. Nagy et al. [5] presents the role and the impact of industry 4.0 in the aspect of using the internet of things on the business strategy of the value chain. In the ear of industrialization, companies are increasingly investment on resources to integrate network to improve value chain and competitive advantage using IoT tools. The main purpose of Industry 4.0 is to achieve improvements in terms of automation and operational efficiency as well as effectiveness. Ray et al. [6] proposes the concept framework of smart manufacturing systems for industry 4.0 consisting of Cyber-Physical Systems (CPS), the Internet of Things (IoT), cloud computing, and big data analytics. The smart manufacturing includes smart design, smart machining, smart monitoring, smart control, and smart scheduling. Wang, L. H. [7] presents machine availability monitoring and machining process planning using cloud manufacturing in order to achieve cost effective, flexible, sharing resources and services. Could manufacturing employs mainly internet and web-based service system. It can show the shop floor in terms of machine status on real time both locally and remotely. Xu, X. [8] explains cloud manufacturing based on cloud computing which is to change the way of industries and business operations to become dynamic and virtualization in the forms of internet services. It can create intelligent factory networks to work as collaboration. CAD systems are interfaced to CAM, CNC machines, ERP systems, SCM system and other systems via intercommunication of Bus protocol. As a result, knowledge-based system and intelligent methods are established and liked to the platform. The systems can interface to the smart shop floor devices such as RFID, sensors, active badge, smart phone, web camera, and smart card. Zhong et al. [9] presents smart manufacturing conceptual framework for industry 4.0 using internet of things (IOT), cyber-physical systems (CPS), cloud computing and big-data analysis as well as AI in order to serve uncertain demands, high quality, and shorter time to market and converting manufacturing system onto a smart level. The components of smart manufacturing consisting of smart design, smart machining, smart monitoring, smart control, and smart scheduling including decision making and the integration. The first goal is to create a new and innovative product design and optimized process design. Smart design tools are CAD, 3D printing whereas smart machine uses intelligent CNC machine with smart robotics and AGVs. Smart monitoring uses cloud server, and IOT linked with data input by sensors and counters. Smart control uses smart phones and mobile applications. Smart scheduling uses the system to link with process planning and control at the shop floor level. Smart control uses cloud based numerical control. The cloud is worked as a mobile server with can access anywhere via HMI panel. The cloud server contains design, production, control, and logistic information. Magesh et al. [10] presents the state of the art of the internet of things and its applications. It can link with embedded system and peripheral hardware via IP address or web API. The embedded system consists of Wi-Fi, Bluetooth, Ethernet, ISDL, whereas the input devices receive data from sensors, actuators, and drives. Henzel and Herzwurm [11] proposes cloud manufacturing which is currently challenged in the manufacturing industry to gain more competitive while delivering value to customers. Dachyar et al. [12] proposes knowledge growth of the IOT applications which is classified into 3 categories; technology, industry, and method. Industry area looks at the manufacturing in various aspects such as logistic, supply chain management, decision support, production process, smart industry, robot, industrial control system, manufacturing system, smart manufacturing, smart factory, etc. Problems need from current manufacturing technology that requires monitoring devices to support distribution of sensors and high-speed wireless networks.

Supply chain management is widely applied to many organizations in different characteristics. A subcontractor strategy is implemented to a manufacturing cluster of automotive part industry. Resources are shared cost and managed by the 1st tier company. The subcontractors are performed as production agent. Purchasing orders are controlled by the 1st tier and distributed to the subcontractors. The 1st tier is also responsible for product planning and marketing, purchasing raw material, production planning and control, engineering and information management. On the other hand, subcontractors take care with quality, delivery, competitive cost and continuous process improvement.

Rubber part process consists of several processes. As the supply chain, purchasing orders are received from the 1st tier company and distributed to different companies within a network chain. The orders contain many items that are produced and delivered on the due date plan. The items are varieties which are needed to make under standard operation time. In addition, different material has a certain period of lifetime. Therefore, the material must be correctly controlled both lifetime and temperature. The raw material is made from a mix of synthetic rubbers because the different material types which is used for different proposes and functions of the applications. The raw material is placed inside a mold which is designed in different sizes and cavities. The size is based on the machine size such as 410 × 410 mm, 500 × 500 mm, 600 × 600 mm etc. Cavities are designed based on the optimization of cycle time quantity, the part's sizes and shapes. The operation begins with preparation of material shape such as barrel, rectangular, cylinder and so on. In general, the raw material is made as a sheet with a certain thickness. Then, the sheet is slit in the shape that fits with cavity's shape and size. The machine is heated under the suitable temperature. The prepared materials are placed in the cavities and hold in the standard time. The compressed parts are picked up out of the mold. The parts are inspected and packed according to the purchasing order. The parts need to deliver every day under the delivery plan. Production planning and control are produced every day for 2 shifts and 20 h a day. Standard process is set by the 1st tier company collaborated with the sub-contractors. Therefore, the sub-contractors must follow the standard operation process and controlling parameters. Quality control is checked both at the factory and at the mother company. Performance is evaluated every month. Fuzzy logic concepts are widely applied to deal with constraints and dynamic situations as well as uncertainty risk assessment modeling for bus body manufacturing supply chain using AHP and fuzzy AHP [13]. Kamchai and Suthep [14] proposed uncertain inventory management using TOPSIS and FAHP method in automotive supply chain in order to select effective subcontractors and suppliers.

3 Development of Smart Manufacturing Platform

The concept of smart manufacturing platform covers the life cycle of manufacturing process which is integrated into a single database as a network linked to several agent bases. Intelligent agent-based system is applied to connect each other by Wi-Fi and internet of things (IoT) and cloud system both internal and external manufacturing process inside the supply chain network.

Figure 1 shows the smart manufacturing platform for automotive subcontractor's environment. It can be divided into two parts: the 1st tier side and the the Subcontractor (SCT) side. Both parts are communicated via cloud computing, IoT, Wi-Fi, and Ethernet. The 1st tier consists of several functions such as SCM, marketing, engineering, raw material, mold making, and inventory. On the other side, the SCT consists of planning, store and inventory, logistics, and production including maintenance. Wi-Fi is used to connect information between functions inside companies. Smart manufacturing is designed to achieve high efficiency, high value of the production outputs. The pain point of the platform is mainly concerned about quality, delivery, and back order.

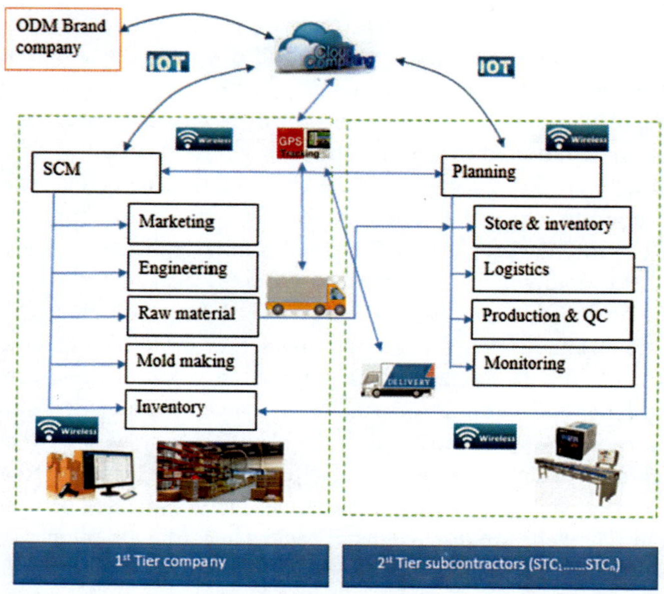

Fig. 1. Smart manufacturing platform for automotive subcontractors.

The assumption is that the quality should be better and more accurate, the delivery should be faster and on time schedule, the back order should be release rapidly. The impact factors influence the performance which are accompanied include quality of raw material control, compression mold, logistics, production planning and control, IT/IoT communication. Moreover, the smart manufacturing platform should integrate with AI decision supporting system such as Fuzzy logic, Neuro fuzzy to deal with uncertainty of demand and supply. Therefore, fuzzy logic is applied to the platform.

4 Fuzzy Logic Modeling

The fuzzy logic modeling assists the smart manufacturing platform to estimate performance of a subcontractor in operation management based on supply chain network. The model concerns mainly on the output of manufacturing performance estimation as shown in the Fig. 2. Three major factors are measured including quality, delivery, and communication. The fuzzy model is designed for 5 ranges of membership input functions depended on the important criteria. The raw material, mold and machine are divided into 5 levels of linguistic language; very high, high, medium, low, and very low. Quality consists of 5 criteria; raw material, mold, machine, logistics and production planning and control. Delivery composes of 2 criteria; logistics and production planning and control. Communication consists of 6 criteria; raw material, mold, machine, logistics, production planning and control, Wi-Fi/IoT/Cloud. All of input criteria are divided into 5 level of the membership functions; very high, high, medium, low, and very low (Table 1).

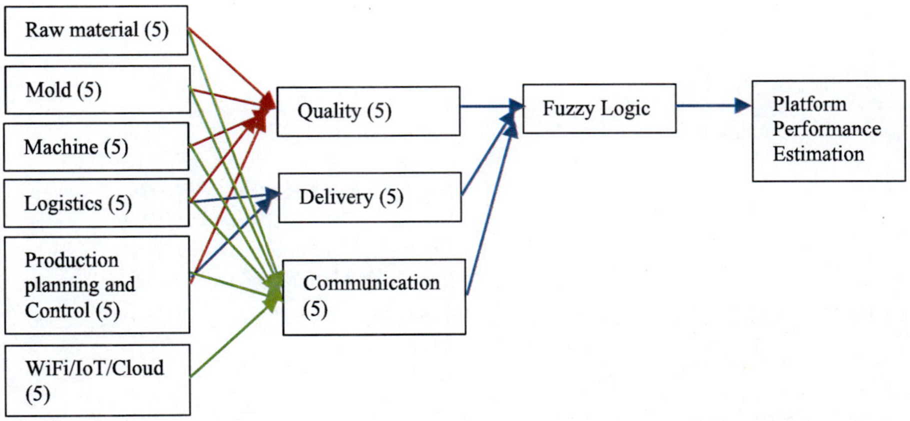

Fig. 2. Fuzzy logic modeling for the smart manufacturing platform to estimate platform performance.

Table 1. Level fuzzy FIS acceptance.

Linguistic variable	Term set	Values
Raw material score	VH, H, M, L, VL [99.5%, 99.0%, 95.0%, 90.0%, 85.0%]	**Very high;** Controlled on-line, Sufficient stock, Safety stock, Well-prepared **High;** Safety stock, Sufficient stock, Well-prepared **Medium;** Sufficient stock, Well-prepared **Low;** Well-prepared **Very low**; Normal-prepared
Mold score	VH, H, M, L, VL [99.0%, 95.0%, 90.0%, 85.0%, 80.0%]	**Very high;** Feed-back control, Sensors, force, counter, Node red IoT **High**; Monitoring execution system **Medium;** Maintained by ISO standard procedure **Low;** Automatic control **Very low;** Manual control
Machine score	VH, H, M, L, VL [95.0%, 90.0%, 85.0%, 80.0%, 75.0%]	**Very high:** Could computing, Fuzzy controller, MES, Node-red IoT **High:** MES, could computing, flexible manufacturing **Medium:** Temp. sensor control **Low:** Mold control **Very low:** Automatic control, Stand alone *MES = Monitoring Execution System
Logistics score	VH, H, M, L, VL [99.5%, 95.0%, 90.0%, 85.0%, 80.0%]	**Very high:** GPS tracking, Cloud server, Optimization route, MES, QR code **High:** GPS tracking, Cloud server, MES **Medium:** GPS tracking, MES **Low:** GPS tracking **Very low:** Route control

(*continued*)

Table 1. (*continued*)

Linguistic variable	Term set	Values
Production planning and control	VH, H, M, L, VL [99.0%, 95.0%, 90.0%, 85.0%, 80.0%]	**Very high:** ERP, MRP, SCM, Cloud computing, Mobile dashboard, MES **High:** ERP, MRP, SCM, MES, Cloud server **Medium:** ERP, MRP, internet Wi-Fi **Low:** Basic program, internet Wi-Fi **Very low:** Manual and basic program
Wi-Fi/IoT/Cloud	VH, H, M, L, VL [99.5%, 95.0%, 90.0%, 85.0%, 80.0%]	**Very high:** NB IoT, GPS tracking, Cloud server **High:** IoT, CPS, Cloud server **Medium:** IoT, QR code, RFID **Low:** Internet IoT **Very low:** LAN, Wi-Fi,

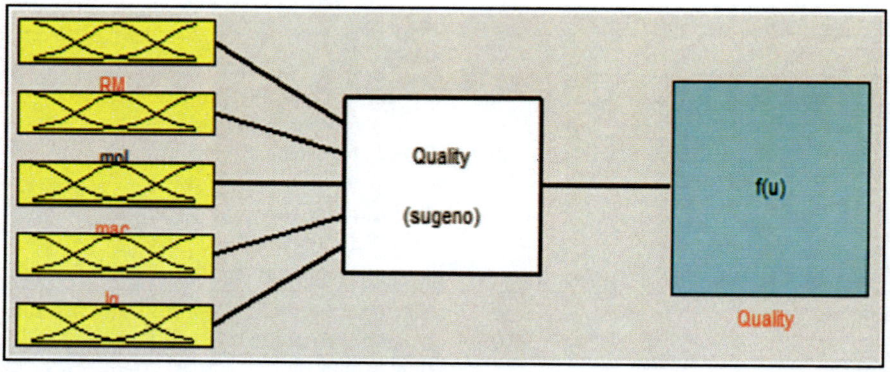

Fig. 3. The FIS membership functions of the platform performance estimation.

Figure 3 shows FIS membership function of the platform performance estimation. The input consists of raw material conditions, mold, machine, logistics and production planning and control conditions. Figure 4 shows the rule viewer of the platform performance estimation result. The example data shows that when the raw material performance is 85%, the mold performance is 80%, the machine performance is 75%, the logistic performance is 80%, the production planning performance is 80%, then the total platform performance estimation is 80%. The input membership functions can be modified according to the actual situation which is automatically come out with that result.

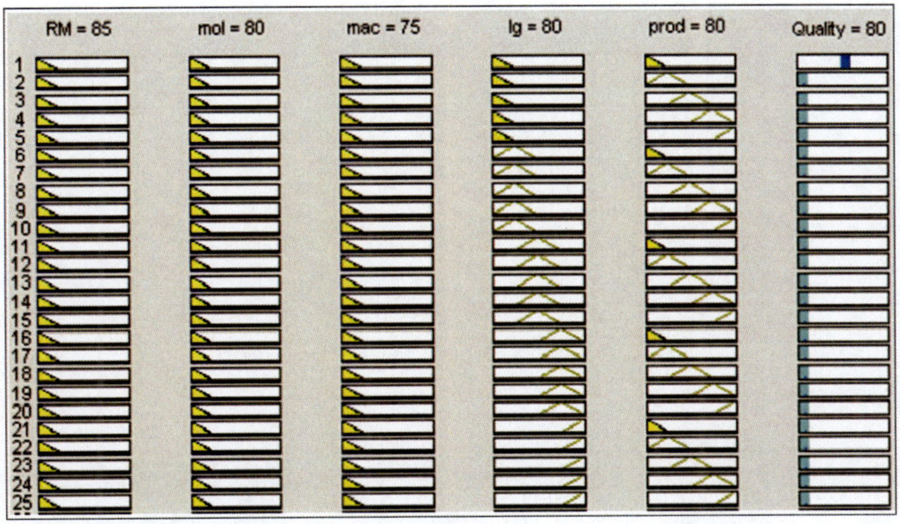

Fig. 4. The rule viewer of the platform performance estimation result.

Figure 5 shows the raw material membership function consisting of 5 levels: VH, H, M, VL, and L. For example, the rank of VH is between 81.3 to 88.62 which is on the triangle base, whereas the top of the triangle is 85 and that is the middle point of the triangle. The Fig. 6 shows the platform performance estimation output function which consists of 3125 rules.

```
    [System]
    Name='Quality'
    Type='sugeno'
    Version=2.0
    NumInputs=5
    NumOutputs=1
    NumRules=3125
    AndMethod='prod'
    OrMethod='probor'
    ImpMethod='prod'
    AggMethod='sum'
DefuzzMethod='wtaver'
Input1]
    Name='RM'
    Range=[85 99.5]
    NumMFs=5
MF1='VL':'trimf',81.375 85 88.625]
MF2='L':'trimf',[85.0019571044774 88.6253894272665 92.2502379833295]
MF3='M':'trimf',[88.6256372446178 92.2499266101254 95.8735454318426]
MF4='H':'trimf',[92.25 95.87 99.5]
MF5='VH':'trimf', [95.8749521503765 99.4997009398531 103.125]
```

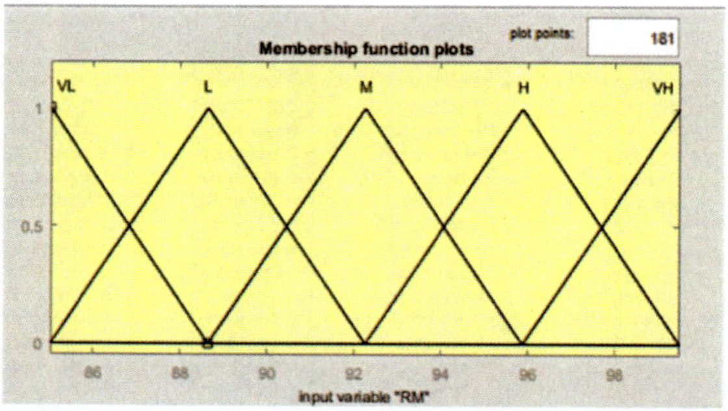

Fig. 5. The membership function for raw material.

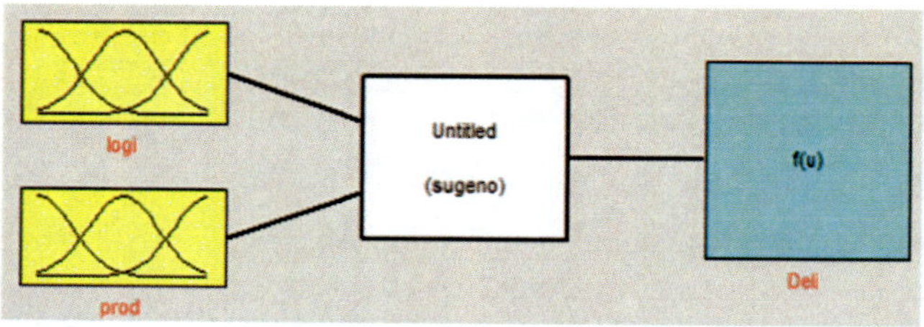

Fig. 6. The FIS membership function of the input to delivery performance estimation.

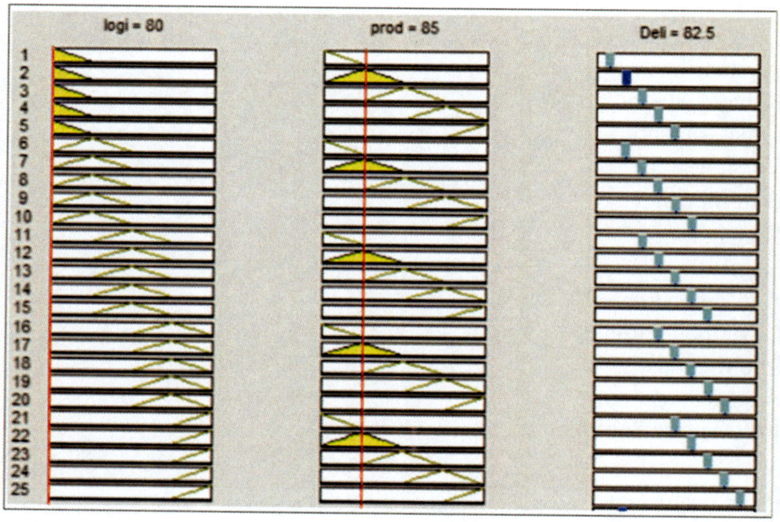

Fig. 7. Rule viewer of the delivery performance

Figure 7 shows the FIS membership function of the input to delivery performance estimation. The input consists of two parameters: logistic system and production planning control. The membership functions concern with the quality of operations and effectiveness. Figure 8 shows rule viewer of the delivery performance estimation. The inputs are logistic condition and production planning and control, and the output is delivery quality. Then, the total quality is evaluated to get the result.

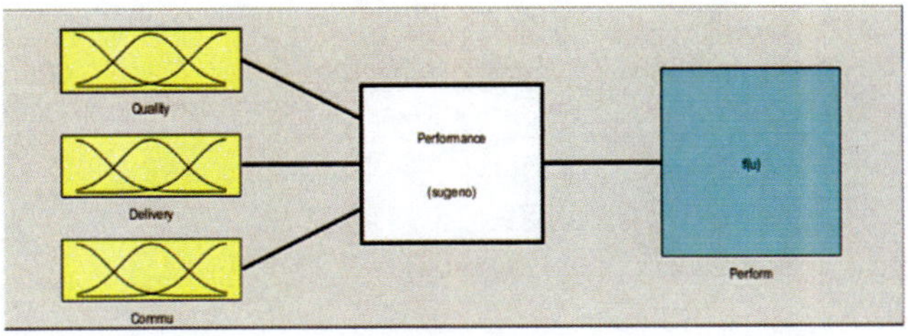

Fig. 8. The FIS membership function of the profile quality estimation.

Figure 8 shows the FIS membership function of the input and output for estimation platform performance from the smart platform. The input membership function consists of quality, delivery, and communication.

[Input1]
Name='Quality'
Range=[80 95.5]
NumMFs=5
MF1='VL':'trimf',[76.125 80 83.875]
MF2='L':'trimf',[80 83.875 87.75]
MF3='M':'trimf',[83.8339947089947 87.7089947089947 91.5839947089947]
MF4='H':'trimf',[87.75 91.625 95.5]
MF5='VH':'trimf',[91.625 95.5 99.375]

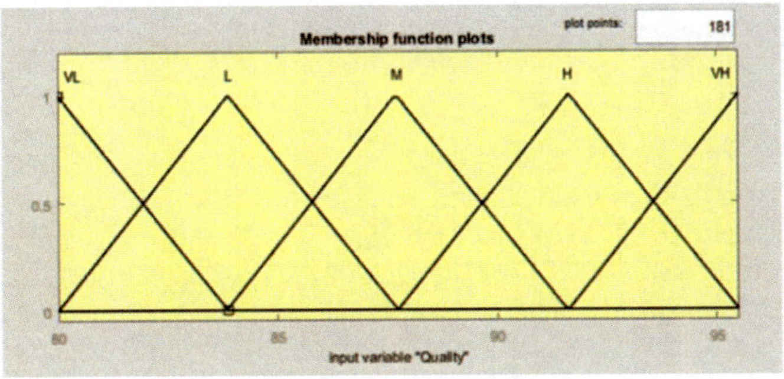

Fig. 9. The input membership function of the quality

Figure 9 shows the input membership function of the quality of the smart manufacturing platform. The quality membership is divided into 5 levels; Vl, L, M, H, and VH. The value set ranks from 80 to 95.5. The Vl value set ranks from 76.125 to 83.875. The VH value ranks from 91.625 to 99.375.

Input2]
Name='Delivery'
Range=[80 99.25]
NumMFs=5
MF1='VL':'trimf',[75.1875 80 84.8125]
MF2='L':'trimf',[80 84.8125 89.625]
MF3='M':'trimf',[84.8125 89.625 94.4375]
MF4='H':'trimf',[89.625 94.4375 99.25]
MF5='VH':'trimf',[94.4375 99.25 104.0625]

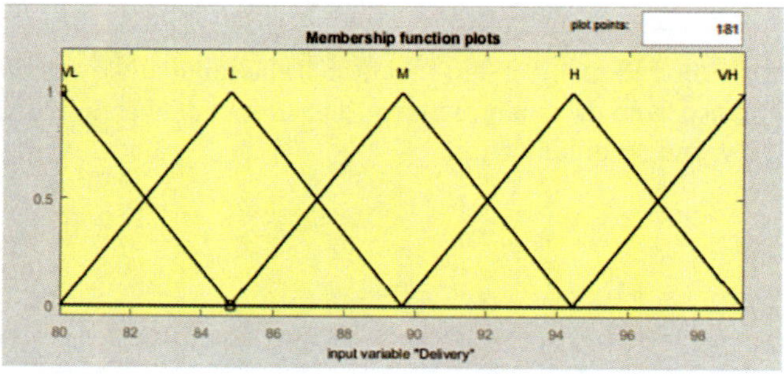

Fig. 10. The membership function of the delivery.

Figure 10 shows the membership function of the delivery. It consists of 5 levels: VL, L, M, H, VH. The total ranks are between 80.0 to 99.25. The VL ranks from 75.1875 to 84.8125. The VH ranks from 94.4375 to 104.0625.

Input3]
Name='Commu'
Range=[80 96.17]
NumMFs=5
MF1='VL':'trimf',[75.9575 80 84.0425]
MF2='L':'trimf',[80 84.0425 88.085]
MF3='M':'trimf',[84.0425 88.085 92.1275]
MF4='H':'trimf',[88.1277777777778 92.1702777777778 96.2127777777778]
MF5='VH':'trimf',[92.1275 96.17 100.2125]

Figure 11 shows the membership function of the communication. It is divided into 5 levels: VL, L, M, H and VH. The total ranks are between 80 to 96.17. The VL ranks from 75.9575 to 84.0425. The VH ranks from 91.625 to 99.375.

Figure 12 shows the FIS output of smart manufacturing platform performance estimation. It consists of 3 parameters: quality, delivery, and communication. The example shows that if the quality is 87.3, the delivery is 89.5, the communication is 88.1, and then the performance is expected is 88.5%.

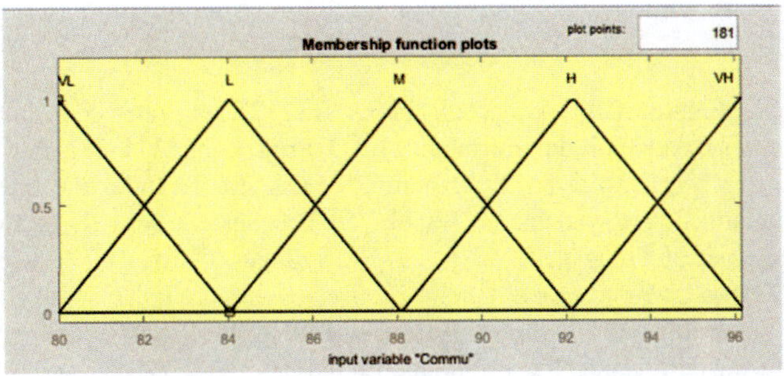

Fig. 11. The membership function of the communication.

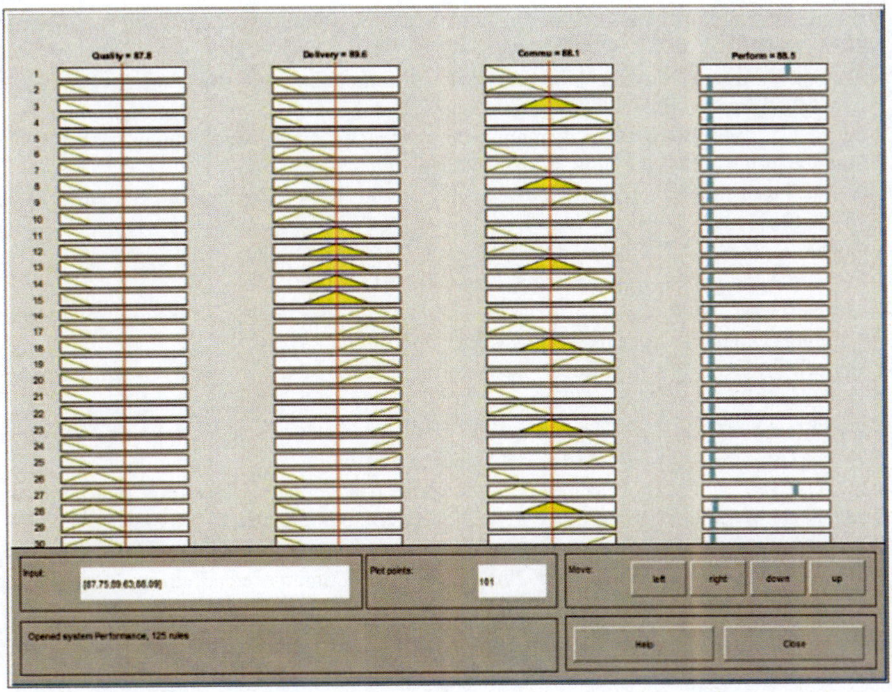

Fig. 12. The FIS output of smart manufacturing platform design performance estimation.

5 Conclusion

The paper has presented the smart manufacturing platform estimation for SME in automotive rubber part supply chain management. Three major criteria are taken into account: quality, delivery, and communication. The modelling is designed by using fuzzy logic. The set of values is collected from expert domain. The output expectation of the platform design can be selected from the modelling. The content of the modelling is based on the subcontractors in the rubber part automotive supply chain manufacturing.

References

1. Dalenogare, L., Benitez, G.B., Ayala, N.F., Frank, A.G.: The expected contribution of Industry 4.0 technologies for industrial performance. Int. J. Prod. Econ. **204**, 383–394 (2018)
2. Jeschke, S., Brecher, C., Meisen, T., Özdemir, D., Eschert, T.: Industrial Internet of Things and cyber manufacturing systems. In: Jeschke, S., Brecher, C., Song, H., Rawat, D.B. (eds.) Industrial Internet of Things. SSWT, pp. 3–19. Springer, Cham (2017). https://doi.org/10.1007/978-3-319-42559-7_1
3. Frank, A.G., Dalenogare, L., Ayala, N.F.: Industry 4.0 technologies: implementation patterns in manufacturing companies. Int. J. Product. Econ. **210**, 15–26 (2019)
4. Ghadge, A., Er Kara, M., Moradlou, H., Goswami, M.: The impact of Industry 4.0 implementation on supply chains. JMTM **31**(4), 669–686 (2020)
5. Nagy, J., Olah, J., Erdei, E., Mate, D., Popp, J.: The role and impact of industry 4.0 and the internet of things on the business strategy of the vale chain – the case of Hungary. Sustainability **10**, 3491 (2018)
6. Zhong, R.Y., Xu, X., Shoshin, A.: Smart manufacturing systems for industry 4.0: a conceptual framework. In: CIE47 Proceedings (2017)
7. Wang, L.H.: Machine availability monitoring and machining process planning towards Cloud manufacturing. CIRP J. Manuf. Sci. Technol. **6**(4), 263–273 (2013)
8. Xu, X.: From cloud computing to cloud manufacturing. Robot. Comput. Integr. Manuf. **28**(1), 75–86 (2012)
9. Zhong, R.Y., Xu, X., Klotz, E., Newman, S.T.: Intelligent manufacturing in the context of industry 4.0: a review. Engineering **3**, 616–630 (2017)
10. Magesh, K.M., Vertripriya, M., Brigetta, A., Akila, A., Keerthana, D.: Analysis on Internet of Things and its application. IJSRSET **2**(2), 1040–1047 (2016)
11. Henzel, R., Herzwurn, G.: Cloud manufacturing: a state of the art survey of current issues. Procedia CIRP **72**, 947–952 (2018)
12. Dachyar, M., Teuku, Y.M., Zagloel, S.L.R.: Knowledge growth and development: Internet of Things (IoT) research 2006–2018. Heliyon **5**(8), e02264 (2019)
13. Suthep, B., Puntiva, P.: Uncertain risk assessment modeling for bus body manufacturing supply chain using AHP and fuzzy AHP. Procedia Manuf. **30**, 663–570 (2019)
14. Kamchi, T., Suthep, B.: Uncertain inventory management using TOPSIS and FAHP method in automotive supply chain. In: IEEE 7th International Conference on Industrial Engineering and Applications, pp. 441–446 (2020)

Allocation of Human Resources in Sustainable Production Context

Agnieszka Tychoniuk[1] (✉) and Ryszard Wyczółkowski[2]

[1] Faculty of Organization and Management, Silesian University of Technology, Zabrze, Poland
agnieszka.tychoniuk@polsl.pl
[2] Faculty of Mechanical Engineering, Silesian University of Technology, Gliwice, Poland

Abstract. The idea of sustainable production, among many others, takes into account the aspect of work, in particular the role of man in the production process. In addition to the need to provide the employee with a sense of well-being, this issue also has economic significance: in many situations, the time and quality of task implementation depends on soft competences, e.g. creativity, learning skills, openness to change, flexibility, knowledge based on experience. In addition, selected skills are important when performing certain jobs, e.g. spatial planning skills, a sense of aesthetics, and the set of such skills needed is different for different jobs. However, the selection of employees for tasks, especially in the case of a large variety of tasks performed (e.g. in services, maintenance provided for many plants) is not easy. It requires prior recognition and assessment of employees' soft competences and proper recognition of the requirements that a specific task presents to the employee. Only then, it is possible to select the teams that carry out the tasks correctly. As the complexity of production processes increases in industrial enterprises and the need to make decisions about the allocation of employees to tasks based on real-time information arises, the problem described is becoming increasingly important. In the article, authors present the original concept of a tool supporting the process of selecting employees for tasks of various nature. The possibility of using selected analytical methods to determine the soft skills of employees and an advisory system supporting the determination of the desired soft skills needed for the task was presented. Solving these two problems will then allow developing a tool to support the scheduling work of employees.

Keywords: Sustainable production · Resource allocation · Ontology

1 Introduction

The term - sustainable production - is defined in many ways, depending on the author and the direction of research carried. Some authors define sustainable production as a strategy or approach, while others define it as a paradigm or a system [1]. Production of goods and services in accordance with the idea of sustainable production:

– implements processes and systems that do not pollute the environment (reducing pollution),

A. Batako et al. (Eds.): GCMM 2021, LNNS 335, pp. 505–516, 2022.
https://doi.org/10.1007/978-3-030-90532-3_38

- preserves energy and raw materials,
- is economically realistic,
- is safe and not endangering human health,
- is socially and creatively profitable for all working people.

The main goal of sustainable production is to strike a balance between the social, environmental, technical and economic dimensions of implemented processes in such a way as to meet the requirements of interested parties and gain a competitive advantage [2–4]. For industrial enterprises, the idea of sustainable production is an important element in building their competitive advantage. Searching for technological and organizational solutions, they try to use effectively the available resources that are at their disposal. One of the most valuable resources of an organization from the point of view of an industrial company providing technical services are people [5]. Employees are an irreplaceable element of the production and service system, and their competences become a key determinant of the efficiency and effectiveness of this system. According to one of the principles of sustainable production [6], employees should be encouraged to develop their talents and opportunities. Their work should be organized to increase their efficiency and creativity and encourage them to participate in making decisions. Moreover, safety and well-being of employees should be a priority. Proper selection and assignment of employees to their tasks are therefore a key issue in enterprises operating based on sustainable production principles.

The purpose of the article is to present the concept of a system supporting the decision-maker in the process of selecting employees to perform specific tasks. The system is to indicate which employees are best suited to the implementation of planned work, taking into account their characteristics and task properties. This is an early stage of work and the article presents some assumptions that allow investigating the practical application of the presented concept.

2 Job Assignment: Linking Employees and Tasks

The most important role in the process of selecting employees for tasks is played by the decision maker, i.e. a person completing the composition of the team, who should know the capabilities and limitations of available employees to the greatest extent and is able to use knowledge and experience in the process of optimal delegation of tasks. Undoubtedly, there are such decision-makers who, based on their own experience, comprehensive knowledge, as well as intuition can make choices in a perfect way. Decisions are difficult to measure; they are often the result of imprecise language information such as thoughts, beliefs, opinions and feelings [7].

The selection of task team members is a complicated process of identifying and assessing available candidates in relation to the requirements arising from the scope of work. The decision to select a specific composition of a team of employees to perform a task requires analysing a number of data and information in order to acquire knowledge of at least: customer requirements (whether or not articulated), legal requirements, competences and capabilities of employees adequately to demand, etc. [8] (Fig. 1).

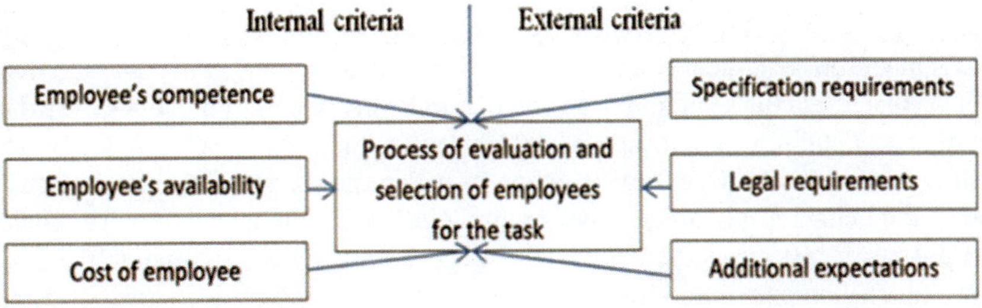

Fig. 1. Basic factors influencing process of assessment and selection of employees to predefined tasks, source: own work

Hence, the basic questions arise when selecting employees for tasks:

1. What competencies of employees are relevant to the task?
2. Which employees have the competences expected from the task point of view?
3. In the case of building several teams to carry out various tasks, in the event of a shortage of employees with specific competences, how to optimize the composition of these teams?

Obtaining the right combination of these elements is a necessary factor in the process of selecting employees for tasks. However, this is not a simple task, because there are many different possible combinations of employee assignment and many conflicting factors that usually need to be considered (e.g. quality, cost, time). At the same time, choosing the right teams to accomplish a given task is important. It may contribute to a better and faster task performance or, if the wrong choice is made, extend the lead time and the quality of the work performed and, as a consequence, expose the company to losses resulting not only from the worse (less effective) use of human resources but also contractual penalties or loss of the client [9]. That is why it is advisable to introduce tools supporting this process.

The paper presents the concept of building a decision support system for the allocation of human resources. This system is dedicated to enterprises operating in the technical services industry, where orders flow in a certain uneven stream, they are most often performed at the clients' facilities, they fit into a specific company profile, but each of them has its own individual character and individual resource needs.

The system's assumption is its universality, but it was developed basing on data and experience gained from a selected enterprise that performs assembly, measurement and service works in the power industry. Typical tasks include preparing and completing the necessary resources, both technical in the form of control and measuring equipment, materials or technical means supporting the implementation of processes, as well as employees envisaged to carry out the works. Preparation for the implementation of tasks also requires the analysis of documentation related to the subject of work as well as legal and internal requirements of the client, which may affect the way they are organized and the entire process of their implementation.

Orders for technical services carried out in the field, often at new clients, at unknown facilities in various technical conditions, require the decision-maker to analyse a number of aspects that affect the selection of a team of employees to carry out work. The problem concerns many conditions that should be analysed, including: task implementation time, technical requirements necessary to meet formal requirements, additional technical, physical and behavioural competences of employees that can have a positive impact on better and faster task implementation. Thus, it is a typical decision-making situation - we have more than one option to choose from, but it is not possible to assess the results of the decision earlier. An additional difficulty is the limited set of information available to the decision maker and often their inaccuracy or approximate nature. Considering that the accuracy of the decisions taken has an impact on the effectiveness of the company's operations, it is appropriate to support the decision-making process, especially in the case of high dynamics of incoming orders.

3 The Framework of an Expert System Supporting Identification of Expected Soft Skills Necessary for Tasks Implementation

The general structure of the proposed system is shown in Fig. 2. The important parts of the system are:

- a knowledge base of implemented tasks, allowing determining what features, skills and formal rights should have employees performing them,
- knowledge base on human resources, containing the description of employees' competences,
- the requesting module, which allows you to assign the desired features of the team carrying out them based on the description of the features of the new task. This module can be expanded with a part that allows you to optimize the assignment of teams to various tasks carried out simultaneously in conditions of limited resources and uncertainty as to the future demand for resources (the appearance of subsequent orders).

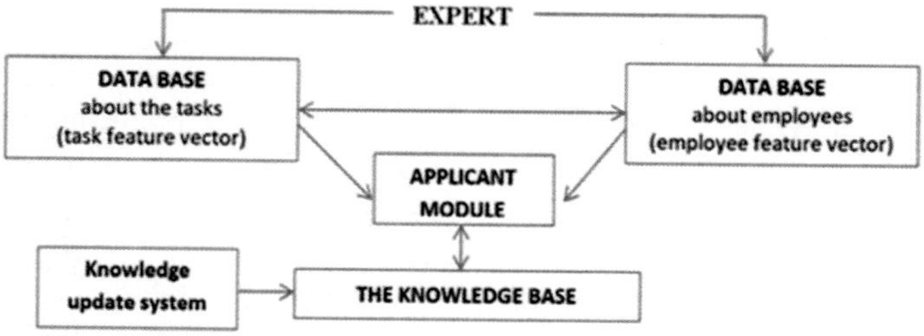

Fig. 2. Fragment of a decision support system dedicated for job assignment problem, source: own work

The starting point for developing such a system is acquiring knowledge, structuring, and formalizing it properly so that it can be saved in the system under construction and used in the application process. Data are usually obtained in various forms and from various sources, e.g. an interview with an expert, using a survey, based on observations, based on collected historical data. Acquired knowledge can be recorded using the different formal notation of knowledge. The rest of the article presents the method of acquiring knowledge and its formalization for the needs of the discussed system.

3.1 Selected Analytical Methods of Defining Soft Competences of Employees

When analysing employees' competences, usually hard and soft competences are identified. Hard competences are knowledge and skills of a routed nature that are necessary to perform a given employee role [10]. Examples of such competences are: specialist knowledge, the ability to operate specific equipment or knowledge of industry-specific software. For manufacturing and service-technical enterprises, technical competence determines the ability to perform specific tasks. They include: knowledge of the process and practical skills to do the job. Certificates, diplomas or certificates confirm hard competences. Soft competences are personal and interpersonal skills, they are tacit knowledge. The literature on the subject lacks a uniform position in relation to methods for determining employee competences [11]. Four groups of methods for assessing soft skills are the most popular and most often cited:

a) tests,
b) interviews with employees,
c) multi-source assessment,
d) Assessment & Development Centre (Table 1).

Table 1. Advantages and disadvantages of selected methods of competences assessment, source: own work

Method	Advantages	Disadvantages
Tests	• Simplicity and speed of use, • Easy analysis of results, • Reliability of the results obtained, • Adaptable to the organization's specifics, • The ability to test a large number of people at the same time	• Limited number of possible interpretation of results, • definition misunderstandings, • Requires completion using a different chosen assessment method
Interviews with employees	• Provides samples of employee behaviour during future tasks	• Due to the subjectivity of the respondent's assessment, it cannot be the only source of assessment

(continued)

Table 1. (*continued*)

Method	Advantages	Disadvantages
Multi-source assessment	• A large variety of methods, • Multi-source opinions, • Higher quality assessment	• Time-consuming and expensive assessment process due to the large number of assessors, • Subjective selection of persons performing the assessment, • Attribution errors in employee assessment
Development centre	• High level of objectivity, • Versatility of assessment, • The employee's potential is assessed in strict relation to key competences	• Process time consumption and high method costs, • Lack of assessment goals, • Used mainly to assess employees in managerial positions

As part of the work carried out, the knowledge of experts was used, which was based on an interview with employees and their own experience. The use of other tools, e.g. Development Centre, was considered but in this case, it was not justified due to the area of application - mainly employees in managerial positions.

In cooperation with industry experts, historical data was analysed about the orders carried out by the company, in particular the scope of works carried out, and they were decomposed into smaller, autonomous tasks. The essence of the decomposition was to identify these works, which on the one hand constitute a minimal, logically coherent whole, and on the other are repetitive components of larger tasks. Then, based on the analysis of the scope and activities included in each decomposed task, the experts assigned to it potentially expected employee characteristics, potentially necessary for their implementation (Fig. 3).

Then, based on a list of competences defined in such a way and in cooperation with experts, individual employees were assessed, assigning them the degree of "possession" of a given competence. The progress of past orders was also assessed and the suitability of individual employees to perform these tasks was determined based on the history of their implementation. A three-level scale was adopted for the assessment (mismatched, moderately suited to the task, very useful for the task). In this way, a set of employees' competences was obtained with a label describing the employee's suitability for the implementation of a given, previously decomposed task (Table 2).

In order to specify the set of competences relevant for a given task and verify the assessments of experts from the first stage, a classifier was determined based on the labelled data set determining the suitability of individual employees, represented by the set of competences they possess, to perform a given job. The algorithm C4.5 of decision tree structure was used in the construction of the classifier. A fragment of the resulting decision tree is shown in Fig. 4.

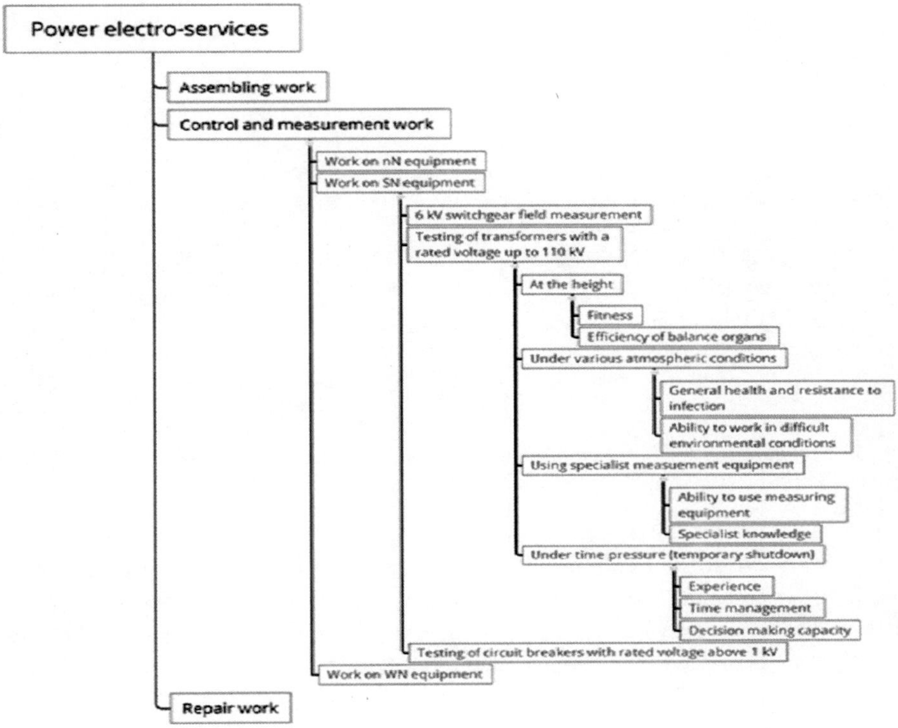

Fig. 3. Decomposition of orders, source: own work

Table 2. Features of employees, source: own work

Employee no.	Abilities	Anxiety	Health	Knowledge	Fitness	Experience	Manage	**Fitting degree**
A	La.	La.	Sm.	La.	Av.	Av.	Sm.	**MF**
B	La.	Av.	La.	Sm.	Av.	Av.	Av.	**MF**
C	Sm.	Av.	La.	La.	Av.	Av.	La.	**MF**
D	Sm.	Sm.	La.	Av.	Av.	Sm.	Sm.	**WF**
E	La.	La.	Av.	La.	La.	La.	La.	**VF**
F	Sm.	Sm.	La.	Av.	Av.	Sm.	Sm.	**WF**
G	La.	La.	La.	La.	Sm.	La.	La.	**VF**
H	Av.	Av.	Av.	La.	Av.	Av.	La.	**MF**
I	Av.	La.	Sm.	La.	Av.	Av.	La.	**MF**
J	Av.	Sm.	Av.	Sm.	Av.	Sm.	Sm.	**WF**

Competences: Sm. – small, Av. – average, La. – large
Fitting degree: WF – worst fitted, MF – medium fitted, VF – very fitted

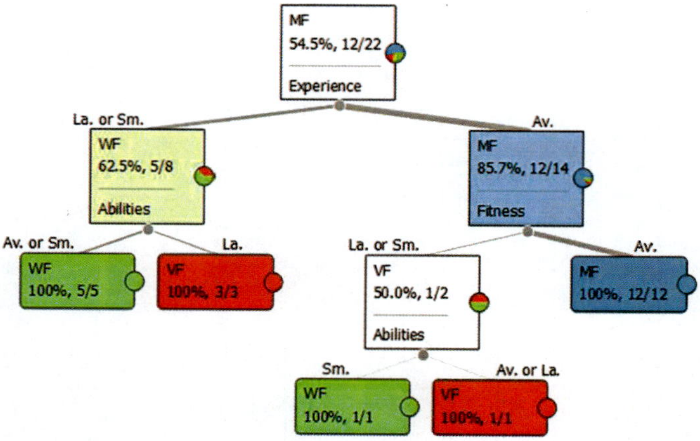

Fig. 4. Decision tree for soft competences assessment, source: own work based on Orange software [12]

Each branch of the tree is a classification question about the level of employee matching for assembly work, and its leaves are part of the data properly classified in terms of degree of matching. At the same time, the features and competences that have actually been included in the classification process are visible.

3.2 Formal Records of Knowledge on Tasks

After analysing the data, clearing out unnecessary information noise, the acquired expert knowledge must be represented and coded in a specific format. In the described system, ontologies were used to formally record knowledge. Ontology is a formal explicit description of terms in a domain of discourse (classes), properties of each concept describing various features and attributes of the concept (slots), and restrictions on slots (facets) [13].

Ontologies are used in order to benefit from their semantics capabilities and their integrated inference mechanism [14]. The authors decided to use the tool Ontorion Fluent Editor [15]. It is a tool what enables editing, manipulating and searching complex ontologies written in OWL, SWRL or RDF. Fluent Editor is compatible with most of the Semantic Web W3C standard but at the same time uses the Ontorion Controlled Natural Language (OCNL) that is a human friendly alternative to XML ontology language. Furthermore, there is rOntorion package, which allows for operation with the ontology from the R environment.

There is no "one and only" methodology for developing an ontology [16]. This closed improvement process can be modified depending on the planned application and the envisaged extensions. When creating an ontology, a series of questions that can be classified into two groups should be formed. The first group includes questions that should be asked before working on the ontology. In the second group, there are questions to which the developed ontology should answer.

In the first group, there are the following questions:

– What area does ontology concern?
– What for will we use ontology?
– Who will use and maintain ontology?
– What kind of questions does the ontology answer?

The extension of the last question is a series of questions detailing further work on the ontology itself:

- How can we group individual tasks?
- What features can we describe depending on the type of task?
- Which of these features are important during their implementation?

The ontology prepared in this way should enable the following questions to be answered (second group of questions):

- What features of the task should be taken into account when assigning an employee to its implementation?
- What do we expect from the employee who is to be assigned to perform the task?
- For which task does each employee have predispositions?
- Which employee competencies disqualify him or her when choosing a task team?

The simple example presents the basic elements of the ontology developed, allowing the collection and processing of knowledge about the tasks performed.

Individual groups of works after their decomposition were written in the form of classes and subclasses, respectively (Fig. 5). It should be noted that the division can be carried out in many ways, e.g. due to the nature of the order, the assembly class can be distinguished, mechanical and electrical assembly subclasses, and due to the location of the place of the order execution, the departure class, with close and distant subclasses.

```
Part-1: 'simple hierarchy of beings'.
Comment: 'Names of instances occurring in the
ontology'.

Every montage is a task.
Every measurement is a task.

Every electrical-montage is a montage.
Every mechanical-montage is a montage.
Every electrical-measurement is a measurement.
Every faraway-trip is a trip.
Every close-trip is a trip.
```

Fig. 5. Class description properties, source: own work

Each of the defined classes and subclasses may be assigned the resources necessary for its implementation, e.g. driving license (a man with a driving license) and a car for long-distance travel, or assembler (as an impersonation of a certain set of features and skills, the detailed calculation of which has been omitted in the article) for mechanical assembly, and a good fit for completing the order, which additionally has a short delivery time (Fig. 6).

```
Every faraway-trip needs one driving-licence.
Every faraway-trip needs a car.
Every mechanical-montage needs an fitter.
```

Fig. 6. Class description properties, source: own work

In addition, ontology can contain rules (knowledge) that allow assigning tasks or resources to specific classes based on their individual characteristics (Fig. 7).

```
Every-single-thing that is a task and has-distance
greater-than 30 is a faraway-trip.
Every-single-thing that is a task and has-distance
lower-than 30 is a close-trip.
Every-single-thing that is a task and has-value
greater-than 1000 and has-deadline lower-than 3 needs
a good-fitter.
Every-single-thing that is a task and has-deadline
lower-than 3 needs a supported-fitter.
Every-single-thing that is an employee and has-
experience greater-than 10 and has manual-skills and
has precision-of-workmanship is a good-fitter.
```

Fig. 7. Class description properties, source: own work

Specific orders can be assigned to a given class directly or by specifying their individual characteristics. For example, order-1 has been explicitly assigned to two classes: measurement and electrical assembly, and at the same time, based on its features, the inference mechanism assigned it to the distant trip class (Fig. 8).

```
Order-1 is an electrical-montage.
Order-1 is an electrical-measurement.
Order-1 has-distance equal-to 50.
Order-1 has-deadline equal-to 2.
Order-1 has-value equal-to 1500.
```

Fig. 8. Class description properties, source: own work

Sample class tree, obtained after the so-called materialization, i.e. the use of information recorded in the defined rules is shown in Fig. 9.

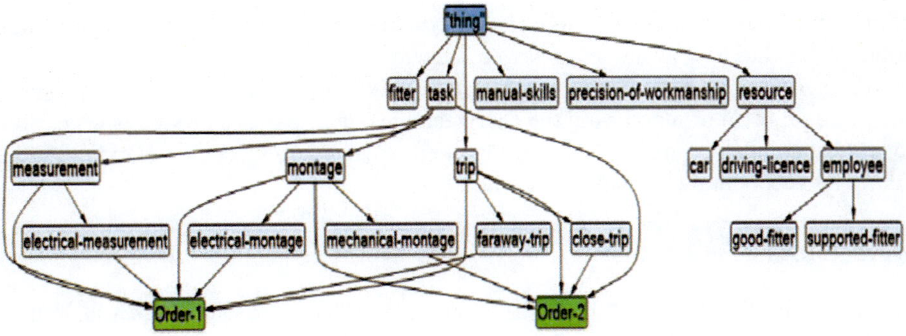

Fig. 9. Classes diagram, source: own work

After materializing the ontology, simple questions can be asked by using the built-in tool - the reasoner (Fig. 10). In this way, e.g. to the question "What resources does order-1 require?" the system will collect and return all the requirements assigned to the classes to which this order belongs.

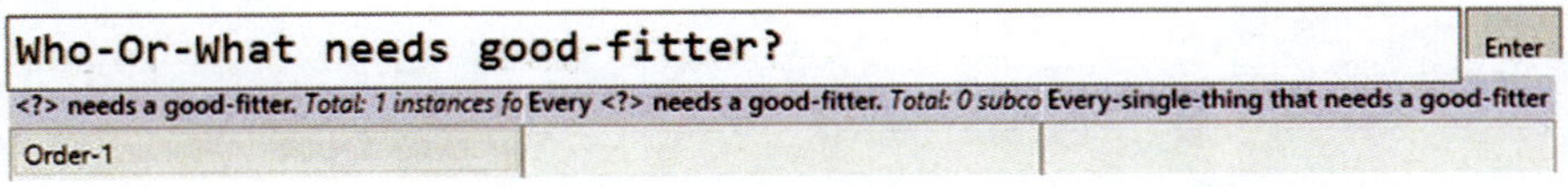

Fig. 10. An example for a query to the database, source: own work

The developed simplified ontology, together with a set of classes, subclasses, their properties and rules, constitutes the knowledge base that is the basis for conducting inference. The advantage of Fluent Editor is the ability to connect to the R computing environment, which allows you to expand the system and introduce your own search and inference mechanisms.

4 Conclusion

The work develops the concept presented by the authors in [8]. Selected analytical methods for determining the soft skills of employees with an analysis of their advantages and limitations were presented. The problem of selecting employees for tasks requires defining a solid competence framework, while it should be assumed that it is a continuous process and the models change and adapt to the requirements of the work. The paper presents examples of formalization of knowledge about tasks in order to select employees responsible for their implementation, taking into account the competences required. Decision trees and knowledge record in the form of domain ontology were used for this purpose. The simulations carried out based on the knowledge accumulated in cooperation with experts in a given field confirmed the possibility of using ontology to record it, creating a knowledge base about tasks and showed that it is possible to search for the best connections of a task - an employee with specific competences. The key problem of preparing data for analysis is to isolate, in particular, technical aspects of knowledge about planned tasks relevant to employee selection.

Further work will be continued towards developing the presented concept. Only when the problem of identifying the characteristics of employees needed is it possible to build a module optimizing the selection of employees from a specific working group, taking into account the stream of tasks flowing in time.

References

1. Jasiulewicz-Kaczmarek, M., Żywica, P.: The concept of maintenance sustainability performance assessment by integrating balanced scorecard with non-additive fuzzy integral. Eksploatacja i Niezawodność – Mainten. Reliab. **20**(4), 650–661 (2018)
2. Jasiulewicz-Kaczmarek, M.: The role and contribution of maintenance in sustainable manufacturing. IFAC Proc. Vol. (IFAC-PapersOnline) **46**(9), 1146–1151 (2013)
3. Hadas, L., Stachowiak, A., Cyplik, P.: Decision making model in integrated assessment of business-environment system - a case study. Environmental Science and Engineering (Subseries: Environmental Science), pp. 419–429 (2011)
4. Junior, A.N., Célia, M., de Oliveira, A., Helleno, L.: Sustainability evaluation model for manufacturing systems based on the correlation between triple bottom line dimensions and balanced scorecard perspectives. J. Clean. Prod. **190**, 84–93 (2018). https://doi.org/10.1016/j.jclepro.2018.04.136
5. Jasiulewicz-Kaczmarek, M., Saniuk, A.: Human factor in sustainable manufacturing. In: Antona, M., Stephanidis, C. (eds.) UAHCI 2015. LNCS, vol. 9178, pp. 444–455. Springer, Cham (2015). https://doi.org/10.1007/978-3-319-20687-5_43
6. Greiner, T.J.: Indicators of Sustainable Production - Tracking Progress, A Case Study on Measuring Eco-Sustainability at Guilford of Maine, Inc. (2020). http://www.sustainableproduction.org/downloads/Guilford%20Case%20Study.pdf
7. Saad, R.M., Ahmad, M.Z., Abu, M.S., Jusoh, M.S.: Some fuzzy techniques for staff selection process: a survey. In: AIP Conference Proceedings, vol. 1522, p. 462 (2013)
8. Tychoniuk, A., Wyczółkowski, R.: Assessment of the possibility of using AHP (Analytical Hierarchical Process) to support the process of selecting employees for the implementation of technical services. In: Wybrane elementy zarządzania procesami produkcyjnymi, Gliwice, Wydawnictwo Politechniki Śląskiej, pp. 115–130 (2019)
9. e Silva, L.C., Costa, A.P.C.S.: Decision model for allocating human resources in information system projects. Int. J. Proj. Manage. **31**. 100–108 (2013)
10. Cimatti, B.: Definition, development, assessment of soft skills and their role for the quality of organizations and enterprises. Int. J. Qual. Res. **10**(1), 97–130 (2016). https://doi.org/10.18421/IJQR10.01-05
11. Bohloulia, M., Mittasb, N., Kakarontzasb, G., Theodosioub, T., Angelisb, L., Fathia, M.: Competence assessment as an expert system for human resource management: a mathematical approach. Expert Syst. Appl. **70**, 83–102 (2017)
12. Orange 3 (2020). https://orange.biolab.si/
13. Noy, N.F., McGuinness, D.L.: Ontology Development 101: A Guide to Creating Your First Ontology. Stanford Medical Informatics Report SMI-2001-0880 (2002)
14. Talhi, A., Fortineau, V., Huet, J.-C., Lamouri, S.: Ontology for cloud manufacturing based product lifecycle management. J. Intell. Manuf. **30**(5), 2171–2192 (2017). https://doi.org/10.1007/s10845-017-1376-5
15. Fluent Editor (2015). https://www.cognitum.eu/semantics/FluentEditor/
16. Li, Z., Yang, M.C., Ramani, K.: A methodology for engineering ontology acquisition and validation. Artif. Intell. Eng. Des. Anal. Manuf. **23**, 37–51 (2009)

The Complexity of Data-Driven in Engineer-To-Order Enterprise Supply-Chains

Richard Addo-Tenkorang[1](✉), Petri Helo[2], Ari Sivula[3], and Norman Gwangwava[4]

[1] Brunel University London, Uxbridge UB8 3PH, UK
raddotenkorang@yahoo.com
[2] Department of Production, University of Vaasa, PL 700, 65101 Vaasa, Finland
petri.helo@uva.fi
[3] Seinäjoki University of Applied Sciences, Kampusranta 9, Seinäjoki, Finland
ari.sivula@seamk.fi
[4] Department of MEIE, Botswana International University of Science and Technology, P.O. Box 16, Palapye, Botswana
gwangwavan@biust.ac.bw

Abstract. The complexity of data-driven engineer-to-order manufacturing enterprise supply-chains for effective and efficient decision making has received a lot of attention both within the original equipment manufacturing industrial research and development circle and supply-chains operations research and management circles. However, despite these complexities, most of the published supply-chains research in operations research and management have neglected the 'engineer-to-order perspective within the original equipment manufacturing supply-chains sector. This research employs a comprehensive study of complex supply-chains management activities to attempt to propose feasible and measurable essential propositions and/or framework for "best practices" in data-driven engineer-to-order supply-chains. There seems to be no specific comprehensive study on the complexity of data-driven engineer-to-order supply-chains within the original equipment manufacturing sectors for complex products such as the aerospace, marine, and/or power plant industries, etc. However, because this area of complexity of data-driven engineer-to-order within enterprise supply-chains have not been much researched or explored; there is an expected challenge of finding enough available literature to draw-on or makes an inference to. Hence, this study will take solace from mostly real-life industrial case(s) and/or activities, etc. Therefore, this paper presents a comprehensive study of the complexity of data-driven engineer-to-order enterprise supply-chains as well as outlining essential propositions and/or framework to enhance effective and efficient resilient complex engineer-to-order supply-chains. This paper will thus, contribute to the development of a more robust and resilient framework when dealing with the complexity of data-driven engineer-to-order enterprise supply-chains.

Keywords: Complexity · Engineer-to-Order · Supply-chains · Original-equipment-manufacturer · Bigdata

A. Batako et al. (Eds.): GCMM 2021, LNNS 335, pp. 517–532, 2022.
https://doi.org/10.1007/978-3-030-90532-3_39

1 Introduction

Engineer-to-Order (ETO) design of original equipment manufacturer (OEM) products includes design for large or huge products; such as aircraft, large electric machine, huge centrifugal pumps, diesel/natural fuel power plant engines, steam turbine, huge boilers, ship power engines, etc. [1]. ETO is a product development process, which starts with a product specification and finishes with engineering design as its final deliverable mostly excluding the actual manufacturing processes of the complex product. One of the notable problems associated with ETO complex products design and delivery is excessive lead-time. Thus, a data-driven ETO enterprise supply chain network is seen as a supply chain (SC) network system where the 'decoupling point' is located at the design stage, so that, customers' orders extends beyond the engineering design phase of a complex/new product [2] where these actives are enhanced by information, communication, and technology (ICT) enablers. Hence, an ETO process is associated with huge or large, complex project environments in sectors including the construction and capital-intensive OEM investment initiatives all generating huge/big data concurrently. This is the era where big data is constantly being generated within industrial settings; because of the huge acceptance of information, communication, and technology (ICT) enablers as the main embodiment of enterprise SCs. This has now been coined in many different terms such as internet of things (IoT); artificial intelligence (AI) aided by smart systems (SS) or internet of services (IoS). All these maps-up to what is now known as "Cloud/Virtual Manufacturing"; etc. Hence, complexity within data-driven ETO SC is rarely researched on or used in existing or trending research in this area: Thus a confusion exists as to the appropriate definitions for this specific type of industrial operation and the appropriate "best practices" employed to enhance competitiveness and resilience within a complex data-driven ETO enterprise supply chain network activities. Therefore, the purpose of this paper is to study the complexity of OEM data-driven ETO enterprise supply chain activities and attempt to introduce an optimum "best practice" framework by outlining effective and efficient operational propositions for a competitive and resilient ETO enterprise supply chain management (SCM).

SC is a complex network of industrial entities networked into the upstream, intermediate-stream and finally downstream flows of products processes and/or services, along with the related finances and information [3, 4]. Supply chain management (SCM) involves the systemic and strategic activity network of these flows both within and across OEM industries' SC intending to reduce costs; improve customer satisfaction and gain a competitive advantage as well as enhance the resilience within both the entity industries and the SC as a whole [3, 4]. The voluminous amount of data along with the dynamic and uncertain or volatile network environment of OEM SCs contributes to their complex nature. Thus, these complexities are inherent in SCs and are observed in different forms and origins, including static complexity, which is related to the interconnectivity and structure of subsystems with the network involved in a specific SC. However, dynamic complexity, on the other hand, consists of results from the operational behaviour of the systems and subsystems within an SC network and its environment. Decision-making form of SCM complexity is that which involves both static and dynamic aspects of complexity [5]. According to [5], the complex nature of SC contributes to the complexity in effectively managing OEM SCs, to the extent that it has almost become literal to say

SCM is about managing the complexity of the SC network. Over a decade and beyond now there have been series of studies or research on complexity in the SC to achieving effective and efficient SC performances [6–15].

This section of the paper will continue with the following: an example of complexity in ETO – SC projects; the problem statement of this research. Then, the rest of the paper is structured as follows: Sect. 2 – Measuring complexity of data-driven ETO enterprise supply chains; Sect. 3 – Complexity in supply chain structures; Sect. 4 – Dynamic aspects of supply chain complexity; Sect. 5 – Product and system-related complexity and finally; Sect. 6 – Concludes the paper along with industrial implication and recommendations.

1.1 Examples of Complexities in Data-Driven ETO Enterprise SC Projects

ETO of OEM industries is characterized and described in terms of their markets, products, services, and the internal and external processes of their organization. The variety of SC activities in data-driven ETO projects are mainly customized and complex products due to the underlying uncertainties of characterizing the processes within the SC and their subsequent markets. Furthermore, complexity in OEMs ETO SC is also imminent in all data-driven ETO enterprise SC project procurements and marketing need in terms of information integration within the SCs as well as the other processes such as tendering and design. These characteristics put constraints and/or complexity on the application of established data-driven ETO enterprise supply chain management methods. Industrial ETO produce projects may require a detailed and specific set of the data-driven item, such as numbers, bills of material (BOM), and routings of scheduled manufacturing processes and activities. Thus, these details are usually complex with huge volumes of data or big data as well as long lead times. Therefore, customers are also normally heavily involved throughout the entire design process for specific ETO project products. They are usually limited to an engineering design process which involves the tasks of engineering analysis, concept design, architectural design, detailed design, and manufacturing process design. Hence, these processes do not include the manufacturing phase of materials acquisition, fabrication, and assembly. The product design process phase takes enormous time for the entire process lead-time. Thus, a new and improved data-driven ETO enterprise SC network effectively and efficiently coordinating the other aspects of the manufacturing phase not included in the traditional ETO process would attempt to reduce the long lead-time associated with ETO processes. This approach would enable all the partners on data-driven ETO enterprise SC to operate and function more effectively in terms of information communication and exchanging of essential ETO project product details on a common platform in near real-time if not in real-time.

The literature on SCM has begun addressing the inter-relationships SC network of suppliers as a supplier–to-supplier relationships [16, 17]. Although buyer–to-supplier networks have received sufficient attention in the literature [18, 19], supplier–to-supplier networks, these are also active in the supply chain networks, but are still trending actively in SCM research [20, 21]. For instance, when the network among suppliers, move beyond a dual-linked context (e.g., one supplier to another supplier) into a trio-linked context (e.g., two suppliers competing against each other to provide products to another supplier in the same industrial OEM); these dynamics become much more complex [21, 22]. For example, in the marine – ship power industry, groups of suppliers who used to compete against each another on a certain supply would now have to link themselves up to cooperate in a specific and/or single OEM ETO project. Thus, utilizing organizing themselves into a common effective and efficient ETO SC network to gain more leverage in their approach and activities, it enhances the effective use and analytics of the huge or big data generated among these data-driven ETO enterprise SC networks. This new improved approach as compared with the traditional ETO process will bring all the stakeholders on the ETO project together both internally within the OEM's SC and globally among their external ETO project suppliers. This interconnectivity could be facilitated and enhanced using information technology enablers. Thus, further to the exchange of product parts, exchange of information also occurs within supplier–to-supplier interaction. Therefore, a typical competitive supplier–to-supplier relationship may still lack certain objective information about the other supplier within the network (such as predictable demand, competitor's product features, and production capacity). This limitation could by means of technology enablers [6, 7, 16, 23, 24].

1.2 Research Problem Statement

Dynamism and volatility seem to have been increasingly dominant in recent times. Thus, globalization has made many industries to experience volatility within the industrial marketplace which has virtually prompted industries to have a rethink of the way business is conducted [25] and the huge data generated among the SCs are efficiently utilized. Hence, this has also led to certain vital trade-offs between for example labour costs, transportation costs, inventory costs and response time to the customer *(supply and/or delivery cost)* are becoming increasingly complex [26]. Therefore, it seems impossible to only focus on one's OEMs to gain a competitive advantage as far as resilience in their SC network activities is concerned. It has become obvious that the success of an individual OEM is dependent on the effective and efficient performance and reliability of its suppliers as well as the customers.

Based on the purpose and/or main objectives of this paper, the research question thus formulated is:

- How could "best practices" complexity be measured in a data-driven ETO enterprise SC network for it to be effectively assessed for competitiveness and resilience?

Industrial organizations engage in value-adding activities in their industrial businesses such as purchasing and supply of products and services mainly from a group of suppliers within their supply chain network [27, 28]. The group of suppliers is thus referred to as "supplier network" of the SC and the manufacturing industry that purchases from this supplier network are referred to as the "original equipment manufacturer – OEM." Operationally, the OEM is seen to be at the centre or have an oversight view of all suppliers in the supply chain network, thus, coordinating and controlling all the supply chain management activities. Therefore, OEMs producing engineered products on a data-driven engineer-to-order (ETO) concurrent engineering approach (i.e. partners from various divisions in the organizations) have traditionally maintained some level of business competitiveness. Through the design and product development expertise and could respond to demands for customization with improved product performance as customers are increasingly seeking for just-in-time (JIT) and affordable products. These competitiveness, responsiveness, or resilience in operational data-driven ETO enterprise SCs could be best achieved through improved OEM manufacturing systems' efficiencies. Thus, these OEMs require manufacturing ETO SC network system that is driven by the improved integration of the huge data generated from industrial designs, manufacturing and production processes and procurement functions of a data-driven ETO enterprise SCM system [29]. This facilitates the improvement of effective and efficient integration of the various stakeholders involved in an ETO SC project. Furthermore, it improves the overall product development lead-time, along with enhanced and better-defined product documentation a concurrent engineering process provides. Concurrent engineering process thus operates in the same vein as a collaborative or simultaneous engineering process. It is a process in which the appropriate partner organizations are committed to integrating their various manufacturing system to work interactively. Concurrent engineering in this sense is a moderate collaborative engineering approach that is much more focused on the engineering product design process of its concurrent cross-functional partner organizations. They come together to create products that are much better, much affordable and with a much-reduced lead-time [1].

2 Measuring the Complexity of Data-Driven ETO SCs

SC complexity fundamentally varies conceptually based on nature or operations processes of OEMs, thus, influencing the absolute direct control of an ETO SC project. This is due to the seeming integration among the stakeholders on the SC network, hence, posing some level of complexity or performance measurement within the organizational operations [30]. Measuring complexity within the SC, where the network coordination is no longer a concern of the SC ownership only, but instead on networking across various partners and stakeholders; the complexity within measurement system may be reflected as an engineering system of measuring the unmeasurable or uncertainties. SC network activities not under the direct coordination of an individual OEM company (i.e. a manufacturer) have to be measured and coordinated (by the manufacturer and its supply chain partners), making the supply chain more transparent, to a level not experienced before and leading the way for performance improvements [23, 30, 31]. According to the [32], there are two kinds of complexity in any industry: the good complexity and bad complexity. Thus, implying that, industries may employ one kind of complexity to their advantage while they seek to eliminate the other. So, what does complexity mean in this context? Some may think that complexity is merely the opposite of simplicity; others may think that complexity is synonymous for complicacy. However, both definitions are said to be incorrect, especially the second one [12]. In that, according to [12] "Complicated" and "complex" both come from Latin words, but the first one originally means "of things knotted, entwined with each other", while the second one means "of things which interact among each other". Complexity is the fusion of three things, which includes: The number and variety of components *(information or data about things, objects, machines or equipment)*, systems and people within an industry and across its entire value-chain as well as the interrelationships among all of them and also the pace at which they vary.

Therefore, complexity could be said to continue to increase exponentially more components, systems and/or people to the value-chain interrelationship [12]. Thus, culminating in a huge voluminous generation of industrial "big data."

As yet, there is no agreed or one definition of industrial big data. Hence, many enterprises industrial SCM network stakeholders and experts predict that big data will have a positive impact on their industrial ETO SC operations and activities [23]. Thus, enabling them to make more strategic data-driven and informed decisions. This then enables industrial big data generated for from industrial operation, product development activities as well as services of data-driven ETO SCs to effectively and efficiently utilize these data to coordinate their network activities in an IoT perspective. Big data generated in the form of data-driven ETO SC network on an enterprise SC platform, operates in the form of IoT which only becomes useful when it is analysed and processed for value-adding. According to [23] findings, IoT came to the forefront as an enabling driver of high growth of data both in terms of quantity and category in industrial data-driven ETO SCs perspective. Thus, providing the opportunity for the application and development of big data for effective and efficient industrial operations management activities. On the other hand, the application of data-driven ETO enterprise SC networks as big data platform to IoT also accelerates the research advances and business models of IoT in industries [33]. Furthermore, IoT can also enable improvement of product development activities as well as services. Also, data analytics for customer experience, security, etc., if it is properly filtered and authenticated and harnessed for such industrial or service purposes. IoT also has the potential to transform traditional business-to-customer interactions in a way previously not thought of [34] when networking sensors such as RFIDs are embedded in a variety of electronic devices and/or machines to communicate and exchange data or information in real-time and in a real-world activity [35]. Data-analytics (e.g. machine learning) and data-driven decision making are an essential part ETO SC projects and processes because of the enormous amount of data. This data can be utilized in several ways to provide extended value for ETO SC. Data can be analysed based on data content, which can be for example, with text, audio, video, or other content. Data analytics always knowledge extraction from the dataset as Fig. 1 is illustrating.

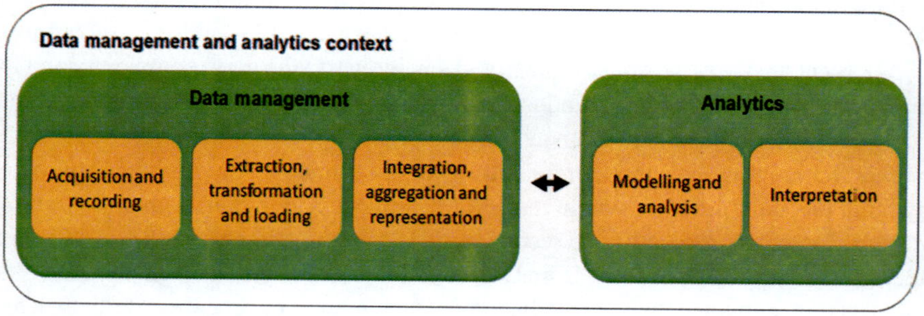

Fig. 1. Processes of knowledge extraction from the data ([36] [adapted])

Process of knowledge extraction from the data can be distributed in data management and analytics processes as Fig. 1 is proposing. First, the data is collected using different methods and devices (e.g. sensors). Data is extracted, transformed, and loaded (ETL) to an appropriate data store where it is integrated, aggregated, and represented in a certain format. After these preparatory processes data is already useful but it can be analysed more with data modelling with machine learning and other analytics. Data analytics does not provide any value without appropriate interpretation which can be done in several formats. Thus, from a manufacturing industry's system engineering point of view, "complexity" surfaces most often when there is variety existing within the boundaries of its SC network. These entities of various existing within the boundaries of these varieties are connected and interlinked with many others: for example, selecting a specific finished product on the global market among a plethora variety of many others may require different and but equally demanding industrial technologies; to control and exploit the variety of different sales channels. Furthermore, the same effort would be required to deliver and/or transport same to a large number of logistic customers; to be able to effectively manage any large variety of different subgroups, components and raw materials connected or interlinked within or among the SC network. Substantial research work has been carried out discussing how the variety of products, of distribution channels, of suppliers, of components, etc. can enhance the competitive strength on one side, but increase coordination and management costs on the other, up to the point of more than counterbalancing those benefits [12].

3 Complexity in Enterprise Supply Chains and Enterprise Structures

Manufacturing systems configuration has extensively been used to effectively address the OEM manufacturing system's specific operations responsiveness by a series of companies dealing with mass ETO project products (mass customization). Several applied types of research have yielded feasible solutions to some of these operational process issues by way of proposing frameworks. These have since been introduced to enable their systematic planning, analyses, development, and implementation data-driven ETO enterprise SC processes. Traditional research has thereby either focused on defining modelling techniques for the configuration model of stable products, on improved configuration algorithms, or the impact of configurators on specific OEM project operations [37]. However, not much focus has been placed on complex issues within data-driven ETO enterprise SC projects in terms of research. Therefore, this paper attempts to outline the critical "best practice" propositions to assess and facilitate these proposals effectively and efficiently. The categorisation of SCs according to specific networks, features and/or clusters that have been widely researched on in previous SC literature and still trending. Therefore, this section of the paper will seek to touch on most of the widely used and adopted manufacturing systems processes and sometimes referred to as 'SC structures.' These SC structures turn to make sense of the diverse range of SC operational processes by outlining how different sections of a data-driven ETO enterprise SC interact with customer purchasing orders, etc. Many research literature within this research area have not adequately addressed the full range of enterprise manufacturing SC or manufacturing systems structures but have, however, focus on just one of these domains or just a selected of these specific structures outlined in Table 1 below. Many of the structures for thinking about effectively connecting ETO SCs to the appropriate marketplace have been organized around the concept of the 'decoupling point' [2].

According to [2], the customer order decoupling point (CODP) is a stockholding point that separates the part of the SC that responds directly to the customer from the part of the SC that uses forecast planning. The various manufacturing system decoupling point structures can act as operational strategies as a buffer against the variable system demand within the SC as well as efficiently schedule an effective way to standardize ETO projects whilst reacting to uncertain orders or complex issues within the SC network. Table 1 below outlines most of the popularly used manufacturing SC system structures. [38–40] identifies with most of the conceptual foundations for SC structures in the different logistics and SC structures for the various decoupling points:

Table 1. Manufacturing SC system structures

#	Manufacturing systems and/or processes	Acronym	System/Process definitions
1	Engineer-to-order	ETO	The decoupling point is located at the end of the detailed engineering design phase with the project SC. Hence, the deliverable for the data-driven ETO enterprise SC is a detailed engineering design of an ETO project
2	Make and ship-to-stock	M/STS	The decoupling point is located at finished goods in a national organization
3	Make to stock	MTS	The decoupling point is located at finished goods in a supply centre
4	Assemble to order	ATO	The decoupling point is located at subassemblies within the assembly process
5	Make to order	MTO	The decoupling point is located at purchased goods
6	Purchase and make to order	P/MTO	The decoupling point is located at the supplier

4 Dynamic Aspects of Data-Driven ETO Supply Chain Complexity

Industrial organizations into data-driven ETO SC projects are facing various challenges in terms of competitiveness and pressures arising from new competitors and emerging competitions [42, 43]. Furthermore, customers are calling for higher flexibility and thereby challenging OEMs to make their ETO SC projects more accessibility in terms of global presence. Detailed engineering design is a core deliverable of the ETO SC project business, defining about 50–60% of the total ETO project life-cycle processes [44]. Figure 1 below, illustrates an optimal framework for ETO SC projects feasibly capable of transforming data-driven ETO SC network activities with the requisite ICT enablers into an improved data-driven ETO enterprise SCM network.

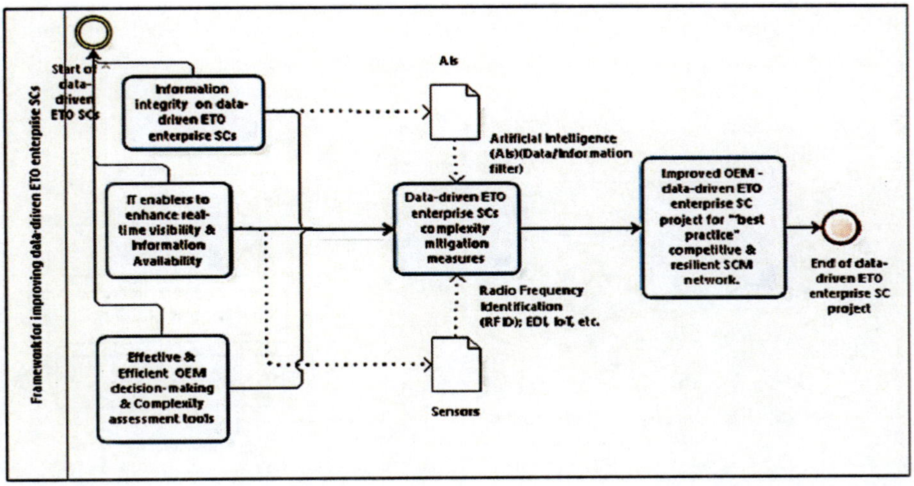

Fig. 2. Framework model for improved data-driven ETO enterprise SCs – *Bizagi Modeler*

5 Product and System-Related Complexities

Complexity in data-driven ETO enterprise SCs as hinted in previous sections of this paper as well as some past literature mainly has to do with ETO projects' product design lead-times which are the main issue of attention in manufacturing systems structure. Another area of attention is the design tools and/or system technologies applied in the product design processes and within other product design domains in terms of design integration or interfacing. Thus, in each of the product design domains, tools and/or technology employed causes some level of complexities which arises mainly due to the numerous data-driven elements within the multitude of relationships or networks involved. These include but not limited to the huge/big data generated between the components of an ETO product being designed as well as the activities or processes employed to design the products, and also among the enterprise SC network stockholders involved in these activities. According to [45] one of the very capable and feasible approaches in handling and/or managing complexity within data-driven ETO enterprise SC is to represent and analyse the various product design domains' structures or enterprise system architectures (ESA). The design structure matrix (DSM) has proved to be a very helpful tool for representing and analysing these complex enterprise system architecture of a specific enterprise SC such as a data-driven ETO enterprise SC [7, 45]. Below is an example of a typical automobile OEM data-driven manufacturing SC project illustrating the complexity of an OEM electrical wiring system (harness) for different customer cars. This electrical wiring system (harness) collects, communicates, and monitors in real-time various data/information inside and outside the car to enable the car functions appropriately for purpose. There are two types of customer products namely, KSK (these are customer-specified concepts/cars) and NKSK (non- customer-specified cars). KSK is made up of individual building blocks which are engine, cock-pit, body and basis whereas NKSK use of the same combination of electrical wires and these include roofs, seats, bumpers, and doors. Figure 3 below, shows a schematic picture showing a different position of harness in a car. JIT and FIFO principles are incorporated from the beginning of the manufacturing process until the harness reach the customer.

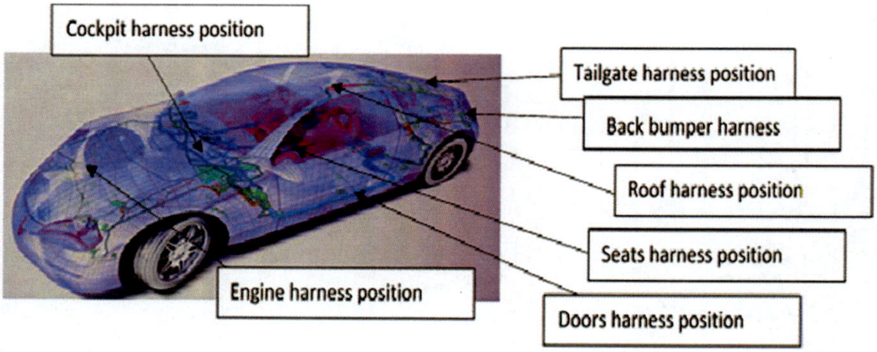

Fig. 3. Harness positions in a car

Figure 4 below, shows a typical example of a complete electrical wire harness and Fig. 5 below systematically illustrates the harness production or manufacturing process and stages. An electrical wire harness is made up of a different combination of wires and components; to mention a few, antennas, wire protectors, housings/connectors, seals, terminals, and grommets.

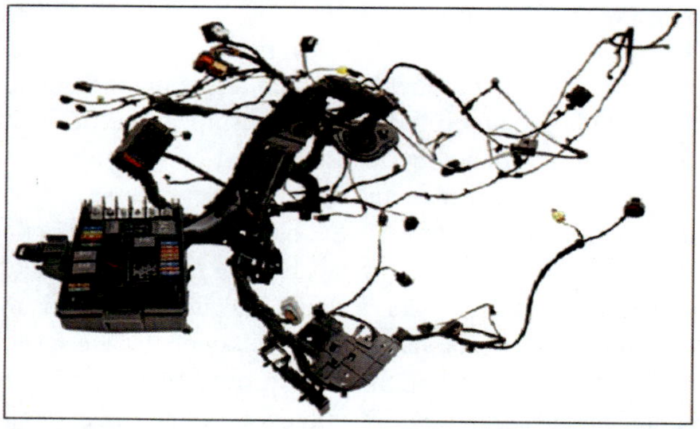

Fig. 4. Typical example of a manufacturing automobile OEM car wiring system harness

5.1 Piloted Real-Life Industrial Case Study Example

Industrial SCM networks are confronted with various challenges. Thus, most of them find themselves struggling to survive operationally. Therefore, most OEMs embarked on high data-driven ETO enterprise SCM activities in complex engineering design and delivery product development activities such as marine – SP engines, aviation – jet engines, automobile manufacturing, etc., are being compelled to improve their SCM network

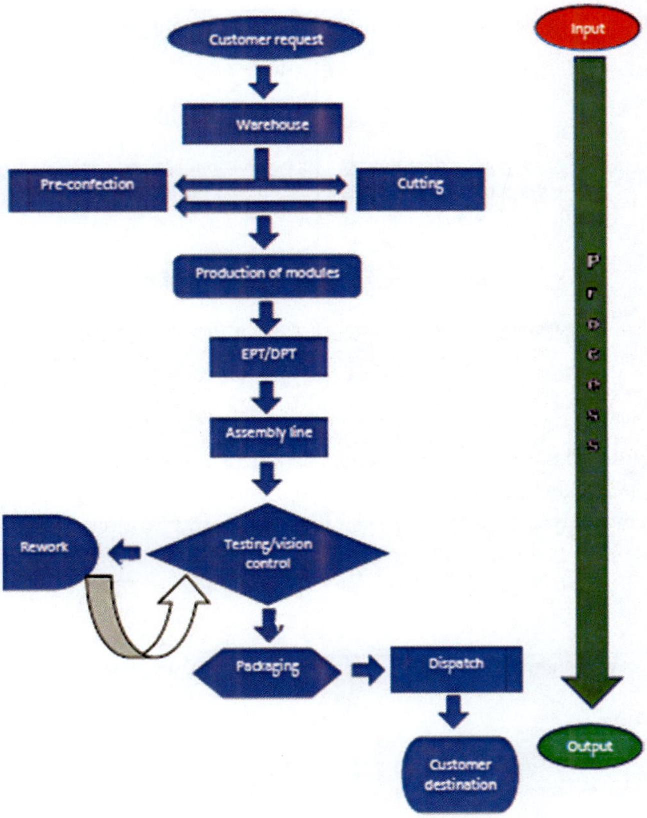

Fig. 5. Harness production process

activities [7]. In this globalized or digital era; most of the real-life challenges facing these OEMs include evolving data-driven actives such as customer demands, stiff global competition market research and the need to improve new or complex products' time to market [46–48]. These persisting operational challenges could be effectively dealt with when the multidisciplinary teams/partners on an enterprise SC network concurrently work together effectively and the information flow and exchange communication network among them are effective and efficient. The marine ship power OEM is one such industry with high data-driven ETO enterprise SCM network approach. Thus, a classical applied industrial case example from [7] on such industrial operation data-driven analysis approach has been adopted as a typical example to illustrate the feasibility and efficacy of the proposed framework in this research paper. In perspective relevance, the case OEM's logistics and the supply chain management division were used for this real-life pilot study. The OEM has over 280 major suppliers globally. The company aims at improving business process efficiency by integrating its industrial operational process with that of its suppliers and customers as well as sharing and exchanging real-time information within its SC network concurrently.

People to People	People to Systems
DSM *(Design Structure Matrix)*	DMM *(Domain Mapping Matrix)*
People to Systems	**System to Systems**
DMM *(Domain Mapping Matrix)*	DSM *(Design Structure Matrix)*

Fig. 6. DSM, DMM data analysis matrix (Source: [27])

According to [7] data-driven analysis matrix illustrated in Fig. 6 above; system-to-system analysis is also best affected by employing the same DSM simulation approach. However, when analysing people to systems, these are two domains that need to be effectively mapped or integrated or interfaced. Therefore, the approach employed is domain mapping matrix (DMM) [49].

Type Of Information		1	2	3	4	5	6	7	8	9	10	11	12	13	14	15	16	17	18	19	20	21	22	23	24	25	26
Delivery Address	1	1																									
Incoterms	2		2									1															
Delivery Times	3	1		3						1	1			1					1			1					
Shipping Marks	4				4																			1		1	1
Type of Package	5					5									1	1							1				
Project Name	6						6																				
Material needs (external & engines) Requirement date for external materials	7							7	1									1									
Materials	8							1	8																		
Requirement date	9			1						9	1																
Pick-up Address and Consolidation Point	10	1		1							10																
Incoterms (with supplier)	11		1									11															
Price	12			1					1	1		1	12										1				
Mode of Transport & Forwarding Agent	13	1		1										13	1					1	1						
Package Type	14					1									14	1							1				
Packages (Content)	15					1									1	15											
Handling Unit No. (Internal & External)	16					1											16										
Dimentions & Weights	17								1									17									
Routes	18	1		1															18		1						
Fowarding Company	19					1														19							
Delivery Type (Planned & Actual)	20	1		1																	20						
Type of Transpotation (Sea, Truck, Air)	21	1		1		1																21					1
Sea, Air, Truck Package Type	22					1																	22				
Tracking No.	23				1																			23	1	1	1
Case Information (Handling Unit No., Project No., Purchase Order No., Stock	24						1																		24		
Case Specification (Pick-Up, Destination Address & Schedule)	25	1																							1	25	
Ship / Vessel [IMO No. - Sea, Project No., Transport Company]	26																										26

Data / Information Dependency by DSM tool

[1] Indicates that the data on "Dimensions & Weights" depends on data / information of the "Material" type

Fig. 7. Design structure matrix (DSM) data/information types relationship entries. Source: [27].

Figures 7 above illustrate some of the graphics from the DSM simulation tool employed in [7] research on the concurrent enterprise: a conceptual framework for enterprise supply-chain network activities. Moreover, several other data-driven models and methods can be employed to support ETO SCM networks. For instance, machine learning provides several algorithms to manage the ETO SCM data which is illustrated in Fig. 5 below (Fig. 8).

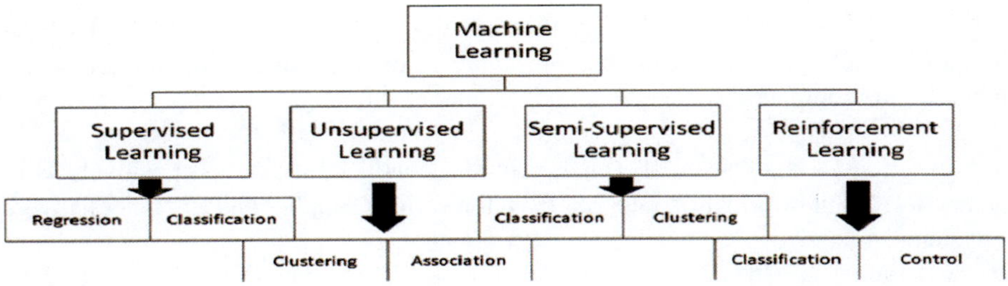

Fig. 8. Machine Learning algorithms. Source: ([50] adapted).

Machine learning can be utilized several ways is ETO SCM depending on the dataset available. Machine learning provides value searching patterns from the data and formulating models which can be utilized in different applications. Machine learning can be combined with vehicle routing systems as computation capacity increases [51]. Complex ETO SCM process can take major advantage of machine learning algorithms as complexity is increasing. Machine learning can be utilized, for instance, in modular structures in the process to estimate process length in a certain configuration. Moreover, digital twins can be utilized to support ETO SCM. Digital twins are replicas of physical objects, which will be developed in the future. Digital twins can be utilized to illustrate the new product in various ways. NASA, for example, is utilizing digital twins to support vehicle development in several ways [52].

6 Conclusions

SC network itself is a complex system. Moreover, integrating enterprise SC complexity management into the SCM system is a necessary action. Therefore, OEMs coming to terms the inherent complexity of their data-driven ETO enterprise SC and taking necessary actions to mitigate, manage and avert these complexities that arises from the voluminous data generated; enable them to perform better and also ensures customer satisfaction. Therefore, this paper defines data-driven ETO enterprise SC complexity in a more practical and system dynamics perspective. It also attempts to categorize the manufacturing system structures of enterprise SC complexity according to the type and origin of the complexity. Further to this, it studies, investigates, and outlines some "best practice" propositions on how data-driven ETO enterprise SCs could be assessed optimally, effectively, and efficiently for industrial competitive advantage in operational complexity. For a competitive, resilient, and successful data-driven ETO enterprise SC network, complexity must be effectively and efficiently managed and/or mitigated. Therefore, it is imperative to identify and understand the causes and drivers of the uncertainties and complexities that arise due to the voluminous amount of data generated during industrial operations within an organization. This paper thus proposes some feasible "best practice" solution strategies to deal with the complexities that arise within OEM – data-driven ETO enterprise SCs to enhance and promote effective and efficient SCM systems network activities [7, 53]. Therefore, below outlines the "best practice" propositions in addition to the framework model illustrated at Fig. 2 above, to enhance OMEs data-driven ETO enterprise SC operations in a more competitive and resilient approach in industrial SC complexities:

- Proposition 1. Data-driven ETO enterprise SC network complexity is positively associated with full information integrity from the various supplier and customer network systems.
- Proposition 2. When OEMs designing and/or developing data-driven ETO enterprise SC projects employs and effectively utilizes enterprise IT enablers to enhance the total visibility of the ETO SC projects and real-time availability of information on the SC network; this approach promotes competitiveness and resilience of a data-driven ETO enterprise SC network.
- Proposition 3. The effective and efficient use of OEM "best practice" operations management tools for decision-making as well as competitive and resilient ETO SC network are essential for data-driven ETO enterprise SC complexity assessment and mitigation measures.

References

1. Addo-Tenkorang, R., Eyob, E.: Engineer-to-order: a maturity concurrent engineering best practice in improving supply chains. In: Customer-Oriented Global Supply Chains: Concepts for Effective Management. Industrial Engineering: Concepts, Methodologies, Tools, and Applications (3 Volumes). IGI Global - Information Resources Management Association (USA) (2013). https://doi.org/10.4018/978-1-4666-1945-6.ch095
2. Gosling, J., Naim, M.M.: Engineer-to-order supply chain management: a literature review and research agenda. Int. J. Prod. Econ. **122**, 741–754 (2009). https://doi.org/10.1016/j.ijpe.2009.07.002
3. Cooper, M.C., Lambert, D.M., Pagh, J.D.: Supply chain management: more than a new name for logistics. Int. J. Logist. Manag. **8**(1), 1–14 (1997)
4. Mentzer, J.T., DeWitt, W., Keebler, J.S., Min, S., Nix, N.W., Smith, C.D.: Defining supply chain management. J. Bus. Logist. **22**(2), 1–25 (2001)
5. Serdaran, S.: A review of supply chain complexity drivers. Comput. Ind. Eng. **66**, 533–540 (2012)
6. Musa, A., Gunasekaran, A., Yusuf, Y.: Supply chain product visibility: methods, systems, and impacts. Expert Syst. Appl. **41**(1), 176–194 (2014)
7. Addo-Tenkorang, R., Helo, P.T., Kantola, J.: Concurrent enterprise: a conceptual framework for enterprise supply-chain network activities. Enterp. Inf. Syst. **11**, 1–38 (2016)
8. Blecker, T., Kersten, W., Meyer, C.: Development of an approach for analyzing supply chain complexity. In Blecker, T., Friedrich, G. (eds.) Mass Customization Concepts – Tools – Realization, pp. 47–59. Gito Verlag, Berlin (2005)
9. Bozarth, C.C., Warsing, D.P., Flynn, B.B., Flynn, E.J.: The impact of supply chain complexity on manufacturing plant performance. J. Oper. Manag. **27**(1), 78–93 (2009)
10. Koudal, P., Engel, D.A.: Globalization and emerging markets the challenge of continuous global network optimization. Build. Supply Chain Excell. Emerg. Econ. **98**, 37–66 (2007)
11. KPMG. Supply chain complexity: Managing constant change. A study of supply chain maturity (2011). http://www.kpmg.com/UK/en/IssuesAndInsights/ArticlesPublications/Documents/PDF/Advisory/Supply-Chain-Survey.pdf:KPMGLLP(UK)

12. Perona, M., Miragliotta, G.: Complexity management and supply chain performance assessment. A field study and a conceptual framework. Int. J. Prod. Econ. **90**(1), 103–115 (2004)
13. Pricewaterhouse Coopers. In 9th Annual Global CEO Survey: Globalization and Complexity (2006). http://www.pwc.com/Extweb/ncevents.nsf/docid/0D298977BF0FBD3380 25716C0035AFB9/$file/Presentation_WB.pdf:PricewaterhouseCoopers
14. Vachon, S., Klassen, R.D.: An exploratory investigation of the effects of supply chain complexity on delivery performance. IEEE Trans. Eng. Manage. **49**(3), 218–230 (2002)
15. Van der Vorst, J.G.A.J., Beulens, A.J.M.: Identifying sources of uncertainty to generate supply chain redesign strategies. Int. J. Phys. Distrib. Logist. Manag. **32**(6), 409–430 (2002)
16. Choi, T.Y., Wu, Z., Ellram, L., Koka, B.R.: Supplier–supplier relationships and their implications for buyer- supplier relationships. IEEE Trans. Eng. Manage. **49**(2), 119–130 (2002)
17. Wu, Z.: Supplier–Supplier Interactions and Their Effects on Supply Performance: An Empirical Study. Unpublished doctoral dissertation, Arizona State University (2003)
18. Cusumano, M., Takeishi, A.: Supplier relations and management: a survey of Japanese, Japanese-transplant, and S. auto plants. Strateg. Manage. J. **12**(8), 563–588 (1991)
19. Helper, S.: Strategy and irreversibility in supplier relations: the case of the U.S. automotive industry. Bus. History Rev. **65**(4), 781–824 (1991)
20. Kamath, R.R., Liker, J.K.: A second look at Japanese product development. Harv. Bus. Rev. **72**(6), 154–170 (1994)
21. Wu, Z., Choi, T.: Supplier–supplier relationships in the buyer-supplier triad. J. Oper. Manage. **24**, 27–52 (2005)
22. Smith, P.C., Laage-Hellman, J.: Small group analysis in industrial networks. In: Axelsson, B., Easton, G. (eds.) Industrial Networks: A New View of Reality. Routledge, London (1992)
23. Addo-Tenkorang, R., Helo, P.T.: Big data applications in operations/supply-chain management: a literature review. Comput. Ind. Eng. **101**, 528–543 (2016)
24. Choi, T.Y., Krause, D.R.: The supply base and its complexity: Implications for transaction costs, risks, responsiveness, and innovation. J. Oper. Manage. **24**(2006), 637–652 (2006)
25. Wilding, R.: The supply chain complexity triangle: uncertainty generation in the supply chain. Int. J. Phys. Distrib. Logist. Manage. **28**(8), 599–616 (1998). https://doi.org/10.1108/09600039810247524
26. Sharma, K.: Managing on the edge of instability. In: Brace, G. (ed.) Logistic Information Management. Sterling Publishing, London (1997)
27. Addo-Tenkorang, R., Helo, P.T., Shamsuzzoha, A.H.M., Ehrs, M., Phuong, D.: Logistics and Supply Chains Management Tracking Networks: Data-Management System Integration/Interfacing Issues. Technology Management for Emerging Technologies (PICMET, 2012). In: Proceedings of PICMET 2012, Portland, OR (2012)
28. Shamsuzzoha, A., Addo-Tenkorang, R., Phuong, D., Helo, P.: Logistics tracking: an implementation issue for delivery network. In: PICMET 2011Conference Technology Management in the Energy Smart World, Portland, Oregon, USA, 31 July 31 – 4 August 2011 (2011)
29. Cameron, N.S., Braiden, P.M.: Using business process re-engineering for the development of production efficiency in companies making engineered to order products. Int. J. Prod. Econ. **89**(3), 261–273 (2004)
30. Van Hoek, R.I.: Measuring the unmeasurable measuring and improving performance in the supply chain. Supply Chain Manage. Int. J. **3**(4), 187–192 (1998)
31. Addo-, R., Helo, P.T.: Analysis of enterprise supply chain communication networks in engineering product development. Int. J. Logist. Manage. **28**(1), 47–74 (2015)
32. Cincom Whitepaper. (2012). The Complex Enterprise. Accessed 26 Jan 2017
33. Min, C., Shiwen, M., Yunhao, L.: Big data: a survey. Mob. Netw. Appl. **19**, 171–209 (2014)

34. Kees, A., Oberländer, A., Röglinger, M., Rosemann, M.: Understanding the internet of things: a conceptualization of business-to-thing (B2T) interactions. In: The proceedings of twenty-third European Conference on Information Systems (ECIS), pp. 1–16, Münster, Germany (2015)
35. Zhong, R.Y., Huang, G.Q., Lan, S., Dai, Q.Y., Xu, C., Zhang, T.: A big data approach for logistics trajectory discovery from RFID-enabled production data. Int. J. Prod. Econ. **165**, 260–272 (2015)
36. Gandomi, M., Haider, M.: Beyond the hype: big data concepts, methods, and analytics. Int. J. Inf. Manage. **35**(2), 137–144 (2015). https://doi.org/10.1016/j.ijinfomgt.2014.10.007
37. Bonev, M., Korell, M., Hvam, L.: New complex product introduction by means of product configuration. In: Proceedings of the 15th International Configuration Workshop, 29–30 August 2013, Vienna, Austria (2013)
38. Christopher, M.: Logistics and Supply Chain Management: Creating Value-Adding Networks. Financial Times Prentice Hall, Dorchester (2005)
39. Mason-Jones, R., Naylor, B., Towill, D.R.: Engineering the leagile supply chain. Int. J. Agil. Manag. Syst. **2**(1), 54–61 (2000)
40. Hoekstra, S., Romme, J.: Integral Logistics Structures: Developing Customer Oriented Goods Flow. McGraw-Hill, London (1992)
41. Olhager, J.: Strategic positioning of the order penetration point. Int. J. Prod. Econ. **85**(3), 319–329 (2003)
42. Gepp, M., et. al.: Assessment of engineering performance in industrial plant business. In: IEEE IEEM 2013 (2013)
43. Large Industrial Plant Manufacturer's Group (VDMA). 'Staying competitive in a volatile environment' Status Report - 2013/2014, Frankfurt (2015)
44. Percivall, G.: Systems Engineering in the automotive industry. In: Proceedings of the 2nd Annual Conference INCOSE, S, pp. 501–508 (1992)
45. Danilovic, M., Browning, R.T.: Managing complex product development projects with design structure matrices and domain mapping matrices. Int. J. Project Manage. **25**, 300–314 (2007)
46. Murman, E., Rebentisch, E., Walton, M.: Challenges in the Better, Faster, Cheaper Era of Aeronautical Design, Engineering and Manufacturing. Lean Aerospace Initiative. Massachusetts Institute of Technology (MIT) (2000). web.mit.edu/lean
47. Molina, A., Aca, J., Wright, P.: Global collaboration engineering environment for integrated product development. Int. J. Comput. Integr. Manuf. **18**(8), 635–651 (2005)
48. De Brentani, U., Kleinschmidt, E.J., Salomo, S.: Success in global new product development: impact of strategy and the behavioral environment of the firm. J. Prod. Innov. Manag. **27**(2), 143–160 (2010). https://doi.org/10.1111/jpim.2010.27.issue-2
49. Yassine, A.A.: An introduction to modeling and analyzing complex product development processes using the Design Structure Matrix (DSM) method. Urbana **51**, 1–17 (2004). users.ipfw.edu
50. Fumo, D.: Types of Machine Learning Algorithms You Should Know. Towards Data Science (2017). https://towardsdatascience.com/types-of-machine-learning-algorithms-you-should-know-953a08248861
51. Rasku, J.: Toward Automatic Customization of Vehicle Routing Systems. JYU dissertations (2019). https://jyx.jyu.fi/handle/123456789/65790
52. Glaessgen, E., Stargel, D.: The digital twin paradigm for future NASA and US Air Force vehicles. In: 53rd Structures, Structural Dynamics, and Materials Conference: Special Session on the Digital Twin (2012)
53. Gunasekaran, A., Ngai, E.W.: Build-to-order supply chain management: a literature review and framework for development. J. Oper. Manag. **23**(5), 423–451 (2005)

Human-Centred Approach in Industry 4.0: Lighting Comfort in the Workplace

Kanisius Karyono[1,2](✉), Badr M. Abdullah[1], Alison J. Cotgrave[1], Ana Bras[1], and Jeff Cullen[1]

[1] Faculty of Engineering and Technology, Liverpool John Moores University, Byrom Street Campus, Liverpool L3 3AF, UK
k.karyono@2019.ljmu.ac.uk

[2] Faculty of Engineering and Informatics, Universitas Multimedia Nusantara, Scientia Garden, Gading Serpong, Tangerang 15810, Indonesia

Abstract. One of the Industry 4.0 aims is enhancing the worker's life by providing a better environment at work. The uncomfortable working environment can decrease productivity. Lighting comfort is one of the comfort factors that should be considered at work. The visually impaired group is one among 15% of the world population living in disability. The lighting comfort standard is different for this group of people. The current standard in lighting comfort, lighting in the workplace standard BS EN 12464-1:2011 only mentions the standard for non-visual impaired people. This work highlights the comfort standard needed for visually impaired people, which is different from the non-visually impaired people. The illumination ranges are considered to be varied and tend to higher illumination values. Comfortable Reading illuminance level, which is defined as 300 to 500 lx in the standard, was not adequate for visually impaired people. They prefer to have 300 to 750 lx for comfort. The preferable font sizes are also different and correlate with the illumination in the room. The use of Bold Typeface font does not have an impact on the reading comfort for the visually impaired respondents. This result highlights the need for an adaptive lighting system to be considered for achieving human comfort and enhance productivity. There are more extensive opportunities for the visually impaired workers, as identified in the Industry 4.0 advancement.

Keywords: Lighting comfort · Productivity · Illumination standard · Visual impaired · Adaptive

1 Introduction

Industry 4.0, which is referred to as the Fourth Industrial Revolution, elaborates on the latest innovative technologies, processes, and products. These advancements produce a leap in the industrial field. The major factors involved are pervasive use of many digital technologies on the massive scale in the forms of cyber-physical systems (CPS) and transform the factories into smart factories. Like the early adopters, European Member States are continuously putting their innovation to develop Industry 4.0. One aspect of

A. Batako et al. (Eds.): GCMM 2021, LNNS 335, pp. 533–546, 2022.
https://doi.org/10.1007/978-3-030-90532-3_40

the ability to implement Industry 4.0 technologies is the employment shifting paradigm in manufacturing [1]. The Relation between Industry 4.0, sustainability manufacturing and circular economy in the integrated framework can become an essential factor in the development of this field [2]. The sustainability issue will push forward the way of traditional manufacturing aspects to the advanced process for all process involved in the manufacturing process like the maintenance 4.0 [3].

Industry 4.0 can give advantages such as faster operations cycle times, shorter delivery times, faster time to market, better quality, customisable products, and more substantial consumer involvement, which can lead to better loyalty. These advantages can help the company to win the new and emerging markets with a differentiation strategy or a new blue ocean strategy. This approach also nurtures the novel human and production organisation systems, which can impact the overall value chain, society, and the environment. The people as one of the critical aspects are the dominant players in the innovation and development, while machines assist with routine tasks. The employees with creativity, decision-making skills, technical and ICT expertise are the company assets. The team member with the right competencies has more autonomy and interesting work [4].

This approach also gives a broader opportunity for disabled people to take part in the team. Another goal of Industry 4.0 is to enhance the worker's life by giving a better environment at work. The uncomfortable working environment can decrease productivity. However, the current standard for the working place still addressed only the non-impaired people. A notable example of this standard is the BS EN 12464–1:2011 light and lighting: Lighting of workplaces: Indoor workplaces [5]. The standard only mentions the standard for non-visual impaired people.

This work complemented the previous work done with the older people as the primary respondent and focused on Age-Related Macular Degeneration (AMD) patient. Our trial focus on the visually impaired people who are about to enter the workforce when the Industry 4.0 are fully implemented at work. All these trials address the need for standards which acknowledge the need of the visually impaired people. This work also addresses the possibility of using a smart system to cope with the employer's comfort needs, which can lead to better productivity.

2 Industry 4.0 and Smart Factory

The advances in manufacturing, services, and information technology are the driver of the next Industrial Revolution. It brings many improvements due to the synergy of all of these aspects. It can lead to increased productivity, both service systems, and manufacturing environment. The coordination and connection become mandatory and supported by the technology, including communication and networking (Industrial Internet), embedded systems (Cyber-Physical Systems), adaptive robotics, cybersecurity, data analytics and artificial intelligence, and additive manufacturing. Industry 4.0 is taking the advantages of the interconnected and smart products.

Industry 4.0 promises more productive and continuous systems if it is appropriately adapted. The design principles should be adequately studied, and the working framework with the technologies is described in detail. The fundamental links between smart products and smart processes in the production control and monitoring change from human

labour centred production to fully automated. The industry requires strategic workforce planning, constructing the right organisation structure, developing partnerships, participating, and sharing the technological standardisation. These are essential factors to drive technological advancements. The primary implementation areas are the real-time supply chain optimisation, human-robot collaboration, smart energy consumption, digital performance management, and predictive maintenance. The next generation of technology involved nanotechnology, self-organised, self-motivated, and self-learning systems powered by artificial intelligence algorithms [6].
Industry 4.0 is driven by:

- Digitisation and integration of vertical and horizontal value chains.
- Digitisation of product and service offering.
- Digital business models and customer access.

The typical components of this new paradigm are the data and analytics. Reaching customer satisfaction is the main goal and a multi-stage, never-ending process [7].

There are four areas of Industry 4.0, productivity, revenue growth, employment, and investment. Productivity improvement varies by industry. Manufacturers for industrial components can achieve among the most significant productivity improvements, which are about 20 to 30% in Germany. Industry 4.0 drives revenue growth due to the manufacturer's demand for enhanced equipment and consumer demand for a wider variety of customised products. The enhancement gives greater flexibility for the manufacturing environment. Industry 4.0 leads to a 6% increase in employment during the next ten years. However, different skills are needed by the industry, which is related to technology. This competency transformation is one of the critical challenges for team member [8].

The smart factory is one of the examples of Industry 4.0. The advantages of smart factories:

- ability to solve problems related to the complex manufacturing process
- the goods produced have a longer lifespan and less defect
- manufacturing resources including people and machines are in synchronous and connected

Human capital becoming essential to support the efficient use of IT. Workplaces are changing and put high value in innovation. The capability to acquire new skills has become far more critical [9]. The smart factories also concern with their workers' health conditions. Advanced automation changes the factories to more ergonomic environments. This transformation creates stress-free and more comfortable places for employees. The challenges for implementing this are the time and spatial flexibility. The threats to the employees are more psychological rather than physical or disorders. The muscle force is becoming less and taken up by robots or factory automation [10].

This condition opens the possibility for the disabled team member with the specific needed skills to get a role in the factory. The employees are knowledgeable in their fields and able to utilise the machines with less muscle force involved. Employees are shifted on doing more strategic, coordinating, and creative activities. Furthermore, the social values of employees and demographics altered regarding this change [11]. Although robots are replacing workers in some jobs, the employment rate remains high. Overall, it raises the demand for labour. Technology is changing the needed skills, especially for skills that cannot be replaced by robots. Cognitive skills, like critical thinking and socio-behavioural skills, are among them [12]. Training will become a vital part of the process of nurturing workers [2].

3 Ergonomic in Industry 4.0

The team member role gets greater importance in Industry 4.0. There are two main focuses which are related to performance and wellbeing. Some of the empowerment for performance is gained by facilitating better communication, creating interactive environments between humans and robots and assisting decision-making processes. The wellbeing support is achieved by reducing physical workload and reducing dangerous and repetitive human tasks. Monitoring of health conditions and physical activities are integrated with the real-time information processing system capable of solving analytical complexities [13].

Their wellbeing factors are increased through the contribution of Ergonomics. Some issues related to the postures of the operator during the practical activities become the centre of the design. The issue can be identified, mapped, and solved through the new technology such as virtual reality. The new technology saves the use of resources spent on the progress of the project. The ergonomic study of the workplace also can play an essential role in the saving of used resources and become a tool for increasing productivity. This method can obtain additional information and different variants of assessments to optimise the result. The key factors, such as the operation of manual work, visibility, accessibility, the usability of equipment, comfort, and risk factors, can be easily mapped. The key factors give the designer the optimal conditions for the careful development procedures of decision-making processes in the preventive phase of industrialisation [14].

The ergonomic approach deals with the human-machine interaction identify and classifies stages, and individually works operations. The ergonomic will analyses the postures and movements of the team member to analyse psychophysical wellbeing and increase the environmental condition. The human factor is the central subject in human-machine systems. The factors which are involved are human activities, environmental, instrumental, and organisational conditions. The factors also cover the safety of manufacturing plants, their needs, protecting life, increasing efficiency, and the reliability of human-machine systems. Increased competitiveness can be done with the combination, ergonomics, and operating participative models. Trough Industry 4.0, the economic boost, which offers many opportunities, can be obtained through the connection between the real and the virtual world. Sensors, machines, real operators, and IT systems are connected to form technical and economic benefits. The smart systems' properties are the presence of technologies linked to each other and generate a change in the production paradigm. The ergonomic analysis allows the identification of innovative methods and procedures to design better general equipment [15].

The significant improvement in line integration and interoperability is achieved using the latest engineering formation. It requires denser cognitive efforts and skills socio-emotional, communication, motivation, autonomy, perseverance, self-control, resilience, collaboration, and creativity. 5C Architecture is used to map the implementation of Industry 4.0. In the Connection level, the primary attribute is communication. The primary attribute of the informational is the attribute for conversion level while the controllable is for the cybernetic level. In the level of cognition, self-cognition is the primary attribute, and predictive maintenance is one of the implementations. For intelligent production, the primary attribute for the Configuration level is the auto-configuration [16].

The collaboration between people and machines can work efficiently and safely. The team member acquires a new role in this revolution environment. Collaborative Robots, for example, can share with people the same space work to carry out tasks, contributing to each part with the best of both worlds [17]. Robotic technology can also be used to give the team member support so that they can have better strength in the form of the exoskeleton. This support is also available for disabled people so that they can have an equal chance to work in the industry.

4 Lighting Standard and Productivity

The British Standard related to light and lighting for indoor workplaces is numbered as BS EN 12464-1:2011 light and lighting: Lighting of workplaces: Indoor workplaces [5]. The lighting condition should be evenly distributed to give comfort to the team member. The light illuminance level should be adequate and no glare, which can distract the team member. Some notable illuminance levels in the workplace, according to the standard, are 200 lx, 300 lx, 500 lx, and 750 lx. Lower illumination leads to lower team member satisfaction, lighting discomfort, and lower productivity.

The location should be illuminated with at least 200 lx: a canteen, pantries, feed preparation, dairy, utensil washing, bakery preparation and baking, laundry ironing, lounges, entrance halls, and library bookshelves area. Location for bakery finishing, glazing, decorating, playroom, nursery, handicraft room, classroom, music practice, and computer practice should be illuminated with at least 300 lx. The illumination of 500 lx is necessary for the sickbay, hairdressing, cutting, gilding, embossing, block engraving, work on stones and platens, printing machines, matrix making, writing, typing, reading, data processing, restaurant's kitchen, library reading area, the classroom for evening classes, lecture hall, demonstration table, practical rooms, and handicraft room. Higher illumination of 750 lx is needed for laundry inspection and repairs, art, and technical drawing.

There has been researched to lower down the energy used for lighting by lowering down the standard illumination. The research focuses on visual performance over the printed material. The study suggested decreasing the illuminance level standard of 250 to become 150 lx and reduce energy consumption by up to 40% [18]. However, the study was only based on the non-visually impaired people as the respondents. Other research also reveals that the display illumination affects the search speed. The recommended e-paper displays illumination is higher than 700 lx to maintain search performance [19]. The ambient illumination also has to be considered for e-paper reading comfort. The minimum ambient illumination requirement for the e-paper display is at least 62 lx. At this value, the reflective-type display started presenting its better legibility, and the subjective visual fatigue started decreasing [20]. In the room that considers the glare and the activity related to the computer, the value for vertical illuminance is 351.6 lx [21].

5 Visual Impairment

According to the World Health Organization, at least 2.2 billion people have a vision impairment or blindness. The visually impaired group is one among 15% of the world population living in disability. This value also consists of those with moderate or severe distance vision impairment or blindness due to unaddressed refractive error and near vision impairment. These impairments are classified into two groups, distance and near presenting vision impairment. The experience of vision impairment varies due to many different factors [22]. Distance vision impairment:

- Mild – presenting visual acuity worse than 6/12
- Moderate – presenting visual acuity worse than 6/18
- Severe – presenting visual acuity worse than 6/60
- Blindness – presenting visual acuity worse than 3/60

Previous experiments found that the illumination level should be adjusted to fulfil the need for low vision occupants. The most significant change in lighting illumination happens in the kitchen around the sink. The standard value is about 300 lx and should be adjusted to 1000 lx while the lowest change is for the dining table. From the initial value of 250 lx, have to be adjusted to reach 400 lx [23].

The other focus for lighting comfort is for the needs of the elder who have the Age-Related Macular Degeneration (AMD) [24]. The work presents the need for higher illumination, which can lead to better reading performance and comfort. From 13 respondents, it is shown that the pattern of reading comforts are varied, unlike the reading speed. The reading comfort can decrease while the reading speed is constant or even increase. Another work-related to AMD also presented with 25 respondents [25]. The paper of the low vision identified the need for home lighting levels which higher for people with ocular disease. The work also gives a good illustration of the AMD respondents needs compared to the healthy control group.

6 Lighting Comfort for Visual Impairment Trial

This work aims to verify the previous research that the visually impaired needs the different illumination values compared to non-visual impaired people in order to perform better in their role. Many of the researches are focus on the AMD or the elder people group. The authors focus on the young, visually impaired group to fill in the gap. The authors conducted experiments with 11 respondents. The respondents who represent visually impaired people are seven students from primary and secondary school, ageing 10 to 14 years old. The reason for choosing this age group is because they have not yet worked so that the current working environment condition will not give a bias to the result.

The respondents are digital native; they are raised in the era of information technology. They are the next generation workforce which will be working with the infrastructure of pure industry 4.0. Five of them suffer Moderate (presenting visual acuity worse than 6/18) and Severe (presenting visual acuity worse than 6/60). Two subjects with blindness (presenting visual acuity worse than 3/60) are excluded from this observation because they could not sense the artificial light used in this trial. The other four people are the non-visual impaired people for the control group.

6.1 Methods and Limitations

The trial uses the lighting illuminance level from 150 lx to 2000 lx. This trial has aimed to define the correct, comfortable illumination level for the visually impaired and non-visual impaired people for reading. In this trial, the respondents are allowed to maintain their reading habits, for example, reading with a very close distance to the text. The preferred illuminance level value for reading then associated with their lighting comfort in their workplace. The preferred illuminance is done personally with the interview method.

In the trial of enhancing productivity, working performance increase or decrease due to the use of proper illumination value is represented by the smallest font size that the subject is feeling comfortable. The use of reading analogy is due to the nature of indoor work which related to reading and working with the objects. The respondents are given the text containing a list of characters with Arial font and Arial Bold font from font size 8 pt up to 26 pt. They must mention the smallest font size that they are comfortable to read, with the font type (regular font or the bold font) at a specific value of illumination. These characters are correctly aligned (not mixed) to avoid the distraction happened when multiple font sizes are being given in the random order. The interview was done personally for each respondent without other respondent comments, so the result was not biased.

Working with this age group is different compare to the more mature respondent. Therefore, the authors use the interview approach and not using the questionnaire. This work also has limitations in term of the sample size. Only limited respondents can participate in this research. Due to this small sample size, the result has to be validated or compared with the previous work mentioned before, in chapter 5 [25].

6.2 The Trial Results

There is no effect on the font type to their reading comfort. The regular font and the bold font do not affect their reading comfort. All participants consider that the bold font does not make them more comfortable. The detailed value chart for the lighting comfort for each respondent is shown in Fig. 1.

- Respondent number 1 can read with the illuminance value of 180 lx, and the smallest font size that s(he) can read is 12 pt. S(he) is more comfortable to read at 600 lx and can read the font of 12 pt.

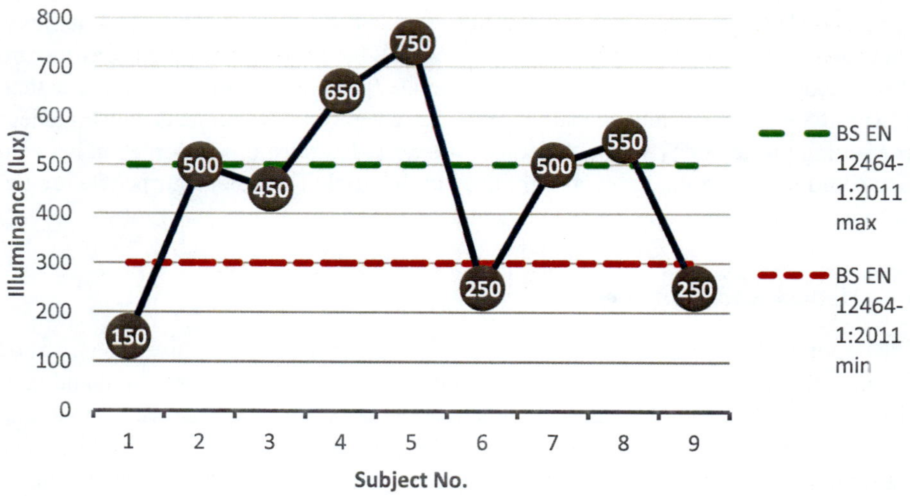

Fig. 1. The comfortable reading illuminance level for each respondent

- Respondent number 2 is able to read with the illuminance value of 150 lx, and the smallest font size that s(he) can read is 10 pt. At 600 lx, s(he) can read the font of 10 pt. At 1200 lx s(he) can sense the font with size 8 pt. The 8 pt font can be read at 1700 lx, but it triggers the uncomfortable reading condition.
- Respondent number 3 at illuminance value of 150 lx can read the 16 pt font as the smallest readable font size. At illuminance 700 lx, s(he) can read 12 pt font size. This value is the smallest font size that can be read even the illuminance level is raised to 1700 lx. This high illuminance triggers an uncomfortable situation.
- Respondent number 4 has a different result. At illuminance 150 lx and 700 lx, s(he) can read the font size 12 pt. The font 10 pt can be read at an illuminance value of 1700 even it triggers the uncomfortable condition.
- Respondent number 5 has a result similar to Respondent number 4. S(he) can read 20 pt font size at the illuminance value of 150 lx. At illuminance value 700 lx, s(he) is comfortable with 14 pt font size. At the illumination value of 1700 s(he) can read the 10 pt font size.
- Respondent numbers 6 to 9 (control group) non-visually impaired can read the font size 8 pt at all illumination value even with the lowest illumination value, but they feel the discomfort. They have their illumination comfort, 250 lx, 500 lx, 550 lx, and 250 lx, respectively.

The illuminance level affects the visually impaired people able to read the small text, but this is not the case with the control group. The chart shows the Relation between the smallest font size and the illumination level for each respondent is shown in Fig. 2.

Using Bold Typeface font does not have an impact on the reading comfort for the visually impaired students. Upon the low-level light, all the visually impaired respondents have the same response. They do the same trick to be able to read the sentences by tilting the reading materials and adjust their viewing distance to search for a better

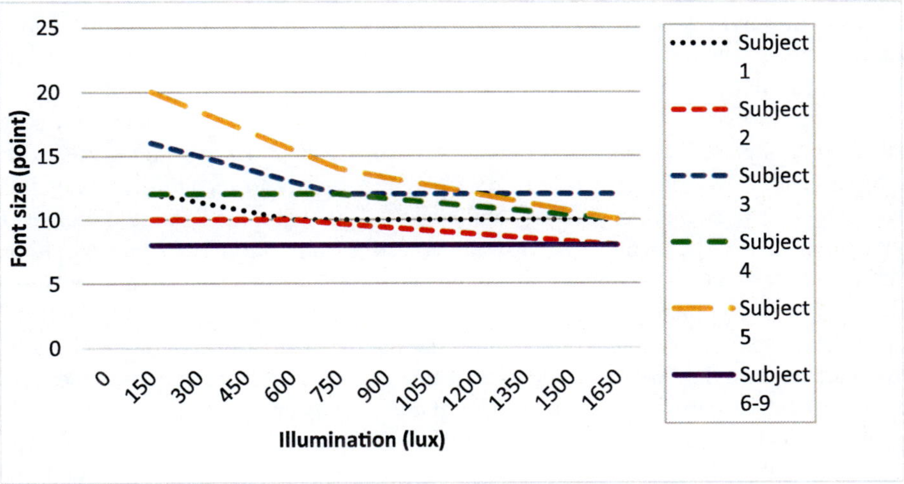

Fig. 2. The relation between the smallest font size and the illumination level

vision. Upon the high level of light, they tried to avoid the lighting source, block the light source, or squint their eyes.

6.3 Analysis and Discussion

According to The British Standard BS EN 12464-1:2011, the standard illuminance for reading is about 300 lx to 500 lx. The previous work discussed in chapter 4 also shows similar illuminance values. Based on the trial result, visually impaired respondents tend to have higher lighting comfort value than those who are non-impaired. They need a brighter environment for their activities, particularly reading activities. The result is like the previous research discussed in chapter 5. The exciting part of these results is the variation of the data captured. The control group shows that the data variations are insignificant. In the intervention group, both trials have very high data variations, even can reach doubled of the values [25].

Our observation also found that the result is not uniform and cannot be generalised to form a visually impaired standard requirement. There is the data that shows fewer illuminance levels for comfort, and the respondent also needs a higher level of illuminance. The difference in comfort is caused by the different levels of visual impairment, a difference in human preferences, and the difference in the reading style/behaviour. The respondent interviewed in this observation is not many, but it already represents the people or employees who use it.

Most of the employees have similar requirements, but there are always employees that require less lighting or more due to their conditions and habit. The value difference is the reason why the system is designed based on the majority preferences. The approach is similar to the case of thermal comfort requirements. At least 80% of the occupants have to be in a comfort situation [26]. The observation result shows that designing the automatic lighting system based on the shared values are not adequate. The lighting comfort is personal, and with the majority design approach, there are 20% or more users feel discomfort. There always be a case that the user has unique properties, which can only be achieved by a personally customised system, like the case of the visually impaired respondents in our trial.

The trial uses the reading ability as the representation of the productivity aspect. All visually impaired respondents experience a similar performance with regards to the lighting level. Low lighting level can decrease their ability to recognise smaller font sizes. Three respondents can read the smaller size of the font if the lighting illuminance is increased. The respondents can improve their results from 12 to 10 pt, from 16 to 12 pt, and from 20 to 14 pt. Two respondents still get the same performance, but they feel more comfortable in completing their task. At their comfortable lighting illuminance level, most of the respondents can read regular font sizes, 10, 12, and 14 pt. The highest lighting level can increase their performance to read smaller font sizes for three respondents (8 and 10 pt), but this also triggers an uncomfortable reading environment.

It is prevalent that the solution for the disabled comes from the assumption from the non-disabled. The solution should be discussed and appropriately tested with the disabled. Without this crucial step, the solution which seems to be right in the planning phase is not suitable for the disabled. This trial also highlights that premise. This trial starts with the underlying assumption that the bold font is more visible to the visually impaired respondents. This observation proves that this is the wrong assumption. Our observation concludes that using Bold Typeface font does not impact the reading comfort for the visually impaired respondents.

The information from [25] supports our observation result, which later drives our intent to use the adaptive techniques.

6.4 User Experience/User Interfaces (UX/UI) Factor

Besides the trial for the illuminance, the authors also assess the possibility of using a smartphone as user interface devices. It seems funny to suggest the use of a smartphone to the visually impaired occupants as their way to control the environmental settings. In fact, from the observation of these young, virtual-impaired respondents, they are all smartphone users, just like the non-disabled. Even the respondents with the highest visual impaired level (can be considered entirely blind) are the WhatsApp users. S(he) use the text chat feature of WhatsApp, like the non-disabled respondents. The typing speed is also considered to be fast. The use of the smartphone is so often that all the users, even visually impaired users, are accustomed to using the virtual keyboard (on-screen keyboard). In this case, the design of the application does not need to be specific. The only thing which is more important is the standardisation of the button located in their smartphone application. Their adaptation is faster on this standards location than introducing them to the new set of specific display designs for the visually impaired users. The standard password is still accessible by the visually impaired group because of the use of the virtual keyboard on their smartphones.

7 The Need of the Adaptive Approach

7.1 Adaptive User Behaviour

People naturally have adaptive behaviour to make a more comfortable environment condition. This action is known as self-adaptation. There are three types of adaptation. The first is a physiological adaptation, which represents the body's reaction because of the change in the surroundings. The second is psychological adaptation. This adaptation is derived from the state of mind of previous experiences. The third adaptation is related to human behaviour [27]. Adaptive behaviours in response to indoor environmental conditions provided by lighting systems are as follows [28]:

1. Electric Lighting category: Covering room surface, Opening or closing operable curtains, adjusting electric lighting operative hours, using desktop USB lamp instead of electrical lighting, adjusting desktop or task surface, switching lighting(s) manually, switching lighting(s) using the feedback system.
2. Natural Lighting category: Changing position or direction of furniture, Covering room surface, Opening or closing operable curtains, Opening or closing curtain(s) by remote control, Opening or closing curtain(s) by a feedback system, Opening or closing the operable window(s), Opening or closing window(s) by a feedback system.

7.2 Adaptive System

Based on our trial, lighting illuminance values are very personal. It can vary from person to person according to their condition and disability. Developing a personally customised system is very expensive. The framework review for personalised control is also presented to make the system perform the automatic task [29]. The automatic control, which has the aim of lower energy usage, also have been studied [30, 31]. These systems focused on the implementation of the on-off system. The other parameter which can support energy saving like blinds can also be controlled automatically. The artificial light can be controlled to provide high contrast to make a better transition to and from night shift work. This method can adjust the circadian clock better for the rapidly rotating shift workers.

The adaptive system is the solution to develop a system which able to cope with personal preference. The system can be part of the Industry 4.0 development for supporting the team member. This system can adapt to personal preferences, increase human comfort, and increase productivity. The main component of this system is the Artificial Intelligence (AI). Neural Network is one of the preferred solutions for the core of the adaptive system. This system can be accustomed to the users within the area. Two identical systems can behave differently under different environments, even that the software and hardware used are precisely the same. The different learning data can lead to different system behaviour under specific circumstances. Using this adaptive system, the real user behaviour, which is unique, can be captured by the system.

8 Conclusion

Industry 4.0 can enhance the team member quality of life by focusing on better roles and better environment at work. A more comfortable environment can increase productivity. Industry 4.0 can give a new role to the team member with the new requirement of acquired skills. This role can be more open to disabled people so that they can have an equal chance to work in the industry.

According to The British Standard BS EN 12464-1:2011, the standard illuminance for reading is about 300 lx to 500 lx, and the standard only refers to the non-visual impaired people. Based on the trial result, visually impaired respondents tend to have higher lighting comfort value compared to non-impaired respondents. They prefer to have 300 to 750 lx for comfort.

Our observation shows that the lighting illuminance level is personal. The result shows the non-uniform data and cannot be generalised to form a standard requirement for the visually impaired people. Low lighting level can decrease their ability to recognise smaller font sizes. The use of Bold Typeface font does not have an impact on the reading comfort for the visually impaired respondents.

This result highlights the need for an adaptive lighting system to be considered for achieving human comfort and enhance productivity. There can be more extensive open opportunities for the visually impaired workers, as identified in the Industry 4.0 aims.

References

1. Muscio, A., Ciffolilli, A.: What drives the capacity to integrate Industry 40 technologies? Evidence from European R&D projects. Econ. Innov. New Technol. **29**(2), 169–183 (2020). https://doi.org/10.1080/10438599.2019.1597413
2. Bag, S., Pretorius, C., Harm, J.: Relationships between industry 4.0, sustainable manufacturing and circular economy: proposal of a research framework, Int. J. Organ. Anal. ahead-of-print, no. ahead-of-print. Accessed 18 Nov 2020. https://doi.org/10.1108/IJOA-04-2020-2120
3. Jasiulewicz-Kaczmarek, M., Legutko, S., Kluk, P.: Maintenance 4.0 technologies–new opportunities for sustainability driven maintenance. Manage. Prod. Eng. Rev. **11**(2), 74–87 (2020). https://doi.org/10.24425/mper.2020.133730
4. Fonseca, E.: Industry 4.0 and the digital society: concepts, dimensions and envisioned benefits, (in English), vol. 12, issue number 1, p. 386 (2018). https://doi.org/10.2478/picbe-2018-0034
5. Light and lighting — Lighting of work places BSI, 30 June 2011
6. Salkin, C., Oner, M., Ustundag, A., Cevikcan, E.: A conceptual framework for Industry 4.0. In: Industry 4.0: Managing The Digital Transformation. SSAM, pp. 3–23. Springer, Cham (2018). https://doi.org/10.1007/978-3-319-57870-5_1
7. Reinhard Geissbauer, J.V., Schrauf, S.: Industry 4.0: building the digital enterprise, in "2016 Global Industry 4.0 Survey," PWC, Germany (2016). https://www.pwc.com/gx/en/industries/industries-4.0/landing-page/industry-4.0-building-your-digital-enterprise-april-2016.pdf
8. Rüßmann, M. L., Gerbert, P., Waldner, M., Justus, J., Engel, P., Harnisch, M.: Industry 4.0: The Future of Productivity and Growth in Manufacturing Industries, 09 April 2015. http://www.inovasyon.org/pdf/bcg.perspectives_Industry.4.0_2015.pdf
9. Sima, V., Gheorghe, I.G., Subić, J., Nancu, D.: Influences of the Industry 4.0 revolution on the human capital development and consumer behavior: a systematic review. Sustainability **12**(10), 4035 (2020)

10. Nuray Çelik, F.Ö.: The upcoming issues of industry 4.0 on occupational health and safety specialized on turkey example. Int. J. Econ. Bus. Manage. Res. **1**(5), 236–256 (2017)
11. Shaw, P., Varghese, R.: Industry 4.0 and Future of HR. J. Manage. **5**(6), 96–103 (2018)
12. Bank, W.: World development report 2019: The changing nature of work. World Bank Group, Washington, DC (2019). https://hubs.worldbank.org/docs
13. Silva, V.L., Kovaleski, J.L., Pagani, R.N., Corsi, A., Gomes, M.A.S.: Human factor in smart industry: a literature review/Fator Humano em Industria Inteligente: Uma Revisao de Literatura, (in English). Future Stud. Res. J. Trends Strategy Rep. **12**, 87 (2020)
14. Laudante, E.: Industry 4.0, innovation and design. a new approach for ergonomic analysis in manufacturing system. Des. J. **20**(sup1), S2724–S2734 (2017). https://doi.org/10.4995/IFDP.2016.3297
15. Laudante, E., Caputo, F.: Design and Digital Manufacturing: an ergonomic approach for Industry 4.0. Syst. Des. Beyond Process. Think. 922–934 (2016)
16. Rocha, M.F.M., de Oliveira, K.F., Polezi, I., Munhoz, A.C.S.A.: Industry 4.0: technology mapping and the importance of cognitive ergonomics. Int. J. Adv. Eng. Manage. Sci. **5**(5), 296–303 (2019). https://doi.org/10.22161/ijaems.5.5.1
17. Munoz, L.: Ergonomics in the industry 40 exoskeletons. J Ergon. **8**, 176 (2018)
18. Atmodipoero, R.T., Pardede, L.: Research on minimum illumination as a function of visual performance. Energy Build. **36**(7), 644–649 (2004). https://doi.org/10.1016/j.enbuild.2004.01.002
19. Lee, D.-S., Shieh, K.-K., Jeng, S.-C., Shen, I.H.: Effect of character size and lighting on legibility of electronic papers. Displays **29**(1), 10–17 (2008). https://doi.org/10.1016/j.displa.2007.06.007
20. Lin, Y.-T., Hwang, S.-L., Jeng, S.-C., Koubek, R.J.: Minimum ambient illumination requirement for legible electronic-paper display. Displays **32**(1), 8–16 (2011). https://doi.org/10.1016/j.displa.2010.09.002
21. Suk, J.Y.: Luminance and vertical eye illuminance thresholds for occupants' visual comfort in daylit office environments. Build. Environ. **148**, 107–115 (2019). https://doi.org/10.1016/j.buildenv.2018.10.058
22. W. H. Organization: Blindness and vision impairment (2019). https://www.who.int/news-room/fact-sheets/detail/blindness-and-visual-impairment. Accessed 15 July 2019
23. Brunnström, G., Sörensen, S., Alsterstad, K., Sjöstrand, J.: Quality of light and quality of life – the effect of lighting adaptation among people with low vision. Ophthalmic Physiol. Opt. **24**(4), 274–280 (2004). https://doi.org/10.1111/j.1475-1313.2004.00192.x
24. Fosse, P., Valberg, A.: Lighting needs and lighting comfort during reading with age-related macular degeneration. J. Vis. Impair. Blindness **98**(7), 389–409 (2004). https://doi.org/10.1177/0145482x0409800702
25. Perlmutter, M.S., Bhorade, A., Gordon, M., Hollingsworth, H., Engsberg, J.E., Baum, M.C.: Home lighting assessment for clients with low vision. Am. J. Occup. Ther. **67**(6), 674–682 (2013). https://doi.org/10.5014/ajot.2013.006692
26. ANSI/ASHRAE Standard 55 Thermal Environmental Conditions for Human Occupancy, ASHRAE (2017). https://www.ashrae.org/technical-resources/bookstore/standard-55-thermal-environmental-conditions-for-human-occupancy
27. Parsons, K.: Human Thermal Comfort. CRC Press, Boca Raton (2020)
28. Keyvanfar, A., et al.: User satisfaction adaptive behaviors for assessing energy efficient building indoor cooling and lighting environment. Renew. Sustain. Energy Rev. **39**, 277–295 (2014). https://doi.org/10.1016/j.rser.2014.07.094
29. O'Brien, W., Gunay, H.B.: The contextual factors contributing to occupants' adaptive comfort behaviors in offices – a review and proposed modeling framework. Build. Environ. **77**, 77–87 (2014). https://doi.org/10.1016/j.buildenv.2014.03.024

30. Nagy, Z., Yong, F.Y., Schlueter, A.: Occupant centered lighting control: a user study on balancing comfort, acceptance, and energy consumption. Energy Build. **126**, 310–322 (2016). https://doi.org/10.1016/j.enbuild.2016.05.075
31. Gunay, H.B., O'Brien, W., Beausoleil-Morrison, I., Gilani, S.: Development and implementation of an adaptive lighting and blinds control algorithm. Build. Environ. **113**, 185–199 (2017). https://doi.org/10.1016/j.buildenv.2016.08.027

Advanced Risk Model for the Safety Evaluation of Food Transport Logistics

Rasheed Onakoya[1], Zhuohua Qu[1], Zaili Yang[2](✉), and Ai Ha Thi Nguyen[3]

[1] Liverpool Business School, Liverpool John Moores University, Liverpool, UK
[2] Liverpool Logistics, Offshore and Marine Research Institute, Liverpool John Moores University, Liverpool, UK
Z.Yang@ljmu.ac.uk
[3] Faculty of Agronomy, Nong Lam University Ho Chi Minh City, Ho Chi Minh City, Vietnam

Abstract. This paper aims to identify and evaluate the risk factors influencing food transport logistics (FTL) chains. Using the combination of Fuzzy Rule-Based and Bayesian Network (FRB-BN) approach and the primary data collected from two major food supply chains in Vietnam, the risk factors of high-risk levels are analysed as "leadership in food safety management", "food supplier transparency", and "quality deterioration and adaptation to food standard regulation", respectively. The findings provide useful insights for transport logistics providers, food risk assessors and the internal/external auditors to handle uncertain and imprecise data to support safety-based decision making in the food and logistic industries. It contributes to the FTL risk analysis from different perspectives and applies an advanced uncertainty modelling technique to evaluate such diversified FTL risks with high uncertainty in data in the same framework.

Keywords: Food transport safety · Food logistics · Hazard · Fuzzy logic · Bayesian network

1 Introduction

The Food supply chain involves the farmers, growers, ingredient suppliers, product processors, product distributors, product importers and product retailers from different regions in the world [1]. In the food industry, food manufacturing, processing and transport logistics are important to ensure global food supply. Specifically, the transport logistics chain is likely the most critical chain throughout the food journey from farm to fork because of its perishable nature, efficiency requirement, safety, sustainability and cost optimization of the food products into the global market [2, 3]. Food transport logistics (FTL) providers play a major role in ensuring that food products after production are transported to the right place, at the right time and in the right condition [4]. However, due to the increasing number of stakeholders involving in the process of FTL chain and their widespread geographical locations, FTL poses a severe and complicated risk challenge that threatens the smooth flow of the food products with a severe consequence to the global economy [5]. Thus, the international food trade agreement has highlighted

A. Batako et al. (Eds.): GCMM 2021, LNNS 335, pp. 547–559, 2022.
https://doi.org/10.1007/978-3-030-90532-3_41

the importance of conducting a risk analysis for the international elaboration of food safety standard [6]. This paper aims to apply an advanced risk analysis model by combining fuzzy rule base (FRB) and Bayesian networks (BN) to evaluate the risk factors affecting FTL chain with data in high uncertainty. To achieve it, this paper is structured as follows. An overview of the applications of FRB and BN in risk studies is described in Sect. 2. Section 3 describes the methodology of modelling FTL risk evaluation. A real case study of seven agri-food transport logistics chains from the Republic of Vietnam is investigated to demonstrate the feasibility and reliability of the proposed model in Sect. 4. Section 5 concludes the paper.

2 Literature Review

A literature review is conducted based on the 171 papers collected by the authors from WoS core journals in food science technology, agriculture, business economics, engineering covering topics on food supply chain and safety published between the period from 1995 to Q4 2020. The start date was set on the basis that there was a significant increase in research on food safety from the said date. To collate the relevant papers for review, search strings and substrings of the study variables - food safety, supply chain safety, food supply, food transport, food logistics, risk factors, risk assessment and risk models were selected as topics and sub-topics and to ensure that the most relevant literature was evaluated. The study-imposed a (stratified) selections by defining the criterion for including or excluding studies, only the articles published in English were considered and included. Conference proceedings, textbooks, working paper, anonymous paper and other unpublished works were excluded. The retained articles were then thoroughly reviewed to ensure their quality and relevance to the research subject. The author carefully screened the papers using the criteria including 1) relevant titles, keywords and abstracts so they can make relevant contributions to the study 2) cover one or more food supply chain risk analysis topics. Articles that do not meet the filtration criteria were excluded. Finally, the reference list of the selected articles was evaluated to ensure that there are no other relevant articles omitted. The result shows that there has been limited research done in.

the assessment of the risk factors in an end to end food supply chain [7, 8], Previous studies on the subject had only focused on the risk assessment of food supply chains from the production [9], post-harvest [10, 11] and processing chain [12], there are very few studies on a whole FTL chain, particularly from quantitative assessment perspective due to imprecise and uncertainty of data [13], revealing a significant research gap to address. If the risk factors present in FTL chain is left unaddressed, it will affect the whole supply chain link of the product and creates a considerable loss to the global economy [14]. In assessing the risks with the uncertainty of data presented in an FTL chain, the advanced quantitative risk assessment models based on uncertain treatment theories such as Bayesian algorithm [15], Fuzzy logic [16], Dempster-Shafer theory [17] have exposed a wider range of applications in the literature in handling data uncertainties. However, a careful review of the literature reveals that the uncertain treatment models are often used in a combined manner to overcome the drawbacks when using them individually in practice [18, 19]. Fuzzy logic has the advantages of tackling uncertainty caused by vagueness in reasoning that influences human judgements due to their lack or insufficient evidence [20]. The framework for the assessment of risk based on fuzzy theory has been found in the studies of [21, 22], although fuzzy theory contributes to the development of precise failure analysis and enables rational risk hazards evaluation and prioritization [23], the model renders itself vulnerable in its practical application. For instance, its application involves a large number of rules and the process of applying the set of rules with domain experts is a tedious task. Other critics on a fuzzy logic model can be found in the research conducted by Yang et al. [18]. Similarly, risk assessment based on a BN model is used to capture the non-linear relationship between the risk factors in the form of prior probabilities [18]. The BN tools also show some drawbacks in its applications - it requires too much data in the form of prior probabilities, which are difficult or impossible to obtain in risk assessment [18]. Hence, to compensate for both models' disadvantages, a combined FRB and BN model is proposed for effective risk evaluation. While the combination of the FRB and BN models facilitates the study risk assessment process, its application in the FTL chain is new and also poses some challenges. For instance, a large FRB belief structure will be required, given the multiple risk parameters employed for FTL chain risk analysis. Section 3 describes a new mechanism for the use of the combined FRB and BN in risk analysis of FTL chain.

3 Methodology to Model FTL Chain Risk Evaluation

The major challenge in the assessment of risks in a FTL chain is that the objective data are limited and often only available to a certain level. For instance, around January to April 2016 more than 1700 tons of Vietnamese rice transported to the United States was rejected by the America officials due to safety concern [24], the information about the subjective probability of the significant risk factors and its consequence that lead to the safety issue was unclear. To address such uncertainty in the assessment of risk, the ensuring section describes how the FRB and BN are combined within the context of FTL.

3.1 Combined FRB and BN in FTL

The concept of the fuzzy theory was introduced by Zadeh [25]. The theory provides a fundamental basis of possibility, allowing the problem of imprecision due to uncertainty and ambiguity associated with human thinking to be solved. The FRB theory is used to model vagueness and uncertainty in the decision-making process due to the lack of complete information, by using linguistics terms to represent decision-maker preferences [26]. BN presents a graphical model to capture the probabilistic interdependency between a set of variables [27]. BN technique is applied to facilitate and synthesize the rule. The combined FRB and BN have the advantage to model the interrelationship between the attribute of the risk parameters and risk status flexibly and to process and transform the information into subjective conditional probabilities to handle the imprecise information challenge in the assessment of risk in the FTL chain. The combination of FRB and BN has been applied successfully to solve many complex real-world problems (e.g. [28]). Figure 1 presents the steps for modelling the FTL chain risks based on the proposed combined FRB and BN approach.

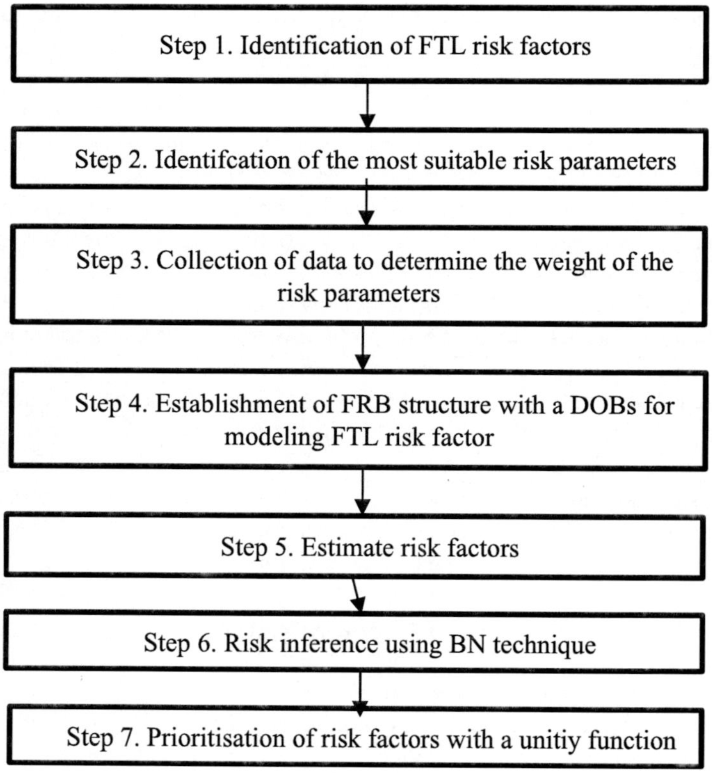

Fig. 1. Steps for modelling the FTL chain risks

Step 1: Identification of the risk factors in the FTL chain. The various risk factors associated with FTL chain were identified through a careful review of the literature, following a Delphi technique process with the industry experts to classify and verify the reviewed risk factors. The preliminary study of classifying the literature review risk factors took place in March 2019 in the UK in a brainstorming exercise with domain experts (five academic and two industry members, each with more than 15 years working experience in the FTL industry) during a workshop until a consensus was reached. Furthermore, in June 2019, five experts in the food industry from the Republic of Vietnam (the world-leading rice and cocoa export country) were selected to pilot test and verify the risk factors in a real-life industry concern. Table 1 represents the verified 46 risk factors influencing FTL chain.

Step 2: Identification of the most suitable risk parameters for FTL chain risk evaluation. In this study, the risk parameters include likelihood probability (L), resilience (R), probability of hazard detection (P), and disintegration of the consequence of the risk factors into three categories based on the risk consequence severity on time delayed in transporting the food products (CT), the severity on the product quantities (CQV) and the severity on the product qualities (CQ). They were identified and verified by scholars and industry experts as the major consideration in the FTL chain risk management. To facilitate data collection, Likert scale assessment grades are allocated to each of the risk

Table 1. The verified 46 FTL chain risk factors

RF1. Flood	RF17. Cross-contamination	RF33. Storage and warehouse
RF2. Government transport policies	RF18. Insufficient holding space	RF34. Lack of sufficient cargo handling equipment's
RF3. External legal issues	RF19. Timely availability of the vehicle	RF35. Risk of applying sensing technology
RF4. Labour strike	RF20.Truck accident	RF36. Humidity monitoring/control
RF5. Worker union relation	RF21. Human error	RF37. Negligently equipment maintenance
RF6. Pilferage	RF22. Improper sanitation and backhauling hazardous material	RF38. Power system
RF7. Cyberattack	RF23. Poor pest control	RF39. Poor packaging
RF8. Changes in the currency exchange rate	RF24. Communication failure among partners	RF40. Poor preservation
RF9. Payment delay from shippers	RF25. Labour skilled personnel	RF41. Inaccurate shipment from the supplier
RF10. Higher transportation cost	RF26. Employee wages	RF42. Low supplier transparency
RF11. Excessive inventory	RF27. Overburden employee	RF43. Low supplier integration
RF12. Food quality deterioration	RF28. Poor motivation among the workforce	RF44. The poor performance of sub-contractor
RF13. Poor handling	RF29. Leadership in food safety management	RF45. Poor logistics contract
RF14. Fire accident	RF30. Adaptation to food standard regulation change	RF46. Global sourcing network
RF15. Product damage in transits	RF31. Poor employee hygiene	
RF16. Temperature abuse	RF32. Obsolete technology	

parameters (Table 2). For example, a set of five linguistic terms such as {Highly unlikely, unlikely, likely, highly likely, definite} for "Likelihood Probability" was defined by the domain expert [29].

Step 3: Collection of data to determine the weight of the risk parameters. A questionnaire is distributed among the relevant experts to compare the relative importance of the risk parameters in the assessment of FTL chain risks. The study applied an AHP technique [30] to obtain the relative weight of the risk parameters.

Step 4: Construction of FRB structure. Based on the six risk parameters and their associated weights, the FRB structure was constructed using Eq. (1).

Table 2. The risk parameters and their grading scales

Parameters	Likelihood probability (L)	Resilience (R)	The probability of risk detection (P)	The consequence of severity (CT, CQV, CQ)
Grading scales	Highly unlikely	Very low	Highly unlikely	Very low
	Unlikely	Low	Unlikely	Low
	Likely	Medium	Likely	Medium
	Highly likely	High	Highly likely	High
	Definite	Very high	Definite	Very high

R_k: IF $A_1{}^k$ and $A_2{}^k$ and …. $A_M{}^k$, THEN risk estimate is $\{(\beta_1{}^k, D_1), (\beta_2{}^k, D_2), (\beta_N{}^k, D_N),\}$

$$\left(\sum_{j=1}^{N} \beta_j^k = 1\right), \tag{1}$$

Where $A_i{}^k$ (i = 1,2,…., M) is the linguistics variables of the risk parameters, $\beta_j{}^k$ (i = 1,2,…, N) is a belief degree distribution to the grades used to describe risk, D_j is consequent in the kth packet rule, M is the number of all the possible risk parameters, N is the number of all the possible risk grades and $\sum_{J=1}^{N} \beta_j^k = 1$ satisfies that the Kth packet rule is considered a complete rule base otherwise, it is incomplete. To rationalise the belief degree distribution in the THEN part of a FRB, a weighted approach is used here [28]. For instance, the FRB for modelling the relationship between L, R, P, C and risk is shown in Table 3.

Table 3. The established L, R, P, C FRB with a belief structure for the assessment of FTL risk factors

Rules	Antecedent attribute (input)				Risk result (output)				
No	L	R	P	C	Very low	Low	Medium	High	Very high
1	Highly unlikely	Very low	Highly unlikely	Very low	1				
2	Highly unlikely	Very low	Highly unlikely	Low	0.71	0.29			
…	…	…	…	…	…	…	…	…	…
624	Definite	Very high	Definite	High				0.29	0.71
625	Definite	Very high	Definite	Very high					1

Step 5: Collection of expert judgement on the risk parameters in terms of an investigated FTL. A questionnaire was distributed among the stakeholders engaging in the FTL to seek their opinion and experience on the investigated risk factors based on the six risk parameters assessment grade. The collected data will be used as the priority probability of each risk parameter and used in BN for risk inference [31].

Step 6: Risk inference using BN Technique. The risk inference for the risk assessment is simplified by calculating the marginal probability of the risk status (NR) using Eq. (2).

$$\mathrm{P(R_H)} = \sum\nolimits_{\mathrm{I=5}}^{5} \sum\nolimits_{\mathrm{J=5}}^{5} \sum\nolimits_{\mathrm{K=5}}^{5}, \mathrm{P(R_H|L_i R_j P_k C_l)(P(L_i) P(R_j) P(P_k)} \sum\nolimits_{\mathrm{C}} \mathrm{P(C)}$$
$$\text{where} \sum\nolimits_{\mathrm{c}} \mathrm{P(Cl)} = \sum \mathrm{P(Cl|C_T C_{cqv} C_{cq}) P(C_T) P(C_{cqv}) P(C_{cq})}.$$

$$(\mathrm{h} = 1, 2, 3, 4, 5). \tag{2}$$

Step 7: Prioritization of risk factors with a utility function. The study assigned a linear numerical utility value as defined in the study of Yang et al. [31], to describe the preference degree of the five utility expression. Thus, the new risk priority index value (RPI) is determined using Eq. (3).

$$\mathrm{RPI} = \sum\nolimits_{h=1}^{5} P(\mathrm{Rh})\mathrm{U_{RH}} \tag{3}$$

4 A Real Case Study on FTL Chains Risk Factors

4.1 Case Description

To demonstrate the feasibility and applicability of the proposed model, seven leading food exporters (i.e. 5 for rice and 2 for cocoa) with own large logistics department or professional logistics sub-contractors from the republic of Vietnam into the global market were invited to participant in the study to conduct the FTL chain risk evaluation. A questionnaire was designed to collect data from the company most senior managers, each with more than 15 years of experience in dealing with agro-food export and logistics service into the global markets. The experts' knowledge and experience are described in Table 4.

In the questionnaire, the seven industry experts were asked to evaluate each of the forty-six FTL risk factors with respect to the six risk parameters, in terms of their associated linguistic grades and DoBs. Their feedback data were first checked, to ensure practical and non-biased belief function, then combined by computing their average to produce the failure input value of the six risk parameters. The ranking priority of the 46 significant FTL risk factors is obtained using Eqs. (2) and Eq. (3).

4.2 Result of Factors of the High Likelihood

For all the 46 significant FTL risk factors, fluctuation in the currency exchange rate, followed by humidity monitoring control, global sourcing of networks and a payment

Table 4. Expert knowledge and experience

1	Trading manager	Had worked as a supervisor and line manager in food supply chain industries with more than 20 years of experience
2	General manager	Had more than 20 years' experience with the global transport and logistics of Agri-food products
3	Director	Had more than 20 years in the production handling, storing and transporting food product from Vietnam to the global market
4	Senior sales executive	Had Worked for more than 15 years for the company, with the responsibility to arrange good transport and logistics to complete the sales contract
5	Sales executive	Had more than 15 years' experience in arranging multi-modal transportation for a food product in and out of Vietnam
6	Production manager	Had worked in Agri-food production and logistic unit for more than 20years
7	Logistic manager	Had more than 20 years of experience in dealing with food transport and logistics

delay from the shippers are the top four risk factors of the high likelihood of occurrence. Based on the average opinion of the seven industry experts and before combining them for a final risk ranking, fluctuation in currency exchange has a DoB value of 0% highly unlikely, 12.86% unlikely, 22.86% likely, 24.29% highly unlikely and 40% definite likelihood of occurrence. Similarly, the humidity monitoring control has an average risk value of 78.57% highly likely and 21.43% definite likelihood of occurrence in the FTL chain (Table 5).

Table 5. Factors of the high likelihood

	Fluctuation in the currency exchange rate	Humidity monitoring control	Global sourcing network	Payment delay from the shipper
Highly unlikely	0.00%	0.00%	7.14%	12.86%
Unlikely	12.86%	0.00%	7.14%	8.57%
Likely	22.86%	0.00%	38.57%	45.71%
Highly likely	24.29%	78.57%	30.00%	17.14%
Definite	40.00%	21.43%	17.14%	15.71%

4.3 Result of Factors of the High Probability of Risk Detection

The probability of risk detection is the ability of a stakeholder to detect the risk factor before its occurrence. based on the average expert opinion, deterioration in service quality, leadership in food safety, adaptation to food standard and storage/warehousing are the four top risk factors with a high detection probability (Table 6). Deterioration in service quality is ranked first among the 46 risk factors in the FTL chain with a failure input value of 21.42% likely, 38.57% highly likely and 40% definite probability of detection.

Table 6. High probability of risk detection

	Deterioration in service quality	Leadership in food safety	Adaptation to food standard	Storage and warehousing
Highly unlikely	0.00%	0.00%	0.00%	38.57%
Unlikely	0.00%	0.00%	0.00%	4.29%
Likely	21.43%	12.86%	12.86%	14.29%
Highly likely	38.57%	64.29%	67.14%	27.14%
Definite	40.00%	22.86%	20.00%	15.71%

4.4 Result of Factors of the High Resilience Impact

The resilience impact assesses the ability of a stakeholder in the FTL chain to recover from the consequence outcome of a risk factor or be in a better state than its initial stage after the risk occurs. Low supplier transparency, leadership in food safety, poor food preservation and communication failure among the partners are the top risk factor with a high resilience impact (Table 7).

Table 7. Resilience

	Low supplier transparency	Leadership in food safety	Poor food preservation	Communication failure among partners
Very low	0.00%	0.00%	0.00%	0.00%
Low	0.00%	0.00%	8.57%	2.86%
Medium	00.00%	12.86%	8.57%	25.71%
High	32.86%	20.00%	18.57%	12.86%
Very high	67.14%	67.14%	64.29%	58.57%

4.5 Result of Factors of the High Consequence

The consequence assesses the effect of a risk factor on the stakeholder after its occurrence. Based on a risk factor effect on the delay (time), volume or quantity of the product (quantity) and the quality of the product (quality) on the arrival to its destination, its risk consequence is qualified and evaluated. The survey result reveals that adaptation to food standard, followed by leadership in food safety, employee wages and fire accident are the top risk factors. Adaptation to food standard has failure input value of 78.57% high and 21.43% very high (Table 8). The average responses from the experts show that adaptation to food standard has the highest severity consequence on the product delivery time. Similarly, employee wages, follow by humidity monitoring control, leadership in food safety standard and lack of sufficient handling equipment are the top four risk factors with a significant consequence on the quantity of the products. The employee wages ranked first with a failure input value of 0% very low, 38.57% low, 4.29% medium, 30.00% high, and 27.14% very high (Table 9). The significant risk factors based on the consequence of product quality include low supplier transparency, followed by humidity monitoring control, fire accident and the deterioration in service quality (Table 10). This result shows that the proposed model is capable of presenting a flexible risk results and can simplify the problem of imprecise data information, to support a safety-based decision on a real situation in the FTL chain.

Table 8. Consequence time based

	Adaptation to food standard	Leadership in food safety	Employee wages	Fire accident
Very low	0.00%	0.00%	0.00%	0.00%
Low	0.00%	0.00%	0.00%	0.00%
Medium	0.00%	7.14%	38.57%	40.00%
High	78.57%	72.86%	42.86%	44.29%
Very high	21.43%	20.00%	18.57%	15.71%

Table 9. Consequence quantity based

	Employee wages	Humidity monitoring control	Leadership in food safety	Lack of sufficient handling equipment
Very low	0.00%	0.00%	0.00%	0.00%
Low	38.57%	0.00%	0.00%	45.71%
Medium	4.29%	0.00%	12.86%	11.43%
High	30.00%	78.57%	67.14%	25.71%
Very high	27.14%	21.43%	20.00%	17.14%

4.6 Result of the Final Priority Ranking of Risk Factor

Estimation of all the risk factors in terms of the combined six risk parameters is obtained and transformed into a form of prior probability using Eq. (2), the priority index of

Table 10. The consequence of quality based

	Low supplier transparency	Humidity monitoring control	Fire accident	Deterioration in service quality
Very low	0.00%	0.00%	25.71%	0.00%
Low	0.00%	0.00%	2.86%	0.00%
Medium	20.00%	0.00%	10.00%	1.43%
High	8.57%	32.86%	22.86%	77.14%
Very high	71.43%	67.14%	38.57%	21.43%

each risk factor is obtained using Eq. (3). Findings show that leadership in food safety management, low supplier transparency, deterioration in service quality and adaptation to changes in food standard are the four top priority risk factors in the FTL chain (Table 11). Hence, this risk factor requires more attention with respect to FTL chain risk management.

Table 11. The Four top priority risk factors in FTL chain

Risk factors	RRI values	Ranking
R29. Leadership in food safety management	2140.0569	1st
R42. Low supplier transparency	2103.0957	2nd
R12. Food quality deterioration	1961.326	3rd
R30. Adaptation to changes in food standard and regulation	1842.0904	4th

5 Conclusion

This paper provides experimental evidence on the successful application of the FRB-BN model to evaluate and priorities risks in FTLs using subjective risk data. It also shows that the model is capable of presenting a sensible and flexible risk result in a real-life situation by employing multiple risk parameters with a set of linguistic grades to simplify the failure information in the assessment of the concerned risk in the FTLs. The new findings reveal that 1) the risk factors of different characteristics are evaluated on the same plane for effective overall risk control from a supply chain perspective, 2) in terms of practical implication, the high-level risk factors against each risk parameter and their combination (from an overall risk angel) are identified, which can provide useful insights for transport logistic managers, food risk assessors and the internal/external auditors for effective risk control of the selected FTLs, and 3) the generic methodology can be tailored and applied to model risks of other food supply chains from, as its theoretical contribution. Although the proposed framework does not focus on modelling and evaluating the causal risk mitigation strategies. A further extension of the study could be developed to provide an advanced model capable of evaluating the relevant root causes influencing the hazards(s) on the significant factor and risk mitigation strategies.

Acknowledgements. This research is financially supported by Liverpool John Moores University, UK and EU H2020 MC RISE programme (Grant No. GOLF-777742).

References

1. Benson, A.P.: Communicating risk to consumers in domestic and internationally traded products. Food Control **22**(9), 1529–1534 (2011)
2. Ho, W., Zheng, T., Yildiz, H., Talluri, S.: Supply chain risk management: a literature review. Int. J. Prod. Res. **53**(16), 5031–5069 (2015)
3. Garvey, M.D., Carnovale, S., Yeniyurt, S.: An analytical framework for supply network risk propagation: a Bayesian network approach. Eur. J. Oper. Res. **243**(2), 618–627 (2015)
4. Ornstein, A.T.C., Tavasszy, L.A., van Damme, D.A.: Factors determining distribution structure decisions in logistics: a literature review and research agenda. Transp. Rev. **39**(2), 243–260 (2019)
5. Tang, O., Nurmaya Musa, S.: Identifying risk issues and research advancements in supply chain risk management. Int. J. Prod. Econ. **133**(1), 25–34 (2011)
6. Food and Agriculture Organization of the United Nations and World Trade Organization: "Trade and food standards" (2017)
7. Marvin, H.J.P., Kleter, G.A., Frewer, L.J., Cope, S., Wentholt, M.T.A., Rowe, G.: A working procedure for identifying emerging food safety issues at an early stage: implications for European and international risk management practices. Food Control **20**(4), 345–356 (2009)
8. Van der Fels-Klerx, H.J., Burgers, S.L.G.E., Booij, C.J.H.: Descriptive modelling to predict deoxynivalenol in winter wheat in the Netherlands. Food Addit. Contam. Part A Chem. Anal. Control Expo Risk Assess. **27**(5), 636–643 (2010)
9. Chen, C., Zhang, J., Delaurentis, T.: Quality control in food supply chain management: an analytical model and case study of the adulterated milk incident in China. Int. J. Prod. Econ. **152**, 188–199 (2014)
10. Hsiao, H.I., van der Vorst, J.G.A.J., Kemp, R.G.M., Omta, S.W.F.O.: Developing a decision-making framework for levels of logistics outsourcing in food supply chain networks. Int. J. Phys. Distrib. Logistics. Manag. **40**(5), 395–414 (2010)
11. Faour-Klingbeil, D., Murtada, M., Kuri, V., Todd, E.C.D.: Understanding the routes of contamination of ready-to-eat vegetables in the Middle East. Food Control **62**, 125–133 (2016)
12. Van der Spiegel, M., Sterrenburg, P., Haasnoot, W., Van der Fels-Klerx, H.J.: Towards a decision support system for control of multiple food safety hazards in raw milk production. Trends Food Sci. Technol. **34**(2), 137–145 (2013)
13. Ackerley, N., Sertkaya, A., Lange, R.: Food transportation safety: characterizing risks and controls by use of expert opinion. Food Prot. Trends **30**(4), 212–222 (2010)
14. Septiani, W., Marimin, M., Herdiyeni, Y., Haditjaroko, L.: Risk dependency chain model of the dairy agro-industry supply chain using the fuzzy logic approach. Supply Chain Forum Int. J. **17**(4), 218–230 (2016)
15. Lee, B.H.: Using Bayes belief networks in industrial FMEA modelling and analysis, pp. 7–15 (2002). ieeexplore.ieee.org
16. Xu, K., Tang, L., Xie, M., Ho, S., Zhu, M.: Fuzzy assessment of FMEA for engine systems. Reliab. Eng. Syst. Saf. **75**(1), 17–29 (2002)

17. Liu, J., Yang, J.-B., Wang, J., Sii, H.-S.: Engineering system safety analysis and synthesis using the fuzzy rule-based evidential reasoning approach. Qual. Reliab. Eng. Int. **21**(4), 387–411 (2005)
18. Yang, Z., Bonsall, S., Wang, J.: Fuzzy rule-based bayesian reasoning approach for prioritization of failures in FMEA. IEEE Trans. Reliab **57**(3), 517–528 (2008)
19. Eleye-Datubo, A.G., Wall, A., Wang, J.: Marine and offshore safety assessment by incorporative risk modeling in a fuzzy-Bayesian network of an induced mass assignment paradigm. Risk Anal. **28**(1), 95–112 (2008)
20. Ung, S.-T.: Development of a weighted probabilistic risk assessment method for offshore engineering systems using fuzzy rule-based Bayesian reasoning approach. Ocean Eng. **147**, 268–276 (2018)
21. Hong, I.H., et al.: An RFID application in the food supply chain: a case study of convenience stores in Taiwan. J. Food Eng. **106**(2), 119–126 (2011)
22. Samvedi, A., Jain, V., Chan, F.T.S.: Quantifying risks in a supply chain through integration of fuzzy AHP and fuzzy TOPSIS. Int. J. Prod. Res. **51**(8), 2433–2442 (2013)
23. Meng Tay, K., Peng Lim, C.: Fuzzy FMEA with a guided rules reduction system for prioritization of failures. Int. J. Qual. Reliab. Manag. **23**(8), 1047–1066 (2006)
24. Doa, T.: Vietnam suspends rice exports to the US after pesticide violations - VnExpress International (2016)
25. Zadeh, L.: Fuzzy sets as a basis for a theory of possibility. Fuzzy Sets Syst. **1**(1), 3–28 (1978)
26. Awasthi, A., Chauhan, S.S., Omrani, H.: Application of fuzzy TOPSIS in evaluating sustainable transportation systems. Expert Syst. Appl. **38**(10), 12270–12280 (2011)
27. Bouzembrak, Y., Marvin, H.J.P.: Impact of drivers of change, including climatic factors, on the occurrence of chemical food safety hazards in fruits and vegetables: a Bayesian network approach. Food Control **97**, 67–76 (2019)
28. Alyami, H., Lee, P.T.-W., Yang, Z., Riahi, R., Bonsall, S., Wang, J.: An advanced risk analysis approach for container port safety evaluation. Marit. Policy Manag. **41**(7), 634–650 (2014)
29. Yang, Z., Ng, A.K.Y., Wang, J.: A new risk quantification approach in port facility security assessment. Transp. Res. Part A Policy Pract. **59**, 72–90 (2014)
30. Joshi, R., Banwet, D.K., Shankar, R.: A Delphi-AHP-TOPSIS based benchmarking framework for performance improvement of a cold chain. Expert Syst. Appl. **38**(8), 10170–10182 (2011)
31. Yang, Y.C.: Risk management of Taiwan's maritime supply chain security. Saf. Sci. **49**(3), 382–393 (2011)

Advanced Bayesian Model to Quantify the Adequacy of Organization for Human Reliability: A Maritime Case

Zhuohua Qu[1], Khalifa Mohamed Abujaafar[2], Jin Wang[2], Salman Nazir[3], Steven C. Mallam[3], Eduardo Blanco-Davis[2], Fangbin Guo[2], Hazel Sivori[2], Stephen Symes[2], and Zaili Yang[2](✉)

[1] Liverpool Business School, Liverpool John Moores University, Liverpool, UK
[2] Liverpool Logistics, Offshore and Marine Research Institute, Liverpool John Moores University, Liverpool, UK
Z.Yang@ljmu.ac.uk
[3] Department of Maritime Technology and Innovation, University College of Southeast Norway, Kongsberg, Norway

Abstract. Adequacy of Organisation (AOO) is a critical contributor in Human Reliability Analysis (HRA). In this paper, the human, organisational and technological Performance Influencing Factors (PIFs) and sub-PIFs affecting AOO are identified. A novel hierarchical process is developed, in which PIFs and sub-PIFs are categorised horizontally according to their cause/effect relationships, and vertically according to their functioning levels. It, for the first time, contributes to the state-of-the-art study on the influential factors of AOO in a new structure dealing with their causal relationship and functionality simultaneously. This process underlies the first step of developing a new Bayesian Network (BN) model for quantifying AOO in HRA. A maritime operation case study is investigated to demonstrate the feasibility of the established AOO-BN model, and reveal its applicability in an accident investigation involving human errors.

Keywords: Human reliability analysis · Adequacy of Organisation · Bayesian network · Performance influencing factors

1 Introduction

Focusing on the human contribution to accidents, well-established methods distinguish erroneous human actions made at different levels of an organisation. The generation and evolution of an accident depend on the level of the organisation at which errors are made: errors made at the highest level of the organisation do not necessarily become immediately visible, but remain in a latent state. They propagate and expand throughout the organisation affecting a large number of subsequent decisions and then become apparent at the level of active operation and control. It is, therefore, becoming evident that Adequacy of Organisation (AOO) is a fundamental, but often "hidden", factor influencing human reliability. The current literature however, reveals that most studies on AOO are qualitative in nature. As a result, from a risk modelling perspective, it is often the case that

A. Batako et al. (Eds.): GCMM 2021, LNNS 335, pp. 560–576, 2022.
https://doi.org/10.1007/978-3-030-90532-3_42

the analysts cannot easily rationalise the resources to address the deficiencies without a clear understanding of the causal relationship among the errors at different levels, and the extent to which they influence each other in a quantitative manner. This paper aims to develop a new Bayesian network (BN) mode to quantify AOO for human reliability.

This paper first identifies and evaluates the human, organisational and technological Performance Influencing Factors (PIFs) affecting AOO, and tests the PIFs, within the context of maritime operations. A hierarchical process is then developed to group PIFs horizontally according to their cause-effect relationships, and categorise them vertically based on their functioning levels in Sect. 2. The findings in the form of qualitative description, are then used to develop a new BN model for the new attempt of quantitative AOO assessment in Sect. 3. A real case study scenario is developed to demonstrate the feasibility of the new BN model in quantifying AOO within the context of maritime operation in Sect. 4. In the same section, based on the response of the model to different sets of evidence, useful implications are generated to guide the improvement of the AOO in marine operations before the conclusion in Sect. 5.

2 Identification and Analysis of PIFs and Sub-PIFs of AOO in Maritime Operations

The identification of PIFs and sub-PIFs is carried out with reference to the literature available on the research of HRA from the nuclear, aviation, health, and chemical industries. To adapt PIFs and sub-PIFs in maritime operations, a hierarchical process is developed to classify the identified PIFs and sub-PIFs in four categories with consultation from three marine chief engineers (each of over ten year working experience at sea). They are designated from the top-down respectively as root cause, sub-condition, condition and functional PIFs. Horizontally, the PIFs and sub-PIFs are classified into three groups, specified from left to right as process, culture, and resources (see Table 1). PIFs and sub-PIFs' relationship as children-parents, descendants-ancestors, can be identified intuitively from their definitions and descriptions in Table 2.

Table 1. AOO PIFs and sub PIFs hierarchical process

PIFs cat	Process	Culture	Resources
Root cause	Management commitment Crew involvement Proactive control measures Reactive control measure Standards Policy Communication Strategy and objectives Maintenance procedures Contingency procedures Operational procedures	Management commitment Crew involvement	Proactive control measures Reactive control measure Availability of equipment and records Quality of equipment and records Availability of condition monitoring and controls Trust of condition monitoring and controls Skill of human resources Knowledge of human resources Motivation of human resources

(continued)

Table 1. (*continued*)

PIFs cat	Process	Culture	Resources
Sub-condition	Controls Safety culture Norm		Controls
Condition	Human resources Organisational culture Management quality Organisational structure Safety management system (SMS)	Safety culture Norm. Organisational structure Operational processes	Operational processes Organisational structure SMS Equipment and records Condition monitoring and performance control Human resources
Functional	Operational processes	Organisational culture	Resources management
End-point effect	Adequacy of organisation		

Table 2. The definitions and descriptions of the PIFs and sub-PIFs

PIFs and sub-PIFs	Definitions	Descriptions	References
Operational process	Operational processes constitute the core functions of an effective and efficient system operation. The competences of operators or supervisors in controlling marine engineering operations are affected by working contexts	The operational process effect described by two states [effective, ineffective] is directly influenced by the state levels of SMS, organisational structure, and management quality	[1, 2]
Management quality	Management quality reflects the follow-up and support of managerial process tools such as standards, polices, and controls used to govern the effectiveness and efficiency of the operational process in correlation with the SMS and organisational structure	Management quality effect [high and low] is directly influenced by the discrete state levels of variable standards deployed, policy followed, and proactive and reactive control measures implemented	[1–3]

(*continued*)

Table 2. (*continued*)

PIFs and sub-PIFs	Definitions	Descriptions	References
Resources management	Resource management [effective and ineffective] is related to the efficient and effective deployment of an organisation's resources when they are needed	Such resources may include human skills, knowledge and motivation, availability and trust of condition monitoring and performance control technology utilised in system operational processes, and quality and availability of equipment and records' inventory	[4]
Organisational culture	Organisational culture [acquired and not acquired] creates the climate that influences the issue as to how well people can communicate, plan and make decisions	Organisational culture consists of norms, values and unwritten rules of conduct as well as management styles, priorities, beliefs and interpersonal behaviours that prevail	[1, 5]
Organisational structure	Organisational structure [effective and ineffective] is the mapping of the formal and informal consistency of operational processes	Formal consistency lays down detailed policies, procedures, and standing orders, so that everybody is aware of his/her duties and obligations towards the effective and ineffective operational processes. Informal organisation has group goals, social roles to play, leader-follower relationship, unwritten behaviour patterns, and a code of conduct. It enhances the effectiveness of organisational communication channels	[1–3]
Norm	Norms are the acceptable standards of behaviour that are shared by a group of members	Norms [committed and not committed] act as a means of influencing the behaviour of its members with a minimum of external control	[1, 2]

(*continued*)

Table 2. (*continued*)

PIFs and sub-PIFs	Definitions	Descriptions	References
Safety culture	The safety culture [acquired and not acquired] is the product of individual and group values, attitudes, perceptions, competencies, and patterns of behaviour that determine the commitment to, and the style and proficiency of, an organisation's health and safety management	Safety culture is an evaluative concept that includes criteria for the operation of sound safety-critical systems	[1, 3, 6–10]
Safety management system	SMS is a comprehensive control of safety and includes the management of methods, procedures and people	It emphasises the role of management as a body that controls and takes charge of safety. The management is responsible for setting goals, providing resources and supervising implementation	[6, 8, 11–13]
Condition monitoring & Performance control	Condition monitoring and performance control [reliable and not reliable] as sophisticated automation systems appear everywhere in marine engineering operations	Poor partnerships between people and automation will become increasingly costly and catastrophic. Such flawed partnerships can be described in terms of misuse and disuse of automation	[14, 15]
Standards	Standards are the acknowledged measure of comparison in a form of rules, regulations and guidelines established by recognised bodies such as classification societies, legislative organisation and system manufacturers	Two states [adopted, not adopted]	[1–3]

(*continued*)

Table 2. (*continued*)

PIFs and sub-PIFs	Definitions	Descriptions	References
Management commitment	Organisational commitment [high and low] to safety refers to the degree to which an organisation's management prioritises safety in decision making, and allocates adequate resources to safety management	Organisational commitment refers to the degree to which those directly involved in supporting work or the supervision of crews are actually committed to safety and reinforce the safety values espoused by upper management	[16]
Human resources	Human resources reside in the knowledge, skill, and motivation of humans, who are under the right conditions, to learn and grow better with age and experience, which no other resources can do	Human resources are the scarcest and most crucially productive resource that creates the largest and longest-lasting advantage for an organisation	[4, 17]
Controls	Controls are a set of proactive and reactive measures and procedures performed by required control bodies	Control effect [effective and ineffective] is directly influenced by the discrete state levels of variables' proactive control measures and reactive control measures	[1–3]
Policies	Policies are mandatory general guidelines adapted from standards to be implemented	Policies [clear and unclear] are required to be considered by operators and supervisors while performing their managerial, operational, maintenance, and contingency tasks	[1–3]
Availability	Availability is determined by reliability (the probability of the system effectively working) and maintainability (the ability to restore the system to service)	Availability [high and low] is the measure of the time at which the system is in an operational state compared to the total life time	[18]

(*continued*)

Table 2. (*continued*)

PIFs and sub-PIFs	Definitions	Descriptions	References
Equipment & records	Equipment [sufficient and insufficient] is the tangible properties characterised by a durable nature which is useful in the operation of systems. Records are documents that memorialise and provide objective evidence of activities performed, events occurred, results achieved, or statements made	The equipment and records of a maritime organisation are directly influenced by their quality and availability	[2]
Operational procedure	Procedures are written instructions as to how tasks are to be done, what they involve, and the sequence to be undertaken	Procedures [performed and not performed] are typically designed in accordance with the constraints and characteristics of the operational process to be controlled	[2, 19]
Communication	Communication [effective and ineffective] is a dynamic force that shapes an organisational behaviour	Communication shifts information of value acquired from within, and between the organisation's environment to various departments, crew and individuals	[2]
Trust	Trust [high and low] can be defined as the attitude that an agent will help achieve an individual's goals in a situation characterised by uncertainty and vulnerability	Trust is a meaningful concept to describe human-automation interaction in both naturalistic and laboratory settings	[14, 20–25]
Proactive control measures	Proactive control measures are dedicated to prevent, detect, protect, recover and contain the events that may combine to initiate an accident	These measures are exploited for probabilistic safety assessment, safety analysis of maximum credible accidents, design of standard and emergency procedures, design of decision support tools, and operator training	[26, 27]

(*continued*)

Table 2. (*continued*)

PIFs and sub-PIFs	Definitions	Descriptions	References
Reactive control measures	Reactive control measures are used to learn the lessons of past experience and to develop appropriate feedback mechanisms	They are exploited for PSA, safety analysis of maximum credible accidents, design of standard and emergency procedures, design of decision support tools, and operator training	[26, 27]
Quality	Quality [high and low] of equipment and records is a measure of fineness or state of being free from defects, deficiencies and significant variations	A standardised definition of quality is described as the totality of features and characteristics of a product or service that bears its ability to satisfy stated or implied needs	[2]
Skill	Skill is the ability and capacity acquired through deliberate, systematic, and sustained effort to smoothly and adaptively carry out complex activities or job functions	The activities associated with skill involve ideas (cognitive skills), things (technical skills), and/or people (interpersonal skill)	[1, 2, 28, 29]
Strategy and objectives	Effective strategic [clear and unclear] management creates a productive alliance between an organisation's culture and the resources it has at its disposal. An organisation's goals [clear and unclear] may be pursued in accordance with organisational principles, which are based on employee beliefs, values and attitudes	Organisational principle determines the culture of an organisation and provides a set of standards that govern the overall conduct of an organisation. Clearly defined and agreed objectives are the first stage in the design of an organisation's structure. The success of a maritime organisation is measured by the progress of its employees towards goals set via such objectives	[30–32]

(*continued*)

Table 2. (*continued*)

PIFs and sub-PIFs	Definitions	Descriptions	References
Motivation	Motivation [high and low] is the internal and external factors that stimulate desire and energy in humans to be continually interested in and committed to a job, role or subject, and to exert persistent effort in attaining a goal	Motivation results from the interactions among conscious and unconscious factors such as the intensity of desire or need, incentive or reward value of the goal, and expectations of the individual	[1–3]
Knowledge	Knowledge [good and poor] in engineering and science can be defined as a body of justified true beliefs (such as laws, models, objects, concepts, know-how, processes, and principles) acquired by human beings about a system of interest	The three knowledge categories of cognitive knowledge, correct reasoning from hypotheses, and intellectual or emotional acceptance of a proposition constitute the human cognition of knowledge that might be different from a future state of knowledge achieved by an evolutionary process	[33]

3 Methodology of Establishing the AOO-BN in Maritime Operations

3.1 Process of Building the AOO-BN Model

The construction of BNs consists of two main parts, qualitative and quantitative networks. The qualitative network refers to the model of causal relationship among the variables and definition of the variable states/grades. In this study, it has been analysed and descried based on the literature review (see Table 2). Such relationship needs to be verified by a d-separation approach [34] to ensure that the model's integrity complies with BN characteristics, matching a perception of a real world, and showing how mutual information is conditionally propagated among model variables [35]. The quantitative network means the assignment of prior probabilities to each variable in the network, including unconditional (to root nodes) and conditional probabilities (to the non-root nodes). Probability values are usually estimated based on certain sources of evidence such as empirical data, expert judgements, and/or literature review. The prior probabilities of all root, sub-condition, condition, functional, and effect nodes are assigned subjectively due to the lack of objectively measured data. Further investigation could be done to measure states' probability based on empirical data and structured surveys in order to reduce data subjectivity. The unconditional probabilities of all root cause nodes of the generic AOO-BN model are equally disrupted on each node state to show the generic case (unknown situation). The aforementioned three marine engineers were interviewed to provide their subjective elicitation of conditional probabilities of non-root nodes of binary states. Elicited conditional probabilities (CPs) are aggregated and normalised by using Eq. 1 [36], to develop the conditional probability tables (CPTs) that are used to establish the AOO-BN model.

$$\mathrm{CP} = \frac{\sum_{e=1}^{n} e(x)}{n} \tag{1}$$

where, CP is the aggregated and normalised value of conditional probabilities elicited by n experts, e(x) is the conditional probability given by the eth expert. Verbal Elicitor (VE) developed by [37] is used in support of the qualitative elicitation of unconditional probabilities of leaf nodes, as listed in Table 3.

Table 3. VE cue words and their associated probabilities

Cue word	Probability
Certain	100%
Probable	90%
Expected	70%
Fifty-fifty	50%
Uncertain	30%
Improbable	15%
Impossible	00%

3.2 CPT Data and Finding of the AOO-BN with Prior Probabilities

Using the VE approach, the CPTs of all the non-root nodes in the AOO-BN are obtained. For instance, Tables 4 and 5 are associated with the CPTs of the nodes of AOO and operational process, respectively. As a result, the AOO-BN with prior probabilities is developed and shown in Fig. 1. The structure of this BN diagram encodes the perception that AOO is affected by the nodes of operational process, organisational culture, and resource management. These, in turn, are affected by the causes of the condition/sub-condition nodes depicted in the model. The other relationships represented by the model can be read similarly. When applying the AOO-BN model for reasoning, the impact of changing any variable based on the real observation for a given scenario is transmitted right through the network following the relationships expressed by the CPTs. Changes in any node simply arise from the combined effect of changes in all the nodes linked to it either directly or indirectly. In formal terms, the BN encodes a joint probability distribution over all the nodes. Every time the state of a node changes, the joint distribution is updated through the iterative application of Bayes' theorem [34]. Changes in the BN are observed as changes in the chance that a node is in a particular state. Due to the uncertainty in the CPTs, it is rare for a node to be definitely in one state or another; it is far more common for probability distributions across all the states of a node to be observed [38].

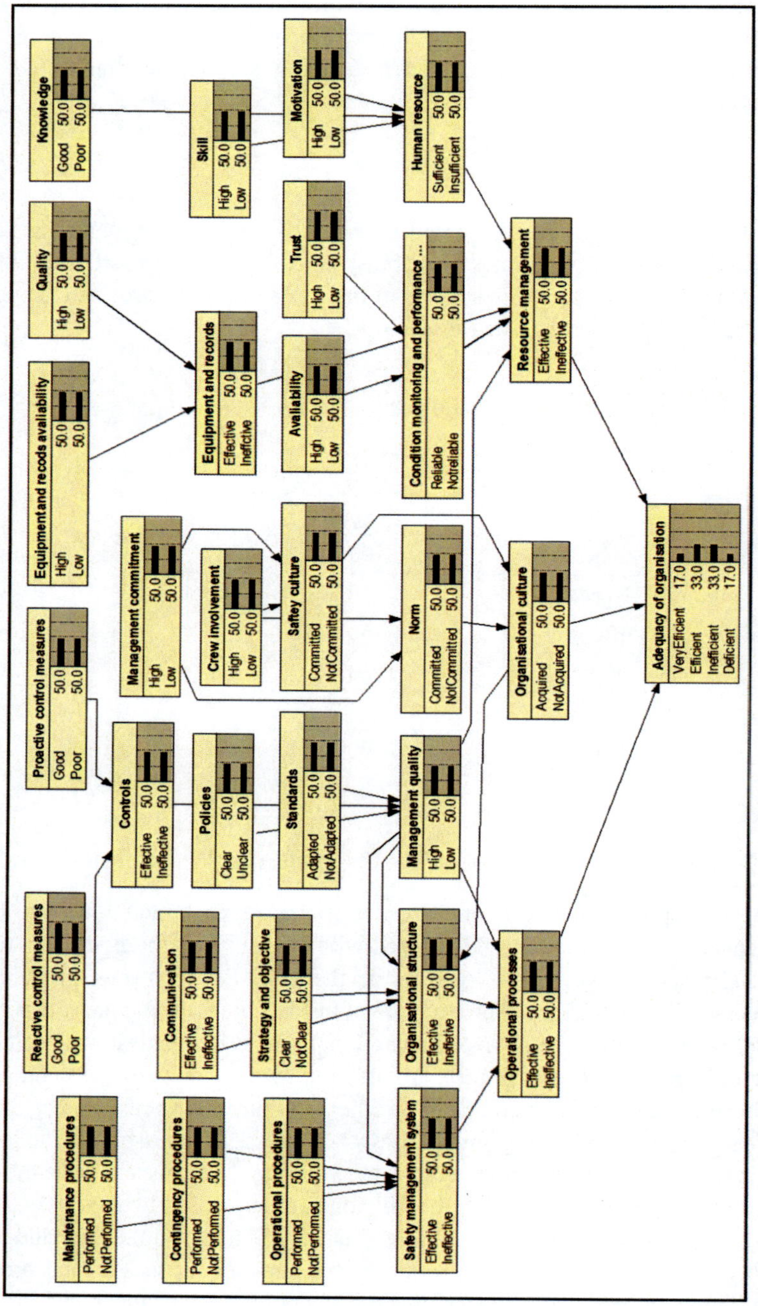

Fig. 1. The generic AOO-BN model

Table 4. AOO CPT

AOO CPT (%)								
Operational processes	Effective				Ineffective			
Organisational culture	Acquired		Not acquired		Acquired		Not acquired	
Resources management	Effective	Ineffective	Effective	Ineffective	Effective	Ineffective	Effective	Ineffective
Very efficient	100	10	10	0	0	0	0	0
Efficient	0	90	85	45	55	5	0	0
Inefficient	0	0	5	55	45	85	90	0
Deficient	0	0	0	0	0	10	10	100

Table 5. Operational processes CPT

Operational processes CPT (%)								
Safety management system	Effective				Ineffective			
Organisational structure	Effective		Ineffective		Effective		Ineffective	
Management quality	High	Low	High	Low	High	Low	High	Low
Effective	100	80	75	20	80	25	20	0
Ineffective	0	20	25	80	20	75	80	100

4 Case Study: The Grounding Accident of the MV HANJIN DAMPIER

4.1 Case Description and Preliminary Analysis

The selection of the MV HANJIN DAMPIER grounding accident as the case study is because the error occurrence was found to be due to inadequacy of organisation including 1) the influence of the management exchanges between the Master, Pilot and Chief Engineer which inhibited competent vessel control, 2) the need for an efficient organisational structure and effective SMS to ensure significant levels of crew competence control, and 3) the vital need for the effective communication of the current risks that would shape the safest departure process.

Based on the Australian Transport Safety Bureau [39], organisationally the Chief Engineer failed to communicate the seriousness of the generator problem to the Master, which directly contributed to the grounding incident. At 1032 on 25 August 2002, the Korean flag bulk carrier "HANJIN DAMPIER" departed from the port of Dampier, Western Australia. A pilot was overseeing the navigation of the ship, which was loaded with iron ore and had a displacement of 233,158 tonnes, with draughts of 17.94 m forward and 18.10 m aft. At 1127, two of the ship's three main generators stopped, leaving only one generator running and connected to the main switchboard. At 1152, the third generator's circuit breaker tripped open. With the total loss of power to the main switchboard the main engine stopped, and the ship lost propulsion. The rudder had stopped at 10° to starboard, which prevented effective steering control of the vessel. As the ship slowed, it started to turn to starboard towards shallow water. The emergency generator failed to start automatically and, as a result, steering was not restored for some four minutes. At 1202, "HANJIN DAMPIER" touched bottom. The ship suffered only minor damage to the bottom shell plating, but the consequences of this incident could have been a lot worse.

In light of the above description and ATSB report [39], the following facts (observations) are derived with respect to the root cause variables states. The ineffective communication between the Chief Engineer and the Master resulted in low crew involvement that led to contingency procedures not being performed. Such an action could be manifested in low crew skill and knowledge capability, which had affected their motivation and stimulated their low trust in interaction with the functionality of control systems. As a result, the crew was not able to infer the main reasons for the tripping of the main power supply circuit breakers. Indeed such uncertainty would reduce the crew's competence and therefore lessen the ability in restoring the power supply system availability. Moreover, the situation was aggravated by the failure of the emergency generator starting system, due to a faulty battery. Also, the duty engineers' failure to drain the settled water in the diesel oil settling and service tanks daily, and to detect the low performance of purification equipment, is considered as nonconformity with the operational procedure. Furthermore, the badly installed gravity disc and faulty starting battery for the emergency generator indicated that maintenance procedures were not followed properly. These conditions reasonably reflect the possibility of unclear maintenance policy aggravated by low management commitment and low crew involvement. Maintenance and operational procedures were not followed due to unclear operational and maintenance policies, which are commonly caused by non-adaptable standards, and the lack of proactive control measures. As a result, the performance of the purification system and the emergency generator starting system had deteriorated.

4.2 Implication Analysis Using the AOO-BN

Based on the preliminary analysis, the prior probability assessment are provided as input observations into the generic AOO-BN model shown in Fig. 2. As a result, the marginal probabilities of AOO in light of new evidence changes. In this process, the new observation probabilities assigned to the prior probabilities of the leaf nodes' crew involvement, availability, contingency procedures, operational procedures, communication, trust, maintenance procedures, skill and knowledge, are based on the probable cue word of VE and its associated probability value in Table 3. On the other hand, the root cause nodes' management commitment, standards, policies, equipment and records availability, proactive control measures, reactive control measures, quality and motivation, are left with no change, as they are not observed or verified in the accident analysis.

Model readout posterior probabilities of the condition, sub-condition, functional nodes, and end-point effect node (AOO) are conditionally inferred based on their associated CPTs and the prior probabilities of their respective influencing parent nodes and the assigned prior probability of root cause nodes. Consequently, the BN model end-point effect node (AOO) inferred marginal probabilities are 6.33% very efficient, 20.9% efficient, 38.2% inefficient and 34.4% deficient. Accordingly, the model inferred quantitative belief certainty on inefficient and deficient states compatible with the qualitative finding of the incident analysis. The new marginal probabilities of AOO are sensitive to the assigned probable observations of nine leaf nodes and the likelihood findings of the other nine root cause nodes.

Deductively, a prognosis scenario as a proactive measure to formulate plans for eliminating the possibility of the occurrence of such accidents in the future could be developed. It is important to note that the AOO-BN serves as an interactive tool that

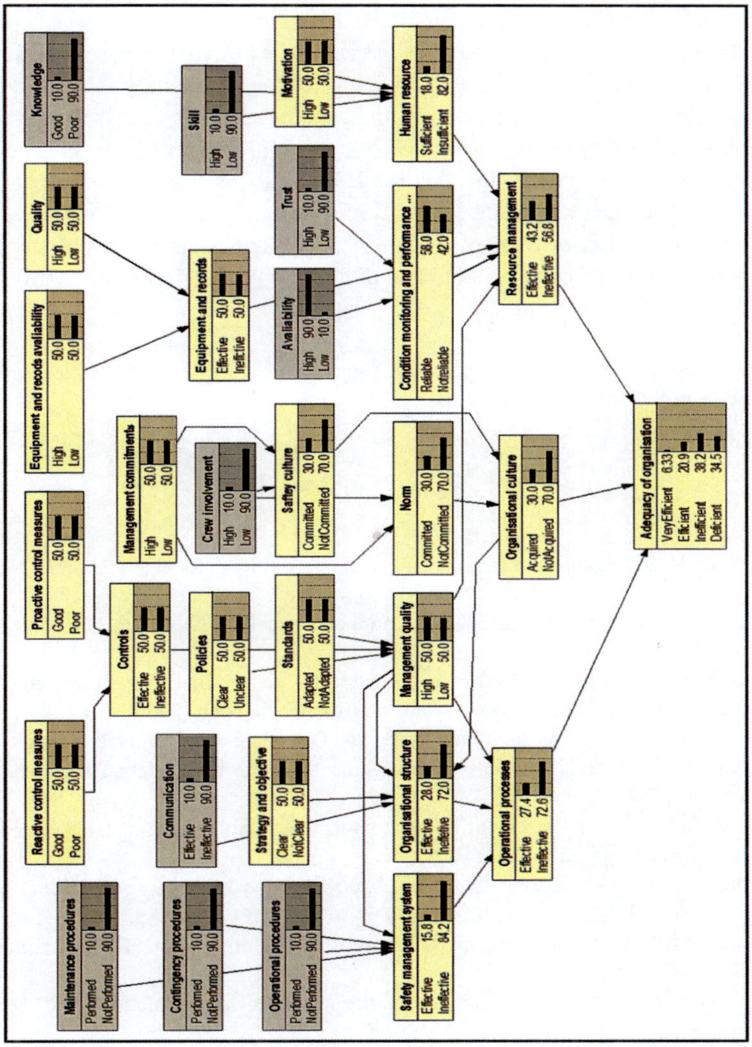

Fig. 2. The Grounding of "MV HANJIN DAMPIER" case analysis based on the AOO-BN model

enables experts to create a probabilistic model, simulate scenarios, and reflect on the results of the inference.

5 Conclusion

This paper first identifies all the relevant PIFs and sub-PIFs influencing AOO in HRA through a literature review. Secondly, it brings new contributions for categorising the identified factors into different groups based on their causal relationship and functionality simultaneously by a rigorous contextual analysis. The outcomes from the first two steps contribute to the development of a pioneering AOO-BN model to quantify the extent of AOO in maritime operations. In this process, both d-separation and VE approaches are used to aid the qualitative and quantitative network development respectively. As a result, the application of the AOO-BN in the HANJIN DAMPIER case reveals that the model can help effectively conduct 1) a backward diagnosis analysis for the accident investigation and 2) a forward prediction analysis for accident prevention in the future when the real observations becomes available in an initial stage of an incident occurrence. It can therefore provide useful insights for HRA analysis focusing on organisational factors in maritime operation accident analysis in specific and HRA in other sectors in general. However as the first attempt to quantify AOO, an objective of strong subjectivity, the binary definitions of each variable and their prior probabilities at a 50:50 ratio are subject to further investigation.

Acknowledgements. This work is partially supported by EU H2020 Marie Curie RISE ENHANCE project (MSCA-RISE 823904).

References

1. Kondalker, V.G.: Organisational Behaviour. New Age International (P) Limited, New Delhi (2007)
2. Mullins, L.J.: Management and Organisational Behaviour. Pearson Education Limited, Harlow (2005)
3. Schein, E.H.: Organizational Culture and Leadership. Jossey-Bass, San Francisco (1985)
4. Kelly, A.: Managing Maintenance Resources. Elsevier, London (2006)
5. Sinn, L.: Corporate culture. In: Management and Organizations. Hund publishing (1991)
6. HSE: Successful Health and Safety Management. Health and Safety Executive, HMSO, London, UK (1997)
7. Booth, R.T., Lee, T.R.: The role of human factors and safety culture in safety management. Eng. Manufact. **209**(5), 393–400 (1995)
8. Kuusisto, A.: Safety Management Systems. Audit Tools and Reliability of Auditing. Technical Research Centre VTT, Finland (2000)
9. Clarke, S.: The contemporary workforce implications for organisational safety culture. Pers. Rev. **32**(1), 40–57 (2003)
10. Lee, T., Harrison, K.: Assessing safety culture in nuclear power stations. Saf. Sci. **34**(1), 61–97 (2000)
11. Hale, A.R., Heming, B.H.J., Carthey, J., Kirwan, B.: Modelling of safety management systems. Saf. Sci. **26**(1), 121–140 (1997)

12. Hale, A.R., Baram, M.: Safety Management. The Challenge of Change. Elsevier, Oxford (1998)
13. EPSC: Safety Performance Measurement. European Process Safety Centre, UK (1996)
14. Parasuraman, R., Riley, V.: Humans and automation: use, misuse, disuse, abuse. Hum. Factors **39**(1), 230–253 (1997)
15. Janssen, C., Donker, S., Brumby, D., Kun, A.: History and future of human-automation interaction. Int. J. Hum. Comput. Stud. **131**, 99–107 (2019)
16. Thaden, T.L., Gibbons, A.M.: The Safety Culture Indicator Scale Measurement System. Federal Aviation Administration, USA (2008)
17. McGregor, D.: The Human Side of Enterprise. Penguin, New York (1987)
18. Benbow, D.W., Broome, H.W.: The Certified Reliability Engineering Hand Book. ASQ, Milwaukee (2008)
19. Dien, Y.: Safety and application of procedures in nuclear power plants. Saf. Sci. **29**(1), 179–187 (1998)
20. Muir, B., Moray, N.: Trust in automation: experimental studies of trust and human intervention in a process control simulation. Ergonomics **39**(1), 429–460 (1996)
21. Lee, J.D., Moray, N.: Trust control strategies and allocation of function in human machine systems. Ergonomics **35**(1), 1243–1270 (1992)
22. Halprin, S., Johnson, E., Thornburry, J.: Cognitive reliability in manned systems. IEEE Trans. Reliab. **22**(1), 165–169 (1973)
23. Lewandowsky, S., Mundy, M., Tan, G.: The dynamics of trust: comparing humans to automation. Exp. Psychol. Appl. **6**(1), 104–123 (2000)
24. Gausdal, A.H., Makarova, J.: Trust and safety onboard. WMU J. Marit. Aff. 1–21 (2017). https://doi.org/10.1007/s13437-017-0126-z
25. Kim, T., Gausdal, A.: Leading for safety: a weighted safety leadership model in shipping. Reliab. Eng. Syst. Saf. **165**, 458–466 (2017)
26. Maurino, D.E., Reason, J., Johnston, N., Lee, R.B.: Beyond Aviation Human Factors. Avebury Aviation, Aldershot (1995)
27. Vervloesem, W.: Ship Survey and Audit Companion, a Practical Guide. The Nautical Institute, London (2000)
28. Flin, R.: CRM (nontechnical) skills: a European perspective. In: Crew Resource Management, pp.185–206 (2019)
29. Flin, R., Maran, N.: Basic concepts for crew resource management and non-technical skills. Best Pract. Res. Clin. Anaesthesiol. **29**(1), 27–39 (2015)
30. Brown, A.: Organizational culture: the key to effective leadership and organizational development. Leadersh. Org. Dev. J. **13**(2), 3–6 (1992)
31. Richardson, B., Thompson, J.: Strategic Competency in the 1990s'. Administrator **1**(1), 2–3 (1994)
32. Tilles, S.: Making strategy explicit. In: Business Strategy. Penguin, New York (1969)
33. Ayyub, B.M., Klir, G.J.: Uncertainty Modelling and Analysis in Engineering and the Sciences. Taylor & Francis, New York (2006)
34. Jensen, F.V.: Bayesian Networks and Decision Graphs. Spring-Verlag, New York (2001). https://doi.org/10.1007/978-1-4757-3502-4
35. Jensen, F.V.: An Introduction to Bayesian Networks. University College London Press, London (1996)
36. Klir, G.R., Yuan, B.: Fuzzy Sets and Fuzzy Logic: Theory and Applications. Prentice-Hall, Upper Saddle River (1995)
37. Hope, L., Nicholson, A., Korb, K.: Knowledge Engineering Tools for Probability Elicitation. Technical report, School of Computer Science and Software Engineering, Monash University (2002)

38. Cain, J.: Guidelines for using Bayesian Networks to Support the Planning and Management of Development Programmes in the Water Sector and Beyond. The Centre for Ecology & Hydrology, UK (2001)
39. ATSB (2003). https://www.atsb.gov.au/publications/investigation_reports/2002/mair/mair184/. Accessed 28 Dec 2003

Fabrication and Applications of Antibacterial Surfaces and Nano Biosensing Platforms

Alka Jaggessar(✉), Jafar Hasan, and Prasad K. D. V. Yarlagadda

School of Mechanical, Medical and Process Engineering, Queensland University of Technology, Brisbane, QLD 4000, Australia
{alka.jaggessar,y.prasad}@qut.edu.au

Abstract. The development of antibiotic resistant superbugs poses significant threat to global healthcare. With the overuse of antibiotic medication over time, these superbugs have evolved to fight the effects of pharmacological therapies causing death to patients with easily treatable infections. For this reason, research has shifted towards investigating ways in which infection can be cured without antibiotic medication. Research has found that some nanostructured materials have been found to exhibit antibacterial behaviours either through antibiofouling or bactericidal (killing) mechanisms. These nanostructures have an inherent property which repels or kills bacteria upon physical contact, without the need for chemical therapy. This review gives a perspective on the current state of research which uses nanotechnology to fabricate antibacterial surfaces. In addition, this review presents various biosensing techniques for human biosensing and bacteria sensing, particularly for food and water borne pathogens. This review also discusses future directions for ways in which these two research areas can be combined to develop a nanosensing platform which detects bacteria in hospital environments, helping to reduce waiting times between sampling and detection, length of patient hospital stays, patient discomfort and burdens to the healthcare system, as well as control and contain the spread of infection in hospital environments.

Keywords: Bactericidal surfaces · Antibiofouling surfaces · Bacteria detection · Nanosensing · Biosensing · Electrochemical anodization

1 Introduction

In recent times, applications of nanotechnology have increased exponentially, revolutionising research in a number of industries, such as disease control, drug delivery, food industries, electronics and advanced manufacturing, to name a few. Nanostructured and nanotextured surfaces in particular have gained attention for their use in biosensing and antibacterial surfaces.

In recent times, nanopillared surfaces, fabricated using various techniques and materials have exhibited antibacterial properties against various bacteria strains [1–6]. These studies aim to replicate the natural nano and microstructure of some insect wings, plant

A. Batako et al. (Eds.): GCMM 2021, LNNS 335, pp. 577–588, 2022.
https://doi.org/10.1007/978-3-030-90532-3_58

leaves and animal skin, which results in antibacterial behaviour [7–10]. These surfaces are being used for a wide range of potential applications, from preventing biofilm formation on medical implants to biofouling reduction on ship hulls.

Since their conception, biosensors have evolved to address and improve sensing techniques, accuracy and application. Biosensors are usually used for detecting DNA of bacteria or viruses, or proteins generated for the immune system of infected or contaminated living organisms [11]. Biosensors have many applications ranging from exercise-based sensors for monitoring heart rate and calories to glucose testing using human tears.

This review presents recent research, fabrication methods and applications of antibacterial surfaces and biosensing platforms which employ nanotechnology and nanostructured surfaces. Suggestions for merging these two concepts to form duel-functional antibacterial surfaces with biosensing capabilities as a future direction of study are also presented.

2 Antibacterial Nanostructures

Antibacterial surfaces, or surfaces that prevent bacterial adhesion are classified as either bactericidal or anti-biofouling surfaces. Antibiofouling surfaces repel and prevent cell attachment due to surface chemistry or unfavourable surface topography, whereas bactericidal surfaces disrupt the cell, causing death [5, 12]. Antibacterial microstructures are a promising strategy for biomaterials that repel microbes or kill them in surrounding areas [13].

There are several species that exhibit antibiofouling properties through micro and nanoscale structures. These structures create superhydrophobic properties, thus reducing biofilm formation and bacterial infection. Species such as lotus and taro leaves exhibit superhydrophobic properties due to nano and micro-scale hierarchal structures that result in a self-cleaning effect [14–17]. In addition, species such as cicada and butterfly wings are active in killing bacteria upon contact, due to the nanotexture present [4, 7, 10, 18–22]. The purpose of replicating these phenomena is to eliminate difficult-to-treat infections, due biofilm-forming microorganisms' increasing resistance to antimicrobial therapies [13].

2.1 Anti-biofouling Surfaces

Biofouling poses a significant problem in many applications, leading to the build-up of bacteria and viruses on a surface. Anti-biofouling implies the ability to prevent accumulation of microorganisms (such as bacteria and viruses) on wetted surfaces [23]. Superhydrophobic (contact angle above 150°) anti-biofouling surfaces are self-cleaning in nature [24], and repel bacterial adhesion by allowing dirt and contaminants to freely roll off surfaces with water [25]. Similarly, superhydrophilic surfaces (contact angle below 15°) with a high degree of hydration, increase the energetic potential for water removal by microorganisms when they try to attach to surfaces [23]. Anti-biofouling surfaces are effective at repelling bacterial adhesion for strains such as *Staphylococcus aureus* (*S. aureus*) [26–29], *Escherichia coli* (*E. coli*) [29, 30], *Staphylococcus mutans* [31] and *Pseudomonas aeruginosa* [26, 28].

Anti-biofouling surfaces have an exceptionally large number of applications from marine applications to implant devices. Superhydrophobic nanomaterials designed to mimic the naturally occurring nano and microstructure of shark skin, have been found to reduce fluid drag in water, air, and oil. These surfaces have application in reducing drag for airplanes, wind turbines, ship hulls (also reducing the attachment of mussels, barnacles and algae [32]) and pipelines [33]. The mechanism of drag reducing properties of these surfaces comes from anisotropic flow, which facilities efficient fluid movement over the surfaces. This minimizes movement of fluid molecules, thus reducing energy losses, leading to lower drag. The combination of liquid repellence from superhydrophobic nanostructures with anisotropic flow results in drag reducing, self-cleaning and antifouling properties [33].

Superhydrophilic coatings have been used for anti-fogging applications to prevent condensation of water on surfaces. These coatings are often applied to glass or plastic to improve their optical characteristics. Intricate nanostructured films fabricated by layer-by-layer assembly of rice hush ash and chitosan nanosilica have shown excellent anti-fogging and anti-biofouling properties applicable for industrial applications [23].

Anti-biofouling surfaces are particularly important for medical devices such as catheters, drug delivery devices [34], stents and grafts, and sutures. The insertion of these artificial materials in the body triggers a foreign response, which results in biofouling. This biofouling can lead to invasive revision procedures and limited device lifetime. Next generation implants are designed to include hydrophobic and hydrophilic materials, which reduce the amount of biofouling [35].

Antifouling properties were exhibited by graphene nanowells (vertically orientated graphene nanosheets) using methane and plasma enabled synthesis techniques. When exposed to *S. aureus* and *E. coli*, both materials exhibited antifouling behaviour. Gram-positive *S. aureus* attachment was lower when exposed to sugarcane bagasse derived graphene nanowells, compared to methane derived, and copper (control surface). Gram-negative *E. coli* attachment was lowered in comparative amounts by both materials compared to copper [36] as seen in Fig. 1.

Hydrophobic and hydrophilic nanostructured surfaces are fabricated using a number of methods, such as layer-by-layer assembly [23, 30], micro-moulding [37, 38], vacuum casting [39], reactive ion etching [40], femtosecond lasering [26], lithography [41–45], chemical vapor deposition [29] and surface functionalisation [27].

2.2 Bactericidal Fabrication

In recent times, there has been heavy research focus towards using inspiration from nature to fabricate surfaces which inhibit the development of bacterial biofilms. Researchers are attempting to bio-mimic the naturally occurring nanostructure found on cicada and dragonfly wings bactericidal surfaces that kill bacteria upon contact. It is postulated that the physical topography of these nanostructured surfaces cause the death of some bacteria strains by stretching and disfiguring the cell wall, thus leading to bacteria death [7, 22]. Studies have found that there is correlation between stretching of the cell wall and geometric parameters of nanotextured surfaces. Geometric parameters found to influence nanostructure bactericidal performance include aspect ratio, structure density [8], structure height [3, 46, 47], hydrophobicity [7, 48] and surface roughness [49, 50].

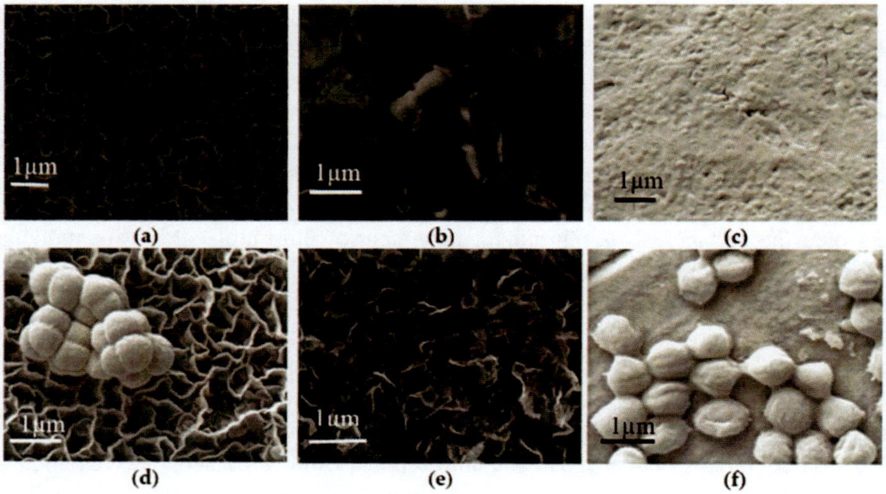

Fig. 1. SEM images of *E. coli* cell attachment on the surfaces of (a) methane-derived (GNW_M) and (b) bagasse-derived (GNW_B) graphene, and (c) pure copper substrate after 4 h of incubation at 22 °C. SEM images of *S. aureus* cell attachment on the surfaces of (d) GNW_M, (e) GNW_B, and (f) copper after incubation under the same conditions. Image from [36], reproduced with permission.

The majority of studies in this area focus on creating antibacterial surfaces for implant applications. The benefit of the proposed antibacterial mechanism is that it is strictly physical [50], eliminating the need for chemical therapies such as antibiotic medication, the use of which contributes towards the global rise of antibiotic resistant superbugs [47].

Mechanically stable titanium dioxide nanostructures have been fabricated using hydrothermal synthesis. These surfaces exhibit bactericidal behaviour against both Gram-negative *P. aeruginosa* and *Gram-positive S.* aureus bacterial cells. In addition, when exposed to these nanostructured surfaces, the cellular metabolic rate of osteoblast cells remained unaffected after 24 h, when compared to flat Ti-6Al-4 V (commonly used implant material) [51, 52]. This work also showed that the bactericidal efficiency of nanostructured surfaces is correlated to nanostructure height, and developed a function predicting nanostructure height from hydrothermal reaction conditions (NaOH concentration, reaction time and reaction temperature) [3, 46, 47].

Highly ordered and consistent titanium nanostructures were fabricated using electron beam lithography (EBL) by mimicking the naturally occurring nanostructure on cicada wings. Investigation was done into three cicada species using various techniques to determine wing topography, structure geometry and bactericidal efficiency. The wing species displaying the highest bactericidal efficiency (*Psaltoda claripennis*) was then replicated using EBL for form titanium nanostructures that cause damage to cell membranes of *S. aureus* and *P. aeruginosa* [1, 53].

Randomly orientated black titanium nanostructures, similar to that of black silicon, was fabricated through a chlorine-based reactive ion etching process. These surfaces were found to have a high bactericidal efficiency against *E. coli*, *P. aeruginosa*, *Mycobacterium smegmatis* and *S. aureus* within 4 h. In addition, these surfaces supported attachment and proliferation of human stem cells and osteogenic lineage, resulting in multi-biofunctional surfaces for potential use of orthopaedic implants [6].

Wet etching was used to fabricate randomly aligned nanoridges on aluminium 6063 surfaces. These surfaces were found to have high surface roughness and caused *S. aureus* and *P. aeruginosa* cells to be physically deformed and rendered non-viable upon attachment, after 3 h of contact. In addition, these surfaces were tested against rhinovirus and respiratory syncytial virus (two common respiratory viruses) and were found to significantly affect virus recovery after 2 h and showing significant reduction in virus viability counts after 24 h. Nanoindentation showed that these surfaces exhibit excellent durability under sustained loads without any damage. This significant, novel study shows that nanotextured surfaces effective against both bacteria and viruses for potential use in hospital environments to stop the spread of infections through surface transmission (Fig. 2) [54].

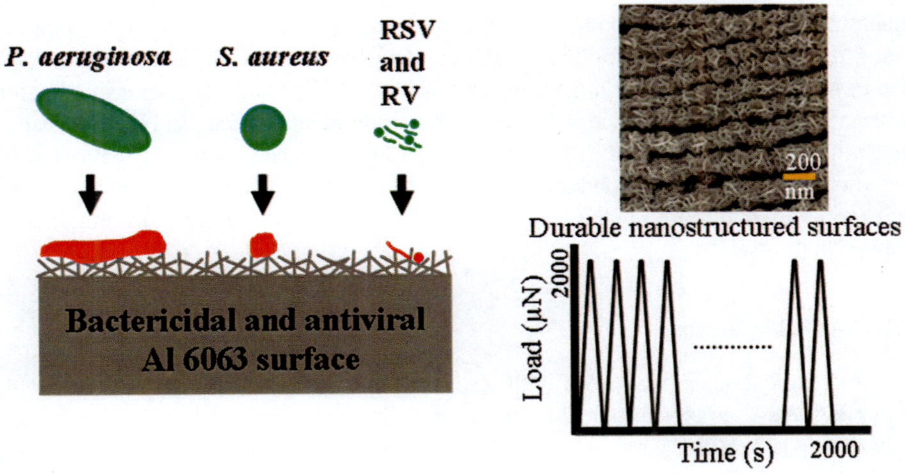

Fig. 2. Nanostructured bactericidal Al 6063 surfaces effective against P. aeruginosa and S. aureus have exhibited antiviral properties against rhinovirus and respiratory syncytial virus. Image from [54], reproduced with permission.

In a study conducted by Prasad et al. 2017, reduced graphene oxide nanosheets with silver nanoparticles were used to form a nanocomposite material which mediated contact killing. Interestingly, this material was just as effective against *S. aureus* and *Proteus mirabilis* as nitrofurantoin antibiotic medication, and when tested against E. coli (multi-drug resistant), was much more effective than nitrofurantoin. In addition, these nanocomposites worked much faster than the antibiotic medication [55].

As it can be seen, there are many studies being conducted within the bactericidal surface landscape with various substrate materials and fabrication techniques being investigated. This keen research interest shows that next generation orthopaedic implants and hospital surfaces are on the horizon to eradicate the spread of bacterial and viral infection through the use of nanostructured materials.

3 Biosensing

A biosensing device is defined by its biological or bioinspired receptor unit that has unique specificities towards corresponding analytes. These analytes are usually biologically based and include bacteria and virus DNAs, proteins, simple molecules (such as glucose) or pollutants [11]. Each biosensor device is made up of two parts: a biological component (enzymes, antibodies, microbial cells, tissue etc.) and the electronic part of the biosensor which transmits and detects physicochemical pathways (e.g. optical, electrochemical, piezoelectrical or thermometric) [56]. Since their discovery, biosensors have been a tremendous addition to the medical industry, providing useful insight into the health and performance of individuals. Initially, biosensors were largely used to detect physical markers such as step counts, calories burned and heart rate [57]. Now however, they have evolved to have applications in cardiovascular health, cancer and diabetes, would healing, and artificial limbs [58]. In addition, nano biosensors are also being researched for their application in detecting bacteria and contaminants in food and beverage industries. Nanomaterials are promising candidates for increasing sensor sensitivity and lowering detection limits down to individual molecules [11]. Biosensing platforms with enhanced stability and sensitivity are produced using a combination of metal or metal oxide with conducting polymers [59]. Biosensors for wearable devices and food and waterborne pathogens are discussed here, particularly those employing nanotechnology.

3.1 Wearable Devices

Electrochemical sensors coupled with wearable devices has revolutionised this field, providing real-time physiological information in a simple, low cost, flexible and compact device [57, 60]. Wearable biosensors come in many forms, from exercise-related devices, epidermal, ocular and oral biosensors, each providing non-invasive, continuous monitoring and real-time physical information [57, 60, 61]. Wearable devices have huge benefit to the general population, but are particularly for useful for vulnerable individuals, such as those with diabetes who need monitor glucose levels, and the elderly. The continuous nature of these devices provides individuals and healthcare professions up-to-date information, alerting them to significant changes in biomarkers.

Since the early 2000's, research has focused on nano biosensors to detect biomarkers in sweat, tears, saliva and interstitial fluid through wearable devices [57]. Sweat and interstitial fluids contain biomarkers such as metabolites (e.g. glucose), electrolytes, trace elements and molecules (such as proteins) [62] which have been found to have a good correlation to concentrations of these biomarkers in the blood stream [63], as these biomarkers are passed directly from the bloodstream into saliva. Sampling of these biomarkers on the skin surface is achieved when these fluids are transferred over the biosensor surface. Such biosensors could be in form of E-skin, printed temporary tattoos, wristbands or patches or even directly embedded in textiles, ensuring close contact with the skin [57].

Human tears can be used to detect glucose levels, as the glucose concentration found in tears correlates well with blood-glucose levels [57]. However, testing human tears directly comes with several issues such as small sample volumes, ease of sample evaporation and sample variation due to variation in tear production among different individuals at different times throughout the day [57]. Interestingly, tear-based wearable biosensors provide a solution to these problems, where the biosensor is embedded into contact lenses worn by the individual [64–66]. However, further research is needed to convert this technology into widely used products.

Within the human body, biomarkers are passed directly from the bloodstream into saliva. Using saliva in biosensing is an ideal, non-invasive technique for detecting biomarker concentration and disease [57]. There have been several studies which investigate the use of oral biosensors to detect biomarker concentrations. For example, graphene-based nanosensors have been printed on water-soluble silk and transferred directly onto tooth enamel. This technology has allowed for passive and wireless detection of bacteria [67]. The major challenge facing this innovative and exciting technology is that the concentration of biomarkers in saliva is significantly lower than in the bloodstream. To combat this issue, some concentration readings have been scaled to reflect correct values [57], however this scaling value may vary from person to person.

Common methods used to fabricate wearable biosensors include screen printing (common for glucose monitoring) [68], roll-to-roll gravure printing (common in the health and fitness industry) [69], stamp transfer electrodes [70], 3D printing [71], and various forms of lithography such as soft lithography, photolithography and electron beam lithography [60].

3.2 Food and Waterborne Pathogens

Prevention and early detection of bacterial infection caused by foodborne pathogens are one of the most significant problems facing the industry. It is estimated that one third of food produced for human consumption globally is wasted during difference stages of the food chain process, due to microbial contamination [56]. Rapidly growing numbers of foodborne illness (for example, food poisoning) has lead researchers to investigate contamination prevention and detection methods of food and waterborne pathogens [72–74]. In addition to this, traditional and standard bacterial detection methods (such as culture and colony counting methods) may take up to 7 to 8 to produce results [75], which is inadequate for timely prevention of pathogen spread. Common pathogens found in food and water include *E. coli*, listeria and salmonella [56].

There have been a number of studies using nanotechnology, such as nanostructured surfaces for sensing bacteria pathogens in food. For example, highly ordered nanopillar arrays produced by soft/photolithography and metal evaporation have been found to detect *E. coli* with high reproducibility and sensitivity (10 × 101 colony forming units (CFU)). These flexible nanopillar-based sensors also exhibit superior electrochemical and mechanical properties [76]. Similarly, zinc oxide coated micropillars were found to detect *E. coli* bacteria in the range of 106–101 CFU/mL. The presence of the micropillar structure increases surfaces area and strongly adheres to bacteria, enhancing bacterial capturability [72].

Carbon nanotubes and graphene based potentiometric biosensors have been used to rapid and ultra-sensitive bacterial detection. This technology is able to detect a single CFU/mL of *S. aureus* in 1–2 min [77]. In addition, magnetic nanoparticles have been used in combination with microfluid chip technology to detect Listeria monocytogenes for a sensitivity of 104–105 CFU/mL in 3 h in milk, lettuce and ground beef [78].

Optical biosensors are very popular in bioanalysis due to its high selectivity and sensitivity capabilities [75]. Gold nanoparticles have been used to develop an immune sensor that detects *Lactobacillus* and *S. aureus* strains in various food samples (including sugar cane) with a minimum detection limit of 105 and 120 CFU/mL, respectively [79]. Optical biosensors have also been proven to detect *E. coli* to a 3 × 103 CFU/mL sensitivity in cucumber and ground beef [80].

Electrochemical sensing has been used for rapid and sensitive bacterial detection in drinking water. Gold nanoparticles functionalized with positively charged quaternary amine headgroups, combined with electrostatic interactions resulted in increased enzyme activity when bound to bacteria. This indicated the presence of *E. coli* and *S. aureus* bacteria in less than 1 h, with high sensitivity (102 CFU/mL). This technology produces rapid quantification with low cost and minimal instrumentation [81]. Multiwall plate microfluidic devices have been designed to visualise multiplex detection of waterborne pathogens such as *E. coli* and rotavirus (common waterborne pathogens). These systems can be used to monitor the quality of drinking water systems [82], however more research is needed in this area.

4 Conclusion and Future Direction

This review shows that there is currently a significant need and wide range of applications for both antibacterial surfaces and biosensors. The inclusion of nanoparticles or nanostructured surfaces for sensing has proven to be a low cost and effective way of improving sensor performance. Our group is currently investigating proven antibacterial nanostructured surfaces for detecting bacteria and viruses in blood and body fluids using electrochemical anodization to develop a nanosensing platform. Using bacteria DNA and a Hoechst electrolyte, a solution will be analysed using the nanostructured surface as the working electrode, in combination with reference and counter electrodes. Surfaces with different chemical and physical properties will be used to determine the dominant factor which maximises detection accuracy. This work may develop a sensor that rapidly detects bacteria, to be used in hospital and healthcare environments, helping to reduce the length of patient hospital stays, patient discomfort and burdens to the healthcare system, as well as control and contain the spread of infection in hospital environments. Particular focus needs to aim at detecting specific strains of bacteria or viruses, such as methicillin-resistant *S. aureus* and SARS-CoV-2.

Acknowledgements. The authors would like to acknowledge the Central Analytical Research Facility at Queensland University of Technology. This work is funded by the Australian Research Council Discovery Grant (DP180101098).

References

1. Shahali, H., et al.: Multi-biofunctional properties of three species of cicada wings and biomimetic fabrication of nanopatterned titanium pillars. J. Mater. Chem. B **7**(8), 1300–1310 (2019)
2. Jaggessar, A., Yarlagadda, P., Qiu, T., Li, T., Tesfamichael, T.: Fabrication of nano pyramid texture on Ti-6Al-4V using nanosphere lithography. Mater. Today Proc. **5**(5, Part 2), 11593–11600 (2018)
3. Jaggessar, A., Mathew, A., Tesfamichael, T., Wang, H., Yan, C., Yarlagadda, P.K.: Bacteria death and osteoblast metabolic activity correlated to hydrothermally synthesised TiO_2 surface properties. Molecules **24**(7), 1201 (2019)
4. Hasan, J., Chatterjee, K.: Recent advances in engineering topography mediated antibacterial surfaces. Nanoscale **7**(38), 15568–15575 (2015)
5. Hasan, J., Crawford, R.J., Ivanova, E.P.: Antibacterial surfaces: the quest for a new generation of biomaterials. Trends Biotechnol. **31**(5), 295–304 (2013)
6. Hasan, J., Jain, S., Chatterjee, K.: Nanoscale topography on black titanium imparts multi-biofunctional properties for orthopedic applications. Sci. Rep. **7**, 41118 (2017)
7. Hasan, J., et al.: Selective bactericidal activity of nanopatterned superhydrophobic cicada Psaltoda claripennis wing surfaces. Appl. Microbiol. Biotechnol. **97**(20), 9257–9262 (2013). https://doi.org/10.1007/s00253-012-4628-5
8. Hazell, G., Fisher, L.E., Murray, W.A., Nobbs, A.H., Su, B.: Bioinspired bactericidal surfaces with polymer nanocone arrays. J. Colloid Interface Sci. **528**, 389–399 (2018)
9. Huang, J., Wang, X., Wang, Z.L.: Controlled replication of butterfly wings for achieving tunable photonic properties. Nano Lett. **6**(10), 2325–2331 (2006)
10. Ivanova, E.P., et al.: Natural bactericidal surfaces: mechanical rupture of pseudomonas aeruginosa cells by cicada wings. Small **8**(16), 2489–2494 (2012)
11. Holzinger, M., Le Goff, A., Cosnier, S.: Nanomaterials for biosensing applications: a review. Front. Chem. **2**, 63 (2014)
12. Jaggessar, A., Shahali, H., Mathew, A., Yarlagadda, P.K.D.V.: Bio-mimicking nano and micro-structured surface fabrication for antibacterial properties in medical implants. J. Nanobiotechnol. **15**(1), 64 (2017). https://doi.org/10.1186/s12951-017-0306-1
13. Francolini, I., Vuotto, C., Piozzi, A., Donelli, G.: Antifouling and antimicrobial biomaterials: an overview. APMIS **125**(4), 392–417 (2017)
14. Balčytis, A., Seniutinas, G., Lapierre, F., Juodkazis, S.: Artificial antibacterial surfaces that are simple to fabricate. In: Ivanova, E.P., Crawford, R.J. (eds.) antibacterial surfaces, pp. 27–40. Springer, Cham (2015). https://doi.org/10.1007/978-3-319-18594-1_3
15. Ensikat, H.J., Ditsche-Kuru, P., Neinhuis, C., Barthlott, W.: Superhydrophobicity in perfection: the outstanding properties of the lotus leaf. Beilstein J. Nanotechnol. **2**(1), 152–161 (2011)
16. Yan, Y.Y., Gao, N., Barthlott, W.: Mimicking natural superhydrophobic surfaces and grasping the wetting process: a review on recent progress in preparing superhydrophobic surfaces. Adv. Colloid Interface Sci. **169**(2), 80–105 (2011)
17. Zhang, C., McAdams, D.A., Grunlan, J.C.: Nano/micro-manufacturing of bioinspired materials: a review of methods to mimic natural structures. Adv. Mater. **28**(30), 6292–6321 (2016)
18. Bandara, C.D., et al.: Bactericidal effects of natural nanotopography of dragonfly wing on Escherichia coli. ACS Appl. Mater. Interfaces **9**(8), 6746–6760 (2017)
19. Bhadra, C.M., et al.: Antibacterial titanium nano-patterned arrays inspired by dragonfly wings. Sci. Rep. **5,** 16817 (2015)
20. Diu, T., et al.: Cicada-inspired cell-instructive nanopatterned arrays. Sci. Rep. **4**, 7122 (2014)

21. Linklater, D.P., Nguyen, H.K.D., Bhadra, C.M., Juodkazis, S., Ivanova, E.P.: Influence of nanoscale topology on bactericidal efficiency of black silicon surfaces. Nanotechnology **28**(24), 9 (2017)
22. Pogodin, S., et al.: Biophysical model of bacterial cell interactions with nanopatterned cicada wing surfaces. Biophys. J. **104**(4), 835–840 (2013)
23. Choi, M., et al.: Superhydrophilic coatings with intricate nanostructure based on biotic materials for antifogging and antibiofouling applications. Chem. Eng. J. **309**, 463–470 (2017)
24. Fang, Y., Sun, G.: Complex wettability and self-cleaning performance of butterfly wing surface. Appl. Mech. Mater. **723**, 943 (2015)
25. Cheng, Y.T., Rodak, D.E., Wong, C.A., Hayden, C.A.: Effects of micro- and nano-structures on the self-cleaning behaviour of lotus leaves. Nanotechnology **17**(5), 1359–1362 (2006)
26. Fadeeva, E., et al.: Bacterial retention on superhydrophobic titanium surfaces fabricated by femtosecond laser ablation. Langmuir **27**(6), 3012–3019 (2011)
27. Tang, P., et al.: Effect of superhydrophobic surface of titanium on staphylococcus aureus adhesion. J. Nanomater. **2011**, 1–8 (2011)
28. Privett, B.J., et al.: Antibacterial fluorinated silica colloid superhydrophobic surfaces. Langmuir **27**(15), 9597–9601 (2011)
29. Crick, C.R., Ismail, S., Pratten, J., Parkin, I.P.: An investigation into bacterial attachment to an elastomeric superhydrophobic surface prepared via aerosol assisted deposition. Thin Solid Films **519**(11), 3722–3727 (2011)
30. Yang, H., Deng, Y.: Preparation and physical properties of superhydrophobic papers. J. Colloid Interface Sci. **325**(2), 588–593 (2008)
31. Okada, A., et al.: Inhibition of biofilm formation using newly developed coating materials with self-cleaning properties. Dent. Mater. J. **27**(4), 565–572 (2008)
32. Kesel, A., Liedert, R.: Learning from nature: non-toxic biofouling control by shark skin effect. Comput. Biochem. Physiol. A Mol. Integr. Physiol. **146**(4), S130 (2007)
33. Bixler, G.D., Bhushan, B.: Rice-and butterfly-wing effect inspired self-cleaning and low drag micro/nanopatterned surfaces in water, oil, and air flow. Nanoscale **6**(1), 76–96 (2013)
34. Damodaran, V.B., Murthy, N.S.: Bio-inspired strategies for designing antifouling biomaterials. Biomat. Res. **20**(1), 18 (2016). https://doi.org/10.1186/s40824-016-0064-4
35. Harding, J.L., Reynolds, M.M.: Combating medical device fouling. Trends Biotechnol. **32**(3), 140–146 (2014)
36. Prasad, K., et al.: Effect of precursor on antifouling efficacy of vertically-oriented graphene nanosheets. Nanomaterials **7**(7), 170 (2017)
37. Kim, T.W.: Assessment of hydro/oleophobicity for shark skin replica with riblets. J. Nanosci. Nanotechnol. **14**(10), 7562–7568 (2014)
38. Li, X., et al.: The nanotipped hairs of gecko skin and biotemplated replicas impair and/or kill pathogenic bacteria with high efficiency. Nanoscale **8**(45), 18860–18869 (2016)
39. Zhao, D.-Y., Huang, Z.-P., Wang, M.-J., Wang, T., Jin, Y.: Vacuum casting replication of micro-riblets on shark skin for drag-reducing applications. J. Mater. Process. Technol. **212**(1), 198–202 (2012)
40. Fisher, L.E., Yang, Y., Yuen, M.-F., Zhang, W., Nobbs, A.H., Su, B.: Bactericidal activity of biomimetic diamond nanocone surfaces. Biointerphases **11**(1), 011014 (2016)
41. Chou, S.Y.: Nanoimprint lithography. In: Cabrini, S., Kawata, S. (eds.) Nanofabrication Handbook, pp. 187–206. CRC Press, Boca Raton (2012)
42. Dickson, M.N., Liang, E.I., Rodriguez, L.A., Vollereaux, N., Yee, A.F.: Nanopatterned polymer surfaces with bactericidal properties. Biointerphases **10**(2), 021010 (2015)
43. Liu, W., et al.: Bioinspired polyethylene terephthalate nanocone arrays with underwater superoleophobicity and anti-bioadhesion properties. Nanoscale **6**(22), 13845–13853 (2014)
44. Zhang, G., Zhang, J., Xie, G., Liu, Z., Shao, H.: Cicada wings: a stamp from nature for nanoimprint lithography. Small **2**(12), 1440–1443 (2006)

45. Cho, J.-Y., Kim, G., Kim, S., Lee, H.: Replication of surface nano-structure of the wing of dragonfly (Pantala Flavescens) using nano-molding and UV nanoimprint lithography. Electron. Mater. Lett. **9**(4), 523–526 (2013). https://doi.org/10.1007/s13391-013-0042-0
46. Jaggessar, A., Yarlagadda, P.K.D.V.: Modelling the growth of hydrothermally synthesised bactericidal nanostructures, as a function of processing conditions. Mater. Sci. Eng. C **108**, 110434 (2020)
47. Jaggessar, A., Yarlagadda, P.K.D.V.: Modelling the height of hydrothermally synthesized titanium dioxide nanostructures. Adv. Mater. Lett. **11**(6), 1–6 (2020)
48. Hasan, J., Raj, S., Yadav, L., Chatterjee, K.: Engineering a nanostructured "super surface" with superhydrophobic and superkilling properties. RSC Adv. **5**(56), 44953–44959 (2015)
49. Hasan, J., Jain, S., Padmarajan, R., Purighalla, S., Sambandamurthy, V.K., Chatterjee, K.: Multi-scale surface topography to minimize adherence and viability of nosocomial drug-resistant bacteria. Mater. Des. **140**, 332–344 (2018)
50. Velic, A., Mathew, A., Hines, P., Yarlagadda, P.K.D.V.: Control of bacterial attachment by fracture topography. J. Mech. Behav. Biomed. Mater. **91**, 416 (2019)
51. Jaggessar, A., Mathew, A., Wang, H., Tesfamichael, T., Yan, C., Yarlagadda, P.K.D.V.: Mechanical, bactericidal and osteogenic behaviours of hydrothermally synthesised TiO_2 nanowire arrays. J. Mech. Behav. Biomed. Mater. **80**, 311–319 (2018)
52. Jaggessar, A., Tesfamicheal, T., Wang, H., Yan, C., Yarlagadda, P.K.D.V.: Investigation of mechanical properties and morphology of hydrothermally manufactured titanium dioxide nanostructured surfaces. Procedia Manuf. **30**, 373–379 (2019)
53. Shahali, H., Hasan, J., Wang, H., Tesfamichael, T., Yan, C., Yarlagadda, P.K.D.V.: Evaluation of particle beam lithography for fabrication of metallic nano-structures. Procedia Manuf. **30**, 261–267 (2019)
54. Hasan, J., Xu, Y., Yarlagadda, T., Schuetz, M., Spann, K., Yarlagadda, P.K.D.V.: Antiviral and antibacterial nanostructured surfaces with excellent mechanical properties for hospital applications. ACS Biomater. Sci. Eng. **6**(6), 3608–3618 (2020)
55. Prasad, K., et al.: Synergic bactericidal effects of reduced graphene oxide and silver nanoparticles against Gram-positive and Gram-negative bacteria. Sci. Rep. **7**(1), 1591 (2017)
56. Hameed, S., Xie, L., Ying, Y.: Conventional and emerging detection techniques for pathogenic bacteria in food science: a review. Trends Food Sci. Technol. **81**, 61–73 (2018)
57. Kim, J., Campbell, A.S., de Ávila, B.E.-F., Wang, J.: Wearable biosensors for healthcare monitoring. Nat. Biotechnol. **37**(4), 389–406 (2019)
58. Shafiee, A., Ghadiri, E., Kassis, J., Zarandi, N.P., Atala, A.: Biosensing technologies for medical applications, manufacturing, and regenerative medicine. Curr. Stem Cell Rep. **4**(2), 105–115 (2018). https://doi.org/10.1007/s40778-018-0123-y
59. El-Said, W.A., Abdelshakour, M., Choi, J.H., Choi, J.W.: Application of conducting polymer nanostructures to electrochemical biosensors. Molecules **25**(2), 307 (2020)
60. Ferreira, P.C., et al.: Wearable electrochemical sensors for forensic and clinical applications. TrAC Trends Anal. Chem. **119**, 115622 (2019)
61. Windmiller, J.R., Wang, J.: Wearable electrochemical sensors and biosensors: a review. Electroanalysis **25**(1), 29–46 (2013)
62. Sonner, Z., et al.: The microfluidics of the eccrine sweat gland, including biomarker partitioning, transport, and biosensing implications. Biomicrofluidics **9**(3), 031301 (2015)
63. Seshadri, D.R., et al.: Wearable sensors for monitoring the physiological and biochemical profile of the athlete. NPJ Digit. Med. **2**(1), 72 (2019)
64. Farandos, N.M., Yetisen, A.K., Monteiro, M.J., Lowe, C.R., Yun, S.H.: Contact lens sensors in ocular diagnostics. Adv. Healthc. Mater. **4**(6), 792–810 (2015)
65. Mitsubayashi, K., Arakawa, T.: Cavitas Sensors: Contact Lens Type Sensors & Mouthguard Sensors, pp. 1170–1187. Wiley-VCH Verlag, Weinheim (2016)

66. Pankratov, D., González-Arribas, E., Blum, Z., Shleev, S.: Tear Based Bioelectronics, pp. 1250–1266. Wiley-VCH Verlag, Weinheim (2016)
67. Mannoor, M.S., et al.: Graphene-based wireless bacteria detection on tooth enamel. Nat. Commun. **3**(1), 763 (2012)
68. Carpini, G., Lucarelli, F., Marrazza, G., Mascini, M.: Oligonucleotide-modified screen-printed gold electrodes for enzyme-amplified sensing of nucleic acids. Biosens. Bioelectron. **20**(2), 167–175 (2004)
69. Bariya, M., et al.: Roll-to-roll gravure printed electrochemical sensors for wearable and medical devices. ACS Nano **12**(7), 6978–6987 (2018)
70. Windmiller, J.R., Bandodkar, A.J., Parkhomovsky, S., Wang, J.: Stamp transfer electrodes for electrochemical sensing on non-planar and oversized surfaces. Analyst **137**(7), 1570–1575 (2012)
71. Nesaei, S., et al.: Micro additive manufacturing of glucose biosensors: a feasibility study. Anal. Chim. Acta **1043**, 142–149 (2018)
72. Lee, K.S., et al.: Development of zinc oxide-based sub-micro pillar arrays for on-site capture and DNA detection of foodborne pathogen. J. Colloid Interface Sci **563**, 54–61 (2020)
73. Gracias, K.S., McKillip, J.L.: A review of conventional detection and enumeration methods for pathogenic bacteria in food. Can. J. Microbiol. **50**(11), 883–890 (2004)
74. Swaminathan, B., Feng, P.: Rapid detection of food-borne pathogenic bacteria. Ann. Rev. Microbiol. **48**(1), 401–426 (1994)
75. Lazcka, O., Campo, F.J.D., Muñoz, F.X.: Pathogen detection: a perspective of traditional methods and biosensors. Biosensors. Bioelectron. **22**(7), 1205–1217 (2007)
76. Park, Y., et al.: Flexible nanopillar-based electrochemical sensors for genetic detection of foodborne pathogens. Nano Converg. **5**(1), 1–8 (2018). https://doi.org/10.1186/s40580-018-0147-0
77. Hernández, R., Vallés, C., Benito, A.M., Maser, W.K., Rius, F.X., Riu, J.: Graphene-based potentiometric biosensor for the immediate detection of living bacteria. Biosensors. Bioelectron. **54**, 553–557 (2014)
78. Kanayeva, D.A., et al.: Efficient separation and sensitive detection of Listeria monocytogenes using an impedance immunosensor based on magnetic nanoparticles, a microfluidic chip, and an interdigitated microelectrode. J. Food Prot. **75**(11), 1951–1959 (2012)
79. Verdoodt, N., Basso, C.R., Rossi, B.F., Pedrosa, V.A.: Development of a rapid and sensitive immunosensor for the detection of bacteria. Food Chem. **221**, 1792–1796 (2017)
80. Wang, Y., Ye, Z., Si, C., Ying, Y.: Monitoring of Escherichia coli O157: H7 in food samples using lectin based surface plasmon resonance biosensor. Food Chem. **136**(3–4), 1303–1308 (2013)
81. Chen, J., et al.: Electrochemical nanoparticleenzyme sensors for screening bacterial contamination in drinking water. Analyst **140**(15), 4991–4996 (2015)
82. Zhao, X., Dong, T., Yang, Z., Pires, N., Høivik, N.: Compatible immuno-NASBA LOC device for quantitative detection of waterborne pathogens: design and validation. Lab Chip **12**(3), 602–612 (2012)

Flexible Manufacturing System with Uncertainty Management Using Fuzzy Logic for Machine Shop

Suthep Butdee and Siridech Kunhirunbawon(✉)

Department of Production Engineering, Faculty of Engineering, King Mongkut's University of Technology North Bangkok, 1518 Pracharaj I, Bangsue, Bangkok 10800, Thailand
siridech_k@psru.ac.th

Abstract. Presently, modern and intelligent machining is needed to achieve more flexibility in order to deal with uncertain demands, a variety of part requirements as well as small-lot customer orders. The flexible manufacturing system is widely applied to deal with the current problems. However, there are a lot of applications of the FMS that are implemented but they are not effective enough to respond to the fast changes and win the competitive advantage in the edge of industry 4.0. This paper proposes fuzzy logic to deal with uncertain circumstances. The customer orders are received by CAD files in the form of rotational and prismatic parts. The 3D objects are then simulated by CAM and determined for an optimal solution. Then, the optimized process plans are created by the CAPP program. The critical factors affected by the manufacturing system are taken into account as fuzzy input. The output of the fuzzy system can predict the level of effective FMS routes with available facilities under shop floor constraints.

Keywords: Flexible manufacturing system · Uncertainty management · Small lots · Fuzzy logic · Machine shop · CAD/CAM/CAPP

1 Introduction

Machine shops are mostly used to make complex machined parts consisting of CNC machining center, CNC turning, CNC Drill, CNC grinding, and so on. The type of production system can be designed by different jobs such as job shop, continuous flow production, or project manufacturing. Nowadays, the machining process needs to respond the rapid change or be more flexible in order to serve small lot quantity size and more product variety. In addition, the manufacturing process route should deal with uncertain demands. The new factory needs to select high-efficiency machine and performs production under maximum capability. In real-world decision-making to produce some products still depends on the experience of the machinist on the shop floor. It is important to improve an existing product to be more effective and more flexible. The flexible production system (FMS) consists of flexible machines linked to a chain workstation and connected to be automated material management systems, which are controlled by

A. Batako et al. (Eds.): GCMM 2021, LNNS 335, pp. 589–600, 2022.
https://doi.org/10.1007/978-3-030-90532-3_43

computers. The benefits of this system are the reduction of change-overtime, shorter production cycle time, reduction of labor demand, better machine utilization, and increased product quality [1]. However, it is not enough to solve the whole production system such as uncertainty of alternative for allocation workpieces to the suitable machines which consisting various factors to be taken into account. In a general machine shop, customer orders are received by CAD files in the form of rotational or prismatic parts. The machining planer starts with thinking back to the process, or route that could be done in the past. However, it is a deficiency and a lack of competitive advantage. Modern planning should be more flexible and deal with uncertain demands. Therefore, this paper proposes a new approach of fuzzy logic modeling to deal with uncertain circumstances in the FMS environment. It can impact production cost, operation time, tools life, and the precision of the product dimension. The paper is organized as follows. The next section briefly reviews literature that has improved the FMS and problem solving used by fuzzy logic as well as the background of FMS. The fuzzy model development is proposed in Sect. 4. In Sect. 5, the results of the model are discussed and compared to each alternative. Finally, conclusions are presented in Sect. 6.

2 Literature Review

The flexible manufacturing system was used to manage machine tools, materials handling systems, storage areas, and work in-process inventory control. The flexible system consists of machine flexibility, process flexibility, product flexibility, routing flexibility, volume flexibility, expansion flexibility, operation flexibility, and production flexibility [2]. Some interesting researchers studied involve scheduling problems in the FMS such as part input sequencing using a robot and machine scheduling by simulation experimentation [3]. They found that the best FMS robot scheduling rules are the shortest remaining processing time (SRPT) and the earliest due date (EDD) with state-dependent part input sequencing. In simulation analysis of part launching for FMS found that part launch rule was capable of good performance regarding minimum mean tardiness and maximum service level [4]. In the part of optimization in process of FMS that has been studied such as solving the machine loading problem of FMS using a discrete firefly algorithm identify the sequence of the optimal jobs [5]. A genetic algorithm (GA) approach was employed to minimize the make span of the scheduling problem [6]. The processing rates have given the FMS throughput target, work-in-process level, part routes that used nonlinear mean value analysis, and Petri net method [7, 8]. In addition, there was a methodology aimed at dealing with ambiguity and uncertain problems that were the fuzzy set theory, which was proposed by Zadeh in 1965 [9]. The fuzzy model of the fuzzy rule-based system was as well-known as the fuzzy inference system (FIS) or fuzzy logic. Its function consists of fuzzy if-then rules, database, decision-making unit, fuzzification, and defuzzification [10]. The researcher applied to generate a fuzzy scheduling model to solve allocation operated, operation scheduling, tool allocation problems [11], machine-tool selection [12, 13], machine load problem of FMS [14, 15], and real-time produce sequence for machines [16–18] that were conducted by assessing the batch size, due date, remaining processing time, essential operation processing time, and optional operation processing time. In addition, it also studies involve order distribution in a cluster manufacturing, which can achieve the appropriate costs and profits [19].

3 Flexible Manufacturing System

The flexible manufacturing system (FMS) emerged in 1975 which utilized an automatic tool change (ATC) of the machining center operated by computer control [20]. The system derived from diverse capabilities in the production of different parts at the same time. The volume of production can adapt to meet uncertain demands. There are three levels of manufacturing flexibility [21] basic flexibilities, system flexibilities, and aggregate flexibilities.

Basic flexibilities consist of three types. Machine flexibility can make various operations supported by computerized programming. Material handling flexibility transports different part types to suitable positions at the various machine tools in a system or section loading/unloading. Operation flexibility can use for alternative operation sequences for manufacturing. System flexibility consists of five types serving with different proposes. Volume flexibility is efficiency that operates profitably at different volumes of the existing part types. Expansion flexibility has the ability to build and expand a larger system. Routing flexibility is an alternative route that can be effectively tracked through a system for the specified process plan. Process flexibility is the volumetric measurement of parts sets that the system can produce without causing any more settings. Product flexibility is a volume of a set of part types that can be produced in the system with setting not much. Aggregate flexibility consists of three types. Program flexibility is the capability of a system to perform long periods without interference from external. Production flexibility can produce without more investment in equipment. Market flexibility can deal with order changing according to market patterns conditions. In summary, the objective of flexibility is to solve complex problems in the production system such as increased machine utilization, decreased work in progress, decreased lead times, increased throughput, and decreased store inventory levels, increased quality. Generally, FMS is a highly automated group technology that consists of a CNC milling machine, CNC lathe which is connected by a material handling system (automated guide vehicle: AGV, robot, conveyor) and controlled by the computer system [22].

4 Methodology

This section considers only the initial allocation part to the machine shop which ignores the workload on each machine. That is divided into two sections. They are manual allocated planning based on experiences and fuzzy logic rule-based methods, which algorithm runs in MATLAB software and the hardware of Intel Core i5-4210U CPU @ 2.40 GHz. The example machine part in this research assumes to begin with production ordered by the customer as shown in Fig. 1. It presents the common example part types to test with the fuzzy process planning development. The parts are mixed with rotational parts, prismatic parts, and some hole operations.

4.1 Manual Allocation Planning

Manual process planning is used to define machine operations. That is considered by an expert process engineer or an experienced supervisor in a machine shop. This research

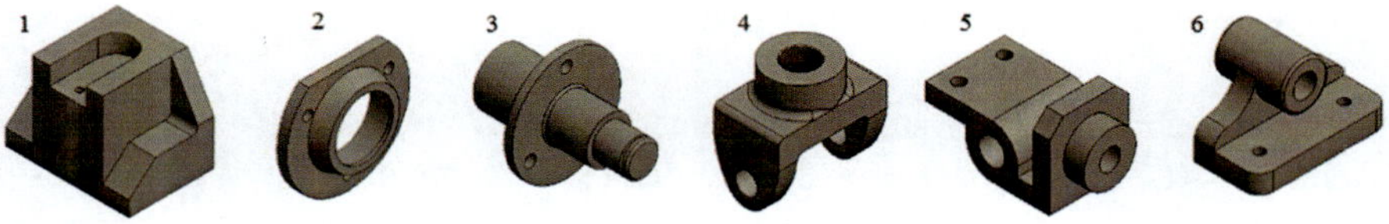

Fig. 1. CAD parts file for machining [23].

is assumed that there are three machines on the shop floor available to produce six-parts which have three alternatives for allocation as shown in Table 1 and the performance was under automated transfer part to machine. The details of cutting parameters for each part do not show in the plan, but it is generated by CAM. An alternative route for allocation to the machine is based on experience. In this case consideration, in every sequence of operations, each machine assumes to produce maximum capacity. This case is defined by general machines for production in the machine shop which consists of CNC lathe (M1), CNC milling (M2), and CNC drill (M3).

Table 1. Alternative part allocated jobs to machines.

Machine		Part1			Part2			Part3			Part4			Part5			Part6		
		Alternative			Alternative			Alternative			Alternative			Alternative			Alternative		
		1	2	3	1	2	3	1	2	3	1	2	3	1	2	3	1	2	3
CNC lathe	M1	0	0	0	1	0	1	1	1	0	1	0	1	1	0	0	1	0	0
CNC milling	M2	1	1	0	1	1	0	0	1	0	1	1	1	1	1	1	1	1	1
CNC drill	M3	1	0	0	1	1	1	1	0	0	1	1	0	1	1	0	1	1	0

Note: 1 = used, 0 = not used.

Table 1 show alternative part allocated by following rules:

1. If a part is rotational, then machine selection is CNC Lathe.
2. If a part is prismatic, then machine selection is CNC Milling.
3. If a prismatic part has a hole, then machine selection is CNC Drill or CNC Milling.
4. If a rotational part has a hole, then machine selection is CNC Drill or CNC Lathe.

4.2 Fuzzy Approach

The fuzzy model is developed to find an efficient allocation process planning in more detail and more flexible, and based on the relation between the membership of machines and different machining characteristics in production and critical factors affected by the manufacturing system. Figure 2shows the model fuzzy in five steps consisting of input, fuzzification, inference engine (FIS), defuzzification, and output in order to find an alternative allocated part type to fit with machine selection. The Mamdani method [24] is used for calculation in a fuzzy inference system. The first step is to collect all information based on the experience of machining sequence which follows the geometry of the workpiece in order to use on the machine. Then, they are allocated to machine type based on a significant level of tools as shown in Fig. 3. After that, the fuzzy rule is created. Next, output criteria are designed concerning efficiency in allocating parts to the machine before starting the production line. The fuzzy modeling aims to reduce cost, short-time operation, long-life tools, and high precision based on flexibility. They are four criteria as shown in Fig. 4.

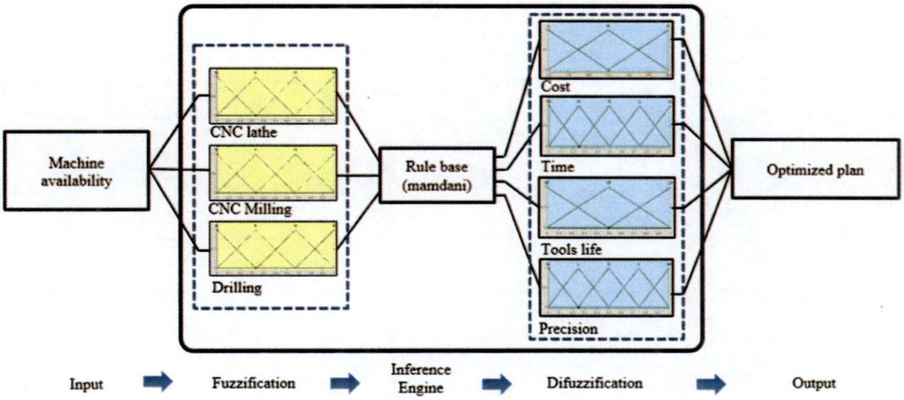

Fig. 2. Fuzzy based model for a part selection.

4.3 FIS Membership Functions

There are three membership functions for input variables of fuzzy inference system consisting of CNC lathe (M1), CNC milling (M2), and CNC drill (M3) as shown in Fig. 3. The output fuzzy variables consist of four membership functions as shown in Fig. 4. The fuzzy control technique uses a specific shape of the fuzzy set, which is depended on the system behavior identified by the knowledge engineers. Therefore, there is no standard method for selecting the suitable shape of fuzzy sets of control variables. A triangular shape was employed to describe the fuzzy sets. The abbreviation of input fuzzy membership is explained in Table 2 and Table 3.

The fuzzy expression in Table 2 contains the levels of the tool used in each machine. They are four-level in each machine and Table 3 shows the abbreviation for output of time, precision, costs, and tools life.

4.4 Fuzzy Rules

A set of fuzzy rules has been constructed for making decision to select a part on suitable machine based on the knowledge and skills of machine tool operators. Rule bases consist of a group of if-then control rule as shown below as an example:

1. If (M1 is MI) and (M2 is MX) and (M3 is ML) then (C is M) (T is M) (TL is MT) (PC is M)

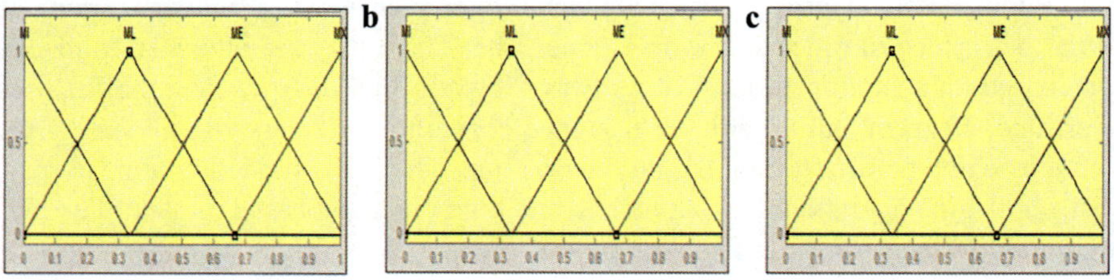

Fig. 3. Membership function input (a) M1; (b) M2; (c) M3.

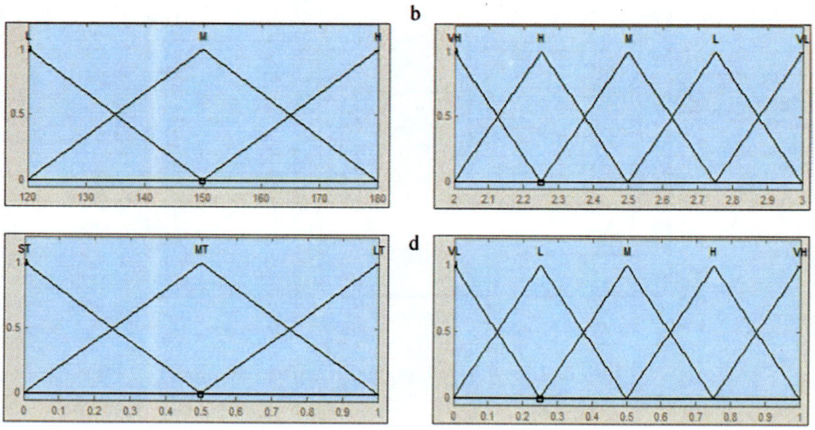

Fig. 4. Membership function output (a) cost; (b) time; (c) tools life; (d) precision.

Table 2. Fuzzy expressions for input.

Machine using tools	
Abbreviation	Expression
MI	Minimum tool changes
ML	Low tool changes
ME	Medium tool changes
MX	Maximum tool changes

2. If (M1 is MI) and (M2 is MX) and (M3 is MI) then (C is M) (T is H) (TL is ST) (PC is VH)
3. If (M1 is ME) and (M2 is MX) and (M3 is ME) then (C is H) (T is L) (TL is LT) (PC is VL)
4. If (M1 is MI) and (M2 is MX) and (M3 is MX) then (C is M) (T is L) (TL is MT) (PC is M)
5. If (M1 is MI) and (M2 is MX) and (M3 is MI) then (C is L) (T is H) (TL is ST) (PC is H)

Table 3. Fuzzy expressions for output.

Time (T) and Precision (PC)		Cost (C)		Tools life (TL)	
Abbreviation	Expression	Abbreviation	Expression	Abbreviation	Expression
VH	Very long time, very high precision	L	Low cost	LT	Long tool life
H	Long time, high precision	M	Medium cost	MT	Medium tool life
M	Medium time, medium precision	H	High cost	ST	Short tool life
L	Short time, low precision				
VL	Very short time, very low precision				

6. If (M1 is MX) and (M2 is ML) and (M3 is ML) then (C is H) (T is M) (TL is MT) (PC is VH)
7. If (M1 is MI) and (M2 is MX) and (M3 is ML) then (C is L) (T is M) (TL is ST) (PC is H)
8. If (M1 is MX) and (M2 is MI) and (M3 is ML) then (C is L) (T is H) (TL is MT) (PC is M)

4.5 Fuzzy Ranks

The efficient and optimized plan is set by fuzzy ranks to imply the efficiency level. Table 4 shows the time rank for fuzzy value. The rank is divided into 5 levels; VH, H, M, L, VL. The highest time efficiency is 100%, whereas low efficiency is 25%. There is assigned maximum 3 h and minimum 2 h according to the CAM simulation.

Table 4. Time rank for fuzzy value.

Abbreviation	Rank (h)	Efficiency
VH	2.000–2.125	100
H	2.126–2.375	80
M	2.376–2.625	60
L	2.626–2.875	40
VL	2.876–3.000	20

Table 4 shows the time ranks for a fuzzy value between 2 h to 3 h, which assumed the overall process operated by CAM simulation (CNC milling, CNC lathe, CNC Drill). In this case, it is divided a range into five-level machining time. These data is used in this paper to demonstrate and verify the application of a fuzzy model.

Table 5. Precision rank for fuzzy value.

Abbreviation	Precision value	Efficiency
VL	0.000–0.125	20
L	0.126–0.375	40
M	0.376–0.625	60
H	0.626–0.875	80
VH	0.876–1.000	100

Table 5 shows the precision rank for a fuzzy value between 0–1, which based on multi axes of machining which can change a clamping work piece method based on the part feature surface positions. However, clamping position changes can affect precision. If there are many changes in positions of a work piece, it infers that precision is very low. In this case, the precision rank is divided into 5 levels; VH, H, M, L, and VL respectively.

Table 6. Cost and tool life rank for fuzzy value.

Abbreviation	Cost (Baht)	Tools life	Efficiency
L, LT	120–135	0.76–1.00	100
M, MT	136–165	0.26–0.75	66.67
H, ST	166–180	0.00–0.25	33.33

Table 6 shows the cost estimation between 120 to 180 Baht (The value, multiplied by 10) and tools life ranked by the time of use between 0 to 1, which based on a number of tools affect to cost and duration of tool operation has an effect on the tools life. In this case, both cost and tools life is divided into 3 levels; L, M, H and LT, MT, ST respectively.

5 Result and Discussion

The result shows that the highest efficiency percentage is an alternative way for allocation work piece to the machine. In the case of maximum operation in each alternative. Figure 5 shows an example rule viewer of the inference system after entering all information which has a level number of fuzzy each membership function in case the first alternative part allocated to the machine of part 1.

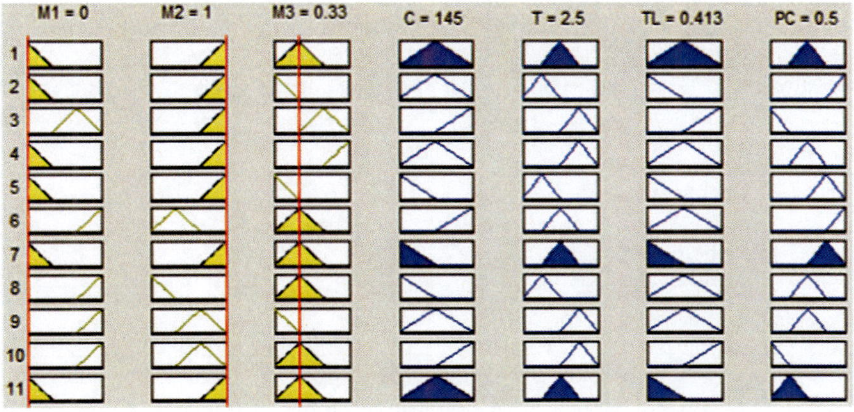

Fig. 5. Rule viewer of fuzzy inference system.

Table 7 shows the fuzzy result and efficiency plan of the alternative allocated parts to a machine shop, that is related efficiency condition of cost, operation time, tools life, and precision which distinguish the result of fuzzy and scale of ranking. The efficiency percentage was shown for the alternative allocated work piece to manufacturing. This fuzzy logic strategy model can simulate an operator's experience and expertise in the decision-making management allocated job to FMS for the machine shop. The allocation decisions can select a work piece for the machine. It is important in terms of input of the production line. In this research, the machine used for general metal machining is selected such as CNC lathe, CNC milling and CNC Drill. These are the effects that cause instability in the production process, such as throughput, due date, and quality of productivity.

Table 7. The efficiency allocate plan in machining process.

Alternative			The impact factor of ranking criteria				Total score	Efficiency percentage
			Cost	Time	Tools life	Precision		
Part 1	1	Fuzzy level	145	2.50	0.413	0.50		
			66.67	60	66.67	60	253.34	63%
	2	Fuzzy level	150	2.25	0.163	0.920		
			66.67	80	33.33	100	280.00	70%
	3	Fuzzy level	0	0	0	0		
			0	0	0	0	0	0

(*continued*)

Table 7. (*continued*)

Alternative			The impact factor of ranking criteria				Total score	Efficiency percentage
			Cost	Time	Tools life	Precision		
Part 2	1	Fuzzy level	170	2.50	0.50	0.905		
			33.33	60	66.67	100	260	65%
	2	Fuzzy level	145	2.50	0.413	0.625		
			66.67	60	66.67	80	273.34	68%
	3	Fuzzy level	130	2.25	0.50	0.50		
			100	80	66.67	60	306.67	77%
Part 3	1	Fuzzy level	130	2.25	0.50	0.50		
			100	80	66.67	60	306.67	77%
	2	Fuzzy level	150	2.75	0.50	0.50		
			66.67	40	66.67	60	233.34	58%
	3	Fuzzy level	0	0	0	0		
			0	0	0	0	0	0
Part 4	1	Fuzzy level	170	2.74	0.814	0.123		
			33.33	40	100	40	213.33	53%
	2	Fuzzy level	145	2.50	0.413	0.50		
			66.67	60	66.67	60	253.34	63%
	3	Fuzzy level	150	2.50	0.50	0.50		
			66.67	60	66.67	60	253.34	63%
Part 5	1	Fuzzy level	170	2.75	0.837	0.08		
			33.33	40	100	20	193.33	48%
	2	Fuzzy level	150	2.75	0.50	0.50		
			66.67	40	66.67	60	233.34	58%
	3	Fuzzy level	145	2.26	0.198	0.785		
			66.67	80	33.33	80	260	65%

(*continued*)

Table 7. (*continued*)

Alternative			The impact factor of ranking criteria				Total score	Efficiency percentage
			Cost	Time	Tools life	Precision		
Part 6	1	Fuzzy level	150	2.75	0.50	0.25		
			66.67	40	66.67	40	213.34	53%
	2	Fuzzy level	145	2.50	0.413	0.50		
			66.67	60	66.67	60	253.34	63%
	3	Fuzzy level	145	2.25	0.163	0.795		
			66.67	80	33.33	100	280	70%

6 Conclusion

The paper has presented the characteristics of the machine shop and the way to improve as modern and more effective. Optimal process planning is done to couple with CAD files to give more flexibility to respond to the rapid change of product design and uncertain demands from the market. Fuzzy logic modeling is developed and created to assist a modern machinist to select the optimal plan rapidly and effectively before the machining process. That are under ranking criteria. The factors that influence the production system consist of cost, time, tool life, and precision. The output of the fuzzy modeling is efficient and more accurate than human experience.

References

1. Mellichamp, J.M., Kwon, O.J., Wahab, A.F.: FMS designer: an expert system for flexible manufacturing system design. Int. J. Prod. Res. **28**(11), 2013–2024 (1990)
2. Browne, J., Dubois, D., Rathmill, K., Sethi, S.P., Stecke, K.E.: Classification of flexible manufacturing systems. FMS Mag. **2**(2), 114–117 (1984)
3. He, Y., Stecke, K.E., Smith, M.L.: Robot and machine scheduling with state-dependent part input sequencing in flexible manufacturing systems. Int. J. Prod. Res. **54**(22), 6736–6746 (2016)
4. Wang, Y.C., Chen, T., Chiang, H., Pan, H.C.: A simulation analysis of part launching and order collection decisions for a flexible manufacturing system. Simul. Model. Pract. Theory **69**, 80–91 (2016)
5. Bottani, E., Centobelli, P., Cerchione, R., Gaudio, L., Murino, T.: Solving machine loading problem of flexible manufacturing systems using a modified discrete firefly algorithm. Int. J. Ind. Eng. Comput. **8**(3), 363–372 (2017)
6. Candan, G., Yazgan, H.R.: Genetic algorithm parameter optimisation using Taguchi method for a flexible manufacturing system scheduling problem. Int. J. Prod. Res. **53**(3), 897–915 (2015)

7. Schweitzer, P.J., Seidmann, A.: Optimizing processing rates for flexible manufacturing systems. Manage. Sci. **37**(4), 454–466 (1991)
8. Aized, T., Takahashi, K., Hagiwara, I.: Modeling and performance optimization of multiple product FMS using colored Petri net and response surface methods. In: 2007 IEEE International Symposium on Assembly and Manufacturing, pp. 94–99. IEEE (2007)
9. Zadeh, L.A.: Fuzzy sets. Inf. Control **8**(3), 338–353 (1965)
10. Passino, K.M., Yurkovich, S.: Fuzzy Control. Addison-Wesley, California (1997)
11. Srinoi, P., Shayan, E., Ghotb, F.: A fuzzy logic modelling of dynamic scheduling in FMS. Int. J. Prod. Res. **44**(11), 2183–2203 (2006)
12. Rai, R., Kameshwaran, S., Tiwari, M.K.: Machine-tool selection and operation allocation in FMS: solving a fuzzy goal-programming model using a genetic algorithm. Int. J. Prod. Res. **40**(3), 641–665 (2002)
13. Kłosowski, G., Gola, A., Świć, A.: Application of fuzzy logic controller for machine load balancing in discrete manufacturing system. In: Jackowski, K., Burduk, R., Walkowiak, K., Woźniak, M., Yin, H. (eds.) IDEAL 2015. LNCS, vol. 9375, pp. 256–263. Springer, Cham (2015). https://doi.org/10.1007/978-3-319-24834-9_31
14. Vidyarthi, N.K., Tiwari, M.K.: Machine loading problem of FMS: a fuzzy-based heuristic approach. Int. J. Prod. Res. **39**(5), 953–979 (2001)
15. Shih, H.M., Sekiguchi, T.: Fuzzy inference-based multiple criteria FMS scheduling. Int. J. Prod. Res. **37**(10), 2315–2333 (1999)
16. Bilkay, O., Anlagan, O., Kilic, S.E.: Job shop scheduling using fuzzy logic. Int. J. Adv. Manufact. Technol. **23**(7–8), 606–619 (2004)
17. Pires, M.G, Castro, P.A., Morandin, O., Camargo, H.A., Kato, E.R.: A real time production sequencing rule for machines of a FMS using fuzzy logic. In: SMC 2003 Conference Proceedings. 2003 IEEE International Conference on Systems, Man and Cybernetics. Conference Theme-System Security and Assurance (Cat. No. 03CH37483), vol. 5, pp. 4381–4386. IEEE (2003)
18. Kong, Y., Gong, X., Wang, J.: Realtime scheduling for FMS based on fuzzy logic. In: First International Conference on Innovative Computing, Information and Control-Volume I (ICICIC 2006), vol. 2, pp. 195–198. IEEE (2006)
19. Butdee, S., Nitnara, C.: A fuzzy logic combined with LP model for performance evaluation to distribute purchase orders in cluster manufacturing. Procedia Manufact. **30**, 19–25 (2019)
20. Ranky, P.G.: FMS in CIM (flexible manufacturing systems in computer integrated manufacturing). Robotica **3**(4), 205–214 (1985)
21. Shivanand, H.K., Benal, M.M., Koti, V.: Flexible Manufacturing Systems. New Age International, New Delhi (2006)
22. Saren, S.K., Tiberiu, V.: Review of flexible manufacturing system based on modeling and simulation. Fascicle Manage. Technol. Eng. **24**(14), 113–118 (2016)
23. Boundy, A.W.: Engineering Drawing, 6th edn. McGrew-Hill, Singapore (2006)
24. Mamdani, E.H.: Advances in the linguistic synthesis of fuzzy controllers. Int. J. Man Mach. Stud. **8**(6), 669–678 (1976)

Adaptation of Methods to Improve the Efficiency of Business Processes Using the BPRPM Method * as a Chance to Gain a Competitive Advantage

Filip Nowak[1] and Piotr Biernacki[2](✉)

[1] Lukasiewicz Research Network—Institute of Logistics and Warehousing, ul. Estkowskiego 6, Poznań, Poland
filip.nowak@ilim.lukasiewicz.gov.pl

[2] European Bank for Reconstruction and Development, MGX Infoservice, ul. Letnia 11, 05-515 Nowa Iwiczna, Poland
piotr.biernacki@mgx.com.pl

Abstract. The aim of the article is to present the needs of managers and to present the method of meeting those needs by means of BPRPM (Business Process Rapid Prototyping Method). The needs were identified by means of a survey. The survey checked which methods are used in management decisions. Based on the results of the survey, a hypothesis was put forward that the BPRPM method proposed by Filip Nowak solves many problems faced by managers and perfectly complements the methods used in decision-making. The method generally consists in creating analyzed business process models in the As-Is and To-Be state, and then prototyping them and simulating business process behavioral variants, so that organizations can experience change before it occurs in reality. A detailed description of this method is included in Filip Nowak's doctoral thesis "Method of applying a simulation experiment for transformation of information and decision-making processes in small and medium-sized enterprises" (Metoda zastosowania eksperymentu symulacyjnego do transformacji procesów informacyjno-decyzyjnych w małych i średnich przedsiębiorstwach). The effectiveness and efficiency of the presented method was confirmed by its implementation in several economic organizations.

Keywords: BPRPM · BPM · Process analysis

1 Introduction

The surveys are of an internal nature and allow to recognize the opinion of respondents on socio-economic phenomena taking place in economic organizations [1]. The main objective of the survey is to aggregate empirical data in order to solve the research problem. Respondents' answers to the questionnaire questions should be consistent and constitute a logical sequence of information whose analysis will lead to solving the scientific problem [2].

The aim of the analysis is to verify the hypothesis that economic organizations need a method of rapid implementation of changes through prototyping. Methods of business

A. Batako et al. (Eds.): GCMM 2021, LNNS 335, pp. 601–613, 2022.
https://doi.org/10.1007/978-3-030-90532-3_44

process simulation have been known and used since the mid-1990s. Nevertheless, there is a lack of awareness among the management of economic organizations that such methods can be used to make accurate management decisions. To verify this hypothesis a survey conducted by Filip Nowak has been used, full research data and a detailed description of the BPRPM method will be published as part of the doctoral dissertation entitled "Method of applying a simulation experiment for transformation of information and decision-making processes in small and medium-sized enterprises" at Faculty of Engineering Management at the Poznan University of Technology.

The surveys are in nature actions that are in one's own, personal interest [3] and allow to recognize the opinion of respondents on socio-economic phenomena taking place in economic organizations [1]. The main objective of the survey is to aggregate empirical data in order to solve the research problem. Respondents' answers to the survey questions should be consistent and constitute a logical sequence of information whose analysis will lead to solving the scientific problem [2].

The need for a scientific approach to the identification of problems related to the implementation of process changes results from the characteristics of management and quality sciences, which are strongly related to the activities of business entities [1, 8].

2 Survey

The above was the reason for choosing the survey method, which made it possible to gather the opinions of respondents with regard to the introduction of process changes. The choice of the survey method was also influenced by the possibility of covering a larger number of business entities, as well as the greater anonymity of respondents. The limitations occurring during the survey should also be noted. The main ones refer to difficulties in interpreting the content of the questions, receiving incorrectly filled in questionnaires and low return rate of filled in forms [4].

In the survey of managers' needs, for getting answers to the survey questions, were used:

- interviewers who play a technical role in the survey [5],
- direct interview method [4] and
- persons who filled in the questionnaire received by e-mail on their own.

The use of various methods of response collection, the so-called methodological triangulation was consciously applied in order to achieve the effect of complementation and synergy of research methods [7].

The research technique used was an electronic questionnaire and the research tool used was a survey questionnaire containing 21 questions. The range of questions allows to recognize expectations and barriers related to the implementation of process changes in Polish enterprises.

The questionnaire form used in the survey was facade validated by two domain experts.

An overview of the databases available on the market does not allow for the purchase or preparation of a list of all companies that could operate the population (time-consuming and high-cost operation). Therefore, the selection of the research sample was carried out using the "snowball" method - recruitment of survey participants by other participants [5, 6]. The limitation of this method of selection of the research population is the possibility that the research sample will include respondents with similar characteristics. This may result from recommending subsequent respondents from the group of people who know the original respondents well, and thus both have similar attitudes. In order to eliminate the effect of the respondents' affinity, various methods of response collection were used (methodological triangulation).

Approximately 30 people were invited to take part in the survey, who declared to provide the questionnaire to at least 150 people in total. 102 correctly completed anonymous forms were received, which were filled in electronically by the respondents between December 2019 and mid-March 2020. The majority of respondents declared a positive (42%) or very positive (46%) attitude towards their current work. Data on respondents and companies where they currently work is presented in Fig. 1, 2, 3, 4 and 5.

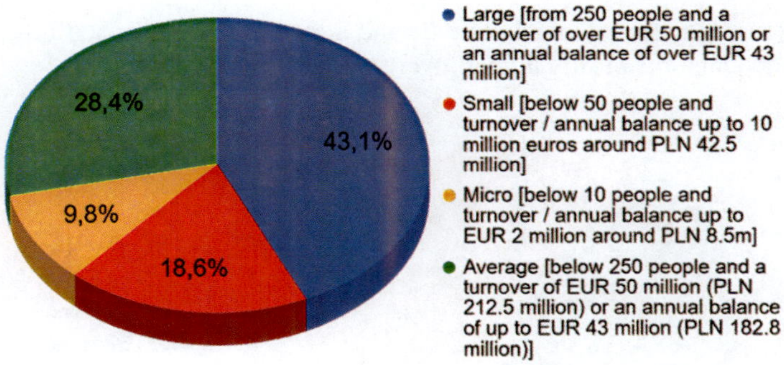

Fig. 1. Company size

Companies with more than 50 employees constitute a vast majority (72%) of the total number of surveyed organizations. That is, those for which business decisions should be based on objective process data.

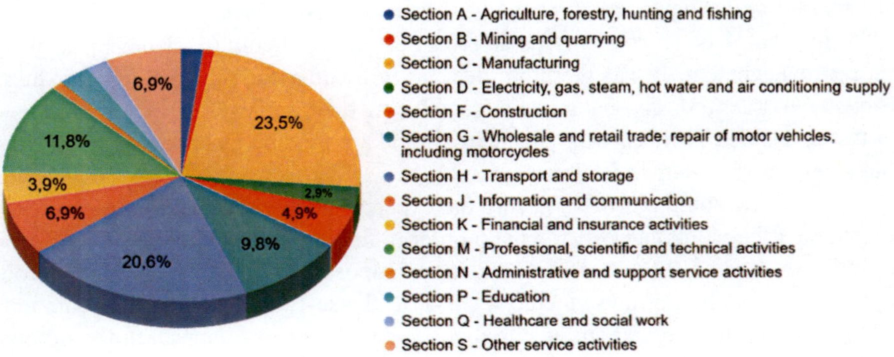

Fig. 2. Industry.

Companies belong to most of the sections of the Polish Classification of Activities (PKD). However, the following sections were most represented: production and logistics.

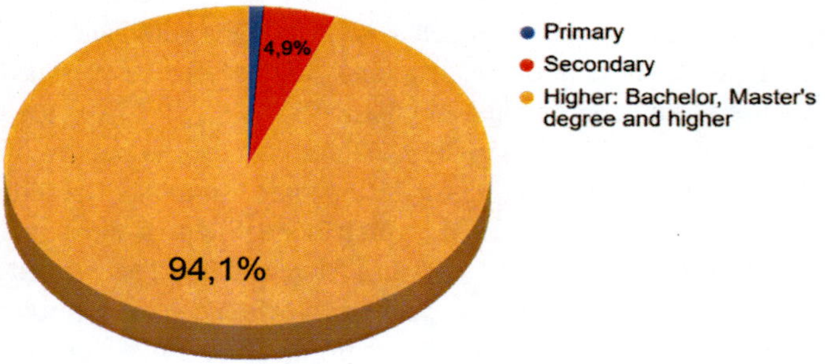

Fig. 3. Education.

As the survey involved mainly middle and higher management staff, the education of the respondents was mostly higher (over 94%).

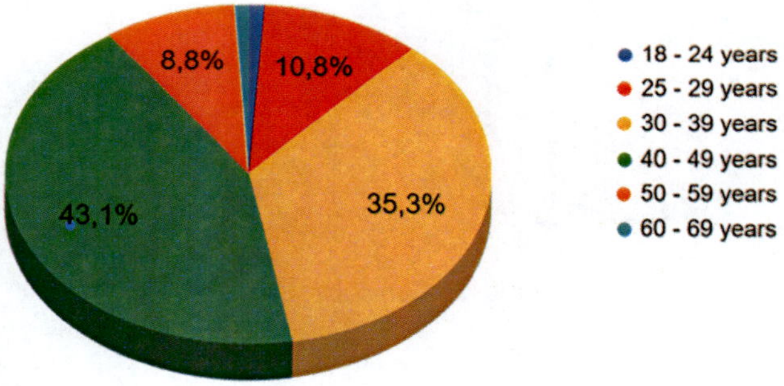

Fig. 4. Age.

The vast majority of respondents were between 30 and 50 years of age. The result is typical for the management of economic organizations in Poland.

More than ¾ of respondents declare barriers and difficulties related to business process management. It was therefore decided to identify the main reasons for these problems in order to propose a method for reducing their impact.

It was decided to check what the main criteria are used by managers when making long-term and short-term decisions.

There are two methods of achieving the expected end result: the iterative method, i.e. making a large number of variants in order to find the most effective solution, and the intuitive method, i.e. choosing a solution proposal that is good for meeting our expectations. The first of these methods concerns the so-called "hard implementation" and entails very high social and material costs, but allows us to approach the optimal solution in the long run. On the other hand, the second method does not guarantee obtaining an optimal solution, but only a sufficient one; its advantage is the very quick

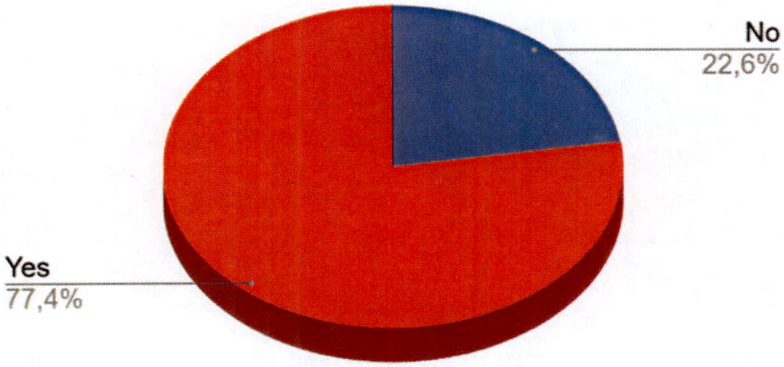

Fig. 5. Do you identify obstacles/difficulties in process management?

experience of change, but it does not necessarily have to be a successful change. What in the case of several unsuccessful attempts may lead to reluctance towards any changes. It would be desirable to be able to apply the first method without high social and material costs, i.e. efficient virtualization of the proposed change. This is the method of rapid prototyping of changes in business processes (BPRPM). It allows not only to identify processes and design the improvement of their efficiency, but also to experience the effects of the process change before its final implementation. Such a procedure will allow to draw important conclusions, which will be used for possible corrections of the target implementation. This allows for multi-variant testing of the target process and its experience in the future environment of its functioning. This ensures maximum benefits with minimum effort (Fig. 6).

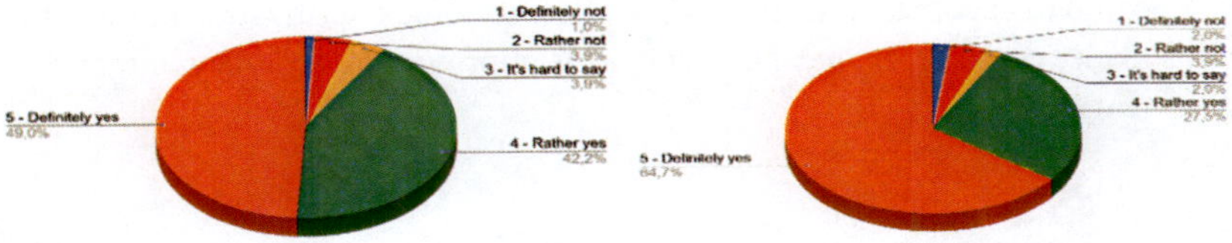

Fig. 6. The importance of experience in long-term and short-term decision-making

The vast majority of respondents (91% for long-term purposes and 92% for short-term purposes) base their decisions on experience, so it seems crucial for them to gain experience before implementing the change. Such opportunities are provided by the BPRPM method to the management (Fig. 7).

The graphs above show that regardless of age, respondents declared high and very high significance of their experience in making decisions concerning long-term and short-term actions. This indicates that their application of the BPRPM method would enable them to make decisions according to their expectations (on the basis of experience) with a significant reduction in the time of acquiring experience.

Figure 8 illustrates that for all age ranges experience is important, but for younger managers, probably because of less experience, it is not a basic decision criterion. The BPRPM method allows for very rapid acquisition of experience, which may ultimately increase confidence in basing decisions on experience.

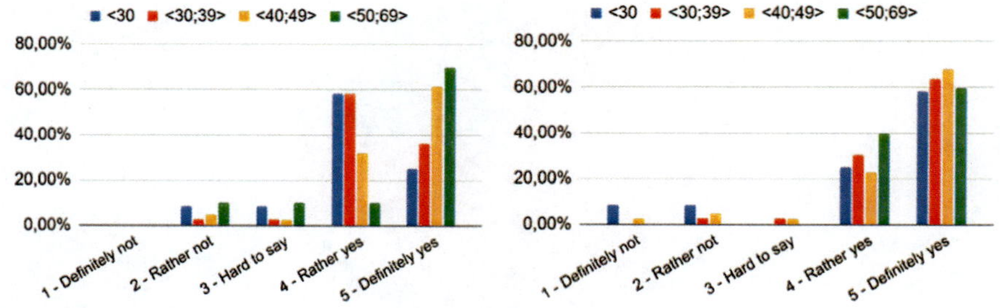

Fig. 7. The importance of experience in long-term and short-term decision-making—age of respondents for particular weights of this criterion

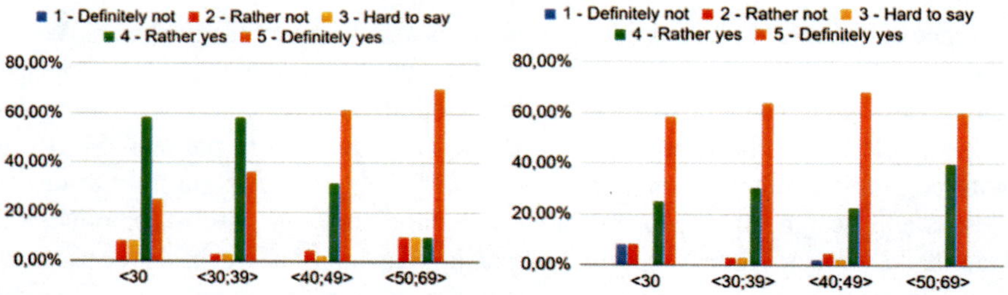

Fig. 8. Aspect 1.4/1.5 The importance of experience in long-term and short-term decision-making—the importance of this criterion in particular age brackets

Experience is even more important when making short-term decisions. Action scenarios prepared in accordance with BPRPM allow for quick verification of the effects of decisions made, even before they are made. This improves the accuracy of management decisions (Fig. 9).

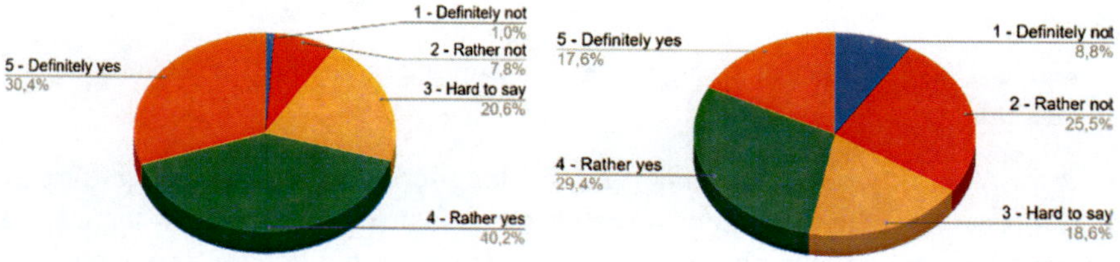

Fig. 9. The importance of analyzing market/industry trends in making long-term and short-term decisions

The majority of respondents are also guided by market/industry trends in their long-term decisions. The use of trends is associated with a high risk of succumbing—"fashion", so it is advisable to assess its long-term effects before making a decision. The BPRPM method allows to verify various scenarios of market behaviors (trends) and examine the sensitivity of processes to changes resulting from market forecasts. The method allows for determining the most convenient date of change and thus initiating the change at the most convenient moment (prediction in order to hit the change on time).

The method also enables to design emergency withdrawal scenarios based on trends that turn out to be wrong (Figs. 10 and 11).

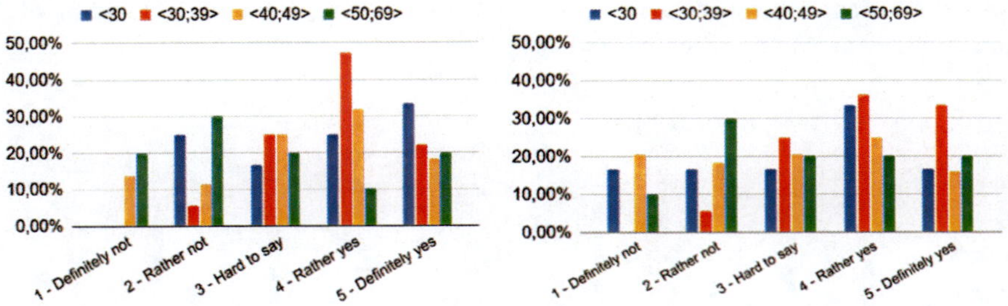

Fig. 10. The importance of analyzing market/industry trends in making long-term and short-term decisions—age of respondents for particular weights of this criterion

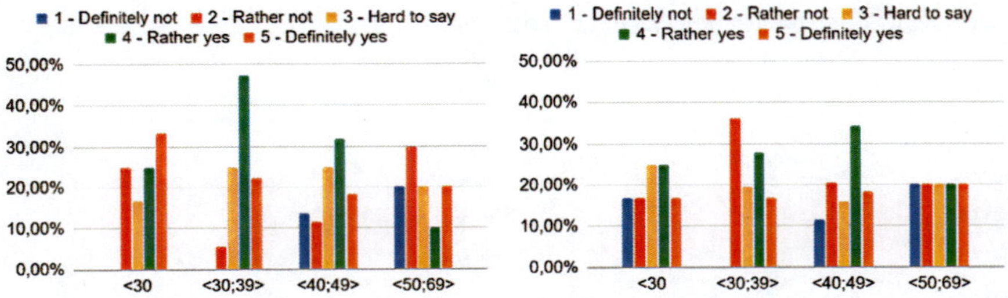

Fig. 11. The importance of analyzing market/industry trends in making long-term and short-term decisions—the importance of this criterion in particular age brackets

High confidence in trends is declared by a group of respondents aged 30–39, especially for them it seems to be effective in making long-term decisions supported by the BPRPM method. In the case of short-term decisions made by survey participants, trends do not play as important a role as for long-term decisions (Fig. 12).

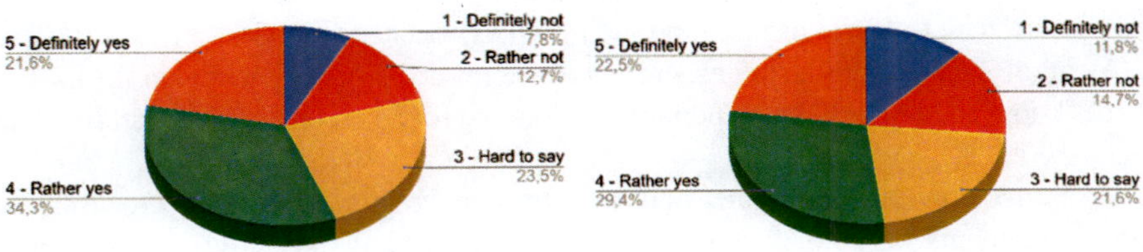

Fig. 12. The importance of process analysis in long-term and short-term decision-making

In the long term, the majority of respondents (56%) see the importance of processes in decision-making, while a relatively small group (21%) does not see the importance of process research. Similarly, in the context of short-term decisions, respondents using process analysis will apply the BPRPM method relatively quickly, as they already know the benefits of process analysis, and quick prototyping will enable them to efficiently verify the effects of planned decisions. For those who have negatively or neutrally referred to supporting their decisions with process knowledge, BPRPM is not only a method of improving the accuracy of decisions, but also a chance for a better understanding of the processes they are doing (Figs. 13 and 14).

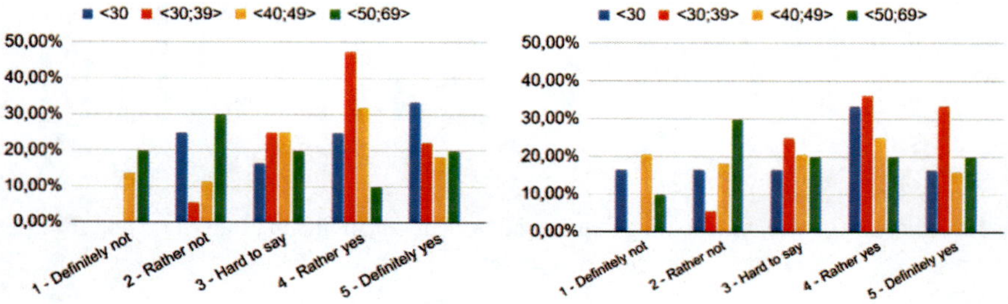

Fig. 13. The importance of process analysis in long-term and short-term decision-making—age of respondents for particular weights of this criterion

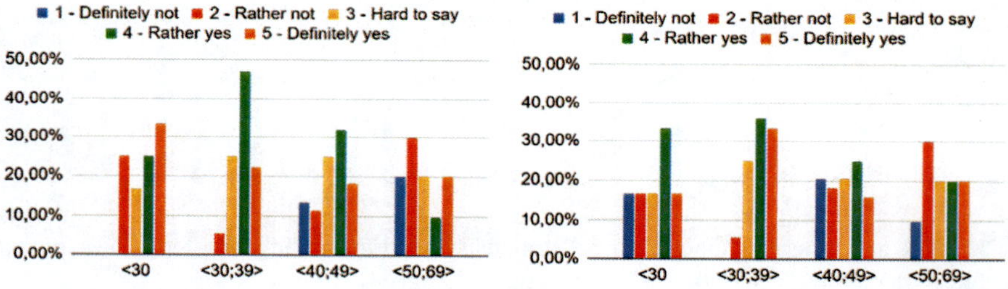

Fig. 14. The importance of process analysis in long-term decision-making and short-term—the importance of this criterion in particular age brackets

Younger respondents (up to 40 years of age) attach higher importance to knowledge gained from business processes than respondents over 50 years of age. This may result from greater dissemination of process analyses in the form of lectures and exercises at universities and numerous trainings in this field. On the other hand, older people attach more importance to social aspects and their own experience (Fig. 8).

As in the case of long-term decisions, short-term decisions are made on the basis of process knowledge mainly by respondents belonging to the second age bracket (between 30 and 39 years old).

Figure 15 shows the importance of observations in the decision-making process. Most of the respondents (73%) make long-term decisions based on the analysis of operational activities. As the BPRPM method is also used for standardized recording and analysis of current activities, it seems that it is particularly dedicated to this group of respondents. The strength of the BPRPM method is its ability to correlate planned future activities with current operational activities and to compare them with the use of Key Performance Indicators (KPIs) (Figs. 16 and 17).

The vast majority of respondents "believe in what they see" and make long-term decisions based on current observations. Only among a small group of experienced

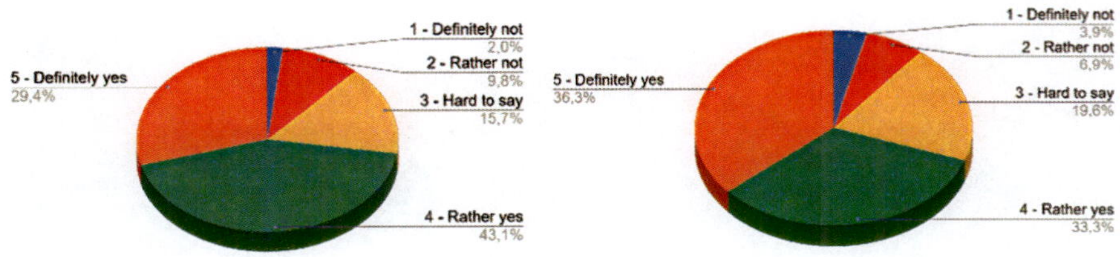

Fig. 15. The importance of observations of activities performed by resources in long-term and short-term decision-making

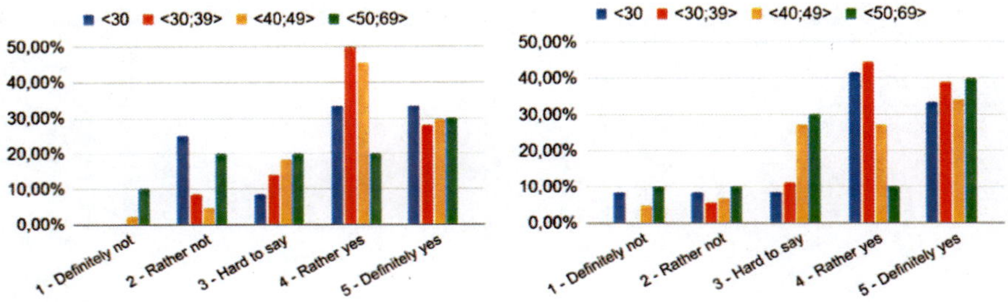

Fig. 16. The importance of observations of activities performed by resources in long-term and short-term decision-making—age of respondents for particular weights of this criterion

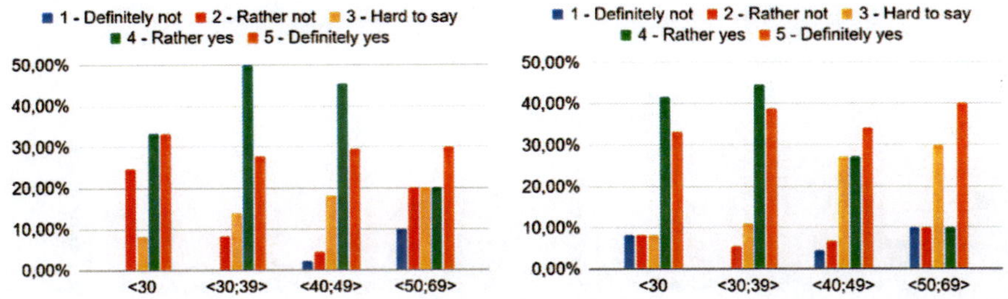

Fig. 17. The importance of observations of activities performed by resources in long-term and short-term decision-making—the importance of this criterion in particular age brackets

people (over 40 years old) there is skepticism to make strategic decisions based on current operational activities, this may result from their experience of dynamic technological development, which has changed traditional business models in recent decades (Fig. 18).

By far the highest percentage of respondents make both long-term (84%) and short-term (82%) decisions based on the analysis of the effects of operational activities. Such an approach requires permanent observation, which will allow to notice the correlation between the availability of resources and the effects of their activities. Since the observation of effects is a process, the BPRPM will be particularly effective in improving its effectiveness (Figs. 19 and 20).

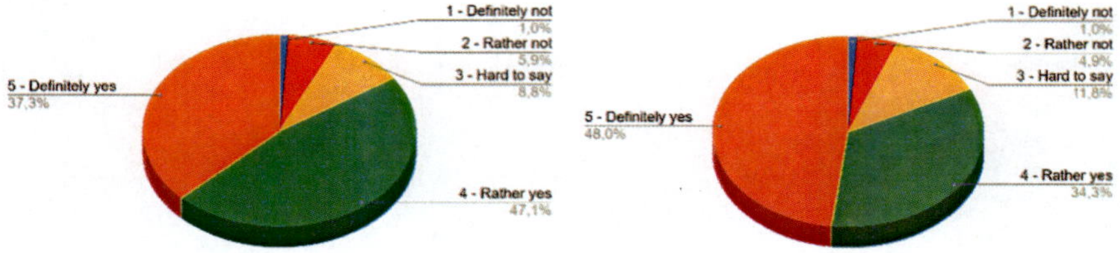

Fig. 18. The importance of analysis of the effects of current operational activities in long-term and short-term decision-making

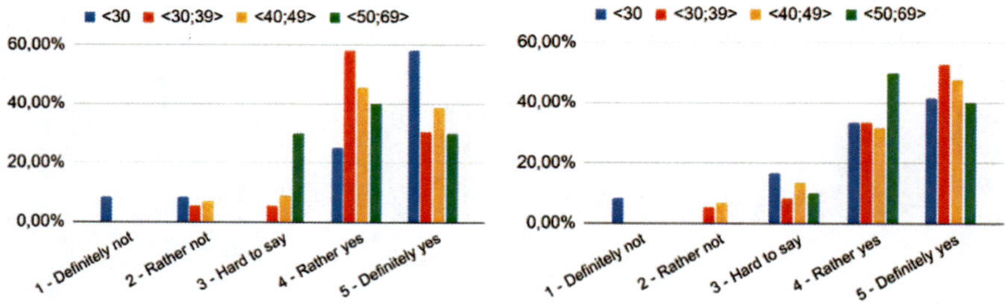

Fig. 19. The importance of analysis of the effects of current operational activities in long-term and short-term decision-making—age of respondents for particular weights of this criterion

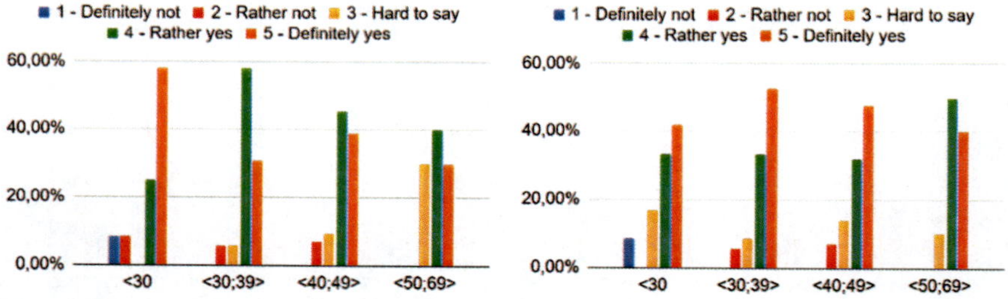

Fig. 20. The importance of analysis of the effects of current operational activities in long-term and short-term decision-making—the importance of this criterion in particular age brackets

Each of the surveyed age groups strongly declares that they make decisions by analyzing the effects of current activities. The BPRPM approach provides a tool for quick identification of process parameters determining the achievement of efficiency acceptable (expected) by decision-makers. Therefore, the aspect of efficient performance monitoring should become the main trigger for BPRPM (Fig. 21).

The lack of analysis of business processes is explained by a lack of tools. This doesn't mean that they are not aware of the existence of tools for process analysis. Lack of time, lack of skills and lack of knowledge are also an important barrier in conducting process analysis. Popularization of knowledge about the importance of processes, the possibility to feel the effects of process change and providing easy and friendly tools to study processes and evaluate the effects of proposed changes would allow the business organizations to develop much more effectively. Significant benefits would be to reduce

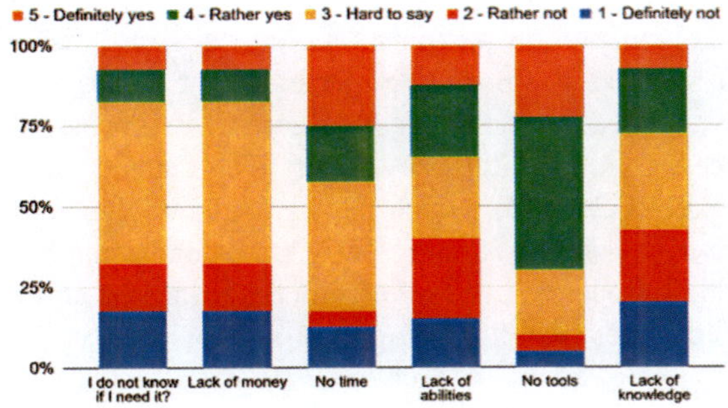

Fig. 21. Reasons for lack of process analysis

the costs of change by eliminating erroneous development decisions already at the stage of making them, shortening the time from the identification of the need to respond to the market situation to the implementation of a solution that responds to it. Such action is part of a Six Sigma process improvement tool DMAIC and refers to a data-driven improvement cycle used for improving, optimizing and stabilizing business processes and designs. DMAIC is not exclusive to Six Sigma and can be used as the framework for other improvement methodologies [9]. The application of the methods of virtualization of selected stages of the DMAIC cycle significantly shortens the time necessary to complete the full cycle (Fig. 22).

Fig. 22. DMAIC cycle

3 BPRPM

The method of rapid prototyping of business processes BPRPM (Business Process Rapid Prototyping Method), can be used regardless of the level of process maturity of the organization. BPRPM supports the "fast track" transformation of functionally managed organization to process managed organization. It allows to examine the impact of process changes even before their implementation. Business organizations, especially small and medium-sized ones, should not act intuitively, and process changes should result from accurate analyses, which are possible in a short period of time and with minimal involvement of the organization's personnel and financial resources. The method consists of three phases: Phase 1—modelling of the current process state (AS IS), Phase 2—transformation and creation of the target version of the process (TO BE), Phase 3—experience of process change (PROTOTYPING). The method, in the semantic layer, uses the notation BPMN 2.0 [10], which allows for unambiguous modelling of processes taking place in the organization in a way that is understandable both for people directly implementing processes, as well as process managers and IT specialists who implement process changes (Fig. 23).

Fig. 23. Phases in BPRPM

The first two phases should focus on building parametrized process models in order to compare their efficiency with the defined KPIs. Parameters are obtained during interviews or direct observations at the place of process execution. To verify the models and process behavior, software for simulating business processes is used. The study was carried out using iGrafx Process. The tool was selected due to the possibility of deep process parameterization, the possibility of process simulation together with subprocesses and many processes running in parallel (examination of mutual influence of processes on each other) and the possibility of creating simulation scenarios [11–13].

In phase 3, it is important to quickly start the process in the target state in order to gather information about its real effectiveness from the people who are implementing it. At this stage, the main functionalities supporting the designed change are delivered in the form of prototypes of mobile applications, web services and dedicated interfaces for information exchange.

The practical application of BPRPM makes it possible to check what would happen if the decisions to implement the simulated changes were taken a month, a quarter or a year earlier. This allows the assessment of their effectiveness in the natural ecosystem of the surveyed entity. The resulting business process models also allow to examine the resilience of processes, both before changes (AS IS) and after their implementation (TO BE), to various threats, e.g. fluctuations in demand for products and services, variability of raw material supply, absenteeism of process operators, interruptions in media supply, increase in labour costs, legislative changes, etc. The ability of BPRPM to predict the behaviour of business processes on the basis of prototypes of these processes and to enable decisions to be made well in advance fits into the overall concept of business process management (BPM) [14] although it can also be applied independently of a comprehensive BPM implementation.

4 Résumé

In making both short-term and strategic decisions, companies use, in most cases, ad hoc methods of analysis of the current situation, supported mainly by experience and opinions, rather than relevant data. Conclusions based on incomplete data are most often explained by a lack of tools, as well as a lack of time, skills and knowledge.

Since a low proportion of real-time (on-line) monitoring of business processes has been observed, there is a great potential for improving the organization through proper identification of processes, their simulation in order to assess the potential effects of change and their ongoing monitoring.

Quick adaptation of techniques included in BPRPM will allow business organizations to build a competitive advantage in a short time.

References

1. Sudoł, S.: Nauki o zarządzaniu (The sciences of management). PWE, Warszawa (2012)
2. Groves, R.M., Fowler, F.J., Jr., Couper, M.P., Lepkowski, J.M., Singer, E., Tourangeau, R.: Survey Methodology. Wiley, Hoboken (2009)
3. Matejun, M.: Metodyka badań ankietowych w naukach o zarządzaniu — ujęcie modelowe (Survey methodology in management science — a model approach). In: Lisiński, M., Ziębicki, B. (eds.) Współczesne problemy rozwoju metodologii zarządzania (Contemporary problems of management methodology development), pp. 341–354. Fundacja Uniwersytetu Ekonomicznego w Krakowie, Kraków (2016)
4. Chybalski, F., Matejun, M.: Organizacja jako obiekt badań — od zbierania danych do analizy wyników (Organisation as a research objective — from data collection to results analysis). In: Adamik, A. (eds.) Nauka o organizacji (Organizational studies), pp. 93–151. Oficyna Wolters Kluwer business, Warszawa (2013)
5. Babbie, E.: The Basics of Social Research. Wadsworth, Belmont (2013)
6. Taherdoost, H.: Sampling methods in research methodology; how to choose a sampling technique for research. Int. J. Acad. Res. Manag. (IJARM) **5**(20), 18–27 (2016)
7. Stańczyk, S.: Triangulacja — łączenie metod badawczych i urzetelnienie badań (Triangulation — combining research methods and making research more reliable). In: Czakon, W. (ed.) Podstawy metodologii badań w naukach o zarządzaniu (Basics of research methodology in management sciences), Oficyna Wolters Kluwer business, Warszawa (2015)
8. Kołodziejczak, M., Sobczyk, J.R.: Metodologiczne aspekty ogólnej charakterystyki organizacji i zarządzania jako odrębnej dyscypliny naukowej (Methodological aspects of the general characteristics of organisation and management as a separate scientific discipline). In: Błaszczyk, W. (ed.) Nurt metodologiczny w naukach o zarządzaniu (Methodological trend in management sciences), Wydawnictwo UŁ, Łódź (2006)
9. Wikipedia — DMAIC — a, 20 November 2020. https://en.wikipedia.org/wiki/DMAIC
10. ISO/IEC 19510 (2013)
11. Melão, N., Pidd, M.: Use of business process simulation: a survey of practitioners. J. Oper. Res. Soc., 2–10 (2003)
12. Alexander, C.W.: Discrete event simulation for batch processing. In: Proceedings of the 2006 Winter Simulation Conference, Monterey, CA, pp. 1929–1934 (2006)
13. Khemiri, A., Amine Hamri, M.E., Frydman C., Pinaton, J.: Improving business process in semiconductor manufacturing by discovering business rules. In: 2018 Winter Simulation Conference (WSC), Gothenburg, Sweden (2018)
14. Elzinga, D.J., Horak, T., Lee, C.-Y., Bruner, C.: Business process management: survey and methodology. IEEE Trans. Eng. Manag. **42**(2), 119–128 (1995)

Intelligent Modelling of Exploitation Systems of Machines and Devices – A Review of Concepts and Methods

Marian Brzezinski[1], Stanislaw Legutko[2(✉)], and Hubert Kędziora[2]

[1] Faculty of Security, Logistics and Management, Military University of Technology, 2nd S. Kaliskiego Street, 00-908 Warsaw, Poland
[2] Faculty of Mechanical Engineering, Poznan University of Technology, 3th Piotrowo Street, 60-695 Poznan, Poland
stanislaw.legutko@put.poznan.pl

Abstract. Modelling the systems of exploitation of machines and devices is a very important and a very complex issue, particularly in relation to new intelligent technical objects. Models can be applied to the assessment of the processes of exploitation, designing systems, forecasting of a system functioning and making exploitation decisions. A survey of literature, as well as the authors' own knowledge and practical experience have allowed to formulate the conclusion that systemic approach can be applied to modelling the system of machines and devices exploitation. Hence, the purpose of the present paper was, firstly, to review the applied methods of modelling the systems of exploitation of technical objects and to indicate their faults and merits; secondly, to indicate the possibilities of applying the systemic approach to modelling to exploitation of technical systems with the use of intelligent tools and techniques of Industry 4.0.

Keywords: Industry 4.0 · Modelling of exploitation systems · Systemic approach · Exploitation system security

1 Introduction

Modelling the systems of exploitation of machines and devices is a very important and a very complex issue. Models can be applied to the assessment of the processes of exploitation, designing systems, forecasting of the system behavior and making exploitation decisions. Although there is a number of works in the field, there still exist the necessity to seek new ways of modelling. A survey of literature, as well as the authors' own knowledge and practical experience have allowed to formulate the conclusion that systemic approach can be applied to modelling the system of machines and devices exploitation.

It has been assumed that the purpose of the elaboration can be achieved by: characterizing the methodological basis of exploitation modelling, presentation of modelling methods and elaboration of the process of modelling the system of exploitation of machines and devices [1–4]. The object of investigation is the systems of exploitation of machines and devices and the processes realized within them. In the investigation, both theoretical and empirical methods have been applied to solve the presented problems.

A. Batako et al. (Eds.): GCMM 2021, LNNS 335, pp. 614–623, 2022.
https://doi.org/10.1007/978-3-030-90532-3_45

2 Methodological Basis of Modelling Exploitation Systems

In the engineering of systems, the basic tool of learning the reality is models. They are recognized as originals because they are the only reality available for us. Learning a definite fragment of reality, which is the system of exploitation, and its modification requires creation of a model [4–8]. In modelling exploitation systems, holistic approach should be applied. It is an epistemological-methodological orientation according to which the exploitation systems should be viewed as wholes from various points of view.

Modelling is a sequence of actions leading to the creation of a model of the system. The elaborated model should correspond to the actual exploitation system. A model is corresponding if there is functional similarity of the system and the model, i.e. similarity of the elements and relations within the system and with its environment. The degree of similarity between the system of exploitation and its model can be projected isomorphically or homomorphically. In the case of isomorphic projection of the exploitation system by its model, mutually unique projection between elements, as well as between the relations of the system under examination and its model takes place. This means that, to each element of the system has a uniquely defined corresponding relation in the model of the system and vice versa. The correspondence is symmetrical and, therefore, acts in both directions. In the case of isomorphic models, projection of functional properties takes place.

Homomorphic models are models of the exploitation system simplified in various ways and, therefore, simplifications of its isomorphic model. Consequently, a given system can have one isomorphic model and many homomorphic ones determining various kinds and degrees of similarity between that system and its models. This means that, if we learn the examined system by examining its homomorphic models, we acquire less information about it. The less degree of similarity between the system and the given model, i.e. the weaker is homomorphic projection of the examined system by the given model. In the case of isomorphic models, we deal with mapping of structural properties. Modelling of exploitation systems can be performed by descriptive, graphic and mathematical modelling [9]. The result of modelling can be semantic, graphic, heuristic and mathematical models.

Semantic models of exploitation systems result from descriptive modelling. They contain descriptions of super system, exploitation system and its subsystems, purposes, tasks and structure, features and parameters of the system, problems to be solved and the system environment. It can also be a system of notions, standards, regulations and principles of exploitation functioning. Conceptualization of a semantic model consists in identification or determination of the set of major features important for the description of the exploitation system, relations between them and ways of its functioning. The exploitation systems can be presented in the form of layouts, schedules, tables, charts of process progress, etc. which are graphic models of them [10, 11].

In graphic presentation of the models of exploitation systems, various symbols, conventional designations, verbal, numerical, letter descriptions and lines of predetermined thickness are used. Graphic models make it possible to find the major links and discontinuities in the exploitation system, subjects and times necessary to perform the individual actions [12].

Heuristic models are most often used for making forecasts and in the processes of decision making. They can be used to determine the state of exploitation, to indicate the directions and dynamics of its development or to select tote best exploitation strategy. They make it possible to discover new facts and relations between them and to discover new truths. Heuristic models are built basing on the opinions of experts taking advantage of their knowledge, experience and intuition. Building of heuristic models is most often effected with the use of a brain storm and the Delphi method. In the process of exploitation modelling, inquiry techniques are applied and the investigation tool is a questionnaire.

Mathematical models reflect exploitation systems by means of mathematical relationships. In the opinion of the authors, presentation of an exploitation system, even in the simplest mathematical form, can bring the best learning effects. Mathematical models are made with predetermined assumptions and limitations and, consequently, they are a kind of simplification of the exploitation system as any other model. Elaboration of a mathematical model of an exploitation system is preceded by elaboration of a semantic model and often a graphic one, too.

Mathematical models are used, first of all, for the assessment of the systems and processes of exploitation of machines and devices There are many practical reasons for which exploitation assessment of machines and devices, as well as exploitation systems, is purposeful. It is a complicated issue, nevertheless it is purposeful to perform a possibly effective verification of the parameters of assessment of machines and devices, as well as exploitation systems. Each assessment contains an element of learning contributing additional information to the set of knowledge at the disposal of the assessing person or puts in order the set of information about technical objects and the exploitation system of machines and devices.

Depending on the character of parameters, investigation of the exploitation systems can performed with the use of the following kinds of models: deterministic ones, probabilistic ones, stochastic ones and simulation ones.

3 Methods of Modelling Exploitation of Machines and Devices

Bases for creation of mathematical assessment models in the field of exploitation can be sought, first of all, in:

- multicriterial comparative analyses;
- theory of reliability;
- mass service theory;
- the field of artificial intelligence.

The methods of multicriterial comparative analyses can be applied to compare the exploitation systems with identical or close functional destination. The systems are compared according to identical features which can be presented quantitatively [1, 5–8].

In the field of multicriterial comparative analyses, there are many methods including: numerical taxonomy, ELECTRE, Belinger's AHP (Analytic Hierarchy Process), PROMETHEE.

The number and variety of the features of an exploitation system which can be assessment criteria is very high and not all of them can be learned. It is enough to distinguish important features (which determine their nature and enable them to be selected from other systems) from all the features of the exploitation systems. The selection of features by means of which a system can be characterised is the most important and the most difficult element of the process of assessment. It depends on the purpose, scope and investigation aspect of the exploitation system. The principal purposes of the assessment of exploitation systems include comparison of two or a number of systems, verification of the degree of meeting the requirements, assessment of the system design, selection of the best solution.

Considering temporal aspects, two kinds of assessments are distinguished: retrospective ones (ex post) formulated for the past states of the exploitation system and prospective ones (ex ante) formulated for the future states of the exploitation system.

In the authors' opinion, the most important features of the exploitation systems are: usefulness, durability, readiness, economy, reliability, elasticity and controllability.

In multicriterial methods of comparative analyses, the comparison criterion of the systems can be the state of quality which is a function of the best essential features of the exploitation systems being compared [5, 21]. The assessment criteria can be quantitatively and qualitatively determined. The quantitatively determined values of a criterion criteria are its parameters, i.e. real denominate numbers measurable in the physical sense. The application of the multicriterial comparative analysis will also require quantification of the qualitative criteria by attributing definite numerical values to them.

According to the authors, assessment of the technique exploitation systems can be performed basing on the indices by means of which the exploitation systems can be characterised. They are, therefore, features of the exploitation systems. Next, by attributing to them weights and application of a selected method of multicriterial analyses, one can determine a collective index of the assessment of exploitation systems [13]. Using the indices, one can also assess the processes of utilization, servicing or exploitation strategies. Assessment of the exploitation systems can be effected with the use of the indices determined as result of the application of models such as: reliability models of the whole system, as a combination of the reliability of the subsystems of utilization, servicing and management, overall equipment effectiveness (OEE), Key Performance Indicators (KPI) [14–17].

Assessment of exploitation systems requires selection of indices and determination of their values. The parameters which determine the values of indices differ in the direction of influencing the complex index of the assessment of the exploitation system. This influence can be positive, negative or neutral. In this aspect, the assessment criteria can be divided into stimulants, destimulants and nominants. A stimulant is such a diagnostic variable whose increase is to be associated with growth of the complex index of assessment of the system (technical object). A destimulants, on the other hand, is such a diagnostic variable whose increase is associated with a drop of the complex index of the assessment of the system; its drop, on the other hand, with a growth of that index. Nominant is a variable which has a definite value, most advantageous (in the aspect of a complex phenomenon assessment) value referred to as the nominal value. The nominant has values lower or higher than the nominal one with the drop or growth of the complex index of the assessment of the exploitation system, respectively. In practice, there are situations in which the nominal values constitute a definite numerical interval.

Any deviations of the nominant from the normal level are a negative phenomenon from the point of view of the investigated index of the assessment of the exploitation system. Nominants can be easily transformed into destimulants by determining the absolute of the given value deviation from the level considered as nominal.

A very important thing to enable correct process of aggregation is unification of the variables character, i.e. putting all variables into the role of stimulants or destimulants. In practice, the first option prevails. Special formulas unifying the variables have been elaborated.

Basing on the knowledge of partial assessment indices, applying a selected method of multicriterial analyses, one can determine the collective index of assessment of, e.g. the quality of the exploitation system. Next, by comparing to another system, requirements or standard model, one can select the exploitation system of the highest quality. The performed assessment can also be a basis for perfecting the exploitation system.

Assessment of the exploitation systems of machines and devices can be performed with the use of models built on the basis of the theory of reliability [5, 8, 18, 19]. The term, "reliability" is used for general, not numerical, description. It characterises the exploitation system's ability to ensure utilization of technical objects and servicing them in definite conditions and period of time. One can distinguish utilization reliability, servicing reliability and management reliability.

Reliability is one of the most important features of each technical object. It is a function of such features as damage proofness, serviceability and provision of servicing media. Provision of servicing media is the organization ability to provide, upon request, the media necessary for servicing the object under the conditions of its utilization and servicing. This feature determines the quality of functioning of the servicing system. Consequently, reliability of military technology depends not only on its features but also on the features of utilization and features of servicing. Reliability of an exploitation system is a collective feature and includes subjectively selected partial properties which can be defined quantitatively by the assessment indices.

The assessment indices should:

- incorporate the influence of various internal and external factors on the technical object or exploitation system and its reliability;
- be useful for engineers' reliability calculations;
- enable formulation of reliability requirements for designed and acquired machines and devices or designed exploitation systems;
- be easy to apply in the exploitation practice.

Basing on the knowledge of the partial indices of assessment of technical objects, one can determine the collective index of the reliability assessment of an object or exploitation system.

Summing up, due to the accidental character of damages of machines and devices, the reliability indices are probabilistic magnitudes. They are determined by means of the methods of mathematical statistics and theory of probability. The reliability indices of technical object can be used for:

- planning of exploitation;
- comparison of reliability of technical objects;
- comparison of exploitation system;
- organization of utilization and servicing processes;
- planning of provision of technical materials.

The basis for building assessment models in the field of exploitation systems can be sought in the mass service theory, also referred to as the queueing theory [1].

The queueing theory is derived from theory of stochastic processes which is a part of the theory of probability. A stochastic process is considered as a mathematical abstraction of certain empirical arrangements randomly changing in time.

Stochastic processes are random ones depending on time. They are defined as a $X(t)$ function depending on the time parameter, whose values are random variables at each moment. Stochastic processes in which there is a finite or calculated set of values to be taken by the realizations of those processes are called stochastic processes of discrete realizations. The values, which the realizations of such stochastic processes can have, are called states, and the whole process is considered as a phenomenon of random changes of the state of a certain system. In the authors' opinion, queue models can be applied for the assessment of the systems of repairs and technical servicing with the use of Markov processes.

Creation of classical mathematical models of exploitation processes is a very complex matter and not always possible. The use of classical mathematics to build models of exploitation requires incorporation of a large number of assumptions, limitations and simplifications in order to bring them to the simplest form and to present them in the form of linear equations. Hence, models of exploitation processes can reflect exploitation reality to a small extent and be not much useful in assessments being bases for decision making. In classic approach, problem solving is effected by means of algorithms, that is a finite series of clearly defined actions. Since not all the parameters considered in the process of creating exploitation models can be linearized, many variables have qualitative character and the relations between them are not always known, modelling can be based on techniques originating from artificial intelligence [20]. In the authors' opinion, in modelling of the exploitation processes, artificial neuron networks can be used [5, 21].

Neuron networks can be used for description of phenomena with imprecise, ambiguous character which cannot be coped with by classical mathematics and two-valued logic. They are used to solve complex problems not by means of algorithms but by the processes of knowledge analysis.

Models of exploitation processes built basing on artificial neuron networks can be used, particularly in conditions involving problems of information quality. The attributes of information quality are, first of all, up-to-dating, completeness, availability, credibility and accuracy. In such conditions, making exploitation decisions is particularly complicated and artificial neuron networks would play the role of a specific expert.

Artificial neuron networks can boost decision making in the field of organization and management of exploitation systems of machines and devices and the processes of utilization and servicing. In the author's opinion, the most important criteria of the assessment of exploitation decisions can be: effectiveness (the degree of the purpose realization) or the cost of the exploitation processes.

One of the known structures of artificial neuron networks to be used for building models of exploitation processes is the one-layer one-direction perceptron with an algorithm of iterative teaching of the network backward propagation). The architecture of an artificial neuron network consists of: input, layers of data processing and output. neutrons. The neutrons are arranged in the following layers: input one, hidden ones and the output one. They are interconnected with the principle "each to each" between the layers but there is no connection between the neutrons within the individual layers. The degree of complexity of the artificial neuron network is determined by the number of hidden layer and the number of neutrons. When using an artificial neuron network to boost the exploitation processes, it is best to apply an algorithm of teaching the network with a teacher which requires indication of the desired input value as a model one. This can be cost minimization or maximum effectiveness.

Input information to the artificial neuron network is a set of data characterizing the exploitation system. When using an artificial neutron network to boost exploitation decisions, the input information can be grouped in the following way: kinds of machines and devices, utilization features of machines and devices, exploitation conditions, features related to the performance of servicing processes, kind, conditions and intensity of the operational processes of the organization, damages of machines and devices and their structure, training and experience of exploitation managers, as well as staff and servicers. Each of the stated pieces of information can be characterized by sets of parameters which should be quantified. Output of the artificial neuron network is, for example, assessment of effectiveness of the realization of the exploitation processes which is a significant part of the assessment of the degree of realization of the operational objectives of the organization.

4 Process of Modelling the Exploitation System of Machines and Devices

There are many practical reasons why it is purposeful to assess the exploitation systems of machines and devices. It is a complicated issue, nevertheless is advantageous to perform an effective verification of the parameters of assessment of the exploitation systems of machines and devices. Each assessment contains a cognitive element contributing additional information to the set of knowledge of the person performing the assessment or puts in order the set of information about the exploitation system. The knowledge is a basis of diagnosis and making exploitation decisions.

Since the exploitation systems are particularly complex social- technical systems combining many various functions, their investigation and description should be performed from various points of view. They can be examined and described in a systemic, economic, management, computer aspect, etc. The essence of modelling exploitation system has been shown in Fig. 1.

Each description of an exploitation system is performed from a definite point of view and, consequently, it is a subjective one. The sum of subjective descriptions from various points of view gives more knowledge about the exploitation system. The knowledge is still subjective, but it contains more information about the system under examination

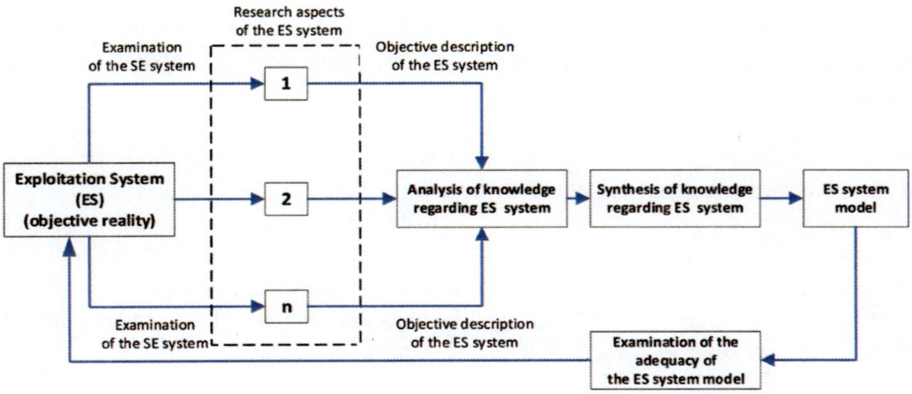

Fig. 1. The essence of modelling of machine and device exploitation systems.

and, as such, it is closer to reality. Hence, the model built on the basis of the acquired knowledge is closer to the original.

The process of assessment of exploitation systems of machines and devices includes the following actions:

- determination of the assessment purposes and description of the assessment object;
- determination of the assessment criteria;
- elaboration of the assessment model;
- elaboration of the assessment algorithm and performance of calculations;
- analysis of the assessment results and verification of them;
- application of the assessment results.

The objectives of the assessment of the exploitation system of machines and devices can be as follows: determination of the system (process) condition, learning its functioning, designing the system, forecasting its development and making decisions.

One of the most important actions in the process of assessment is determination of the criteria and creation of an assessment model. The process of assessment should start with recognition of the object of assessment, i.e. acquisition of knowledge about it making it possible to determine its features [22]. The number of system features is infinite and they exist objectively. From the whole variety of features, we are usually interested in those which depend on the purpose, scope and aspect of investigation. Analysing the required assumptions and limitations concerning the assessment, one can determine a subset of assessment criteria in the set of features. The number of criteria is finite and they are subjectively determined. Sometimes one criterion is sufficient for the assessment of a military technical object; the criterion may be, for example reliability.

The set of assessment indices should meet the following requirements [23, 24]:

- completeness of the assessment consisting in incorporation of all possible aspects of the problem under consideration;
- coherence of the assessment consisting in that each criterion properly forms the global preferences of the decider;
- no redundancy, i.e. unrepeatability of the semantic scopes of the criteria.

It is recommended that the set of criteria contains no more than 7 ± 2 criteria due to the limitations of the decider's perception and, consequently, the quality of the decision making process [23].

Determination of the features of assessment criteria and, particularly, quantification of them and creation of quantitative models of assessment of exploitation systems of machines and devices is a very complex issue.

The elaborated model should be examined in the aspect of adequacy to the real object or exploitation system. This consists in verification of its design in accordance with the limitations and requirements formulated prior to the modelling process. A model is adequate if there is functional similarity of the object or system and their model, i.e. the elements and relations inside the system and with its environment.

5 Summary

The principal purpose of the paper was theoretical elaboration of the issue of modelling exploitation systems of machines and devices. It has been indicated that systemic approach can be applied to modelling the exploitation system of machines and devices. It is a complicated issue; nevertheless it is purposeful to perform a possibly effective verification of the quality of the exploitation systems. There are many practical reasons for which examination of exploitation systems is purposeful. Firstly, models can be used to evaluate and analyse technical systems and the environment in which they perform their functions. Secondly, by modelling the exploitation system, one can identify its weaknesses and thus determine the direction of improvement activities. Thirdly, the exploitation models are a tool for decision makers, thanks to which they can make rational and safe decisions.

References

1. Brzeziński, M.: Modelling the System of Military Equipment Repairs of Land Forces. Military University of Technology, Warsaw (2010). (in Polish)
2. Brzeziński, M., Kijek, M., Owczarek, P., Głodowska, K., Zelkowski, J., Bartosiak, P.: Aspects of improvement in exploitation process of passenger means of transport. J. Adv. Transp. **2018**, 13 pages (2018). Article ID 5062165
3. Legutko, S.: Development trends in machines operation maintenance. Eksploatacja i Niezawodnosc. Maint. Reliab. **2**(42), 8–16 (2009)
4. Jasiulewicz-Kaczmarek, M., Legutko, S., Kluk, P.: Maintenance 4.0 technologies – new opportunities for sustainability driven maintenance. Manag. Prod. Eng. Rev. **11**(2), 74–87 (2020)
5. Brzeziński, M.: Assessment of Land Forces Technology Exploitation Systems. Military University of Technology, Warsaw (2018). (in Polish)
6. Brzeziński, M.: Logistics systems engineering. Military University of Technology, Warsaw (2015). (in Polish)
7. Loska, A.: Scenario modelling exploitation decision-making process in technical network systems. Eksploatacja i Niezawodnosc Maint. Reliab. **19**(2), 268–278 (2017)
8. Loska, A.: Methodology of variant assessment of exploitation policy using numerial taxonomy tools. Manag. Syst. Prod. Eng. Eksploatacja i Niezawodnosc Maint. Reliab. **18**(2), 96–104 (2015)

9. Krolczyk, J.B., et al.: Material flow optimization - a case study in automotive industry. Tehnicki Vjesnik-Tech. Gazette **22**(6), 1447–1456 (2015)
10. Kotowska, J., Burduk, A.: Optimization of production support processes with the use of simulation tools. In: Wilimowska, Z., Borzemski, L., Świątek, J. (eds.) ISAT 2017. AISC, vol. 657, pp. 275–284. Springer, Cham (2018). https://doi.org/10.1007/978-3-319-67223-6_26
11. Górnicka, D., Burduk, A.: Improvement of production processes with the use of simulation models. In: Wilimowska, Z., Borzemski, L., Świątek, J. (eds.) ISAT 2017. AISC, vol. 657, pp. 265–274. Springer, Cham (2018). https://doi.org/10.1007/978-3-319-67223-6_25
12. Wyczółkowski, R., Piechowski, M., Gładysiak, V., Jasiulewicz-Kaczmarek, M.: The concept of intelligent chlorine dosing system in water supply distribution networks. Adv. Intell. Syst. Comput. **835**, 350–359 (2019)
13. Jasiulewicz-Kaczmarek, M., Żywica, P.: The concept of maintenance sustainability performance assessment by integrating balanced scorecard with non-additive fuzzy integral. Eksploatacja i Niezawodnosc. Maint. Reliab. **20**(4), 650–661 (2018)
14. Jasiulewicz-Kaczmarek, M., Bartkowiak T.: Improving the performance of a filling line based on simulation. In: ModTech International Conference - Modern Technologies in Industrial Engineering IV (2016). IOP Conference Series: Materials Science and Engineering 145, 042024
15. Drożyner, P., Jasiulewicz-Kaczmarek, M.: Maintenance stakeholders from OEE perspective – case study. In: Yue, X.G., Mc Aleer, M. (eds.) International Conference on Economics and Management Innovations (ICEMI 2017), vol. 1, no. 1, pp. 397–400 (2017)
16. Domingo, R., Aguado, S.: Overall environmental equipment effectiveness as a metric of a lean and green manufacturing system. Sustainability **7**, 9031–9047 (2015)
17. Stadnicka, D., Antosz, K.: Overall equipment effectiveness: analysis of different ways of calculations and improvements. In: Hamrol, A., Ciszak, O., Legutko, S., Jurczyk, M. (eds.) Advances in Manufacturing. LNME, pp. 45–55. Springer, Cham (2018). https://doi.org/10.1007/978-3-319-68619-6_5
18. Krolczyk, G., Legutko, S., Krolczyk, J., Tama, E.: Materials flow analysis in the production process - case study. In: Velisek, K., Kostal, P., Nad, M. (eds.) Novel Trends in Production Devices and Systems, Book Series: Applied Mechanics and Materials, vol. 474, pp. 97–102 (2014)
19. Antosz, K., Ratnayake, R.M.C.: Machinery classification and prioritization: empirical models and AHP based approach for effective preventive maintenance. In: Proceedings of the 2016 IEEE IEEM, pp. 1380–1386 (2016)
20. Antosz, K., Pasko, L., Gola, A.: The use of intelligent systems to support the decision-making process in lean maintenance management. IFAC PapersOnLine **52–10**, 148–153 (2019)
21. Brzeziński, M., Kijek, M., Gontarczyk, M., Rykała, Ł., Zelkowski, J.: Fuzzy modelling of evaluation logistic systems. In: Proceedings of 21st International Scientific Conference. Transport Means 2017, Part II, Kaunas, pp. 377–382 (2017)
22. Jasiulewicz-Kaczmarek, M.: Identification of maintenance factors influencing the development of sustainable production processes–a pilot study. IOP Conf. Ser. Mater. Sci. Eng. **400**, 062014 (2018)
23. Żak, J.: The Methodology of multiple criteria decision making/aiding in public transportation. J. Adv. Transp. **45**, 1–20 (2011)
24. Franciosi, C., Voisin, A., Miranda, S., Riemma, S., Iung, B.: Measuring maintenance impacts on sustainability of manufacturing industries: from a systematic literature review to a framework proposal. J. Clean. Prod. **260** (2020). Article 121065

Development of New Vitrified Bond for Diamond Grinding of Polycrystalline Diamond Compacts

Xiao-Fu Zhang[1], An-Xian Lu[2], Yu Wang[2], and Andre D. L. Batako[3](✉)

[1] Research for Abrasives and Grinding, 121 Wutong Street, Zhengzhou 450001, Henan, China
[2] School of Materials Science and Engineering, Central South University, Changsha 410083, China
[3] General Engineering Research Institute, Liverpool John Moores University, Byrom Street, Liverpool L3 3AF, UK
a.d.batako@ljmu.ac.uk

Abstract. It is well-known that diamond, be it natural or industrially man-made is the hardest material. Therefore, machining diamond parts such as polycrystalline diamond compacts (PDC) is a serious challenge because there is second to none in terms of hardness. This work presents an investigation and the development of a new vitrified material based on Li2O-Al2O3-SiO2 glass ceramics, which was subsequently used to manufacture a grinding wheel. The performance of the developed diamond grinding wheel is studied here in a cylindrical grinding process of PDCs. The factors that influence the properties of grinding wheel such as amount of vitrified bond and the types and the amount of stuffing and additives in grinding wheel matrix were also investigated. It was found that additives such as Al2O3 or SiC reduced the grinding wheel life. Importantly, it was noticed that the new vitrified bond firmly held the diamond grains providing an extended wheel life of about 2.5–3 times of that of resin bond. The precision of the PDCs ground parts was improved from 0.03 μm to 0.01 μm. The grinding time was reduced by a factor of two, and the process was quieter than grinding with resin bonded wheels.

Keywords: Polycrystalline diamond compact (PDC) · Vitrified bond · Diamond grinding wheel

1 Introduction

A polycrystalline diamond compact (PDC) is made of a polycrystalline diamond layer and a tungsten carbide substrate. The polycrystalline diamond layer possesses both extremely high hardness and abrasion resistance, whereas the tungsten carbide substrate greatly improves the toughness and the weldability of PDC. Therefore, PDC is widely used as the aiguilles in the petroleum industry, the geology exploration tools or the cutting tools for processing nonferrous metals and nonmetal materials [1, 2].

The diamond layer has a larger cylindrical diameter than the tungsten substrate layer, and it is very irregular after high pressure and high temperature processing [3]. For their applications, PDC cylinders are ground to a finished size. However, this is

A. Batako et al. (Eds.): GCMM 2021, LNNS 335, pp. 624–631, 2022.
https://doi.org/10.1007/978-3-030-90532-3_46

a challenging task because both PDC and tungsten carbide are hard to machine [4]. Commonly resin bond diamond wheels are used to grind the cylinder of PDC, with the associated drawbacks such as high tools wear, low grinding efficiency, high process cost, poor process precision, high noise, and smell which is harmful to the worker's health [5].

Therefore, there is a need in developing new bonds, which is the purpose of the paper, where a new vitrified bond is engineered and investigated. Vitrified bonds diamond grinding wheels are sintered at temperatures over 1273k, and the diamond grains bind strongly with the bond to secure higher wheel strength with reference to resin bond wheels, which is solidified about 473k. To secure a high efficiency during the grinding process, pores are introduced in the structure of vitrified bond. Since the vitrified bond is of a glassy structure, it is rigid almost non deformable during the grinding and this rigidity lead to high precision grinding with a very good size holding for the PDCs.

Though vitrified bond diamond grinding wheels are suitable for grinding the cylinder of PDC [6], diamond grits oxidize at temperatures over 1023k [7]. However, existing techniques of producing vitrified bond grinding wheels use a sintering process above 1473k so that to achieve adequate bonding strength. This affected the performance of conventional vitrified diamond wheels; consequently, seldom these wheels are used for grinding the cylinder of PDC [8, 9]. In this paper, a new vitrified bond diamond grinding wheel for grinding PDC is developed, and the main factors which influence the performance of grinding wheel and the process are investigated.

2 Experimental Work

The vitrified bond was developed on the basis of the Li_2O-Al_2O_3-SiO_2 compound glass ceramic. Li_2O–Al_2O_3—SiO_2 glass ceramic is widely used for its high mechanical strength along with a wide adjustable range of thermal expansion coefficient [10], however the Li_2O –Al_2O_3—SiO_2 does not bind firmly with diamond grains at the sintering temperatures lower than 1273K because of its high melting temperature and high viscosity. Therefore, some ZnO, B_2O_3 and P_2O_5 were added to the compound to decrease its melting temperature [11]. The resulting chemical composition of the compound is listed in Table 1.

Table 1. Composition of the ceramics cement (Mole fraction, %)

Li_2O	ZnO	Al_2O_3	SiO_2	P_2O_5	B_2O_3
2.1–5.0	14.5–19.5	15.5–22.0	48–65	1.3–1.5	4.5–10.8

The compound is melted at the requested temperature, held up for a specific time, then cooled in water with subsequent drying and crushing into powder, and then mixed with the diamond grains. There are six grades of synthetic diamonds on the market, these are in order of lowest to the highest, RVD1, RVD2, MBD4, MBD6, MBD6, SMD, with their commercial price being proportional to the grade. To secure a good wheel life along with desired bonding strength with the diamond grains the RVD1 grade was used in the experiment. To overcome diamond instability at high temperature, the diamond grains were coated with a thin film of Titanium. This provided a reinforced bonding strength between titanium and diamond and the vitrified bond itself. The titanium coat on the surface of diamond crystal oxidized during sintering resulting into Ti_3O_5 an inert substance that protects diamond from oxidization during the sintering process [12].

In general grinding wheel are specified by a standard marking and in this investigation the following wheel was designed: 1A1 400 * 40 * 203 * 10. This reads outer diameter ~400 mm, inner diameter ~203 mm, thickness of diamond abrasive layer ~40 mm, and wheel width ~10 mm grinding. The vitrified diamond abrasive was pressed into segments that were glued onto an aluminum disc as shown in Fig. 1a. Here, the grains size were 90–107 μm with 21% (wt %) of additives to the bond, which resulted into a density of segments being around $2.35 * 10^3$ kg/m^3.

The grinding tests were undertaken on a cylindrical grinder with a wheel speed of 45 m/s and the PDC workpiece was ground from a diameter of ϕ14.6 mm down to ϕ13.4 mm. The workpiece had a tungsten carbide of 0.8 mm in thickness. The coolant used was a 2% soda water.

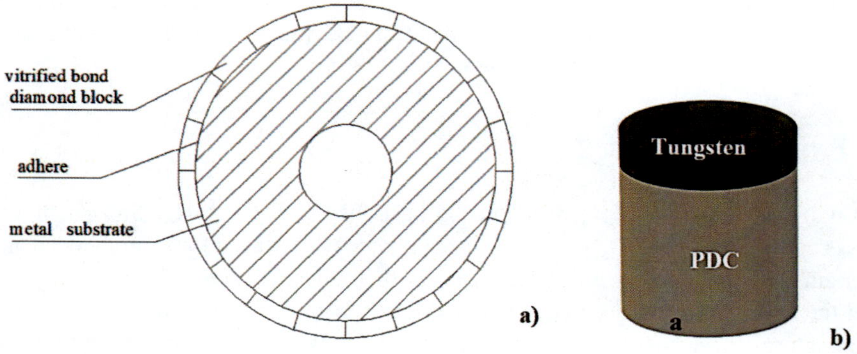

Fig. 1. Outlines, **a)** Grinding wheel; **b)** PDC cylinder.

3 Results and Discussion

To contrast the performance of the newly developed bond, set of tests were undertaken in real high volume production plant and were compared with the reference test grinding using conventional resin bond.

3.1 Wheel Life

The performance of the new wheel was characterized though a longitudinal study whereby new grinding wheels were utilized to grind workpieces until a full and complete wear out of the abrasive layers. The initial experimental results these studies are given in Table 2.

Table 2. Comparison of the effects for processing PDCs by using different bond wheels

Bond type	Time/part, sec	Precision, μm	Grinding ratio	Wheel life, parts/wheel	Noise level
Resin	200	±30	~0.17	750	**High**
Vitrified	**112**	**±10**	**~0.43**	**~2200**	**none**

It is observed from Table 2 that the new bond outperformed the resin bond across all parameters. This is because the rigidity of the grinding wheel plays an important role and the resin bond wheel has a young modulus of 500 MPa, while that of vitrified bond is 46.10^3 MPa [13, 14]. Therefore, for the same grinding conditions, the cutting force would be the same; however, the wheel deflection would be larger on the resin bond wheel. This explains the low precision on the ground parts, whereas, the deflection of the vitrified bond wheel was about 1% of that of resin bond grinding wheel.

The 40% lead on grinding time is due to the sharpness of the vitrified wheel and less deflection, which contribute to shorter machining time per part. In addition, the bond itself that is Al_2O_3 is an abrasive, which potentially contributes to the cutting process.

In terms of wheel life, the rigidity has a major effect along with the strong bonding achieved by introducing the titanium coating that generated a Ti_3O_5 crystal lattice leading to a structure like $Al_2O_3{\cdot}TiO_2$ in vitrified bond [15]. In the resin bond, there is no chemical reaction between the elements, but a simple mechanical force between diamond grains and bond, therefore, the diamond grains are easily dislodged during grinding. In addition, the temperature in the cutting zone is about 923k–1123k that affect the ability of the resin bond hold the grain [16–19]. Conversely this temperature does not affect the vitrified bond, and a grinding wheel with pure vitrified bond (no additives) has a life of 2.5–3.0 times of that of resin bond.

Figure 2 illustrates the life performance of vitrified bonded wheel as a function of the bond content. Here the test done in a continuous cutting process until the entire layer of the abrasives was stripped off the aluminium disc. It is observed that 21% of vitrified bond provided the maximum number of parts ground per wheel.

3.2 Effects of Additives on Wheel Life

To control the grinding process performance, workpiece integrity and wheel live it necessary to include in the bond a relatively small amount of additives, [20]. In general, corundum and carborundunm are the two main additives, which are widely used in diamond grinding wheels. In this study, a controlled proportion of corundum and carborundunm was added to the vitrified bond and grinding tests were undertaken to quantify the effect of these additives on the wheel life. Figure 3 shows a relative performance of vitrified bond over resin bond as a function of additives percentage. Here, it observed that when there is no additive, i.e. 100% pure vitrified bond, the outperformed the resin bond by a factor of 3 in terms of wheel life. However, the additives led to a reduction of life with 35% of additive giving a good wheel life. The carborundum had an acceptable performance compared to the corundum.

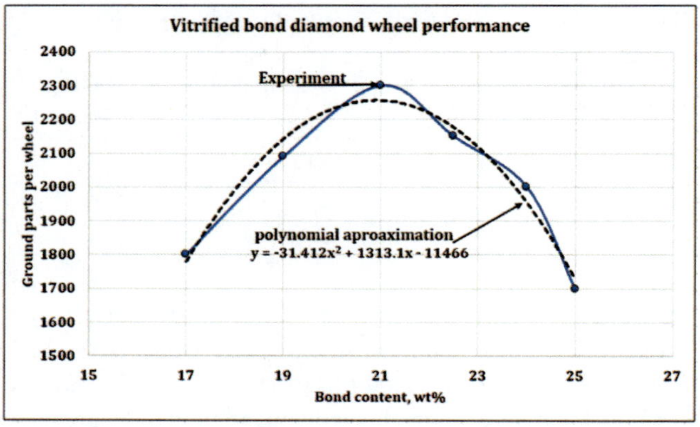

Fig. 2. The longevity change with the amount of vitrified bond in the grinding wheel

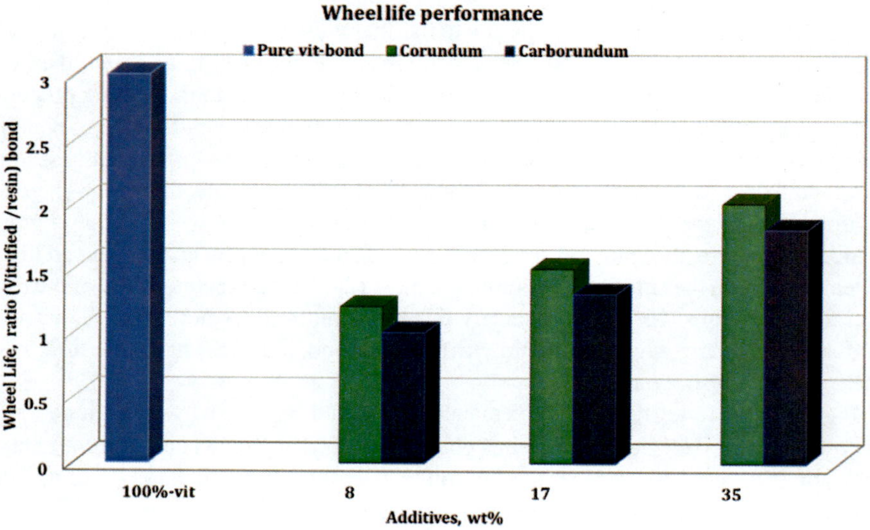

Fig. 3. Wheel life ratio as a function of additives

Fig. 4. SEM photograph of diamond-ceramics cement additives, **a**) Carborundum, **b**) Corundum.

Further investigation was carried out into the structure of the bond matrix. Figure 4 depicts the micro-structure of the bonds with both corundum and carborundum additives. The bond with carborundum had relatively larger pores, thus it was more friable than its counterpart with corundum that was relatively compact and homogenous. This is because the chemical elements in carborundum react with O_2 at high temperature creating a CO_2 that is trapped in the structure inducing large pores which contributes to the friability [21]. Consequently, the wheel life was not very high compared to corundum, this is because the carborundum does not bond easily with the vitrified bond at temperatures of 1073–1123k therefore may form isolated compounds in the bonded structure.

In Fig. 2 it is seen that after 21% of bond the wheel life deteriorated, therefore, some 24% bond segments were fabricated and investigated under the Energy Dispersive Analysis System. Here a large number of pores were identified, though the glassy vitrified bond are produced by smelting the entire compound. The chemical composition was tested at points 'e' and 'f' as marked in Fig. 5a.

The close up to the pore in Fig. 5 and the elemental count at the edges of the pore did not reveal the existence of O_2 or C and or their compound CO_2 that could generate the voids. Therefore, further analysis is required to understand the origin of the voids. However, this investigation showed that, any further increase in bond content led to the distortion of the segment at the sintering stage.

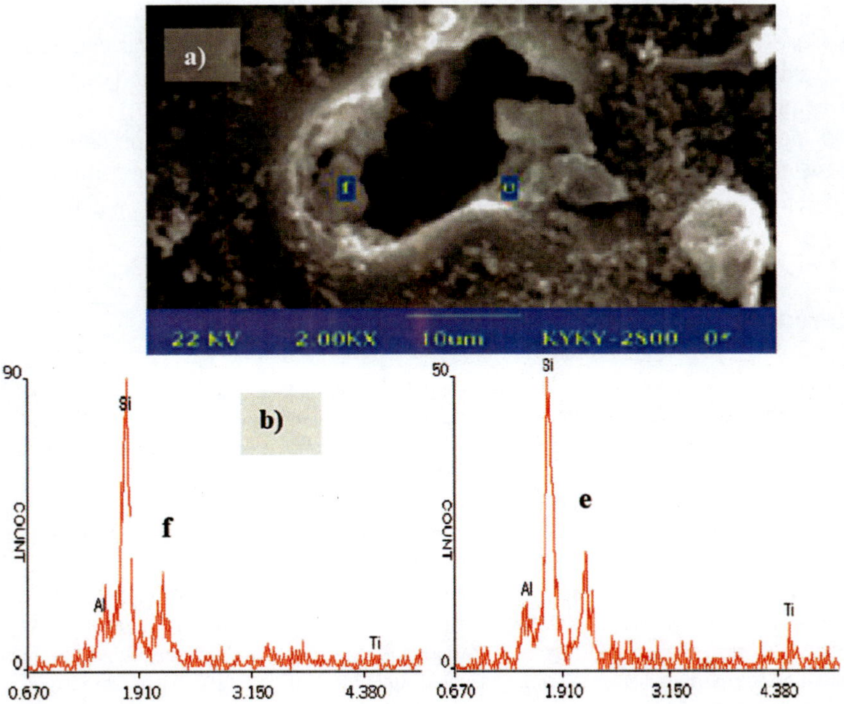

Fig. 5. Close-up to pores, (**a**) SEM picture of diamond grinding segment, (**b**) Elemental count at point "**f**" and "**e**"

4 Conclusion

A new vitrified bond was engineered and thoroughly investigated in terms of components constitution, their interaction and their effects on the performance of the grinding wheel. The performance of the new bond was compared with a standard resin bond diamond wheel with the same specification process parameters in grinding PDC parts.

- It was shown that the constituting elements have a direct effect on the diamond grinding wheel life. Adding carborundum and corundum into the binder led to a reduction of wheel life of about one third of that of pure vitrified bond developed.
- Using pure vitrified bond newly developed secured a wheel life of 3 fold of the life of a standard commercially available resin bond diamond grinding wheel.
- It was identified that 21% of binder content provided the best performance achieving over 2300 PDCs parts ground per grinding wheel.
- It was shown that increasing binder content over 21% led to a decrease of wheel life as pores were generated in the abrasive compound matrix at the sintering stage, along with distortion of the segments.
- The coating of the diamond grain with titanium provided the required strong bond between the grain and the binder due to a similarity of the substance of titanium thin film and the aluminium oxide. This secured the required rigidity for precision grinding and long life.
- It was established that the reduction in wheel life due to additives was caused by the excessive pores formation in the carborundum, as CO_2 was formed and trapped in the structure.
- The poor performance of corundum was attributed to the weak binding strength between the corundum matter and the diamond grains. However, the formation of small pores in the glassy binder requires further investigation and this will be studied.

References

1. Mamalis, A.G., Horvath, M., Grabchenko, A.I.: Diamond grinding of super—hard materials. J. Mater. Process. Technol. **97**, 120–125 (2000)
2. Li, Y., Han, X., Xiaoi, W.: Research the mechanical properties of polycrystalline diamond compact. Diamond Abras. Eng. **125**(5), 27–30 (2001). (in Chinese)
3. Akaishi, M., Yamaoka, S., Ueda, F., Ohashi, T.: Synthesis of polycrystalline diamond compact with magnesium carbonate and its physical properties. Diamond Relat. Mater. **5**(1), 2–7 (1996). https://doi.org/10.1016/0925-9635(95)00332-0
4. Li, D.-P., Yang, T.-L., Zheng, C.: Cylindrical grinding machaining of cubic boron nitride polycrystalline compact. Jewel. Sci. Technol. **13**(3), 15–18 (2003). (in Chinese)
5. Qiu, Z., Zhang, F., Luan, D., Gao, D.: Research grinding alumina ceramic with different bond diamond grinding wheels. Diamond Abras. Eng. **120**(6), 49–51 (2000). (in Chinese)
6. Zeng, J., Wa, C.-L.: The vitrified bond of diamond wheel in process PCD tools have superiority. Super Hard Mater. Eng. **17**(8), 18–20 (2005). (in Chinese)
7. Zhang, H.-X., Wang, G.-M., Hua, Y.: Research vitrified bond for diamond grinding wheels. China Ceramic **39**(1), 38–39 (2003). (in Chinese)

8. Jackson, M.J., Mills, B.: Microscale wear of vitrified abrasive materials. J. Mater. Sci. **39**, 2131–2143 (2004)
9. Shimamaoka, H.: New machine and wheel system for grinding sintered diamond tools. Ind. Diamond Rev. **42**, 155–160 (1982)
10. Wang, C., Xu, W.: Study of crystallisation of the Li2O-ZnO-Al2O3-SiO2 system glasses. J. Non-Crystall. Solids **80**(1–3), 237–242 (1986). https://doi.org/10.1016/0022-3093(86)90401-1
11. Davkova, K., Zafirovskis, Pocev, S.: Preparation of precipitated batch composition for glass ceramic in the system MgO. Glass Technol. **41**, 197–198 (2000)
12. Wang, Y.H., Zang, J.B., Wang, H.X., Wang, M.Z.: Effect of Ti coating on the wear stages of diamond in tools. Key Eng. Mater. **202–203**, 199–202 (2001)
13. Fu, J.: The Principle and Technics of Grinding, pp. 26–38. Hunan University Press, Changsha (1986).(in Chinese)
14. Wen, B., Huang, Y.: Handbook of Nonmetal Materials. Electronic Industry Press, Beijing (2006).(in Chinese)
15. Lu, P.: Science Base of Inorganic Materials, pp. 39–57. Wuhan University of Technology Press, Wuhan (1996).(in Chinese)
16. Feng, K., Cui, H.: Research of grinding polycrystalline diamond. Abras. Grinding **4**, 22–24 (1993). (in Chinese)
17. Huang, B.: Manufacture of Vitrified Bond Grinding Wheel. Committee of Chinese Machinery and Bureau of Machine Tool, Beijing, pp. 215–223 (1987). (in Chinese)
18. Grabchenko, A.: Role of wheel diamond concentration in grinding Polycrystalline of super hard materials. Soveit. J. Super. Hard. Mater. **6**(1), 58–62 (1984)
19. Takeshi Tanaka, S., Esaki, K., Nishida, T., Nakajima, K.U.: Development and application of porous vitrified-bonded wheel with ultra-fine diamond abrasives. Key Eng. Mater. **257–258**, 251–256 (2004). https://doi.org/10.4028/www.scientific.net/KEM.257-258.251
20. Grabchenko, A.: Role of wheel diamond concentration in grinding polycrystalline of super hard materials. Soveit J. Super Hard Mater. **1**, 58–62 (1984)
21. Liu, F., Fan, W.: Research the oxidation of carborundum in vitrified bond abrasive. Diamond Abras. Eng. **149**(5), 45–47 (2005). (in Chinese)

Flame Intensity Analysis for Hot Molten Metal Pouring in the Steel Industry by Applying Image Segmentation

Ivan Popov and Grazia Todeschini(✉)

Faculty of Science and Engineering, Swansea University, S1 8EP Swansea, UK
{708967,grazia.todeschini}@swansea.ac.uk

Abstract. Pouring large quantities of hot metal (HM) can release substantial amounts of flame. This problem is frequently encountered within the Basis Oxygen Furnace (BOF) steelmaking process where large quantities of HM (frequently exceeding 300 t) are poured into the converter vessels, generally loaded with scrap metal. The HM is contained in ladles and poured using overhead girder cranes. Excess release of flame may damage surrounding components (such as crane ropes) and consequently reduce their lifecycle. Limiting the release of flame during pouring allows extending the lifetime of the components located in proximity of the ladle. The scope of this paper is to characterise flame generation during pouring operations at a BOF steelmaking plant and to relate the amount of flame generated to some of the process factors. Due to the complexity of the process, this paper does not aim at providing solutions to eliminate flame generation, but rather to identify approaches to its mitigation. The methodology adopted in this work utilises a standard CCTV camera to record videos of pours. Colour based image segmentation analysis is then performed and flame intensity is calculated. This process is partially automated for each video by using functions in MATLAB. A total of 167 videos are analysed and the pours that cause higher flame intensity generation are identified. In the second part of the paper, analysis is performed on two process factors that impact flame generation: pouring rate and scrap metal composition.

Keywords: Flame intensity analysis · Image segmentation · Process control · Steel industry · Lifecycle management

1 Introduction

The BOF process involves three main stages: charging, blowing and tapping. During charging phase, the scrap and the hot liquid iron are loaded into the converter vessel. In the blowing stage, fluxes are added to aid the oxidation process and a water-cooled lance, through which oxygen is supplied, is inserted into the vessel and oxygen blowing is initiated. The last stage involves tapping of the converter vessel, to allow liquid steel to exit through a tapping nozzle, separating molten steel from carbon heavy slag.

A. Batako et al. (Eds.): GCMM 2021, LNNS 335, pp. 632–646, 2022.
https://doi.org/10.1007/978-3-030-90532-3_47

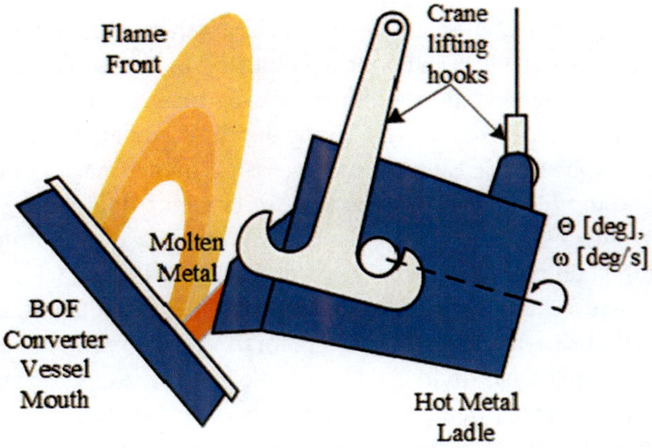

Fig. 1. Diagram of HM pouring into BOF converter vessel

A diagram of flame release during HM pouring into a BOF converter vessel is displayed in Fig. 1, where the main components involved in the process are shown: the ladle, the converter vessel and the hooks used to lift and tilt the ladle.

Scrap metals, varying between 15 and 25% of the total metal weight in the converter vessel, are loaded during the charging phase and vary by size and composition depending on the grade of steel being manufactured [1]. Charging with hot metal (HM) then takes place where hot iron is poured into the converter vessel from purpose-made ladles. In modern Basic Oxygen Steelmaking (BOS) plants, the ladles have a capacity to hold over 300 t of HM. Specialised overhead girder charging cranes are used to transport the ladles and pour their contents in to the converter vessel. Heat, in the form of flame, is released during pouring, reaching temperatures of 1300 °C, and causing thermal stress to the crane and surrounding structures [2]. To minimise damage to these components, the release of flame during HM charging needs to be minimised. The flame intensity and propagation vary with each consecutive charge and depends on several process factors.

The above concerns are well-known in the steelmaking industry and several research groups have worked on methods to improve this process. The heat front from HM pouring into a converter vessel was previously recorded in [3] using infrared imaging. The aim of this work was to analyse the thermal damage caused to charging crane cables and to develop a cable condition monitoring system.

Along with flames, metallic fumes and dust are released while pouring the HM. In [4], several video recordings were used to analyse the amount of fumes generated during charging. The average fume flow rate and total mass were then evaluated and used in a Computational Fluid Dynamic (CFD) model. Similarly in [5], CFD simulations were carried out based on emissions recorded during HM pouring. The following data was used as an input to the model: flame temperature, recorded using a thermocouple device positioned in the vicinity of the flame front, gas concentration and flow velocity. For a separate model, the volume of smoke plume released was determined using structural particle image velocimetry.

The use of flame segmentation has been applied in other industrial processes, specifically those related to the metal manufacturing industry. In [6], the flame from an iron ore sintering furnace was segmented by first pre-treating images containing furnace flame using discrete Fourier transformation (DFT) and then by applying Red Green Blue (RGB) colour space decomposition. Examples of processing flame images in rotary sintering kilns can be found in [7, 8], providing designs for complex flame recognition systems. A system for recognising flame regions in a Blast Furnace raceway for the estimation of temperature using pixel colour information is presented in [9].

Flame segmentation has also been applied to study the behaviour of flame in industrial boilers. In [10], a flame segmentation algorithm for use with infrared images is presented, where the shape of the flame is divided into multiple geometries. A Bayesian approach is used to estimate the boundary of each geometry. This segmentation algorithm was tested on a flame generated by a boiler burner and characterised against process conditions. An algorithm based on visible light spectrum images has also been developed for this application [11]. This boiler flame segmentation is based on YCbCr colour space with automatic thresholding of luma (Y) and red chroma (Cr) components. The blue chroma (Cb) component showed to provide limited colour information for identification of flame.

According to the literature review carried out, no previous studies have been performed where image segmentation was used to analyse flame produced during HM charging in BOF converter vessel. In this paper, a method of analysing the causes of high flame release is proposed based on image segmentation. The approach consists in performing a colour segmentation analysis on video recordings of HM pouring into a converter vessel to estimate the intensity of the produced flame. Flame intensity is then related to two process factors: pouring rate and scrap metal amount and composition. A case study was performed at the Tata Steel Port Talbot BOS plant, where a CCTV camera (Pelco Sarix Enhanced IME322 [12]) was positioned in proximity to the converter mouth to record the HM charging operation. Using MATLAB, each video was transformed into CIE L*a*b* colour space and a segmentation algorithm was applied to calculate flame intensity. An evaluation was then made against known process conditions. The results from this case study, where 167 pouring videos were analysed, are presented in this paper.

2 CIE L*a*b* Colour Space

Most commercially available video acquisition devices, including the CCTV camera used for this project, output video in an RGB format. For each pixel, this method encodes a value ranging from 0–255 for the individual R (red), G (green) and B (blue) colour component. Although colour segmentation in images is possible using the RGB format, in this paper the CIE L*a*b* colour space was utilised since it is more uniform than other approaches, allowing for better colour representation of the flame [13]. The CIE L*a*b* colour space consists of 3 components: L* - Luminance (or lightness), which ranges from 0 (black) to 100 (white), a* - ranging from -128 (green) to 128 (red), and b* - ranging from -128 (blue) and 128 (yellow). As a result, the CIE L*a*b* colour space provides a spherical 3D representation of the colour spectrum, where each individual colour has its own cartesian coordinate [14]. This representation is displayed in Figure 2.

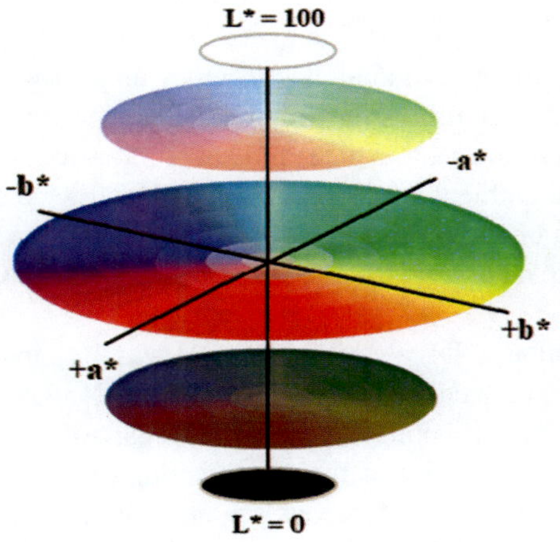

Fig. 2. CIE L*a*b* 3D coordinate representation.

Colour space conversion from RGB to CIE L*a*b* can be performed using the following equations [15]:

$$\begin{bmatrix} X \\ Y \\ Z \end{bmatrix} = [M] \begin{bmatrix} R \\ G \\ B \end{bmatrix} \tag{1}$$

$$L^* = \begin{cases} 116 \times \sqrt[3]{Y/Y_r} - 16, \text{for}(Y/Y_r) > 0.008856, \\ 903.3 \times (Y/Y_r), \text{otherwise}, \end{cases} \tag{2}$$

$$a^* = 500 \times (f(X/X_r) - f(X/X_r)), \tag{3}$$

$$b^* = 200 \times (f(Y/Y_r) - f(Z/Z_r)), \tag{4}$$

$$f(t) = \begin{cases} \sqrt[3]{t}, \text{for} t > 0.008856, \\ 7.787 \times t + 16/116 \text{otherwise}, \end{cases}$$

Where $[M]$ is the RGB to XYZ conversion matrix (in this case $[M]$ is the conversion matrix for standard RGB (sRGB) [16]), R, G and B are the data channels from RGB encoding ranging from 0–255 and X_r, Y_r and Z_r are reference values of colour white (in this case taken as 0.9504, 1.0000 and 1.0888 respectively [17]).

3 Nearest Neighbour Colour Segmentation

Image segmentation based on the analysis of colour spectrum, falls under two categories: supervised and unsupervised. In supervised classification, the segmentation region classifiers are typically pre-identified by the user. In unsupervised classification, these regions are estimated using the spectral colour information of a given image

[18]. Although requiring prior knowledge to establish classifiers by the user, the use of supervised classification may lead to higher accuracy image segmentation results [19].

In the videos recorded for this work, colour variation was caused only by the appearance of flame. Since the camera was placed in a fixed location, and an indoor environment, variations due to ambient conditions were excluded. Due to the lack of external factors influencing colour variation, a supervised classification method was chosen as it was deemed possible to determine the classification parameters using a single example video.

The Nearest Neighbour (NN) algorithm [20, 21] was used in this analysis due to its simple implementation and accurate results [19, 22]. The NN algorithm works by evaluating a distance metric between a point of interest x and classifiers $y = \{y_1, y_2, y_3, \ldots y_n\}$, where the smallest distance indicates the classification of x. When applied in colour-based image segmentation, points of interest are the individual pixels and the classifiers are colours to undergo segmentation. The distance calculation is performed on the colour space encoding values. Typically, the Euclidean distance metric is applied to find the minimum distance according to the following expression:

$$d(a, b) = \sqrt{\sum\nolimits_{i=1}^{n} (a_i - b_i)^2} \tag{5}$$

where $d(a, b)$ is the calculated distance, a and b are the cartesian coordinates of x and y respectively and n is the dimensionality of x and y.

In this analysis, the CIE L*a*b* colour space was utilised for the classification of pixels and the distance metric $d(a, b)$ for each pixel was calculated using the L*, a* and b* encoding values [23]. Therefore, the dimensionality value n in (5) was set to 3 where the difference of a_i and b_i was calculated as the difference between the values of L*, a* and b* for pixel x and classifier y, respectively.

For this analysis, four classifiers were selected, and their corresponding L*, a* and b* values were determined from a single video recording of a HM charge. These four colours were further sub-classified as the primary colours of the flame and the secondary colours of the background image. The primary colours were white, and a yellow shade taken as a cumulative average from sampling the recorded flame images at different points. The secondary colours are brown and grey: the first used for the identification of background with flame reflection and the second without reflection. The L*, a* and b* values corresponding to the four classifiers are displayed in Table 1.

Table 1. L*, a* and b* classifiers

	L* (b_1)	a* (b_2)	b*(b_3)	Colour
y_1 - White	99	0	0	
y_2 - Yellow	90	-8	39	
y_3 - Brown	70	0	30	
y_4 - Grey	70	0	0	

4 Methodology

Image segmentation was performed using MATLAB ver. R2018a with Image Processing Toolbox [24] and Parallel Computing Toolbox [25]. A NN segmentation MATLAB code was developed based on the example provided in [23]. The analysis was performed on each frame of the recorded videos. For example, with the recording set at 30 fps (frames per second), a 2:00 min recording resulted in the analysis of 3600 frames. Both 30 fps and 20 fps videos were used in this work (154 30 fps and 13 20 fps videos). To reduce the computational time, a region of interest (ROI) was selected: each frame was cropped to display only the region where the flame is most prominent, resulting in an analysis window measuring 670 × 1100 pixels. Each pixel in this window was then transformed from RGB to the CEI L*a*b* colour space. The NN classification was then performed on each pixel using classifiers displayed in Table 1 and subsequently the image of flame was segmented from the background image. The last step in this analysis was the estimation of flame intensity, which was taken as the total count of the segmented pixels displaying flame. The flow diagram for the analysis of a single frame is displayed in Fig. 3.

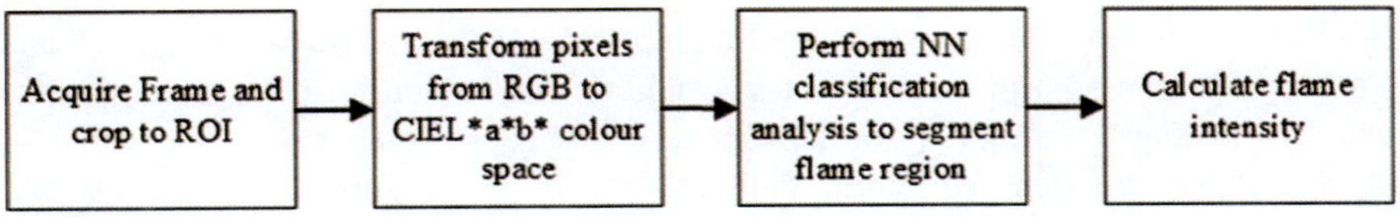

Fig. 3. NN Flame segmentation flow chart

Fig. 4. Example of flame segmentation: (a) original image and (b) after flame segmentation is applied.

The analysis described above was repeated for each frame. The flame intensity for the entire video was taken as the average value of calculated flame intensity for the individual frames. Figure 4a displays a frame prior to the application of flame segmentation and Fig. 4b displays the identical frame with the flame segmented from the background. For this image, the flame intensity was 37754. The overall average flame intensity for 167 recorded videos ranged from 7626 (Low flame) and 101523 (High flame).

5 Analysis of Results

In this section the results obtained from flame segmentation are compared against process factors. Pouring rate, and scrap amount and type are the two factors investigated in this work as main contributors to the production of flame. Other factors that may impact the amount of flame generated are not considered in this paper, although they may be taken into the account in future works.

At first, the relationships between 1. flame intensity and pouring rate and 2. flame intensity and total amount of scrap used are presented. Then, the contribution of different types and amount of scrap to flame generation is assessed. Lastly, further analysis is carried out to determine the combined effect of pouring rate and scrap type and weight on flame generation. This section is concluded by providing suggestions on modifications of the pouring process to reduce flame intensity.

5.1 Average Pouring Rate and Flame Intensity

The first factor considered for the impact on flame intensity was average pouring rate. This parameter was determined by dividing the HM weight (ranging between 275 and 315 t) by the total pouring time, obtained from the video recordings of each individual pour. The average pouring rate for the videos considered varied between 1.79 and 5.43 t/s. The actual rate differs at different stages of a pour due to the geometry of the ladle and individual pouring behaviours of the crane drivers. Figure 5 displays a scattered graph where pouring rate is plotted against the average flame intensity for each recording.

With the application of a trend line, a linear relationship is identified, suggesting that flame intensity increases with higher pouring rates. The coefficient of determination (R^2) is calculated to quantify the relationship between average flame intensity and pouring rate. This parameter is a statistical measure of how close the data is to the fitted regression line [26]. In theory, the value of R^2 ranges between 0 and 1, where values close to 1 indicate that the regression line provides a good description of the data. For the case in Fig. 5, the R^2 value is close to 0.11, indicating substantial scattering of the results. Therefore, a relationship between pouring rate and flame intensity could not be concluded based on this graph alone. This result suggests that there are other factors contributing to the generation of flame.

5.2 Scrap Amount and Flame Intensity

The relationship between the total scrap weight and average flame intensity is shown in Fig. 6. The slope of the trend line in this case suggests that flame intensity increases with the increase of scrap weight. However, the R^2 value in this case is close to 0.06, being even lower than the one obtained in Fig. 5 and indicating that scrap weight alone is not the only contributing factor in flame generation. Six pours where no scrap was used were discarded from the results in Fig. 6.

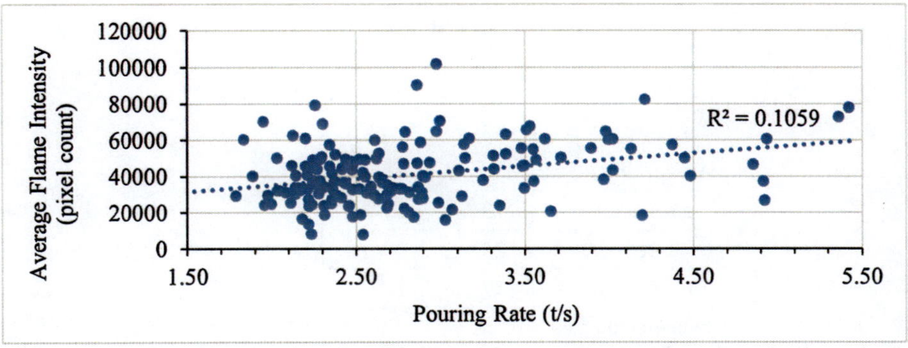

Fig. 5. Plot showing pouring rate vs average flame intensity

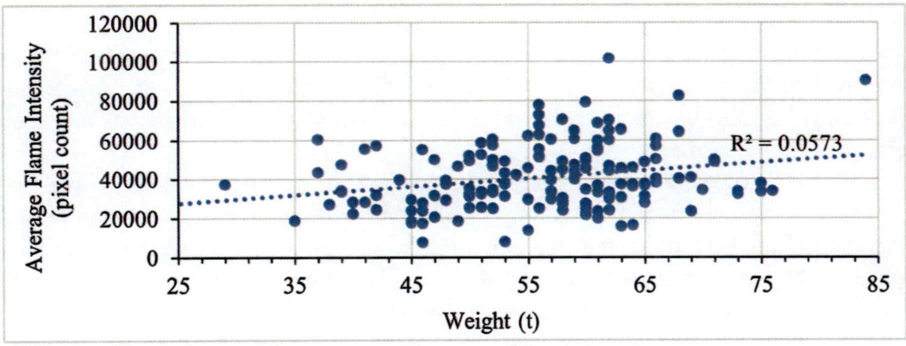

Fig. 6. Plot showing scrap weight vs average flame intensity

5.3 Scrap Type – Frequency of Occurrence Analysis

At the Port Talbot BOS plant scrap is classified in to 9 groups: Turnings, A Steel Skull, C Steel Skull, Cold Iron, Tin/Steel Cans, Incinerated Bales, Mill Products/Slabs, Tundish Skull and A0/Demo. These scrap types can be further sub-categorised into internally (A and C Steel Skulls, Cold Iron, Mill Products/Slabs and Tundish Skull) and externally (Turning, Tin/Steel Cans, Incinerated Bales and AO/Demo) sourced scrap types [1]. Internal scrap arises from leftover solidified iron and steel and trimmings of steel strips/slabs. External scrap incorporates a variety of commercial and industrial equipment, incinerator scrap and used tin coated steel cans [1]. Each scrap type varies in size and composition and the use of different scrap types and amounts depends on the grade of steel being manufactured. Internally sourced scrap is typically quite homogenous. Compositional variations are more prominent with externally sourced scrap types, varying in iron and impurity contents from batch to batch.

Based on detailed process data on scrap weight and composition for each pour, further analysis was carried out to identify a pattern between the use of different scrap

Table 2. Maximum, minimum and classification high/low limit value of scrap weight

Scrap type	Max (t)	Min (t)	Median (t)
Turnings	13	1	7
A steel skull	16	1	9
C steel skull	13	1	7
Cold iron	25	3	14
Tin/steel cans	17	1	9
Incinerated bales	15	2	9
Mill products/slabs	39	1	20
Tundish skull	23	1	12
A0/demo	3	2	2

types and the generated flame intensity. In this analysis, the top and bottom quartile of pours with the highest and lowest recorded flame intensity, respectively, were considered. Excluding the pours where no scrap was used, a total of 40 pours were included in each top and bottom quartile.

Each scrap type was categorised into Low amount and High amount. These two categories were determined by taking the median value between the maximum and minimum amounts recorded for each individual scrap type, where the weights above the median value categorised as High and values equal to or below as Low. The maximum, minimum and median weights for each scrap type are displayed in Table 2.

The frequency of occurrence of each scrap type was determined for the pours included in the top and bottom quartiles using the Low and High scrap weight classifiers. Figure 7 shows the results for Low scrap: the frequency of occurrence is plotted for each scrap type. Two bars are shown: the orange corresponding to the top quartile and the blue to the bottom quartile. The frequency of occurrence appears to be uniform, with a maximum difference observed for Mill products/Slabs. Figure 8 shows the same results for the case of High scrap: in this case, the frequency of occurrence varies more visibly, where the uses of High A Steel Skull, C Steel Skull, Cold Iron, Tin/Steel Cans and Incinerated Bales are more frequent in the Top quartile. The uses of High Mill products/Slabs and Tundish Skull are more frequent in the Bottom quartile.

These results suggest that individual scrap types, when used in small amounts, have little to no influence on the intensity of generated flame. For high scrap amounts, a possible correlation between certain scrap types and flame intensity could be established, since A Steel Skull, C Steel Skull, Cold Iron, Tin/Steel Cans and Incinerated Bales are used in the highest amount in the top quartile. In turn, this suggests that the use of these scrap types, in high amounts, contributes to the production of flame.

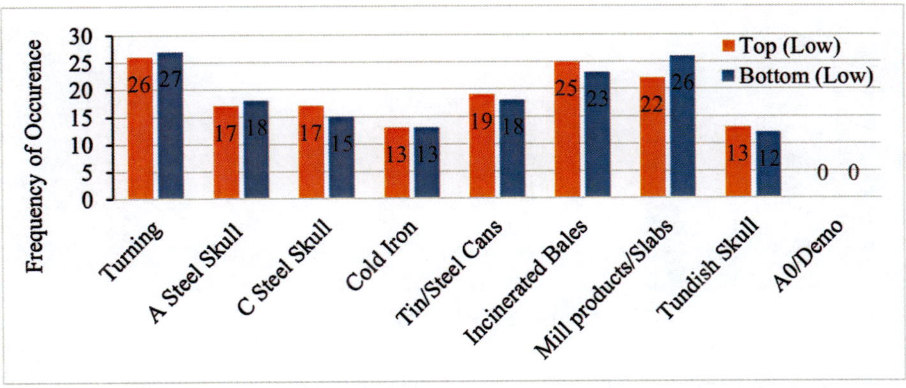

Fig. 7. Bar chart of top vs bottom quartiles – low scrap weight frequency

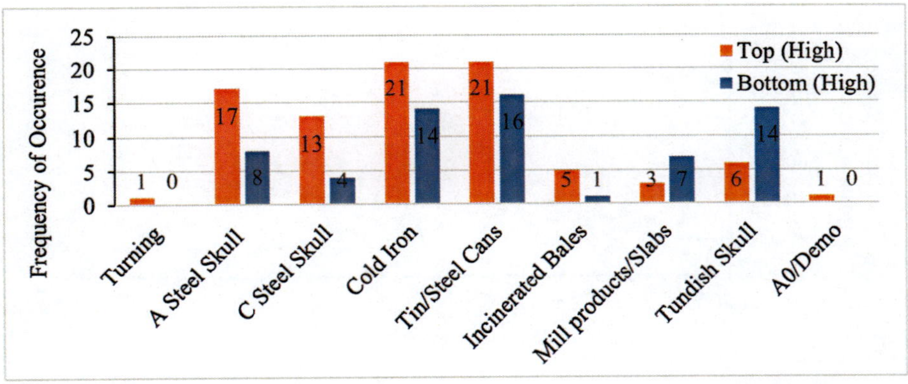

Fig. 8. Bar chart of top vs bottom quartiles – high scrap weight frequency

5.4 Pouring Rate and Scrap Classification – High Amount, High Frequency Scrap

The results from Sect. 5.3 showed that there is a potential correlation between the use of high amounts of A Steel Skull, C Steel Skull, Cold Iron, Tin/Steel Cans and Incinerated Bales and high flame generation. The weights these five scrap types were summed for each recorded charge, when the weight was above the median value (weight displayed in Table 2). Charges with no scrap were not considered in this analysis. Five groups were then identified, arranged by increasing value of the sum. These groups were: <10 t, 10–19 t, 20–29 t, 30–39 t and >39 t. The frequency of occurrence for each weight group were: 28, 54, 35, 18 and 26, respectively. Figure 9 displays the average value of flame intensities in each weight group. An increase of average flame intensity with increase of scrap amounts is observed, further indicating a correlation between the use of these scrap types and the generation of flame.

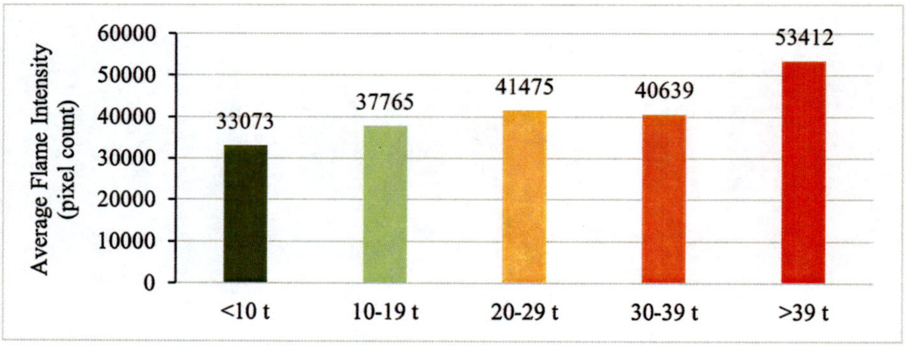

Fig. 9. Bar chart of high freq. scrap weight groups vs average flame intensity

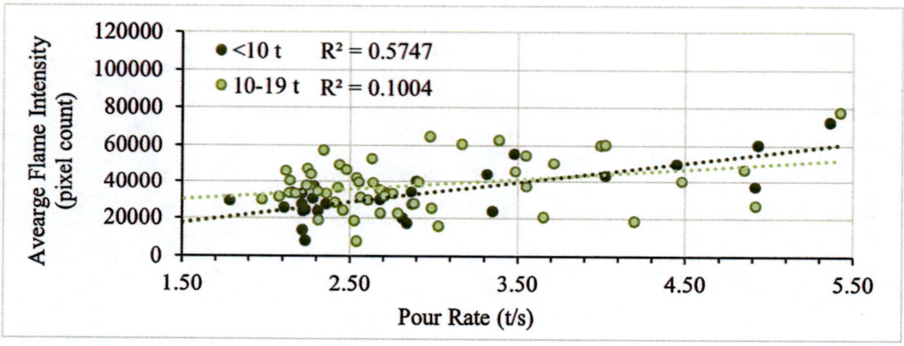

Fig. 10. Pouring rate vs average flame intensity by high freq. weight, <10 t and 10–19 t

For each scrap weight group, the relationship between pouring rates and flame intensity is visualised via scatter graphs in Figs. 10 (<10 t and 10–19 t), Fig. 11 (20–29 t and 30–39 t) and Fig. 12 (>39 t). By interpolating a trend line, a linear relationship between pouring rate and flame intensity is established. For each scrap group, an increase in flame generation is expected with faster pouring rates. Referring to the R2 value for each scrap group, the best result is displayed for group < 10 t, where R2 is calculated to be 0.57. This value is deemed as acceptable in industrial process that are not automated and where the human factor contributes to scattering of the results. In this context, the value is high enough to allow predicting the amount of flame generated with pouring rate to an acceptable level of accuracy. For groups 10–19 t, 20–29 t, 30–39 t and > 39 t, the R2 value is much lower, approximately measuring 0.1, 0.12, 0.16 and 0.24, respectively. This result is visually confirmed by the increased scattering of points in these weight groups. In addition to the approach of individual crane drivers, scattering of the results can be also justified by the relatively low number of videos analysed. It is expected that the analysis of more videos will provide further insights into the process factors thus leading to establishing better prediction of the flame generated by each pour.

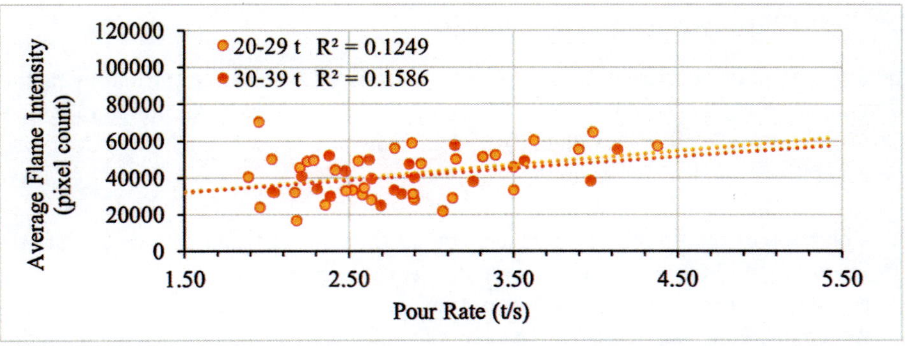

Fig. 11. Pouring rate vs average flame intensity by high freq. weight, 20–29 t and 30–39 t.

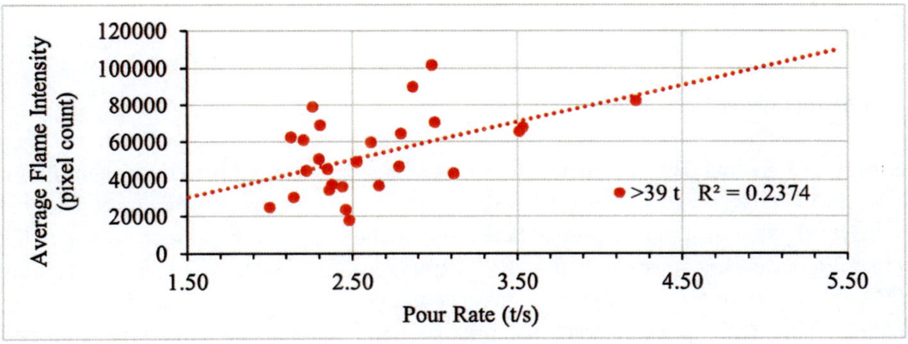

Fig. 12. Pouring rate vs average flame intensity by high freq. Weight, > 39 t.

5.5 Suggestions on Process Improvement

Based on the results obtained in this paper, suggestions can be made to improve the pouring process with the aim of reducing the amount of flame generated.

Firstly, from the results in Fig. 9, by reducing the amount of scrap types that were identified to be major contributors to flame generation it is possible to reduce the generation of flame. However, this might not be a financially viable option since different scrap types have different costs and substituting them with scrap identified to have a lower impact could increase manufacturing costs. Additionally, the type of scrap used for each pour depends on the grade of steel being manufactured.

Secondly, it may be possible to reduce the generation of flame by adjusting the pouring rates in accordance to scrap weight groups and to the trend lines in Figs. 10, 11 and 12. For example, to keep the intensity of flame below an arbitrary value of 40000, for group <10 t, the pouring should be below 3.5 t/s and for 30–39 t pouring rate should be kept below approximately 2.5 t/s.

5.6 Other Process Factors and Conditions

In this paper, pouring rate and scrap type were considered as the two contributing factors to flame generation. These two factors have been identified based on literature review, experience of operators and availability of data. However, other factors may also impact the generation of flame, for example:

1. Operation of the fume extraction system: At the time of video data collection, the operation of the fume extraction system was not known.
2. Moisture content in the scrap: Water, trapped in small crevices, may be present in the scrap metal due to rainfall. The presence of moisture could increase generation of flame.

The above factors may be considered in future works on this topic, given availability of data.

6 Conclusion

In this paper an analysis was carried out to correlate process factors to the generation of flame during charging of a converter vessel with HM. The two process factors considered in this work were the pouring rate and the amount and composition of metal scrap in the converter vessel. Image segmentation techniques were applied to 167 videos recording HM charging, in order to establish the average flame intensity for each pour. Regression analysis was then applied to find a correlation between the process factors described above and flame intensity. It was established that fast pouring and high amounts of total scrap contributed to the generation of flame. However, it was difficult to establish a linear relationship due to high spreading of results using these two factors alone.

By looking at the frequency of occurrence and at the weight of individual scrap types in the bottom and top quartiles, arranged by their flame intensity, scrap types that have a major contribution to the generation of flame were established. Then, using the weight information of the identified high frequency scrap types, a refined relationship between the use of scrap, the pouring rate and the recorded flame intensity was established.

Through this analysis, possible process improvement suggestions were identified: specifically, using lower amounts of identified scrap and pouring liquid metal at a lower rate. Although it was difficult to establish a defined linear relationship between pouring rate, scrap weight and flame intensity, these results can be used as a basis for further analysis, to establish other process factors contributing to the generation of flame.

Acknowledgments. The authors would like to acknowledge the M2A funding from the European Social Fund via the Welsh Government (c80816) and Tata Steel Strip Products UK that has made this research possible. Additionally, the authors would like to thank John Madill, Alison Tuling and Ben Jenkins of Tata Steel Strip Products UK.

References

1. Short, J., Grosvenor, R.I., Prickett, P.W.: The development of a process charge expert system for a basic oxygen steelmaking plant. In: 18th International Conference on Knowledge-Based and Intelligent Information and Engineering Systems, Gdynia, Poland (2014). http://orca.cf.ac.uk/81264/. (Unpublished)
2. Popov, V., Gabtykaev, D.: Cables with a steel core for hot-metal charging cranes. Metallurgist **52**(9–10), 588–591 (2008)
3. Maslov, V.: Vlijanie termociklicheskih vozdejstvij rasplavlennogo metalla na sistemu "konverter-kran" i sovershenstvovanie ee jelementov (Influence of Thermocyclic Actions of Molten Metal on the "Converter–Crane" System and Improvements of its Elements). Cherepovets State University, Thesis (2004). (in Russian)
4. Plikas, T., Woloshyn, J., Johnson, D.: Application of CFD modeling to the design of fume control systems in the steel industry. In: AIST Proceedings: AISTech 2007, vol. 1, pp. 745–756. Association for Iron and Steel Technology, Warrendale, PA (2007)
5. Marx, K., Roedl, S.: Efficient optimization of steel plant dedusting. Stahl. Eisen. **132**(5), 61 (2016)
6. Chen W., et al.: RGB color decomposition and image feature extraction of flame image in rear of sintering machine. In: Lui T., Zhao Q., (eds.) Proceedings of the 36th Chinese Control Conference (CCC), pp. 5460–5463. IEEE, Piscataway, NJ (2017)
7. Li, W., Mao, K., Chai, T., Zhang, H., Wang, H.: Gabor filter and eigen-flame image-based burning state recognition for sintering process of rotary kiln. In: Proceedings of 50th IEEE Conference on Decision and Control and European Control Conference, pp. 3216–3221. IEEE, Piscataway, NJ (2011)
8. Wang, Q., Sheng, Y., Chai, L.: Local entropy based pre-processing for flame images of alumina rotary kiln. In: Proceedings of 9th IEEE Conference on Industrial Electronics and Applications, pp. 1476–1480. IEEE, Piscataway, NJ (2014)
9. Zhang, R., Cheng, Y., Li, Y., Zhou, D., Cheng, S.: Image-based flame detection and combustion analysis for blast furnace raceway. IEEE Trans. Instrum. Meas. **68**(4), 1120–1131 (2019)
10. Marques, J.S., Jorge, P.M.: Visual inspection of a combustion process in a thermoelectric plant. Signal Process. **80**(8), 1577–1589 (2000)
11. Xie, M., Wu, J., Zhang, L., Li, C.: A novel boiler flame image segmentation and tracking algorithm based on YCbCr color space. In: Proceedings of 2009 International Conference on Information and Automation, pp. 138–143. IEEE, Piscataway, NJ (2009)
12. Pelco. Sarix IME Indoor and Environmental Mini Domes. Product Specification (2019). https://www.pelco.com/docs/
13. Celik, T.: Fast and efficient method for fire detection using image processing. ETRI J. **32**(6), 881–890 (2010)
14. Sappi Fine Paper North America. Defining and Communicating Color: The CIELAB System (2013). https://www.sappi.com/files/defining-and-communicating-colorpdf
15. Lindbloom, B.: Useful Color Equations. http://www.brucelindbloom.com/index.html?Equations.html. Accessed 22 Apr 2020
16. Lindbloom, B.: RGB/XYZ Matrices. http://www.brucelindbloom.com/index.html?Eqn_RGB_XYZ_Matrix.html. Accessed 22 Apr 2020
17. MathWorks: rgb2lab. https://uk.mathworks.com/help/images/ref/rgb2lab.html. Accessed 22 Apr 2020
18. Song, X., Fan, G.: A study of supervised, semi-supervised and unsupervised multiscale Bayesian image segmentation. In: Proceedings of the 2002 45th Midwest Symposium on Circuits and Systems. IEEE, Piscataway, NJ (2002)

19. Surlakar, P., Araujo, S., Sundaram, K.M.: Comparative analysis of k-means and k-nearest neighbor image segmentation techniques. In: Proceedings of the 2016 IEEE 6th International Conference on Advanced Computing, pp. 96–100. IEEE, New York, NY (2016)
20. Cover, T., Hart, P.: Nearest neighbor pattern classification. IEEE Trans. Inf. Theory **13**(1), 21–27 (1967)
21. Duda, R.O., Hart, P.E.: Pattern Classification and Scene Analysis. Wiley, New York (1973)
22. Harini, R., Chandrasekar, C.: Image segmentation using nearest neighbor classifiers based on kernel formation for medical images. In: Proceeding of the International Conference on Pattern Recognition, Informatics and Medical Engineering (PRIME-2012), pp. 261–265. IEEE, Piscataway, NJ (2012)
23. MathWorks: Color-Based Segmentation Using the L*a*b* Color Space. https://uk.mathworks.com/help/images/color-based-segmentation-using-the-l-a-b-color-space.html. Accessed 13 May 2020
24. MathWorks: Image Processing Toolbox. https://uk.mathworks.com/products/image.html. Accessed 22 Apr 2020
25. MathWorks: Parallel Computing Toolbox. https://uk.mathworks.com/products/parallel-computing.html. Accessed 22 Apr 2020
26. Hocking, R.R.: Methods and Applications of Linear Models: Regression and The Analysis of Variance, 3rd edn. John Wiley & Sons, Hoboken (2013)

Intelligent Implant System for Bone Lengthening

Joachim Strittmatter[1,2](✉), Nikolas Hör[1], and Lazar Bošković[1,2]

[1] Materials Engineering Laboratory, HTWG Konstanz University of Applied Sciences, Alfred-Wachtel-Str. 8, 78462 Konstanz, Germany
joachim.strittmatter@htwg-konstanz.de

[2] WITg – Institute for Material Systems Technology Thurgau, Konstanzerstrasse 19, 8274 Tägerwilen, Switzerland

Abstract. A novel implant system for bone elongation will be presented. With this technique, the body's own bone material, so-called callus; can be formed by gradual distraction of the tubular bones, thus achieving an extension of femur and tibia bones. The driving principle of this fully implantable bone lengthening system is based on a shape memory element. During the surgical treatment, the intramedullary nail serves to stabilize the severed bone and enables the formation of new, endogenous bone material to lengthen the limbs or to bridge bone defects. The intramedullary nail is implanted into the medullary cavity and fixed at both ends with locking bolts. A receiver coil implanted under the skin receives the necessary energy twice a day through high-frequency energy transport to activate the thermal phase transformation of the shape memory element. This gradually increases the bone gap by 0.5 mm each time and stimulates callus formation. Consequently, osteoblasts or osteocytes are formed in the area of the desired bone extension and load-bearing bone material is formed. Three nail prototypes have already been tested for their functionality in a cadaver study in a German clinic. Currently a redesign of this intelligent implant system is underway, focusing on a novel coil geometry, a monitoring sensor system and control technology and a novel connection technology for the drive components. With this intelligent implant system, it will be possible for the first time to lengthen the bones in a patient-friendly manner and to continuously monitor, document and evaluate the entire lengthening process.

Keywords: Bone elongation · Limb lengthening · NiTi · Shape memory · Joining technology · Monitoring sensor system

1 Introduction

Today, medical applications of shape memory alloys (SMAs) can be found in cardiovascular devices, gastroenterology and urology as well as in the area of orthopedic implants, orthodontic devices and clinical instrumentation. Their functional properties combined with excellent biocompatibility increase the possibility and the performance of minimally invasive surgeries. Overviews of existing applications can be found in [1, 2].

A. Batako et al. (Eds.): GCMM 2021, LNNS 335, pp. 647–655, 2022.
https://doi.org/10.1007/978-3-030-90532-3_48

Within the medical field, most of the applications with shape memory (SM) material take advantage of the superelasticity of Nickel-Titanium (NiTi) SMAs. In contradiction to the superelastic or mechanical SM effect, the application described in this study uses the thermal SM effect for a new medical implant. Before explaining the SM driven intramedullary bone nail in detail, a short introduction to the bone elongation technique is given. With regards to the SMAs, their effects and extraordinary and outstanding properties, in this paper reference is made to the literature [3, 4].

2 Medical Background

The background of this application is based on the medical fact that normally any tissue reacts to an injury with repair and healing processes through multiplication of cells. If after a transverse osteotomy a strain stimulus is activated, for example by tensile stress, this multiplication of cells and new formation of tissue may be continued for any length of time. Due to this mechanism, even considerable loss of bone caused by fractures or congenital defective positions, may be compensated without bone grafts. The technique of callus distraction by means of external fixation or intramedullary nail stimulates the formation of callus in the bone gap. Callus is the repair tissue of the bone generated in the fracture gap in case of bone fracture or osteotomy. The gap to be bridged should not be wider than 1 mm per day [5]. The process starts with the exudation of callus around the ends of the broken bone. At first, callus is more like a fibrous tissue. Later it hardens due to deposition of calcium and eventually it is converted into true bone. Three weeks after severance, the vascular system is formed. An overview of current bone lengthening techniques, also called callus distraction, can be found in [5]. External systems are normally used for the extension of bones, the bone fragments being fixed on rings by wires. The decisive disadvantages of those external systems are primarily the considerable risk of infection due to protruding wires, noticeable discomfort for the patient because of the external rings, a coarse cosmetic result because of scarring, as well as rather long hospitalization. Therefore, internal bone extension systems are of great interest to orthopedic surgery.

Several internal bone extension systems for tubular bones were developed in the past but are no longer available on the market. These are for example: The Intramedullary Skeletal Kinetic Distractor, ISKD® is mentioned in [6]. The nail can be totally implanted into the medullary canal of a long bone. The extension of the ISKD-nail is based on a ratchet mechanism and a threaded rod. The ratchets allow a rotational movement of the parts from three to nine degrees against each other and cause an elongation of approximately 1 mm after 160 rotational movements. A control of the extension can be achieved with an external monitor that detects an integrated magnet [7]. A similar internal extension system is the so-called Albizzia-nail [8] that needs rotational movements from 20 to 30° for the bone lengthening. A further development of the Albizzia-nail is realized in the Guichet-Nail [9]. An extension of 1 mm is achieved by 15 rotations of the leg by 10°.

At present three systems for internal bone extension are available: The fitbone® is an internal lengthening system for tubular bones that was introduced into market by the German company Wittenstein intens in the year 1997 [10]. The distraction is obtained by means of an integrated motor and gear that drives a threaded rod. The required energy is supplied by induction via a receiver that is implanted under the skin. The further developed intramedullary limb-lengthening system, also called Fitbone II, is distributed by Orthofix since 2020 [11].

The Precise® system of the NuVasive company in San Diego, CA, [12] consists of a fully implantable nail with a telescoping metal rod and a magnetic motor inside of it. The activation is realized via an External Remote Controller (ERC) that has to be placed from outside against the leg or arm to achieve the elongation process.

Synoste Oy in Espoo, Finland, developed an implantable nail for the treatment of bone length discrepancy. The system is based on smart materials such as SMA and PEEK. Globus Medical has acquired Synoste in 2020 and is selling the system now [13].

3 Experimental Results

The newly developed intramedullary implant for bone elongation and bone segment transport [14] that is shown in this paper is driven by a SM element. During surgical treatment, the intramedullary nail serves to stabilize and enables the body's own formation of new bone material for limb lengthening or bridging of defects according to the "Betz-Baumgart method" [15]. After dissection of the tubular bone, the medullary nail is inserted into the medullary cavity and fixed by means of locking bolts. A subcutaneously implanted receiver coil is activated through high-frequency energy from an outside generator coil, converts the electrical current and feeds a heating element that triggers the thermal effect of the SM element. The temperature at the interface to the body during activation is a maximum of 48 °C to prevent tissue damage. In combination with a bias spring the fracture gap increases twice a day by 0.5 mm each. This leads to the stimulation of osteocyte formation and the desired bone length.

3.1 Prototypes Tested in a Cadaver Trial

The functional principle of the SM linear driver is partially shown in Fig. 1. The SM element consists of a NiTi tube (f) that is connected with two sliding elements (e). The sliders are under mechanical stress through a bias spring (d) and equipped with catches (c), partially formed with a saw tooth profile. Inside of the NiTi tube, a heating element is placed that can be activated via a high-frequency energy coupling. Heating above the austenite finish temperature of the SM element causes the shortening of the driver mechanism: the SM tube contracts, because the catches in the rear slider shift one notch forward and its teeth engage in the internal saw tooth profile of the external tube (a). Additionally, the bias spring becomes further tensioned. For this step, a heating time of less than 3 min at very low power supply is sufficient. After the interruption of the high-frequency induction the SM tube will be cooled to body temperature within a longer period of time. When the SM element passes the phase change temperature to the martensite structure, the tension in the bias spring will overbalance the reset force of the SM tube and causes the elongation of the driver mechanism: the SM tube will be elongated into its initial position when the catches from the front slider shift one notch foreword. Since the front slider is fixated to the connection rod of the intramedullary nail, the bone lengthening will be executed.

In Fig. 1 it can also be observed that at the end of the nail a switch (b) is provided in order to interrupt the heating of the NiTi tube when the teeth of the rear catch engage in the saw tooth profile and the first step of the elongation is executed. Three such nail prototypes were tested in a cadaver trial on a human corpse in a German clinic [16]. These tests revealed some vulnerability in the safe functioning of the switch and the connection of the SM tube to the nail parts. In the following some improvements of the ongoing re-design are presented.

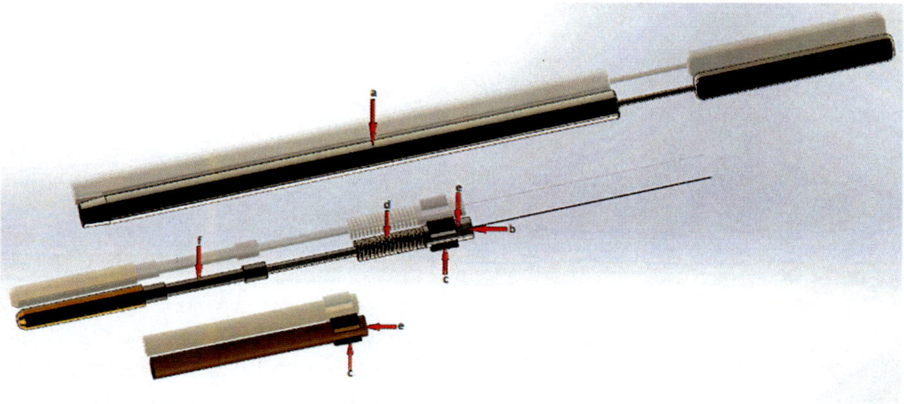

Fig. 1. The shape memory alloy limb lengthening nail. (a) external tube, (b) switch, (c) catches, (d) spring element, (e) slider, (f) NiTi tube.

3.2 New Connection Technology

The joining of the NiTi tube to the other parts of the implant is of fundamental importance for a reliable and safe operation of the implant system. Based on in-situ measurements [17] and the requirements of the surgeons, displacement forces well above 1.000 N can occur during bone lengthening. These high forces generally take place in the end of the lengthening procedure, when the implant has to operate against the highest resistivity forces of the muscles and tendons. Due of the working principle of the NiTi tube in combination with a spring element as bias spring, forces up to 3.000 N are required at the NiTi tube. Taking into account a safety factor of two, this results in transmitting forces in the order of 6.000 N. In the past, different fastening techniques were investigated, ranging from simple threaded connections to more complex crimp technologies. To further miniaturize the implant, a new connection technology was developed that uses shape memory heat shrinkable rings.

In Fig. 2 the new design is visualized: The NiTi tube (1) is joined over the pins of the inner shaft tip (2) and the transmitting shaft (5) and two heat shrinkable metal rings (3) are positioned over the two ends of the NiTi tube. A simple heating up to 165 °C causes the shrinking of the metal rings due to their thermal SM effect. With optimum geometry of the pin outer diameter and the tube inner bore, holding forces up to 8.000 N can be achieved in this way [18]. The used shrink rings consist of a NiTiNb SMA that exhibit a significant enlarged hysteresis. Using this approach, fasteners made of this NiTiNb SMA remain martensitic to 50 °C, recover to austenite and build full clamping stress by 165 °C. This clamping stress maintain on cooling to −65 °C and the fasteners do not become fully martensitic until cooled below −120 °C.

In this view also the heating element (4) can be seen that is positioned in the center of the NiTi tube.

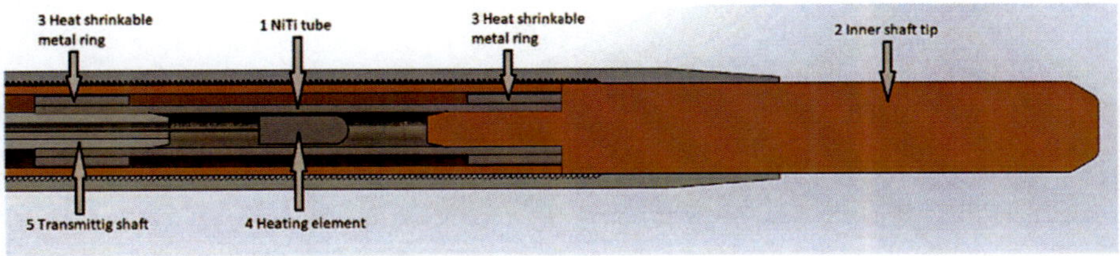

Fig. 2. Design of the new connection technology: NiTi tube (1), inner shaft tip (2), heat shrinkable metal rings (3), heating element (4) and transmitting shaft (5).

In order to achieve the highest transmitting force of the connection, several tests were carried out in the Materials Engineering Laboratory on dummy pairs in which not only the diameters of the pins, but also the geometry and surface conditions were varied. For the pin profiles conical and cylindrical variants with and without shoulders were investigated. In addition, different special electrochemical surface treatments were used. For the NiTi tubes different surface treatments on the inner diameter were performed by the supplier [19], whereby also a special slurry-cleaning process was applied. Last but not least, also for the shrinkable metal rings some variations concerning their inner diameter and their width was investigated.

The dummy pairs were assembled using a self-developed joining device, the shrinking process was activated in a thermal cabinet and the maximum pull-out forces were tested using a universal tensile testing machine. Figure 3 reveals the maximum pull-out forces of the four combinations with the highest values, all slightly above 7 kN with little variation. In this configuration the inner surface of the NiTi tubes were prepared using the slurry-cleaning process for reducing the oxide layers [19] and the pins of the inner shaft and the transmission shaft were electro polished. It can be stated that by using heat shrinkable metal rings a reliable connection of the NiTi tube to the other parts of the implant can be realized.

3.3 Novel Coil Geometry

The illustration in Fig. 4 shows the intramedullary nail and the connection cable to the new receiver coil. This coil is embedded in a silicone body and is positioned subcutaneously in the leg. The old design consisted in round coil resulting in a total diameter of about 5 cm of the silicone part. In laboratory tests and in the cadaver test this old design worked very well, but was not favorable for a future patient. First of all such a large part in the leg is not comfortable and leads to the patient constantly working on it

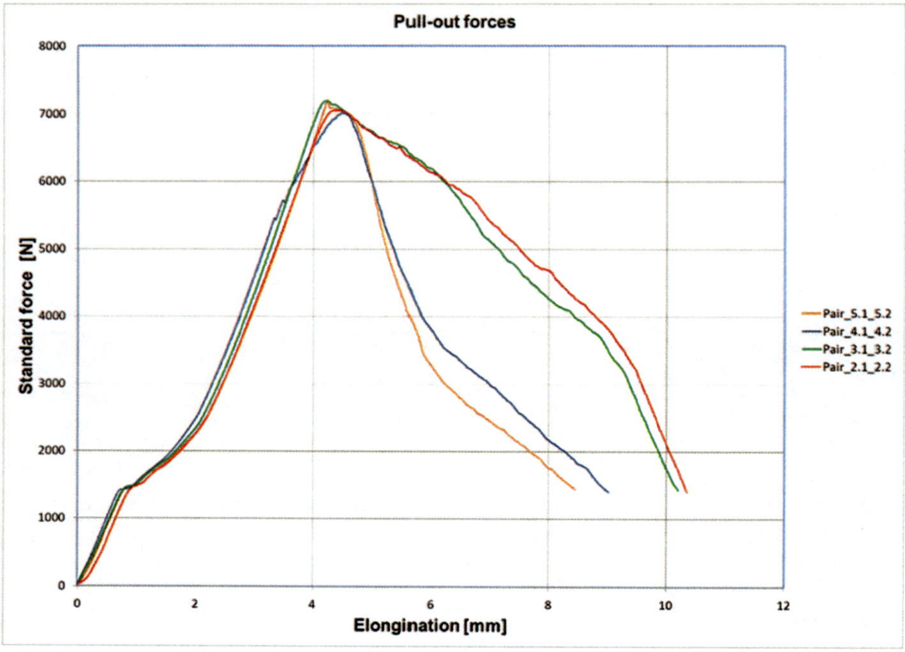

Fig. 3. Pull-out forces of the four combinations with the highest values.

with his fingers. Another disadvantage could be that the tissue above the center of the silicone implant will not be sufficiently supplied with blood during treatment. In consultation with the surgeon and a coil specialist, the new coil design was manufactured and tested. Thanks to the elongated design, the novel coil geometry is now very slim and flat and allows the use of an embedded RFID element in the center. This radio frequency identification (RFID) is one part of the new monitoring system.

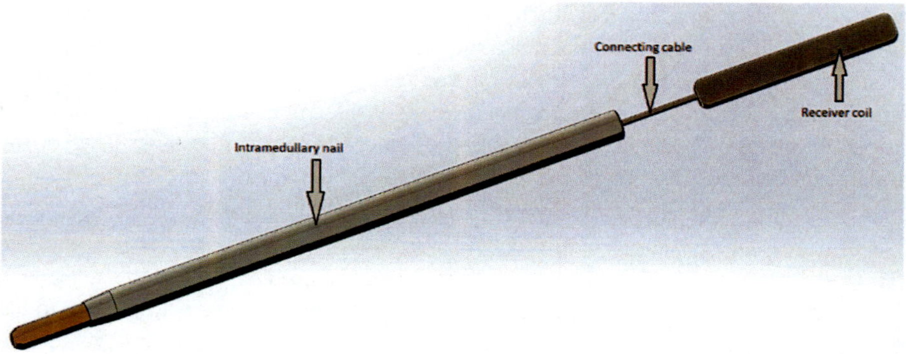

Fig. 4. Novel coil geometry of the Intelligent Implant System.

3.4 New Switch Mechanism

Redesign and investigation was undertaken to find a better solution of the switch mechanism. First of all, a switch is necessary to stop the electrical activation of the heating element after the movement of the catches into the next step of the saw tooth profile. In the last prototype generations a switch was positioned at the end of the inner nail. Although it worked very well in most of the tests, this technique was not reliable for the cases where the accumulation of the tolerances exceeded a threshold value. Due to the increasing manufacturing costs to maintain the required tolerances, a new switch mechanism was developed that is positioned directly under the catches. The Fig. 5 gives some ideas of this new mechanism.

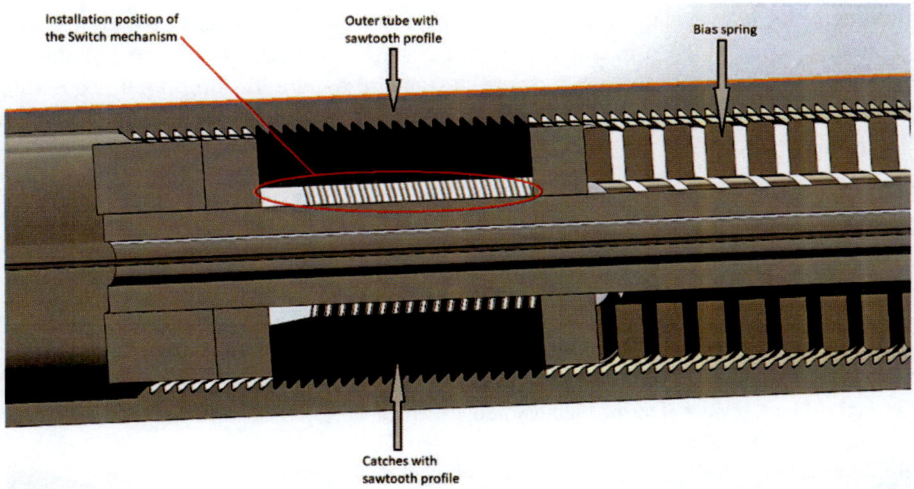

Fig. 5. Views of the new switch mechanism of the Intelligent Implant System.

3.5 New Control Technology

The innovative part of this new intelligent implant system for bone lengthening is the possibility to monitor and document the elongation process for the first time. For this a resonant circuit consisting of a capacitor, receiver coil and heating resistor for the secondary circuit offers the possibility to detect the steps of the SMA limb lengthening nail when the additional capacitors are activated/deactivated by the new switching system. Background is an evaluable phase shift that is generated by the oscillation heating element in the parallel resonant circuit when the resonance frequency changes. In combination with RFID elements implemented in the center of the new coils a reliable monitoring system is available.

The new switching mechanism and this new control technology are in the test phase of the ongoing re-engineering and have so far produced very encouraging results.

4 Conclusion

A new implant system for bone elongation is under development. With this technique, the body's own bone material, so-called callus, can be formed by gradual distraction of the tubular bones, thus achieving an extension of femur and tibia bones. The driving principle of this fully implantable bone lengthening system is based on a SM element in combination with a bias spring. Latest design improvements in development include a new switch mechanism, novel coil geometry and control technology, and a new connection technology. In particular, the approach to a reliable connection technology that can withstand tensile forces in the range of 6.000 to 7.000 N between the active SM element and the implant parts in this small installation is presented. For this shrinkable metal rings of NiTiNb are used. The excellent performance of the SM elements requiring minimal space allows large forces to be transferred with this innovative solution.

With this intelligent implant system for bone lengthening, it will be possible to lengthen bones more gently than the current standard of care. In addition, it is possible for the first time to continuously monitor, document and evaluate the whole elongation process.

References

1. Duerig, T.M.: An overview of Nitinol medical applications. Mater. Sci. Eng., A **273–275**, 149–160 (1999)
2. Fischer, H.: Applications of shape-memory alloys in medical instruments. Mater. Sci. Forum **394–395**, 9–16 (2002)
3. Jani, J.M.: A review of shape memory alloy research, applications and opportunities. Mater. Des. **56**, 1078–1113 (2014)
4. Strittmatter, J., Gümpel, P., Hiefer, M.: Intelligent materials in modern production - current trends for thermal shape memory alloys. In: GCMM-2018, Procedia Manufacturing, vol. 30, pp. 347–356. Elsevier Ltd., Amsterdam (2019)
5. Rüter, A., et al.: Kallusdistraktion. Urban und Schwarzenberg, München (1998)
6. Cole, J.D., Paley, D., Dahl, M.: Intramedullärer skelettaler kinetischer Distraktor ISKD – Chriurgisches Verfahren für das Femur. Orthofix GmbH, (IS-03002(B)-OPT-G2 08/08) (2008)
7. Pip, S.: Intramedulläre Beinverlängerung mit dem ISKD®. Dissertation an der Medizinischen Fakultät der Westfälischen Wilhelms-Universität Münster (2014)
8. Guichet, J.-M.: Beinverlängerung und Deformitätenkorrektur mit dem Femur-Albizzia-Nagel. Orthopade **28**, 1066–1077 (1999)
9. Toren, V.: Zur Entwicklung invasiver Behandlungsmethoden der Beinlängendifferenz. Dissertation an der Medizinischen Fakultät der Ludwig-Maximilians-Universität zu München (2002)
10. FITBONE® Product Information: Wittenstein intens GmbH. https://www.orthofix.com/wp-content/uploads/2020/03/2015_Produktbroschure_FITBONE_Intens_8Seiter_OF.pdf. Accessed 10 Dec 2020
11. Maier, S.: Wittenstein verkauft seine Medizinsparte an das Medtec-Unternehmen Orthofix. Pressemitteilung Wittenstein SE 04.02.2020. https://www.wittenstein.de/fileadmin/04-3-Presse/2020/2020-02-04/pm-win-orthofix-04-02-2020-de.pdf. Accessed 10 Dec 2020
12. Landge, V., Shabtai, L., Gesheff, M., et al.: Patient satisfaction after limb lengthening with internal and external devices. J Surg. Orthop. Adv. **24**(3), 174–179 (2015)

13. N.N.: Evonik und HThtGF verkaufen Synoste-Anteile. medtech zwo. Technologie. Wirtschaft. Gesundheit, Heft 2.2020. 4. Jahrgang BIOCOM, 50 (2020)
14. Strittmatter, J., Storz-Irion, R, Hör, N., Bošković, L.: Shape memory driven intramedullary nail for bone elongation. In: Conference Proceedings from the International Conference on Shape Memory and Superelastic Technologies (SMSTTM2019), pp. 264–265. ASM International® (2019)
15. Baumgart, R.: A fully implantable motorized intramedullary nail for limb lengthening and bone transport. Clin. Orthop. Relat. Res. **343**, 135–143 (1997)
16. Dünnweber, L.W.: Evaluation of the SMALL nail: drive technology and behavior in situ. Med. Eng. Phys. **38**, 1518–1523 (2016)
17. Glaser, M.: Entwicklung eines instrumentierten Distraktionsmarknagels. Dissertation an der Universität Konstanz (2010)
18. Borden T.: Intrinsic Devices, Inc.: Homepage. https://www.intrinsicdevices.com/. Accessed 10 Dec 2020
19. Slurry Cleaning Inner Diameter Surface: Euroflex GmbH homepage. https://www.euroflex.de/en/downloads.html. Accessed 10 Dec 2020

Risk Assessment of a Production System with Series-Parallel Reliability Structure

A. Burduk(✉)

Faculty of Mechanical Engineering, Wrocław University of Science and Technology, Wybrzeże Wyspiańskiego 27 Street, 50-370 Wrocław, Poland

anna.burduk@pwr.edu.pl

Abstract. Each economic activity is accompanied with a risk influencing financial results of companies. Risk cannot be eliminated, but, with use of specially designed methods and tools, its size can be estimated and the actions aimed at its reduction can be undertaken. Risk management by systematic pursuing the policy that includes analysis, assessment and actions leading to a reduction of losses caused by the risk, is more and more often employed by production companies. Reduction of the risk negatively affecting financial results of companies is perceived as one of the possibilities to improve effectiveness and a chance to obtain the competitive advantage. In the paper, a concept of the reliability method of risk assessment is presented and verified on a practical example. The suggested method is composed of two stages. At the first stage, the risks of selected phases of the production process and production lines were determined by the FMEA method (Failure Mode and Effects Analysis). In the FMEA, the data from observations of the manufacturing process and the reports generated in the maintenance system were used. Application of FMEA made it possible to determine risks of selected elements of the entire manufacturing system. At the second stage, reliability structure of the production system was determined and, on that ground, the risk of non-performance of production orders for the entire plant was calculated.

Keywords: Risk in production systems · Methods of risk assessment · Risk factors

1 Introduction

Recently, increasing interest of manufacturing companies in risk management can be observed. This is caused by successively increasing economic risk and by introduction of risk management to many management standards. An example can be the ISO standards used in most manufacturing companies, like ISO 9001, ISO 14001, ISO 45001, ISO/IEC 27001, ISO 23301, IATF 16949, ISO 22000, ISO 17025 and other ones considering the risk management aspects. The newest ISO standard series 9001:2015 emphasizes the risk management much more than it was in the previous version 9001:2008. Another cause of increasing interest in the risk management is the fact that the modern production systems and their manufactured goods are more and more complex and complicated. In

A. Batako et al. (Eds.): GCMM 2021, LNNS 335, pp. 656–666, 2022.
https://doi.org/10.1007/978-3-030-90532-3_49

the case of both the production system and its product, the question often appears about the failure-free period of their future operation [1–5].

From the viewpoint of management of production systems, the risk management means making a decision on the ground of the information about disturbances of the production system, reflected in production losses, and suitably responding to these disturbances [6–10]. The term "risk management" can be only used, when it is possible to quantify the risk, since then it becomes an important element of planning, control, inspection and organization in the company. The purpose of implementing the risk management in a company is reduction of costs caused by occurrence of risk factors.

Because of the occurring risk, production systems are not deterministic, but stochastic. A stochastic process can be completely assessed by the theory of probability only. However, the probability theory alone is unable to predict either reliability or safety of the system. Here, required is analysis of the system operation, its structure, way of functioning, environmental influences and many other factors affecting the system. The probability theory can be only one of the tools accessible for an engineer in order to convert his/her knowledge about the system in a prediction of its probable future behavior. It this case, the reliability theory is much more required.

In literature several approaches have gained popularity including: Bayesian networks [11], Petri Nets [11] Hazard and Operability Analysis (HAZOP) [12], Fault Tree Analysis (FTA) [13], Failure Modes and Effect Analysis (FMEA) [14]. These approaches often simplify the system safety and reliability analysis processes by making assumptions about system behaviour. Such assumptions include: focusing only on the technical part of the system, availability of failure data, single mode of operation for system, binary states (failed or non-failed) of components, non-repairability of components, and. However, in practice all these assumptions are not valid. Unfortunately, classical approaches are not capable of accounting for these aspects. Modern systems frequently operate in multiple modes or phases, making them dynamic systems [11, 15].

2 Reliability Method of Risk Assessment for Production Systems

In the operational sense, reliability is most often defined as probability that the system, product or component will carry out its assigned jobs for the scheduled period in the determined work conditions [2, 17, 18] This probability is determined by the reliability function, which can be written in the following way:

$$N(t) = P\{t \leq T\}. \tag{1}$$

The expression (1) should be understood as the probability that the object that started its work at the moment $t = 0$ will not be operationally incapable before the moment $t \leq T$, where T is the durability. Durability is understood as the time during that the object maintains its usable properties in the determined operation conditions [17, 20]. Any economic or technical system is designed and built in order to fulfill the determined functions. Most often, productivity is introduced as the measure of a system ability to fulfill the given function [16, 17].

Opposite to the reliability function is the failure probability function $Z(t)$. Assuming that a failure of an object or system (till the moment t or later) is a certain event, the following can be written:

$$Z(t) + N(t) = 1. \tag{2}$$

In an organizational or production system, risk will be probability of the fact that the system will not fulfill the functions for that it was designed. In other words, risk can be determined as probability of occurrence of losses in the system, caused by potential hazards (so-called risk factors). Therefore, identification of a risk refers to location of hazards to the system, which can cause losses. As the probability of these losses, the risk (R) is a synonym of failure probability (Z), that is [16, 21]:

$$R = Z. \tag{3}$$

For the so accepted interpretation and using the formula (2), the following equation should be true [16, 21]:

$$R(t) = 1 - N(t). \tag{4}$$

So, analysis and assessment of risk make it possible to determine reliability of the system, as well as its failure probability. The reliability approach at planning and assessment of risk gives more possibilities and makes it possible to localize risk factors in the production system.

As a result of differentiated reliability levels of individual elements of a production system, reliability of the system as a whole is significantly dependent on the way how these elements are linked, i.e. on its reliability structure. This means that various reliability structures of the system, built of the same numbers of identical, independent elements, result in various reliability levels of the system. Analysis of the system reliability structure should be preceded by dividing the system into individual components, i.e. decomposition of the system, reflecting the logic links in the system in such a way that its individual parts are statistically independent and possibly largest. Depending on types of links between the components in the system, basic structures (series and parallel) and mixed structures (series-parallel and parallel-series) can be generally distinguished. In practice, the systems with mixed series-parallel and parallel-series structures are most often encountered [16].

A system has the series reliability structure, when unserviceability of any element results in ineffectiveness of the entire system. It results from this definition, that the system is efficient then and only then, when all its elements are serviceable and that its reliability will decrease with increasing number of the elements. Reliability of a system with series structure (N_S) will be determined by the formula [20]:

$$N_S = N_1 \cdot N_2 \cdot \ldots \cdot N_n = \prod\nolimits_{i=1}^{n} N_i, \tag{5}$$

where N_1, N_2, N_n – reliability of individual elements of the system.

Using the formula (4), total risk (R_C) of the system with series structure can be determined as follows:

$$R_C = 1 - [(1 - R_1) \cdot (1 - R_2) \cdot \ldots \cdot (1 - R_n)], \tag{6}$$

where R_1, R_2, R_n – risk occurring in individual elements of the system.

A system with parallel reliability structure is capable, when at least one of its elements is serviceable. The parallel structure is characteristic for the systems in that the elements execute the same task. Reliability of such a system will increase along with increasing number of its elements and will be determined by the formula [20, 21]:

$$N_S = 1 - [(1 - N_1) \cdot (1 - N_2) \cdot \ldots \cdot (1 - N_n)] = 1 - \prod_{i=1}^{n} (1 - N_i). \quad (7)$$

where $N_1, N_2, \ldots, N_n$ – reliability of individual elements of the system.

Total risk (R_C) of the system with parallel structure can be determined using the formula:

$$R_c = R_1 \cdot R_2 \cdot \ldots \cdot R_n = \prod_{=1}^{n} R_i, \quad (8)$$

where R_1, R_2, R_n – risk occurring in individual areas/elements of the system.

Production systems with the so defined parallel reliability structure do not appear in practice or can be met very rarely, since excess of the elements (e.g. machines, workers, means of transport etc.) results in unused resources, which involves increasing production costs [16, 19, 21, 22]. In this case, it is much more appropriate to determine the risks for individual production components. An example of such a structure is shown in Fig. 1.

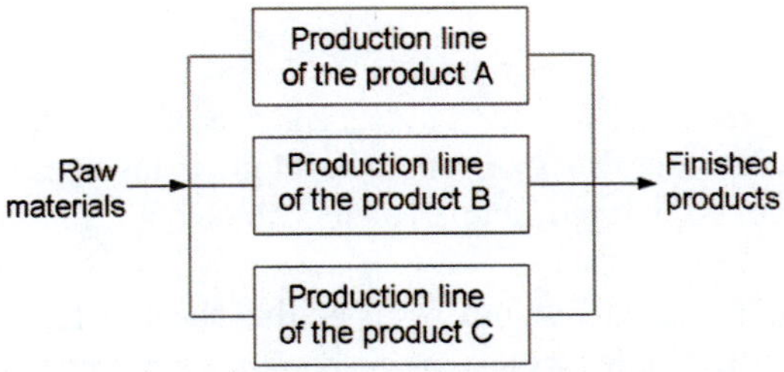

Fig. 1. Exemplary parallel production structure

When treating the system shown in Fig. 1 as a parallel arrangement of production components and taking into account the nature and specificity of production systems, the risk considering weights of the system components should be expressed by the formula (9) The weights are assigned depending on productivity of the line and reflect the influence of individual lines on total risk. These lines operate independently of each other and feed the output in different degrees.

$$R_{SC} = \alpha_1 R_1 + \alpha_2 R_2 + \ldots + \alpha_n R_n, \ where \ \alpha_1 + \alpha_2 + \ldots + \alpha_n = 1. \quad (9)$$

where $\alpha_1 + \alpha_2 + \ldots + \alpha_n = 1$.

Interpretation of risk as a synonym of failure probability, suggested in the above-presented method, makes it possible to identify the system elements most exposed to action of risk factors, or the elements in that action of risk factors will most strongly affect functioning of the entire production system. Certainly, real production systems have most often mixed series-parallel and parallel-series structures. In such a case, the above-described basic structures should be suitably modified to the needs of risk assessment of the analyzed system.

3 Characteristics of Production System of Steel Profiles

The data for verification of the suggested methodology were acquired in the company manufacturing steel profiles, mostly for the furniture and window industry. Manufacture is carried-out on individual orders and individual production orders significantly differ from each other in sizes of products and their complexity degrees. In the production floor of the company, shown in Fig. 2, four independent production lines are located, differing in sizes, ages of machinery, productivities and numbers of rolling stations. During production planning, individual orders are assigned for execution on individual production lines depending on sizes of products and their complexity degree.

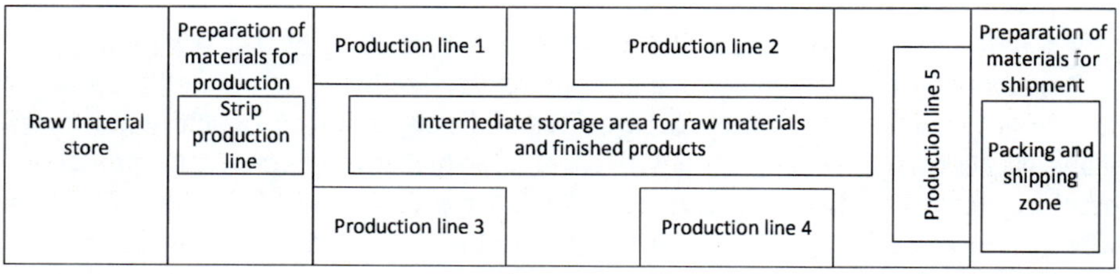

Fig. 2. Simplified layout of the plant

The production process is preceded by acceptance of an order and requisition of materials for production. It is also necessary to develop a detailed production technology in that, apart from process and product parameters, die blocks, rolls and shapes are also designed.

The first stage of the production process is cutting sheet metal from coils 1250 mm wide to strips of that the profiles will be produced. The strip production line consists of an uncoiler for longitudinal slitting. After uncoiling, sheet metal is moved to the section equipped with cutting knives. Up to 30 strips can be made from one coil. Next, strips are transported to intermediate storage areas in the production floor, where they are installed on the roll forming machines.

Steel profiles are manufactured on 4 production lines with various technical parameters presented in Table 1.

Table 1. Basic parameters of production lines

Line number	Power [kW]	Max. sheet thickness [mm]	Max. sheet width [mm]	Line age [years]	Number of rolls	Punching facility
LP 1	4 × 7	4	10–350	12	46	1
LP 2	4 × 4	3	10–350	10	28	1
LP 3	4 × 4	3	10–350	5	16	1
LP 4	4 × 7	4	10–350	3	11	0

The most technically advanced, although the oldest, is the production line LP1. This line is 3 times larger than a standard roll forming machine and is characterized by the highest power and productivity. Thanks to large number of rolls, most complicated and longest profiles can be produced on this machine. It is suitable for processing sheet metal up to 4 mm thick. Except the LP4 line, all the forming machines are equipped in punching modules.

4 Analysis of Risk Factors

In order to identify risk factors in the considered plant, all the 4 production lines were subjected to detailed observation. During 2 months, on specially prepared forms, the employees acquired data on random factors disturbing the production process, recording the information concerning kind of the risk, time of its occurrence, duration and size of losses caused in the production process. In addition to the observations, daily reports in form of diagrams generated by the program monitoring the production process were considered in the analysis of risk factors. Thanks to the reports, it was possible to determine in details numbers and duration times of stoppages. This information was compared with the data on numbers, sizes, complexity degrees and materials of the profiles.

Results of the observations showed that the following factors are crucial for the processes running in the plant:

- frequent failures of machines in the production lines LP1 and LP2,
- material specificity making a necessity to adjust or re-tool the machines, and n
- errors at packing the products

Results of the observations were not surprising, since the LP1 and LP2 lines are the oldest ones in the plant. The most frequent failures related to all the production lines included a rupture of the driving belt in the saw, wear or a fracture of the saw-blade disk, damages of bearings in the side rolls and too small amount of the cooling agent or its incorrect composition. Steel is a uniquely difficult material for cold-rolling and, in this connection, dimensions and shapes of the processed material require continuous control and adjustment. Moreover, under the influence of temperature and friction forces, such elements like threads, washers and clamps lose their properties, which can make it necessary to replace them or correct the machine settings. In consequence, the production process is interrupted. A noticeable factor delaying execution of orders was also high number of errors in packing the finished products. Incorrect selection or application of special separators securing products during transport caused damages and disintegration of packages or damages of the products themselves.

4.1 Assessment of Impact of Risk Factors on Production Lines with Use of the FMEA Method

In the next step, parameters related to severity for the process (S), frequency (probability) of occurrences (O) and detectability (D) were assigned to the identified risk factors. Then, values of the parameters were estimated according to the scale shown in Fig. 3.

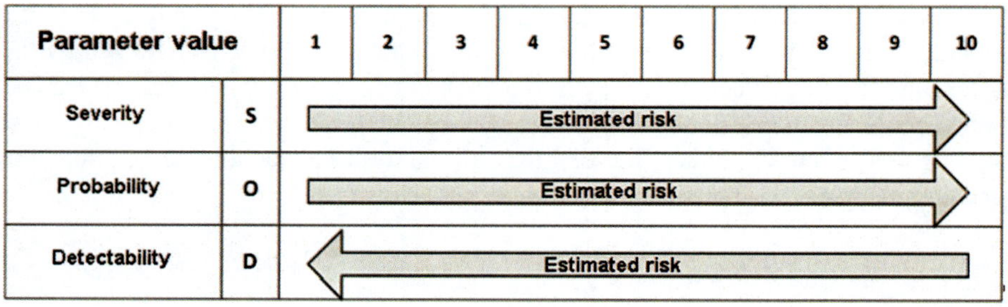

Fig. 3. Accepted scale of risk indices

Next, risk values for individual production lines, defined as the Risk Priority Number (RPN) were determined from the formula:

$$RPN = (S) \cdot (O) \cdot (D). \tag{10}$$

The RPN value ranges between 1 and 1000. High RPN value is an equivalent of high risk in the process. The S, O and D values were assigned by the specially designed team composed of the most experienced employees. The team decided that the RPN values below 100 would be considered as the acceptable risk. In the case of the RPN values over 100, corrective actions leading to reduction of the risk to the acceptable level will be carried-out subsequently.

Each of the lines is characterized by different technical parameters, so the FMEA tables were prepared for each line separately. The FMEA table includes also other risk factors related to packing finished products. This risk factor was considered separately, since packing of the products manufactured on all the production lines takes place in one separate zone in the production floor, see figure. The brief version of the FMEA table for all the 4 analyzed production lines is shown in Table 2.

Table 2. Abbreviated version of FMEA tables for 4 analyzed production lines

		S	O	D	RPN
LP1	Failures of machines	8	8	5	320
	Additional re-toolings	9	6	4	216
LP2	Failures of machines	7	8	5	280
	Additional re-toolings	8	5	4	160
LP3	Failures of machines	4	2	6	48
	Additional re-toolings	6	5	4	120
LP4	Failures of machines	2	2	6	24
	Additional re-toolings	5	6	4	120
	Packing errors	5	4	6	120

The maximum RPN value for each production line is 8000 and for the packing operation it is 1000. So, values of the risk (R_{LPi}) occurring in individual lines will be:

$$R_{LP1} = \frac{320 + 216}{8000} = 0.067$$

$$R_{LP2} = \frac{280 + 160}{8000} = 0.055$$

$$R_{LP3} = \frac{120}{8000} = 0.015$$

$$R_{LP4} = \frac{120}{8000} = 0.015$$

$$R_{pack} = \frac{120}{1000} = 0.12$$

The performed FMEA analysis confirmed the observation results. The highest risk occurs in the production lines LP1 and LP2 and this is the risk of a failure, which is probably related to the age of these lines. For the LP3 and LP4 lines, the risk of a failure is small, so, according to recommendations of the FMEA method, they were omitted at determination of losses. Like in the case of a failure, the highest risk of additional re-toolings occurs in the LP1 and LP2 lines. This factor occurs in all the production lines, but it causes smaller losses in the LP3 and LP4 lines, which is related to shorter ages of the machines and to the fact that less complicated profiles are produces on these lines. The analysis confirmed that the risk of packing errors is relatively high, amounting to 12%.

5 Characteristics of Production System of Steel Profiles

The data for verification of the suggested methodology were acquired in the company manufacturing steel profiles, mostly for the furniture and window industry. Manufacture is carried-out on individual orders and individual production orders significantly differ from each other in sizes of products and their complexity degrees. In the production floor of the company, shown in Fig. 2, four independent production lines are located, differing in sizes, ages of machinery, productivities and numbers of rolling stations. During production planning, individual orders are assigned for execution on individual production lines depending on sizes of products and their complexity degree.

When building the reliability structure, considered were only those elements, in that no risk factors occurred. Figure 4 shows the reliability structure of the analyzed production system.

With regard to specificity of production orders based on individual customers' orders, it is impossible to determine the average daily production for individual production lines. Complexity levels of the products can be also highly different. In consequence, potential losses of production volume, caused by occurrence of the risk factors, can be also diversified. Therefore, a profile with medium complexity degree, 6 m long, was selected. The following productivity (W) values were established for this profile on individual lines:

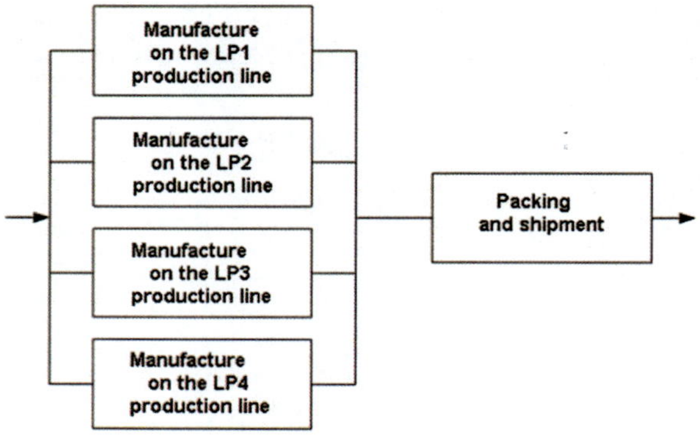

Fig. 4. Reliability structure of the analyzed production system

- productivity on LP1: WLP1 = 300 pcs. per shift,
- productivity on LP2: WLP2 = 200 pcs. per shift,
- productivity on LP3: WLP3 = 200 pcs. per shift,
- productivity on LP4: WLP4 = 300 pcs. per shift.

According to the formula (9), in the parallel-series structure of the analyzed production system shown in Fig. 4, the weights α_{LP1}, α_{LP2}, α_{LP3} and α_{LP4} corresponding to daily productivities should be assigned to individual production lines. It is assumed that the production lines operate independently on each other. This means that a stoppage of one line does not cause any breaks in functioning of the other lines.

Considering first the parallel structure of the analyzed production system and designating the risk for individual lines by RLPi, total risk of the system R_{SC} will be expressed by the formula:

$$RSC = 1 - [1 - \alpha PL1RLP1 + \alpha PL2RLP2 + \alpha PL3RLP3 + \alpha PL4RLP4.\ 1 - RPak] \tag{11}$$

where:

$$\alpha_{PL1} + \alpha_{LP2} + \alpha_{LP3} + \alpha_{LP4} = 1$$

R_{pack} – packing risk.

It was determined from the formula (9) and on the grounds of daily productivity of the lines, that $\alpha_{PL1} = \alpha_{LP4} = 0.3$ and $\alpha_{PL2} = \alpha_{PL3} = 0.2$. According to the formulae (4), (6) and (9), total risk R_C for the production system from Fig. 4 will take the form:

$$R_C = 1 - \left[\left(1 - \sum_{i=1}^{4} \alpha_i R_i\right) \cdot \left(1 - R_{pack}\right)\right], \text{ and so}$$

$$R_C = 1 - [(1 - (0.3 \cdot 0.067 + 0.2 \cdot 0.055 + 0.2 \cdot 0.015 + 0.3 \cdot 0.015))(1 - 0.12)] = 0.15$$

The obtained total risk value for the entire production system (R_C) is small, amounting to 8% only. This value means that the scheduled production will not be accomplished with the probability of 8%. The risk value for the entire system is much smaller than the risk values for the individual elements. In fact, in the event of a failure of one production line, production is executed on the other lines. This value is also consistent with the actual values obtained from the company, in that the annual risk of delays in shipments to customers amounts to 15%.

6 Summary

Nowadays, in order to obtain a competitive advantage, companies are searching for improvement of efficiency of their processes in various ways. One of them is implementation of risk management preceded by analysis and assessment of risk. Implementation of the risk management is more and more often regarded by production companies as one of possibilities to improve their production efficiency. In this article, a method of risk assessment is suggested and verified on a practical example. The method makes it possible to assess the risk of a complex technical system like a production system composed of several independently functioning production lines.

Verification of the method was carried-out on the grounds of the data acquired in the company manufacturing steel profiles. Risk was defined as probability of non-fulfilment of the production schedule. For the needs of the analysis and assessment, 2-month observation was carried-out and the data about the occurring risk factors were acquired. The risk was assessed in two stages. At the first stage, the FMEA method was used, in order to determine the risks of individual process stages and production lines. At the second stage, the reliability structure of the entire system was determined and, on this ground, total risk was determined. The results showed that the risks of failures and additional re-toolings are high for individual production lines, but the risk for the entire system, considering its reliability structure, is relatively low, amounting to 8%.

References

1. Aabo, T., Fraser, J.R.S., Simkins, B.J.: The rise and transformation of the chief risk officer: a success story on enterprise risk management. J. Appl. Corp. Fin. **17**(3), 62–75 (2005)
2. Antosz, K., Stadnicka, D., Ratnayake, R.C.: Development of a risk matrix for the assessment of maintenance suppliers: a study based on empirical knowledge. IFAC-PapersOnLine **50**(1), 9026–9031 (2017)
3. Aven, T.: Risk Analysis. Wiley, New York (2015)
4. Bagowsky, I.: Reliability Theory and Practice. Prentice-Hall, Englewood Cliffs (1961)
5. Billinton, R., Ronald, N.A.: Reliability Valuation of Engineering Systems. Plenum Press, New York (1992)
6. Burduk, A., Musiał, K., Kochańska, J., Górnicka, D., Stetsenko, A.: Tabu search and genetic algorithm for production process scheduling problem. LogForum **15** (2019)
7. Burduk, A., Chlebus, E.: Evaluation of the risk in production systems with a parallel reliability structure, Eksploatacja i Niezawodnosc-Maint. Reliab. (2), 84–95 (2009)
8. Cholewa, M., Helman, J., Molasy, M., Rosienkiewicz, M.: Identification of challenges to be overcome in the process of enhancing innovativeness based on implementation of central European projects funded from Interreg Programme. In: Burduk, A., Chlebus, E., Nowakowski, T., Tubis, A. (eds.) ISPEM 2018. AISC, vol. 835, pp. 185–194. Springer, Cham (2019). https://doi.org/10.1007/978-3-319-97490-3_18

9. Cwikla, G., Grabowik, C., Kalinowski, K., Paprocka, I., Banas, W.: The initial considerations and tests on the use of real time locating system in manufacturing processes improvement. IOP Conf. Ser. Mater. Sci. Eng. **400**(4), 042013 (2018)
10. Cwikla, G., Gwiazda, A., Banas, W., Monica, Z., Foit, K.: Assessment of the efficiency of the continuous improvement system based on Kaizen in an example company. IOP Conf. Ser. Mater. Sci. Eng. **400**(6), 062008 (2018)
11. Dunjó, J., Fthenakis, V., Vílchez, J.A., Arnaldos, J.: Hazard and operability (HAZOP) analysis. a literature review. J. Hazard. Mater. **173**(1–3), 19–32 (2010)
12. Gajdzik, B., Grzybowska, K.: Internal sources of risk in building employees' attitudes of safe work in metallurgical enterprise. Metalurgija **5**(2), 285–288 (2013)
13. Hoffmann, R, Kiedrowicz, M., Stanik, J.: Risk management system as the basic paradigm of the information security management system in an organization. In: MATEC Web of Conferences, vol. 76, p. 04010. EDP Sciences (2016)
14. IEEE Std 610.12-1990: IEEE Standard Glossary of Software Engineering Terminology, Institute of Electrical and Electronics Engineers, New York (1990)
15. Jasiulewicz-Kaczmarek, M., Legutko, S., Kluk, P.: Maintenance 4.0 technologies – new opportunities for sustainability driven maintenance. Manag. Prod. Eng. Rev. **11**(2), 74–87 (2020)
16. Kabi, S., Papadopoulos, Y.: Applications of Bayesian networks and Petri nets in safety, reliability, and risk assessments: a review. Saf. Sci. **115**, 154–175 (2019)
17. Kłos, S., Patalas-Maliszewska, J.: An Analysis of the efficiency of a parallel-serial manufacturing system using simulation. In: Burduk, A., Chlebus, E., Nowakowski, T., Tubis, A. (eds.) Intelligent Systems in Production Engineering and Maintenance, ISPEM 2018. Advances in Intelligent Systems and Computing, vol. 835, pp. 32–43. Springer, Cham (2018). https://doi.org/10.1007/978-3-319-97490-3_4
18. Kłosowski, G., Gola, A.: Risk-based estimation of manufacturing order costs with artificial intelligence. In: 2016 Federated Conference on Computer Science and Information Systems (FedCSIS), pp. 729–732. IEEE (2016)
19. Kuric, I., Císar, M., Tlach, V., Zajačko, I., Gál, T., Więcek, D.: Technical diagnostics at the department of automation and production systems. In: Burduk, A., Chlebus, E., Nowakowski, T., Tubis, A. (eds.) Intelligent Systems in Production Engineering and Maintenance. ISPEM 2018. Advances in Intelligent Systems and Computing, vol. 835, pp. 474–484. Springer, Cham (2019). https://doi.org/10.1007/978-3-319-97490-3_46
20. Ratnayake, R, Antosz, K.: Risk-based maintenance assessment in the manufacturing industry: minimisation of suboptimal prioritisation. Manag. Prod. Eng. Rev. **8**, 38–45 (2017)
21. Shakya, R., Prashant, C.: Modelling of risk analysis in production system. IOP Conf. Ser. Mater. Sci. Eng. **691**(1). IOP Publishing (2019)
22. Sawhney, R, Subburaman, K, Sonntag, C., Rao, P.R.V., Capizzi, C.: A modified FMEA approach to enhance reliability of lean systems. Int. J. Qual. Reliab. Manag. **27**(7), 832–855 (2010)
23. Stanik, J., Kiedrowicz, M., Waszkowski, R.: Security and risk as a primary feature of the production process. In: International Conference on Intelligent Systems in Production Engineering and Maintenance, pp. 701–709. Springer, Cham (2018). https://doi.org/10.1007/978-3-319-64465-3
24. Yazdi, M., Kabir, S., Walker, M.: Uncertainty handling in fault tree based risk assessment: state of the art and future perspectives. Process Saf. Environ. Prot. **131**, 89–104 (2019)

A Manufacturing Cost Estimation by Utilizing a Novel Sensor Based Cost Model in a Knowledge Management System

Hande Tokucoglu, Xun Chen(✉), Abdennour El Rhalibi, and Tahsin T. Opoz

Faculty of Engineering and Technology,
Liverpool John Moores University, Liverpool L3 3AF, UK
H.Tokucoglu@2016.ljmu.ac.uk, {X.Chen,A.Elrhalibi,
T.T.Opoz}@ljmu.ac.uk

Abstract. Nowadays, many small or medium size manufacturing companies are struggling to identify the right solution to tackle the problems of long production cycle time, poor quality and expensive cost in their manufacturing processes. An attractive method to solve these problems is to introduce knowledge management system in their processes. However, most available knowledge management systems are not feasible for them, as they are either too expensive or lack of necessary functions to fit their specific requirements. In order to generate a flexible and effective knowledge management system for small or medium size manufacturing companies, it is important to facilitate suitable sensors to observe and monitor manufacturing processes in a company, so that the efficiency of the manufacturing processes can be assessed in real time and the cost of the manufacturing processes can be calculated accurately. In this research, the manufacturing processes are observed via a power meter that is assembled on the machine. The efficiency measurement and calculation of the manufacturing processes are gathered in a knowledge management system database that is created by utilising Microsoft Access Software. One of the key beneficial feature of the database is the calculation of the annual manufacturing cost taking account of the real utilization rate of the machines. In this paper, how the real utilisation rate is monitored and integrated into general cost model is presented together with the application cases of two industry companies.

Keywords: Manufacturing costing · Optimisation · Knowledge support system · Industry 4.0

1 Introduction

Nowadays, it is important to serve products to customers responsively with 100% satisfactory in the full product life cycle. However, from time to time, companies have difficulties in fulfilling customer needs in a cost effective way. According to the meetings with the industrial partners, one of the considerable reasons of this problem is that the companies cannot recognise cost related troubles in a manufacturing process effectively because of utilizing conventional costing methods. Since the conventional costing

A. Batako et al. (Eds.): GCMM 2021, LNNS 335, pp. 667–682, 2022.
https://doi.org/10.1007/978-3-030-90532-3_50

methods do not consider the influence of the process variation, the inefficiency in the production do not get penalty in the conventional costing. Sometimes, lack of effective machine maintenance and process control could cause problems in a manufactured product, which may need manufacturing a new one or return for repair. The consequences of these deficiencies are longer cycle time of production leading to higher manufacturing cost of the products. The longer cycle time delivery process, order filling and all kind of services of a facility introduce instability, generate increasing cost and decrease customer satisfaction [1]. On the other hand, higher capital and labour costs in manufacturing that combined with high level of technological and competitive alterations, create difficulties for production companies in developed countries [2].

This paper presents an investigation, which combines a database with sensor based cost models that are capable for calculating unit manufacturing cost more accurately and presenting the efficiency of the manufacturing process more clearly to the user, in order to improve the performance of a manufacturing process and customer dissatisfaction. Here, the paper presents initial studies of the development of a knowledge support system framework of a generic cost model for small or medium size companies in order to calculate unit manufacturing cost of a product accurately in a database system that links to sensors in manufacturing systems. Besides, the system can select the most suitable machine in a facility for a specific manufacturing process. Moreover, it can give necessary information for the operator to choose manufacturing process steps and manufacturing techniques. In the next section, there will be a review of the related work, followed by thee methodology of the research, which covers knowledge framework of the database, general cost model for calculating unit manufacturing cost, the experiment about obtaining data for the cost model with a power meter and another experiment for testing WIFI technology of the power meter. Afterwards, the system that is created to calculate the utilization rate of a machine as an efficiency rate of a manufacturing process in the cost model will be explained. Subsequently, the benefits of the research will be explained and lastly, there will be a conclusion and future works of the research.

1.1 Related Works

In literature, there are some examples of similar works to this research. For instance, there are a few researches reporting how a smart knowledge based system is developed in order to forecast the manufacturing cost modelling of a product at the cognitive design level of product life cycle. Such a smart knowledge based system contains a CAD solid modelling system, user interface, material selection, manufacturing process/machine selection, and cost forecasting methods as well [3]. Moreover, other systems encompassing a supplier selection and assessment system that depends on knowledge management system with case-based reasoning decision support system for outsourcing processes at Honeywell Consumer Products Limited in China is proposed in [4]. On the other hand, a research takes a step to enhance a hybrid system through combined Case-Based Reasoning (CBR) and Artificial Neural Networks (ANN) as a Product Unit Cost (PUC) estimation model for a mobile phone company. A group of qualitative and quantitative aspects are examined and decided based on the cost formula of mobile phones and the opinions of the experts. CBR retrieves similar cases of phone products from case base through using qualitative factors and ANN is utilized for searching relationship between the quantitative factors and product unit cost [5].

Another paper utilizes Case-Based Reasoning method that includes modelling, case-based building and reasoning levels in order to schedule semiconductor manufacturing systems. Through the case-based reasoning system, a schedule for a new case with different manufacturing necessities with initial setup status and number of setup alteration operators is gathered by sequentially retrieving and reusing the previous cases that are most analogue to the new case come across during schedule creation in the reasoning level [6]. Besides, Potential Support Vector Machine (PSVM) is utilized to generate a model in order to optimize surface roughness during milling aluminium alloys (AA6061-T6) with carbide coated inserts. The validity test of the fit and suitability of the offered model has been done via analysis of variance. It is seen that the generated model is in the limits of the acceptable fault (about 2–9%) compared to experimental results [7].

There are also some researches in which some sensors are being used for quality control of a process: Some cost-effective sensors, including an accelerometer and GPS, were utilized in order to create an accelerometer-based system for gauging road roughness based on the vibrational response of roads for road maintenance [8]. Moreover, a thermoresistive strain sensor was generated in order to detect the strains that were distributed in a random area of a transparent substrate in roll-to-roll manufacturing. Besides, sequential operations of sensor patterning by inkjet printing, material stabilization by flashlight sintering, thermal creation by Joule heating, and strain analysing by IR inspection can be successfully indicated through thermoresistive strain sensor [9].

As reviewed on available literatures, we can see that sensors are used in order to measure efficiency of the manufacturing process as well. For example, a machine maintenance policy, which is called a sensory-updated degradation-based predictive maintenance policy, is presented. In this maintenance policy, contemporary degradation models, which join component based real-time degradation signals with degradation and reliability features of the population of the component, are used for estimating and updating the residual life distribution (RLD). After the generation of the machine maintenance policy, the performance of the policy will be assessed by using a simulation model of the basic manufacturing cell [10]. On the other hand, multi-sensor based tool condition monitoring system, which can provide wireless communication and is a cost effective system, is proposed. In multi-sensor based tool condition monitoring system, several sensors data are collected and utilized, for example vibration, cutting force and power data along with machining parameters, in order to facilitate effective tool condition observation and life prediction [11].

After conducting some investigation in literature, support vector machine (svm) method is decided to be utilized as a machine learning method for classification of the milling machine status as idle or working according to power values of the machine. Besides, case based reasoning method will be used for the validation of machine status classification and calculation of the utilization rate of the milling machine.

2 Methodology

The investigation of sensor based costing were conducted in two small size-manufacturing companies, which will be expressed as Company 1 and Company 2, in order to make the research more reliable and accurate. The brief explanation about the content of the knowledge support system framework is given below in 3 steps:

- The annual cost of the manufacturing process is calculated through a generic cost model that includes direct and overhead cost of the process. The utilization rate of the machine is added to the cost model during the calculation of the overhead costs.
- The utilization rate of the machine, which is utilized in the manufacturing process, is calculated based on the time and power data that are collected through using the power meter. The utilization rate refers to a percentage of the time at which the machine actually works and produces a product. The utilization rate of the machine gives a clear idea about the efficiency of the manufacturing process.
- Cased Based Reasoning method is applied in order to calculate utilization rate of a machine of a manufacturing process of a product based on similar previous cases.

2.1 Knowledge Framework of the Database

Knowledge framework of the database, which will be utilized for annual unit manufacturing cost calculation, production process and scheduling optimization, has been generated through using Microsoft Access software. Structure of the knowledge support system is presented as a flowchart in Fig. 1.

According to the figures in Fig. 1, the flow is built on manufacturing system structure. In the creation level of the knowledge support system, the elements of a manufacturing system, which are manufacturing processes of a product, input of the system that is the material of the product, resource of the system that is knowledge management of the manufacturing system and control mechanism of the user of the database or the operator of the manufacturing system, are considered. By courtesy of the functions of the elements of the manufacturing system, the output is the new product. According to the meetings with Company 1 and 2, it is apparent that the resources of the manufacturing system affect the quality of the product significantly. Therefore, two of the resources of the manufacturing system, - cost calculation system of the product and artificial intelligence, are added to the knowledge support system.

2.2 Generic Cost Model for Calculating Unit Manufacturing Cost

The development of generic cost model for which calculating annual unit manufacturing cost will be explained as following:

- Step 1: An initial time and sensor based generic cost model was generated through Segmenting Cost Estimation Method (SCEM). The generic cost model should include direct cost parameters as well as overhead cost parameters in order to give a reliable result to the user of the knowledge support system database. Therefore, the parameters were determined through having meetings with Company 1 and Company 2. After SCEM was considered, sub-formulas of the cost model were generated as the segments of the cost model. Subsequently, sub-formulas and other cost parameters were gathered under four group: Fixed Cost, Variable Cost, Fixed Overhead and Variable Overhead

Cost. Lastly, the related sub-formulas were combined with other cost parameters and initial general manufacturing cost model was obtained.

- Step 2: According to the survey with Companies, a core model, which are intermediary cost models between initial cost model and final cost model, were created based on initial general cost model and needs and requests of the companies. The core model is shown in Table 1.

  The core model can be tailored for individual company with some small modification.

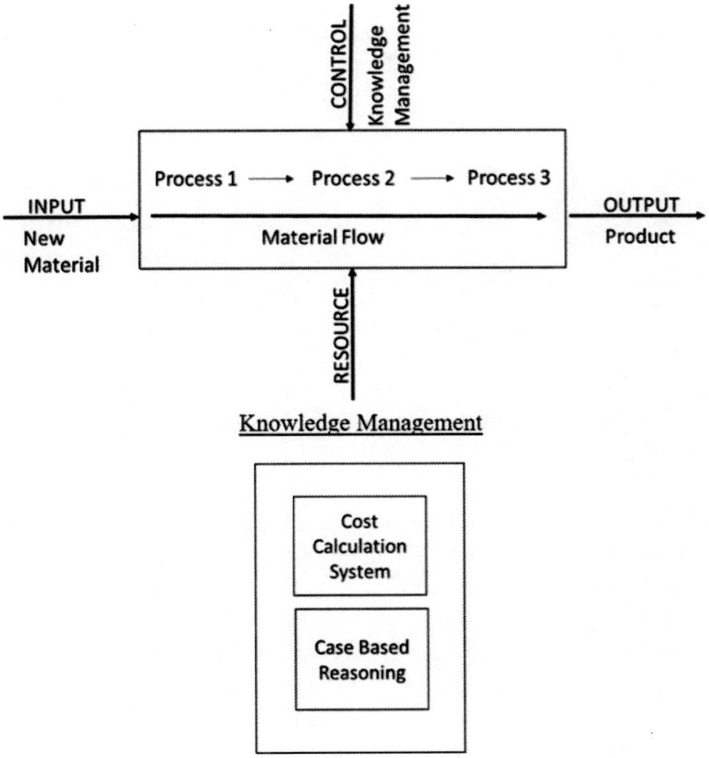

Fig. 1. The structure of the knowledge management system.

- Step 3: After the core model was created, the parameters of the model were combined under four group that were presented in Table 1 in order to generate the general cost model. It is seen that some parameters need a sub-formula for an accurate calculation subsequent to combination of two core models. The sub-formulas were given in following:

$$Electricity\ Cost\ per\ Process = Total\ active\ energy\ *\ electricity\ price\ *\ cycle\ time \quad (1)$$

$$\textit{Fluid Cost per Process} = \textit{Total fluid consumption} * \textit{unit fluid price} * \textit{cycle time} \quad (2)$$

$$\textit{Gas cost per Process}\ [12] = (\textit{Current Reading}(\textit{Cubic meter}) * \textit{Calorific value of the gas}) * 1.02264)/3.6 \quad (3)$$

$$\textit{Direct Cost of Sales} = \textit{Hourly salary of the operator} * \textit{case time} \quad (4)$$

$$\textit{Packaging Cost} = \textit{Packaging Equipment Price} * \textit{Packaging Time} \quad (5)$$

$$\textit{Hire Purchase and Leasing Commitments} = \textit{Insurance} + \textit{Estate taxes} + \textit{Rent of the facility} \quad (6)$$

$$\textit{Indirect Cost of Sales} = \textit{Hourly non-direct labour salaries} * \textit{Administrate personnel working hours} \quad (7)$$

$$\textit{Storage Cost}\ [13] = \textit{Total Storage area cost}/(\textit{total number of units shipped during previous}\ 12\ \textit{months} + \textit{total number of units that is stored currently}) \quad (8)$$

- Step 4: Based on the information of the previous steps, the general cost model for calculating annual unit manufacturing cost was generated. The parameters of the cost model were showed in Table 2.

As it can be seen in Table 1 and Table 2, when the core model and initial cost model were joined, with the difference of the parameters of the core model of Company 1, "Work Expenses" is added into fixed costs and "Standing Electricity Charge" is attached to fixed overhead costs. The system allows users to modify or select parameters according to real situation of individual company, which gives flexibility for the user.

After determining the parameters of the general cost model, the general cost model, which can provide flexibility to small or medium size companies through its sensor based data gathering and human-machine interface that is ensured with a database in Microsoft Access, is generated.

The formula of the general cost model is given as Eq. (9):

$$\textit{Unit Manufacturing Cost} = \textit{Fixed Cost} + \textit{Variable Cost} + [\textit{Fixed Overhead Cost} + \textit{Variable Overhead Cost}] \quad (9)$$

Based on the general cost model, this research can make difference in literature with its calculation of the electricity energy cost. The electricity cost of the manufacturing process will be considered as a variable cost, while overall electricity cost is being involved in variable overhead costs. Therefore, with this research, the electricity cost of a manufacturing process will be separated from overall electricity cost of a facility.

Table 1. The core model parameters.

Fixed cost	Variable cost	Fixed overhead cost	Variable overhead cost
Computer software cost	Electricity cost per process	Tangible fixed assets	Storage cost
	Fluid cost per process	Debtors	Light and heat
	Gas cost per process	Hire purchase and leasing commitments	Post and stationery
	Raw material cost	Pension costs and other post-retirement benefits	Telephone
	Direct cost of sales	Research and development	Travelling
	Packaging cost	Indirect cost of sales	Staff training
		Security charges	
		Staff welfare	
		Repairs & maintenance	
		Trade subscriptions	
		Courier services	
		Computer costs	
		Medical insurance	
		Accountancy	
		Legal and professional fees	
		Advertising	

Obtaining Data from Company 1 for Creating a Core Model. The experiment, for which is transferring and analysing data from power meter on a machine to a computer in order to obtain data for calculating electricity energy cost of the manufacturing process, has five stages:

- Observation of individual pressing process steps
- Measuring cycle time
- Calculation the cost of electrical energy consumption of the process.
- Calculating utilization rate of the machine during the experiment time.
- Analysing the power graph of the machine.

Table 2. The parameters of general cost model.

Fixed cost	Variable cost	Fixed overhead cost	Variable overhead cost
Computer software	Electricity cost per process	Tangible fixed assets	Storage cost
Work expenses	Fluid cost per process	Debtors	Light and heat
	Gas cost per process	Hire purchase and leasing commitments	Post and stationery
	Raw material cost	Pension costs and other post-retirement benefits	Telephone
	Direct cost of sales	Research and development	Travelling
	Packaging cost	Indirect cost of sales	Staff training
		Security charges	
		Staff welfare	
		Repairs & maintenance	
		Trade subscriptions	
		Courier services	
		Computer costs	
		Medical insurance	
		Accountancy	
		Legal and professional fees	
		Advertising	
		Standing electricity charge	

The experiment was done through considering laser sintering process and laser sintering machine in Company 1 The aim of the experiment is to observe the laser sintering process, measuring the cycle time of the process and calculating the electrical energy consumption of the laser sintering machine during the time of the experiment in the facility of Company 1. As a result of this, electrical energy cost of the machine, which is one of the parameters of the cost model, can be calculated according to three types of power that are given in the following according to current and voltage data:

- Active (Real) Power: It is power that is utilized or spread essentially in a circuit.
- Reactive (Non-Active) Power: It is known that reactive loads, for instance, inductors and capacitors, do not spread any power. However, they decrease the voltage and draw

current. This situation gives a misleading effect that power is actually spread. This "phantom power" is called reactive power.

- Apparent Power: It is the combination of active power and reactive power [14].

The minimum, maximum and average values of active, reactive and apparent power of laser sintering machine are presented in Table 3:

After the value of power types were certain, the cost of electrical energy consumption for laser sintering process was calculated based on the Eq. (10):

$$\begin{aligned} \textit{Cost of Electrical Energy} &= (\textit{Power Consumption } (kW) * \textit{Electricity Price } (£/kWh) \\ &* \textit{Machine On Time } (h)) \end{aligned} \tag{10}$$

Table 3. Values of active, reactive and apparent power.

Power type	Minimum	Maximum	Average
Active power (kW)	−0.000	2.085	0.851
Reactive power (var)	0.000	7.257	1.020
Apparent power (VA)	0.000	7.296	1.328

Based on power meter calculations, total cycle time is 167.97 h and electricity price of Company 1 is 0.127 £/kWh. Therefore, considering the value of active energy in calculations, total cost of electricity energy consumption for laser sintering process in six days is £18.13.

The next aim of the experiment is calculating the utilization rate of the laser sintering machine for six days according to Eq. (11):

$$\begin{aligned} \textit{Utilization Rate of the Laser Sintering Machine} &= \big[(\textit{Number of Hours the machine} \\ &\textit{works}) \backslash (\textit{Total Working Hours})\big] * 100 \end{aligned} \tag{11}$$

Considering Eq. (11), the utilization rate of laser sintering machine for six days is 53.37%. Lastly, active power graph of the laser sintering machine was analysed based on job scheduling document and was showed in Fig. 2.

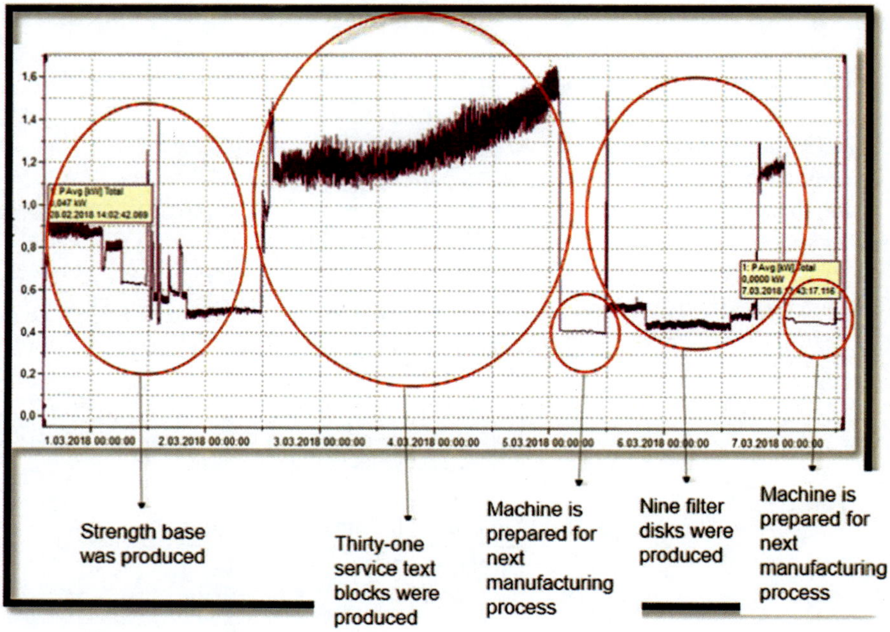

Fig. 2. Active power graph of company [15].

As it can be seen in Fig. 2, one component and two types of products were manufactured: strength base as the component, service text blocks and filter disks as the products. Moreover, between the production processes of the products, the machine was prepared for the next process.

It is aimed that the evaluation process of the power graph will be automated through a suitable rule based expert system in the future.

Same experiment were done for Company 2 through assembling the power meter on cutting machine. As a result of the experiment, total electricity energy cost for Company 2 is £82.15 and the utilization rate of cutting machine is 42.15% for six days.

It is aimed that utilization rate will be calculated automatically and more accurately by using appropriate machine learning methods.

2.3 The Experiment with WIFI Technology of the Power Meter

The aim of the experiment is testing WI-FI technology of the power meter in order to provide transferring data from the sensors on the machines to the database via internet; when this aim will be fulfilled, Industry 4.0. approach can be achieved in a facility. The steps of the experiment are given in following:

- The power meter was assembled to the milling machine in the laboratory of Liverpool John Moores University and machine started working.
- It started logging and saving the voltage and current data of the milling machine.

- The data were transferred from the power meter to the computer every two hours between 9:36 am and 15:36 pm without interfering the power meter while the power meter was still logging.
- When it was 15:36 pm, the power meter and the machine were shut down and the power meter disassembled from the machine
- The cost of milling process during the experiment time was calculated through general cost model that was explained in Sect. 2.2.

After the steps of the experiment were performed, it is seen that the experiment proceeded for 6 h 14 min 46 s and the number of current and voltage data that were collected in 6 h 14 min is 375.

As a conclusion of the experiment, it is seen that data transfer can be accomplished from power meter that is assembled to the machine to the computer with WI-FI technology without interfering the power meter while it was working. This kind of data transfer can provide Industry 4.0 approach in a manufacturing facility.

2.4 Calculation of the Utilization Rate of the Machine via Machine Learning Method as a Parameter of the Cost Model

In order to obtain more reliable results from the cost model and track the manufacturing process more efficiently, a manufacturing process efficiency parameter, which is the utilization rate of the machine, will be calculated and added to the cost model. The parameter will show the rate of the actual usage of the machine for manufacturing a product during an observation of the process through a sensor.

Utilization rate of the machine is planned to be calculated in 4 steps:

- Classification of the machine status according to its power level values that were obtained from the power meter.
- Calculation of times of the target status, total time and utilization rate of the machine
- The Implementation of Utilization Rate in the Cost Model
- Validation of the system.

Step 1: Classification. The aim of this step is to classifying the status of milling machine of Liverpool John Moores University Laboratory according to its power level values. As a result of the experiment, the power meter had been logging on the milling machine for three weeks and 30244 power level value data was collected. In order to achieve the classification of machine status, multiclass model for support vector machine (svm) is decided to utilize in Matlab. Subsequently, every week is analysed separately and in the first week, the code is trained and made to predict the status. In second and third week, the code only made prediction based on the trained model of svm. The machine has two status: Idle (the machine is on, however it does not do milling. Power level should be between 0 and 600) and working (the machine does milling. The power level is between 600 and 1000). Support Vector Machine code of week 1 of the experiment of milling machine is given in following as an example (Fig. 3):

```
filename = 'milling machine power data v4.xlsm';
[Anum, Btxt] = xlsread(filename,'Shifting','a5575:f6000');
n0 = 1;
n=424;
X = Anum (n0:n,1:5);
Y = Btxt (n0:n);
L1= false(n,1);
Y1= zeros(n,1);
for i=1:n
   L1(i)= strncmp(Y(i),'working',7);
if L1(i)
   Y1(i)= 1;
else
Y1(i)= 2;
end
end
Mdl=fitcecoc(X,Y1,'OptimizeHyperparameters','auto',...'Hyperpar
ameterOptimizationOptions',struct('AcquisitionFunctionName',...
'expected-improvement-plus'));
filename = 'milling machine power data v2.xlsx';
[Anum,Craw] = xlsread(filename,'Shifting','a2200:e2445');
n0 = 1;
n = 244;
XX = Anum (n0:n,1:5);
yfit = predict(Mdl,XX);
zzz=[XX,yfit];
plot(zzz);
```

Fig. 3. Support Vector Machine code of week 1 of the experiment of milling machine.

As it can be seen in the code, first of all, 424 data was selected from the dataset and power values was identified as input (X) and status (idle or working), which was expressed as text, defined as output (Y). Then, output data was converted from text to numbers. (working is 1 and idle is 2.) in order to ease the calculation of "working" and total machine-on time. In the next step, the code was trained through generating and optimizing a model that is named as "Mdl" via using multiclass svm. Finally, 244 input data was selected from the dataset and the code predicted the status (output) that is shown as "yfit".

In the next step, total "working" and total machine-on time is calculated according to time (it includes date and time) and status (the result of the codes of week 1, week 2 and week 3 that are called as yfit, yfit2 and yfit3) of the power level values.

Step 2: Calculation of Time of the Status. Calculation of total "working" that is the time that the machine actually producing a product and machine-on time that is the sum of idle and working time of the machine is required for the calculation of the utilization rate of the machine. The formula of the utilization rate is given in following:

$$\textit{Utilization Rate} = (\textit{Total ``Working'' Time} \,/\, \textit{Total Machine-On Time}) * 100\,\% \quad (12)$$

In order to calculate total working time, total machine on time and the utilization rate of the milling machine, a program code was written in Matlab and week 1, week 2 and week 3 were calculated individually. The results of the Matlab codes are given in following (Table 4):

Table 4. The results of utilization rate of milling machine.

	Week 1	Week 2	Week 3
Number of records	10,491	10,080	9670
Total "working" status	24	279	279
Total machine-on-time	174 h 49 m 58.06 s	167 h 59 m 0.067 s	161 h 9 m 0.065 s
Total working time	0 h 23 m 59.673 s	4 h 39 m 0.283 s	4 h 39 m 0.283 s
Missing records time	0 h 0 m 0 s	0 h 0 m 0 s	0 h 0 m 0 s
Utilization rate	0.2287%	2.7682%	2.8856%

Step 3: The Implementation of Utilization Rate in the Cost Model. The formula of the cost model is given as Eq. (9). The formula is changed a little bit in order to add the contribution of the utilization rate of the machine that is calculated through using Matlab Software. The new equations of the cost model are given below:

$$\begin{aligned} \textit{Total Process Overhead Cost} &= \textit{Depreciation of the machine} + \textit{Computer Costs} \\ &+ \textit{Post and Stationary} \end{aligned} \tag{13}$$

$$\textit{Total Process Overhead Cost After Considered Utilization Rate} = (\textit{Total Process Overhead Cost That Does Not Count Machine}) + (\textit{Machine Overhead Cost}) * (1/\textit{Utilization Rate}). \tag{14}$$

$$\textit{Company Overhead Cost} = \textit{Debtors} + \textit{Pension costs and other post} - \textit{retirement benefits} + \textit{Security Charges} + \textit{Staff Welfare} + \textit{Trade subscriptions} + \textit{Courier Services} + \textit{Medical insurance} + \textit{Accountancy} + \textit{Legal and Professional fees} + \textit{Total non-direct labour cost} + \textit{Advertising} + \textit{Telephone} + \textit{Travelling} + \textit{Insurance} + \textit{Total Gas Cost} + \textit{Standing Electricity Charges} + \textit{Estate Taxes} + \textit{Rent of the facility} + \textit{Maintenance Cost} \tag{15}$$

$$\textit{Total Annual Manufacturing Cost} = \textit{Fixed Cost} + \textit{Variable Cost} + \textit{Total Process Overhead Cost After Considered Utilization Rate} + \textit{Company Overhead Cost} \tag{16}$$

As it can be seen in Eq. (14), total process overhead cost and utilization rate of the machine are inversely proportional. When the percentage of the utilization rate of the machine increases, total process overhead cost of the company decreases. This decline in process overhead cost reverberates to total annual manufacturing cost of the product.

Step 4: Validation of the System. In order to validate the system of the calculation of utilization rate of a machine, Case Based Reasoning System will be utilized as a future work. The basic principle of the validation is that the parameters of the previous cases will be compared with the parameters of the current problem and the solution, which is the utilization rate of the machine, will be adapted to the current problem.

3 Discussion of Potential Benefits of Sensor Based Costing Method

Potential beneficial outcomes of the research are:

- The knowledge support system can calculate unit manufacturing cost of a product through a novel generic sensor based cost model for SMEs in order to add flexibility to the knowledge support system. The novelty of the cost model comes from the one of its cost parameters that is an indicator of the efficiency of the manufacturing process of a product. This cost parameter is the utilization rate of the machine of the manufacturing process.
- The knowledge support system database can give information about steps of the manufacturing process of the current product and previous products (cases) that were produced via the same manufacturing process.
- The knowledge support system can estimate the manufacturing cost of the new product as well, before the beginning of the manufacturing process.
- In order to estimate the utilization rate of the machine of the new product, the Case Based Reasoning (CBR) System, which chooses the previous product that has the most similar manufacturing system to the new product among the previous cases of the facility and gives the utilization rate of the machine of the previous product, was established inside the database.
- A base for Industry 4.0 approach can be established in a manufacturing system through the sensor based cost model.
- The sensors that will be applied on machines in a facility show the efficiency of production and the power level of the machine, which is one of the outcomes of the sensors, indicates potential quality problems and productivity problems of the manufacturing process.

4 Conclusion and Future Works

This research presents the studies of the development of a knowledge support system framework with a generic cost model for small or medium size companies in order to calculate annual unit manufacturing cost of a product accurately in a database system.

The novelties of the project are given in the following:

- The utilization rate of a machine during a manufacturing process will be calculated through the data, which are utilized inside of a suitable machine learning technique for the calculations of the utilization rate, from the sensor on the machine. Receiving the value of the utilization rate of the machine has two benefits:

- The efficiency of a manufacturing process can be seen and measured effectively through the value of the utilization rate of the machine.
- Overhead costs of the manufacturing process can be calculated more effectively by using the value of the utilization rate of the machine.
- The cost of the electricity energy consumption of a manufacturing process is considered and added to the Direct Process Level Cost of the generic cost model through calculating the electricity energy consumption of a machine of a manufacturing process via the data of the power meter that is assembled on the machine.
- In the future, the validation of the system of the calculation of utilization rate of a machine, which is an efficiency parameter of the cost model, will be completed through utilizing case based reasoning method. Subsequently, the parameter will be added to the cost model and knowledge support system database, which includes the cost model, will be tested in industry.

References

1. Meyer, C.: Fast Cycle Time: How to Align Purpose, Strategy, and Structure for Speed. The Free Press, New York (2007)
2. Yin, Y., Stecke, K.E., Swink, M., Kaku, I.: Lessons from seru production on manufacturing competitively in a high cost environment. J. Oper. Manag. **49**(51), 67–76 (2017)
3. Shehab, E., Abdalla, H.: Manufacturing cost modelling for concurrent product development. Robot. Comput. Integr. Manufact. **17**(4), 341–353 (2001)
4. Choy, K., Lee, W., Henry, C.L., Choy, L.: A knowledge-based supplier intelligence retrieval system for outsource manufacturing. Knowl. Based Syst. **18**(1), 1–17 (2005)
5. Chang, P.-C., Lin, J.-J., Dzan, W.-Y.: Forecasting of manufacturing cost in mobile phone products by case-based reasoning and artificial neural network models. J. Intell. Manufact. **23**(3), 517–531 (2012). https://doi.org/10.1007/s10845-010-0390-7
6. Lim, J., Chae, M.-J., Yang, Y., Park, I.-B., Lee, J., Park, J.: Fast scheduling of semiconductor manufacturing facilities using case-based reasoning. IEEE Trans. Semicond. Manuf. **29**(1), 22–32 (2016)
7. Kadirgama, K., Noor, M.M., Rahman, M.M.: Optimization of surface roughness in end milling using potential support vector machine. Arab. J. Sci. Eng. **37**(8), 2269–2275 (2012). https://doi.org/10.1007/s13369-012-0314-2
8. Bidgoli, M.A., Golroo, A., Nadjar, H.S., Rashidabad, A.G., Ganji, M.R.: Road roughness measurement using a cost-effective sensor-based monitoring system. Autom. Constr. **104**, 140–152 (2019)
9. Liao, K.-H., An, C.-H., Lo, C.-Y.: Continuous inkjet-patterned and flashlight-sintered strain sensor for in-line off-axis detection in Roll-to-Roll manufacturing. Mechatronics **59**, 95–103 (2019)
10. Kaiser, K.A., Gebraeel, N.Z.: Predictive maintenance management using sensor-based degradation models. IEEE Trans. Syst. Man Cybern. Part A Syst. Hum. **39**(4), 840–849 (2009)
11. Zhang, X., Lu, X., Wang, S., Wang, W., Li, W.: A multi-sensor based online tool condition monitoring system for milling process. Procedia CIRP **72**, 1136–1141 (2018)
12. Your Energy Bill - How Energy Bills Calculated? British Gas. https://www.britishgas.co.uk/products-and-services/gas-and-electricity/tips-and-advice/energy-bill-cost.html. Accessed 18 Apr 2019

13. Raley, T.: How to Calculate Warehouse Storage Costs. Sciencing. https://sciencing.com/how-5869978-calculate-warehouse-storage-costs.html. Accessed 18 Apr 2019
14. Electronics Tutorials: Power in AC Circuits. Electronics Tutorials. https://www.electronics-tutorials.ws/accircuits/power-in-ac-circuits.html. Accessed 18 Apr 2019
15. Tokucoglu, H., Chen, X., El Rhalibi, A., Opoz, T.: Sensor based cost modelling for a knowledge support system development. In: 25th International Conference on Automation and Computing (ICAC), Lancaster, United Kingdom, pp. 1–6 (2019)

High-Tech Enterprise Value Evaluation Based on Real Options

GuangLei Xu and HongBing Jiang(✉)

Zhengzhou University, Zhengzhou 450001, China
jhbymx@foxmail.com

Abstract. Because high-tech enterprises have the characteristics of high risk, high investment and dynamic sequence of decision-making, and their intangible assets account for a relatively large part of the enterprise value, the traditional valuation theory has been unable to meet the value evaluation requirements of high-tech enterprises. In order to solve this problem, this article uses the Schwartz-Moon model, which is more applicable in the real option method according to the characteristics of high-tech enterprises, to evaluate the value of high-tech enterprises, And then conducts a valuation analysis of listed high-tech enterprise A, and the model is verified according to the enterprise's stock valuation, which enriches the valuation method of high-tech enterprises to a certain extent.

Keywords: Real options · High-tech enterprise · Enterprise valuation

1 Introduction

High-tech enterprises are faced with many risks in technology, market and production when carrying out product research and development, and they have great uncertainty. Moreover, due to their unique growth, high-tech enterprises often invest hugely in the early stage and lack stable cash flow. The particularity of high-tech enterprises makes it difficult for traditional valuation methods, such as discounted cash flow method and cost method, to accurately assess the intangible assets and human capital of high-tech enterprises, nor can they predict the risk value and growth of high-tech enterprises ability. However, the real option method regards the enterprise activity as a random process influenced by uncertain factors. It can segment the enterprise activity and analyze it more flexibly according to the information of different stages, which can more accurately quantify the current value of high-tech enterprises.

By comparing the cash flow discount method and the real option valuation method, Chengyan Yang and Jie Zhang [1] think that the real option evaluation method considers the volatility of the company's operation and the flexibility of management, and reasonably explains the "invisible" value of the enterprise in the period of strategic transformation. In 2000, Schwartz [2] established a continuous real option pricing model based on the assumption that the future revenue and cost of the network companies follow a random process, which is far ahead in the evaluation of the overall value of high-tech enterprises by using the continuous time pricing model. After that, Kellogg and Charnes

A. Batako et al. (Eds.): GCMM 2021, LNNS 335, pp. 683–689, 2022.
https://doi.org/10.1007/978-3-030-90532-3_51

[3] used the decision tree algorithm to quantify the stock price of agouon.Inc, a high-tech biochemical company, and found that the binomial lattice method was more accurate in the early value evaluation of the company.

Domestic scholars have gradually applied the real option method to all aspects. He Liu and Lili Fan [4] applied the real option method to the valuation of enterprise carbon intangible assets, pointing out that the value of carbon intangible assets is not the asset itself, but the flexible value derived from management. Honghai Wang, Yuqing Wu et al. [5] used Monte Carlo simulation to evaluate the value of enterprises based on the characteristics of the value evaluation interval of high-tech enterprises. According to the characteristics of Internet enterprises, Jianfeng Guo and Dan Wang et al. [6] use the BS model of catastrophe progression method and real option method to evaluate the value of Internet enterprises. Xiaohua Zhou and Yajun Tan et al. [7] tried to combine the real option method with the traditional valuation method for the IPO pricing of high-tech companies. Domestic and foreign researches on the use of real option methods in specific industries are gradually emerging, but their methods mainly use methods such as BS model and binary tree. This article will use the Schwartz-Moon model to estimate a high-tech enterprise based on the characteristics of China's high-tech enterprises value. The model was modified according to the evaluation situation, which simplified the calculation amount to a certain extent, verified the reliability of the model, and then enriched the value evaluation method of high-tech enterprises.

2 Construction of Valuation Model of High-Tech Enterprises Based on Real Options

2.1 The Rationality of Introducing Real Option Method to the Valuation of High-Tech Enterprises

High-tech enterprises generally have the following characteristics: First, high risks and high returns. High-tech enterprises face many risks such as market, finance, production, technology, etc., but high-tech enterprises that successfully overcome the risks have a high return on investment. Second, asset composition softened. Traditional enterprises account for a large proportion of tangible assets, while high-tech enterprises tend to account for a large proportion of intangible assets, but intangible assets in the economic sense are very broad, and the accounting confirmation of the balance sheet reflects very little, which has a great impact on the value evaluation of high-tech enterprises. Third, there is a lack of basic information required for value evaluation. High-tech enterprises are often different from traditional industries, most of which lack of relevant information such as risk coefficient and cash flow for reference, which makes the parameter prediction foundation for value evaluation relatively weak and difficult to make convincing evaluation data.

Because of the above characteristics of high-tech enterprises, traditional enterprise valuation methods can not accurately evaluate them. Real option is the expansion of financial option in the real field. Since the 1970s, real option theory has been widely used to solve the investment decision-making problem of irreversible investment projects under uncertain conditions. Compared with the traditional cost-benefit analysis of investment decision-making methods, the real option method has the characteristics of managing investment and controlling risk flexibility. It is very suitable for analyzing investment projects with the characteristics of high investment risk, huge investment amount, long investment time and irreversible investment. Because high-tech enterprises have these characteristics, it is suitable to use the real option method to solve the problem of enterprise valuations.

2.2 Valuation Model of High-Tech Enterprises Based on Real Option

From the perspective of real options, an enterprise can be regarded as a financial asset system. The equity held by investors is regarded as a call option. The value of the company is regarded as the target. The execution price is the debt price. On the maturity date, when the company's value is greater than the debt price, the investor can choose to exercise options or waive. Once the debt price is lower than the debt price, the investor can renounce the execution of the contract, and then the company will liquidate the debt into bankruptcy proceedings.

Schwartz and Moon proposed the Schwartz-Moon model when evaluating the value of Amazon in 2000. The model is randomly related to revenue, revenue growth rate, and cost rate, The specific concepts are as follows: First, a continuous-time model is constructed. and then the random path of the main variables of the model is simulated using the Monte Carlo simulation method.

The Schwartz-Moon model has three uncertainties. These are the uncertainty of revenue, expected growth rate and cost rate. By analyzing the cost and revenue data of Company A from 2010 to 2016, the correlation coefficient between the two is calculated to be 0.999, which means that cost and revenue are highly correlated, so only two uncertainties, Uncertainty of revenue and uncertainty of the expected growth rate.

The model is constructed as follows:

2.2.1 Continuous Time Model

Suppose that the operating income of company A at time t is R(t), and its changing process follows geometric Brownian motion.

$$\frac{dR(t)}{R(t)} = \mu(t)d(t) + \sigma(t)dZ_1 \quad (1)$$

Among them, $\mu(t)$ is used to describe the expected growth rate of income, and $\sigma(t)$ is the volatility of income. dZ_1 represents the impact of unforeseen circumstances on the company's revenue. Since the operating environment facing high-tech companies is highly uncertain and competitive, it is assumed that the initial high growth rate will randomly converge to the industry average. That is, $\mu(t)$ follows the mean recovery process and converges to the industry's long-term average growth rate $\bar{\mu}$.

$$d\mu(t) = \kappa(\bar{\mu} - \mu(t))dt + \eta(t)dZ_2 \quad (2)$$

$\eta(t)$ represents the volatility of income growth rate. dZ_2 represents the uncertainty of income growth rate. The mean recovery coefficient κ represents the speed at which the expected income growth rate converges to the average, and $\ln 2/\kappa$ can be regarded as half the time that the growth rate returns to the long-term average. In addition, assuming that there is no correlation between the two variables Z_1 and Z_2, it is assumed that the volatility $\sigma(t)$ of the expected income growth rate will converge to 0, which means that the income change will return to the industry normal level over time.

$$d\sigma(t) = \kappa 1(\sigma - \sigma(t))dt \tag{3}$$

$$d\eta(t) = -\kappa 2\eta(t)dt \tag{4}$$

The company's operating costs include fixed costs and variable costs. Variable costs are related to revenue, and are $\gamma(t)R(t)$, and fixed costs are expressed as F.

$$\mathrm{Cost}(t) = \gamma(t)R(t) + F \tag{5}$$

Let the net profit after tax be Y(t), the company's fixed tax rate is τ, and ignore other factors, such as deferred income tax and capital expenditure:

$$Y(t) = [R(t) - \mathrm{Cost}(t)](1 - \tau) \tag{6}$$

Without considering depreciation, r is the risk-free rate of return, assuming that the company's available cash at time t is X(t):

$$dX(t) = \{rX(t) + Y(t)\}dt \tag{7}$$

At time t, the company's value is composed of two parts: available cash and continuing operating value. Under the neutral probability measure, the company value can be expressed by the following formula:

$$V(0) = E_Q\{X(t) + M \times [R(t) - \mathrm{Cost}(t)]\} \cdot e^{-rt} \tag{8}$$

e^{-rt} is the continuous compound interest discount factor, M is set to 10.

2.2.2 Discrete Time Model

Considering the path dependence of the parameter variables in the above model, when calculating the company value, the discrete time model is used to approximate the continuous model, and the Monte Carlo simulation method is used for simulation. Assuming that all the response coefficients are the same, the model is as follows:

$$R(t + \Delta t) = R(t)exp\left\{\left[\mu(t) - \frac{\sigma(t)^2}{2}\right]\Delta t + \sigma(t)\Delta t^{\frac{1}{2}}\varepsilon_1\right\} \tag{9}$$

$$\mu(t + \Delta t) = e^{-\kappa\Delta t}\mu(t) + \left(1 - e^{-\kappa\Delta t}\right)\bar{\mu} + \left(\frac{1 - e^{-2\kappa\Delta t}}{2\kappa}\right)^{\frac{1}{2}}\eta(t)\varepsilon_2 \tag{10}$$

$$\sigma(t) = \sigma_0 e^{-\kappa t}\mu(t) + \bar{\sigma}\left(1 - e^{-\kappa t}\right) \quad (11)$$

$$\eta(t) = \eta_0 e^{-\kappa t} \quad (12)$$

ε_1 and ε_2 are independent normal distribution variables.

When using Monte Carlo simulation, you need to calculate the various parameter values in the model, then use a random number generator to generate ε_1 and ε_2, and finally calculate the company value.

3 The Output of Company A by Monte Carlo Simulation

3.1 Parameter Estimation

The relevant parameters are analyzed and estimated by relevant industry analysis and company A's financial statements from 2009 to 2016.

According to the company's financial statements, we can directly obtain the following parameters: the initial income, R = 1589210000; the cash available at the beginning of the period comes from the company's monetary funds and cash equivalents, X = 311829000; the fixed cost is obtained through regression analysis of operating cost to operating income, F = 47267210; the average income growth rate, μ = 27.92%; the standard deviation of the change in the beginning income, σ = 26.16%; The standard deviation of the percentage change in income growth rate η = 27.51%.

Then we use some specific methods to calculate other parameters. We select 20 listed companies of similar nature in the industry, calculate and average their operating income growth rates from 2010 to 2016, and obtain the long-term revenue growth rate μ = 24.90%. The standard deviation of the long-term revenue growth rate fluctuation percentage, $\bar{\sigma}$ = 17.97%, is got by calculating the fluctuations in the revenue growth rate of 20 companies. According to the regression analysis of operating cost to operating income, we obtain γ = 0.5851 and the goodness of fit is 0.999. Since Company A obtained the high-tech enterprise certification, it enjoys tax incentives, so τ = 15%. It is generally believed that treasury bonds have no default risk, so the yield to maturity of treasury bonds can be regarded as a risk-free interest rate. We select 7 treasury bonds with a term of more than 10 years to obtain a long-term average yield of 3.47%, r = 3.47%. The κ indicates the speed at which the growth rate returns to the industry average. The growth rate of the company's revenue will decrease in an exponential form $e^{-\kappa t}$, and ln2/κ can be regarded as the time for the company's growth rate to return to half of the long-term average growth rate. Suppose that company A leads the industry average for 0.5 years, κ = ln2/(0.5 * 0.5) = 2.773. According to the research done by Schwartz and Moon in 2001, we assume that the company can earn more than the industry profit for 10 years, and then return to the industry average ten years later, T = 10. Since the financial data used are all from the company's annual report, we set Δt = 1.

3.2 The Output of Monte Carlo Simulation

Using MATLAB simulation software for programming, the path is simulated 1000 times, and 1000 values are obtained. The 1000 values are averaged and discounted using a continuous compounding factor. The theoretical value of Company A on December 31, 2016 is V = 15608739146. It can be seen from the balance sheet of Company A that year that the total liabilities at the end of the year were 588810000. According to the accounting identity, the theoretical equity capital of Company A in the current year is about 150.193 million. Looking at the company's share capital holdings, the number of share capital at the end of 2016 was approximately 184.02 million shares. From the above data, it can be estimated that on December 31, 2016, the share price of Company A was about 64.62. There is a certain difference from the company's actual share price of 50 per share. This shows that the company's stock price is undervalued.

After the analyze of the main parameters in the valuation model that affect the company's value at the beginning of the income growth rate, the average recovery coefficient, the variable cost rate, the valuation period and the beginning income volatility, we can get a conclusion that the valuation period has the greatest impact on the company's value. The longer the valuation period, the longer the company can earn profits that exceed the industry's average level, and the higher the company's value. The mean recovery factor is an important factor that affects the value of the company. $\ln 2/\kappa$ means that half of the time it takes to recover from the initial growth rate to the industry average. The smaller the κ, the slower the average recovery rate, indicates that the longer the company can maintain its competitive advantage, the greater the value of the company. In addition, a slight change in the variable cost rate γ will bring about a huge change in the company's value. As the cost increases, the company's value will become significantly smaller. The initial income growth rate and the initial income volatility determine the initial shape of the model, which has an impact on income volatility and income growth rate, and has little effect on stock prices.

From the analysis, we can understand the important factors that affect the company's value, so as to help companies make targeted strategic development decisions. Enterprises must intensify their technological innovation in order to maintain the competitiveness of the industry and maintain their competitive advantage, so that they can stay ahead of the industry for a long time and obtain excess profits. At the same time, cost control management is also an important means for enterprises to maintain their own operating advantages.

4 Conclusion and Prospect

4.1 The Purpose of This Paper Is to Overcome the Limitations of Traditional

Valuation methods in the valuation of high-tech enterprises. and based on the characteristics of management flexibility values, intangible assets and human capital in the operation of high-tech enterprises, the Schwartz-Moon model is used to make quantitative valuation analysis of a listed high-tech enterprise A, taking better account of the uncertainty in the operation, and through the sensitivity of the main parameters of the model, we have obtained important factors that affect the value of the company, which enriches the existing research results of high-tech enterprise value evaluation. However, there is a certain deviation between the valuation results of the model and the real values. On the one hand, the choice of model parameters has a certain subjectivity, which does not fully reflect the real situation of the market. On the other hand, because the model has too many assumptions, it is not completely consistent with the actual situation. In addition, most of the parameter data are obtained through the financial statements of listed companies. For non listed companies, the model is not applicable. These problems need further study.

References

1. Yang, C., Zhang, J.: Comparison and application of discounted cash flow method and real option valuation method. Financ. Acc. Monthly **36**(19), 26–31 (2016)
2. Schwartz, E.S., Moon, M.: Rational pricing of internet companies revisited. Financ. Anal. J. **36**(3), 62–75 (2000)
3. Kellogg, D., Charnes, J.M.: Real-options valuation for a biotechnology company. Financ. Anal. J. **56**(3), 76–84 (2019)
4. Liu, H., Fan, L.: Value evaluation of enterprise carbon intangible assets from the perspective of real options. Forum Sci. Technol. China **31**(8), 116–121 (2016)
5. Wang, H., Yuqing, W., Li, S.: Research on interval characteristics and numerical simulation of high tech enterprise value evaluation. Commun. Financ. Acc. **33**(20), 50–51 (2013)
6. Guo, J., Wang, D., Fan, Y., Liu, Y.: Research on the value evaluation system of Internet enterprises – based on the analysis of real option model. Price Theory Pract. **36**(7), 153–156 (2017)
7. Zhou, X., Tan, Y., Zhou, S.: Research on IPO pricing of high tech enterprises based on multi variable causal compound option. Commun. Financ. Acc. **32**(30), 133–135 (2012)

The Enterprise Organization Reform Under the Industrial Internet Situation

Lin Zhang[1], Shuaijie Shi[2], and Jinfa Li[1](✉)

[1] School of Management Engineering, Zhengzhou University, Zhengzhou, China
lijinfa@zzu.edu.cn
[2] Haier Group, Beijing, China

Abstract. The wide application of information technology and Internet technology has some influence on the traditional organizational structure, forcing enterprises to establish a new organizational structure that is suitable for the industrial Internet. Based on lewin's model of organizational change, this paper explores the problems related to organizational change in industrial Internet, and constructs a structure model of "force field" of organizational change in industrial Internet. The model shows that the reformed organization is a stable organization form that adapts to the industrial Internet after balancing the driving force and the resistance force. Finally, the feasibility of the model is illustrated with the case of haier's organizational change.

Keywords: Industrial Internet · Organizational change · Force field structure model

1 Introduction

Since the birth of enterprises, the research on their organizational structure has never stopped. Since the industrial revolution, pyramid organization has been the main form of enterprise organization all over the world. Pyramid organization was founded by the father of scientific management – Frederick wirger tirol. Pyramid structure plays a great role in the context of relatively backward productivity and relatively closed information, and its stability brings great benefits to enterprises. However, Silva Ivanovitch et al. [1] found through the analysis of a large number of enterprise cases that the traditional pyramid organizational structure is no longer applicable in today's advanced information technology, and the organizational rigidity of pyramid organizations limits the enterprise's adaptability, so the hierarchical system must be changed. Yenni Tim et al. [2] explained from different perspectives that the hierarchical structure was adopted in the old era because of its insufficient management range. However, Internet technology makes the way of transferring tasks between superiors and subordinates become standardized and stereotypical, and the message is delivered more quickly, on time and correctly. The upper control over the lower has been strengthened, and the management range has been greatly increased. Oliver Lyons et al. [3] analyzed from the perspective of effective communication and interaction among employees. They believed that

A. Batako et al. (Eds.): GCMM 2021, LNNS 335, pp. 690–697, 2022.
https://doi.org/10.1007/978-3-030-90532-3_52

the use of Internet technology enables direct communication between team members, reduces the possibility of losing authenticity in the process of information transmission, and makes decisions more accurate. Kauffman, Weber and others [4] also believe that the elimination of intermediate management level is the structure of an efficient new enterprise.

Questions about what should be the new enterprise, Torgeir Welo et al. [5] believed that the new enterprise organization should be composed of many teams with cross functions. Each team is user-centered and integrates multiple capabilities such as product innovation, process management and service innovation. The enterprise no longer adopts a fixed operation mode, but adopts flexible operation. The emphasis of flexible operation is to adopt a large number of new technologies, including NC, GT, CAD, CAM, ERP and CIMS. These technologies can make rational use of resources and improve the enterprise's adaptability and competitiveness. The flexibility of products, innovations, modifications, batches, processes and materials constitutes a flexible operating system.

Although the research on new enterprise organizations has not stopped, most of them remain in the conceptualization stage and lack a systematic and specific model. Therefore, this paper will systematically analyze the influence factors of Internet on organizational change and build an organizational change model in the context of industrial Internet in combination with lewin's organizational change model for reference and research.

2 Analysis and Model Construction of Enterprise Organizational Change in the Context of Industrial Internet

2.1 Analysis of the Causes of Organizational Change

The driving force of organizational change is a force that pushes the organization to change. The most common driving force of organizational change is the change of environment. The driving factors of organizational change in the industrial Internet era include:

- Change of enterprise strategy: In the era of industrial Internet, the enterprise is no longer a traditional and simple product manufacturer, it has become a service provider for users, providing value and benefits for users. The traditional enterprise takes management and production as the center, while the new enterprise takes service for users as the center. (The first: The incorrect statement has been deleted here 'The text should be set to single line spacing').
- Production mode change: After the emergence of industrial Internet, new concepts such as industry 4.0, intelligent manufacturing, flexible production, big data, artificial intelligence, collaborative manufacturing and cloud manufacturing have caught the public's eye. Production mode is facing changes, and the fourth industrial revolution themed by intelligent manufacturing is taking place slowly but firmly.
- Organizational structure reform: The traditional hierarchical system and centralized control show the shortcomings of high consumption, low efficiency and rigidity. The technology and management ideas brought by the industrial Internet enable enterprises

to conduct management and production in parallel and decentralized way, which makes it more open and convenient for communication, and has the advantages of low consumption, high efficiency and flexibility. In the context of industrial Internet, the main forms of organizational structure change are: team structure, virtual structure, and borderless organization. The general characteristics of enterprise organizational structure reform are: flat, flexible, discrete and network.

2.2 Advanced Concepts and Organizational Structure in the Context of Industrial Internet Subsection Sample

The rise of makers. Maker is defined as a person who has a knowledge base, is good at innovation and practice, and can turn innovation into reality. Its emergence shows that invention and creation are not only exclusive activities of high-end laboratories, but also can be invented and innovated by ordinary people. Now that the Internet has made it even easier for ordinary people to become makers, innovation that once required a large institution can now be achieved by a single person. Applying the concept of maker to the enterprise organization and enabling each maker group to realize self-innovation can not only flatten and network the organizational structure, but also increase the flexibility, innovation ability, communication ability and practice ability of the organizational structure, so that the individual ability, especially the innovation ability, can be realized. The organizational structure of makers is shown in the figure below (see Fig. 1).

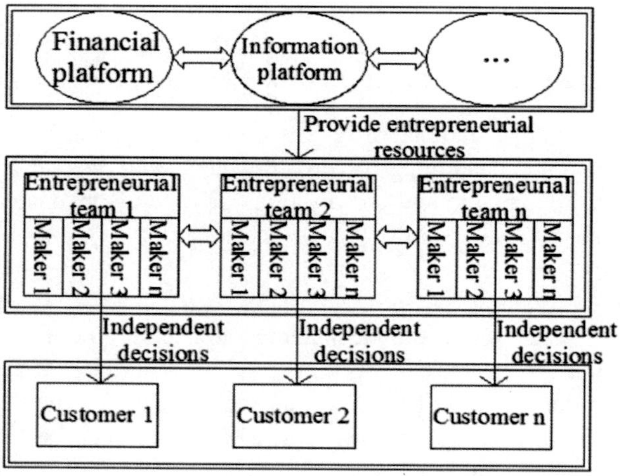

Fig. 1. Maker structure. The image is derived from: industrial Internet

Pull type organization. Different from the previous organizational structure of separation of production and marketing, the pull type organization is characterized by the fact that each department of the enterprise can communicate directly with the market and use the market demand to drive the production and operation of the organization. The group headquarters, main functional departments and subordinate business units arrange their own development, marketing and production departments according to the market demand and their own process characteristics. The advantage of pull production is to reduce a large amount of inventory, it can make the enterprise waste greatly reduced. Pull type organization is the expansion of the production, it emphasizes that embarks from the market to understand the needs of the users and according to the requirements to determine to provide users with the amount and type of services or products. Each department of a pull organization can communicate directly with the market, thereby increasing the competitive advantage of its products and services, which are provided entirely by the user. (The second: Here, increased the descriptions of the advantage of pull type).

The marketing department is in direct contact with the market. They collect and process the market information for their own business division, make the sales plan according to the processed information, and pass the information to the design department and the production department in a timely manner. The design department improves the product after receiving the information collected by the marketing department. The production department shall organize production after receiving the market demand information. Various information systems can realize the timely communication and feedback of information between the group headquarters and subordinate business departments. The drawing organization structure is shown in the figure below (see Fig. 2).

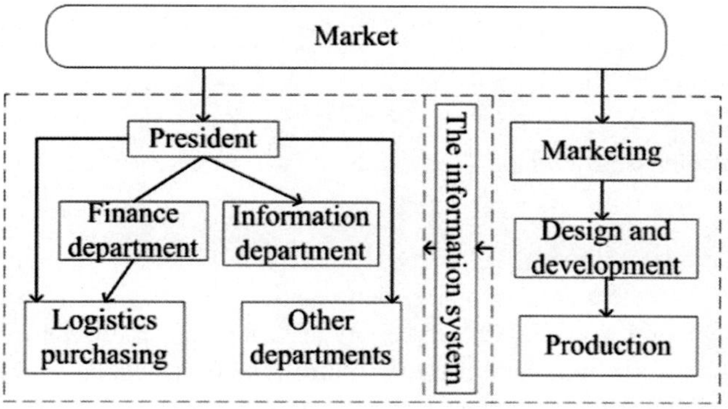

Fig. 2. Pull type organization. The image is derived from: the web (The third: indicate the source of the picture.)

Platform organizational structure. Under the thinking mode of Internet, isolated competition with other enterprises will no longer be an effective way for enterprises to compete, and enterprises will no longer be closed, and the boundary between enterprises will become more and more blurred. Resource sharing will make the future mainstream, and resources in the market will be fully integrated on the platform. Each enterprise will obtain any resources on the platform, and individual competition will become systematic competition. Enterprises obtain more kinds of resources, thus can create more customer value and obtain more profits [6]. With the help of the industrial Internet, enterprises can interact with customers in a timely manner. Consumers can participate in the production process of products, and finally buy the products they really need. Mass manufacturing becomes mass customization, so as to realize personalized production. In order to adapt to the Internet era, the platform enterprises have several basic structural characteristics:

canceling the centralization, canceling the intermediation and canceling the sideways trend.

2.3 Organizational Change Model in the Context of Industrial Internet

In the era of industrial Internet, enterprises need to strike a balance under various forces and re-establish an open, low-consumption, high-efficiency and flexible organization. Therefore, based on the influence of the Internet on organizational change, a model of organizational change in the context of industrial Internet is constructed by combining the current advanced organizational forms and lewin's organizational change model. The model is shown in the figure below (see Fig. 3).

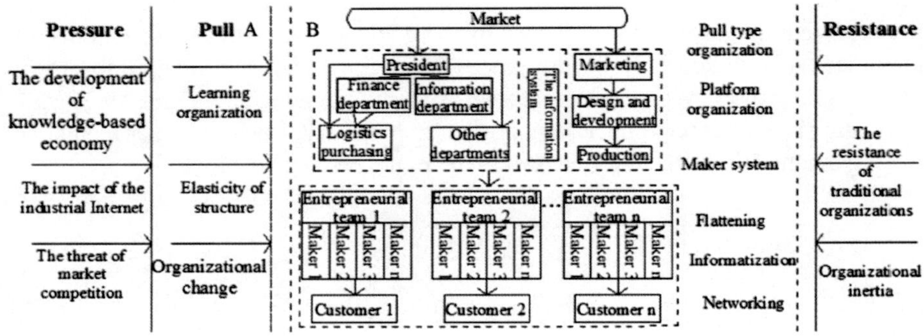

Fig. 3. Pull type organization. (The fourth: 'A' and 'B' were added to the picture)

The whole model is a concept of 'force field' in physics, and the organization is under various forces. 'A' is the organizational structure before the reform. 'B' in the dotted box represents the general result model of the organization reform driven by the Internet. It represents a platform type enterprise with pull type organization and makerization. It can effectively break the organizational rigidity of the pyramid structure, organically combine the advantages of various modern advanced organizations. This model has the characteristics of openness, flexibility, efficiency and science, and is very competitive. (The fifth: Here, it describes why the new model is competitive).

In the figure, 'pressure' represents the driving force for organizational change. The rapidly emerging knowledge economy in the Internet era, the rapidly changing market environment, the threat of new enterprises and the limitations of its own structure are all the driving forces for organizational change. 'Pull' refers to the motivation of enterprise's own reform, which is an internal motivation of enterprise's desire for change. Learning organization represents a typical 'pull', which has the ability to constantly learn, optimize and improve itself. The innovation of organizational culture, the renewal of managers' ideas and the elasticity of organizational structure include: employee inertia, employee thought resistance, employee benefit reduction, employee insecurity, information transmission distortion, organizational inertia, departmental resistance, threats to professionals, and organizational culture. Organizational change also faces great resistance factors, which need to be overcome.

3 Case Study of Haier's 'Platform + Small and Micro Enterprises' Organizational Structure

3.1 Problems Existing in the Organizational Structure Before Haier's Reform

Before the reform, Haier's awareness of rights and responsibilities was not clear enough, its regulatory mechanism was backward, its management was unreasonable, and it did not form a strong market concept. The efficiency of the enterprise was not ideal, and the company suffered serious losses for several years in a row and was on the verge of bankruptcy. Before the reform, Haier adopted the traditional pyramid organizational structure, which was widely used in enterprises all over the world at that time. The problems in Haier's organizational structure are as follows:

- The decision-making mechanism of brief concentration: Haier's decision-making right at that time only existed in the hands of a few people, the decision-making process was very closed and brief, and it was not well combined with the market. This decision-making mechanism put haier in a state of hysteresis and closure, and the production plan of the enterprise often lagged far behind the changes of the market. The ability of employees is limited, the overall response of the organization is sluggish, the products and services provided are fixed and single, the information transmission is slow and there is the risk of information distortion, the decision-making cycle is long and the efficiency is low.
- Organizational efficiency is limited by the traditional structure: In the traditional management mode, the overly complex intermediate management layer restricts the overall efficiency of the organization. The structure is complex, the response is stiff, the individual ability of employees is limited, there are fewer professional and technical personnel but more managers, the thinking and quality of internal personnel are backward, and the enterprise lays too much emphasis on management instead of providing better products and services for users.
- Unable to meet the personalized needs of customers: The Internet era requires enterprises to take users as the center, achieve the personalized and diversified needs of users, and provide better services. However, haier cannot flexibly carry out r&d, manufacturing and service, nor can it efficiently coordinate various resources to provide users with personalized services and products.

3.2 Haier's Reform

In order to build an organizational structure that can better serve customers, Haier overturns the traditional bureaucratic system and chooses a platform-based organizational structure. The traditional employment system was abolished and the employee maker system was adopted, which greatly increased the power of employees and facilitated the provision of personalized services for users [7]. The use of 'platform + small and micro enterprises' to achieve a win-win situation between enterprises and users. Haier's organizational change model is established based on the organizational change model established in Sect. 3, as shown in the figure below (see Fig. 4).

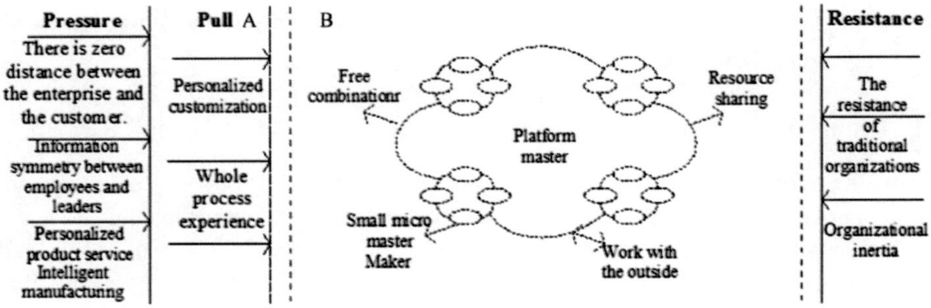

Fig. 4. Haier's organizational change model. (The last: 'A' and 'B' were added to the picture)

'A' refers to the organizational structure before the reform of Haier; 'B' refers to the new organizational structure after the reform; 'pressure' refers to the factors that promote the reform of haier under the context of the Internet; 'pull' refers to the strategic decision of organizational reform carried out by the organization under the stimulation of external forces; 'resistance' refers to the forces that resist the reform of Haier.

The new Haier has no hierarchy of traditional organizations, only the platform owner, small micro and makers. In the 'platform + small and micro enterprises', Haier's platform is a resource provider, while xiao wei is a service provider providing personalized services to users. Xiao wei's resources come from the platform, and the platform can integrate a wide range of resources. Platform + small and micro enterprise is a flat organization, the new type of organization is flexible and fast. Each micro is independent and interrelated, they can freely combine and provide users with different products and services. Enterprise work is no longer just about strategies and management organization, but rather to become a server, establish a platform to make rich products, services, ideas, thoughts and suggestions to get integration and gathering, attract entrepreneurs to become a maker of their enterprises, and provide all kinds of support for makers, including capital, technology and power. Each small micro maker selects a small micro master, who can manage and make decisions by himself to provide users with personalized products and services. All employees in the enterprise organization have become self-entrepreneurs. Self-employment enables each employee to give full play to his/her abilities without being restricted by the organization. Employees have higher freedom, so their sense of happiness is stronger. The new 'platform + small and micro enterprises' carry out the internal operation through the way of tracking after sales service, gathering and distributing benefits according to orders, and cooperating with external commitments. The mode carries out external operation by means of customer communication and interaction platform, maker training platform, industrial chain platform and third-party supervision platform.

4 Conclusions

This paper sets up a general model of enterprise organization reform under the industrial Internet situation from the perspective of external pressure, internal tension and resistance. And it combined with the specific case of haier organizational change to explain its specific use.

The factors affecting organizational change can be divided into three forces: pressure, tension and resistance. The enterprise is regarded as a balance body under various forces. In order to achieve the ideal balance state, the enterprise has to carry out organizational change. Enterprises in the process of change should know clearly the stress, and strengthen the stress and strain and weaken resistance or change the resistance to pressure and tension to make organizational change more smoothly.

This paper discusses the limitations of enterprise organization reform without technical support in the context of industrial Internet. In the future, the method of technical analysis can be used for in-depth research: to establish a mathematical model of organizational change and analyze organizational change from the mathematical perspective. In addition, an attempt can be made to quantify the pressures, strains, and resistances of an organization in order to accurately study the impact of external forces on organizational change.

References

1. de Barros Filho, I.E., Silva, I., Viegas, C.M.D.: An effective extension of anti-collision protocol for RFID in the industrial Internet of Things (IIoT). Sensors **18**(12), 4426 (2018)
2. Tim, Y., Hallikainen, P., Pan, S.L., Tamm, T.: Actualizing business analytics for organizational transformation: a case study of Rovio Entertainment. Eur. J. Oper. Res. **281**(3), 642–655 (2018)
3. Lyons, O., Timmons, J., Cohen-Hall, A., LeBlois, S.: The essential characteristics of successful organizational transformation: findings from a Delphi panel of experts. J. Vocat. Rehabil. **49**(2), 205–216 (2018)
4. Kauffman, R.J., Weber, T.A.: Special section: the digital transformation of vertical organizational relationships. J. Manag. Inf. Syst. **35**(3), 837–839 (2018)
5. Welo, T., Ringen, G.: Investigating organizational knowledge transformation capabilities in integrated manufacturing and product development companies. Procedia CIRP **70**, 150–155 (2018)
6. Kiel, D., Arnold, C., Voigt, K.-I.: The influence of the Industrial Internet of Things on business models of established manufacturing companies – a business level perspective. Technovation **68**, 4–19 (2017)
7. Huang, Y.: Haier's organizational and business model changes in the Internet Era. Mod. Bus. (27), 98–99 (2017). (in Chinese)

The Significance of Coordinated Research Against SARS-CoV-2

Jafar Hasan[1], Amar Velic[1], Alka Jaggessar[1], Asha Mathew[1], Tejasri Yarlagadda[2], Kirsten Spann[2], Seeram Ramakrishna[3], and Prasad K. D. V. Yarlagadda[1](✉)

[1] Science and Engineering Faculty, Queensland University of Technology, Brisbane, QLD 4000, Australia
y.prasad@qut.edu.au

[2] Centre for Immunology and Infection Control, Faculty of Health, Queensland University of Technology, Brisbane, QLD 4000, Australia

[3] Center for Nanofibers and Nanotechnology, National University of Singapore, Singapore, Singapore

Abstract. With many unanswered questions about SARS-CoV-2, the research community needs to immediately prioritize research areas in the fight against the current COVID-19 pandemic. An accelerated and coordinated multidisciplinary research effort is needed from the wider scientific community in many sectors from vaccine and antiviral development to digital technology. This mini-review highlights key research opportunities such as vaccine development, diagnostics, antiviral masks, environmental transmission and antiviral treatments, and outlines how systematic efforts can be focused in these key areas. Some alternate strategies such as the use of novel nanostructured surfaces to slow and prevent the spread of COVID-19 are presented.

1 Introduction

Coronavirus disease 2019 (COVID-19) is caused by the transmission of the novel virus named 'severe acute respiratory syndrome coronavirus 2' (SARS-CoV-2). Genome analysis revealed that the virus is 79.6% similar to the SARS-CoV virus and 96% identical to a coronavirus observed in bats [1]. SARS-CoV-2 was initially detected in December 2019 in Wuhan city, the capital of Hubei province in China, which became the epicenter of the outbreak in China. The virus was later declared as a pandemic on March 11, 2020, causing global ramifications. SARS-CoV-2 may have likely started as a zoonotic transmission from the wet animal market in Wuhan city, which was then followed by wide spread human-to-human transmissions. SARS-CoV-2 is the seventh coronavirus known to infect humans, with the other six causing severe (SARS-CoV, MERS-CoV and SARS-CoV-2) to mild illnesses (HKU1, NL63, OC43 and 229E) [2]. As of 13th Nov 2020, there have been over 52 million affected individuals and over 1.2 million deaths worldwide due to the COVID-19 pandemic, with numbers rapidly rising. To control the pandemic, it is important to understand the mechanisms of the spread, including key characteristics such as timing, mode of exposure, and risk factors for infection and disease progression [3]. In addition to epidemiology and surveillance activities worldwide, exploration and

A. Batako et al. (Eds.): GCMM 2021, LNNS 335, pp. 698–713, 2022.
https://doi.org/10.1007/978-3-030-90532-3_53

development in the areas of diagnostics, vaccines, alternative therapeutic approaches, antiviral personal protection equipment, PPE and the use of digital technology, must be promptly initiated or expedited to find efficient ways of overcoming the epidemic (Fig. 1) [4]. This research article presents a holistic view on COVID-19 by collating insights on important research areas and solution strategies relevant for the current crisis and its long-term prevention.

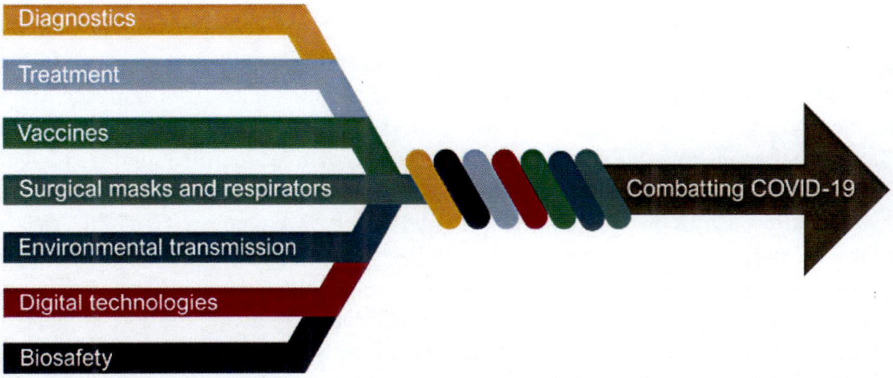

Fig. 1. Key research areas in the fight against COVID-19 pandemic. Parallel advances in these areas intertwine to produce a synergized response to COVID-19 that is more efficient and effective. Combatting COVID-19 requires diagnostics that are specific, sensitive and readily available; vaccines that give long-term protection; therapeutic drugs that work in the majority of the population; and additional physical and digital technologies that further support these efforts.

2 Methodology

For this review, an intensive literature search was performed using tools such as Scopus, Google Scholar and Web of Science. Most of the articles were published from 2019 onwards with some exceptions such as in case of explaining a concept. Since there are a vast number of reports present on COVID-19 in various research databases, only peer-reviewed articles were selected and none of the preprints were included in the review.

3 Diagnostics

The key element in control of an outbreak is the identification of positive cases that can then be isolated and traced effectively. Early detection is critical in isolating cases, identifying contacts of known cases and understanding the rate of community transmission. Detection also helps in allocation of medical resources and staff to areas of high need and enables preparedness wherever transmission forecast is on the rise. Countries such as Iceland, Egypt, Estonia, Norway and Italy have performed a higher coverage of tests. This has not, however, translated to lower rates of acquisition or mortality. A shortage of testing kits globally has hindered efforts to understand the full extent of community transmission and has necessitated the use of precautionary, broad sweeping measures such as social distancing and home isolation for entire communities regardless of unknown infection status [5].

Research and development of diagnostic tests has been on the rise since the pandemic was declared [6]. Highly sensitive and specific diagnostic methods are essential to identify positive cases and understand how the virus is transmitted. In the case of SARS-CoV, which caused a pandemic in 2003–2004, the virus was readily transmitted via respiratory droplet and oral-faecal contamination [7, 8]. By contrast, SARS-CoV-2 appears to be transmitted primarily by respiratory droplets [8–11]. Although viral RNA has been rarely detected in faeces [12] and blood [13], samples from both the upper and lower respiratory tract, such as nasopharyngeal swab, bronchoalveolar lavage and sputum are the most reliable specimens to collect for positive identification [11, 14–17]. Urine has been tested, although no positive cases have been identified [7, 15, 18].

Reverse transcription, quantitative polymerase chain reaction (RT-qPCR) assays are the most accurate and rapid means of identifying SARS-CoV-2 virus. Different assays have been developed to detect the virus. Some PCR assays have been modified for SARS-CoV-2 so that primers target different regions of the virus's genetic sequence. The relative positions of the targets on the SARS-CoV-2 sequence include RNA-dependent RNA polymerase (RdRp gene) with a gene position of 15,361–15,460, Envelope protein (E-gene) with a gene position of 26,141–26,253, and Nucleocapsid protein (N-gene) with a gene position of 28,555–28,682 according to SARS-CoV-2 GenBank NC_004718 [19]. Some other targets such as Spike (S gene) with a gene position of 21,563–25,384 and orf1ab with a gene position of 13,791–14,290 from GenBank Accession No. MN908947.3, and Membrane (M gene) with a gene position of 26,523–27,191 from Accession No. EPI_ISL_412026, have also been identified for detection of SARS-CoV-2 [15, 20]. For the coronavirus, various kinds of PCR assays have been established, namely RT-PCR, real-time reverse transcription PCR (rRT-PCR), reverse transcription loop-mediated isothermal amplification (RT-LAMP), as well as real-time RT-LAMP [21, 22]. Computed tomography (CT) scans are another way of clinical diagnosis of COVID-19, involving the acquisition of cross-sectional chest X-ray images taken at different angles, which are analysed by radiologists to diagnose abnormal features [23]. However, the specificity of CT scans is low (25%) due to overlapping of the images with viral pneumonia [24].

All the tests have demonstrated varying performance characteristics, likely underpinned by the inherent nature of the present crisis which is novel, complex and ever-evolving. As a result, the sensitivity and specificity on various sample types is also unclear [25]. For nucleic-acid based tests, swabs from the nasopharynx and/or oropharynx are the most common sample types, and it appear that those involving the nasopharynx are more sensitive [26]. The true sensitivity and specificity of the clinical molecular diagnostic test is, however, unknown and not likely to be 100%.

Previously for other pathogens, RT qPCR has shown higher sensitivity and specificity when compared to RT- Reverse Transcription Loop-mediated Isothermal Amplification (RT LAMP) and RT PCR techniques [27–29]. For SARS-COV-2, no comparative studies have been conducted yet to compare the sensitivities and specificities of the molecular tests. However, in a recent comparison between CT scan and RT-PCR, only 59% of patients had positive RT-PCR results, whereas 88% had positive chest CT scans [24]. Furthermore, performance of the RT-PCR tests targeting different RNA-dependent RNA polymerase (RdRp)/helicase (Hel), spike (S), and nucleocapsid (N) genes in comparison to the RdRp-P2 assay was evaluated [30]. It was found that the sensitivity and specificity of the novel COVID-19-RdRp/Hel assay (limit of detection, LOD of 11.2 RNA copies/reaction in vitro transcripts) was much higher (43.6%) in comparison to RdRp-P2 (28.2%). In another test, RT-PCR assays were found to be highly sensitive with best results obtained for the E gene and RdRp gene assays (LOD of 3.9 and 3.6 copies per reaction respectively) [19].

The limitation with viral diagnostics is the reliance on the reverse transcriptase PCR methods. Some novel methods are needed such as detection of single molecules of RNA. Using CRISPR techniques, nucleic acid detection can be developed as shown in some of the recent studies [31, 32]. CRISPR associated enzymes such as Cas13a can be reprogrammed for specific RNA sensing [31, 33, 34]. Such methods have earlier been used to detect Dengue or Zika virus single stranded RNA and may provide good reference for SARS-CoV-2 detection [32]. Other techniques such as isothermal nucleic acid amplification based methods like LAMP based, and microarray based methods have the potential to detect viruses such as SARS-CoV-2 [32].

Serological tests help with identification of the source of infection, surveillance, epidemiology and symptoms but currently remain a big challenge [35]. In these types of tests, previous infections are tested by characterizing the antibodies for the virus. If these antibodies are found, they can be used by the immune system to neutralize the pathogenic virus. Serology tests are primarily performed from enzyme-linked immunosorbent assay (ELISA) followed by microneutralization test for confirmation. They are complement to the nucleic-acid based detection methods.

There have been a few studies involving serological tests for SARS-CoV-2 [36, 37]. In a study during the SARS-CoV infection in 2015, 13% of the healthcare workers in Singapore showed positive asymptomatic infections by SARS serology [38]. The results were later confirmed through neutralization test. For SARS-CoV-2, serological testing may help in diagnosing acute and past infections, although the assay kinetics, specificity and sensitivity require more exploration [39].

Worldwide, there are several companies that have launched diagnostic products (FDA and CE Mark approved) for detection of SARS-CoV-2 using PCR assays with result time ranging between less than 1 to more than 12 h [40]. There has been a surge in the number of interesting research articles both in PCR and serological driven diagnostics.

A concern with SARS-CoV-2 is its rapid person to person transmission which leads to high rate of infection worldwide. More importantly, some infected people do not show any trace of illness for a specific time span and they become the main source of transmission chain. Therefore, it becomes vital to develop diagnostic kits that can rapidly detect the trace of coronavirus in any biological media [41]. Currently most of the countries and health authorities are using RT-PCR assay to detect the trace of coronavirus in nasal or throat swabs. As of now, there is no consensus on the accuracy of the testing, but understanding the limitations of the test becomes extremely important for accurate diagnosis.

4 Treatment

The main aims of treating a disease from a pandemic like COVID-19 are to alleviate symptoms among those infected and to lower the overall mortality rate. As such, several companies are working towards drug development of COVID-19 with the final goal of treating symptoms, including anti-inflammation and anti-viral effects [40]. A number of these drugs are in different phases of the clinical trial stages [40]. The idea is to repurpose existing antiviral agents which already have proven therapeutic experience such as antibodies, oligonucleotide- and interferon-based therapies [42]. Hence, existing drugs known to treat diseases such as Ebola, malaria and HIV are also being tested against SARS-CoV-2 [42–46]. The list includes chloroquine, arbidol, ritonavir, lopinavir, remdesivir, galidesivir, ivermectin, ribavirin and favipiravir, to name a few.

These drugs can be classified as RNA inhibitor or nucleoside analogs in the form of adenine or guanine derivatives. The nucleoside analogues target the RNA dependent RNA polymerase and block the RNA synthesis in viruses. One such analog is Remdesivir which was developed to treat Ebola, and has shown positive results against MERS, SARS-CoV and SARS-CoV-2 [42, 47, 48]. Despite early anecdotal evidence that Remdesivir may have a strong action against COVID-19, recent, smaller controlled trails suggest that the drug is not effective in cases of severe disease. That being said, studies on this drug are in their initial stages and current publicly available results in the case of mild disease are inconclusive [49]. There are also other nucleoside analogs such as ribavirin (for HCV, RSV), favipiravir (for Influenza, Ebola, chikungunya, yellow fever) and R-1479 (for Dengue) that are potential candidates to inhibit SARS-CoV-2 (Fig. 2) [50–52].

Hydroxychloroquine has been primarily used as an antimalarial drug and also shows antiviral properties as a potential treatment against COVID-19 [45, 47]. Currently, it is the most common therapeutic agent being trialed for the disease. In the current scenario, it is critical to fast-track *in-vitro* research as well as pre-clinical animal and clinical trials, and to find a common platform between researchers and clinicians.

In a recent study, it has been found that angiotensin converting enzyme 2 (ACE2) is the critical receptor for SARS-CoV-2 and inhibition of the interaction has been targeted for the treatment of COVID-19 [53]. The study shows that a trial drug named "human recombinant soluble ACE2" can be used to block SARS-CoV-2 infections [53]. Another recent development has been the design of promising lead compounds that target the main protease, M^{pro}, which mediates RNA replication and transcription [54–56].

In a small clinical study, convalescent plasma containing neutralizing antibody such as IgG and IgM anti-SARS-CoV-19 showed improvement in infected patients [57]. The use of convalescent plasma has been earlier employed for SARS-CoV, pandemic 2009 influenza A (H1N1), avian influenza A (H5N1), several hemorrhagic fevers such as Ebola, and other viral infection. Therefore, human convalescent serum maybe a significant treatment option for prevention of the COVID-19 disease [58, 59]. This is still in very early stages and maybe more available when there are sufficient numbers of recovered people who can donate the immunoglobulin-containing serum [60].

Prescription of previously developed anti-viral drugs is a fast and essential route towards effective treatment approaches for novel viral infections. In this matter, hydroxychloroquine has been widely used as an effective anti-SARS-CoV-2 agent in-vitro [61]

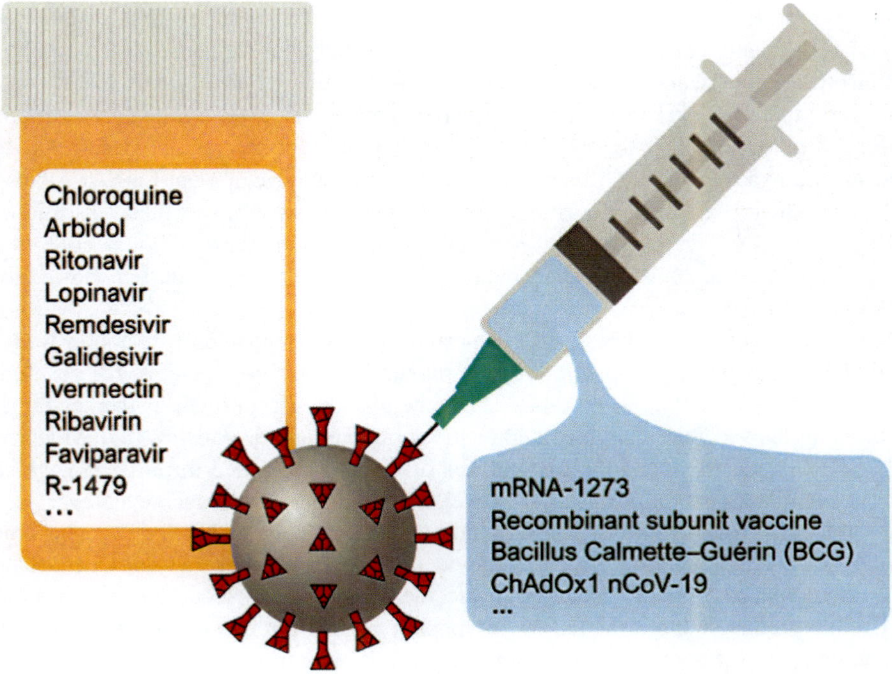

Fig. 2. Promising therapies and vaccines in the COVID-19 pipeline. Various drugs and vaccines have been initiated since the pandemic. Listed here are some potential candidates which have received the most attention and hold promise toward completed development.

which can reduce the drug-drug interaction [62]. Likewise, azithromycin is another effective drug that was previously used against Zika, Ebola and patients with severe respiratory tract infections [63, 64]. Combinatorial usage of hydroxychloroquine and azithromycin has been found to effective for the treatment of COVID-19 which was approved by the French government. In a work by Philippe Gautreta et al. (Marseille, France), they reported that infected patients with COVID-19 were treated with receipt of hydroxychloroquine and azithromycin during 6 days that fully inhibited the virus [65, 66].

Despite the fact that some anti-viral drugs have been successfully used to treat people infected with COVID-19, however, in a majority of cases the drugs are found to be ineffective and for some undiscovered reasons they cannot present a uniform performance against the illness in diverse parts of the world. Thereby, it is vital to develop a specific antiviral drug for treating the COVID-19 based on its mechanism of action and reproduction route.

5 Vaccines

For the SARS-CoV-2, vaccine development remains undoubtedly the most highly anticipated and the prioritized research area for scientists and medical companies. The first step towards vaccine development is antigen selection. Different antigens include the (i) whole cell antigens (WCA) which contain lipids, polysaccharide, nucleic acids, and some other components, (ii) spike protein (S protein), (iii) nucleocapsid protein (N protein), (iv) membrane protein (M protein) and (v) envelope protein (E protein). Among these, S protein is likely the most promising candidate for the vaccine design due to the surface exposure [67]. Different types of vaccines in development for SARS-CoV-2 can be categorized as (i) live-attenuated vaccines, (ii) subunit vaccines, (iii) mRNA vaccines, (iv) DNA vaccines, (v) live-vector vaccines and/or (vi) synthetic peptide vaccines. The different antigen and vaccine types sections have been recently reviewed for SARS-CoV-2 [68].

Several vaccine candidates have been established by companies, research institutes and universities which are undertaking the maximum efforts to accelerate the development of the vaccine (Fig. 2) [40, 44]. The trimeric spike glycoprotein is the key target protein for the vaccines including therapeutic antibodies and diagnostics [67]. Moderna Therapeutics has produced the first batches of the vaccine called mRNA-1273 which has begun its phase-1 clinical trial. However, no previous mRNA vaccine has ever been approved and therefore, the vaccine will have to overcome several challenges to become available for immunization against the pandemic [44]. The University of Queensland has also produced a recombinant subunit vaccine of SARS-CoV-2 S protein which is stabilized in prefusion conformation by polypeptide moiety through the molecular clamp technology [69]. Some research groups will also initiate trials with the BCG vaccine which is originally used for tuberculosis [70]. Recently, a phase 1 clinical trial has begun by University of Oxford researchers and around 1,110 people will take part in the trial [71]. The new vaccine called ChAdOx1 nCoV-19 is based on an adenovirus vaccine vector and the SARS-CoV-2 spike protein.

Although there are many promising candidates, vaccine development is afar due to several challenges. These include vaccine related safety issues, assay requirements and manufacturing logistics, all of which are amplified by the need to rapidly upscale for worldwide availability in the current scenario. Moreover, in the last 2 decades, there has been an emergence of pathogenic beta coronaviruses (SARS, MERS, and SARS-2) affecting the global population. A generic vaccine would be a better solution as it could provide a wide coverage against this pathogen and thus also prevent similar pandemics in the future. The most promising vaccine candidates must be developed such that they are ready for trials in case of an emergency. A recent article has recommended establishing a global finance system to facilitate large-scale vaccine manufacturing, delivery and deployment. This could also help ensure fair allocation, and protects private-sector partners from financial losses [72].

During the SARS-CoV, though two vaccines completed the phase-1 trials, the SARS epidemic halted before the products could be developed and tested under natural challenge [44, 73]. Therefore, it must be ensured that vaccines are developed and end-to-end deployed before another similar pathogen is encountered, an opportunity which was missed out in the outbreak of SARS in 2002–2003.

Researchers should also investigate alternatives such as nanotechnology that may directly or indirectly assist in vaccine development and treatment. Novel experimental methods such as small interfering RNAs (siRNAs) and virus-like nanoparticle vaccines also hold potential in the development of immunization against SARS-CoV-2 [44, 50].

The outbreak of this novel virus in diverse parts of the world has raised the requirement for reliable vaccines through development of antibodies. Development of viral antibodies is a great challenge because of the rapid spread of the pathogenic viral disease. Therefore, establishment of a reliable and fast procedure is required to generate antibodies that can react with responsible viruses for infections and protect people from such pandemic [74–76]. Development of vaccines based on effective antibodies could boost the immune system and protect vulnerable/sensitive people from such viral infectious disease. In the current scenario, treatment using various drug candidates and vaccine development remain one of the top priorities as researchers and companies put their respective efforts in both the sectors (Fig. 2). From the environmental transmission side, however, some novel strategies maybe needed to combat the SARS-CoV-2.

6 Surface Transmission

SARS-CoV-2 virus remains viable on surfaces such as steel for up to 72 h and in aerosols for up to 3 h [77]. The titer count reduces with time in all the conditions. Some common biochemical disinfectants such as ethanol, alkyldimethylbenzylammonium chloride, EDTA and hydrogen peroxide have been shown to effectively disinfect the coronavirus [78]. However, that involves application of disinfectant on the surfaces continuously or intermittently, which is neither affordable nor practical. Therefore, novel surface based engineering strategies are required that can effectively inactivate the virus on the surface or that inhibit the surface transmission. These strategies include but are not limited to surface treatment, chemical coating and surface engineering. Physical inactivation of the virus elicited by topography driven nanomaterials may show some promise. These nanomaterials are presently effective against bacterial cells which are physically ruptured by surface topography shortly after contact (Fig. 3) [79, 80]. Nanomaterials with characteristic dimensions of the surfaces smaller than the size of SARS-CoV-2 (60–140 nm) may be able to inactivate and hinder the virus upon exposure (Fig. 3). Our research group is currently working toward this aim by designing nanostructured antiviral surfaces with tuned surface topography. We have demonstrated previously that such nanostructured surfaces have excellent bactericidal property against a range of bacterial cells [81–83]. Additionally, our recent results have shown for the first time that nanostructured surface topography can inactivate respiratory viruses such as rhinovirus (RV) and respiratory syncytial virus (RSV) upon exposure on etched aluminium [84]. We used a scalable fabrication technique to generate nanostructures on the surfaces of aluminium alloys. Exposure of RV and RSV on the nanostructured surfaces showed that there is a rapid significant reduction of the virus counts compared to the control surfaces. There was approximately 4 log reduction for RV after 24 h on the nanostructured surface compared to the control surface. Although the mechanism of inactivation is still unknown, but it may be possible that there is either entrapment of the viruses or envelope disruption that would inactivate virus particles on the nanostructured surfaces. Nevertheless the antiviral mechanism is different from the bacterial rupture on same surfaces. Based on these results, it would be a feasible option to design surfaces with specific nanoscale features or design nanoscale materials that would have the potential to inactivate the SARS-CoV-2 virus.

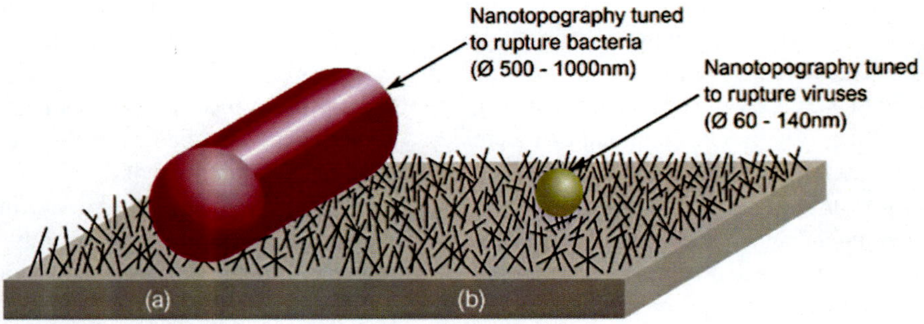

Fig. 3. A schematic of nanostructured decontaminant surface to help minimize surface transmission of disease. (a) A typical nanostructured surface effective against bacteria. (b) A nanostructured surface effective against viruses [84].

Plasma processing has also been investigated for decontamination against viruses [85]. Plasma is a neutral ionized gas with chemical species such as electrons, ions, neutral molecules, and atoms, as well as charged species. Antiviral treatment is dependent on the duration of the plasma. Treatment for 1 min by nitrogen gas plasma showed an approximate two-log reduction in influenza virus titer [86]. Treating for higher time intervals showed better results as performed using dielectric barrier discharge (DBD) plasma torch treatment against the feline calicivirus (FCV F9 strain) [87]. From several studies, it is known that the reactive species from the plasma may damage the RNA and the protein coat of the viruses [85–87]. Plasma can also be used as sterilization process as the reactive plasma species can reach all surfaces and penetrate small openings for antiviral effect. This treatment in particular can be used for sterilization of PPE and medical equipment in the hospitals. Thus, there are a number of opportunities that can be realized with plasma processing to fight COVID-19.

7 Surgical Face Masks and Climate Change

Due to the outbreak of COVID-19 virus pandemic, citizens' public health awareness is unprecedented. Wearing a surgical facemask is ubiquitously followed to prevent the spread of virus and to provide a sense of security and well-being to the general public. In addition to the general public, they are used by the healthcare professionals as well as all those in essential services to keep the society functional and secure. Most people use non-medical grade surgical masks while the healthcare professionals use medical grade surgical masks and respirators. Due to shortages of surgical masks and respirators, government regulatory bodies are considering to approve sterilization and sanitization of them to be reused [88]. This is almost unheard of in the past, and many rules and regulations were 'bent' so as to accommodate the decreased availability of supplies. Reusing surgical masks is not the most efficient nor economic way, but as COVID-19 is confirmed to be aerosolized [77], it is vital for the healthcare professionals in the front line, essential services workers and general public to be well protected for this insidious threat. All the current masks are designed for single use and disposed of safely. They just capture aerosolized viruses but do not deactivate or kill them. Hence there is a great need to design and develop antiviral and virucidal surgical face masks and respirators. Viruses will be deactivated or killed upon coming in contact with them. This requires multidisciplinary efforts involving materials engineers, nanotechnologists, chemists, biologists and virologists. Benefits of developing antiviral or anti-COVID-19 masks include a) the masks not only filter but kill COVID-19 upon contact thus reducing or preventing cross-infection upon accidental contact with the mask, b) the virucidal and self-sterilizing abilities extend the duration of usage of masks thus alleviating the problem of shortages, c) masks with antimicrobial properties are suitable for immune-compromised patients, seniors and children, and d) bi-directional anti-COVID-19 effects protect infected user from spreading and non-mask user being infected. Currently, billions of masks are used per day and thrown away causing environmental pollution. They are made from non-degradable non-woven polypropylene fabric material [89]. Antiviral and self-sterilizing masks made from degradable bioplastics would prevent the long-term adverse effects of petrochemical derived single use plastics.

Despite the severe impact that COVID-19 is having on the global economy due to individual countries imposing lockdowns and restricting ground and air travel, there has been significant improvement to greenhouse gas emissions (particularly NO_2 concentration [90]), air quality and the ozone layer. COVID-19 epicenters such as USA, Italy, Spain and Wuhan have reported a reduction pollution levels between 20–30% [91]. COVID-19 has revealed how quickly these important environmental factors can improve, and has given us insights into a low-carbon emission future, for which governments and businesses could strive.

Through the pandemic, opportunities for decarburization without compromising economies have emerged. Sustainable efforts towards climate change have the potential to now create new areas of research as well as create new jobs and industry sectors. Due to social distancing measures and large percentages of people working from home, the consumption of environmental resources is reducing, yet digital transformation and use of artificial intelligence (AI) and automation has increased. Therefore, digital technology has become a necessity which can now also be used to fight COVID-19.

8 Digital Technology and AI

Though vaccine development and treatment strategies against COVID-19 remain the foremost challenges, their facilitation incurs a number of cumbersome issues such as conventional clinical and drug discovery pathways [44]. Fortunately, these issues can be assisted by digital technologies such as big data, blockchain, Internet of Things (IoT), predictive analysis and machine learning models [92]. For example, AI could provide solutions for the prediction of broad-spectrum antiviral agents, which requires advanced knowledge of data, medicinal chemistry and computational drug design. AI can also offer prediction and forecasting of the spread of the disease, patient management in hospitals and general mitigation of the pandemic. Although AI has great potential in many aspects of COVID-19, it is dependent on accurate as well as robust data. Big data will provide assistance in modeling the disease and other related aspects such as preparedness and geographical factors. IoT can assist in live tracking of COVID-19 cases and also help in operation of virtual clinics. Blockchain can facilitate manufacturing and distribution logistics of vaccines, once developed. Overall, digital technology stands to provide immense assistance in surveillance, detection, and prevention as well as mitigation of the overall impact of COVID-19 [92, 93]. Using health data, several contact tracing apps have been launched to track COVID-19 patients [94–96]. It is suggested that governments, mobile network companies, technology companies and researchers work collaboratively on the most recent data to provide digital solutions for the pandemic [97]. Accordingly, whilst molecular biologists, microbiologist, virologists, physicians and healthcare staff continue to focus on drug or vaccine development, they should also network with digital technology experts to develop more permanent solutions for the long term.

9 Biosafety

Due to the highly transmissible nature of SARS-CoV-2 and the lack of vaccines to protect scientists and other workers from contracting an infection, the WHO has released guidelines on handling SARS-CoV-2 [98]. Standard diagnostics and any testing or handling of killed virus, viral RNA or protein can be conducting in a BSL/PC-2 laboratory where standard PPE, such as gloves, gowns, and eyewear, is used. Where the live virus is being isolated, cultured or handled in any way the risk of infection is higher and this work should be conducted in a BSL/PC3 laboratory using N93/P2 facemasks, one-way airflow, correct donning and doffing of PPE and work separation from routine lab-work. Any work that may generate aerosols, either at BSL-2 or 3 levels must be conducted in a class II biosafety cabinet.

10 Conclusion

Although progress has been made in several research areas, such as vaccine, diagnostic and treatment development, any significant leap forward will ultimately require understanding mechanisms of the disease and its viral replication. This is the fundamental basis of effective vaccine design and antiviral treatment. To support this, we can draw on a number of aspects, such as our knowledge of the previous coronavirus outbreak, SARS-CoV, as well as evolving understandings of the SARS-CoV-2 crisis presently at hand. Most importantly, we can draw on our cooperative and collaborative capabilities. This multifaceted problem requires a multifaceted solution. Accordingly, synergies between scientific and clinical groups from diverse backgrounds such as medicine, microbiology, chemistry, manufacturing and digital technologies are the only way to efficiently bring about significant advancements.

There are many interesting research outputs archived in medical or biology preprint servers which have not been included in this review due to their pending peer review. This attests to the need for the scientific community to fast track the publishing processes of research related to SARS-CoV-2 with the assistance of publishers, editors, peer-reviewers and others.

With the current technology of the healthcare system, the world is better equipped to handle COVID-19 or any other pandemic than ever before. Though our progress so far in the fight against SARS-CoV-2 has been initiated quite swiftly by researchers all over the globe, a long journey ahead still looms. This rests on the ability of research centers, medical companies and health organizations to muster a concerted and accelerated effort to enhance clinical productivity in the chaotic and uncertain context of the COVID-19 pandemic.

References

1. Zhou, P., et al.: A pneumonia outbreak associated with a new coronavirus of probable bat origin. Nature **579**(7798), 270–273 (2020)
2. Corman, V.M., et al.: Hosts and sources of endemic human coronaviruses. In: Advances in Virus Research, pp. 163–188. Elsevier (2018)
3. Lipsitch, M., Swerdlow, D.L., Finelli, L.: Defining the epidemiology of Covid-19 — studies needed. N. Engl. J. Med. **382**(13), 1194–1196 (2020)
4. Fauci, A.S., Lane, H.C., Redfield, R.R.: Covid-19—navigating the uncharted. N. Engl. J. Med. **382**(13), 1268–1269 (2020)
5. Roser, M., et al.: To understand the global pandemic, we need global testing – the Our World in Data COVID-19 Testing dataset. OurWorldInData.org (2020). https://ourworldindata.org/coronavirus
6. Udugama, B., et al.: Diagnosing COVID-19: the disease and tools for detection. ACS Nano **14**(4), 3822–3835 (2020)
7. Chan, K.H., et al.: Detection of SARS coronavirus in patients with suspected SARS. Emerg. Infect. Dis. **10**(2), 294 (2004)
8. Seto, W., et al.: Effectiveness of precautions against droplets and contact in prevention of nosocomial transmission of severe acute respiratory syndrome (SARS). Lancet **361**(9368), 1519–1520 (2003)
9. Otter, J., et al.: Transmission of SARS and MERS coronaviruses and influenza virus in healthcare settings: the possible role of dry surface contamination. J. Hosp. Infect. **92**(3), 235–250 (2016)
10. Poutanen, S.M., et al.: Identification of severe acute respiratory syndrome in Canada. N. Engl. J. Med. **348**(20), 1995–2005 (2003)
11. Ong, S.W.X., et al.: Air, surface environmental, and personal protective equipment contamination by severe acute respiratory syndrome coronavirus 2 (SARS-CoV-2) from a symptomatic patient. JAMA. **323**, 1610–1612 (2020)
12. Yeo, C., Kaushal, S., Yeo, D.: Enteric involvement of coronaviruses: is faecal–oral transmission of SARS-CoV-2 possible? Lancet Gastroenterol. Hepatol. **5**(4), 335–337 (2020)
13. Ng, L.F., et al.: Detection of severe acute respiratory syndrome coronavirus in blood of infected patients. J. Clin. Microbiol. **42**(1), 347–350 (2004)
14. To, K.K.-W., et al.: Temporal profiles of viral load in posterior oropharyngeal saliva samples and serum antibody responses during infection by SARS-CoV-2: an observational cohort study. Lancet. Infect. Diseases **25**, 565–574 (2020)
15. Wang, W., et al.: Detection of SARS-CoV-2 in different types of clinical specimens. JAMA. **323**, 1843–1844 (2020)
16. Gu, J., Han, B., Wang, J.: COVID-19: gastrointestinal manifestations and potential fecal-oral transmission. Gastroenterology **58**, 1518–1519 (2020)

17. Xiao, F., et al.: Evidence for gastrointestinal infection of SARS-CoV-2. Gastroenterology **158**, 1831–1833 (2020)
18. Pan, Y., et al.: Viral load of SARS-CoV-2 in clinical samples. Lancet. Infect. Dis **20**(4), 411–412 (2020)
19. Corman, V.M., et al.: Detection of 2019 novel coronavirus (2019-nCoV) by real-time RT-PCR. Eurosurveillance **25**(3), 2000045 (2020)
20. Wang, C., et al.: The establishment of reference sequence for SARS-CoV-2 and variation analysis. J. Med. Virol. **92**, 667–674 (2020)
21. Zhang, N., et al.: Recent advances in the detection of respiratory virus infection in humans. J. Med. Virol. **92**(4), 408–417 (2020)
22. Cordes, A.K., Heim, A.: Rapid random access detection of the novel SARS-coronavirus-2 (SARS-CoV-2, previously 2019-nCoV) using an open access protocol for the Panther fusion. J. Clin. Virol. **125**, 1043405 (2020)
23. Lee, E.Y., Ng, M.-Y., Khong, P.-L.: COVID-19 pneumonia: what has CT taught us? Lancet. Infect. Dis **20**(4), 384–385 (2020)
24. Ai, T., et al.: Correlation of chest CT and RT-PCR testing in coronavirus disease 2019 (COVID-19) in China: a report of 1014 cases. Radiology. **296**, 200642 (2020)
25. Patel, R., et al.: Report from the American society for microbiology COVID-19 international summit, 23 March 2020: value of diagnostic testing for SARS-CoV-2/COVID-19. mBio. **11**(2), e00722–20 (2020)
26. Zou, L., et al.: SARS-CoV-2 viral load in upper respiratory specimens of infected patients. N. Engl. J. Med. **382**(12), 1177–1179 (2020)
27. Sugiyama, H., et al.: Comparison of loop-mediated isothermal amplification, real-time PCR, and virus isolation for the detection of herpes simplex virus in genital lesions. J. Med. Virol. **75**(4), 583–587 (2005)
28. Khan, M., et al.: Comparative evaluation of the LAMP assay and PCR-based assays for the rapid detection of Alternaria solani. Front. Microbiol. **9**, 2089 (2018)
29. Lau, S.K.P., et al.: SARS coronavirus detection methods. Emerg. Infect. Dis. **11**(7), 1108–1111 (2005)
30. Chan, J.F.-W., et al.: Improved molecular diagnosis of COVID-19 by the novel, highly sensitive and specific COVID-19-RdRp/Hel real-time reverse transcription-PCR assay validated in vitro and with clinical specimens. J. Clin. Microbiol. **58**(5), e00310-e320 (2020)
31. Freije, C.A., et al.: Programmable inhibition and detection of RNA viruses using Cas13. Mol. Cell **76**(5), 826-837.e11 (2019)
32. Shen, M., et al.: Recent advances and perspectives of nucleic acid detection for coronavirus. J. Pharmaceut. Anal. **10**, 97–101 (2020)
33. Gootenberg, J.S., et al.: Multiplexed and portable nucleic acid detection platform with Cas13, Cas12a, and Csm6. Science **360**(6387), 439 (2018)
34. Gootenberg, J.S., et al.: Nucleic acid detection with CRISPR-Cas13a/C2c2. Science **356**(6336), 438 (2017)
35. Meyer, B., Drosten, C., Müller, M.A.: Serological assays for emerging coronaviruses: challenges and pitfalls. Virus Res. **194**, 175–183 (2014)
36. Bai, S.L., et al.: Analysis of the first cluster of cases in a family of novel coronavirus pneumonia in Gansu province. Zhonghua Yu Fang Yi Xue Za Zhi **54**, E005 (2020)
37. Xiao, S.-Y., Wu, Y., Liu, H.: Evolving status of the 2019 novel coronavirus infection: proposal of conventional serologic assays for disease diagnosis and infection monitoring. J. Med. Virol. **92**(5), 464–467 (2020)
38. Wilder-Smith, A., et al.: Asymptomatic SARS coronavirus infection among healthcare workers. Singapore. Emerg. Infect. Diseases **11**(7), 1142–1145 (2005)
39. Haveri, A., et al.: Serological and molecular findings during SARS-CoV-2 infection: the first case study in Finland, January to February 2020. Eurosurveillance **25**(11), 2000266 (2020)

40. Ventures, A.: Fight the pandemic (2020). https://www.av.co/covid
41. Chan, J.F.-W., et al.: A familial cluster of pneumonia associated with the 2019 novel coronavirus indicating person-to-person transmission: a study of a family cluster. Lancet **395**(10223), 514–523 (2020)
42. Li, G., De Clercq, E.: Therapeutic options for the 2019 novel coronavirus (2019-nCoV). Nat. Drug. Rev. Disc. **19**, 149-150 (2020). (Nature Publishing Group)
43. Dong, L., Hu, S., Gao, J.: Discovering drugs to treat coronavirus disease 2019 (COVID-19). Drug Disc. Ther. **14**(1), 58–60 (2020)
44. Hodgson, J.: The pandemic pipeline. Nat. Biotechnol. **38**, 523–532 (2020)
45. Hu, T.Y., Frieman, M., Wolfram, J.: Insights from nanomedicine into chloroquine efficacy against COVID-19. Nat. Nanotechnol. **15**, 247–249 (2020)
46. Caly, L., et al.: The FDA-approved drug ivermectin inhibits the replication of SARS-CoV-2 in vitro. Antiviral. Res. **178**, 104787 (2020)
47. Wang, M., et al.: Remdesivir and chloroquine effectively inhibit the recently emerged novel coronavirus (2019-nCoV) in vitro. Cell Res. **30**(3), 269–271 (2020)
48. Grein, J., et al.: Compassionate use of remdesivir for patients with severe Covid-19. New England J. Med. **382**, 2327–2330 (2020)
49. Pandemic, C.: Hopes dashed as coronavirus drug remdesivir 'fails first trial'. In: BBC (2020)
50. Carter, D.C., et al.: A unique protein self-assembling nanoparticle with significant advantages in vaccine development and production. J. Nanomater. **2020**, 1–10 (2020)
51. Choy, K.-T., et al.: Remdesivir, lopinavir, emetine, and homoharringtonine inhibit SARS-CoV-2 replication in vitro. Antiviral. Res. **178**, 104786 (2020)
52. Elfiky, A.A.: Anti-HCV, nucleotide inhibitors, repurposing against COVID-19. Life. Sci. **248**, 117477 (2020)
53. Monteil, V., et al.: Inhibition of SARS-CoV-2 infections in engineered human tissues using clinical-grade soluble human ACE2. Cell. **181**, 905–913 (2020)
54. Dai, W., et al.: Structure-based design of antiviral drug candidates targeting the SARS-CoV-2 main protease. Science. **368**, eabb4489 (2020)
55. Zhang, L., et al.: Crystal structure of SARS-CoV-2 main protease provides a basis for design of improved α-ketoamide inhibitors. Science **368**(6489), 409–412 (2020)
56. Jin, Z., et al.: Structure of Mpro from COVID-19 virus and discovery of its inhibitors. Nature. **582**, 289–293 (2020)
57. Shen, C., et al.: Treatment of 5 critically Ill patients with COVID-19 With convalescent plasma. JAMA. **323**, 1582–1589 (2020)
58. Roback, J.D., Guarner, J.: Convalescent plasma to treat COVID-19: possibilities and challenges. JAMA. **323**, 1561–1562 (2020)
59. Chen, L., et al.: Convalescent plasma as a potential therapy for COVID-19. Lancet. Infect. Dis **20**(4), 398–400 (2020)
60. Casadevall, A., Pirofski, L.-A.: The convalescent sera option for containing COVID-19. J. Clin. Investig. **130**(4), 1545–1548 (2020)
61. Biot, C., et al.: Design and synthesis of hydroxyferroquine derivatives with antimalarial and antiviral activities. J. Med. Chem. **49**(9), 2845–2849 (2006)
62. Yao, X., et al.: In vitro antiviral activity and projection of optimized dosing design of hydroxychloroquine for the treatment of severe acute respiratory syndrome coronavirus 2 (SARS-CoV-2). Clin. Infect. Diseases **71**, 732–739 (2020)
63. Bosseboeuf, E., et al.: Azithromycin inhibits the replication of Zika virus. J. Antivir. Antiretrovir **10**, 6–11 (2018)
64. Bacharier, L.B., et al.: Early administration of azithromycin and prevention of severe lower respiratory tract illnesses in preschool children with a history of such illnesses: a randomized clinical trial. JAMA **314**(19), 2034–2044 (2015)

65. Gautret, P., et al.: Hydroxychloroquine and azithromycin as a treatment of COVID-19: results of an open-label non-randomized clinical trial. Int. J. Antimicrob. Agents. **56**, 105949 (2020)
66. Gautret, P., et al.: Clinical and microbiological effect of a combination of hydroxychloroquine and azithromycin in 80 COVID-19 patients with at least a six-day follow up: a pilot observational study. Travel. Med. Infect. Disease. **34**, 101663 (2020)
67. Wrapp, D., et al.: Cryo-EM structure of the 2019-nCoV spike in the prefusion conformation. Science **367**(6483), 1260 (2020)
68. Zhang, J., et al.: Progress and prospects on vaccine development against SARS-CoV-2. Vaccines **8**(2), 153 (2020)
69. Takashima, Y., et al.: Artificial molecular clamp: a novel device for synthetic polymerases. Angew. Chem. Int. Ed. **50**(33), 7524–7528 (2011)
70. Organization, W.H.: Bacille Calmette-Guérin (BCG) vaccination and COVID-19: scientific brief, 12 April 2020. World Health Organization (2020)
71. Office, U.O.O.N: Oxford COVID-19 vaccine begins human trial stage (2020). http://www.ox.ac.uk/news/2020-04-23-oxford-covid-19-vaccine-begins-human-trial-stage
72. Lurie, N., et al.: Developing Covid-19 vaccines at pandemic speed. New. England. J. Med. **382**, 1969–1973 (2020)
73. Peeples, L.: News feature: avoiding pitfalls in the pursuit of a COVID-19 vaccine. Proc. Nat. Acad. Sci. **117**, 202005456 (2020)
74. Maslow, J.N.: Vaccine development for emerging virulent infectious diseases. Vaccine **35**(41), 5437–5443 (2017)
75. Modjarrad, K.: MERS-CoV vaccine candidates in development: the current landscape. Vaccine **34**(26), 2982–2987 (2016)
76. Coleman, C.M., et al.: Purified coronavirus spike protein nanoparticles induce coronavirus neutralizing antibodies in mice. Vaccine **32**(26), 3169–3174 (2014)
77. van Doremalen, N., et al.: Aerosol and surface stability of SARS-CoV-2 as compared with SARS-CoV-1. New England J. Med. **382**, 1564–1567 (2020)
78. Kampf, G.: Potential role of inanimate surfaces for the spread of coronaviruses and their inactivation with disinfectant agents. Infect. Prevent. Pract. **2**(2), 100044 (2020)
79. Ivanova, E.P., et al.: Natural bactericidal surfaces: mechanical rupture of pseudomonas aeruginosa cells by cicada wings. Small **8**(16), 2489–2494 (2012)
80. Ivanova, E.P., et al.: Bactericidal activity of black silicon. Nat. Commun. **4**, 1–7 (2013)
81. Jaggessar, A., et al.: Bio-mimicking nano and micro-structured surface fabrication for antibacterial properties in medical implants. J. Nanobiotechnol. **15**(1), 64 (2017)
82. Shahali, H., et al.: Multi-biofunctional properties of three species of cicada wings and biomimetic fabrication of nanopatterned titanium pillars. J. Mater. Chem. B. **7**, 1300–1310 (2019)
83. Hasan, J., et al.: Multi-scale surface topography to minimize adherence and viability of nosocomial drug-resistant bacteria. Mater. Des. **140**, 332–344 (2018)
84. Hasan, J., et al.: Antiviral and antibacterial nanostructured surfaces with excellent mechanical properties for hospital applications. ACS Biomater. Sci. Eng. **6**, 3608–3618 (2020)
85. Sakudo, A., Yagyu, Y., Onodera, T.: Disinfection and sterilization using plasma technology: fundamentals and future perspectives for biological applications. Int. J. Mol. Sci. **20**(20), 5216 (2019)
86. Sakudo, A., et al.: N2 gas plasma inactivates influenza virus by inducing changes in viral surface morphology, protein, and genomic RNA. BioMed. Res. Int. **2013**, 694269 (2013)
87. Yamashiro, R., Misawa, T., Sakudo, A.: Key role of singlet oxygen and peroxynitrite in viral RNA damage during virucidal effect of plasma torch on feline calicivirus. Sci. Rep. **8**(1), 17947 (2018)
88. Branley, J.M.: Can we reuse P2/N95 respirators? Current evidence and urgent research questions. University of Sydney, Sydney (2020)

89. Kern, J.J., Stein, R.L., Wildrick, J.P.: Disposable face mask. Google Patents (1997)
90. Zambrano-Monserrate, M.A., Ruano, M.A., Sanchez-Alcalde, L.: Indirect effects of COVID-19 on the environment. Sci. Total. Environ. **728**, 138813 (2020)
91. Muhammad, S., Long, X., Salman, M.: COVID-19 pandemic and environmental pollution: A blessing in disguise? Sci. Total Environ. **728**, 138820 (2020)
92. Ting, D.S.W., et al.: Digital technology and COVID-19. Nat. Med. **25**, 459–461 (2020)
93. Keesara, S., Jonas, A., Schulman, K.: Covid-19 and health care's digital revolution. New England J. Med. **382**, e82 (2020)
94. Reeves, J.J., et al.: Rapid response to COVID-19: health informatics support for outbreak management in an academic health system. J. Am. Med. Inform. Assoc. **27**, 853–859 (2020)
95. Calvo, R.A., Deterding, S., Ryan, R.M.: Health surveillance during covid-19 pandemic. Br. Med. J. **369**, 1–10 (2020)
96. Ferretti, L., et al.: Quantifying SARS-CoV-2 transmission suggests epidemic control with digital contact tracing. Science. **368**, eabb6936 (2020)
97. Oliver, N., et al.: Mobile phone data for informing public health actions across the COVID-19 pandemic life cycle. Sci. Adv. **6**, eabc0764 (2020)
98. Organization, W.H: Laboratory biosafety guidance related to coronavirus disease 2019 (COVID-19): interim guidance, 12 February 2020. World Health Organization (2020)

Operations Management Processes in the Continuous Production of Renewable Energy: The Basic Social Process of Proficient Amalgamating

Jan Green[1](✉) and Alison Mcmillan[2]

[1] Liverpool Hope University, Hope Park, Liverpool L16 9JD, UK
greenj2@hope.ac.uk

[2] Wrexham Glyndŵr University, Plas Coch Campus, Wrexham LL11 2AW, UK

Abstract. Operations management processes are embedded throughout and across business and society, becoming attention-seeking when there is a malfunction or breakdown. Over time they have become increasingly sophisticated and essential to economic success through value chains, quality management, logistics, supply chain and inventory control. The operations concern and priority is a replicable process that matches and meets the continuous expectations of contemporary organisations and requires a fusion between engineers, management and technology. The research question posed in this study is: what are the concern resolving behaviours required to acquire and maintain fusion and inter-connectedness within an operations management department of a renewable energy facility. The objective being the identification and integration of behaviours and practices into revised operational procedures to ensure the everyday processes and procedures are functioning correctly. The research component is classic grounded theory, based on a purposive sample. A new conceptual theory illustrated in the model of proficient amalgamating was the outcome and is evidenced when the replicable processes contributing to the operation of an organization are functioning in a manner which is integrative, effective and efficient. This status supports and contributes to the subsequent amalgamating of operations management techniques and procedures into the background of an organizational environment and therefore emphasises the relevance of research into the use and application of operations management with clear links with maintenance practices.

Keywords: Operations management · Maintenance · Processes · Systems · Grounded theory

1 Introduction

The operations management of a renewable energy company, which consists of an anaerobic digester that feeds energy into the national grid is a central and critical function of the organization in order to generate continuous output and maintain the strategic production targets. A single product/single machine set up is the scope of the system,

A. Batako et al. (Eds.): GCMM 2021, LNNS 335, pp. 714–728, 2022.
https://doi.org/10.1007/978-3-030-90532-3_54

based on the widely used transformation system of changing inputs into outputs through a transformation process. In addition, resource constraints are also an operations management issue requiring attention due to both storage capacity and availability to provide a mix that complies with the measurement instruments and the national grid regulations.

A precise recipe of chicken manure and cow slurry are combined in a tank to achieve a consistency that enables the mix to be pumped into the digester without causing blockages. Once in the digester the contents heat up through the anaerobic process and effective production of gas is dependent on maintaining a consistent temperature in order to produce biogas. The output is pumped through a carbon filter and creates electricity which is subsequently fed into the national grid. Input is metered in order to calculate the amount of energy fed into the grid and generate records of production capacity and the required documentation related to invoicing and payment.

Critical operations management practices for the organization include a reliable supply chain to ensure the raw materials are available in the required quantities and age, to provide the requisite calorific and consistency mix. The production process is continuous and requires up to 24-h for the digester to reach optimum temperatures, therefore maintenance is planned, with a back-up facility via a second pump which was been put into place to limit stoppages that had been experienced in the early stages of the project.

Funding of the enterprise was achieved via a call for shares, purchased by private individuals. The majority having an interest in sustainable businesses and also keen to support local initiatives. When a secure financial foundation had been achieved, planning and building commenced and the senior management team focused activities on the detail of the production process, making rapid progress to the commissioning stage. Early output was satisfactory and stabilized at a level that enabled the interest payments and share capital repayment schedules to be met. Operational costs included corrective and preventative maintenance in addition to production costs and the digester was run in a manner intended to minimize long-run average costs, however the benefits of predictive maintenance were given limited consideration in the early euphoria of the project.

As the replaceable parts of the digester reached and passed warranty periods, maintenance issues began to arise with faults in valves and pumps which resulted in periods of down time and loss of energy production. There were delays in engineers attending the site to undertake diagnosis and carry out the subsequent repairs and renewals. Due to the smooth and uninterrupted honeymoon period insufficient attention had been given to continuity management and the associated operations. It was admitted by the Board that initial plans had not incorporated sufficient contingency detail and adequate plans to support authorized decision making. As the digester roles were part-time, with those involved having other business demands, the focus of running the digester had become sporadic, intermittent and a background priority. A series of interruptions changed this perception, bringing the operation of the digester into the foreground, due to the interruptions.

Therefore, the research problem being addressed by this paper is to establish the concern resolving behaviours that were required in order to design and put into place effective operations management techniques that would result in uninterrupted generating and revert to the status quo situation in the initial stages of the project which are conceptualized as proficient amalgamating.

To identify the management behaviours, classic grounded theory, a stand-alone methodology was used to collect data via a purposeful sample of semi-structured interviews which underwent comparative coding to identify the core issues and emergent concepts that contribute to the ongoing notion of proficient amalgamating within an operations management setting.

The structure of the paper, following the abstract and introduction is to review the literature of relevance within the operations management body of knowledge, acknowledging that this section is limited to an introduction to the topics covered. A summary of classic grounded theory procedures then provides an insight into the selected methodology prior to presenting and analyzing the data. The outcome of the objective of the paper, to revise operational procedures and support continuous production is discussed in the final section. Recommendations for practitioners in a similar situation, to maintain continuous production, are proposed and have value to academics and practitioners working within an operations management field or specialism.

2 Literature Review

Operations management is a discipline that originated to solve management problems in a factory environment [24]. Operations occur within simple and complex organizations and the interaction between operations and organizational systems is a key issue when companies need to ensure processes are [26]. Operations Management involves creating, designing and coordinating business operations that serve to develop and produce goods and/or services using the best available means from a growing and continually changing stock of technologies [18]. Reliability is a central Operations Management objective alongside efficiency, flexibility and responsiveness.

A view posed by [18] is that operations management research is, in the main, carried out by practitioners to introduce novel approaches in order to derive operational benefits. The academic operations management researcher subsequently attempts to explain the outcomes of the practitioners work through theoretical capture. This is a reversal of the usual practice and has the potential to decrease the value of academic research, within a practical sense. To address this a change in emphasis from explanation to exploration is proposed, which aligns with the selected methodology in this paper.

2.1 Transformation Process

The transformation process is the foundation of achieving the desired product or service, according to [1], based on measured inputs and outputs taken at various stages throughout the process of transformation, weighted against a standard objective, which may require cross checking and corrective actions to maintain quality and consistency. Inputs vary according to sector but recognized practice within operations management is to group inputs into broad categories which include people, materials, energy, capital and data, drawing on the overall resources of the organization [17]. The internal transformation is made up of the reoccurring operational processes that result in an output which may be a service, a product or information.

An insightful and critical view into the complexities of the transformation process is provided by [14] who go beyond the notion of a traditional operations management view of sequential sub-processes or activities to recognise the interactive nature of a transformation process where there is partial overlap with other areas of the business and external stakeholders, with an emphasis on the customer. There remains clear acknowledgement that the transformation process is a powerful template for analysis which is governed, coordinated and enhanced by communication. The sub-processes are broadly categorised as request, agreements and commitments which, to be effective, require management preparation and negotiation followed by actions linked to accepted performance measures which relate to the process in question. Throughout there is an emphasis on the requirement for asking, agreeing and actions, resulting in an illustration which mirrors the nature of the transformation process in action. Authors of [8] endorse this view, using the term 'functional deliberation' to indicate the proportion of time allocated to routine operational aspects. However, overall performance is enhanced when the deliberation process is expanded to incorporate top level activities, particularly where the environment lacks stability.

2.2 Continuous Production

In [21] authors comment that many production assets run continuously. Any instances where there is a requirement to shut down and ramp up assets entails costs should be prohibited unless absolutely necessary, for example maintenance or cleaning purposes. Whilst this point is endorsed by [29], it is clear that automated continuous processes may require more effort than batch production to identify deviations which arise from the steady state set points. Therefore, clear process parameters are essential and typically will involve temperature, pressure and fill levels. Sensors are the tradeoff between deviation and delay in reacting to deviations in order to resolve the disturbances. Where the risk is significant when a deviation occurs, the use and application of technology is viewed as the required response to ensure there is automatic shutdown.

To achieve steady state process analytical technology is the core enabler for continuity, and process measurements should be incorporated within the design stage [27]. Steady state is the fundamental property of processes which are continuously operated and incorporate fault detection and data reconciliation, concluding with stopping criteria. The goal of steady state according to [27] is unsupervised process optimization which are established using time series techniques where samples are certified and input variables are run to establish steady state, which is automatically detected.

2.3 Inventory and Decisions

In [9] ones propose that the objective of inventory management is to minimize the average cost per time unit (in the long run) incurred by the inventory system, whilst at the same time guaranteeing a level of service that is pre-specified. To achieve this co-operation is required between the various actors participating in holding stock for onward transmission and effective communication of delivery schedules which is partially dependent on decision making and choice to achieve alignment and co-ordination, across the elements of the supply chain. Costs per order and costs of shortages are elements requiring consideration in the decision-making process which depend on the sharing of knowledge.

Many aspects of operations management are founded on predictions related to inventory where consistent capture of all operational activity is required. The outcome of decisions within this domain are business critical and often made within a short-time frame. According to [7] support systems decisions in many organizations are dependent on subject matter experts with 'look ahead' knowledge and may involve intuitive judgement and the application of external signals.

Aspects of knowledge sharing networks have been drawn together in an extensive review conducted by [2], citing trust as the most important determinant. They can be developed through social relations and network ties which have the potential to contribute to a sense of collegiality. The notion of absorptive capacity is partially dependent on previous related knowledge and therefore effective communication is a priority to address this point, which is enhanced by the use of technology.

2.4 Systems Management

Studies evaluating the improvements generated by operations management practices suggest positive effects. In [4] ones propose the combined use of different systematic approaches such as Total Quality Management, Just In Time, Total Productive Maintenance and Supply Chain Management to achieve increased performance across an enterprise, including financial performance. All of these procedures require effective planning, recording, action, communication and overall optimization of internal resources, working across an organization rather than being restricted to one function or area.

Ones [22] review aspects of the interface between operations management and systems. The authors emphasize the need for additional understanding related to the boundaries between the two practices and the blending that occurs between technology and people, whilst stressing that systems are operational as they are repetitive in nature. The scope for the notion of blending this interface has potential for multiple viewpoints, encompassing most organizational functions. Information sharing is the key aspect to encourage cost savings through lead times based on accurate and timely information. This paper also refers to the growing need for cyber security within operations management and draws on the potential improvements through the increasing use of the Internet of Things. Increasingly, seamless integration and inter-operability via various devices [3] contributes to asset performance and smooth the developments of human-machine interactions through touch improvements.

2.5 Training and Decisions

Effective operations management training includes various elements, frequently commencing with familiarization via demonstration, to introduce the underlying principles, together with cause and effect aspects. From this foundation the incorporation of practice and knowledge of where to go decisions with knowledge of indicators and symbols should be practised. Ones in [28] add that control decisions are frequently made in a matter of seconds and should encompass arrivals and completions to reflect the entire operations management processes. In [4] authors refers to the frequent push which results in people becoming very operative and into taking ideas from practical issues to access information and to develop specific skills. This can cause internal knowledge to be very specialized and strongly connected to the real world. Such an approach is clearly extremely important because it allows for the quick analysis and solution of operative problems and also creates an awareness of the knowledge gap that needs to be filled Links with professional bodies are strongly recommended, in this specific case the Waste Management Industry Training and Advisory Board provides relevant training.

3 Methodology

Classic grounded theory principles and techniques have been applied ~~in order~~ to answer the research problem or concern, which is the maintenance of continuous production. In doing so, an appreciation of 'what counts is only what is data-related' [6] and aligns with the view that "Grounded theory is one of the best research methods to answer the question: 'what is going on here?" [20] as the data collection embraces opinions, actions and context. In [10] Author summarizes grounded theory as the study of abstract problems and their processes to discover emergent problems and the subsequent resolutions to manage the situation. In [5] Author states that grounded theory is a systematic method of analysing and collecting data with an aim of developing theories and is usually approached from an inductive perspective. The stand-alone methodology is comparative and interactive.

Throughout the grounded theory process, the aim is to analyse data using constant comparison methods to inform subsequent data collection. It is also noted by [5] that grounded theory is a method for studying processes; this view justifies the selection of the method for this particular study. In [16] Author provides further endorsement for the choice of method, as the emphasis is on individuals as unique living entities. The research focus is on the world as it is experienced by those contributing to the data collection. It is real world research that generates theory which provides a guide to action using concepts that emerge from statements. An integrated framework is formed to explain an event which is empirically grounded. Further support for the use of grounded theory is the commentary provided in [11] that grounded theory helps the researcher to see a situation as it is, not from preconceived viewpoint. The methodology provides an orientation to see behaviours as data which are then applied using the stages of grounded theory to see what is going on in the context to which it is being applied and help effect a change to the context problem.

The first stage of grounded theory is data collection, which is carried out via semi-structured interviews. Where the emphasis is on introductory open questions and encouraging the participants to provide a picture of the research situation as it is, not as a theoretical preconception, which has the potential to force the data. After each data set is collected, coding is carried out and when subsequent data is collected, constant comparison is undertaken. Summarizing constant comparison, in [12] Author notes the requirement to code incidents and compare the incidents to other data sets, noting similarities and differences and the conditions in which they arise. Codes are abstracted from the data as the integrated framework is developed.

In order to illustrate ideas and take apart the data story, the researcher writes memos which include a title or reference so they are identifiable. The purpose of memos is to capture and develop ideas, the production of memos is continuous and should be written immediately following data collection. Memos raise the description of properties emerging in the data to a theoretical and conceptual level [11]. Finally, memos should be sortable to create the integrated framework.

The basic social process illustrates emergent stages that have individual characteristics and accounts for the evident variations in problematic patterns of behaviour. A basic social process is not static, it occurs over time, has identifiable break points and involves change [13]. Embedded within the basic social process are forms of contact and communication [15] resulting in a clear perception by the social actors.

3.1 Sampling

The Board consists of ten directors with an involvement in the enterprise, the intention of achieving a census sample being to interview all members. Despite an awareness that the operations management of the project had shortcomings, resulted in reduced production and, in more extreme cases, complete shutdowns, with the potential to delay capital repayments to shareholders, one director declined to participate. The rationale for this decision was due to geographical location and therefore limited insight into the core research problem. In this research, the setting was expert and judgmental and was clearly representative of the Board population to acquire an in-depth insight to the problem.

Therefore, purposive sampling was used, ensuring each participant had knowledge and insight of relevance to the problematic processes. Purposive sampling is intentional selection of informants based on their ability to elucidate a specific theme, concept, or phenomenon [25], which is a type of nonprobability sample [23]. This approach enables discourse to take place with peers and experts to provide a crucial contribution resulting in a wealth of ideas [19] as an authentic foundation for the theory.

3.2 Data Collection

In compliance with grounded theory techniques, data was collected via semi-structured interviews which involved asking the participants to answer two questions, of relevance to the research problem, they were:

1. In your opinion, what is/are the causes of the digester breakdowns?
2. What requires attention/change in order to reduce or eliminate the breakdowns?

These questions ensured that the research problem was clearly identified, from the perspective of those who reported breakdowns and secondly the research problem was addressed in line with grounded theory principles by establishing the concern resolving behaviours in an abstract and conceptual manner. Interviews were not recorded, however extensive notes were taken and reviewed at the end of each interview. Memos were written at this stage, followed by initial coding. The interviews took between 20 and 40 min to complete, reflecting the degree of involvement each individual had with the operational side of the enterprise.

The data provided referred to sequences of tasks that repeatedly take place to maintain operations and rapidly patterned out during constant comparison. This outcome indicated the notion of a basic social process as there was a change in the raw materials from slurry and manure during the anaerobic digestion, creating energy in a useable form, which is electricity fed into the grid. Similar incidents were described that resulted in an unscheduled operational break, whilst the perspectives of the participants differed saturation was rapidly achieved. However, for completeness all interviews were conducted.

4 Discussion

The descriptive codes appearing in the data were clustered into three themes, listed in the left-hand column of Table 1. The descriptive components of each theme are listed in the right-hand column. Adhering to grounded theory principles the themes are conceptualized to generate the basic social process of proficient amalgamating and moves the theory beyond description. Relevant data extracts provide in vivo comments and illustrations of the concepts are reproduced in the discussion, providing a degree of contextualization and added insight which go beyond the stated codes. A series of written memos capturing ideas, actions and progression which indicated ways in which the operation of the digester was improved were written to emphasize how processes were changed, recorded and refined to resolve the concerns.

Data was subsequently conceptualized in an abstract manner ensuring the theory is devoid of time, place and people, in compliance with grounded theory principles. Table 1 illustrates the stages which were taken in adapting management behaviours to achieve Proficient Amalgamating of widely recognized operations management practices to support continuous production.

Operational management deficiencies identified in the data pattern out into three main themes beginning with practices and procedures that would benefit from considerable revision, they are listed under the theme Initial Practices. To address these deficiencies the integration of technical tools is indicated in several areas and forms the central section of the basic social process.

Finally, the resolution of the problem requires an overall approach which is themed as systematic functioning across a range of operational management tasks to achieve the fusion between the individuals and related tasks. To clarify the findings, Table 2 groups the three themes under revised headings to illustrate the recommended behaviours and operations management theories to support Proficient Amalgamating of tasks associated with the day-to-day functioning of the digester to achieve prolonged steady state production.

4.1 Precise Checking

It is clear from the data that the early management of the digester was imprecise, and an ad hoc approach was adopted. Measures in terms of inventory control were largely absent, resulting in stock-out and therefore loss of production. Casual observation, without experience, back up data and planning all contributed to this aspect of operations management. In addition, there was no documented purchase order system in place

Table 1 Integrated framework of the problem

Category themes	Codes within the categories
Initial practices	Casual management Consumables and replacement schedules Ineffective delivery/stock control procedure Inadequate back-up/ contingency procedures No scheduling
Technical integrating	Limited documentation of processes Cursory knowledge management, sharing and communication Digital tasks Controls location and scope Using and applying technology Access to measurement app Contrasting sounds/noise as insight
Systematic functioning	Implementation of Rotas/Systems Maintaining Records and logs. Access Undertaking training and ongoing updates Centralised systems management Decision making and parameters Identifying and Defining Scope of authority

Table 2 Operations management contributions to the basic social process of Proficient Amalgamating

Precise checking	Technical integrating	Acquiring competence
→	→	→
Inventory control	Knowledge management	Levels of access
Scheduling	Communication	Training
audit documentation	measures and parameters	systems management
Contingencies	Compliance	Discretionary decisions
	Digital procedures	Authority
	Steady state	

and expectations that a random phone call would result in prompt action proved to be inaccurate. Without an audit trail linked to orders there were accounting issues when invoices were being approved at a later date. This concept readily indicates the context and incidents which occurred and were coded from the data. Precision was largely absent and based on somewhat casual observation and measures. Therefore, attending to these aspects removed several problematic occurrences to improve the operation.

Relevant data extracts are:

"there was no stock control, the approach was too casual in the early days and there were occasions when there was a delivery gap. Product availability is seasonal, this wasn't factored in."

"it sounds basic, but we all needed to know when there had been a delivery – and how much. A load is an imprecise measure! We needed a daily stock check – which was recorded. Not everyone wanted to do that!"

"Production ceased when a pump broke down after three years of operation, it took six weeks to get a replacement. This was another example of no contingency planning or system. The loss of income far exceeded the cost of the replacement pump, this situation has changed. We agreed to a more formal and precise purchasing system"

"Basic maintenance needed to be shared, lubrication, diaphragms and carbon were consumables, we needed a schedule that we could all access which linked to levels of stock and delivery timescales. This was amateurish"

"I was criticized for being too corporate! It took time, I provided a template and explained a pattern of consumption would emerge"

"the rubber diaphragms just go, there is no warning and no pattern developed to indicate how long they would last. We didn't have sufficient stock of low cost consumables. We learned that the cheaper makes were not substantial enough for our processes"

"When the diaphragm ruptured, production stopped – priorities not sorted, all we needed was stock control – and a checking system. These are not expensive items"

The data emphasizes a lack of precision and documentation at many stages of the operation. Visual checks and ongoing records contribute to confidence and reflects the suggestions and actions taken which then blend into the second category of technical integration to progress and refine the operations.

4.2 Technical Integrating

When the digester is fed and up to temperature, replaceable parts are working constantly, production is continuous, and electricity generated from the process is fed into the national grid. This was the required and anticipated status quo. The transfer and sharing of technical knowledge was initially very limited, and is indicated in several of the comments linked to casual management. The result was access delays and limitations related to data recording and interpretation. To improve these aspects attention and training was required to move them into a scheduled background check which took very little time to complete and note. They became integrated into the routine, and as a result no longer a problem when implemented. Using the monitors effectively also acted as an early warning system to take corrective action and added further positive contributions to resolving the overall problem. Data extracts provide additional details.

"The control panel is located at the top of the digester – I don't know why! Access required wearing protective gear and climbing and outside set of stairs. They get wet when it rains and slippy – my interpretation is that the panel was poorly located at the design stage"

"The order schedule referred to earlier – which was introduced - was maintained using technology – I just refused to have a manual paper-based system when we had state of the art equipment measuring the digester output."

"We eventually got to share and use the app so the gas levels and temperature can be checked via a mobile device – anywhere, any time. Poor integration – and knowledge sharing, this took several months to rectify."

"Initial training on the technical side would have prevented so many operational issues arising in the first place"

"I've been using on-line accounting for several years. There was considerable reluctance to integrate the finances of the organization to a cloud base. The problem was two-fold – a lack of confidence and a lack of knowledge linked to technology, which the Finance Director needed to address."

"A three-dimensional view based on input of raw materials, repeated references to technology output to check functions was where we eventually got to. The scope and importance of what we were doing was under-estimated"

"I learned to listen to the digester when everything was working there was an audible hum, when something was wrong the sound was more broken and mechanical. I should have recorded the two contrasting sounds and played them to my colleagues!"

Responses from the sample were both reflective and practical. The operational tasks requiring scheduled attention are not in themselves overly complex and it was increasingly clear that knowledge transfer from the commissioning engineers had not included all staff subsequently tasked with managing routine operations. Knowledge sharing demonstration of technical issues assisted in gaining competence and confidence, prior to progressing to the final stage of the process.

4.3 Acquiring Competence

Anaerobic digestion is a natural function, the variable inputs being the mix which is transformed into electricity. Repetition of loading and mixing is required at regular scheduled intervals to maintain continuous processing. This is a widely recognized method of managing operations. Those involved readily grasped the need for gaining expertise and taking part in relevant training to address avoidable production gaps. An appreciation of this is clear in the data extracts for the final element of the basic social process, viz:

"It sounds basic – but that's operations management – a sequence, a rota, shared records and access to data, that is what supports the continuous function of digestion"

"I didn't feel proficient in the early stages, reading the instruction manual about the range of functions was a priority – but I had to find it first. Then we needed the review the updates"

"Contingencies – no one wanted to think about that – answering the what if question – when we could do that we became effective and we do practice operations management in the way I know it from previous roles – but with more up to date tools and devices"

"an effective operating system is like wallpaper, the breaks between rolls should not be visible, so you don't really notice it – initially our breaks were cracks"

"the digester temperature and output rate can be read continuously – from a mobile phone. Why didn't we know about this?"

"we should have received some guidance – no more than that, formal training – all of us – when the digester was commissioned – about operating it!"

Proficiency is rapidly improved with relevant training, as noted in the data and dovetails with the various functions within operations management. Formal training is referred to and introducing specialized and qualified providers would ensures both neutrality and professional content with precise learning outcomes. The acquisition of relevant and recognized qualifications is data has the potential to contribute to confidence and making better informed decisions.

4.4 Theoretical Sensitivity

The data extracts explain several tasks involved in operating the digester and behaviours of the operators, which form the stages of the conceptual basic social process providing indications related to the concern resolving behaviours of the problem. To conclude the grounded theory, theoretical sensitivity is applied by drawing on the literature related to the findings.

This research is based on operations management and it is clearly evident from the sample of data extracts provided that the literature which defines operational management is both accurate and relevant for this study. Reference being made to solving problems in [18] and to co-ordination, and interaction [15] reliability, flexibility and a stock of technology. All elements which arise in the anaerobic digestion transformation process provide the foundation of successful continuous production. As this did not always occur the design precision of the digester and supporting systems [27] were not given sufficient prominence. This was particularly evident when transferring the managerial authority from the engineering build company following commissioning to the Co-operative and was, in effect, the cause of the majority of the problems subsequently experienced. This created several situations which were the opposite of the problem resolution, namely rhythmic and continuous production.

Initial measures were vague and therefore required considerable attention and revision to incorporate multiple viewpoints. This contributed to ensuring potential stock-outs, maintenance and renewals were addressed in a proactive manner to introduce a clear interface [22] between the technology and those responsible for managing the operations. The formal creation of a system is a clear recommendation, and this may require additional sections as the plant ages. There is the potential for an engine rebuild or panel replacements which would involve capital expenditure and a further recommendation is a systematic and integrated approach to the financial function which depreciates the plant and also introduces the provision of a progressive contingency account to build reserves. The reference to cloud-based procedures requires management attention to further raise the levels of proficiency and strengthen the links within and between the various functions, finance also being an operational procedure. Incorporating a knowledge sharing network [2] is a solution to ensure communication and action are timely and minimize problems as they arise.

Additional literature was reviewed following the data coding and comparison exercises as the emergent theory, expanded beyond the traditional aspects of operations management to encompass ways in which decisions are made via systems [4]. This identified the clear need for targeted training to support the ongoing operation, in this case it was specifically related to technology. Specialist knowledge is developed to support prompt and effective decision making which is illustrated in the behaviour, to become intuitive, steady state experts [7]. This aspect could be strengthened through the acquisition of formal qualifications which are widely recognized in the sector. It is usual practice to widen the initial scope of the literature review when generating grounded theory as there is no attempt to verify previous theories. There is negligible coverage which links both of these aspects together in previous studies and therefore reveals a gap which the concept of proficient amalgamation of operations management procedures addresses. The findings are potentially of particular use to practitioners.

5 Conclusion

In conclusion, operations management should be raised to the strategic level and be discussed at Board level on a regular basis. This change would enhance the prominence given to what may be viewed as routine functions, which have the potential to limit or prevent a business from achieving strategic goals. As technology changes, develops and improves it is also essential that operational staff receive sufficient training and updates as required. Decision-making is a further area which this study has identified as a short-coming and therefore a knowledge of what are the control parameters for decision making should be clear, together with an appreciation of the potential consequences and shared via an integrated decision making policy.

The concept of proficient amalgamation within operations management has grab for operations practitioners. It strongly suggests an intuitive awareness of normality which does not require ongoing attention but is always present through systematic monitoring and, therefore effectively conceptualizes the notion of steady state continuous production as a successful outcome of resolving the stated research problem. This is largely achieved by the fusion of engineering, management and technology skills.

Further research is suggested to embed rigorous and formally recognized operations management training when developing repetitive systems, with a particular emphasis on technological advances to include a timebound element for updating and evaluation.

References

1. Akinola, A.: Transformation Processes in Operations Management: Dangote Cement, Nigeria (2018). https://www.researchgate.net/publication/327752290
2. Asrar-ul-Haq, M., Anwar, S.: A systematic review of knowledge management and knowledge sharing: trends, issues, and challenges. Cogent. Bus. Manage. **3**, 1 (2016). https://doi.org/10.1080/23311975.2015.1127744
3. Atzori, L., Iera, A., Morabito, G.: Understanding the internet of things: definition, potentials, and societal role of a fast evolving paradigm. Ad Hoc Netw. **56**, 122–140 (2017). https://doi.org/10.1016/j.adhoc.2016.12.004
4. Battistoni, E., Bonacelli, A., Colladon, A.F., Schiraldi, M.M.: An analysis of the effect of operations management practices on performance. Int. J. Eng. Bus. Manage. (2013). https://doi.org/10.5772/56919
5. Charmaz, K.: The Power and Potential of Grounded Theory (2020). https://www.semanticscholar.org/paper/The-Power-and-Potential-of-Grounded-Theory-Charmaz/93d28c60474e31cedd4464c5b24ae0af2efbc090?p2df
6. Christiansen, O.: A simpler understanding of classic GT: how it is a fundamentally different methodology. Grounded Theory Rev. **6**(3), 39–58 (2007)
7. Daly, M.: Decision support: a matter of information supply and demand. J. Decis. Syst. **25**, 216–227 (2016). https://doi.org/10.1080/12460125.2016.1187423
8. Demeester, L., De Meyer, A., Grahovac, J.: The role of operations executives in strategy making. J. Oper. Manag. **32**(7–8), 403–413 (2014). https://onlinelibrary.wiley.com/doi/abs/10.1016/j.jom.2009.06.001
9. Fiestras-Janeiro, M.G., Garcia-Jurado, I., Meca, A., Mosquera, M.A.: Cooperative game theory and inventory management. Eur. J. Oper. Res. **210**(3), 459–466 (2011). https://doi.org/10.1016/j.ejor.2010.06.025
10. Glaser, B.: Getting started. Ground. Theory. Rev. **17**(1), 3–6 (2018)
11. Glaser, B.G.: Memoing: A Vital Grounded Theory Procedure. Sociology Press, Mill Valley (2014)
12. Glaser, B.: The constant comparative method of qualitative analysis. Ground. Theory Rev. **7**(1), 1–6 (2008)
13. Glaser, B.: Basic social processes. Ground. Theory. Rev. **4**(1), 1–28 (2005)
14. Goldkuhl, G., Lind, M.: Coordination and transformation in business processes: towards an integrated view. Bus. Process Manage. J. **14**(6), 761–777 (2008). https://doi.org/10.1108/14637150810915964
15. Green, J.: Frontier Rescinding. In: RC33 Conference Proceedings of 9th International Conference on Social Science Methodology, University of Leicester, 11–16 September 2016 (2016). https://www2.le.ac.uk/departments/sociology/research/rc33-conference/3_Green.pdf
16. Hallberg, L.R.-M.: The "core category" of grounded theory: making constant comparisons. Int. J. Qual. Stud. Health Well Being **1**(3), 141–148 (2006). https://doi.org/10.1080/17482620600858399
17. Hill, A., Hill, T.: Essential Operations Management. Palgrave, London (2018)
18. Holmström, J., Romme, A.G.L.: Five steps towards exploring the future of operations management. Oper. Manag. Res. **5**, 37–42 (2012). https://doi.org/10.1007/s12063-011-0060-8
19. Johnson, B., Holness, K., Porter, W., Hernandex, A.: Complex adaptive systems of systems: a grounded theory approach. Ground. Theory. Rev. **17**(1), 52–69 (2018)
20. Jovanovi, M., Mas, A., Mesquida, A.-L., Lali, B.: Transition of organizational roles in Agile transformation process: a grounded theory approach. J. Syst. Softw. **133**, 174–194 (2017)
21. Klosterhalfen, S.T., Holzhauer, F., Fleischmann, M.: Control of a continuous production inventory system with production quantity restrictions. Eur. J. Oper. Res. **268**(2), 569–581 (2018). https://doi.org/10.1016/j.ejor.2018.02.001

22. Kumar, S., Mookerjee, V., Shubham, A.: Research in operations management and information systems interface. Prod. Oper. Manage. **27**(11), 1893–1905 (2018). https://doi.org/10.1111/poms.12961
23. Lavrakas, P.: Encyclopedia of Survey Research Methods. Sage Publications, London (2008). https://doi.org/10.4135/9781412963947.n419
24. Peinado, J., Graeml, A.R., Vianna, F.: Operations management body of knowledge and its relevance to manufacturing and service organizations. Revista de Gestão **25**(4), 373–389 (2018). https://doi.org/10.1108/REGE-03-2018-0049
25. Robinson, R.S.: Purposive sampling. In: Michalos, A.C. (ed.) Encyclopedia of Quality of Life and Well-Being Research, pp. 5243–5245. Springer Netherlands, Dordrecht (2014). https://doi.org/10.1007/978-94-007-0753-5_2337
26. Samson, D., Kalchschmidt, M.: Looking forward in operations management research. Oper. Manag. Res. **12**(1–2), 1–3 (2019). https://doi.org/10.1007/s12063-019-00138-8
27. Simon, L.L.: Continuous manufacturing: is the process mean stationary? Am. Inst. Chem. Eng. **64**(7), 2426–2437 (2018). https://doi.org/10.1002/aic.16125
28. Van Der Zee, D., Slomp, J.: Simulation as a tool for gaming and training in operations management–a case study. J. Simul. **3**(1), 17–28 (2009)
29. Wengerter, M., Li, Y., Nieder, H., Brandner, J.J., Schoenitz, M., Scholl, S.: Energy and resource efficient continuous production of a binder emulsion using microstructured devices. Chem. Eng. Process. **122**, 319–332 (2017)

A Review of Satisfaction Theories and Implications for STEM Teaching

Janani Ganapathi[1(✉)], Jyoti Sharma[2], and Prasad K. D. V. Yarlagadda[1]

[1] Queensland University of Technology, 1 George Street, Brisbane 4000, Australia
janani.ganapathi@qut.edu.au, jananiganapathi14@gmail.com
[2] Department of Science and Technology, Government of India, New Mehrauli Road, New Delhi 110016, India

Abstract. This paper reviews the existing literature to understand the correlation between job satisfaction and STEM teaching, by adopting a best evidence review methodology. Four key job satisfaction theories are elaborated, namely Herzberg's motivator-hygiene theory, Maslow's needs hierarchy theory, job characteristics model, and the dispositional approach to job satisfaction. Considering the limited amount of research conducted in examining job satisfaction theories in relation to teachers, these theories are utilized in this paper to draw parallels between school teachers and STEM education delivery, through description and interpretation. This review presents the possible implications for STEM teaching and the need for further research assessing teachers' satisfaction and the impact it has on STEM delivery.

Keywords: STEM education · Teaching quality · Job satisfaction · School teachers · Satisfaction theories

1 Introduction

More and more nations are recognizing the necessity for human resources development, as a requisite to advance their capacities in science and technology (Sanders 2009). With the increasing complexity of societies, education systems are challenged with the responsibility of preparing the new generation to meet the complexities, sooner than later (Graziano et al. 2012). Creating a new generation of STEM professionals who are innovative and critical thinkers starts at school and would necessitate "a significant focus on the role of the teacher and how to develop teachers' knowledge, pedagogical practices, and dispositions to re-engage students" with these disciplines (Bissaker 2014, p. 56).

Research through the years has consistently underlined the critical role played by teachers in student achievement (Kirchhoff and Lawrenz 2011). Governments that are increasingly prioritizing STEM education also acknowledge that the quality of STEM school teachers would determine the success of STEM outcomes in students (Mervis 2009). Yet, scholars claim that a large number of school teachers and teachers-to-be are unprepared for STEM teaching: most tend to cling to teaching content that is safe

A. Batako et al. (Eds.): GCMM 2021, LNNS 335, pp. 729–742, 2022.
https://doi.org/10.1007/978-3-030-90532-3_55

and familiar as they do not have sufficient training, skills, or knowledge of the sciences, amongst other (Mintzes et al. 2013). While there are countless studies separately analyzing STEM education and the factors contributing to or affecting teaching quality in general, the number of research works focusing on the relationship between job satisfaction of STEM teachers and teaching delivery are very scarce. Keeping this gap in mind, this article attempts to review the existing literature to comprehend the equation between job satisfaction and STEM teaching.

2 Methodology

This article explores scholarly perspectives and criticisms on the topic and the most relevant and prominent theories, using the theoretical literature review approach. In particular, the most discussed job satisfaction theories from the literature are reviewed to better understand their affective nature on employee or teacher and performance at work. The method chosen focuses on four key job satisfaction theories are elaborated in this paper, namely Herzberg's motivator-hygiene theory, Maslow's needs hierarchy theory, job characteristics model, and dispositional approach. The theoretical literature review method not only helps in identifying the existing theories and their relationships, but they also allow the assessment of the extent to which the existing theories have been explored and develop new hypotheses to be examined (Turner 2018).

Considering the limited amount of studies conducted in examining job satisfaction theories in relation to teachers, these theories will be utilized to draw parallels to school teachers and STEM education, through evaluation and description. The works referred to in this paper consist of both recent and old sources, due to the mentioned scarcity of discussions around job satisfaction theories and STEM school teaching, as well as the agedness of each theory and the original literature by the theorists and early scholars.

This review is organized into two key sections; the first section of which is dedicated to the four selected prominent job satisfaction theories in the literature, followed by a discussion section, wherein the implications of STEM teachers' job satisfaction are discussed.

3 Job Satisfaction Theories

Defining a social sciences construct is often times challenging, as in the case of "job satisfaction", due to the non-existence of a universally accepted definition. A rather subjective term, job satisfaction is a broad term since it not only includes elements of the profession but also that of the work environment, both of which impact employees' satisfaction levels (Oswald et al. 2005; Ford et al. 1974). Locke formulated one of the most referenced definitions of job satisfaction in 1976, defining the term as "a pleasant or positive emotional state deriving from an individual's appreciation of his or her experience or profession" (p. 1304). Most scholars seem to be in acceptance of this definition such as Miller et al. (2009) who state that this definition covers the required component necessary to illustrate what the subjective term of job satisfaction could entail.

In each sub-section of this first section on theories, the main approaches to job satisfaction will be outlined, alongside the various scholarly definitions and interpretations, and debates surrounding their efficacy.

3.1 Herzberg's Motivator-Hygiene Theory

The motivator-hygiene theory emphasizes that contrary to what one may think, satisfaction and dissatisfaction at work are two completely different concepts that are distinct and unrelated. Herzberg's theory explains job satisfaction and dissatisfaction as comprising of two key components: motivation and hygiene. Motivation is generally determined by and reliant upon inducing factors such as professional achievement, salary and associated benefits. Herzberg suggests that the fulfillment of these motivation factors by employers is crucial for employees to remain motivated and satisfied of their work and workplace. Contrary to motivation, hygiene is heavily associated with dissatisfaction at work. Hygiene factors in a teaching context can include the structure and policies of an institution, work stability and security, workplace environment and conditions, management or administration standard, and nature of relationship with colleagues.

Apart from satisfaction and dissatisfaction, there are a range of other feelings or perspectives a teacher or an employee in general may have. For instance, since both the motivation and hygiene elements can be considered as separate and independent, it is possible for teachers to feel satisfied about one and dissatisfied about the other; indicating an added state of being known commonly as "neutral" (see Fig. 1 below).

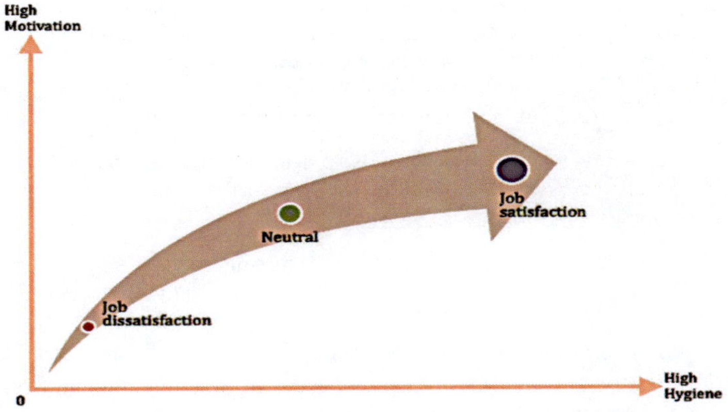

Fig. 1. Representation of Herzberg's motivator hygiene theory.

In some cases, teachers could feel neither satisfied nor dissatisfied, or partially satisfied and partially dissatisfied at the same time. Human emotions in general are rather complex, hence the feelings of teachers can be influenced by various causes, both external and internal such as one's own expectations and opinions.

Several researchers such as Collins (2002); Maher (2002) and Ewen (1964) have criticized Herzberg's work for a number of reasons such as the lack of a rigorous research methodology. Despite the theory's significant contribution to the understanding of human satisfaction and dissatisfaction, it received very limited empirical backing, with some scholars agreeing (Schmidt 1976; Manisera et al. 2005), while more seemed to disagree (Hill 1986; King 1970). Nevertheless, this theory remains one of the most discussed, debated and cited theories till date, the reason for which it has been included in this review article.

However, making one distinction can largely help in understanding teacher satisfaction. Herzberg's theory specifies that employees tend to be dissatisfied when the hygiene conditions are poor, whereas when they are high, employees either display dissatisfaction or a sense of neutrality. In this case, employees do not necessarily display satisfaction, since satisfaction is usually more heavily dependent on the motivation factors. Therefore, for teacher motivation to be achieved, it is essential for the necessary motivating elements be available and fulfilled.

3.2 Maslow's Hierarchy of Needs Theory

While Maslow's hierarchy of needs (Maslow 1995) is more commonly recognized in the area of human motivation, it is one of the key theoretical constructs that explain the factors that influence job satisfaction. Maslow's theory can be illustrated as a five-level hierarchy of human necessities, as shown in Fig. 2 below.

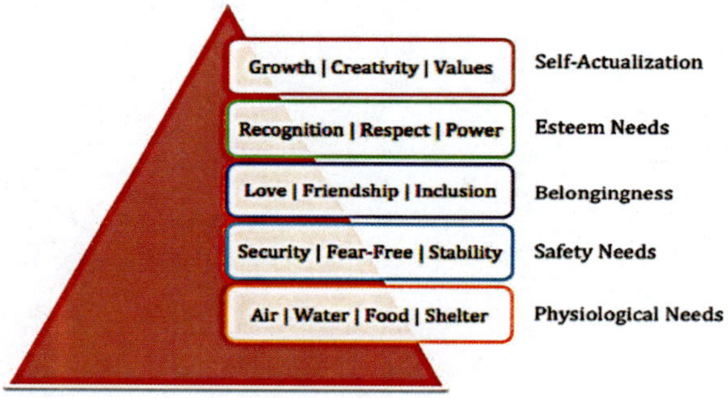

Fig. 2. Representation of Maslow's five-tier needs theory.

This theory not only displays the different and most significant needs of humans, but also underlines the chronology of needs in order of priority; starting from the most basic needs which are physiological, followed by safety, belonging, esteem, and finally self-actualization. Although the hierarchy of needs explains the fundamental day-to-day requirements for human motivation, it can also be used to understand teachers' needs in work environments and thus explain job satisfaction. While food, water and shelter are basic physiological needs for all humans, from any employee's perspective, healthcare benefits and financial compensation are primordial to fulfil those needs. The second important need in the hierarchy is safety (see Fig. 2 above), which includes fearlessness, stability and security. In the workplace, safety can mean a number of things to employees: a work environment free from physical or mental threats, job security and stability, and favorable policies and organizational structure, which all contribute to an employee's sense of belonging. Belonging is the third most important human need and in the context of teachers, which can entail the feeling of being accepted within their organization, team or amongst colleagues. It can also involve the sentiment of being in a cordial relationship with colleagues and supervisors. Following the sense of belonging comes the need for esteem. All humans expect and hold factors such as respect, recognition and approval in high regard in their daily lives. For a teacher, esteem comes from being valued in their workplace, either through appreciation or reward. The last need on Maslow's hierarchy is self-actualization, which is a stage where humans realize their full potential and work on developing it in order to fully utilize their capabilities to achieve their life goals. In the case of a teacher, self-actualization is a key stage where one develops and uses one's full potential to achieve his or her career goals.

Each stage on Maslow's hierarchy can either be understood as a separate set of needs or as chronological and interrelated stages leading to the final stage of self-actualization. Educational institutions seeking the improvement of teacher satisfaction and output should ideally focus on fulfilling these basic human necessities in the workplace. Some scholars criticize this approach for three critical reasons: first and foremost, there seems to be very limited empirical data that supports its success or effects; secondly, this method does not seem to take into consideration employees' cognitive process (Spector 1997); and lastly, the concept of self-actualization is subjective, immeasurable and difficult to be determined in a person (Maher 2002).

3.3 Dispositional Approach

The dispositional approach offers a different explanation for job satisfaction, wherein it describes the phenomenon as being closely associated with an individual's character. This approach, which is also known as Trait Theory, proposes two things: firstly, each person's work satisfaction tends to differ and be fixed at a personal pre-established level; secondly, this level does not fluctuate or change drastically over time (Durham et al. 1997). Figure 3 here below presents a visual representation to help understand this theory.

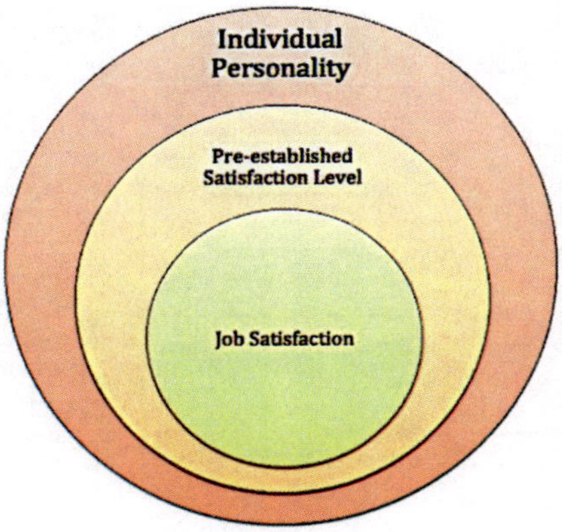

Fig. 3. Representation of the dispositional approach to job satisfaction.

In the literature, the most significant works examining this area in depth are of two kinds: the first that study human character with a direct focus and the second that study other aspects of human behavior and employment, from which human and job satisfaction inferences can be made (Klinger and Judge 2007; Lanen and Judge 2001). The latter includes studies such as those that measure work performance, teacher attitudes, and changes in profession or workplace, amongst others, which are all factors that could explain job satisfaction.

3.4 Job Characteristics Model

Unlike Herzberg and Maslow's theories, the Job Characteristics Model has greater empirical support. Three key review studies from early literature provide immense support to this model, with all three emphasizing the model's influence and effect on job satisfaction (Loher et al. 1985; Glick and Roberts 1981; Ferris and Fried 1987). A significant social psychology study published years later in 2000 (Behson et al. 2000) further supported the Job Characteristics Model, wherein the researchers presented results of meta-analysis of over a dozen separate research works that investigated the function of critical psychological states. This critical research study revealed that within the Job Characteristics Model, these psychological states played an important, theoretical as well as a practical role.

According to the Job Characteristics Model (Oldham 1975), an employee achieves satisfaction in his or her job in a work setting that ideally provides five core job dimensions, such as autonomy, feedback, task identity, task significance and skill variety, which are key employee motivating factors as well as influencers of necessary employee psychological states.

Figure 4 below illustrates how the five core job dimensions contribute to the formation of an employee's critical psychological states including responsibility of work outcomes, work meaningfulness and the knowledge of results.

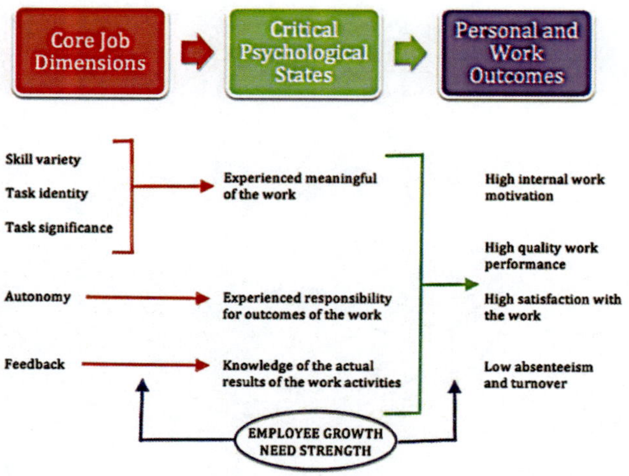

Fig. 4. Job characteristics model adapted from (Hackman and Oldham 1975, p. 161)

In consequence, the psychological states influence a number of outcomes, such as high quality work performance, low absenteeism and turnover, increased internal work motivation, and most importantly high job satisfaction levels. The Job Characteristics

Model therefore seems to imply that workplaces that focus on the improvement of core job dimensions can ensure employees' work satisfaction and create an ideal work environment.

Despite the multiple works in support of the Job Characteristics Model, there are few scholars that have criticized this model. Although it allows the measurement of the impact of core work dimensions on personal and work outcomes, critical psychological states are generally ignored or omitted during research studies (Behson 2000).

4 Discussion

This final section discusses the above theories in relation to and in the particular context of STEM school teachers.

4.1 Defining STEM Education

STEM education is a widely discussed topic across scholarly works. While the acronym STEM commonly refers to a set of fields or subjects relating to science, a clear and accepted definition is inexistent (Xie et al. 2015). In recent times, there is considerable debate around why only the four domains of science, technology, engineering and mathematics are provided exclusivity (Gonzalez and Kuenzi 2012). As a consequence, there have been arguments for the inclusion of other subjects such as arts by numerous scholars such as Wynn and Harris (2012) and Payton et al. (2017), making the definition of STEM education complex. There are also significant variations in STEM teaching across the primary, secondary and tertiary levels (Breiner et al. 2012). For instance, "at the foundational K-6 level, STEM education is synonymous with the math and science curriculum that is required for all students, so research on STEM education at the elementary school level focuses on participation and performance in science and math in general" (Xie et al. 2015, p. 332). In the middle and high school years, STEM subjects become more specialized in nature, with its complexity peaking in the university years (Killewald and Xie 2012; NGSS 2015). The lack of a standard and fixed definition, as well as the vast differences across the learning years could explain why teacher preparedness and delivery in schools are problematic. In accordance with the difference in curriculum and the level of difficulty of a subject, the required skills and qualifications of STEM teachers are non-linear and therefore highly variable.

4.2 STEM Teaching Quality

Several scholars argue that STEM education is tremendously affected by the shortage of quality STEM teachers (Polizzi Jaggernauth et al. n. d.; Asghar et al. 2012; Lesseig et al. n.d.) and attribute student performance and interest in STEM subjects to teaching quality (Tytler et al. 2011; McKenzie 2008; Akiba et al. 2007; Zhang and Campbell 2015). Despite being deeply rooted in a student-centric and constructivist philosophy (as theorized by Vygotsky and Piaget) (Thomas and Watters 2015; Appleton et al. 1999), some studies claim that there has been improvement in learning and teaching at schools over the years (Mitchell and Mitchell 2009). However, some others suggest that the last decades have been witnessing an increasing pressure placed upon teachers to drastically alter their teaching pedagogies and practices, adding new challenges to effective teaching delivery (Hackling et al. 2001; Hackling and Prain 2005). Wieman (2012, p.28) describes some of the key expectations that the STEM system demands of its teachers, in his research:

> *"The effective STEM teacher must understand expert thinking and design suitable practice tasks, (…) target student thinking and learning needs, (…) motivate the student to put in the extensive effort that is required for learning, (…) provide effective feedback that is timely and directly addresses the student's thinking, (…) understand how learning works and use that to guide all of their activities, (…) have deep mastery of the content (…) and pedagogical content knowledge."*

Although teaching quality is dependent on teacher preparedness and ability to design and deliver optimal STEM lessons (Teo and Osborne 2014; Wieman 2012), it demands teachers' time and effort in developing added skills, interdisciplinary capabilities and in specializing further in the subject they are teaching (Hudson et al. 2015; Asghar et al. 2012). The increase in reforms also gives rise to new constraints, such as the need for teachers to further qualify themselves to meet the new professional standards (Ching and Jie 2018; Rushton et al. 2014). Some scholars attribute these continuing challenges to teaching quality as a consequence of the failure of the existing STEM system. They argue that "discourses of teaching quality and teacher effectiveness mirror the straightforward linear conceptions and simplify tendencies of school education policy" (Skourdoumbis 2017, p. 206), ignoring the poor school structures and pedagogical practices in the class (Li and Rawolle 2015).

4.3 Job Satisfaction: Implications for STEM Teaching

Scholars reveal that job satisfaction comprises of multiple facets such as salary, opportunities for development and progress, the nature of the job, the nature and relationship with one's work supervisor and colleagues, and customers (Snipes et al. 2005). Two distinct sets of predictors of satisfaction are identified in Blegan's research (1993): organizational variables and personality traits or personal attribute variables. The former includes professionalism, communication, autonomy, commitment, stress, routinization, autonomy, fairness and peer communication; while the latter includes education, age, locus of control and years of experience (as cited in Guleryuz et al. 2008). Therefore, it can be understood that job satisfaction depends on various factors. For instance, job satisfaction is often greatly influenced by how satisfied an individual is about his or her life (Au and Ho 2008), while performance (Luthans 1992), quality of service (Schneider and Bowen 1985; Ferrell and Hartline 1996), and individual and related characteristics (Miller et al. 2009). Moreover, scholarly research reveals that job satisfaction plays a vital role in mediating between organizational commitment and emotional intelligence (Guleryuz et al. 2008).

It can be observed that job satisfaction is a widely researched topic, particularly by scholars involved in the fields of organizational and occupational psychology. Whilst being a popular topic of study, the accepted and reoccurring definitions offered by practitioners and researchers are of two kinds: as either the pleasure derived from an employee's appraisal and job progression or as the liking an employee has towards his or her profession. This illustrates the fact that existing definitions cover individuals' emotion or feeling towards their profession, which could include work environment conditions, salary, relationship with managers and colleagues, amongst other (Lu et al. 2005). Additionally, job satisfaction can also be determined by measuring how much work outcome expectations are met or surpassed (Luthens 1992). Nevertheless, scholars such as Alliger and Taber (1995) seem to suggest that job satisfaction should not be solely understood as enjoyability of work. Their study revealed that the correlation between enjoyment of individual work tasks and overall job satisfaction were fairly moderate, while there was a reasonable association with job satisfaction at the international scale. The research further presented evidence that contrary to what one may believe, factors such as supervision level, focus and concentration on work, and task prioritization have little to no effect on job satisfaction. This implies that factors other than enjoyment can influence an individual's job satisfaction.

To improve STEM teaching, scholars propose numerous solutions. For instance, a number of scholars who critique the inadequacies of STEM teachers (Polizzi et al. n. d.; Redman 2017) either propose stricter teacher recruitment criteria, competitive examinations, the toughening of teacher license requirements (Epstein and Miller n. d.), added certifications and rigorous high-level science training and pedagogical delivery (Stohlmann et al. 2012) or the development of a more robust professional partnership and collaboration system (Bissaker 2014). Some studies advocate for the necessity for university-based interdisciplinary professional programs, which could enhance teachers' knowledge and skills in interdisciplinary STEM teaching, while equipping them with context-specific problem-solving capabilities (Asghar et al. 2012; Fakayode et al. n. d.; Goodnough et al. 2014), reducing stress and difficulties in teaching subjects. Yet other studies emphasized on the need for more pedagogical tools (Lesseig et al. 2016) and introduction of new technologies and usage of devices such as iPads (Hughes et al. 2017) to improve and facilitate teachers' efforts in teaching STEM subjects in modern classrooms.

Scholars like Hudley and Mallinson (2017) argue that going through trainings and other programs to improve classroom delivery make limited difference in helping STEM teachers become more efficient, since most feel that they do not have enough time to assimilate the new content or practice the process of integration into lessons. Instead, those increase their workload, with teachers having to work an additional 4 to 5 h every evening (Tofel-Grehl and Callahan 2014). Therefore, increased expectations not only put teachers under immense pressure, but also jeopardize their dedication and effectiveness. Scholars like Redman (2017) thus argue that teachers need sufficient space and time to understand themselves and their responsibilities, followed by structured support that handholds them in practically carrying out STEM teaching.

Key scholarly works prescribe the monitoring and regular assessment of job satisfaction in managers, employees or teachers in order to identify any issues early on. Researchers such as Ho and Au (2008) emphasize that it does not suffice to solely evaluate how external factors affect teachers to determine teacher job satisfaction. Instead, as Locke (1976) postulates in his definition, the evaluation of teaching satisfaction would ideally also involve a cognitive and judgmental process (Demirtas 2010). Teacher dissatisfaction can have adverse effects on teaching quality and performance, because of which ensuring high quality education largely depends on ensuring teachers' job satisfaction (Demirtas 2010).

5 Conclusion

This paper examined the literature to decode the correlation between job satisfaction and STEM education delivery by school teachers. Four of the most prominent job satisfaction theories in the literature were selected for this purpose and discussed, alongside the various scholarly definitions, interpretations, and debates surrounding their efficiency. Due to the scarce availability of scholarly research on teacher satisfaction and STEM education delivery, the theoretical understanding gained from the review was further used to form an association with the job satisfaction of school teachers.

This review found that there is a clear lack of scholarly evidence on teacher satisfaction and its measurement. Nor is there an agreed singular solution to addressing the stagnant STEM teaching quality across schools. Instead, existing research works propose a vast range of solutions, leading the dilemma to persist. While job satisfaction theories remain well discussed in various work contexts, their application and effect on school teachers and STEM education in particular are hard to be found in the present literature. Future studies focusing on measuring teacher satisfaction could perhaps unveil potential barriers to STEM teaching quality.

References

Ajzen, I.: Nature and operation of attitudes. Ann. Rev. Psychol. **52**(1), 27–58 (2001)

Alliger, G.M., Taber, T.D.: A task-level assessment of job satisfaction. J. Organisational Behav. **16**(2), 101–121 (1995). https://doi.org/10.1002/job.4030160202

Al Salami, M.K., Makela, C.J., de Miranda, M.A.: Assessing changes in teachers' attitudes toward interdisciplinary STEM teaching. Int. J. Technol. Des. Educ. **27**(1), 63–88 (2015). https://doi.org/10.1007/s10798-015-9341-0

Anderson, R., Greene, M., Loewen, P.: Relationships among teachers' and students' thinking skills, sense of efficacy, and student achievement. Alberta J. Educ. Res. **34**(2), 148–165 (1988)

Appleton, K., Kindt, I.: Why teach primary science? Influences on beginning teachers' practices. Int. J. Sci. Educ. **21**(2), 155–168 (1999)

Bandura, A.: Self-efficacy: Toward a unifying theory of behavioral change. Psychol. Rev. **84**(2), 191–215 (1977)

Bandura, A.: Social Foundations of Thought and Action: A Social Cognitive Theory. Prentice-Hall Inc., Englewood Cliffs (1986)

Bandura, A.: Human agency in social cognitive theory. Am. Psychol. **44**(9), 1175–1184 (1989)

Bandura, A.: Self-Efficacy: The Exercise of Control. Freeman, New York (1997)

Behson, S.J., Eddy, E.R., Lorenzet, S.J.: The importance of the critical psychological states in the job characteristics model: a meta-analytic and structural equations modeling examination. Curr. Res. Soc. Psychol. **5**(9), 170 (2000)

Bissaker, K.: Transforming STEM education in an innovative Australian school: the role of teachers' and academics' professional partnerships. Theory Pract. **53**(1), 55–63 (2014). https://doi.org/10.1080/00405841.2014.862124

Borg, S.: The impact of in-service teacher education on language teachers' beliefs. System **39**(3), 370–380 (2011)

Bragow, D., Gragow, K.A., Smith, E.: Back to the future: toward curriculum integration. Middle Sch. J. **27**, 39–46 (1995)

Capps, D.K., Crawford, B.A.: Inquiry-based instruction and teaching about nature of science: are they happening? J. Sci. Teacher Educ. **24**(3), 497–526 (2013)

Christensen, R., Knezek, G., Tyler-Wood, T.: Alignment of hands-on STEM engagement activities with positive STEM dispositions in secondary school students. J. Sci. Educ. Technol. **24**(6), 898–909 (2015)

Collins, C.: Job satisfaction of CNMs: luxury or necessity? J. Nurse Midwifery **35**(4), 237 (2002)

Collinson, V.: Sources of teachers' values and attitudes. Teach. Dev. **16**(3), 321–344 (2012)

Czerniak, C.M., Weber, W.B., Sandmann, A., Ahern, J.: A literature review of science and mathematics integration. Sch. Sci. Math. **99**(8), 421–430 (1999)

DeChenne, S.E., Koziol, N., Needham, M., Enochs, L.: Modeling sources of teaching self-efficacy for science, technology, engineering, and mathematics graduate teaching assistants. CBE Life Sci. Educ. **14**(3), 1–14 (2015)

Demirtas, Z.: Teachers' job satisfaction levels. Procedia Soc. Behav. Sci. **9**(2010), 1069–1073 (2010)

Diggs, V.: Ask–think–create: the process of inquiry. Knowl. Quest **37**, 30–33 (2009)

Donaghue, H.: An instrument to elicit teachers' beliefs and assumptions. ELT J. **57**(4), 344–351 (2003)

Drake, S.M., Burns, R.C.: Meeting Standards through Integrated Curriculum. ASCD, Alexandria (2004)

Durham, C.C., Judge, T.A., Locke, E.A.: The dispositional causes of job satisfaction: a core evaluations approach. Res. Organisational Behav. **19**, 151 (1997)

Eagly, A.H., Chaiken, S.: The Psychology of Attitudes. Harcourt Brace Jovanovich College Publishers, Fort Worth (1993)

Eccles, J.S.: Understanding women's educational and occupational choices. Psychol. Women Q. **18**(4), 585–609 (1994)

Engeln, K., Euler, M., Maass, K.: Inquiry-based learning in mathematics and science: a comparative baseline study of teachers' beliefs and practices across 12 European countries. ZDM **45**(6), 823–836 (2013)

Ewen, R.B.: Some determinants of job satisfaction: a study of the generalisability of Herzberg's theory. J. Appl. Psychol. **48**, 161 (1964)

Finsterwald, M., Wagner, P., Schober, B., Luftenegger, M., Spiel, C.: Fostering lifelong learning: evaluation of a teacher education program for professional teachers. Teach. Teach. Educ. **29**, 144–155 (2013)

Fox, M.A., Hackerman, N. (eds.): Evaluating and Improving Undergraduate Teaching in Science, Technology, Engineering, and Mathematics. National Academies Press, Washington, DC (2002)

Fried, Y., Ferris, G.R.: The validity of the job characteristics model: a review and meta-analysis. Pers. Psychol. **40**, 287 (1987)

Ginns, I.S., Tulip, D.F., Watters, J.J., Lucas, K.B.: Changes in preservice elementary teachers' sense of efficacy in teaching science. Sch. Sci. Math. **95**(8), 394–400 (1995)

Gottfried, M.A.: The influence of applied STEM course taking on advanced mathematics and science course taking. J. Educ. Res. **108**(5), 382–399 (2015)

Gregoire, M.: Is it a challenge or a threat? A dual-process model of teachers' cognition and appraisal processes during conceptual change. Educ. Psychol. Rev. **15**(2), 147–179 (2003)

Guleryuz, G., Guney, S., Aydin, E.M., Asan, O.: The mediating effect of job satisfaction between emotional intelligence and organisational commitment of nurses: a questionnaire survey. Int. J. Nurs. Stud. **45**(11), 1625–1635 (2008). https://doi.org/10.1016/j.ijnurstu.2008.02.004

Hackett, G., Betz, N.E.: A self-efficacy approach to the career development of women. J. Vocat. Behav. **18**(3), 326–339 (1981)
Hackman, J.R., Oldham, G.R.: Development of the job diagnostic survey. J. Appl. Psychol. **60**, 161 (1975)
Hall, C., Dickerson, J., Batts, D., Kauffmann, P., Bosse, M.: Are we missing opportunities to encourage interest in STEM fields? J. Technol. Educ. **23**(1), 32–46 (2011)
Herzberg, F.: Work and the Nature of Man. World Publishing Company, Cleveland (1966)
Hill, M.D.: A theoretical analysis of faculty job satisfaction/dissatisfaction. Educ. Res. Q. **10**, 36 (1986)
Hoover-Dempsey, K.V., Bassler, O.C., Brissie, J.S.: Parent involvement: contribution of teacher efficacy, school socioeconomic status, and other school characteristics. Am. Educ. Res. J. **24**(3), 417–435 (1987)
Hudson, P., English, L., Dawes, L., King, D., Baker, S.: Exploring links between pedagogical knowledge practices and student outcomes in STEM education for primary schools. Aust. J. Teach. Educ. **40**(6), 134–151 (2015)
Jonassen, D.H.: Designing constructivist learning environments. In: Reigeluth, C.M. (ed.) Instructional Design Theories and Models: A New Paradigm of Instructional Theory. Lawrence Erlbaum Associates Publishers, Mahwah (1999)
Kafai, Y.B., Resnick, M.: Constructionism in Practice: Designing, Thinking, and Learning in a Digital World. Routledge, Mahwah (1996)
Kagan, D.M.: Implication of research on teacher belief. Educ. Psychol. **27**(1), 65–90 (1992)
Kennedy, S.Y., Smith, J.B.: The relationship between school collective reflective practice and teacher physiological efficacy sources. Teach. Teach. Educ. **29**, 132–143 (2013)
Kent, S.I.: Problems of beginning teachers: comparing graduates of bachelor's and master's level teacher preparation programs. Teach. Educ. Q. **35**(4), 83–96 (2000)
King, N.: Clarification and evaluation of the two factor theory of job satisfaction. Psychol. Bull. **74**, 18 (1970)
Klassen, R.M., Chiu, M.M.: Effects on teachers' self-efficacy and job satisfaction: teacher gender, years of experience, and job stress. J. Educ. Psychol. **102**(3), 741–756 (2010)
Klinger, R., Judge, T.A.: Job satisfaction: subjective well-being at work. In: Eid, M., Larsen, R. (eds.) The Science of Subjective Wellbeing, pp. 393–413. Guildford Publication, New York (2007)
Lanen, R.J., Judge, T.A.: Dispositional affect and job satisfaction: a review and theoretical extension. Organ. Behav. Hum. Decis. Process. **8**(6), 67 (2001)
Lee, V., Dedrick, R., Smith, J.: The effect of social organization of schools on teachers' efficacy and satisfaction. Sociol. Educ. **64**(3), 190–208 (1991)
Levin, B., He, Y.: Investigating the content and sources of teacher candidates' personal practical theories (PPTs). J. Teach. Educ. **59**(1), 55–68 (2008)
Loher, B.T., Noe, R.A., Moeller, N.L., Fitzgerald, M.P.: A meta-analysis of the relation of job characteristics to job satisfaction. J. Appl. Psychol. **70**, 280 (1985)
Lortie, D.C., Clement, D.: Schoolteacher: A Sociological Study. University of Chicago Press, Chicago (1975)
Lumpe, A.T., Haney, J.J., Czerniak, C.M.: Assessing teachers' beliefs about their science teaching context. J. Res. Sci. Teach. **37**(3), 275–292 (2000)
Luthans, F.: Organisational Behaviour, 6th edn. McGraw-Hill, New York (1992)
Manisera, M., Dusseldorp, E., Van der Kooij, A.J.: Component structure of job satisfaction based on Herzberg's theory. Rapporti di Ricerca del Dipartimento di Metodi Quantitativi. Brescia University: Working Paper, vol. 253 (2005). http://www.datatheory.nl/pages/fullmanuscript_final_epm.pdf
Mansour, N.: Science teachers' beliefs and practices: issues, implications and research agenda. Int. J. Environ. Sci. Educ. **4**(1), 25–48 (2009)

Maslow, A.: Motivation and Personality. Harper, New York (1995)
Maher, E.: Overcoming Controllable and Uncontrollable Work Difficulties: Change Environment or Self? [PhD Thesis]. Deakin University, Victoria (2002). http://dro.deakin.edu.au/view/DU:30026637
McComas, W.F.: STS education and the affective domain. In: Yager, R.E. (ed.) What Research says to the Science Teacher. The Science, Technology, and Society Movement. National Science Teachers Association, Washington, pp. 161–168 (1993).
McComas, W.R., Wang, H.A.: Blended science: the rewards and challenges of integrating the science disciplines for instruction. Sch. Sci. Math. **98**(6), 340–348 (1998)
Mellati, M., Khademi, M.: The relationships among sources of teacher pedagogical beliefs, teaching experiences, and student outcomes. Int. J. Appl. Linguist. Engl. Lit. **4**(2), 177–184 (2015)
Nespor, J.: The role of beliefs in the practice of teaching. J. Curric. Stud. **19**(4), 317–328 (1987)
Pinnell, M., Rowly, J., Preiss, S., Franco, S., Blust, R., Beach, R.: Bridging the gap between engineering design and PK-12 curriculum development through use of the STEM education quality framework. J. STEM Educ. Innov. Res. **14**(4), 28–35 (2013)
Pinto, R.: Introducing curriculum innovations in science: identifying teachers' transformations and the design of related teacher education. Sci. Educ. **89**(1), 1–12 (2005)
Polat, N.: A comparative analysis of pre-and in-service teacher beliefs about readiness and self-competency: revisiting teacher education for ELLs. System **38**(2), 228–244 (2010)
Powell-Moman, A.D., Brown-Schild, V.B.: The influence of a two-year professional development institute on teacher self-efficacy and use of inquiry-based instruction. Sci. Educ. **20**(2), 47–53 (2011)
Pratkanis, A.R., Breckler, S.J., Greenwald, A.G.: Attitude Structure and Function. Erlbaum, Hillsdale (2014)
Ramey-Gassert, L., Shroyer, M.G., Staver, J.R.: A qualitative study of factors influencing science teaching self-efficacy of elementary level teachers. Sci. Educ. **80**(3), 283–315 (1996)
Raudenbush, S.W., Rowan, B., Cheong, Y.F.: Contextual effects on the self-perceived efficacy of high school teachers. Sociol. Educ. **65**(2), 150–167 (1992)
Redman, C.: Would increasing engineering literacies enable untapped opportunities for STEM education? Theory Pract. **56**(4), 318–326 (2017). https://doi.org/10.1080/00405841.2017.1350493
Rieber, L.P.: A pragmatic view of instructional technology. In: Tobin, K. (ed.) The Practice of Constructivism in Science Education, pp. 193–212. AAAS Press, Washington (1993)
Roehrig, G.H., Kruse, R.A., Kern, A.: Teacher and school characteristics and their influence on curriculum implementation. J. Res. Sci. Teach. **44**(7), 883–907 (2007)
Roberts, K.H., Glick, W.: The job characteristics approach to task design: a critical review. J. Appl. Psychol. **66**, 193 (1981)
Ross, J., Bruce, C.: Professional development effects on teacher efficacy: results of randomized field trial. J. Educ. Res. **101**(1), 50–60 (2007)
Sade, D., Coll, R.: Technology and technology education: views of some Solomon Island primary teachers and curriculum development officers. Int. J. Sci. Math. Educ. **1**(1), 87–114 (2003)
Schmidt, G.L.: Job satisfaction among secondary school administrators. Educ. Adm. Q. **12**, 68 (1976)
Spector, P.E.: Job Satisfaction: Application, Assessment, Causes, and Consequences. Sage, London (1997)
Thibaut, L., Knipprath, H., Dehaene, W., Depaepe, F.: How school context and personal factors relate to teachers' attitudes toward teaching integrated STEM. Int. J. Technol. Des. Educ. **28**(3), 631–651 (2017). https://doi.org/10.1007/s10798-017-9416-1

Tschannen-Moran, M., Woolfolk Hoy, A.: The influence of resources and support on teachers' efficacy beliefs. In: Annual meeting of the American Educational Research Association, New Orleans, LA, April 2002

Turner, J.: Theoretical literature review: tracing the life cycle of a theory and its verified and falsified statements. Hum. Resour. Dev. Rev. **17**(1), 34–61 (2018). https://doi.org/10.1177/1534484317749680

Van Aalderen-Smeets, S.I., Walma van der Molen, J.H.: Improving primary teachers' attitudes toward science by attitude-focused professional development. J. Res. Sci. Teach. **52**(5), 710–734 (2015)

Van Aalderen-Smeets, S.I., Walma van der Molen, J., Asma, L.J.: Primary teachers' attitudes toward science: a new theoretical framework. Sci. Educ. **96**(1), 158–182 (2012)

Webb, D.L.: Engineering professional development: Elementary teachers' self-efficacy and sources of self-efficacy. (Doctoral dissertation). Portland State University, Portland (2015)

Yager, R.E. (ed.). What research says to the science teacher? The science, technology, and society movement. National Science Teachers Association, Washington, DC, pp. 161–168 (N.d.)

Zeichner, K.M., Tabachnick, B.R.: Are the effects of university teacher education. J. Teach. Educ. **32**(3), 7–11 (1981)

Zemelman, S., Daniels, H., Hyde, A.: Best practice in mathematics. In: Zemelman, S., Daniels, H., Hyde, A. (eds.) Best Practice: Today's Standards for Teaching and Learning in America's Schools, pp. 106–118. Heinemann, Portsmouth (2005)

Industry 4.0 Mapping Strategic Decision Making for Seaport Operations Management

Scott Caldwell(✉) and Rob Darlington

Liverpool John Moores University, Liverpool, UK
s.a.caldwell@2017.ljmu.ac.uk

Abstract. Seaports are regarded as vital nodes within maritime supply chain operations that accounted for 10.7 Billion tons in 2017 and represent 80% of globalised trade by volume when compared to other modes of transportation. By their operating nature, seaports and container terminals are characterised by fragmentation, complexity, and uncertainty making them ideal candidates for the potential competitive advantages offered by Industry 4.0. This research proposes a framework for mapping existing operations and the potential opportunities for the implementation of Industry 4.0 technologies to underpin a Decision Support Model (DSM) for strategic investment as part of a Port Community System (PCS). A range of data mapping techniques have been surveyed and the authors propose an observational modelling approach adapted from an existing visualisation technique (Value Stream Mapping) as the route to improved understanding of the system under study. This approach values practitioner-led, direct measurement and data collection with a process of considered planning for potential future states focused upon digital technologies and automation. The paper integrates academic and practitioner literature, facilitating the development of a new mapping technique integrated as part of a DSM framework.

Keywords: Supply chain management (SCM) · Big data analytics (BDA) · Industry 4.0 · Port community systems (PCS) · Smart seaports · Value stream mapping (VSM)

1 Introduction and Motivation

The exploitation of information in seaport operations is regarded as a fundamental strategic asset in seaport management [1]. Information is data that leads to less uncertainty increasing understanding of a process. It is timely, accurate, and presented in a relevant and meaningful format. Information is fundamentally important since it influences actions, decisions, and operating strategies of an organisation [2]. It is also suggested that information systems are indispensable to the overall competitive advantage of a seaport, facilitating the formation of informed decision-making that enhances visibility, efficiency, reliability, and security of both the value and supply chain [1]. Fundamentally, seaports need to develop a knowledge management culture that provides value-added services to their customers, in the form of traceability, supply chain visibility,

A. Batako et al. (Eds.): GCMM 2021, LNNS 335, pp. 743–756, 2022.
https://doi.org/10.1007/978-3-030-90532-3_56

and connectivity. However, seaports are not optimised for the implementation of new digital technology. Many seaports still operate traditional forms of EDI that were first introduced in the 1960s, resulting in scalability and interoperability issues between heterogeneous systems [2]. It is inferred that the maritime sector is lagging behind in the implementation and exploitation of Industry 4.0 technology when compared to the manufacturing, automotive, and financial sectors [3]. The maritime supply chain generates a large volume of data sets, although the majority remain un-utilised in the form of strategic operating decisions [4].

This paper presents a fundamental redesign of the VSM technique that is applied within the paradigm of Lean Management (LM). This approach will utilise the concepts of a current and future state visual representation that enables data capture from embedded Industry 4.0 ready sensors, facilitating a culture of continuous improvement. The mapping process will utilise the successfully applied techniques found within the robust and simplistic methodology of VSM, focusing on the flow, capture, exploitation of information generated within the value stream.

It is suggested that a PCS is well suited to fulfil the function of a DSM to understand and exploit the application of Industry 4.0 ready technological platforms in a maritime value and supply chain. A PCS is a neutral, open, and electronic platform that securely integrates the diverse actors who operate within the confines of the supply chain [5]. Recent literature has concentrated on discussing the importance of actual and high-quality information integration through the functionalities of dedicated systems and DSM [1]. A new mapping technique as part of an integrated DSM framework will contribute to the existing knowledge and offer new insight on PCS and DSM frameworks as enablers of Industry 4.0.

The methodology of LM is based on the continued development of the Toyota Production System (TPS) introduced in the 1950s within the automotive value chain [1, 6]. LM is inferred as the most prominent manufacturing paradigm of recent times, gaining significant popularity in both academic and practitioner-led research [6]. Both Industry 4.0 and LM facilitate the concept of decentralised control, with the overall aim of enhancing productivity and flexibility to leverage a market competitive advantage. However, there is limited research that investigates the close relationship between these two domains [6]. VSM may be defined in generic terms as a simultaneous visual management tool that records, analyses, and attempts to understand the sequence of a single product family, including the flow of information and materials throughout a value stream in the context of graphical representation [6, 7]. VSM was initially devised in 1995 and it was popularised by the research of Rother and Shook in their publication 'Learning to See' [8]. VSM is a robust and proven method that enables the systematic mapping and analysis of processes and supply chains, facilitating a drive for continuous improvement by identifying waste [7].

Industry 4.0 has gained significant research focus from both academia and the manufacturing sector. The concept of Industry 4.0 was first announced at the Hannover Messe in 2011 and it was proclaimed as the fourth industrial revolution [6]. The advent of the digital age in the form of Industry 4.0 has revolutionised the manufacturing sector and the maritime sectors are considered as lagging behind in the exploitation of Industry 4.0 to deliver a market competitive advantage. Industry 4.0 refers to the digital manufacturing systems provided by the successful integration of production processes, information technologies, and techniques [6]. Digitalisation and optimisation of logistical functions are achieved by the utilisation of emerging Industry 4.0 enablers, such as the Internet of Things (IoT), Cloud Computing (CC), BDA, Artificial Intelligence (AI), and Cyber-Physical Systems (CPS) [9].

2 Adaptation of VSM

The literature on the application of VSM is rather extensive and covers a variety of diverse sectors, such as architecture, automotive, construction, healthcare, manufacturing, mining, product development, seaport management, and tourism. This underpins the unique flexibility of the VSM technique and the adaptability of its methodology to integrate emerging Industry 4.0 technology.

Traditional VSM involved mapping the current state and the future state of a product or a family of products. It is essentially a paper exercise that involves the participation of the entire workforce. The re-engineering of the VSM framework will enable seaport management to visualise the flow of information in their value and supply chain. VSM is regarded in the literature as one of the most powerful and versatile LM techniques [7].

3 The Emergence of Industry 4.0

Nurasimham and Parthasarathy devised a new approach that integrated the concept of simulation to VSM (simulation-aided/ sa VSM) to increase the flow of data analysis and to facilitate the process of continuous mapping [12]. McDonald et al. developed simulation models of manufacturing systems to analyse and visualise the dynamic states of both unit processes and the entire process, enabling a holistic overview of the manufacturing sector [13]. The need for research into the practical application of Industry 4.0 is stressed by the literature, relating to the effects of new technologies on existing operations management [4].

Considerable research has also focused on addressing the issues associated with the application of VSM within a dynamic and complex operating environment. Roh, Kunz, and Wegener advocated a structured approach to analyse and improve the functions of information streams, mitigating the issues of inaccurate, incomplete, and delayed information flows. This method enables the characteristics of the information stream to be analysed within complex and sophisticated manufacturing processes [14]. The performance of the information streams is assessed in relation to five indicators who map the current state. A successful validation revealed an increase in overall productivity, based on a reduction of lead-time and operating costs [14]. The static collection of data within the method of VSM is often perceived as a significant drawback that hinders its implementation in dynamic and complex environments. It is argued by Buer, Standhagen, and Chan that the utilisation of Industry 4.0 will enhance VSM by the function of real-time data collection, enabling informed strategic decision making in relation to unexpected fluctuations in the operation and market conditions [7]. The principle of real-time VSM is also addressed by Chen and Chen. They advocate wireless monitoring via the functionality of Radio Frequency Identification (RFID) to automatically produce a real-time VSM that is represented by computer added programming [15].

Kemble, Gunasekaran, and Dhome investigated the direct effects of Industry 4.0 on the concepts of Sustainable Organisational Performance (SOP) and Leam Manufacturing Practices (LMP) [16]. They suggest that the accuracy of VSM is enhanced by the utilisation of Industry 4.0 in the collection of real-time information, in the initial stage of LMP [16]. However, their sample population was limited to 205 Indian manufacturing practitioners [16]. A comprehensive review of information logistics waste (redundant systems, memory media, or media breaks) is provided by Meudt, Metternich, and Abele, based on the functionalities of recording, handling, processing, analysing, and optimising information processes (VSM 4.0) [10]. It is also suggested that this approach will extend the current scope of practitioner understanding with regards to the market opportunities offered by digitalisation, whilst simultaneously enhancing the application of lean production methodology (VSM) [4]. The VSM 4.0 Model was utilised within a case study method at well-established organisations, consisting of mechanical engineering, electronic, transportation, and manufacturing process operations, enabling a robust test of its validity and reliability [10]. The results obtained from the case studies suggest that VSM 4.0 is well suited to generate a current state visualisation of both material and information flow. This concept is also applicable for administrative processes, extending its current scope within the confines of the manufacturing sector [10].

4 Seaport Management

Seaports by their very nature are conservative in their operating strategy, resistant to change, and operate in a very dynamic, complex, fragmented, and uncertain environment [14–17]. Many diverse actors are involved in the operation of the maritime supply chain, such as freight forwarders, terminal operators, harbour pilots, shipping companies, and shipping agents [13]. Seaport actors differ in size, scale, type of operation, and experience in complex inter-organisational information systems (PCS) [18].

Seaports are regarded as vital nodes in the maritime supply chain that facilitate international trade and globalisation, providing significant economic growth (GDP) both regionally and nationally [18]. In 2017, international seaborne trade reached an unprecedented figure of 10.7 billion tons [3]. There is a general consensus in the literature that seaports represent an exciting opportunity for further research as they are lacking in the scope of the empirical and theoretical study [13, 20–22]. Digitisation of seaports is still in its initial stages [13].

The application of VSM techniques within seaports and container terminals was first advocated by Marlow and Paixão – Casaca due to its capacity to identify waste in the value stream [21]. Lloyd et al. utilised a VSM technique to measure and improve the throughput rate at the McDuffie Coal Terminal located within the Port of Mobile (Alabama, USA). The project concluded that LM can be successfully applied to seaports in order to elevate throughput limitations, based on infrastructure, geographical locations, and cargo diversity [23]. VSM techniques were also applied by Franzén and Streling in seaport container terminal operations to identify potential inefficiencies [24].

5 Methodological Approach

A literature review was primarily conducted to determine the current state of VSM re-engineering in relation to the application of Industry 4.0. This review also identified gaps in the literature involving the utilisation of observational mapping techniques in determining strategic investment in the existing PCS. The mapping process will be underpinned by an Action Research methodology (Table 1) that will involve practitioner participation in site selection, data collection (semi-structured interviews, questionnaires, and case studies), process observation, direct measurement, and process mapping (Fig. 3) [24]. The complex operation of seaports is also reflected by the increase in multi-disciplinary techniques in both data collection and knowledge creation [25]. Kurt Lewin (1890–1947) was a Social Psychologist with a passion for investigating the dynamics of human group interaction and he is widely perceived as being the founding father of Action Research methodology [24]. Action Research is dependent on close collaboration between the researcher and the practitioner, with a shared objective for implementing a measure of organisational change and the generation of new knowledge that is both significant and positive [24].

The justification for Action Research methodology is regarded as a *"change agent"*, based on an organisation with a problem to solve [26]. Action Research enhances the relevance of academic enquiry by linking research into operational practice within the real world of the practitioner (Fig. 1). Essentially, a win-win situation increases the likelihood of successful completion of the research project [27]. Therefore, addressing a clearly defined gap in the literature that relates to the empirical study of the practitioner's processes. Action Research is ideally suited to the diverse requirements of multidisciplinary approaches and is associated with the philosophical paradigm of Positivism [24]. A pilot study was recently been conducted at a local SME that was based on the application of a traditional VSM process.

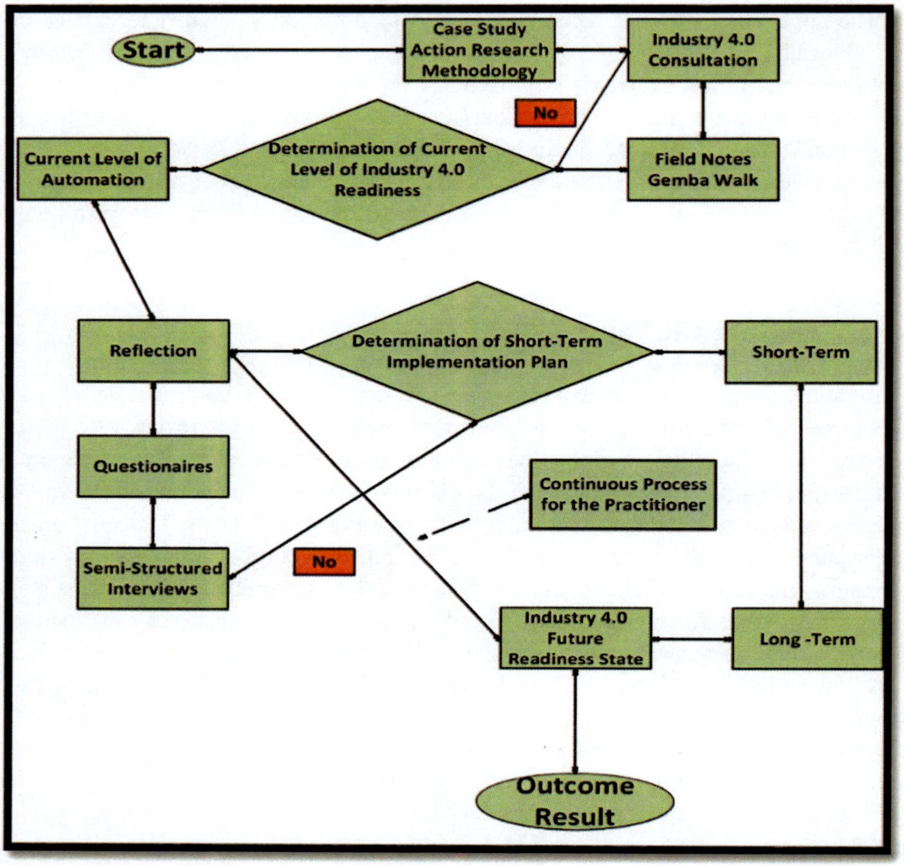

Fig. 1. Research methodology

Table 1. Proposed research methodology

Methodological overview	
Philosophy	Positivism
Research Approach	Inductive – Deductive
Research Strategy	Multiple Case Studies
Research Design	Action Research
Length of Study	Longitudinal
Data Collection Methods	Semi-Structured Interviews (Qualitative), Questionnaires, Case Studies, Industry 4.0 Mapping Technique
Data Analysis	Qualitative Data Analysis

6 Industry 4.0 Mapping

As with all form of VSM, it is first necessary to visit the seaport to observe the process and gain insight and knowledge from the workforce that operate the process. This is also an opportunity to offer justification to the workforce and allay any concerns relating to the mapping and research. Rother and Shook [29] infer that this process is similar to walking along the physical pathways of material and information flow [29]. This process is identified as Gemba walking and fulfils the principle of "go and see", walking the process with concerned people to share knowledge. Gemba walking is a powerful way for an organisation to openly support a culture of continuous improvement and process standardisation [30].

The literature stresses the need for the workforce to take ownership of the LM project, increasing the likelihood of successful completion for the research and the practitioner. This collaborative approach between researcher and practitioner underpins the applied methodology of Action Research. The current state will be identified as the current level of Automation. Industry 4.0 Mapping will interact with the information flow as it transits from processes, people, materials, and CPS. The objective of the Industry 4.0 Map is to improve the performance level of the seaport by mapping: The Current Level of Automation, Information Flows, Integration in the Decision Hierarchy, and Industry 4.0 Readiness State.

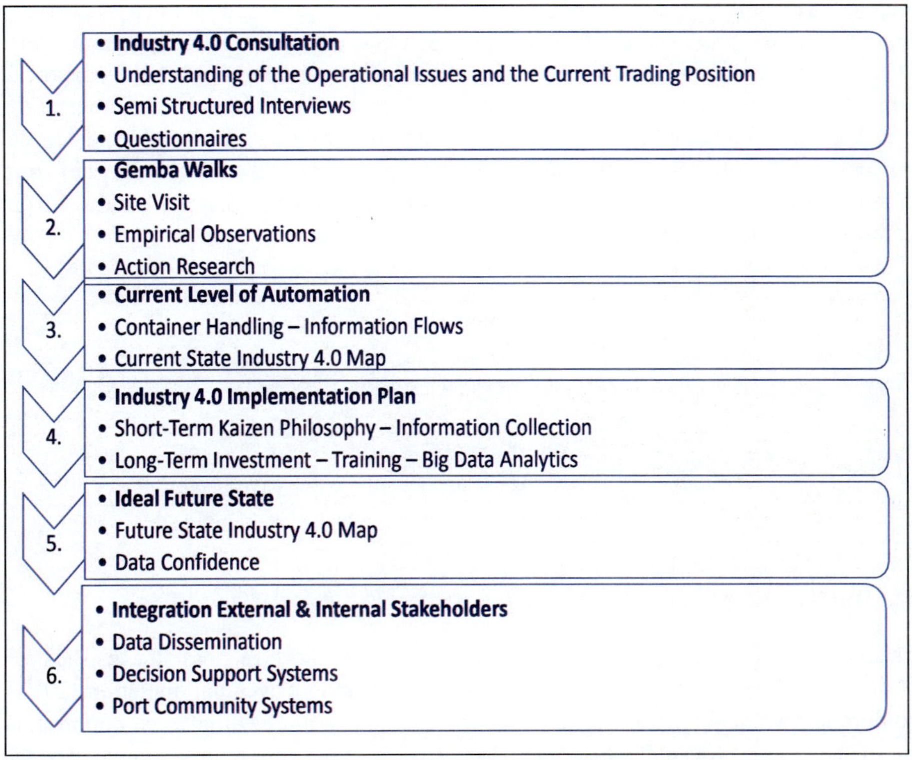

Fig. 2. Overview of the processes of industry 4.0 mapping

7 Data Collection

7.1 Current Level of Automation

The current level of automation (LoA) is the initial stage in the Industry 4.0 Mapping technique. There are many definitions utilised to offer any explanation of the concept of LOA. Most noticeably by Amber and Amber who conclude that LoA relates to the "*extent of which* human *energy and control over the production process are replaced by machines*" [31]. Therefore, the current LoA map needs to accurately record the level of manual integration in the flow of information from the sensors to the interface (PCS) (Table 2).

Table 2. The development of automation

Year of application	Generated data sets per operating days	Total data per year	Total accumulated data
2019	10 GB (average day)	365 days × 10 GB	3,650 GB

The level of LoA concerns the utilisation of intelligent Industry 4.0 sensors that are embedded in the digitalised value chain. To ascertain the rate of automation it is necessary to document the previous implementation of sensors, within the following parameters: year, format, function, location, and cost. This will also determine if the seaport is currently operating at an automation level that is comparable to the functionalities of the fourth industrial revolution. The digitalisation and integration of horizontal and vertical processes. Data utilised in the Industry 4.0 Mapping tool will also be collected from on-site monitoring of the following generic variables (Table 1) that are obtained from archival and current data collection (Table 3).

Table 3. The current state of automation

Year	Sensors	Total sensors	Dowload rate	Cost	Maual int	KPI	Op time
2019	5 × Terminal	5	500 MB Per Hour	£40	PCS	Sea Port	24 h

7.2 Information Flow

The main obstacle to the principle of full automation in seaports relates to the justification of costs in both its research and development stages and its eventual operation (29). A comprehensive mapping of the operation will facilitate a more enlightened opportunity for management to evaluate the potential improvement in performance compared to the issues surrounding cost. The permutations of the current LoA map in relation to the future state map may be analysed via the process of simulation, transforming the static VSM into a dynamic version that records real-time information flow (Fig. 2) [11]. Product customisation demands and value-added services (flexibility) in addition to increased complexity (efficiency), may not always result in improved performance [29].

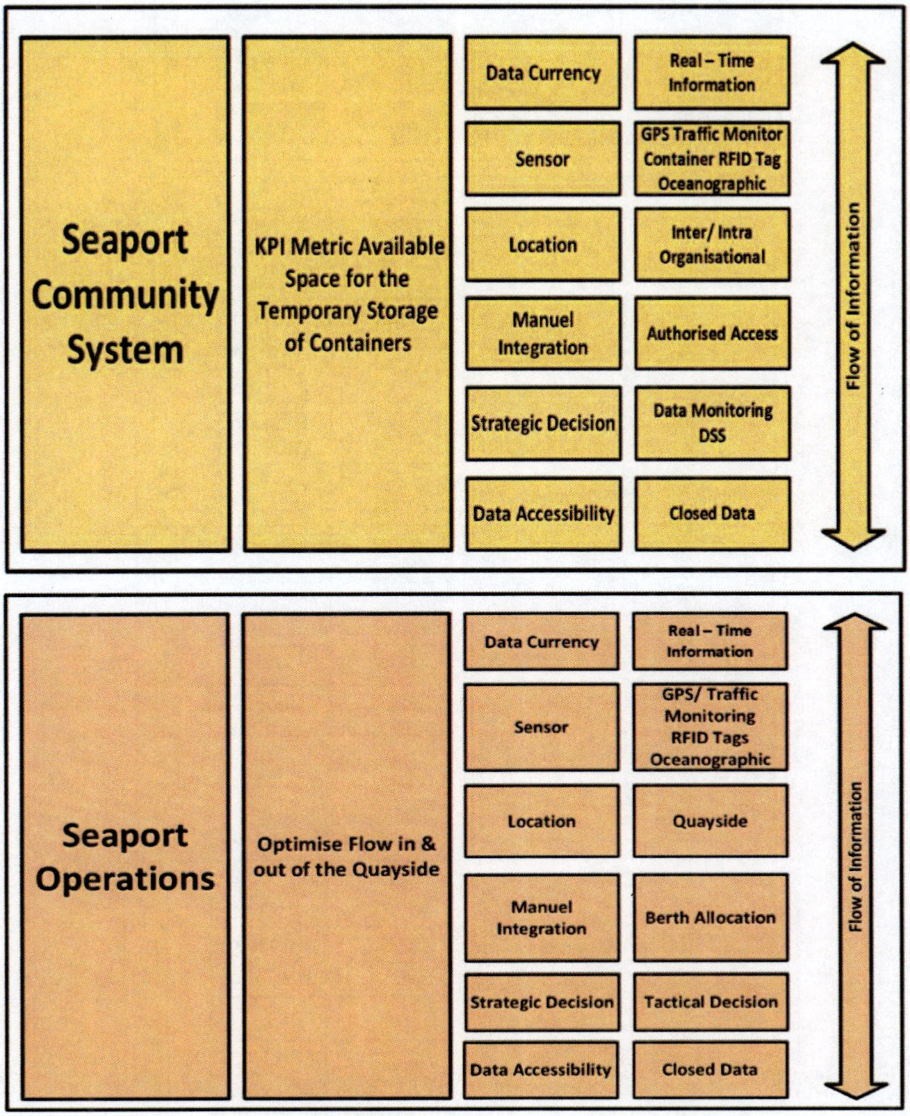

Fig. 3. Information flow of the container terminal handling

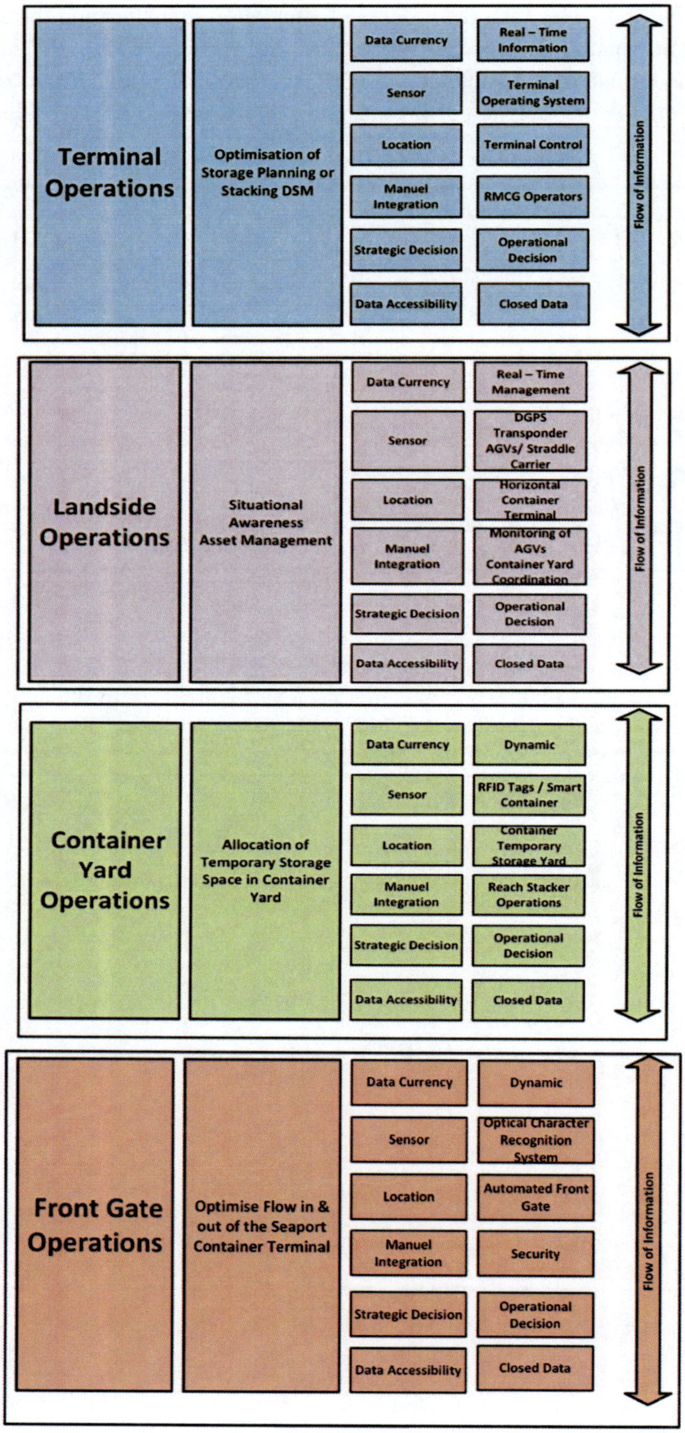

Fig. 4. Information flow of the container terminal handling

Industry 4.0 will be integrated into VSM as a real data collection tool to facilitate insight between the physical throughput of cargo and the generation of KPIs, utilised in determining the performance of the operation. A digitalised map of the seaport will be divided into squares to establish the accurate location of the sensors and it will also record the flow of both open and closed information utilised into the generation of KPI measurements. The flow of information may also be recorded in a graphical format. The mapping of the flow of information within the seaport's value chain will enable the determination of the optimal location of the interoperable Industry 4.0 sensors to generate quality data sets, providing enhanced data confidence. This will enable a continuous monitoring system to display the performance consistency of the seaport.

7.3 Decision Hierarchy

The flow of information will also be mapped in relation to the decision hierarchy of the seaport. This offers a unique opportunity to determine the data confidence of the information created by the Industry 4.0 integrated, enhanced, and connected sensors within the management hierarchical structure at the seaport. To determine this concept, it is advantageous to map the frequency of information access, through the network established by the PCS. The sharing of operational seaport data (Fig. 2) is fundamental in improving the overall effectiveness of the decision-making strategy and it is regarded as a tangible asset in maintaining a competitive advantage [32].

7.4 Industry 4.0 Readiness Mapping

Data confidence will be mapped in the Industry 4.0 Readiness Mapping state in accordance with the following variables (Table 4):

Table 4. Variables and mapping constraints

Variable	Mapping constraint
Quality	Accuracy of Data Sets to Populate KPI metrics
System Integrity	Number of integrated, enhanced, and connected Industry 4.0 Sensors in Operation within the context of a Digitalised Value Chain Facilitating a Sustainable Level of Data Confidence
Completeness	Percentage of Data Sets Without Missing Fields
Currency	Timeliness of Data Sets to Address Market Fluctuations and Uncertainty
Lineage	Data Sets sourced from an Externally Validated/ PCS enabled actor
Security	Mapping of Data Access Points for Open and Closed data within the PCS Security breaches
Governance	ISPS IMO Regulations on Data Accessibility

The Industry 4.0 Readiness Mapping will assist the practitioners in assessing their current level of automation in anticipation that Industry 4.0 will deliver a competitive advantage in the form of product customisation, market flexibility, and increased efficiency [19]. A comprehensive mapping technique would also populate a systematic implementation plan for the continuous development of ICT infrastructure in conjunction with the necessary training and upskilling required by the employees to successfully exploit the fourth industrial revolution. This will also increase the body of knowledge by adding to the limited numbers of empirical studies on the operation of the maritime logistical supply and value chain.

8 Conclusion

The throughput volume handled by seaports is growing at a rapid rate with little evidence of a future decline, this provides a sound justification for continued academic and practitioner research. The Industry 4.0 Mapping project presents an exciting opportunity to study the implications of fully integrating the logistical supply chain from raw material storage, order picking, loading, and production of documentation to final voyage management. This extends to both horizontal and vertical integration across the value/supply chains, linking the managerial information, capture of sales, marketing and technological development, producing accurate KPIs; enabling the various departments within a seaport supply chain to model their solution to meet exact customer requirements, whilst maintaining sustainable rates of throughput. The results of these investigations (Fig. 3) will form the basis of essential decision support for significant investments in automation and cyber-physical systems. Appropriate mechanisms of decision support need to be created, tested and validated to fully realise the potential of this work (Fig. 4).

References

1. Heilig, L., Voß, S.: Information systems in seaports: a categorisation and overview. Inform. Technol. Manage. **18**, 179–220 (2017)
2. Bohari, A.M.: Computerisation of seaport operations management: competitive issues and need of ICT advancement. Inform. Manage. Bus. Rev. **5**(5), 217–224 (2013)
3. Sanchez-Gonzalez, P.L., Díaz-Gutiérrez, D., Leo, J.T., Núñez-Rivas, L.R.: Towards digitalisation of maritime transport. Sensors **19**, 1–22 (2019)
4. Fruth, M., Teuteberg, F.: Digitization in maritime logistics – what is there and what is missing? Cogent Bus. Manage. **4**(1), 1–40 (2017)
5. Srour, F., Oosterhout, M., Baalen, P., Zuidwijk, R.: Port Community System implementation lessons learned from international scan. In: Proceedings of Transportation Research Board 87th Annual Meeting, pp. 1–16. Transportation Research Board, Washington DC (2008)
6. de Lasa, I.S., Castro, R., Laburu, C.O.: Extent of the use of lean concepts proposed for a value stream mapping application. Prod. Plan. Control **20**(1), 82–98 (2009)
7. Buer, S.V., Strandhagen, J.O., Chan, F.T.S.: The kink between Industry 4.0 and lean manufacturing: mapping current research and establishing a research agenda. Int. J. Prod. Res. **56**(8), 2924–2940 (2018)
8. Romanowski, M., Nadolny, K., Sutowski, P.: Value stream mapping of the production process of tankers for the transport of liquid food products. J. Mech. Energy Eng. **1**(41), 123–128 (2017)

9. Liao, Y., Deschamps, F., Loures, E., Ramos, L.: Past, present and future of Industry 4.0 – a systematic literature review and research agenda proposal. Int. J. Prod. Res. **55**(12), 3609–3629 (2017)
10. Meudt, T., Metternich, J., Abele, E.: Value stream mapping 4.0: Holistic examination of value stream and information logistics in production. CIRP Ann. Manuf. Technol. **66**, 413–416 (2017)
11. Huang, Z., Kim, J., Sadri, A., Dowey, S., Dargusch, M.: Industry 4.0 development of a multi-agent system for dynamic value stream mapping in SME's. J. Manuf. Syst. **52**, 1–12 (2019)
12. .Vikraman, P., Kumar, P.S.: Dynamic value stream mapping in the realm of green manufacturing. GRD J. – Global Res. Dev. J. Eng. **2**(3), 4–9 (2017)
13. McDonald, T., Van-Aken, E.M., Rentes, A.F.: Utilising simulation to enhance value steam mapping: a manufacturing case application. Int. J. Logist.: Res. Appl. **5**(2), 213–232 (2002)
14. Roh, P., Kunz, A., Wegener, K.: Information stream mapping analysing and improving the efficiency of information streams in manufacturing streams. CIRP J. Manuf. Sci. Technol. **25**, 1–13 (2019)
15. Chen, J.C., Chen, K.M.: Application of ORFPM system for lean implementation an industrial case study. Int. J. Adv. Manuf. Technol. **72**, 839–852 (2014)
16. Kamble, S., Gunasekaran, A., Dhone, N.C.: Industry 4.0 and lean manufacturing practices for sustainable organisational performance in Indian manufacturing companies. Int. J. Prod. Res. **58**(5), 1319–1337 (2019)
17. Cepolina, S., Ghiara, H.: New trends in port strategies: emerging role for ICT infrastructures. Res. Transp. Bus. Manag. **8**(10), 195–205 (2013)
18. Tijan, E., Aksentijević, S., Čišić, D.: Seaport cluster information systems – a foundation for Port Community System's architecture. In: Proceedings of 2014 Information and Communication Technology, pp. 1557–1562. Electronics and Microelectronics (MIPRO), Opatija Croatia (2014)
19. Heilig, L., Lalla-Ruiz, E., Voβ, S.: Digital transformation in maritime ports: analysis and a name theoretic framework. Netnomics **18**, 227–254 (2017)
20. Vanselslander, T., Sys, C., Lam, J.S.L., Ferrari, C., Roumboutsos, A., Acciaro, M.: A serving innovation typology: mapping port related innovations. Transp. Rev. **39**(5), 611–629 (2019)
21. Marlow, P., Paixão - Casaca, A.: Measuring lean ports performance. Int. J. Transp. Manage. **1**, 189–202 (2003)
22. Lee, P., Lam, J., Lin, C., Hu, K., Cheong, I.: Developing the fifth-generation port concept model: an empirical test. Int. J. Logist. Manage. **29**(3), 1098–1120 (2018)
23. Loyd, N., Jennings, L., Siniard, J., Spayd, M., Holden, A., Rittenhouse, G.: Application of lean enterprise to improve seaport operations. Transp. Res. Rec.: J. Transp. Res. Board **2100**, 29–37 (2009)
24. Franzén, S., Streling, L.: Value stream mapping of container flows at seaports – a case study of four seaport container terminals. Chalmers University of Technology (2017)
25. MacDonald, C.: Understanding participatory action research methodology option. Can. J. Action Res. **13**(2), 34–50 (2012)
26. Woo, S.H., Pettit, S.J., Kwak, D.W., Beresford, A.: Seaport research: a structured literature review on methodological issues since the 1980s. Transp. Res. Part A: Policy Pract. **45**(7), 667–685 (2011)
27. Rapoport, R.: Three dilemmas of action research. Hum. Relat. **23**(6), 419–513 (1970)
28. Wong, L., Davison, R.: Knowledge sharing in a global logistics provider: an action research project. Inform. Manage. **55**, 547–557 (2018)
29. Rother, M., Shook, J.: Learning too See. The Lean Enterprise Institute, Brookline, MA (2003)
30. Tyagi, S., Choudhary, A., Cai, X., Yang, K.: Value stream mapping to reduce the lead-time of a product development process. Int. J. Prod. Econ. **160**, 202–217 (2015)

31. Frohm, J., Lindström, V., Winroth, M., Strahre, J.: Levels of automation in manufacturing. Ergon. Int. J. Ergon. Hum. Fact. **30**(3), 1–28 (2008)
32. Heilig, L., Voβ, S., Stahlbock, R.: 'From digitalisation to data-driven decision making in container terminals', Computer Science (2019). https://arxiv.org/abs/1904.13251. Accessed on 05 Dec 2019

Women in Engineering: Myths, Measures and Policies

Jyoti Sharma[1](✉) and Prasad K. D. V. Yarlagadda[2]

[1] International Cooperation Division (ICD), Department of Science and Technology, Ministry of Science and Technology, Government of India, New Delhi 110016, India
jyotisharma.dst@gmail.com

[2] Smart Systems, Science and Engineering Faculty, Queensland University of Technology, Brisbane 4000, Australia
y.prasad@qut.edu.au

Abstract. Quality Education and Gender Equality are among the 17 Sustainable Development Goals (SDGs) 2030 adopted by all United Nations Member States in 2015. Girls and women are the greatest untapped population who would contribute enormously to the growth of innovation. With evolving policies and institutional efforts to promote women scientists globally, there is a significant gap still lies in Science, Technology, Engineering, and Mathematics (STEM) education. A recent ground-breaking report of UNESCO reveals that only 35% of STEM students in higher education are women who result in 28% of the STEM workforce across all the world's researchers. Differences are observed much wider in the hard sciences like mathematics, physics, engineering, and computer science. Other international research also shows that women make up just 12–15% of all engineers around the world. However, a recent survey report by the Royal Academy of Engineering revealed that over 80% of female engineers are happy in their job and 98% of women engineers find their job rewarding. Despite these facts, most of the nations are losing potential women engineers at every stage of life, and it starts from a young age because of bias and misconceptions in media. There are societal, cultural, institutional factors, and belief systems which are impediments to women in STEM. Gender stereotypes are part of societal belief and govern the attitudes towards female and male roles, their occupations, and their perception of masculine and feminine work. There are many myths and stereotype in a society which push women back in engineering career like girls are bad in Mathematics, engineering is 'too difficult' and require physical strength. Divergent attitudes formed by girls and boys in childhood and systemic biases limit the range of career options and opportunities available to them in adulthood. This study engages with women engineers to record the impact of available myths, unconscious biases, stereotypes, ongoing policies & programs and compare the findings to unfold some of their struggles which is linked to retaining women in engineering professions. Though no single strategy can entirely plug the leak, there is a need for a multi-faceted approach for debunking the myths and spreading the facts. Strengthen the capacity of countries to deliver gender-responsive education, including through the involvement of women role models, their photographs in textbooks, teachers' training, and enhance awareness of the importance of engineering education for girls and women. To achieve all targets of Sustainable Development Goals (SDGs) and future wellbeing, the engineering workforce must equally represent by women.

A. Batako et al. (Eds.): GCMM 2021, LNNS 335, pp. 757–765, 2022.
https://doi.org/10.1007/978-3-030-90532-3_57

Keywords: STEM · Girls · Women · Engineering

1 Introduction

Quality Education and Gender Equality are among the 17 Sustainable Development Goals 2030 adopted by all United Nations Member States in 2015 [1]. Every nation is chasing and working hard to achieve them through the implementation of new policies and innovative programs. Gender issues are significant and influence every aspect of society. Girls and women are the untapped populations who would contribute enormously to the growth of innovation. The inclusion of women in science and Engineering is interlinked with the nation's composition of science & technology (S&T) workforce, market, and overall development. Women constitute half of humanity and any nation cannot become a superpower till the full and equal participation of women.

In recent year's women's participation and representation in Science, Technology, Engineering, and Mathematics (STEM) professions are increasing at entry-level, but a wide gap exists between the percentage of women studying science and the percentage of women doing science. A recent ground-breaking report of UNESCO reveals that only 35% of STEM students in higher education are women who result in 28% of the STEM workforce across all the world's researchers. By the time women reach Senior Lecturer (roughly equivalent to a middle management position), this percentage falls dramatically, resulting in just 20%. Women remain poorly represented in STEMM – just 16% of top-level science and technology researchers and professionals are women [2]. Differences are observed much in the hard sciences like information and communication technologies (ICT), physics, and engineering – where only 3–15% of female students are there. Engineering skills are important and part of every aspect of our life. In Australia, the high demand for qualified engineers relies on skilled workers from overseas could leave Australia vulnerable to factors outside the control. In vocational training, the number is less than one in ten; at universities, it is around one in six. This enormous disparity may exclude women from designing the future. It also means that engineering challenges are being tackled from a narrow set of perspectives [3].

There are cultural, systemic, and societal reasons for the historical underrepresentation of women in engineering areas, such as the impact of discriminatory practices in professional and academic settings. Gender stereotypes are part of societal beliefs and govern the attitudes towards female and male roles, their occupations, and their perception of masculine and feminine work. There are many myths and stereotype in a society which push women back in an engineering career. This paper will analyse the secondary data reported from government and institutional websites with a particular focus on programs designed to support women engineers in Australia.

2 Myths and Stereotypes

2.1 Family Belief

This a general perception of society that genius and eminence are the domain of men. Recent studies affirm that parents play a critical role in determining children's career choices. Many surveys and reports stated that more than 60% of the respondents currently studying STEM subjects at school had been influenced by their parents, teachers, or elder siblings in STEM-related fields. These studies indicate that parents hold a critical role in shaping their children's career choices. The outcome of a Queensland survey conducted in 2016 also revealed that almost 80% of parents would encourage science classes for their children, but less than 60% would encourage a career in science [4]. Parents think girls are less likely than boys in the fields that value "giftedness" such as mathematics and philosophy.

Another observation was observed during the career counselling sessions at the University of Wollongong when parents telling girls not to "waste their ATAR" by enrolling in a STEM course (ATAR is the Australian Tertiary Admission Rank) and being concerned that a STEM course could lead to an insecure career in academia [5]. It was also observed that even if their daughters demonstrated a passion for science or engineering, parents pushing them towards law, medicine, or finance courses, believing these were safer options for future employment. Perhaps parents could picture their daughter as a lawyer, doctor, or accountant, but they could not imagine a successful future for them as a scientist or engineer. It seemed that parents did not appreciate the range of high influence, inspiring, and financially rewarding jobs to which a STEM degree could lead, including in government and industry.

2.2 Girls Are Bad in Mathematics

A meta-analysis of Trends in International Mathematics and Science Study (TIMMS) 2015 which is an international assessment of mathematics and science at the fourth and eighth grades shows that boys had higher achievements in 18 countries, with average deference of 9 points as compared to girls who had higher achievement in 8 countries, with an average difference of 18 points. However, 23 countries showed no difference between boys and girls in average mathematics achievement [6]. The Organisation for Economic Co-operation and Development (OECD) 2015 report also stated that Australian girls are less confident than boys for studying mathematics. The Office of the Chief Scientist Australia also reports that, at the age of fifteen, girls are more likely than boys to believe that mathematical skills will not help them with later study or getting a job. By year twelve, the ratio of boy and girls are 3:1 in and 1.9:1 in physics and advanced mathematics subsequently. In 2017, only 26.7% year 12 female students enrolled in Information and Communication Technology (ICT) and Design & Technology subjects as compared to 39.4% male students [7]. These figures contribute to the broader perception that STEM is a difficult, masculine field, make girls more fearful and they perceive that they are less capable to achieve higher grades or perform in science and mathematics as compared to boys [8]. Though, a 2011 Psychological Bulletin meta-analysis of more than 240 studies published between 1990 and 2007 shows no statistically significant difference in mathematics performance when boys and girls are given the same standardized tests. The Singaporean school system is also challenging stereotyped views where girls are achieving high results in mathematics as boys. Another report described that math self-concept, stronger math–gender stereotypes, and socio-cultural factors are positively related to math achievement on a standardized test [9].

2.3 Women Just Are Not Interested in Engineering Career

Another Myth is that most women are disinterested in careers in Engineering, Physics, and ICT. However, rates of female participation in engineering fields correlate not with 'interest', but with inclusivity and the creation of positive cultural environments. The reason for 50 percent of women engineers in Muslim dominating countries like Jorden, Iran, Malaysia, and Tunisia, and even Saudi Arabia is the result of the vision of their parents [10]. Lyrics from a popular 1950s Arabic TV show in which mothers had sung to their baby girls showed the vision and envision her future in engineering, "And I will say 'My girl has grown up, she will be an engineer'/Oh people, oh people! I love her!/She's her mother's lovely girl."

2.4 There is an Equal Pay for Equal Work

According to the Australian Bureau of Statistics, women's average weekly ordinary full-time earnings across all industries and occupations was $1,558.40 compared to men's average weekly ordinary full-time earnings of $1,812.00 in May 2020. It is approx. 24% in Professional, Scientific, and Technical Services as compared to the national gender pay gap which is 16.2%. There are equal pay rates for entry-level and early-career roles indicates that most organizations provide competitive remuneration at this stage. However, it widens as experience increases. The gender pay gap is higher among leadership positions (28.8%) and not dependent on their age, experience, and whether their highest degree is a bachelor's or Ph.D. As reported by the Office of the Chief Scientists, 32% of males STEM graduates in the highest income bracket earn above $104,000, compared with just 12% of females. There is no pay difference between full-time and part-time workers, and women with, and without children [11].

2.5 The Battle for Gender Equality in STEM Disciplines is Over

Despite 52% of women at undergraduate and 50% at postgraduate levels, only 17% of professors are women in engineering and computer disciplines. Unconscious bias, job insecurity, masculine work culture, and limited career prospects have contributed to this underrepresentation [2]. A significant gender gap in hard science contributes to women's overall underrepresentation in STEM. In Australia, women account for only 12% of the STEM workforce in construction, 17% in mining, 19% in utilities and ICT, and 21% in scientific and technical services [12].

Apart from cultural, pay gap, and other societal factors, it is more about engineering culture that forces women to leave the profession. Women who leave this profession, go, and perform well in other intellectual demanding STEM professions. The stereotype culture of the engineering profession and sexual harassment makes women insecure and vulnerable as compared to other professions like medicine or law. The latest report from Professional Australia reveals that 26.8% of respondents reported that they had been sexually harassed in the course of their employment [13]. Scientists working in the field are at a higher risk to face sexual assault. Approximate two-thirds (64%) of researchers surveyed internationally experiencing sexual harassment, mostly at the hands of a senior researcher [14].

3 Policies and Programs

The Australian government designed and initiated more than 400 separate initiatives across Australia to facilitate women and girls in STEM. These include grants, mentoring & networking programs, work and industry experience, and other aspects of professional life. The Women in STEM Decadal Program, STEM Women, Professionals Australia, the Science in Australia Gender Equity (SAGE), Association for Women in Science (AWIS), a special committee for women in science and engineering, and Women & Leadership Australia (WLA) are few of them [15]. The Science in Australia Gender Equity (SAGE) was set up through a partnership between the Australian Academy of Science (AAS) and the Australian Academy of Technology and Engineering (ATSE) and implemented in thirty-three (33) out of 38 universities, plus a higher education research institution (representing approximately 87% of Australian universities), six (6) out of 50 of Australia's medical research institutes (representing approximately 12% of Australian MRIs) and five (5) out of approximately 12 publicly funded science research agencies (representing 40% of these agencies) till date [16]. The SAGE is based on the Athena SWAN Charter of the United Kingdom that follows the ten principles to support gender parity.

The Women in STEM Decadal Plan, developed by the Australian Academy of Science in collaboration with the Australian Academy of Technology and Engineering offers a vision and opportunities to 2030 to build a strong STEM workforce [17]. The Women in STEM Decadal Plan aims to create a 10-year roadmap in consultation with all stakeholder for achieving sustainable participation and retention of women from school through to careers. A branch of Engineers Australia, Women in Engineering (WIE) also funds a range of initiatives designed to support working female engineers. The Grants for women sponsored by L'Oreal, the 'For Women in Science', Google's Women Techmakers Scholars, and Superstars of STEM programs award financial support to women scientists and technologists. These programs aim to identify, support, and mentor female undergraduates and graduates who excel in the fields of engineering and technology.

3.1 New Initiatives: The Engineering for Australia Taskforce

The Engineering for Australia Taskforce was created by Monash University, the Australian National University, and the University of New South Wales in 2019 that aims to address the stereotyped nature at workplaces. The goal of the task force is to find ways to boost the numbers of women applying for university engineering programs [18]. The new report launched by this taskforce explores the factors that affect a girl's participation in engineering. The report identifies three key actions: create an inclusive vision for STEM and engineering to address the gender disparity; work with the education sector to create a STEM and engineering identity in schools through hands-on activities, and evaluate the ongoing engineering intervention programs [18].

3.2 National Committee for Women in Engineering (NCWIE)

The National Committee for Women in Engineering contributes to the achievement of Engineers Australia's Strategic Priorities by identifying strategies and recommending to the Board actions for inclusion of diversity in the engineering profession. The Committee aims to attract, support, and retain women engineers throughout their engineering career. [19]. As the peak body for the engineering profession in Australia, Engineers Australia understands that there is a need to create a diverse professional workforce and set a target of 30% female representation across Boards and Committees by 2020.

4 Discussion and Conclusion

Women in STEM face hurdles that have nothing to do with their abilities, such as stereotypes, criticism, discrimination, and harassment [20]. Women are not inherently less capable or less interested in engineering than men, and the problem of gender imbalance is not impossible to solve. There is nothing inevitable about inequality, but not only Australia but most of the nation requires a multi-faceted approach across all levels of education and the workforce. There are many drivers impacting girls' interest in STEMM subjects. In some places, having faith in their capabilities is a significant barrier, while in others, encouragement and mentorship or lack of role models are holding them back most. These factors forced the female students less interested and less confident in studying STEM subjects specially engineering and mathematics (Table 1).

The lack of visible, and accessible role models is one of the factors that influenced the choice of career of girls at an early age. A recent study of women between the ages of 11 and 30 across Europe found that most girls become interested in STEMM under the age of 11, but that their interest wanes by 15 [21]. This responsibility lies in the government and media to show the visibility of diverse women in STEMM. The Science and Technology Australia's Superstars of STEM program aims to support up-and-coming female scientists to create role models for younger women and girls who might be considering a STEM career [22]. The Women in STEM Ambassador also works to increase the participation of women and girls in STEM by engaging with stakeholders across Australia to build visibility and drive needed cultural and social change for gender equity in STEM.

The next step is to support and encourage women to pursue their paths in hard-core science such as physics, engineering, and ICT where they make up only around 15 percent of graduates. The identity of the engineering profession among schoolgirls is a persistent problem. Girls are handling sophisticated engineering projects at the school level. However, they are not confident to opt for engineering as a profession. The lessons may be learned from China, Malaysia, and the former Soviet Union. The programs like the Male Champions of Change program also contribute and facilitate the advancement of this work through the involvement of influential and powerful men [23].

Apart from these, the national guidelines and programs must include support for childbearing responsibilities. The Australian Bureau of Statistics reveals that motherhood is one of the main barriers for women to enter the workforce or return after the break [24]. It leads to a sharp drop in the career progression of women once they enter the prime childbearing years of the mid-20s to mid-30s. Apart from motherhood responsibilities, women also share a major responsibility for informal care of family and friends with disabilities, mental illness, chronic illness. It is essential to promote gender-neutral schemes and attitudes to overcome these challenges.

Though, the programs like SAGE are found to be effective to raise awareness and encourages an individual institution for creating a conducive environment for everyone. There is a need to monitor all ongoing 'women in STEM' programs through a data-driven approach to measure their impact on society. The new 'Advancing Women in STEM 2020' Action Plan, released in April 2019, also recommends data collection and analysis, evaluation, and changing institutional practices as per the present need [17]. The federal government also launched a *'STEM Equity Monitor'*, an interactive portal to collect and track data on a range of indicators including girls' attitudes to STEM in schools, engagement rates, qualifications, and workforce participation, pay rates and working conditions [25].

Cultural, social, educational, and unconscious bias in the professional field are various barriers to women's participation in STEM and there is a need to go for the multi-faceted approach. Gender inequality in STEM is not just a problem for women – it is something that should concern all for moral, civic, cultural, and even economic reasons.

Table 1. Percentage of female students interested and confident in STEM subjects

Level of interest in STEM subjects			Confidence in getting good results in STEM subjects		
Discipline	Male	Female	Discipline	Male	Female
Science	68%	61%	Science	64%	60%
Technology	75%	54%	Technology	73%	56%
Engineering	55%	28%	Engineering	50%	26%
Mathematics	56%	45%	Mathematics	65%	60%

Reference: [7]

References

1. Sustainable Development Goals. The United Nations Development Programme. https://www.undp.org/content/undp/en/home/sustainable-development-goals.html. Accessed 21 Sept 2020
2. Gender Equity in STEMM. Science in Australia Gender Equity. Accessed 5 Oct 2020
3. Australia needs more engineers. And more of them need to be women. https://theconversation.com/australia-needs-more-engineers-and-more-of-them-need-to-be-women-130282. Accessed 21 Sept 2020
4. Queenslanders' Perceptions and Attitudes to Science Research Report. Office of the Queensland Chief Scientist. Kantar Public. April (2018). https://www.chiefscientist.qld.gov.au/__data/assets/pdf_file/0018/50148/queenslanders-perceptions-attitudes-science-research-report.pdf. Accessed 22 Oct 2020
5. Calling all parents – Australia's future female scientists need your support now. The Conversation (2017). https://theconversation.com/calling-all-parents-australias-future-female-scientists-need-your-support-now-89025. Accessed 11 Aug 2020

6. Ina, V.S., Mullis Michael, O.: Martin Pierre Foy Martin Hooper. TIMSS 2015 International Results in Mathematics (2015). http://timssandpirls.bc.edu/timss2015/international-results/wp-content/uploads/filebase/full%20pdfs/T15-International-Results-in-Mathematics.pdf. Accessed 11 July 2020
7. Advancing women in STEM strategy. Department of Industry, Science, Energy, and Resources. Australian Government. https://www.industry.gov.au/data-and-publications/advancing-women-in-stem-strategy/snapshot-of-disparity-in-stem/women-in-stem-at-a-glance. Accessed 15 Oct 2020
8. Jaymes Carr. Women in STEM: myths, mentoring and more. Gradaustralia. https://gradaustralia.com.au/career-planning/women-in-stem-myths-mentoring-and-more. Accessed 21 Aug 2020
9. Cvencek, D., Kapur, M., Meltzoff, A.N.: Math achievement, stereotypes, and math self-concepts among elementary-school students in Singapore. Learn. Instr. **39**, 1–10 (2015)
10. Sharma, J., Yarlagadda, T., Sharma, S., Yarlagadda, P.K.: Vertical segregation: issues and challenges of women engineers in Australia. Procedia Manuf. **30**, 671–676 (2019)
11. Australia's Gender Pay Gap Statistics. Workplace Gender Equality Agency. Australian Government (2020). https://www.wgea.gov.au/data/fact-sheets/australias-gender-pay-gap-statistics-2020#:~:text=Australia's%20national%20gender%20pay%20gap%20has%20hovered%20between%2013.9%25%20and,in%20November%202019%2C%20at%2013.9%25. Accessed 15 Aug 2020
12. The gender pay gap in Engineering and Science (2016). Professional Engineers and Professional Scientists Australia. http://www.professionalsaustralia.org.au/professional-women/wp-content/uploads/sites/48/2014/03/Gender-pay-gap-in-Engineering-and-Science.pdf. Accessed 12 Feb 2020
13. Women in STEM Professions Survey Report: Professional Australia. (2018). Accessed 12 Oct 2020
14. Bohannon, J.: Survey of peers in fieldwork highlights an unspoken risk. Science **340**(6130), 265–265 (2013). https://doi.org/10.1126/science.340.6130.265
15. Gender equity. Australian Academy of Science. https://www.science.org.au/supporting-science/diversity-and-inclusion/gender-equity. Accessed 11 July 2020
16. Sharma, J., Yarlagadda, P.: Gender equality in India's sciences: will the Athena Swan Charter help? Nat. India (2020). https://doi.org/10.1038/nindia.2020.41
17. Women in STEM Decadal Plan. Australian Academy of Science. https://www.science.org.au/support/analysis/decadal-plans-science/women-in-stem-decadal-plan. Accessed 22 Nov 2020
18. New national taskforce releases report to tackle gender disparity in engineering. Monash University (2020). https://www.monash.edu/news/articles/new-national-taskforce-releases-report-to-tackle-gender-disparity-in-engineering#:~:text=The%20Engineering%20for%20Australia%20Taskforce%20was%20created%20by%20the%20Deans,tackling%20the%20barriers%20to%20girls. Accessed 11 Aug 2020
19. The National Committee for Women in Engineering. Engineers Australia. https://www.engineersaustralia.org.au/Communities-And-Groups/Special-Interest-Groups/Women-In-Engineering#:~:text=National%20Committee%20for%20Women%20in%20Engineering%20(NCWIE),engineering%20diverse%20and%20inclusive%20profession. Accessed 24 Nov 2020
20. Sharma, J., Dhal, S.: Perspectives, policy and programmes: an empirical analysis of 'women in science' in India. Int. J. Gend. Sci. Technol. **8**(2), 258–278 (2016)
21. Andrew Trotman. Why don't European girls like science or technology? https://news.microsoft.com/europe/features/dont-european-girls-like-science-technology/. Accessed 2 Oct 2020
22. Science and Technology Australia. https://scienceandtechnologyaustralia.org.au/. Accessed 24 Nov 2020

23. Male Champions of Change. https://malechampionsofchange.com/message-to-our-colleagues/
24. Barriers and Incentives to Labour Force Participation, Australia. Australian Bureau of Statistics. (2018–19). https://www.abs.gov.au/statistics/labour/employment-and-unemployment/barriers-and-incentives-labour-force-participation-australia/latest-release. Accessed 19 Nov 2020
25. STEM Equity Monitor. Department of Industry, Science, Energy, and Resources. Australian Government. https://www.industry.gov.au/data-and-publications/stem-equity-monitor. Accessed 22 Nov 2020

Author Index

A. Batako et al. (Eds.): GCMM 2021, LNNS 335, pp. 767–769, 2022.
https://doi.org/10.1007/978-3-030-90532-3